Edited by
Se-Kwon Kim

Marine Microbiology

Related Titles

Hanessian, S. (ed.)

Natural Products in Medicinal Chemistry

2014

ISBN: 978-3-527-33218-2

de Bruijn, F. J.

Handbook of Molecular Microbial Ecology

Set

2011

ISBN: 978-0-470-92418-1

Kornprobst, J.-M.

Encyclopedia of Marine Natural Products

3 Volume Set

2010

ISBN: 978-3-527-32703-4

Brahmachari, G.

Handbook of Pharmaceutical Natural Products

2010

ISBN: 978-3-527-32148-3

Edited by Se-Kwon Kim

Marine Microbiology

Bioactive Compounds and Biotechnological Applications

Verlag GmbH & Co. KGaA

The Editor

Prof. Se-Kwon Kim
Pukyong National University
Marine Bioprocess Research Center
Daeyeon-Dong, Nam-Gu 599-1
Busan 608–737
Republic of Korea

Library of Congress Card No.: applied for

British Library Cataloguing-in-Publication Data
A catalogue record for this book is available from the British Library.

Bibliographic information published by the Deutsche Nationalbibliothek
The Deutsche Nationalbibliothek lists this publication in the Deutsche Nationalbibliografie; detailed bibliographic data are available on the Internet at <http://dnb.d-nb.de>.

Composition Thomson Digital, Noida

Printing and Binding Markono Print Media Pte Ltd, Singapore

Cover Design Schulz Grafik-Design, Fußgönheim

Print ISBN: 978-3-527-33327-1
ePDF ISBN: 978-3-527-66528-0
ePub ISBN: 978-3-527-66527-3
mobi ISBN: 978-3-527-66526-6
oBook ISBN: 978-3-527-66525-9

Printed in Singapore

Printed on acid-free paper

Contents

Preface

The study of the microorganisms in the sea is called marine microbiology. Microbiology is an exceptionally broad discipline encompassing specialties as diverse as biochemistry, cell biology, genetics, taxonomy, pathogenic bacteriology, food and industrial microbiology, and ecology. Marine microbes include bacteria, virus and fungi widely available in marine arena. In the recent years, much attention has been laid on marine derived bioactive compounds for various biological and biomedical applications. Marine microbial bioactives are one of them and much potent from the ancient days. For an example, penicillin is the first drug isolated from *Penicillium* fungi and used to treat several diseases until today. The class of Marine actinobacteria are the most valuable prokaryotes both economically as well as biotechnologically. Among the actinobacteria, Streptomycetes group is considered economically important because out of the approximately more than 10 000 known antibiotics, 50–55% are produced by this genus. There is still need to develop several potent bioactive compounds from marine microorganisms. I had a long-standing interest to edit a book on marine microbial compounds and their biotechnological applications for the drug discovery. Only limited books are available on marine microbiology and many microbiologists and biotechnologists around the globe have written several chapters in various editions. However, full weightage has been given in the present book related to marine microbes. This book will be useful for both novice and experts in the field of marine microbiology, natural product science and biotechnology. By reading this book, scholars and scientists working in the field of marine microbiology can improve their practical knowledge towards its biotechnological applications.

Part I of this book covers the introduction about the microorganism from the deep sea and also global ecosystems. Part II of this book reveals the taxonomic study of marine actinobacteria, potential source for pharmaceutical agents, and usage of microorganism in aquaculture, production of antibiotic and pharmaceutical oriented compounds. Part III of the book extensively compiles the information regarding biological and biomedical activity of marine microbe derived compounds for different applications such as antifungal, anti-mycotoxin, anti-tuberculosis and antimicrobial activities.

I am grateful to all the contributors, my students and colleagues. I am also thankful to Wiley publishers for their constant support. I believe this book will bring new ideas for various biotechnological applications of marine microorganisms.

Busan, South Korea *Prof. Se-Kwon Kim*

FBIO

Prof. Se-Kwon Kim, PhD, is Senior Professor of Marine Biochemistry, Department of Chemistry and Director, Marine Bioprocess Research Center (MBPRC), Pukyong National University, South Korea. He received his BS, MS, and PhD degrees from Pukyong National University and joined the same as a faculty member. He has previously served as a scientist in the University of Illinois, Urbana-Champaign, Illinois (1988–1989), and was a visiting scientist at the Memorial University of Newfoundland, Canada (1999–2000).

Prof. Kim served as the first president of the Korean Society of Chitin and Chitosan (1986–1990) and the Korean Society of Marine Biotechnology (2006–2007). He was also the chairman for the 7th Asia-Pacific Chitin & Chitosan Symposium, held in South Korea in 2006. He is the board member of the International Society of Marine Biotechnology and the International Society for Nutraceuticals and Functional Foods. He has also served as the editor-in-chief of the *Korean Journal of Life Sciences* (1995–1997), the *Korean Journal of Fisheries Science and Technology* (2006–2007), and the *Korean Journal of Marine Bioscience and Biotechnology* (2006–present). He has won several awards, including the Best Paper Award from the American Oil Chemists' Society (AOCS) and the Korean Society of Fisheries Science and Technology in 2002.

Prof. Kim's major research interests include investigation and development of bioactive substances derived from marine organisms and their application in oriental medicine, nutraceuticals, and cosmeceuticals via marine bioprocessing and mass production technologies. Furthermore, he has expanded his research fields to the development of bioactive materials from marine organisms for applications in oriental medicine, cosmeceuticals, and nutraceuticals. To date, he has authored over 520 research papers and has 112 patents to his credit. In addition, he has written and edited more than 45 books.

List of Contributors

Visamsetti Amarendra
SASTRA University
School of Chemical and Biotechnology
Tirumalaisamudram
Thanjavur 613401
Tamil Nadu
India

and

SASTRA University
Genetic Engineering Laboratory
SASTRA's Hub for Research &
Innovation (SHRI)
ASK 302, Anusandhan Kendra
Tirumalaisamudram
Thanjavur 613401
Tamil Nadu
India

Nahara E. Ayala-Sánchez
Universidad Autónoma de Baja
California
Facultad de Ciencias
Km. 103 Tijuana-Ensenada, Highway
Ensenada, BC 22830
Mexico

Sanat K. Basu
Gupta College of Technological Sciences
Department of Pharmaceutics
Ashram More, G.T. Road
Asansol 713301
West Bengal
India

Ira Bhatnagar
Pukyong National University
Department of Chemistry
Marine Biochemistry Laboratory
Busan 608-737
South Korea

and

Centre for Cellular and Molecular
Biology (CCMB)
Laboratory of Infectious Diseases
Hyderabad 500007
Andhra Pradesh
India

Li-Xin Cao
Harbin Institute of Technology
at Weihai
School of the Ocean
West Culture Road 2
Weihai, Shandong 264209
China

Samrat Chakraborty
Gupta College of Technological Sciences
Department of Pharmaceutics
Ashram More, G.T. Road
Asansol 713301
West Bengal
India

Pranjal Chandra
Pusan National University
Department of Chemistry
BioMEMS & Nanoelectrochemistry
Lab
Busan 609-735
South Korea

Surajit Das
National Institute of Technology
Department of Life Science
Laboratory of Environmental
Microbiology and Ecology (LEnME)
Rourkela 769008
Odisha
India

Hirak R. Dash
National Institute of Technology
Department of Life Science
Laboratory of Environmental
Microbiology and Ecology (LEnME)
Rourkela 769008
Odisha
India

Julieta R. de Oliveira
Universidade de São Paulo
Instituto de Química de São Carlos
Av. Trabalhador São-carlense, 400
13560-970 São Carlos, SP
Brazil

Andrew P. Desbois
University of Stirling
School of Natural Sciences
Institute of Aquaculture
Marine Biotechnology Research Group
Stirlingshire FK9 4LA
UK

Pradeep Dewapriya
Pukyong National University
Department of Chemistry
Marine Biochemistry Laboratory
Busan 608-737
South Korea

Mahanama De Zoysa
Chungnam National University
College of Veterinary Medicine
Laboratory of Aquatic Animal
Diseases
Yuseong-gu
Daejeon 305-764
South Korea

Kandasamy Dhevendaran
SASTRA University
School of Chemical and Biotechnology
Tirumalaisamudram
Thanjavur 613401
Tamil Nadu
India

Sung-Hwan Eom
Pukyong National University
Department of Food Science and
Technology
599-1 Daeyeon 3-dong, Nam-gu
Busan 608-737
South Korea

Arijit Gandhi
Gupta College of Technological Sciences
Department of Pharmaceutics
Ashram More, G.T. Road
Asansol 713301
West Bengal
India

Rajendra N. Goyal
Indian Institute of Technology Roorkee
Department of Chemistry
Roorkee 247667
Uttarakhand
India

Lone Gram
Technical University of Denmark
Department of Systems Biology
Søltofts Plads 221
2800 Kongens Lyngby
Denmark

Graciela Guerra-Rivas
Universidad Autónoma de Baja
California
Facultad de Ciencias Marinas
Km. 103 Tijuana-Ensenada, Highway
Ensenada, BC 22830
Mexico

S.W.A. Himaya
Pukyong National University
Marine Bioprocess Research Center
Busan 608-737
South Korea

Le Minh Hoang
Nha Trang University
Faculty of Aquaculture
02 Nguyen Dinh Chieu, Nha Trang
Khanh Hoa 65000
Vietnam

Chiaki Imada
Tokyo University of Marine Science
and Technology
Graduate School of Marine Science
and Technology
4-5-7, Konam
Minato-ku
Tokyo 108-8477
Japan

Ana M. Íñiguez-Martínez
Universidad de Guadalajara
Centro Universitario de la Costa
Av. Universidad de
Guadalajara No. 203
Puerto Vallarta, Jal. 48280
Mexico

Sougata Jana
Gupta College of Technological Sciences
Department of Pharmaceutics
Ashram More, G.T. Road
Asansol 713301
West Bengal
India

Wence Jiao
Dalian University of Technology
School of Life Science and
Biotechnology
Linggong Road 2
Dalian 116024
China

Se-Kwon Kim
Pukyong National University
Department of Chemistry and Marine
Bioprocess Research Center
Marine Biotechnology Laboratory
599-1 Daeyeon 3-dong, Nam-gu
Busan 608-737
South Korea

Young-Mog Kim
Pukyong National University
Department of Food Science and
Technology
599-1 Daeyeon 3-dong, Nam-gu
Busan 608-737
South Korea

Priyanka Kishore
National Institute of Technology
Department of Life Science
Laboratory of Environmental
Microbiology and Ecology (LEnME)
Rourkela 769008
Odisha
India

Dae-Sung Lee
Pukyong National University
Department of Microbiology
599-1 Daeyeon 3-dong, Nam-gu
Busan 608-737
South Korea

Myung-Suk Lee
Pukyong National University
Department of Microbiology
599-1 Daeyeon 3-dong, Nam-gu
Busan 608-737
South Korea

Ulrike Lindequist
Ernst-Moritz-Arndt-University of Greifswald
Institute of Pharmacy
Department of Pharmaceutical Biology
Friedrich-Ludwig-Jahn-Strasse 17
17487 Greifswald
Germany

Neelam Mangwani
National Institute of Technology
Department of Life Science
Laboratory of Environmental Microbiology and Ecology (LEnME)
Rourkela 769008
Odisha
India

Panchanathan Manivasagan
Pukyong National University
Department of Chemistry and Marine Bioprocess Research Center
Marine Biotechnology Laboratory
599-1 Daeyeon 3-dong, Nam-gu
Busan 608-737
South Korea

Maria Månsson
Technical University of Denmark
Department of Systems Biology
Søltofts Plads 221
2800 Kongens Lyngby
Denmark

André L. Meleiro Porto
Universidade de São Paulo
Instituto de Química de São Carlos
Av. Trabalhador São-carlense, 400
13560-970 São Carlos, SP
Brazil

Sabine Mundt
Ernst-Moritz-Arndt-University of Greifswald
Institute of Pharmacy
Department of Pharmaceutical Biology
Friedrich-Ludwig-Jahn-Strasse 17
17487 Greifswald
Germany

Chamilani Nikapitiya
Chonnam National University
Department of Aqualife Medicine
College of Fisheries and Ocean Science
Yeosu
Jeollanamdo 550749
South Korea

Arnab Pramanik
Jadavpur University
School of Environmental Studies
Kolkata 70032
West Bengal
India

Ocky K. Radjasa
Diponegoro University
Department of Marine Science
Soedarto, SH St. 1
Semarang 50275, Central Java
Indonesia

Mirna H. Regali Seleghim
Universidade Federal de São Carlos
Departamento de Ecologia e Biologia
Evolutiva
Via Washington Luís, Km 235
13565-905 São Carlos, SP
Brazil

Jing Ren
Harbin Institute of Technology
at Weihai
School of the Ocean
West Culture Road 2
Weihai, Shandong 264209
China

Richa
Banaras Hindu University
Centre of Advanced Study in Botany
Laboratory of Photobiology and
Molecular Microbiology
Varanasi 221005
Uttar Pradesh
India

Lenilson C. Rocha
Universidade de São Paulo
Instituto de Química de São Carlos
Av. Trabalhador São-carlense, 400
13560-970 São Carlos, SP
Brazil

Gisele N. Rodrigues
Universidade Federal de São Carlos
Departamento de Ecologia e Biologia
Evolutiva
Via Washington Luís, Km 235
13565-905 São Carlos, SP
Brazil

Malay Saha
Sovarani Memorial College
Department of Botany
Jagatballavpur
Howrah 711408
West Bengal
India

Barindra Sana
Nanyang Technological University
School of Chemical and Biomedical
Engineering
Division of Bioengineering
Singapore 637457
Singapore

Ramachandran S. Santhosh
SASTRA University
School of Chemical and Biotechnology
Tirumalaisamudram
Thanjavur 613401
Tamil Nadu
India

and

SASTRA University
Genetic Engineering Laboratory
SASTRA's Hub for Research &
Innovation (SHRI)
ASK 302, Anusandhan Kendra
Tirumalaisamudram
Thanjavur 613401
Tamil Nadu
India

Kalyan K. Sen
Gupta College of Technological
Sciences
Department of Pharmaceutics
Ashram More, G.T. Road
Asansol 713301
West Bengal
India

Amardeep Singh
Pusan National University
Department of Chemistry
BioMEMS & Nanoelectrochemistry
Lab
Busan 609-735
South Korea

Rajeshwar P. Sinha
Banaras Hindu University
Centre of Advanced Study in Botany
Laboratory of Photobiology and
Molecular Microbiology
Varanasi 221005
Uttar Pradesh
India

Irma E. Soria-Mercado
Universidad Autónoma de Baja
California
Facultad de Ciencias Marinas
Km. 103 Tijuana-Ensenada, Highway
Ensenada, BC 22830
Mexico

M.L. Arvinda swamy
Centre for Cellular and Molecular
Biology
Laboratory of Infectious diseases
Uppal Road
Hyderabad, 500007
Andhra Pradesh
India

Lik Tong Tan
Nanyang Technological University
National Institute of Education
Natural Sciences and Science
Education
1 Nanyang Walk
Singapore, 637616
Singapore

Kustiariyah Tarman
Bogor Agricultural University
Faculty of Fisheries and Marine
Sciences
Department of Aquatic Product
Technology
Jl. Agathis 1 Kampus IPB Darmaga
16680 Bogor
Indonesia

and

Bogor Agricultural University
Center for Coastal and Marine
Resources Studies
Division of Marine Biotechnology
Jl. Raya Pajajaran 1 Kampus IPB
Baranangsiang
16144 Bogor
Indonesia

Manoj Trivedi
Pusan National University
Department of Chemistry
BioMEMS & Nanoelectrochemistry
Lab
Busan 609-735
South Korea

Trang Sy Trung
Nha Trang University
Department of External Affairs
02 Nguyen Dinh Chieu, Nha Trang
Khanh Hoa 650000
Vietnam

Bruna Vacondio
Universidade Federal de São Carlos
Departamento de Ecologia e Biologia
Evolutiva
Via Washington Luís, Km 235
13565-905 São Carlos, SP
Brazil

Nguyen Van Duy
Nha Trang University
Institute of Biotechnology and Environment
02 Nguyen Dinh Chieu, Nha Trang
Khanh Hoa 650000
Vietnam

Quang Van Ta
Pukyong National University
Department of Chemistry
Marine Biochemistry Laboratory
Building C13, Room 201
599-1 Daeyeon 3-dong, Nam-gu
Busan 608-737
South Korea

Jayachandran Venkatesan
Pukyong National University
Department of Chemistry and Marine Bioprocess Research Center
599-1 Daeyeon 3-dong, Nam-gu
Busan 608-737
South Korea

Nikolaj G. Vynne
Technical University of Denmark
Department of Systems Biology
Søltofts Plads 221
2800 Kongens Lyngby
Denmark

Matthias Wietz
Technical University of Denmark
National Food Institute
Søltofts Plads 221
2800 Kongens Lyngby
Denmark

and

University of California, San Diego
Center for Marine Biotechnology and Biomedicine
Scripps Institution of Oceanography, La Jolla
San Diego, CA 92093
USA

Xiaona Xu
Dalian University of Technology
School of Life Science and Biotechnology
Linggong Road 2
Dalian 116024
China

Pei-Sheng Yan
Harbin Institute of Technology at Weihai
School of the Ocean
West Culture Road 2
Weihai, Shandong 264209
China

Xinqing Zhao
Dalian University of Technology
School of Life Science and Biotechnology
Linggong Road 2
Dalian 116024
China

1
Introduction to Marine Actinobacteria

Panchanathan Manivasagan, Jayachandran Venkatesan, and Se-Kwon Kim

1.1
Introduction

Marine microbiology is developing strongly in several countries with a distinct focus on bioactive compounds. Analysis of the geographical origins of compounds, extracts, bioactivities, and Actinobacteria up to 2003 indicates that 67% of marine natural products were sourced from Australia, the Caribbean, the Indian Ocean, Japan, the Mediterranean, and the Western Pacific Ocean sites [1].

Marine Actinobacteria have been looked upon as potential sources of bioactive compounds, and the work done earlier has shown that these microbes are the richest sources of secondary metabolites. They hold a prominent position as targets in screening programs due to their diversity and their proven ability to produce novel metabolites and other molecules of pharmaceutical importance [2]. Since the discovery of actinomycin [3], Actinobacteria have been found to produce many commercially bioactive compounds and antitumor agents in addition to enzymes of industrial interest [4]. Approximately, two-third of the thousands of naturally occurring antibiotics have been isolated from these organisms [5]. Of them, many have been obtained from *Streptomyces* [6] and these natural products have been an extraordinary source for lead structures in the development of new drugs [7].

Although the diversity of life in the terrestrial environment is extraordinary, the greatest biodiversity is in the oceans [8]. More than 70% of our planet's surface is covered by oceans and life on Earth originated from the sea. In some marine ecosystems, such as the deep sea floor and coral reefs, experts estimate that the biological diversity is higher than that in the tropical rainforests [9]. As marine environmental conditions are extremely different from the terrestrial ones, it is surmised that marine Actinobacteria have characteristics different from those of terrestrial counterparts and, therefore, might produce different types of bioactive compounds. The living conditions to which marine Actinobacteria had to adapt during evolution range from extremely high pressures (with a maximum of 1100 atmospheres) and anaerobic conditions at temperatures just below 0 °C on the deep sea floor to high acidic conditions (pH as low as 2.8) at temperatures of over 100 °C

Marine Microbiology: Bioactive Compounds and Biotechnological Applications, First Edition.
Edited by Se-Kwon Kim

near hydrothermal vents at the mid-ocean ridges. It is likely that this is reflected in the genetic and metabolic diversity of marine actinomycetes, which remain largely unknown. Indeed, the marine environment is virtually an untapped source of novel Actinobacteria diversity [10,11] and, therefore, of new metabolites [12–14].

However, the distribution of Actinobacteria in the sea is largely unexplored and the presence of indigenous marine Actinobacteria in the oceans remains elusive. This is partly caused by the insufficient effort put into exploring marine Actinobacteria, whereas terrestrial Actinobacteria have been, until recently, a successful source of novel bioactive metabolites. Furthermore, skepticism regarding the existence of indigenous populations of marine Actinobacteria arises from the fact that the terrestrial bacteria produce resistant spores that are known to be transported from land into sea, where they can remain available but dormant for many years [15–17]. In this chapter, we evaluate the current state of research on the biology and biotechnology of marine Actinobacteria. The topics covered include the abundance, diversity, novelty and biogeographic distribution of marine Actinobacteria, ecosystem function, bioprospecting, and a new approach to the exploration of actinobacterial taxonomic space.

1.2 Actinobacteria

Actinobacteria are aerobic, nonmotile, and Gram-positive bacteria with high guanosine–cytosine (GC) content in their DNA (70–80%) and are phylogenetically related to the bacteria based on the evidence of 16S ribosomal RNA cataloging studies [18]. Although originally considered an intermediate group between bacteria and fungi, they are now recognized as prokaryotic organisms. Actinomycetales is an order of Actinobacteria, which have substrate hyphae and form aerial mycelia and spores. Aerial hyphae of Actinobacteria give rise to sporophores that differ greatly in structure. The spore-bearing hyphae of the aerial mycelium have somewhat greater diameter than the substrate mycelium. The spores are resistant to desiccation and can survive in soil in a viable state for long periods. This stage of the life cycle imparts resistance to adverse environmental conditions in the soil such as low nutrients and water availability. These microorganisms are phenotypically highly diverse and found in most natural environments [18].

1.3 Origin and Distribution of Marine Actinobacteria

Actinobacteria are mostly considered as terrigenous bacteria because of their wide occurrence and abundance in soil. Their distribution in the aquatic environment remained largely undescribed for many years. Most of the workers questioned the indigenous nature of aquatic Actinobacteria because these produce resistant spores that are known to be transported from land into sea and other aquatic bodies where they can remain dormant for many years. In fact, they were considered to originate from dormant spores that were washed from land [18].

It is now clear that specific populations of marine-adapted Actinobacteria not only exist in the marine environment but also significantly add to diversity within a broad range of *Actinomycetes taxa* [19,20]. Recent studies have also shown that Actinobacteria can be isolated from mangrove swamps, other coastal environments, and even deep ocean sediments [21,22].

Despite the fact that the selective methods used to culture Actinobacteria targeted only the mycelium-producing strains, thereby omitting the important marine groups such as the mycolate Actinobacteria [23], it can be seen that marine Actinobacteria include new phenotypes that have clearly diverged from those known to occur on land.

Although the ecological roles of marine Actinobacteria remain undefined, it is possible that like their terrestrial counterparts, they are involved in the decomposition of recalcitrant organic materials such as chitin, a biopolymer that is particularly abundant in the sea [21]. Given that Actinobacteria living in the ocean experience a dramatically different set of environmental conditions compared to their terrestrial relatives, it is not surprising that speciation has occurred and unique marine taxa are now being recognized. Not only the extent of marine actinobacterial diversity is yet to be determined, but also the adaptations of these microbes in the sea resulting in the production of secondary metabolites are to be studied.

1.4 Isolation and Identification of Marine Actinobacteria

Actinobacteria are ubiquitous in marine environment and there are several techniques for their isolation. In the conventional isolation techniques, several factors must be considered, namely, choice of screening source, selective medium, culture conditions, and recognition of candidate colonies in the primary isolation. Some of the researchers employ pretreatments of sediments by drying and heating to stimulate the isolation of rare Actinobacteria [24]. An alternative approach would be to make the isolation procedure more selective by adding chemicals such as phenol to the sediment suspension. Many media have been recommended for isolation of Actinobacteria from marine samples. Specialized growth media have been developed to isolate specific actinomycete genera with macromolecules such as casein, chitin, hair hydrolysate, and humic acid that are carbon and nitrogen sources for obtaining rare Actinobacteria. Several antibiotic molecules are also used in selective media to inhibit unwanted microbes, including fast-growing bacteria and fungi.

Strains are preliminarily indentified according to their morphological criteria, including characteristics of colonies on the plate, morphology of substrate mycelium and aerial hyphae, morphology of spores, pigments produced, cell wall chemo type, whole-cell sugar pattern, and so on, and their identification is confirmed by 16s rDNA analysis [25–32].

1.5 Indigenous Marine Actinobacteria

The indigenous deep sea Actinobacteria warrant some specific consideration because if we can define some or all of the features of the deep sea Actinobacterial physiology, this should lead to greater efficacy of isolation. Although an obligate requirement for Na+ and the obligate requirements or tolerance of oligotrophic substrate concentrations, low temperatures, and elevated pressures for growth would provide prima facie evidence of indigenicity, to our knowledge no systematic testing of this hypothesis with respect to deep sea Actinobacteria has been made. In addition, demonstration of growth or metabolic activity *in situ* should be made. We believe that physiological understanding of this type could enable more precise ecosimulation or microcosm approaches to targeting the recovery of a greater diversity of deep sea Actinobacteria [33].

Early evidences supporting the existence of marine Actinobacteria came from the description of *Rhodococcus marinonascene*, the first marine actinomycete species to be characterized [34]. Further support has come from the discovery that some strains display specific marine adaptations [35], whereas others appear to be metabolically active in marine sediments [36]. However, these early findings did not generate enough excitement to stimulate the search for novel Actinobacteria in the marine environment. Recent data from culture-dependent studies have shown that indigenous marine Actinobacteria indeed exist in the oceans. These include members of the genera *Dietzia*, *Rhodococcus*, *Streptomyces*, *Salinispora*, *Marinophilus*, *Solwaraspora*, *Salinibacterium*, *Aeromicrobium marinum*, *Williamsia maris*, and *Verrucosispora* [10–12,14,37]. Among these, the most exciting finding is the discovery of the first obligate new marine Actinobacteria genus *Salinispora* (formerly known as *Salinospora*) and the demonstration of the widespread populations of this genus in ocean sediments by Fenical's research group [19,38]. Subsequently, *Salinispora* strains were also isolated from the Great Barrier Reef marine sponge *Pseudoceratina clavata* [39]. The formal description of *Salinispora*, with two types of species – *Salinispora tropica* and *Salinispora arenicola*, has recently been published [40]. Furthermore, Mincer *et al.* [38] have demonstrated that *Salinispora* strains are actively growing in some sediment samples indicating that these bacteria are metabolically active in the natural marine environment. In this context, Grossart *et al.* [41] have illustrated that actinomycetes account for ~10% of the bacteria colonizing marine organic aggregates (marine snow) [42] and that their antagonistic activity might be highly significant in maintaining their presence that affects the degradation and mineralization of organic matter. Therefore, Actinobacteria are active components of marine microbial communities. They form a stable, persistent population in various marine ecosystems. The discovery of numerous new marine actinomycetes taxa, their metabolic activity demonstrated in their natural environments, and their ability to form stable populations in different habitats clearly illustrate that indigenous marine Actinobacteria indeed exist in the oceans. Another important observation is that novel compounds with biological activities have been isolated from these marine Actinobacteria [13,14,37], indicating that marine actinomycetes are an important source for the discovery of novel secondary metabolites.

1.6 Role of Actinobacteria in the Marine Environment

Actinobacteria have a profound role in the marine environment. The degradation and turnover of various materials are a continuous process mediated by the action of a variety of microorganisms. There is a speculation that the increase or decrease of a particular enzyme-producing microorganism may indicate the concentration of natural substrate and conditions of the environment [43]. Actinobacteria are also reported to contribute to the breakdown and recycling of organic compounds [44].

1.7 Importance of Marine Actinobacteria

1.7.1 Antibiotics

Marine Actinobacteria constitute an important and potential source of novel bioactive compounds [45]. Since environmental conditions of the sea are extremely different from the terrestrial conditions, they produce different types of antibiotics. Several antibiotics have been isolated from marine Actinobacteria by many researchers [46–56]. These isolated antibiotics are entirely new and unique compared to those isolated from the terrestrial ones [57].

The discovery of new molecules from Actinobacteria marked an epoch in antibiotic research and subsequent developments in antibiotic chemotherapy. Since the discovery of streptomycin, a large number of antibiotics, including major therapeutic agents such as amino glycosides, chloramphenicol, tetracyclines, and macrolides, and more recently β-lactam cephamycin group, have been isolated from cultures of *Streptomyces* and *Streptoverticillium* (Atlas of Actinomycetes, The Society for Actinomycetes, Japan, 1997). As more new antibiotics were discovered, the chances of finding novel antimicrobial leads among conventional Actinobacteria dwindled. The focus of industrial screening has therefore moved to markers of less exploited genera of rare Actinobacteria such as *Actinomadura, Actinoplanes, Amycolatopsis, Dactylosporangium, Kibdelosporangium, Microbispora, Micromonospora, Planobispora, Streptosporangium*, and *Planomonospora* [58].

Screening of microorganisms for the production of novel antibiotics has been intensively pursued by scientists for many years as they are used in many fields, including agriculture, veterinary, and pharmaceutical industry. Actinobacteria have the capability to synthesize many different biologically active secondary metabolites: antibiotics, herbicides, pesticides, antiparasitic substances, and enzymes such as cellulase and xylanase that are used in waste treatment. Of these compounds, antibiotics predominate in therapeutic and commercial uses [59–64].

1.7.2 Melanins

Melanins are complex natural pigments, widely dispersed in animals, plants, and microorganisms. They have several biological functions, including photoprotection, thermoregulation, action as free radical sinks, cation chelators, and antibiotics. Plants and insects incorporate melanins as cell wall and cuticle strengtheners, respectively [65]. The function of melanin in microbes is believed to be associated with protection against environmental stress. For example, bacteria producing melanins are more resistant to antibiotics [66], and melanins in fungi are involved in fungal pathogenesis of plants [67]. In mammals, two types of melanin can be distinguished: a dark eumelanin and a yellow to red pheomelanin [65]. Eumelanin, the more ubiquitous mammalian melanin type, is found in different regions of the human body, including the skin, hair, eye, inner ear, and brain [68].

Marine Actinobacteria also synthesizes and excrete dark pigments, melanin or melanoid, which are considered to be useful criteria for taxonomical studies [69,70]. Melanin compounds are irregular, dark brown polymers that are produced by various microorganisms by the fermentative oxidation, and they have radioprotective and antioxidant properties that can effectively protect the living organisms from ultraviolet radiation. Melanins are frequently used in medicine, pharmacology, and cosmetics preparations [71].

Biosynthesis of melanin with tyrosinase transforms the tyrosine into L-DOPA (3,4-dihydroxy phenyl-L-alanine), which is further converted into dopachrome and autoxidized to indol-5,6-quinone. The latter is polymerized spontaneously into DOPA-melanin that gives a dark brown pigment until the further examination [72].

1.7.3 Enzymes

Marine Actinobacteria have a diverse range of enzyme activities and are capable of catalyzing various biochemical reactions [43].

1.7.3.1 α-Amylase

Amylases are enzymes that hydrolyze starch molecules to give diverse products, including dextrins and progressively smaller polymers composed of glucose units [73]. These enzymes are of great significance in the present-day biotechnology with applications in food, fermentation, textile, and paper industries [74]. Although amylases can be derived from several sources, including plants, animals, and microorganisms, microbial enzymes generally meet industrial demands. Today, a large number of microbial amylases are available commercially and they have almost completely replaced chemical hydrolysis of starch in starch processing industry [74].

The history of amylases begins with the discovery of first starch-degrading enzyme in 1811 by Kirchhoff. This was followed by several reports of digestive amylases and malt amylases. It was much later in 1930 that Ohlsson suggested the classification of starch digestive enzymes in malt as α- and β-amylases, according to the anomeric type of sugars produced by the enzyme reaction. α-Amylase (1,4-α-D-glucan glucanohydrolase, EC. 3.2.1.1) is a widely distributed secretary enzyme. α-Amylases of different origins have been extensively studied. Amylases can be divided into two categories: endoamylases and exoamylases. Endoamylases catalyze hydrolysis in a random manner in the interior of the starch molecule. This action causes the formation of linear and branched oligosaccharides of various chain lengths. Exoamylases hydrolyze from the nonreducing end, successively resulting in short end products. Today, a large number of enzymes are known that can hydrolyze starch molecules into different products, and a combined action of various enzymes is required to hydrolyze starch completely [75]. Occurrence of amylases in Actinobacteria is a characteristic commonly observed in *Streptomyces* [76], a genus that is considered a potential source of amylolytic enzymes.

1.7.3.2 **Proteases**

Proteases, also known as peptidyl–peptide hydrolases, are important industrial enzymes accounting for ~60% of all enzyme sales and are utilized extensively in a variety of industries, including detergents, meat tenderization, cheese-making, dehairing, baking, and brewery, in the production of digestive aids, and in the recovery of silver from photographic film. The use of these enzymes as detergent additives stimulated their commercial development and resulted in a considerable expansion of fundamental research into these enzymes [77]. In addition to detergents and food additives, alkaline proteases have substantial utilization in other industrial sectors such as leather, textile, organic synthesis, and wastewater treatment [78,79]. Consequently, alkaline proteases with novel properties have become the focus of recent researches. Alkaline proteases are generated by a wide range of organisms, including bacteria, Actinobacteria molds, yeasts, and mammalian tissues.

Several studies have been made on the proteolytic enzymes of mesophilic actinomycetes [80–84]. Recently, alkaline protease from *Nocardiopsis* sp. NCIM 5124 [85] has been purified and characterized.

1.7.3.3 **Cellulases**

Cellulose, the most abundant organic source of feed, fuel, and chemicals [86], consists of glucose units linked by β-1,4-glycosidic bonds in a linear mode. The difference in the type of bond and the highly ordered crystalline forms of the compound between starch and cellulose make cellulose more resistant to digest and hydrolyze. The enzymes required for the hydrolysis of cellulose include endoglucanases, exoglucanases, and β-glucosidases [87]. While endoglucanase randomly hydrolyzes cellulose, producing oligosaccharides, cellobiose, and glucose, the exoglucanase hydrolyzes β-1,4-D-glucosidic linkages in cellulose, releasing cellobiose from the nonreducing end. On the other hand, β-glycosidases of thermophilic origin, which have received renewed attention in the pharmaceutical industry, hydrolyze cellobiose to glucose.

In the current industrial processes, cellulolytic enzymes are employed in the color extraction from juices, detergents causing color brightening and softening, biostoning of jeans, pretreatment of biomass that contains cellulose to improve nutritional quality of forage, and pretreatment of industrial wastes [88–93]. To date, the alkaline- or alkali-tolerant cellulase producers have mainly been found in the genera *Streptomyces* and *Thermoactinomyces* [94].

1.7.3.4 Chitinase

Chitin, an insoluble linear β-1,4-linked polymer of *N*-acetylglucosamine (GlcNAc), is the second most abundant polymer in nature. This polysaccharide is found in the cell walls of fungi and exoskeleton of insects and the shells of crustaceans. Chitinases (EC 3.2.1.14) are produced by many organisms such as viruses, bacteria, Actinobacteria, and higher plants and animals and they play important physiological and ecological roles [95]. Chitinases hydrolyze the β-1,4 linkages in chitin, yielding predominantly *N*-*N'*-diacetylchitobiose that is further degraded by *N*-acetylglucosaminidases to the GlcNAc monomer [96].

Chitinase is involved in the process of producing mono- and oligosaccharides from chitin. Furthermore, chitinase is a potential antifungal agent because of its chitin degradation activity [97–100]. Among Actinobacteria, the genus *Streptomyces* is the best studied for chitinases [101,102] and is mainly responsible for the recycling of chitinous matter in nature.

Different chitinous substrates that have been reported in the literature for chitinase production include fungal cell walls, crab and shrimp shells, prawn wastes, and flake chitin [103–105]. Production of inexpensive chitinase will be important, if the use of chitinous wastes (shrimp shells, chitin from seafood industry, etc.) will solve environmental problems [106].

1.7.3.5 Keratinase

Keratinase is a specific protease, hydrolyzing keratin, which is a protein found in feathers, wool, and hair. Keratins as well as other insoluble proteins are generally not recognized as a substrate for common proteases. Hydrolysis of bacteria is however affected by specific proteases (keratinases) that have been found in some species of *Bacillus* [107,108], saprophytic and parasitic fungi [109], Actinobacteria [80,84], *Fervidobacterium pennavorans,* and some other microorganisms [110,111]. The microbial degradation of insoluble macromolecules such as cellulose, lignin, chitin, and keratin depends on the secretion of extracellular enzymes with the ability to act on compact substrate surfaces.

Feathers contain about 90% crude proteins in the form of β-keratin [112]. The poultry processing industry produces several millions of tonnes of feathers per year as by-products worldwide [113]. Hydrolysis of keratin-containing wastes by microorganisms possessing keratinolytic activity represents an attractive alternative method for efficient bioconversion and improving the nutritional value of keratin wastes, compared to currently used methods, through the development of economical and environment-friendly technologies [114].

Keratinolytic proteinases play an important role in biotechnological applications like enzymatic improvement of feather meal and production of amino acids or peptides from high molecular weight substrates or in the leather industry [115–118]. These enzymes, keratinases, could be applied for wastewater treatment, textile, medicine, cosmetic, and feed and poultry processing industries, as well as leather industry [119].

1.7.3.6 Xylanases

Xylan, which is the dominating component of hemicelluloses, is one of the most abundant organic substances on Earth. It has a great application in the pulp and paper industry [120–122]. The wood used for the production of the pulp is treated at high temperatures and basic pH, which implies that the enzymatic procedures require proteins exhibiting a high thermostability and activity in a broad pH range [123]. Treatment with xylanase at elevated temperatures disrupts the cell wall structure. As a result, this facilitates lignin removal in various stages of bleaching. For such purposes, (i) xylanases must lack cellulytic activity to avoid hydrolysis of the cellulose fibers and (ii) need to be of low molecular mass to facilitate their diffusion in the pulp fibers. Most importantly, high yields of enzyme must be obtained at a very low cost [89]. Alkaliphilic and cellulase-free xylanases with an optimum temperature of 65 °C from *Thermoactinomyces thalophilus* subgroup C [124] were also reported. Thermostable xylanase were isolated from a number of Actinobacteria [125]. *Streptomyces* sp. have been reported to produce xylanases that are active at temperatures between 50 and 80 °C [125].

1.7.4 Enzyme Inhibitors

Enzyme inhibitors have received increasing attention as useful tools, not only for the study of enzyme structures and reaction mechanisms but also for potential utilization in pharmacology [126]. Marine Actinobacteria are the potential source for production of enzyme inhibitors [127,128]. Imade [127] reported different types of enzyme inhibitors: β-glucosidase, *N*-acetyl-β-D-glucosaminidase, pyroglutamyl peptidase, and α-amylase inhibitors from marine Actinobacteria.

1.7.5 Anticancer Compounds

Cancer is a term that refers to a large group of over a hundred different diseases that arise when defects in physiological regulation cause unrestrained proliferation of abnormal cells [129]. In most cases, these clonal cells accumulate and multiply, forming tumors that may compress, invade, and destroy normal tissue, weakening the vital functions of the body with devastating consequences, including loss of quality of life and mortality. Nowadays, cancer is the second cause of death in the developed world, affecting one out of three individuals and resulting in one out of five deaths worldwide [129]. Diversified groups of marine Actinobacteria are known to produce different types of anticancer compounds. Several kinds of cytotoxic compounds have been reported from marine Actinobacteria [130–137]. The isolated compounds showed significant activity against different cancer cell lines.

1.8 Symbioses

The association of bacteria with marine sponges, bryozoans, tunicates, and holothurians has long been known, and sponge systems have attracted much attention [138–140]. Interest in such animals has been excited by their diversity of bioactive products that most probably are secondary metabolites of their bacterial partners. Actinobacteria are often components of these symbiotic communities, and because of their pedigree as natural product sources, they are increasingly targeted in biodiscovery programs. Actinobacteria are found in reef and deepwater sponges and evidence for sponge-specific symbioses exists [139]. In at least one case, Actinobacterial symbionts such as species of *Micromonospora* have been shown to produce bioactive compounds (manzamines) that have no terrestrial equivalents. Of considerable interest is the reported isolation of *Salinispora* strains (only known previously from marine sediments) from the Great Barrier Reef sponge *P. clavata*, and their activity against other bacterial symbionts [141]. These *Salinispora* isolates possess a polyketide synthase (PKS) gene that is most closely related to the rifamycin B synthase of *Amycolatopsis* [141], and hence might provide a novel marine source of this antibiotic. Based on the greatly conserved but nevertheless distinctive PKS genes for rifamycin found in these two actinomycete genera, the authors consider the system to be a propitious one for recombinant antibiotic synthesis. Reports of Actinobacterial symbionts of sponges have also appeared from Chinese groups. Sponges in the South China Sea harbor a large diversity of Actinobacteria and show evidence of host specificity [142]. The greatest Actinobacterial diversity was found in *Craniella australiensis*, with many of the strains having broad-spectrum antibacterial activities [143]. Similarly, Actinobacteria associated with the Yellow Sea sponge *Hymeniacidon perleve* showed broad taxonomic diversity and included *Actinoalloteichus, Micromonospora, Nocardia, Nocardiopsis, Pseudonocardia, Rhodococcus,* and *Streptomyces* [144]; the value of deploying a wide range of isolation media was again emphasized by this study. Rather less research has been focused on coral-associated Actinobacteria, but two recent reports have alerted our interest: First, a culture-independent study of the recently discovered deep water Mediterranean corals revealed several abundant bacterial phylotypes, one of which was the Actinobacteria [145]. Second, a culture-based study of the symbionts of *Fungia scutaria* [146], a Red Sea species, revealed a large proportion (23%) of Actinobacteria in the mucus layer. Although the cultivation efficiency was low, this was the first account of Actinobacteria being isolated from corals. Continued isolation and screening of coral-associated

Actinobacteria seem entirely warranted, given the early success in discovering valuable bioactive compounds such as thiocoraline [54].

1.9 Bioinformatics

Bioinformatics and its component "-omic" elements have created a paradigm shift in our approach to natural product discovery. Much of the relevant information is contained in Ref. [147], so here we refer only to recent developments that have implications for marine Actinobacteria. Taxonomic databases could provide predictive road maps to chemical diversity, and there is some support for such a relationship at coarse (order Actinomycetales), intermediate (family, e.g., Streptomycetaceae), and fine (genus, e.g., the *Streptomyces violaceusniger* clade) taxonomic ranks within the Actinobacteria; all members of the latter clade produce eliaophylin, geldanamycin, nigericin, and polyene. In some microbial groups, the relationship between taxonomy and the ability to synthesize particular types of natural product is stringent (e.g., in terverticillate penicillia) [148]. Although such patterns have not been demonstrated unequivocally in marine Actinobacteria, pursuit of the relationship is encouraged by recent chemodiversity analyses of the genus *Salinispora* [149], which returns us to the need for further charting of marine Actinobacterial phylogenetic or taxonomic space. Display of 16S rDNA phylogenetic distances in three-dimensional principal coordinate space illustrates dramatically the extensive regions of unexplored Actinobacterial taxonomy [10,150]. Recently discovered marine taxa *Marinospora*, *Verrucocispora maris*, and alkalitolerant *Streptomyces* occupy distinct new regions of phylogenetic space [150] and synthesize exciting new chemical entities (NCE). The prospect of massive sequencing of 16S rRNA and other diagnostic genes (e.g., by means of the 454 pyrosequencing platform) [151] will enable a more representative inventory of marine Actinobacteria to be achieved. Such capacity is crucial for discovering low-abundance marine organisms, including Actinobacteria. Elegant support for this approach has come from the Ref. [152].

1.10 Conclusions

The study of marine Actinobacteria is at an early stage, but the developments in molecular biology and genomics will greatly enhance our capacity to clarify their systematics, to understand their ecology and evolution, and to inform bioprospecting programs. Research programs will need to be focused at the levels of individuals, species, metapopulations, and those communities of which Actinobacteria are components. Furthermore, we have reiterated our belief that natural product search and discovery in marine Actinobacteria shows exceptional promise. Such optimism is based on the spectacular technological armamentarium that is now available and on a fuller, but slower, understanding of marine biology. Optimism is also encouraged by a wide range of natural products and diversity of their applications (biocatalysts, biomaterials, agrichemicals, etc.); however, in this chapter, the focus has been on bioactive metabolites.

Marine actinobacterial search and discovery is one thing, development of discoveries to end products is another. We conclude with a few reflections on this dilemma and although in this context they relate to antibiotics, almost identical arguments are apposite to orphan drugs in general and to "neglected" diseases. There has been much recent comments about the scarcity of new antibiotic entities, why their need has reached alarming proportions, and the reason for the withdrawal of many big pharmaceutical companies from this field. The analysis made by Projan [153] remains true, although there are encouraging signs of newer biotechnology companies filling the innovation gap and, in some cases, focusing on marine organisms. Ultimately, medical necessity as much as business opportunity could be the driver for investment in natural product drugs [154] and this, as Projan has cautioned, will call for urgent changes in public and social policy, and will come at some cost!

References

1 Blunt, J.W. *et al.* (2007) Marine natural products. *Nat. Prod. Rep.*, **24**, 31–86.

2 Ellaiah, P., Ramana, T., Bapi Raju, K.V. V.S.N., Sujatha, P., and Uma Sankar, A. (2004) Investigation on marine actinomycetes from Bay of Bengal near Kakinada coast of Andhra Pradesh. *Asian J. Microbiol. Biotechnol. Environ. Sci.*, **6**, 53–56.

3 Lechevalier, H. (1982) *The Development of Applied Microbiology at Rutgers*, The State University of New Jersey, p. 3.

4 Tanaka, Y. and Omura, O. (1990) Metabolisms and products of actinomycetes: an introduction. *Actinomycetologica*, **4**, 13–14.

5 Takaizawa, M., Colwell, W., and Hill, R.T. (1993) Isolation and diversity of actinomycetes in the Chesapeake Bay. *Appl. Environ. Microbiol.*, **59**, 997–1002.

6 Goodfellow, M. and O'Donnell, A.G. (1993) Roots of bacterial systematic, in *Handbook of New Bacterial Systematics* (eds M. Goodfellow, and A.G. O'Donnell), Academic Press, London, pp. 3–54.

7 Sivakumar, K., Sahu, M.K., Thangaradjou, T., and Kannan, L. (2007) Research on marine Actinobacteria in India. *Indian J. Microbiol.*, **47**, 186–196.

8 Donia, M. and Hamann, M.T. (2003) Marine natural products and their potential applications as anti-infective agents. *Lancet Infect. Dis.*, **3**, 338–348.

9 Haefner, B. (2003) Drugs from the deep: marine natural products as drug candidates. *Drug Discov. Today*, **8**, 536–544.

10 Bull, A.T., Stach, J.E.M., Ward, A.C., and Goodfellow, M. (2005) Marine Actinobacteria: perspectives, challenges, future directions. *Antonie Van Leeuwenhoek*, **87**, 65–79.

11 Stach, J.E.M., Maldonado, L.A., Ward, A. C., Goodfellow, M., and Bull, A.T. (2003) New primers for the class Actinobacteria: application to marine and terrestrial environments. *Environ. Microbiol.*, **5**, 828–841.

12 Jensen, P.R., Gontang, E., Mafnas, C., Mincer, T.J., and Fenical, W. (2005) Culturable marine actinomycete diversity from tropical Pacific Ocean sediments. *Environ. Microbiol.*, **7**, 1039–1048.

13 Fiedler, H.P., Bruntner, C., Bull, A.T., Ward, A.C., Goodfellow, M., Potterat, O., Puder, C., and Mihm, G. (2005) Marine actinomycetes as a source of novel secondary metabolites. *Antonie Van Leeuwenhoek*, **87**, 37–42.

14 Magarvey, N.A., Keller, J.M., Bernan, V., Dworkin, M., and Sherman, D.H. (2004) Isolation and characterization of novel marine-derived actinomycete taxa rich in bioactive metabolites. *Appl. Environ. Microbiol.*, **70**, 7520–7529.

15 Bull, A.T., Ward, A.C., and Goodfellow, M. (2000) Search and discovery strategies for biotechnology: the paradigm shift. *Microbiol. Mol. Biol. Rev.*, **64**, 573–606.

16 Cross, T. (1981) Aquatic actinomycetes: a critical survey of the occurrence, growth and role of actinomycetes in aquatic habitats. *J. Appl. Bacteriol.*, **50**, 397–423.

17 Goodfellow, M. and Haynes, J.A. (1984) Actinomycetes in marine sediments, in *Biological, Biochemical, and Biomedical Aspects of Actinomycetes* (eds L. Ortiz-Ortiz, L.F. Bojalil, and V. Yakoleff), Academic Press, New York, pp. 453–472.

18 Goodfellow, M. and Williams, S.T. (1983) Ecology of actinomycetes. *Annu. Rev. Microbiol.*, **37**, 189–216.

19 Mincer, T.J., Jensen, P.R., Kauffman, C.A., and Fenical, W. (2002) Widespread and persistent populations of a major new marine actinomycetes taxon in ocean sediments. *Appl. Environ. Microbiol.*, **68**, 5005–5011.

20 Stach, J.E.M., Maldonado, L.A., Ward, A.C., Bull, A.T., and Goodfellow, M. (2004) *Williamsia maris* sp. nov., a novel actinomycete isolated from the Sea of Japan. *Int. J. Syst. Evol. Microbiol.*, **54**, 191–194.

21 Sivakumar, K. (2001) Actinomycetes of an Indian mangrove (Pichavaram) environment: an inventory, PhD Thesis, Annamalai University, India, p. 91.

22 Tae, K.K., Garson, M.J., and Fuerst, J.A. (2005) Marine actinomycetes related to the '*Salinospora*' group from the Great Barrier Reef sponge *Pseudoceratina clavata*. *Environ. Microbiol.*, **7** (4), 509–518.

23 Colquhoun, J.A., Mexson, J., Goodfellow, M., Ward, A.C., Horikoshi, K., and Bull, A.T. (1998) Novel *Rhodococci* and other mycolata actinomycetes from the deep sea. *Antonie Van Leeuwenhoek*, **74**, 27–40.

24 Sahu, M.K., Murugan, M., Sivakumar, K., Thangaradjou, T., and Kannan, L. (2007) Occurrence and distribution of actinomycetes in marine environs and their antagonistic activity against bacteria that is pathogenic to shrimps. *Isr. J. Aquacult. Bamid.*, **59** (3), 155–161.

25 Okami, Y., Okazaki, T., Kitahara, T., and Umezawa, H. (1976) A new antibiotic apasmomycin, produced by a streptomycete isolated from shallow sea mud. *J. Anibiot.*, **28**, 176–184.

26 Hayakawa, M. and Nonomura, H. (1987a) Humic acid–vitamin agar, a new medium for the selective isolation of soil actinomycetes. *J. Ferment. Technol.*, **65**, 501–509.

27 Hayakawa, M. and Nonomura, H. (1987b) Efficacy of artificial humic acid as a selective nutrient in HV agar used for the isolation of soil actinomycetes. *J. Ferment. Technol.*, **65**, 609–616.

28 Nonomura, H. (1988) Isolation, taxonomy and ecology of soil actinomycetes. *Actinomycetologica*, **3**, 45–54.

29 Hayakawa, M., Sadaka, T., Kajiura, T., and Nonomura, H. (1991) New methods for the highly selective isolation of *Micromonospora* and *Micronbispora*. *J. Ferment. Technol.*, **72**, 320–326.

30 Cho, S.H., Hwang, C.W., Chung, H.K., and Yang, C.S. (1994) A new medium for the selective isolation of soil actinomycetes. *J. Appl. Microbiol. Biotechnol.*, **22**, 561–563.

31 Kim, C.J., Lee, K.H., Shimazu, A., Kwon, O.S., and Park, D.J. (1995) Isolation of rare actinomycetes from various types of soil. *J. Appl. Microbiol. Biotechnol.*, **23**, 36–42.

32 Nolan, R.D. and Cross, T. (1988) Isolation and screening of actinomycetes, in *Actinomycetes in Biotechnology* (eds M. Goodfellow, S.T. Willams, and M. Mordarski), Academic Press, San Diego, CA, pp. 1–32.

33 Bull, A.T. and Stach, J.E.M. (2007) Marine Actinobacteria: new opportunities for natural product search and discovery. *Trends Microbiol.*, **15** (11), 491–499.

34 Helmke, E. and Weyland, H. (1984) *Rhodococcus marinonascens* sp. nov., an actinomycete from the sea. *Int. J. Syst. Bacteriol.*, **34**, 127–138.

35 Jensen, P.R., Dwight, R., and Fenical, W. (1991) Distribution of actinomycetes in

near-shore tropical marine sediments. *Appl. Environ. Microbiol.*, **57**, 1102–1108.
36 Moran, M.A., Rutherford, L.T., and Hodson, R.E. (1995) Evidence for indigenous *Streptomyces* populations in a marine environment determined with a 16S rRNA probe. *Appl. Environ. Microbiol.*, **61**, 3695–3700.
37 Jensen, P.R., Mincer, T.J., Williams, P.G., and Fenical, W. (2005) Marine actinomycete diversity and natural product discovery. *Antonie Van Leeuwenhoek*, **87**, 43–48.
38 Mincer, T.J., Fenical, W., and Jensen, P.R. (2005) Culture-dependent and culture-independent diversity within the obligate marine actinomycete genus *Salinispora*. *Appl. Environ. Microbiol.*, **71**, 7019–7028.
39 Kim, T.K., Garson, M.J., and Fuerst, J.A. (2005) Marine actinomycetes related to the 'Salinospora' group from the Great Barrier Reef sponge *Pseudoceratina clavata*. *Environ. Microbiol.*, **7**, 509–518.
40 Maldonado, L.A., Fenical, W., Jensen, P.R., Kauffman, C.A., Mincer, T.J., Ward, A.C., Bull, A.T., and Goodfellow, M. (2005) *Salinispora arenicola* gen. nov., sp. nov. and *Salinispora tropica* sp. nov., obligate marine actinomycetes belonging to the family Micromonosporaceae. *Int. J Syst. Evol. Microbiol.*, **55**, 1759–1766.
41 Grossart, H.P., Schlingloff, A., Bernhard, M., Simon, M., and Brinkhoff, T. (2004) Antagonistic activity of bacteria isolated from organic aggregates of the German Wadden Sea. *FEMS Microbiol. Ecol.*, **47**, 387–396.
42 Alldrege, A.L. and Silver, M.W. (1988) Characteristics, dynamics and significance of marine snow. *Prog. Oceanogr.*, **20**, 41–82.
43 Das, S., Lyla, P.S., and Khan, S.A. (2006) Marine microbial diversity and ecology: importance and future perspectives. *Curr. Sci.*, **90**, 1325–1335.
44 Weyland, H. (1969) Actinomycetes in North Sea and Atlantic Ocean sediments. *Nature*, **223**, 858.
45 Colwell, R.R. and Hill, R.T. (1992) Microbial diversity, in *Diversity of Oceanic Life: An Evaluative Review* (ed. M.N.A. Peterson), The Centre for Strategic and International Studies, Washington, DC, pp. 100–106.
46 Bernan, V.S., Montenegro, D.A., Maiese, W.M., Steinberg, D.A., and Greenstein, M. (1994) Bioxalomycins, new antibiotics produced by the marine *Streptomyces* sp. LL-31F508: taxonomy and fermentation. *J. Antibiot.*, **47**, 1417–1424.
47 Biabani, M.A., Laatsch, H., Helmke, E., and Weyland, H. (1997) delta-Indomycinone: a new member of pluramycin class of antibiotics isolated from marine *Streptomyces* sp. *J. Antibiot.*, **50**, 874–877.
48 Woo, J.H., Kitamura, E., Myouga, H., and Kamei, Y. (2002) An antifungal protein from the marine bacterium *Streptomyces* sp. strain AP77 is specific for *Pythium porphyrae*, a causative agent of red rot disease in *Porphyra* spp. *Appl. Environ. Microbiol.*, **68**, 2666–2675.
49 Maskey, R.P., Helmke, E., Fiebig, H.H., and Laatsch, H. (2002) Parimycin: isolation and structure elucidation of a novel cytotoxic 2,3-dihydroquinizarin analogue of gamma-indomycinone from a marine streptomycete isolate. *J. Antbiot.*, **55**, 1031–1035.
50 Maskey, R.P., Helmke, E., and Laastsch, H. (2003) Himalomycin A and B: isolation and structure elucidation of new fridamycin type antibiotics from a marine *Streptomyces* isolate. *J. Antibiot.*, **56**, 942–949.
51 Charan, R.D., Schlingmann, G., Janso, J., Bernan, V., Feng, X., and Carter, G.T. (2004) Diazepinomicin, a new antimicrobial alkaloid from a marine *Micromonospora* sp. *J. Nat. Prod.*, **67**, 1431–1433.
52 Li, F., Maskev, R.P., Qin, S., Sattler, I., Fiebig, H.H., Maier, A., Zeeck, A., and Laatsch, H. (2005) Chinikomycins A and B: isolation, structure elucidation and biological activity of novel antibiotics from a marine *Streptomyces* sp. isolate M045. *J. Nat. Prod.*, **68**, 349–353.
53 Buchanan, G.O., Williams, P.G., Feling, R.H., Kauffman, C.A., Jensen, P.R., and Fenical, W. (2005) Sporolides A and B: structurally unprecedented halogenated macrolides from the marine actinomycete *Salinispora tropica*. *Org. Lett.*, **7**, 2731–2734.
54 Lombo, F., Velasco, A., de laCalle, F., Brana, A.F., Sanchez- Pulles, J.M.,

Mendez, C., and Salas, J.A. (2006) Deciphering the biosynthesis of the antitumor thiocoraline from a marine actinomycetes and its expression in two *Streptomyces* species. *ChemBioChem*, **7**, 366–376.
55 Adinarayana, G., Venkateshan, M.R., Bapiraju, V.V., Sujatha, P., Premkumar, J., Ellaiah, P., and Zeeck, A. (2006) Cytotoxic compounds from the marine actinobacterium. *Bioorg. Khim.*, **32**, 328–334.
56 Lam, S.K. (2006) Discovery of novel metabolites from marine actinomycetes. *Curr. Opin. Microbiol.*, **9**, 245–251.
57 Meiying, Z. and Zhicheng, Z. (1998) Identification of marine actinomycetes S-216 strain and its biosynthetic conditions of antifungal antibiotic. *J. Xiamen Univ. Nat. Sci.*, **37**, 109–114.
58 Lazzarini, A., Cavaletti, L., Toppo, G., and Marinelli, F. (2000) Rare genera of actinomycetes as potential producers of new antibiotics. *Antonie Van Leeuwenhoek*, **78**, 399–405.
59 Lacey, J. (1973) *Actinomycetales: Characteristics and Practical Importance, Society for Applied Bacteriology Symposium Series*, vol. 2 (eds G. Sykes and F. Skinner), Academic Press, London.
60 McCarthy, A.J. and Williams, S.T. (1990) Methods for studying the ecology of Actinomycetes, in *Methods in Microbiology*, vol. 22 (eds R. Grigorova and J.R. Norris), Academic Press, London, pp. 533–363.
61 Ouhdouch, Y., Barakate, M., and Finanse, C. (2001) Actinomycetes of Moroccan habitats: isolation and screening for antifungal activities. *Eur. J. Soil Biol.*, **37**, 69–74.
62 Saadoun, I. and Gharaibeh, R. (2003) The *Streptomyces* flora of Badia region of Jordan and its potential as a source of antibiotics active against antibiotic-reistant bacteria. *J. Arid Environ.*, **53**, 365–371.
63 Waksman, S.A. (1961) *The Actinomycetes, Classification, Identification and Description of Genera and Species*, vol. 2, The Williams and Wilkins Company, Baltimore, pp. 61–292.
64 Oskay, M., Tamer, A.U., and Azeri, C. (2004) Antibacterial activity of some actinomycetes isolated from farming soils of Turkey. *Afr. J. Biotechnol.*, **3** (9), 441–446.
65 Riley, P.A. (1997) Melanin. *Int. J. Biochem. Cell Biol.*, **29**, 1235–1239.
66 Lin, W.P., Lai, H.L., Liu, Y.L., Chiung, Y.M., Shiau, C.Y., Han, J.M., Yang, C.M., and Liu, Y.T. (2005) Effect of melanin produced by a recombinant *Escherichia coli* on antibacterial activity of antibiotics. *J. Microbiol. Immunol. Infect.*, **38**, 320–326.
67 Butler, M.J. and Day, A.W. (1998) Fungal melanins: a review. *Can. J. Microbiol.*, **44**, 1115–1136.
68 Clancy, C.M.R. and Simon, J.D. (2001) Ultrastructural organization of eumelanin from *Sepia officinalis* measured by atomic force microscopy. *Biochemistry*, **40**, 13353–13360.
69 Zonova, G.M. (1965) Melanoid pigments of Actinomycetes. *Mikrobiologiya*, **34**, 278–283.
70 Arai, T. and Mikami, Y. (1972) Choromogenecity of *Streptomyces*. *Appl. Microbiol.*, **23**, 402–406.
71 Dastager, S.G., Li, W.J., Dayanand, A., Tang, S.K., Tian, X.P., Zhi, X.Y., Xu, L.H., and Jiang, C.L. (2006) Separation, identification and analysis of pigment (melanin) production in *Streptomyces*. *Afr. J. Biotechnol.*, **5** (8), 1131–1134.
72 Mencher, J.R. and Heim, A.H. (1962) Melanin biosynthesis by *Streptomyces lavendulae*. *J. Gen. Microbiol.*, **28**, 665–670.
73 Windish, W.W. and Mhatre, N.S. (1965) Microbial amylases, in *Advances in Applied Microbiology*, vol. 7 (ed. W.U. Wayne), Academic Press, New York, pp 273–304.
74 Pandey, A., Shukla, A., and Majumdar, S.K. (2005) Utilization of carbon and nitrogen sources by *Streptomyces kanamyceticus* M 27 for the production of an antibacterial antibiotic. *Afr. J. Biotechnol.*, **4** (9), 909–910.
75 Gupta, R., Gigras, P., Mohapatra, H., Goswami, V.K., and Chauhan, B. (2003) Microbial α-amylases: a biotechnological perspective. *Process Biochem.*, **38**, 1599–1616.
76 Vigal, T., Gil, J.F., Daza, A., Garcia-Gonzalez, M.D., and Martin, J.F. (1991) Cloning characterization and expression of an alpha amylase gene from *Streptomyces griseus* IMRU 3570. *Mol. Gen. Genet.*, **225**, 278–288.

77 Germano, S., Pandey, A., Osaku, C.A., Rocha, S.N., and Soccol, C.R. (2003) Characterization and stability of protease from *Penicillium* sp. produced by solid-state fermentation. *Enzyme Microb. Technol.*, **32**, 246–251.
78 Kalisz, H.M. (1988) Microbial proteinases. *Adv. Biochem. Eng. Biotechnol.*, **36**, 1–65.
79 Kumar, C.G. and Takagi, H. (1999) Microbial alkaline proteases from a bioindustrial viewpoint. *Biotechnol. Adv.*, **17**, 561–594.
80 Noval, J.J. and Nickerson, W.J. (1958) Decomposition of native keratin by *Streptomyces fradiae. J. Bacteriol.*, **77**, 251–263.
81 Nakanishi, T., Matsumura, Y., Minamiura, N., and Yamamoto, T. (1973) Purification and some properties of an alkalophilic proteinase of a *Streptomyces* species. *Agric. Biol. Chem.*, **38**, 37–44.
82 Moormann, M., Schlochtermeier, A., and Schrempf, H. (1993) Biochemical characterization of a protease involved in the processing of a *Streptomyces reticuli* cellulase (Avicellase). *Appl. Environ. Microbiol.*, **59**, 1573–1578.
83 Abd El-Nasser, N.H. (1995) Proteolytic activities by some mesophilic *Streptomyces* species. *J. Agric. Sci.*, **20**, 2913–2923.
84 Bockle, B., Galunsky, B., and Muller, R. (1995) Characterization of a keratinolytic serine proteinase from *Streptomyces pactum* DSM 40530. *Appl. Environ. Microbiol.*, **61**, 3705–3710.
85 Dixit, V.S. and Pant, A. (2000) Comparative characterization of two serine endopeptidases from *Nocardiopsis* sp. NCIM 5124. *Biochim. Biophys. Acta*, **1523**, 261–268.
86 Spano, L., Medeiros, J., and Mandels, M. (1975) Enzymatic hydrolysis of cellulosic waste to glucose. *Resour. Recover. Conserv.*, **1**, 279–294.
87 Matsui, I., Sakai, Y., Matsui, E., Kikuchi, H., Kawarabayasi, Y., and Honda, K. (2000) Novel substrate specificity of a membrane-bound β-glycosidase from the hyperthermophilic archaeon *Pyrococcus horikoshi. FEBS Lett.*, **467**, 195–200.
88 Buchert, J., Pere, J., Oijusluoma, L., Rahkamo, L., and Viikari, L. (1997) Cellulases tools for modification of cellulosic materials, in *Niches in the World of Textiles* (eds J. Buchert, J. Pere, L. Oijusluoma, L. Rahkamo, and L. Viikari), World conference of the Textile Institute, Manchester, England, pp. 284–290.
89 Niehaus, F., Bertoldo, C., Kahler, M., and Antranikian, G. (1999) Extremophiles as a source of novel enzymes for industrial applications. *Appl. Microbiol. Biotechnol.*, **51**, 711–729.
90 Bhat, M. (2000) Cellulases and related enzymes in biotechnology. *Biotechnol. Adv.*, **18**, 355–383.
91 Nakamura, H., Kubota, H., Kono, T., Isogai, A., and Onabe, F. (2001) Modification of pulp properties by cellulase treatment and application of cellulase to wastepaper deinking and mechanical pulp refining. Proceedings of the 68th Pulp and Paper Research Conference, 18-19/06, Japan, pp. 2–5.
92 Vanwyk, J., Mogale, A., and Seseng, T. (2001) Bioconversion of wastepaper to sugars by cellulase from *Aspergillus niger, Trichoderma viride* and *Penicillium funiculosum. J. Solid Waste Technol. Manage.*, **27**, 82–86.
93 Zhou, L., Yeung, K., and Yuen, C. (2001) Combined cellulase and wrinkle free treatment on cotton fabric. *J. Dong Hua Univ.*, **18**, 11–15.
94 Techapun, C., Poosaran, N., Watanabe, M., and Sasaki, K. (2003) Thermostable and alkaline-tolerant microbial cellulase free xylanases produced from agricultural wastes and the properties required for use in pulp bleaching bioprocess: a review. *Process Biochem.*, **38**, 1327–1340.
95 Gooday, G.W. (1990) The ecology of chitin decomposition. *Adv. Microb. Ecol.*, **11**, 387–430.
96 Tsujibo, H., Kubota, T., Yamamoto, M., Miyamoto, K., and Inamori, Y. (2003) Characterization of chitinase genes from an alkaliphilic actinomycete, *Nocardiopsis prasina* OPC-131. *Appl. Environ. Microbiol.*, **69** (2), 894–900.
97 Kunz, C., Ludwig, A., Bertheau, Y., and Boller, T. (1992) Evaluation of the antifungal activity of the purified chitinase I from the filamentous fungus *Aphanocladium album. FEMS Microbiol. Lett.*, **90**, 105–109.
98 Mathivanan, N., Kabilan, V., and Murugesan, K. (1998) Purification,

characterization, and antifungal activity of chitinase from *Fusarium chlamydosporum*, a mycoparasite to groundnut rust, *Puccina arachidis*. *Can. J. Microbiol.*, **44**, 646–651.

99 Ordenlich, A., Elad, Y., and Chet, I. (1988) The role of chitinase of *Serratia marcescens* in biocontrol of *Sclerotium rolfsii*. *Phytophathology*, **78**, 84–88.

100 Roberts, W.K. and Selitrennikoff, C.P. (1988) Plant and bacterial chitinases differ in antifungal activity. *J. Gen. Microbiol.*, **134**, 169–176.

101 Robbins, P.W., Albright, C., and Benfield, B. (1988) Cloning and expression of a *Streptomyces plicatus* chitinase (Chitinase-63) in *Escherichia coli*. *J. Biol. Chem.*, **263**, 443–447.

102 Miyashita, K., Fujii, T., and Sawada, Y. (1991) Molecular cloning and characterization of chitinase genes from *Streptomyces lividans* 66. *J. Gen. Microbiol.*, **137**, 2065–2072.

103 Dahiya, N., Tewari, R., Tiwari, R.P., and Hoondal, G.S. (2005) Chitinase from *Enterobacter* sp. NRG4: its purification, characterization and reaction pattern. *Electron. J. Biotechnol.*, **8** (2), 134–145.

104 Chang, W.T., Chen, C.S., and Wang, S.L. (2003) An antifungal chitinase produced by *Bacillus cereus* with shrimp and crab shell powder as carbon source. *Curr. Microbiol.*, **47**, 102–108.

105 Singh, G., Dahiya, N., and Hoondal, G.S. (2005) Optimization of chitinase production by *Serratia marcescens* GG5. *Asian J. Microbiol. Biotechnol. Environ. Sci.*, **7**, 383–385.

106 Mukherjee, G. and Sen, S.K. (2006) Characterization and identification of chitinase producing *Streptomyces venezuelae* P10. *Indian J. Exp. Biol.*, **42**, 541–544.

107 Williams, C.M., Richter, C.S., MacKenzie, J.M., and Shih, J.C.H. (1990) Isolation, identification, and characterization of a feather-degrading bacterium. *Appl. Environ. Microbiol.*, **56**, 1509–1515.

108 Kim, J.M., Lim, W.J., and Suh, H.J. (2001) Feather-degrading *Bacillus* species from poultry waste. *Process Biochem.*, **37**, 287–291.

109 Dozie, I.N.S., Okeke, C.N., and Unaeze, N.C. (1994) A thermostable alkaline active keratinolytic proteinase from *Crysosporium keratinophylum*. *World J. Microbiol. Biotechnol.*, **10**, 563–567.

110 Wawrzkiewicz, K., Wolsky, T., and Lobarzewsky, J. (1991) Screening the keratinolytic activity of dermatophytes *in vitro*. *Mycopathologia*, **14**, 1–8.

111 Friedrich, A.B. and Antranikian, G. (1996) Keratin degradation by *Fervidobacterium pennavorans*, a novel thermophilic anaerobic species of the order Thermatogales. *Appl. Environ. Microbiol.*, **62**, 2875–2882.

112 Lee, H., Suh, D.B., Hwang, J.H., and Suh, H.J. (2002) Characterization of a keratinolytic metalloprotease from *Bacillus* sp. SCB-3. *Appl. Biochem. Biotechnol.*, **97**, 123–133.

113 Williams, C.M., Lee, C.G., Garlich, J.D., and Shih, J.C.H. (1991) Evaluation of bacterial feather fermentation product, feather-lysate as a feed protein. *Poult. Sci.*, **70**, 85–94.

114 Bertsch, A. and Coello, N. (2005) A biotechnological process for treatment and recycling poultry feathers as a feed ingredient. *Biores. Technol.*, **96**, 1703–1708.

115 Chandrasekaran, S. and Dhar, S.C. (1986) Utilization of a multiple proteinase concentrate to improve the nutritive value of chicken feather meal. *J. Leather Res.*, **4**, 23–30.

116 Dhar, S.C. and Sreenivasulu, S. (1984) Studies on the use of dehairing enzyme for its suitability in the preparation of improved animal feed. *Leather Sci.*, **31**, 261–267.

117 Mukhopadhyay, R.P. and Chandra, A.L. (1990) Keratinase of a *Streptomycete*. *Indian J. Exp. Biol.*, **28**, 575–577.

118 Pfleiderer, E. and Reiner, R. (1988) Microorganisms in processing of leather, in *Biotechnology, Special Microbial Processes*, vol. 6b (ed. H.J. Rehm), Wiley-VCH Verlag GmbH, Weinheim, Germany, pp. 730–739.

119 Mukhopadhyay, R.P. and Chandra, A.L. (1993) Protease of keratinolytic *Streptomycetes* to unhair goat skin. *Indian J. Exp. Biol.*, **31**, 557–558.

120 Dekker, R. and Linder, W. (1979) Bioconversion of hemicellulose: aspects of hemicellulase production by *Trichoderma reesi* QM9414 and enzymic

saccharification of hemicellulose. *S. Afr. J. Sci.*, **75**, 65–71.

121 Chen, C., Adolphson, R., Dean, F., Eriksson, K., Adamas, M., and Westpheling, J. (1997) Release of lignin from kraft pulp by a hyper thermophilic xylanase from *Thermotoga maritima*. *Enzyme Microb. Technol.*, **20**, 39–45.

122 (a) Lee, Y., Lowe, S., and Zeikus, J. (1998) Molecular biology and biochemistry of xylan degradation by thermoanaerobes. *Prog. Biotechnol*, **7**, 275–288; (b) Kalegoris, E., Christakopoulos, D., Kekos, D., Macris, B. (1998) Studies on the solid-state production of thermostable endoxylanases from *Thermoascus aurantiacus*: characterization of two isoenzymes. *J. Biotechnol.*, **60**, 155–163.

123 Jacques, G., Frederic, D.L., Joste, L.B., Viviane, B., Bart, D., Fabrizio, G., Benoit, G., and Jean-marie, F. (2000) An additional aromatic interaction improves the thermostability and thermophilicity of a mesophilic family 11 xylanase: structural basis and molecular study. *Protein Sci.*, **9**, 466–475.

124 Kohilu, U., Nigam, P., Singh, D., and Chaudhary, K. (2001) Thermostable, alkaliphilic and cellulase free xylanases production by *Thermoactinomyces thalophilus* subgroup C. *Enzyme Microb. Technol.*, **28**, 606–610.

125 Bode, W. and Huber, R. (1992) Natural protein proteinase inhibitors and their interaction with proteinases. *Eur. J. Biochem.*, **204**, 433–451.

126 Imade, C. (2004) Enzyme inhibitors of marine microbial origin with pharmaceutical importance. *Mar. Biotechnol.*, **6**, 193–198.

127 Imade, C. (2005) Enzyme inhibitors and other bioactive compounds from marine actinomycetes. *Antonie Van Leeuwenhoek*, **587**, 59–63.

128 Fernandez, L.F.G., Reyes, F., and Puelles, J.M.S. (2002) The marine pharmacy: new antitumor compounds from the sea. *Pharmaceut. News*, **9**, 495–501.

129 Capon, R.J., Skene, C., Lacey, E., Gill, J.H., Wicker, J., Heiland, K., and Friedel, T. (2000) Lorneamides A and B: two new aromatic amides from a southern Australian marine actinomycete. *J. Nat. Prod.*, **63**, 1682–1683.

130 Maskey, R.P., Halmke, E., Kayser, O., Fiebig, H.H., Maier, A., Busche, A., and Laatsch, H. (2004) Anticancer and antibacterial trioxacarcins with antimalaria activity from a marine streptomycete and their absolute stereochemistry. *J. Nat. Prod.*, **57**, 771–779.

131 Georis, J., Giannotta, F., DeBuyl, E., Granier, B., and Frere, J. (2000) Purification and properties of three *endo*-beta-1,4-xylanases produced by *Streptomyces* sp. strain S38 which differ in their capacity to enhance the bleaching of kraft pulp. *Enzyme Microb. Technol.*, **26**, 177–183.

132 Stritzke, K., Schulz, S., Laatsch, H., Helmke, E., and Beil, W. (2004) Novel caprolactones from a marine streptomycete. *J. Nat. Prod.*, **67**, 395–401.

133 Lang, S., Beli, W., Tokuda, H., Wicke, C., and Lurtz, V. (2004) Improved production of bioactive glycosylmannosyl-glycerolipid by sponge-associated *Microbacterium* species. *Mar Biotechnol.*, **6**, 152–156.

134 Liu, R., Cui, C.B., Duan, L., Gu, Q.Q., and Zhu, W.M. (2005) Potent *in vitro* anticancer activity of metacycloprodigiosin and undecyprodigiosin from a sponge-derived actinomycete *Saccharopolyspora* sp. nov. *Arch. Pharm. Res.*, **28**, 1641–1344.

135 Manam, R.R., Teisan, S., White, D.J., Nicholson, B., Grodberg, J., Neuteboom, S.T., Lam, K.S., Mosca, D.A., Lloyd, G.K., and Potts, B.C. (2005) Lajollamycin, a nitro-tetraene spiro-beta-lactonegamma-lactam antibiotic from the marine actinomycetes *Streptomyces nodosus*. *J. Nat. Prod.*, **68**, 240–243.

136 Soria-Mercado, I.E., Prieto-Davo, A., Jensen, P.R., and Fenical, W. (2005) Antibiotic terpenoid chloro-dihydroquinones from a new marine actinomycete. *J. Nat. Prod.*, **68**, 904–910.

137 Jeong, S.Y., Shin, H.J., Kim, T.S., Lee, H.S., Park, S.K., and Kim, H.M. (2006) Streptokordin, a new cytotoxic compound of the methylpyridine class from a marine-derived *Streptomyces* sp. KORDI-3238. *J. Antibiot. (Tokyo)*, **59**, 234–240.

138 Montalvo, N.F. *et al.* (2005) Novel Actinobacteria from marine sponges. *Antonie Van Leeuwenhoek*, **87**, 229–236.

139 Hentschel, U. *et al.* (2006) Marine sponges as microbial fermenters. *FEMS Microb. Ecol.*, **55**, 167–177.

140 Piel, J. (2006) Bacterial symbionts: prospects for the sustainable production of invertebrate-derived pharmaceuticals. *Curr. Med. Chem.*, **13**, 39–50.

141 Kim, T.K. *et al.* (2006) Discovery of a new source of rifamycin antibiotics in marine sponge Actinobacteria by phylogenetic prediction. *Appl. Environ. Microbiol.*, **72**, 2118–2125.

142 Li, Z.Y. *et al.* (2006) Bacterial community diversity associated with four marine sponges from the south China Sea based on 16S rDNA-DGGE fingerprinting. *J. Exp. Mar. Biol. Ecol.*, **329**, 75–85.

143 Li, Z.Y. and Liu, Y. (2006) Marine sponge *Craniella australiensis* associated bacterial diversity revelation based on 16S rDNA library and biologically active actinomycetes screening, phylogenetic analysis. *Lett. Appl. Microbiol.*, **43**, 410–416.

144 Zhang, H. *et al.* (2006) Culturable Actinobacteria from the marine sponge *Hymeniacidon perleve*: isolation and phylogenetic diversity by 16S rRNA gene-RFLP analysis. *Antonie Van Leeuwenhoek*, **90**, 159–169.

145 Yakimov, M.M. *et al.* (2006) Phylogenetic survey of metabolically active communities associated with the deep-sea coral *Lophelia pertusa* from the *Apulian plateau*, Central Mediterranean Sea. *Deep-sea Res.*, **53**, 62–75.

146 Lampert, Y. *et al.* (2006) Diversity of culturable bacteria in the mucus of the Red Sea coral *Fungia scutaria*. *FEMS Microb. Ecol.*, **58**, 99–108.

147 Bull, A.T. (ed.) (2004) *Microbial Diversity and Bioprospecting*, ASM Press.

148 Smedsgaard, J. and Nielsen, J. (2005) Metabolite profiling of fungi and yeast: from phenotype to metabolome by MS and informatics. *J. Exp. Bot.*, **56**, 273–286.

149 Fenical, W. and Jensen, P.R. (2006) Developing a new resource for drug discovery: marine actinomycete bacteria. *Nat. Chem. Biol.*, **2**, 666–673.

150 Ward, A.C. and Bora, N. (2006) Diversity and biogeography of marine Actinobacteria. *Curr. Opin. Microbiol.*, **9**, 1–8.

151 Goldberg, S.M.D. *et al.* (2006) A Sanger/pyrosequencing hybrid approach to the generation of high-quality draft assemblies of marine microbial genomes. *Proc. Natl. Acad. Sci. USA*, **103**, 11240–11245.

152 Sogin, M.L. *et al.* (2006) Microbial diversity in the deep sea and the underexplored "rare biosphere". *Proc. Natl. Acad. Sci. USA*, **103**, 12115–12120.

153 Projan, S.J. (2003) Why is big pharma getting out of antibacterial drug discovery? *Curr. Opin. Microbiol.*, **6**, 427–430.

154 Fox, J.L. (2006) The business of developing antibacterials. *Nat. Biotechnol.*, **24**, 1521–1528.

2
Treasure Hunting for Useful Microorganisms in the Marine Environment

Chiaki Imada

2.1
Introduction

Since the discovery of penicillin, much attention has been given to natural products from terrestrial microbial metabolites. However, research concerning novel beneficial microorganisms on the terrestrial environment has become more and more difficult in recent years as different kinds of microorganisms have been found to produce the same or similar compounds with respect to chemical structure.

Compared to the terrestrial environment, the marine environment has several unique characteristics: high salinity, high hydrostatic pressure, low temperature, and low concentrations of organic matter. The microorganisms living in marine environments are metabolically and physiologically different from the terrestrial ones. More than 90% of marine bacteria are Gram-negative psychrophiles, and nearly half of them require high concentrations of NaCl for their growth. To date, marine microorganisms have been isolated from a neritic environment. A number of natural products from marine microorganisms have recently been reported, and many of these compounds have shown interesting chemical structures and bioactivities.

This chapter aims to introduce useful marine microorganisms that produce various bioactive compounds.

2.2
Microorganisms Living in the Marine Environment

2.2.1
Protease Inhibitor Produced by Marine Bacterium

Protease inhibitors (PIs) have been widely used for the treatment of various diseases caused by abnormally active protease [1]. To date, they have been isolated from terrestrial organisms, although they are important in the field of agriculture and

Marine Microbiology: Bioactive Compounds and Biotechnological Applications, First Edition.
Edited by Se-Kwon Kim

Table 2.1 Comparison of various characteristics of protease inhibitors marinostatins and monastatin [4].

Properties	Marinostatin C-1, C-2*	Monastatin
Structure	Simple peptide	Glycoprotein
Molecular weight	1418, 1644	20 000 (approx.)
pH stability	4–8	2–12
Inhibited protease	Serine protease (excluding trypsin)	Cysteine protease (including fish disease protease)

C-1: Phe—Ala—Thr—Met—Arg—Tyr—Pro—Ser—Asp—Ser—Asp—Glu.
*C-2: Gln—Pro—Phe—Ala—Thr—Met—Arg—Tyr—Pro—Ser—Asp—Ser—Asp—Glu.

fisheries [2]. Marine bacterium producing PIs were herein isolated from neritic seawater of Japan and characterized. The inhibitor-producing strain was a novel bacterium *Pseudoalteromonas sagamiensis* [3]. The strain produced PI only in the presence of seawater. As shown in Table 2.1, the strain simultaneously produced two types of novel PIs whose characteristics are considerably different (simple peptides named marinostatin and glycoprotein named monastatin) [4]. Recent studies on the amino acid sequences of naturally occurring protease inhibitors revealed that extensive homologies in inhibitors were obtained from species of very different phylogenetic origin, that is, mammals and microorganisms [5]. However, marinostatins show no similarity to the known amino acid sequences of these terrestrial inhibitors [6], and they have a unique primary structure in the sequences [7]. As shown in Figure 2.1, monastatin exhibited inhibitory activity against crude protease of sardines. Therefore, it is applicable to the food industry as the inhibitor increases the breaking strength and elasticity of cooked fish meat gel Kamaboko (Table 2.2) [8].

2.2.2
Chitinase Inhibitor Produced by Marine Bacterium

Chitin is the earth's second most abundant biomass after cellulose, and a large quantity is found in the marine environment. Some strains of marine microorganisms have the ability to degrade chitin by chitinase and use it for their growth. However, no report is yet available concerning chitinase inhibitor-producing marine microorganisms. The inhibitor is expected to be applicable to pesticides because it has shown inhibition in insects' ecdysis and growth [9]. Several

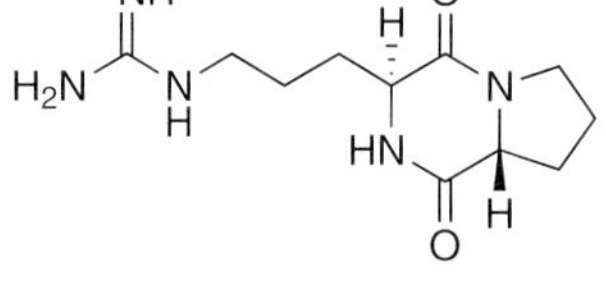

Figure 2.1 Chemical structure of chitinase inhibitor 2,4-dibromo-6-(3,4,5-tribromo-2-pyrrolyl)-1,3-cyclohexadiene-1-ol.

Table 2.2 Inhibitory activity of marinostatin and monastatin against protease on minced spot lined sardines [8].

Inhibitor concentration (%)	Breaking strength (g)	Breaking strain (mm)
Marinostatin		
0	247	7.2
0.075	238	7.0
0.150	227	6.6
0.225	229	6.8
0.300	220	6.6
Monostatin		
0	206	6.6
0.04	287	7.7
0.08	369	8.4
0.12	352	8.3
0.16	368	8.4

inhibitors have been isolated from terrestrial microorganisms [10]. However, no inhibitors have been isolated from marine microorganisms. An attempt was made to isolate a novel inhibitor-producing strain from the marine environment. From about 300 marine bacterial strains, an active strain was isolated from seawater of the coastal area in Shizuoka Prefecture, Japan [11]. The strain was identified as *Pseudomonas* sp. and produced a novel peptide inhibitor CI-4 (Figure 2.2). The substance exhibited no antimicrobial activity or cytotoxicity of various cells [12].

2.2.3
Antibiotic Produced by Marine Bacteria

In 1966, the first antibiotic of marine microbial origin was isolated [13]. The compound has a unique chemical structure containing bromine atoms in the molecule, 2,4-dibromo-6-(3,4,5-tribromo-2-pyrrolyl)-1,3-cyclohexadiene-1-ol. The structure was never seen in the products of terrestrial origin (Figure 2.1). The antibiotic-producing strain was identified as *Pseudomonas bromoutilis.*

As shown in Figure 2.3, marine bacterium isolated from an alga in neritic seawater of the Republic of Belau was found to produce a novel antimicrobial substance korormicin [14]. Korormicin has inhibitory activity against all Gram-negative marine bacteria, whereas no activity is observed against Gram-negative terrestrial bacteria, such as *Escherichia coli,* or all Gram-positive bacteria (Table 2.3).

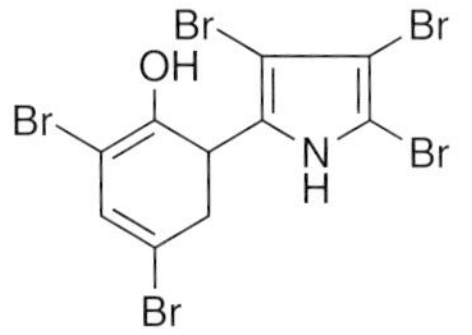

Figure 2.2 Chemical structure of antibiotic.

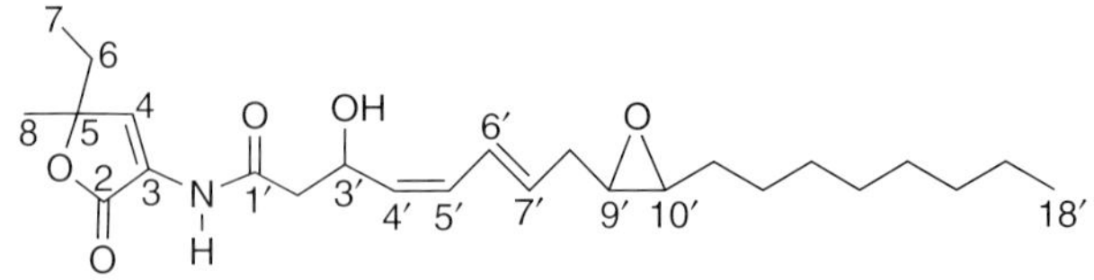

Figure 2.3 Chemical structure of antibiotic korormicin.

Since 90% of the marine bacteria are Gram-negative, the antibiotic had specific inhibitory activity against them.

2.2.4 Antibiotics Produced by Marine Actinomycetes

Most actinomycetes living in the marine environment are considered to be of terrestrial origin because they do not require seawater. *Streptomyces griseus* is an

Table 2.3 Antimicrobial activity spectra of korormicin isolated from marine bacteria.

Bacteria tested	NaCl requirement	Korormicin (μg/ml)					Polymyxin B (μg/ml)
		1000	200	40	10	2	100
Alteromonas macleodii ATCC27126	+	20[a]	17	11	—	—	10
Pseudoalteromonas haloplanktis ATCC14393	+	22	18	15	13	—	08
Halomonas aquamarina ATCC33127	+	17	14	12	9	—	9
Pseudomonas nautical ATCC27132	+	21	18	14	9	—	10
Shewanella putrefaciens ATCC8071	+	26	19	9	—	—	12
Vibrio alginolyticus ATCC17749	+	17	15	—	—	—	8.5
Salinivibrio costicola ATCC33508	+	36	30	24	16	—	12
Pelagiobacter variabilis Ni-2088	+	15	13	9	—	—	9
Oceanospirillum beijerinckii ATCC12754	+	32	25	18	—	—[a]	Low activity
Photobacterium phosphoreum IAM* 12085	+	22	17	14	—	—	Low activity
Marinomonas communis ATCC27118	+	14	12	—	—	—	16
E. coli IFO3301	—	—	—	—	—	—	12

a) Circle diameter (mm) of growth inhibition measured by paper disk method.

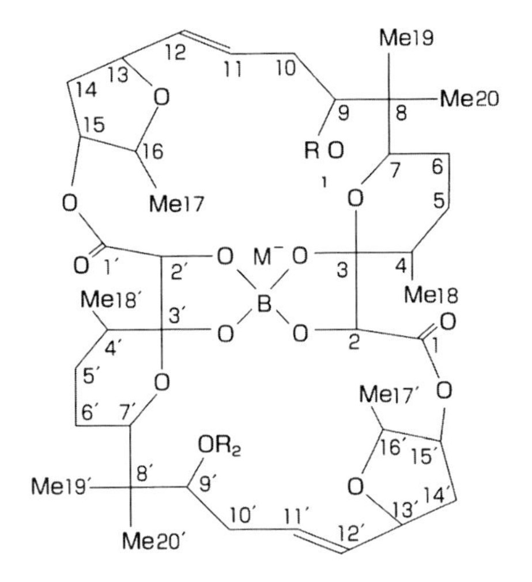

Me : CH_3
I Aplasmomycin $R_1,R_2=H$
II Aplasmomycin B $R_1=Ac$, $R_2=H$
III Aplasmomycin C $R_1,R_2=Ac$

Figure 2.4 Chemical structure of antibiotic aplasmomycin.

actinomycete known for producing streptomycin (SM); however, one strain of *S. griseus* that was isolated from the shallow sea sediment at the Koajiro Inlet in Sagami Bay, Japan did not produce SM in normal culture medium without seawater. When the strain was cultivated in a medium containing seawater and a low concentration of nutrients, it produced a novel antibiotic that has a chemical structure completely different from that of SM [15]. This antibiotic had a boron-containing polyether ionophore structure (Figure 2.4). The compound has anti-plasmodial activity against *Plasmodium berghei* and therefore named aplasmomycin.

Another novel antibiotic SS-228Y was produced by *Chainia* sp. isolated from shallow sea sediment [16] in a medium containing seawater and commercial seaweed powder Kobu-cha from *Laminaria* sp. [17].

Novel enzyme inhibitors such as *N*-acetylglucosaminidase and pyroglutamyl peptidase have also been isolated from marine actinomycetes. These compounds are effective for the treatment of diabetes and other diseases caused by pathogenic bacteria, respectively [18,19].

2.2.5
Antibiotic Produced by Marine Fungi

A marine fungi isolated from the neritic sea of Sardinia in the Mediterranean Sea were found to produce a substance having potent inhibition against pathogenic bacteria [20]. The strain was identified as *Cephalosporium acremonium* based on its taxonomic characteristics. The active compound was purified from the cultured broth, and the chemical structure was determined. Thereafter, the compound was named

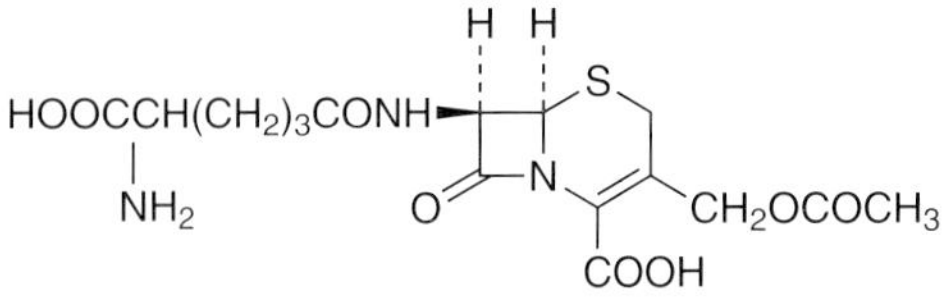

Figure 2.5 Chemical structure of antibiotic cephalosporin C.

cephalosporin C because it was isolated from *C. acremonium* (Figure 2.5). Further study revealed that the antibiotic-producing strain was of marine origin, although it was believed to be a terrestrial bacterium that was transported by river water to marine environment because its isolation point was near the land area. Cephalosporin C is a highly effective antibiotic; therefore, its derivatives are widely applied in the pharmaceutical industry to treat various diseases caused by pathogenic bacteria.

2.2.6
Tyrosinase Inhibitor Produced by Marine Fungi

Several fungi have been isolated from the marine environment; however, only a few reports on the active substance-producing strains are available. A tyrosinase inhibitor (TI)-producing strain was isolated from a sediment sample collected from Niijima Island in the Pacific Ocean [21]. This strain (H1–7) was considered to be a relative species of *Trichoderma viride* based on the sequence analysis of 28S rDNA [22]. The active substance was purified from the culture supernatant, and the chemical structure was determined. As shown in Figure 2.6, the compound was identical to 5-hydroxymethyl-3-isocyano-5-vinyl-cyclopento-2-enone by NMR and mass spectral analysis [23]. The substance showed inhibitory activity against melanization of human melanoma cells; therefore, it is expected to be applied as a skin whitener in the cosmetic industry [24]. The culture supernatant of the strain suppressed black discoloration of various shellfish such as Kuruma prawn (Figure 2.7) and perishable food during storage. It also effectively prevented wine browning in red wine making. This inhibitor is also applicable to various food industries [25,26].

2.3
Microorganisms Living in Deep Sea Water

Since deep sea water (DSW) is very pure, the microbial population is relatively low compared to that of surface seawater (SSW). Figure 2.8 shows the result of a viable

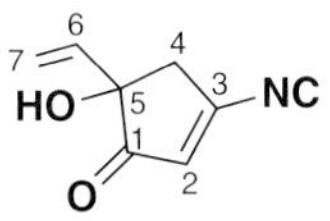

Figure 2.6 Chemical structure of tyrosinase inhibitor [23].

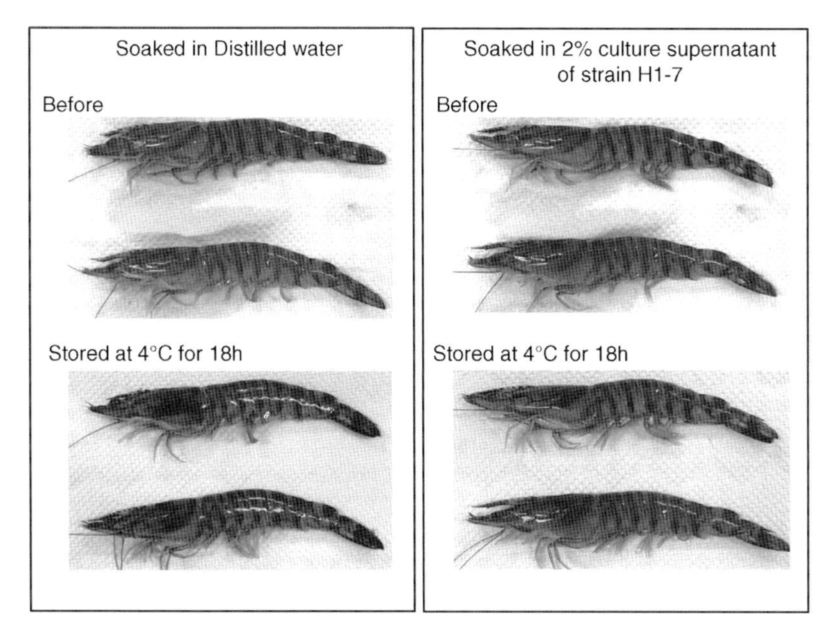

Figure 2.7 Prevention of black discoloration of Kuruma prawn by culture supernatant of strain H1-7.

count of marine bacteria in DSW at a depth of 800 m at Izu-Akazawa, Shizuoka Prefecture, Japan. It is apparent that the viable count in DSW is approximately 1000 times lower than that of SSW. However, a recent research indicates that the specific microbial population existing in DSW is different from those in SSW [27].

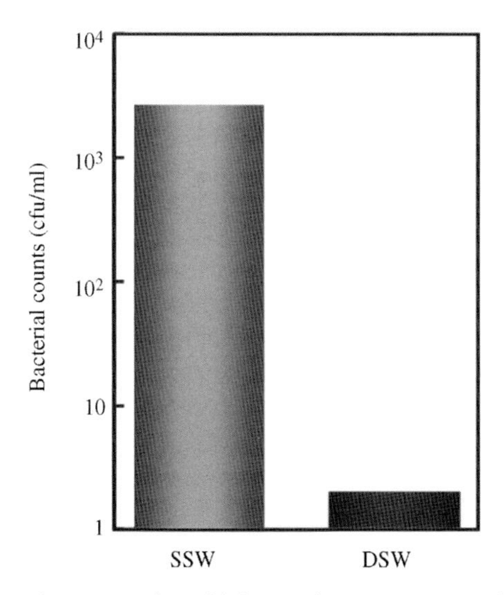

Figure 2.8 The viable bacterial counts in DSW and SSW collected from Izu Akazawa in Shizuoka Prefecture.

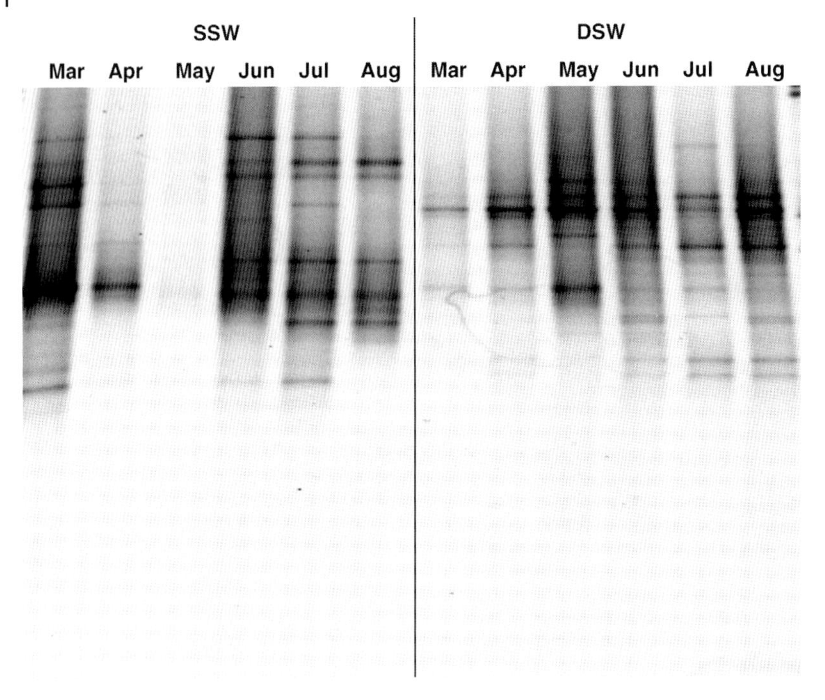

Figure 2.9 The seasonal fluctuation of the community structures of actinomycetes in Izu Akazawa Shizuoka Prefecture (2010).

DSW (800 m) and SSW were collected in Izu Akazawa every month since 2009, and the microorganisms *in situ* were trapped by a sterilized membrane filter (0.2 μm). The microbial DNA was purified and the 16S rDNAs were amplified by polymerase chain reaction (PCR) using universal primers. The community structures of useful microorganisms, such as actinomycetes and lactic acid bacteria, were performed by denaturing gradient gel electrophoresis (DGGE). As shown in Figure 2.9, no seasonal fluctuation was observed in the actinomycetal community structure. Several actinomycetal DNA bands were observed in DSW, whereas a number of bands showed seasonal fluctuation in SSW. On the other hand, the community structure of lactic acid bacteria in DSW did not show as much of a seasonal fluctuation as that of actinomycetes, whereas the number of bands and their patterns showed seasonal fluctuation in SSW (Figure 2.10).

2.3.1 Isolation and Incubation of Lactic Acid Bacteria from Deep Sea Water

To obtain industrially useful lactic acid bacteria from DSW, a bag-type filter (pore size 0.5 μm) that retained deep sea microbes was aseptically cut into 3 cm squares. The filter was rinsed in a small amount of sterilized DSW and an aliquot of the suspension was spread on MRS selection medium [28] and incubated at 27 °C.

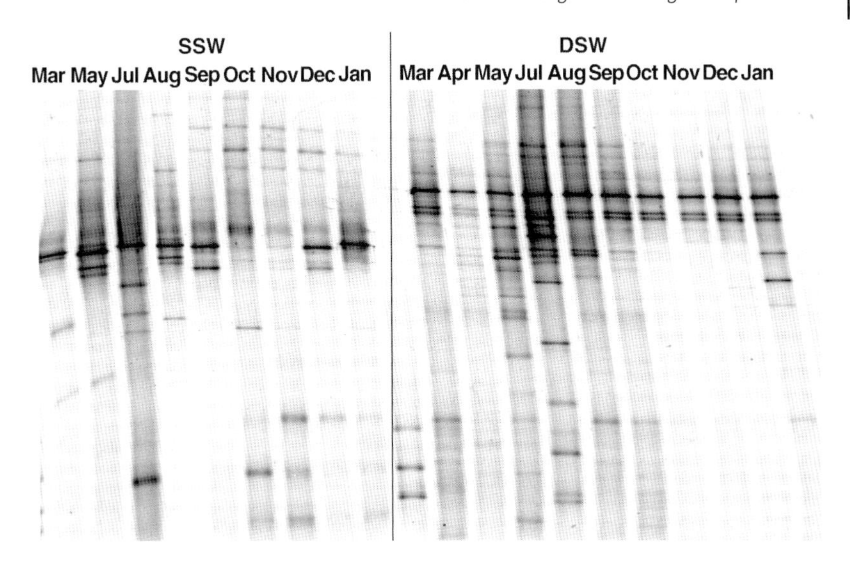

Figure 2.10 The seasonal fluctuation of the community structures of lactic acid bacteria in Izu Akazawa Shizuoka Prefecture (2010–2011).

After incubation, developed colonies on the plate were isolated and were confirmed to be lactic acid bacteria by various taxonomical characterizations. Basic studies were made in the fields of cosmetics and foods using bacteria. Thirty-nine strains were isolated from the bag-type filter. From 16S rDNA sequence analyses, all the strains isolated from DSW were identified as *Lactobacillus plantarum* or *Pediococcus pentosaceus*. These bacteria are well known as terrestrial plant lactic acid bacteria (unpublished data). One of the *L. plantarum* strains collected in June 2009, strain BF-13, showed inhibitory activity against tyrosinase. The activity was much higher than that of arbutin and kojic acid (Figure 2.11). These compounds are well known as skin whiteners in the cosmetic industry. Isolation of the active substance from

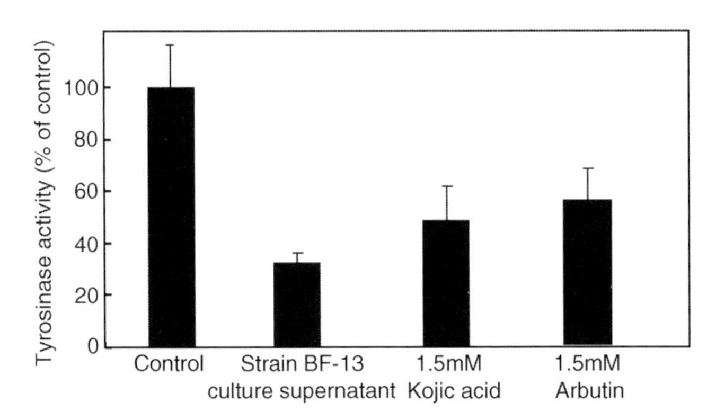

Figure 2.11 Tyrosinase inhibitory activity of lactic acid bacterium (*L. plantarum* strain BF-13) isolated from DSW (unpublished data).

the culture supernatant of the strain and application to various industries will be performed in the near future.

Until recently, not so many useful microorganisms have been isolated from DSW. However, novel bioactive compounds have been isolated and applied to various industries.

References

1 Demuth, H.H. (1990) Recent developments in inhibiting cysteine and serine protease. *J. Enzyme Inhib.*, **3**, 249–278.

2 Terashita, T., Oda, K., Kono, M., and Murao, S. (1981) Promoting effect of acid proteinase inhibitor (S-PI) on fruiting of *Pleurotus ostreatus*. *Hakkokogaku*, **59**, 55–57 (in Japanese).

3 Imada, C., Simidu, U., and Taga, N. (1986) Isolation and characterization of marine bacteria producing alkaline protease inhibitor. *Bull. Jpn. Soc. Sci. Fish.*, **51**, 799–803.

4 Imada, C., Maeda, M., Hara, S., Taga, N., and Simidu, U. (1986) Purification and characterization of subtilisin inhibitors "marinostatin" produced by marine *Alteromonas* sp. *J. Appl. Bacteriol.*, **60**, 469–476.

5 Sugino, H. and Kakinuma, A. (1978) Plasminostreptin, a protein proteinase inhibitor produced by *Streptomyces antifibrinolyticus*: III. Elucidation of the primary structure. *J. Biol. Chem.*, **253**, 1546–1555.

6 Imada, C., Hara, S., Maeda, M., and Simidu, U. (1986c) Amino acid sequences of marinostatins C-1 and C-2 from marine *Alteromonas* sp. *Bull. Jpn. Soc. Sci. Fish.*, **52**, 1455–1459.

7 Kanaori, K., Kamei, K., Taguchi, M., Koyama, T., Yasui, T., Takano, R., Imada, C., Tajima, K., and Hara, S. (2005) Solution structure of marinostatin, a novel ester-linked protein protease inhibitor. *Biochemistry*, **44**, 2462–2468.

8 Imada, C., Nishimoto, S., and Hara, S. (2001) The effect of addition of protease inhibitor from marine bacterium on the strength of gel formation of sardine meat gel (Kamaboko). *Nippon Suisan Gakkaishi*, **67**, 85–89.

9 Sakuda, S., Isogai, A., Matsumoto, S., and Suzuki, A. (1987) Search for microbial insects growth regulators: II. Allosamidin, a novel insect chitinase inhibitor. *J. Antibiot.*, **40**, 296–300.

10 Sakuda, S., Isogai, A., Matsumoto, S., and Suzuki, A. (1986) The structure allosamidine, a novel insect chitinase inhibitor produced by *Streptomyces* sp. *Tetrahedron Lett.*, **27**, 2475–2478.

11 Izumida, H., Miki, W., Sano, H., and Endo, M. (1995) Agar plate method, a new assay for chitinase inhibitors using a chitin-degrading bacterium. *J. Mar. Biotechnol.*, **2**, 163–166.

12 Izumida, H., Imamura, N., and Sano, H. (1996) A novel chitinase inhibitor from a marine bacterium, *Pseudomonas* sp. *J. Antibiot.*, **49**, 76–80.

13 Burkholder, P.R., Pfister, R.W., and Leitz, F. (1966) Production of pyrrole antibiotic by a marine bacterium. *Appl. Microbiol.*, **14**, 649–653.

14 Yoshikawa, K., Takadera, T., Adachi, K., Nishijima, M., and Sano, H. (1997) Korormicin, a novel antibiotic specifically active against marine Gram-negative bacteria, produced by a marine bacterium. *J. Antibiot.*, **50**, 949–953.

15 Sato, K., Okazaki, T., Maeda, K., and Okami, Y. (1978) New antibiotics, aplasmomycin B and C. *J. Antibiot.*, **26**, 632–635.

16 Kitahara, T., Naganawa, H., Okazaki, T., Okami, Y., and Umezawa, H. (1975) The structure of SS-228Y, an antibiotic from *Chainia* sp. *J. Antibiot.*, **28**, 280–285.

17 Okazaki, T., Kitahara, T., and Okami, Y. (1975) Studies on marine microorganisms: IV. A new antibiotic SS-228Y produced by *Chainia* isolated from shallow sea mud. *J. Antibiot.*, **28**, 176–185.

18 Aoyama, T., Kojima, F., Imada, C., Muraoka, Y., Naganawa, H., Okami, Y., Takeuchi, T., and Aoyagi, T. (1995) Pyrostatin A and B, new inhibitors of *N*-acetyl-D-glucosaminidase, produced by

Streptomyces sp. SA-3501. *J. Enzyme Inhib.*, 8, 223–232.
19 Aoyagi, T., Hatsu, M., Imada, C., Naganawa, H., Okami, Y., and Takeuchi, T. (1992) Pyrizinostatin: a new inhibitor of pyroglutamyl peptidase. *J. Antibiot.*, **45**, 1795–1796.
20 Hodgkin, D.C. and Malsen, E.N. (1960) The X-ray analysis of the structure of cephalosporin C. *Biochem. J.*, **9**, 393–402.
21 Imada, C., Sugimoto, Y., Makimura, T., Kobayashi, T., Hamada-Sato, N., and Watanabe, E. (2001) Isolation and characterization of tyrosinase inhibitor-producing microorganisms from marine environment. *Fish. Sci.*, **67**, 1151–1156.
22 Yamada, K., Imada, C., Sato, M., Kobayashi, T., and Hamada-Sato, N. (2008) Prevention of wine browning with culture supernatant of fungus isolated from marine environment. *J. ASEV. Jpn.*, **19**, 2–9.
23 Tsuchiya, T., Yamada, K., Minoura, K., Miyamoto, K., Usami, Y., Kobayashi, T., Hamada-Sato, N., Imada, C., and Tsujibo, H. (2008) Purification and determination of the chemical structure of the tyrosinase inhibitor produced by *Trichoderma viride* strain H1-7 from a marine environment. *Biol. Pharm. Bull.*, **31**, 1618–1620.
24 Yamada, K., Imada, C., Tsuchiya, T., Miyamoto, K., Tsujibo, H., Kobayashi, T., and Hamada-Sato, N. (2007) An application study of culture supernatant of a filamentous fungus isolated from marine environment as an ingredient for whitening cosmetics. *J. Soc. Cosmet. Chem. Jpn.*, **41**, 254–261.
25 Yamada, K., Imada, C., Kobayashi, T., and Hamada-Sato, N. (2007) Prevention of black discoloration of perishable food by fungus isolated from marine sediment. *Nippon Shokuhin Kagaku Kogaku Kaishi*, **54**, 274–279.
26 Yamada, K., Imada, C., Uchino, W., Kobayashi, T., Hamada-Sato, N., and Takano, K. (2008) Phenotypic characterization and cultivation conditions of inhibitor-producing fungus isolated from marine environment. *Fish. Sci.*, **74**, 662–669.
27 Hayashi, T., Shimada, T., Onaka, H., and Furumai, T. (2007) Isolation and phenotypic characterization of enterococci from the deep-seawater samples collected in Toyama Bay. *Jpn J. Lactic Acid Bacteria*, **18**, 58–64 (in Japanese).
28 Joborn, A., Dorsch, M., Olsson, J.C., Westerdal, A., and Kjelleberg, S. (1999) *Carnobacterium inhibens* sp. nov., isolated from the intestine of Atlantic salmon (*Salmo salar*). *Int. J. Syst. Bacteriol.*, **49**, 1891–1898.

3
Strategy of Marine Viruses in Global Ecosystem

Amardeep Singh, Manoj Trivedi, Pranjal Chandra, and Rajendra N. Goyal

3.1
Introduction

In marine microbiology, the microbial processes in the sea have played an important role over the last decades. It is clear that the total flux of matter and energy in marine food webs passes through such organisms by means of dissolved organic matter [1,2]. The concept of the microbial loop in marine food webs has been developed and refined in current years, but viruses were ignored in such studies. However, literature survey reports showed that viruses are not only extraordinarily abundant in marine plankton but are also likely to be significant agents in the control of bacteria and phytoplankton [3–6].

3.2
Reproductive Strategies of Viruses

Naturally, viruses are the most common biological entities in the sea. Viruses are very small, usually 20–200 nm long, and consists of genetic material (DNA or RNA, single- or double-stranded) surrounded by a protein coat. Some of them also have lipids. Viruses also appear many times during the evolution of cellular organisms. These viruses have no intrinsic metabolism and they function only as parasites through the cellular machinery of a host organism. Hence, all cellular organisms are susceptible to infection, often by more than one type of virus, stating that viruses are probably the most diverse creatures on universe. Viruses usually have a restricted range of hosts – often a single species, with some infecting only certain subspecies, whereas others infecting more than one related species or even genus [7]. Viruses attach to their hosts by passive diffusion and use their host's exposed cellular structures, often transporter proteins, as entry points to the cell (Figure 3.1). There are three basic types of virus reproduction: (i) lytic infections, (ii) chronic infections, and (iii) lysogeny infections.

Marine Microbiology: Bioactive Compounds and Biotechnological Applications, First Edition.
Edited by Se-Kwon Kim

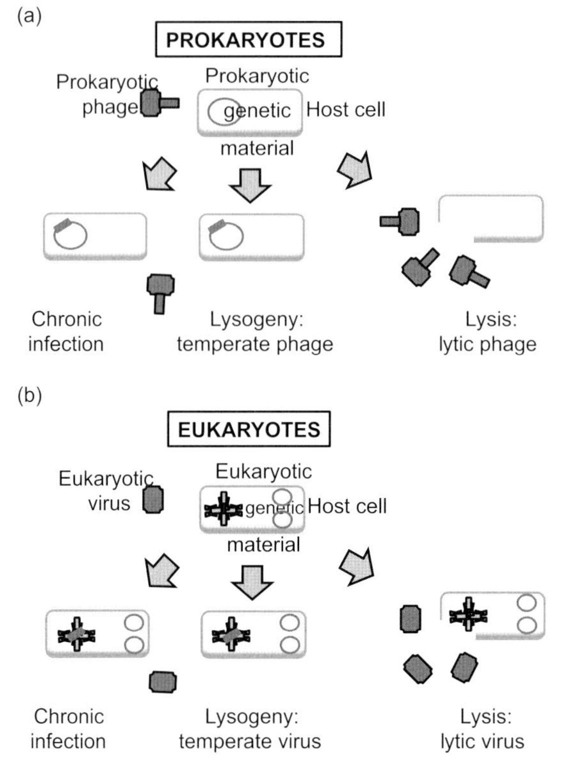

Figure 3.1 (a) Prokaryotic and (b) eukaryotic infection cycles.

3.2.1
Lytic Infection

The virus attach to a host cell injects its nucleic acid into the cell, and then directs the host to produce numerous progeny viruses; these are released by lethal bursting of the cell, allowing the cycle to begin again.

3.2.2
Chronic Infection

The progeny virus release is nonlethal, and the host cell releases them by extrusion or budding over several generations.

3.2.3
Lysogeny Infection

The nucleic acid of the viral genome becomes part of the genome of the host cell and reproduces as genetic material (called a prophage or provirus) in the host cell line. In this case, an induction event, such as stress to the host, can trigger a switch to lytic infection. A less well-defined type of virus–host interaction is termed pseudolysogeny, in which the viral nucleic acid may remain within a host cell for some time (possibly for a few generations) [8] before lysis or cell destruction occurs. Pseudolysogeny may be related to host starvation, in which the virus adopts an inactive state, unable to initiate viral gene expression owing to the low-energy state of the cell; normal viral activity returns only when the cell is fed [9]. Alternatively, pseudolysogeny may be called a transient state of host immunity, apparently induced by an immunizing agent (perhaps a polysaccharide depolymerase) released from infected cells and helping to foster coexistence of host and virus [10,11]. Viruses or virus-like particles may also be involved in killing cells by mechanisms that do not result in virus reproduction [12]. Here, we focus on the accumulated evidences regarding the nature of marine viruses and their ecological effects. The principal conclusion is that viruses can use significant control on marine bacterial and phytoplankton communities, with respect to both production and species composition, influencing the pathways of matter and energy transfer in the system.

3.3
Abundance of Marine Viruses

Marine viruses play a vital role in global ecosystems, as they are the most abundant biological entities in oceans. The mechanism for viruses infecting specific marine organisms has been studied for several decades [13], beginning from studies of pure cultures that focused on organisms rather than on ecological systems. Viruses were not considered as essential components of marine food webs until they were shown by direct counts to be highly abundant. Their small size renders them invisible to ordinary light microscopy. However, they can be visualized by transmission electron microscopy (TEM). Special procedures are required to collect them from seawater. Currently, the ultracentrifugation and TEM methods are employed to collect and visualize marine viruses and prokaryotes, which are slightly different from the original technique published in 1949 [14]. This technique has been extended to natural waters to discover highly abundant prokaryotes and viruses in aquatic habitats. This could lead to the advancement of modern biological oceanography and limnology that are only recently considered significant toward microbiological studies. The first reports of high viral abundance, exceeding the typical bacterial abundance of 109/l [3–5], awakened interest in this area. Many subsequent studies [15–24] have shown that viruses are always the most abundant biological entities in the sea: nearshore to offshore, tropical to polar, sea surface to sea floor, and sea ice to sediment pore water. Viruses are typically 10^{10}/l in surface waters (about 5–25 times the bacteria), and follow the same general abundance patterns as bacteria. These patterns include a decrease of about one order of magnitude between rich coastal waters and oligotrophic (nutrient-poor) open ocean, a decrease of about 5–10-fold from the euphotic zone to the upper midwaters (e.g., 500 m depth), and a further decrease from severalfold to abyssal depths. Sea ice in comparison to the water beneath it is relatively richer in viruses than bacteria [25], while sediment pore waters are highly rich in viruses than the overlying water [18,23]. Viral abundances are dynamic, being particularly responsible for changes in ecological conditions such as algal blooms [6,26]. This provides evidence that viruses are active members of the community rather than inert particles. Viral abundance is usually strongly correlated with bacterial abundance on scales of hundreds of kilometers and less so and with particulate chlorophyll *a* [17–19], suggesting that most of the marine viruses infect bacteria. The term bacteria used throughout refers to prokaryotes in general because most studies do not distinguish true Bacteria (also called Eubacteria) from Archaea. In some environments (particularly the deep sea), half of the prokaryotes may belong to the Archaea [27]. Epifluorescence microscopy with nucleic acid stains such as 4,6-diamidino-2-phenylindole (DAPI), YoPro, and SYBR Green has been used to count viruses accurately, rapidly, and inexpensively compared to TEM [24,27–30]. Viruses stained with SYBR Green can be observed and counted by ordinary epifluorescence microscopy in the field (e.g., on board ship) within an hour of sampling. The counts obtained [24] are similar to or slightly higher (by ~1–1.5 times) than those obtained by TEM. Viruses stained with SYBR Green may also be counted by flow cytometry [31], further expediting aquatic virus studies. Here, again the counts are higher than TEM values. The reasons for the higher fluorescence counts compared to those obtained from TEM preparations are unknown, but fluorescence counts may include viruses that are otherwise hidden by other dark stained particles in TEM preparations. Fluorescence also enables the counting of filamentous viruses or other types that are not recognized easily by TEM. The new stains that permit fluorescence imaging or the flow cytometric detection of a diverse array of viruses may prove useful in biomedical studies. Besides the counting of viruses, marine viral diversity has also been examined by morphology and size distribution. Cultures and natural samples show a broad array of shapes and sizes [32]. A new technique, pulsed field electrophoresis [33,34], has shown the size diversity of viral genomes extracted directly from marine samples. The method gives a minimum estimate of the diversity of the most abundant viruses. There are typically 15–40 visibly distinct genome sizes in a given sample, and the composition is variable in space and dynamic over seasons. Thus, many studies related to viral abundance and distribution have been done worldwide by changes in physiochemical characteristics of the water masses [35,36].

3.4 Viral Activities in Ecosystems

3.4.1 Diversity Regulation

Marine prokaryote diversity regulation began with the development of 16S ribosomal RNA sequence-based techniques that facilitate analysis of most organisms that are resistant to conventional cultivation [37]. As evident from their effect on algal blooms and cyanobacteria, viruses are in a unique position to influence community species compositions. Theoretical analysis [38] indicates that the maximum possible number of bacterial populations in a spatially homogeneous environment is equal to the number of unique resources plus the number of unique virus types. Despite that viruses are responsible for only a small proportion of the mortality of a group of organisms, they can however have a profound effect on the relative proportions of different species or strains in the community [39,40]. Similar ideas have been known from experimental systems for some time [8]. An important postulate is that viral infection is both density dependent and species specific (or nearly so). Because viruses diffuse randomly from host to host, rare hosts are less susceptible to the spread of infection than the more common ones. Lytic viruses can only increase in abundance when the average time to diffuse from host to host is shorter than the average time they remain infectious. Thus, when a particular species or strain becomes more densely populated, it is more susceptible to infection. This may have direct relevance to solving Hutchinson's "paradox of plankton" [41], which asks how so many different kinds of phytoplankton can coexist on only a few potentially limiting resources, when competition theory predicts just one or a few competitive winners. There may be several possible explanations to this question [42], with the most probable being that viral activity assists because the competitive dominants become particularly susceptible to infection, whereas the rare species are relatively protected [13]. Models [43] of the potential factors that control the biomass and species composition of microbial systems include growth limitation by organic carbon, inorganic phosphate or nitrogen (inorganic or organic), and cell losses due to grazing by protists and viral lysis. Even when bacterial abundance is assumed to be controlled by protist grazing, these models consistently show that viruses control the steady-state diversity of the bacterial community, irrespective of whether bacterial growth rate limitation is by organic or inorganic nutrients. Consistent with this picture are studies [33,44] of viral genome sizes and hybridization analyses of total viral communities, which show episodic and spatial changes in viral community composition. Thus, both empirical and theoretical analyses indicate that viruses are important in regulating patterns of diversity.

3.4.2
Rate of Resistance

The resistance to viral infection is well known from nonmarine experimental studies [8] and it is an important postulation in models [42,45] telling about the regulation of diversity by viruses. Resistance has been shown [46] to occur in marine cyanobacteria. The rate of resistance in marine heterotrophic bacterial populations has not been evaluated directly, and doing so would be difficult because most have not yet been grown in culture. However, the high rates of virus production in many marine locations seem to be inconsistent with a community dominated by virus-resistant organisms. Why is it that viral resistance does not appear to be a dominant factor, even though theory and some field data indicate that it should be? Unfortunately, there are no obvious explanations. One option is that resistance is dominant and virus production is not as high as it seems to be, but this would require explanations for the various observations that are consistent with high virus production. Another option is that resistance costs too much physiologically, as it often occurs through the loss of some important receptor and thus confers a competitive disadvantage [8]. Theory suggests that if the cost is too high, resistant organisms cannot persist in competition with sensitive ones, even when viruses are present [38]. Chemostat experiments [47] indicate that even with a cost that allows persistence of resistant cells, they may still be rare unless the cost is low or limiting resources are abundant. However, laboratory experiments indicate that low-cost or no-cost resistance is possible [8], so uncertainty remains. It seems likely that the explanation lies outside the realm of simplified theoretical or laboratory systems. The real ocean has many species of bacteria and viruses living together, with a variety of alternative mortality mechanisms and low, but variable, nutrient conditions. Perhaps there are advantages of being sensitive to infection. In a typical (oligotrophic) marine environment, bacterial growth may be limited by nitrogen, phosphorus, or organic carbon. Unsuccessful viral infection (perhaps stopped intracellularly by a restriction enzyme or with a genetic incompatibility) is then of significant nutritional benefit [5] to the host organism, because the virus injection of DNA is rich in C, N, and P. Even the viral protein coat, left attached to the outside of the cell, is probably digestible by cell-associated exoenzymes [46]. In this situation, one might even imagine bacteria using decoy virus receptors for viral strains that cannot successfully infect them; if an infectious one occasionally gains entry, the strategy might still confer a net benefit to the cell line. Can a model with so many aborted infections still be consistent with high virus production rates? It would seem so given that in the steady state, only one virus out of every burst from cell lysis (typically producing 20–50 progeny viruses) successfully infects another host (Figure 3.2). A large proportion of the remaining 95–98% unsuccessful viruses could end up infecting the "wrong" hosts, giving these hosts an incentive to continue permitting virus attachment and injection. This is a consistent model, but highly speculative.

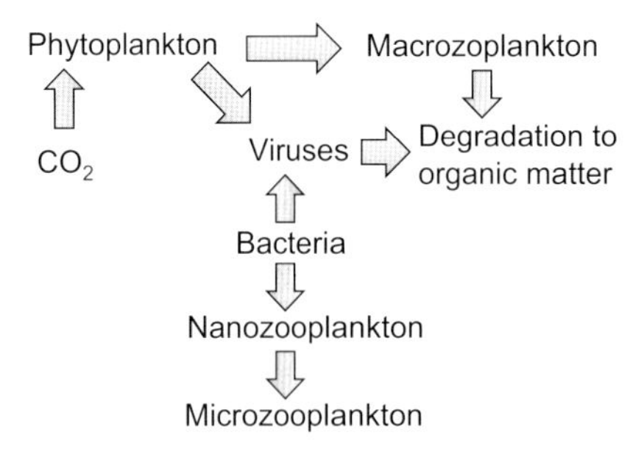

Figure 3.2 Steady-state model of carbon flow for virus's production.

Another consideration that argues against resistance relates to the results of system models that show that, as a group, the heterotrophic bacteria benefit greatly from viral infection (as infection boosts their production significantly, essentially by taking carbon and energy away from large organisms). Viruses also boost the entire system biomass and production by helping to maintain nutrients in the lighted surface waters. However, these explanations would additionally require some sort of group selection theory to explain how individuals would benefit from not developing resistance (e.g., why not cheat by developing resistance and letting all the other organisms give the group benefits of infection?). It seems that most explanations of why resistance is not a dominant factor are either unsatisfactory or highly speculative. Whether lack of comprehensive resistance is due to frequent growth of new virulent strains, rapid dynamics or patchiness in species compositions, or a stable coexistence of viruses and their hosts is yet to be determined. As a final note, the extent of resistance may somehow be related to pseudolysogeny (in the sense of a transient immunity to infection induced by infected hosts) [10,11], but their mechanism is not fully understood.

3.4.3
Lysogeny

Lysogeny has been considered as a survival strategy of viruses living at low host density. At the same time, it may confer advantages to the host, including immunity from infection by related viruses and the acquisition of new functions coded by the virus genome (known as conversion) [48]. In seawater and lake water samples [49,50], lysogens (bacteria that harbor prophage and can be induced to produce lytic viruses) were found to be common, with variable abundances ranging from undetectable to almost 40% of the total bacteria, and this variability can be seasonal [51]. About 40% of cultured marine bacteria are lysogens [52]. There may be certain natural events that induce such lysogens to enter the lytic cycle, and reports [49,53] show that common pollutants such as hydrocarbons, polychlorinated biphenyls (PCBs), and Aroclor can do so. However, growth experiments with native mixed bacterial communities grown in filtered seawater indicate that under typical natural conditions, including exposure to bright sunlight, induction of lysogens is rare and the vast majority (97% or more) of viruses observed in seawater are probably the result of successive lytic infection [36]. The same conclusion was reached from a quantitative interpretation of induction experiments [52,54]. Lysogenic induction may be occurring at a low level all the time or may occur at irregular intervals. Although induction may be rare, the widespread existence of lysogeny probably has major implications for genetic exchange and evolution.

3.4.4
The Exchange of Genetic Material

The most uncertain aspects of marine viruses are probably their role in genetic exchange among microorganisms and its effect on short-term adaptation, population genetics, and evolution. There can be direct effects such as transduction [55,56], in which a virus takes DNA from one host and transfers it to another (often a lysogen). The overall effect over large scales of space and time would be to homogenize genes among the susceptible host populations. Although the conventional view is that transduction usually operates within a highly restricted host range, one report [12] indicates that some marine viruses are capable of nonspecific horizontal gene transfer. Another concern is that viral lysis causes release of DNA from host organisms, which could be transferred to another organism through natural transformation. Dissolved DNA is readily found in seawater, and it has been reported [18] that viral lysis may be a major source mechanism. The latter two processes (generalized viral horizontal gene transfer and transformation) would have the effect of mixing genes among a broad variety of species, with wide-ranging effects on adaptation and evolution. Although these sorts of transfers may be extremely rare, the typical viral abundances of 10^9/l in the euphotic zone and the huge volume of the sea ($3.6 \times 10^7\,\text{km}^3$ in the top 100 m) [57], coupled with the generation times on the order of a day, imply that an event with a probability of only 10^{-20} per generation would be occurring about a million times per day. Thus, this process may have major effects on the genetic structure and evolution of the global population of marine viruses. They should also be considered in the evaluation of the potential spread of genetically engineered microbial genes or of antibiotic resistance introduced by intensive fish farming.

3.5 Recent Advancement of Viruses versus Diseases

Viruses play an important role in ocean ecology. They are not only capable of affecting the biogeochemical and ecological cycles but are also progenitors of diseases in higher organisms. We have very less understanding of the natural reservoirs of viruses that cause disease in marine animals and know even less about their potential to spread to terrestrial systems. They kill the cells and generate about 10^{23} virus infections per second. After killing the cells, they also liberate enough iron to supply the needs of phytoplankton and lead to the production of dimethyl sulfoxide that can influence the climate of the universe. Therefore, due to these activities, marine viruses have a significant impact on global microbial communities and geothermal cycles [58].

Most of the marine viruses are bacteriophages. Many other viruses also infect marine mammals and even cause disease in humans, including adenoviruses, herpesviruses, parvoviruses, and caliciviruses. Recently, novel viruses have been discovered, such as white spot syndrome virus of penaeid shrimp. Many ocean viruses cause disease in marine mammals. Phocid distemper virus is a morbilli kind of virus that killed thousands of harbor seals in Europe. Similar viruses kill dolphins and other cetaceans.

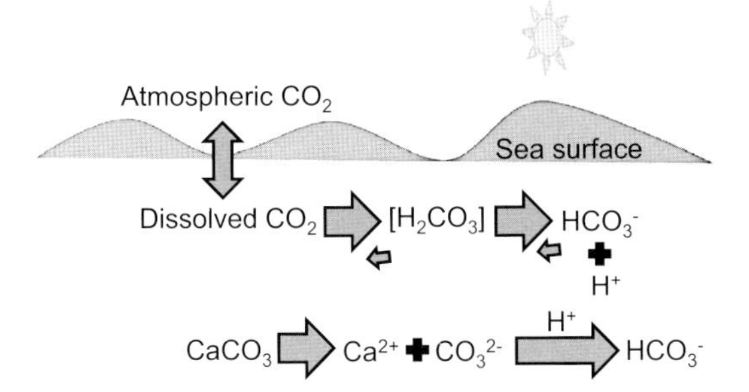

Figure 3.3 Invasion of CO_2 into the ocean causes an acidification.

3.6 The Effect of Ocean Acidification on Marine Viruses

The ocean has a major sink of anthropogenic carbon emitted from fossil fuel use and cement production [59,60]. Besides, the carbonate system also contributes to the large volume of anthropogenic carbon that changes partial pressure of CO_2 (Figure 3.3). Moreover, the uptake of anthropogenic CO_2 induces changes in the pH [61]. The invasion of CO_2 into the ocean causes an acidification of the surface ocean [62]. The effect of pH changes on marine viruses is a big concern and necessary to predict. Published report indicated that some viruses remain unaffected at $pH = 3$ or even less, whereas others are labile at $pH = 7.0$ [63–65]. Weinbauer [66] stated that pH can affect adsorption of phages in freshwater, while marine phages in seawater are typically only affected by pH values.

3.7 Further Aspects

Although there are many reports that have been published on marine viruses to understand the marine communities, a lot of research is still needed to understand the diversity of marine viruses, their role, and also the complex relationships between viruses and the rest of the ecosystem.

Acknowledgments

We would like to gratefully acknowledge Prof. Fed A. Ferhman, University of Southern California, USA and Prof. C.A. Suttle, University of British Columbia, Canada for their immense work published on marine viruses. We are also grateful to Prof. Markus G. Weinbauer and Prof. W. Wilhelm whose contributions on marine viruses have been referred to. Without their published reports, this chapter would not have been possible.

References

1 Fuhrman, J.A. (1992) In *Primary Productivity and Biogeochemical Cycles in the Sea* (eds P.G. Falkowski and A.D. Woodhead), Plenum, New York, pp. 361–383.

2 Fuhrman, J.A. (1999) Marine viruses and their biogeochemical and ecological effects. *Nature*, **399**, 541–548.

3 Sullivan, M.B., Huang, K.H., Ignacio-Espinoza, J.C., Berlin, A.M., Kelly, L., Weigele, P.R., DeFrancesco, A.S., Kern, S.E., Thompson, L.R., and Young, S. (2010) Genomic analysis of oceanic cyanobacterial myoviruses compared with T4-like myoviruses from diverse hosts and environments. *Environ. Microbiol.*, **12**, 3035–3056.

4 Suttle, C.A. (2005) Viruses in the sea. *Nature*, **437**, 356–361.

5 Proctor, L.M. and Fuhrman, J.A. (1990) Viral mortality of marine bacteria and cyanobacteria. *Nature*, **343**, 60–62.

6 Larsen, J.B., Larsen, A., Bratbak, G., and Sandaa, R.-A. (2008) Phylogenetic analysis of members of the Phycodnaviridae virus family using amplified fragments of the major capsid protein gene. *Appl. Environ. Microbiol.*, **74**, 3048–3057.

7 Weinbauer, M.G. (2004) Ecology of prokaryotic viruses. *FEMS Microbiol. Rev.*, **28**, 127–181.

8 Lythgoe, K.A. and Chao, L. (2003) Mechanisms of coexistence of a bacteria and a bacteriophage in a spatially homogeneous environment. *Ecol. Lett.*, **6**, 326–334.

9 Ripp, S. and Miller, R.V. (1997) The role of pseudolysogeny in bacteriophage–host interactions in a natural freshwater environment. *Microbiology*, **143**, 2065–2070.

10 Bohannan, B.J.M. and Lenski, R.E. (2000) Linking genetic change to community evolution: insights from studies of bacteria and bacteriophage. *Ecol. Lett.*, **3**, 362–377.

11 Moebus, K. (1997) Investigations of the marine lysogenic bacterium H24: II. Development of pseudolysogeny in nutrient rich broth. *Mar. Ecol. Prog. Ser.*, **148**, 229–240.

12 Chiura, H.X. (1997) Generalized gene transfer by virus-like particles from marine bacteria. *Aquat. Microb. Ecol.*, **13**, 75–83.

13 Fuhrman, J.A. and Suttle, C.A. (1993) Viruses in marine planktonic systems. *Oceanography*, **6**, 51–63.

14 Sharp, G.D. (1949) Enumeration of virus particles by electron micrography. *Proc. Soc. Exp. Biol. Med.*, **70**, 54–59.

15 Wommack, K.E., Hill, R.T., Kessel, M., Russek-Cohen, E., and Colwell, R.R. (1992) Distribution of viruses in the Chesapeake Bay. *Appl. Environ. Microbiol.*, **58**, 2965–2970.

16 Børsheim, K.Y. (1993) Native marine bacteriophages. *FEMS Microbiol. Ecol.*, **102**, 141–159.

17 Cochlan, W.P., Wikner, J., Steward, G.F., Smith, D.C., and Azam, F. (1993) Spatial distribution of viruses, bacteria and chlorophyll *a* in neritic, oceanic and estuarine environments. *Mar. Ecol. Prog. Ser.*, **92**, 77–87.

18 Paul, J.H., Rose, J.B., Jiang, S.C., Kellogg, C.A., and Dickson, L. (1993) Distribution of viral abundance in the reef environment of Key Largo, Florida. *Appl. Environ. Microbiol.*, **59**, 718–724.

19 Boehme, J. (1993) Viruses, bacterioplankton, and phytoplankton in the southeastern Gulf of Mexico: distribution and contribution to oceanic DNA pools. *Mar. Ecol. Prog. Ser.*, **97**, 1–10.

20 Maranger, R., Bird, D.F., and Juniper, S.K. (1994) Viral and bacterial dynamics in arctic sea ice during the spring algal bloom near Resolute, NWT, Canada. *Mar. Ecol. Prog. Ser.*, **111**, 121–127.

21 Sherr, E.B., Sherr, B.F., and Longnecker, K. (2006) Distribution of bacterial abundance and cell-specific nucleic acid content in the Northeast Pacific Ocean. *Deep-Sea Res. I*, **53**, 713–725.

22 Maranger, R. and Bird, D.E. (1996) High concentrations of viruses in the sediments of Lac Gilbert, Quebec. *Microb. Ecol.*, **31**, 141–151.

23 Steward, G.F., Smith, D.C., and Azam, F. (1996) Abundance and production of bacteria and viruses in the Bering and Chukchi Sea. *Mar. Ecol. Prog. Ser.*, **131**, 287–300.

24 Weinbauer, M.G., Winter, C., and Hofle, M.G. (2002) Reconsidering

transmission electron microscopy based estimates of viral infection of bacterioplankton using conversion factors derived from natural communities. *Aquat. Microb. Ecol.*, **27**, 103–110.

25 Maranger, R., Bird, D.F., and Juniper, S.K. (1994) Viral and bacterial dynamics in Arctic sea ice during the spring algal bloom near Resolute, NWT, Canada. *Mar. Ecol. Prog. Ser.*, **111**, 121–127.

26 Thompson, J.R., Pacocha, S., Pharino, C., Klepac-Ceraj, V., Hunt, D.E., Benoit, J., Sarma-Rupavtarm, R., Distel, D.L., and Olz, M.F. (2005) Genotypic diversity within a natural coastal bacterioplankton population. *Science*, **307**, 1311–1313.

27 Pommier, T., Canbäck, B., Riemann, L., Boström, K.H., Simu, K., Lundberg, P., Tunlid, A., and Hagström, A. (2007) Global patterns of diversity and community structure in marine bacterioplankton. *Mol. Ecol.*, **16**, 867–880.

28 Hara, S., Terauchi, K., and Koike, I. (1991) Abundance of viruses in marine waters: assessment by epifluorescence and transmission electron microscopy. *Appl. Environ. Microbiol.*, **57**, 2731–2734.

29 Hennes, K.P. and Suttle, C.A. (1995) Direct counts of viruses in natural waters and laboratory cultures by epifluorescence microscopy. *Limnol. Oceanogr.*, **40**, 1050–1055.

30 Weinbauer, M.G. and Suttle, C.A. (1997) Comparison of epifluorescence and transmission electron microscopy for counting viruses in natural marine waters. *Aquat. Microb. Ecol.*, **13**, 225–232.

31 Xenopoulos, M.A. and Bird, D.F. (1997) Microwave enhanced staining for counting viruses by epifluorescence microscopy. *Limnol. Oceanogr.*, **42**, 1648–1650.

32 Brussaard, C.P.D. (2004) Optimization of procedures for counting viruses by flow cytometry. *Appl. Environ. Microbiol.*, **70**, 1506–1513.

33 Proctor, L.M. (1997) Advances in the study of marine viruses. *Microsc. Res. Tech.*, **37**, 136–161.

34 Wommack, K.E., Ravel, J., Hill, R.T., Chun, J.S., and Colwell, R.R. (1999) Population dynamics of Chesapeake Bay virioplankton: total-community analysis by pulsed-field gel electrophoresis. *Appl. Environ. Microbiol.*, **65**, 231–240.

35 Steward, G.F. and Azam, F. (1999) Microbial biosystems: new frontiers, in *Proceedings of the 8th International Symposium on Microbial Ecology* (eds C.R. Bell, M. Brylinsky, and P. Johnson-Green), Atlantic Canada Society for Microbial Ecology, Halifax, Canada.

36 Weinbauer, M., Brettar, I., and Hofle, M. (2003) Lysogeny and virus-induced mortality of bacterioplankton in surface, deep, and anoxic waters. *Limnol. Oceanogr.*, **48**, 1457–1465.

37 Winter, C., Moeseneder, M., Herndl, G.J., and Weinbauer, M.G. (2008) Relationship of geographic distance, depth, temperature, and viruses with prokaryotic communities in the Eastern tropical Atlantic Ocean. *Microb. Ecol.*, **56**, 383–389.

38 Fuhrman, J.A. (1997) *Manual of Environmental Microbiology* (eds C.J. Hurst, C.R. Knudsen, M.J. McInerney, L.D. Stetzenbach, and M.V. Walter), American Society for Microbiology, Washington DC, pp. 1–894.

39 Sullivan, M.B., Waterbury, J.B., and Chisholm, S.W. (2003) Cyanophages infecting the oceanic cyanobacterium Prochlorococcus. *Nature*, **424**, 1047–1051.

40 Short, C.M. and Suttle, C.A. (2005) Nearly identical bacteriophage structural gene sequences are widely distributed in marine and freshwater environments. *Appl. Environ. Microbiol.*, **71**, 480–486.

41 Hennes, K.P., Suttle, C.A., and Chan, A.M. (1995) Fluorescently labeled virus probes show that natural virus populations can control the structure of marine microbial communities. *Appl. Environ. Microbiol.*, **61**, 3623–3627.

42 Hutchinson, G.E. (1961) The paradox of the plankton. *Am. Nat.*, **45**, 137–145.

43 Siegel, D.A. (1998) Resource competition in a discrete environment: why are plankton distributions paradoxical? *Limno. Oceanogr.*, **43**, 1133–1146.

44 Thingstad, T.F. and Lignell, R. (1997) Theoretical models for the control of bacterial growth rate, abundance, diversity and carbon demand. *Aquat. Microb. Ecol.*, **13**, 19–27.

45 Helton, R.R., Cottrell, M.T., Kirchman, D.L., and Wommack, K.E. (2005) Evaluation of incubation-based methods for estimating

virioplankton production in estuaries. *Aquat. Microb. Ecol.*, **41**, 209–219.

46 Levin, B.R., Steward, F.M., and Chao, L. (1977) Resource-limited growth, competition, and predation: a model and experimental studies with bacteria and bacteriophage. *Am. Nat.*, **1**, 113.

47 Waterbury, J.B. and Valois, F.W. (1993) Resistance to co-occurring phages enables marine *Synechococcus* communities to coexist with cyanophages abundant in seawater. *Appl. Environ. Microbiol.*, **59**, 3393–3399.

48 Bohannan, B.J.M. and Lenski, R.E. (1997) Effect of resource enrichment on a chemostat community of bacteria and bacteriophage. *Ecology*, **78**, 2303–2315.

49 Hollibaugh, J.T. and Azam, F. (1983) Microbial degradation of dissolved proteins in the seawater. *Limnol. Oceanogr.*, **28**, 1104–1116.

50 Chen, F. and Lu, J.R. (2002) Genomic sequence and evolution of marine cyanophage P60: a new insight on lytic and lysogenic phages. *Appl. Environ. Microbiol.*, **68**, 2589–2594.

51 Thompson, J.R., Pacocha, S., Pharino, C., Klepac-Ceraj, V., Hunt, D.E., Benoit, J., Sarma- Rupavtarm, R., Distel, D.L., and Polz, M.F. (2005) Genotypic diversity within a natural coastal bacterioplankton community. *Science*, **307**, 1311–1313.

52 Cochran, P.K. and Paul, J.H. (1998) Seasonal abundance of lysogenic bacteria in a subtropical estuary. *Appl. Environ. Microbiol.*, **64**, 2308–2312.

53 Jiang, S.C. and Paul, J.H. (1998) Significance of lysogeny in the marine-environment: studies with isolates and a model of lysogenic phage production. *Microb. Ecol.*, **35**, 235–243.

54 Wilcox, R.M. and Fuhrman, J.A. (1994) Bacterial viruses in coastal seawater: lytic rather than lysogenic production. *Mar. Ecol. Prog. Ser.*, **114**, 35–45.

55 Weinbauer, M.G. and Suttle, C.A. (1996) Potential significance of lysogeny to bacteriophage production and bacterial mortality in coastal waters of the Gulf-of-Mexico. *Appl. Environ. Microbiol.*, **62**, 4374–4380.

56 Ripp, S., Ogunseitan, O.A., and Miller, R.V. (1994) Transduction of a freshwater microbial community by a new *Pseudomonas aeruginosa* generalized transducing phage, UT1. *Mol. Ecol.*, **3**, 121–126.

57 Jiang, S.C. and Paul, J.H. (1998) Gene transfer by transduction in the marine environment. *Appl. Environ. Microbiol.*, **64**, 2780–2787.

58 Sverdrup, H.U., Johnson, M.W., and Fleming, R.H. (1942) *The Oceans*, Prentice Hall, Englewood Cliffs, NJ.

59 Goedele, N.M., Stephen, H., and Peter, C. (2010) The mechanism of retroviral integration from X-ray structures of its key intermediates. *Nature*, **468**, 326–329.

60 Riebesell, U., Kortzinger, A., and Oschlies, A. (2009) Sensitivities of marine carbon fluxes to ocean change. *Proc. Natl. Acad. Sci. USA*, **106**, 20602–20609.

61 Sabine, C.L. and Tanhua, T. (2010) Estimation of anthropogenic CO_2 inventories in the ocean. *Annu. Rev. Mar. Sci.*, **2**, 175–198.

62 Doney, S.C., Fabry, V.J., Feely, R.A., and Kleypas, J.A. (2009) Ocean acidification: the other CO_2 problem. *Annu. Rev. Mar. Sci.*, **1**, 169–192.

63 Caldeira, K. and Wickett, M.E. (2003) Anthropogenic carbon and ocean pH. *Nature*, **425**, 365.

64 Krueger, A.P. and Fong, J. (1937) The relationship between bacterial growth and phage production. *J. Gen. Physiol.*, **21**, 137–150.

65 Jin, S., Zhang, B., Weisz, O.A., and Montelaro, R.C. (2005) Receptor-mediated entry by equine infectious anemia virus utilizes a pH-dependent endocytic pathway. *J. Virol.*, **79**, 14489–14497.

66 Weinbauer, M.G. (2004) Ecology of prokaryotic viruses. *FEMS Microbiol. Rev.*, **28**, 127–181.

4
Taxonomic Study of Antibiotic-Producing Marine Actinobacteria

Priyanka Kishore, Neelam Mangwani, Hirak R. Dash, and Surajit Das

4.1
Introduction

Actinomycetes are Gram-positive bacteria, with a high guanine (G) and cytosine (C) ratio in their DNA (>55 mol%), which are phylogenetically related to the evidence of 16S ribosomal cataloging and DNA–rDNA pairing studies [1]. Actinomycetes are soil organisms that have characteristics similar to bacteria and fungi, and yet possess sufficient distinctive features to delimit them into a distinct category in the strict taxonomic sense. Actinomycetes are clubbed with bacteria in the same class of Schizomycetes, but are confined to the order Actinomycetales [2]. Actinomycetes possess many important and interesting features. They are of considerable importance as producers of antibiotics and other therapeutically useful compounds. Antibiotic crams have been a productive research topic in academia in terms of its mode of action and mechanisms of resistance among the microorganisms [3]. They provide challenges and surprises in respect to its elemental scenery, biosynthetic pathways, evolution, and biochemical mode of action as a natural product [4,5]. Being complex in its functionality and chirality, the total synthesis of antibiotic in laboratory will be an intricating task to achieve [6]. Although the antibiotic penicillin was discovered in 1928, it took years to elucidate complete structure of this relatively simple molecule by Dorothy Crowfoot Hodgkin with the help of X-ray crystallographic studies [7] and it was later confirmed by total synthesis in 1959 by Sheehan [8]. The history of antibiotics provides detailed information about the mode of action on target ligands of this tiny vital complex molecule [9,10].

Actinomycetes exhibit a range of life cycles that are unique among the prokaryotes and appear to play a major role in the cycling of organic matter in the soil ecosystem [11]. Therefore, actinomycetes are important due to both their diversity and proven ability to produce new compounds and the discovery of novel antibiotic and nonantibiotic lead molecules through microbial secondary metabolite screening. Actinomycetes are numerous and widely distributed in soil and are next to bacteria in abundance. They are sensitive to acidity or low pH (optimum pH range 6.5–8.0) and waterlogged soil conditions. The population of actinomycetes increases with depth of soil. They are heterotrophic, aerobic, and mesophilic (25–30 °C) organisms,

Marine Microbiology: Bioactive Compounds and Biotechnological Applications, First Edition.
Edited by Se-Kwon Kim

and some species commonly present in compost and manures are thermophilic, growing at 55–65 °C (e.g., *Thermoactinomycetes* and *Streptomyces*). In the order of abundance in soils, the common genera of actinomycetes are *Streptomyces* (nearly 70%), *Nocardia*, and *Micromonospora*, although actinomycetes, *Actinoplanes*, *Micromonospora*, and *Streptosporangium* are also generally encountered.

Marine environments are largely untapped sources for the isolation of new microorganisms with the potential to produce active secondary metabolites [12]. Among such microorganisms, actinomycetes are of special interest, since they are known to produce chemically diverse compounds with a wide range of biological activities [13]. The demand for new antibiotics continues to grow due to the rapid emergence of multiple antibiotic-resistant pathogens causing life-threatening infection. Nowadays, considerable progress has been made in the fields of chemical synthesis and engineered biosynthesis of antibacterial compounds. So, Nature still remains the richest and the most versatile source of new antibiotics [14–16]. Traditionally, actinomycetes have been isolated from the terrestrial sources only, and the first report on mycelium-forming actinomycetes recovered from marine sediments appeared several decades ago [17]. Recently, the marine-derived actinomycetes are recognized as a source of novel antibiotic and anticancer agents with unusual structure and properties [18].

Actinomycetes are the most valuable prokaryotes from the economic and biotechnology viewpoints. Half of the bioactive compounds have been discovered from the actinomycetes. However, marine actinomycetes may generate different compounds than their terrestrial counterparts as they inhabit in an environment completely different from that of the terrestrial actinomycetes, and the marine actinomycetes diversity is largely unknown and is an untapped source of new metabolites [19]. Marine actinomycetes are not only present in the oceans but also in other marine ecosystems.

Actinomycetes population is considered as one of the major groups of soil population, and it is isolated from a diverse range of marine samples, including sediments obtained from the deep sea [20], from the greatest depth of Mariana Trench [21,22], and also from the vicinity of hydrothermal vents [23]. It is now accepted that actinomycetes can be indigenous to the marine environment and that this environment is likely to yield many unusual actinomycetes that have great potential as producers of novel antibiotics and other compounds. These marine actinomycetes play important ecological roles, similar to their saprophytic counterparts in soils, perhaps substantially impacting the cycling of complex carbon substrates in benthic ocean habitats [24]. However, a well-defined biodiversity and taxonomic study of actinomycetes is important to understand actinomycetes from the marine environment [25]. On agar plates, actinomycetes can be easily distinguished from true bacteria. Unlike slimy distinct colonies of true bacteria that grow quickly, they appear slowly and show powdery consistency and also stick firmly to agar surface. Actinomycetes have gained prominence in recent years because of their potential for producing antibiotics [2]. Antibiotics such as streptomycin, gentamicin, and rifamycin that are in use today are the products of actinomycetes. The actinomycetes are important in the fields of pharmaceutical industries and agriculture. Earlier studies showed that actinomycetes isolated from the soil of Malaysia have the potential to inhibit the growth of several plant pathogens [26]. Oskay *et al.* [27] have reported about the ability of actinomycetes isolated from Turkey's farming soil inhibiting *Erwinia amylovora*, a bacteria that cause fireblight to apple, and *Agrobacterium tumefaciens*, a casual agent of Crown Gall disease [28].

Novel bioactive compounds have been reported from marine actinomycetes that have been proved to be valuable sources. Discovery of abbysomicins, a potent polycyclic polyketide active against methicillin-resistant *Staphylococcus aureus*, and salinosporamide A, an anticancer compound produced by *Salinispora tropica*, is the major landmark in the history of production of bioactive compounds from marine actinomycetes [29,30]. In addition, marineosins, which are related to the prodigiosin class of polypyrrole bacterial pigments produced by *Streptomyces* isolated from marine sediment sample, have significant inhibition properties toward human colon carcinoma cell line and leukemia cell line [31]. Salinosporamide A, the bioactive compound secreted by a marine actinomycete *S. tropica*, has been reported to have antimalarial properties [32].

Mangrove forests are among the world's most productive ecosystem that enrich coastal waters, yield commercial forest products, protect coastlines, and support coastal fisheries. However, mangroves exist under the conditions of high salinity, extreme tides, strong winds, high temperature, and muddy anaerobic soils. Mangrove ecosystem symbolizes an untapped source for the isolation of new microorganisms with the potential to produce active secondary metabolites [12]. Among such microorganisms, Actinobacteria are of special interest, since they are known to produce chemically diverse compounds with a wide range of biological activities [13]. Rapid emergence of multiple antibiotic-resistant pathogens causing life-threatening infections has led to the demand for new antibiotics. Of late, considerable progress has been made in the fields of chemical synthesis and engineered biosynthesis of antibacterial compounds [14–16]. Subsequently, marine *Streptomyces* were recognized as a source of novel antibiotic and anticancer agent with unusual structure and properties [18]. Actinomycetales (commonly known as actinomycetes), an order of Actinobacteria, represent a ubiquitous group of microbes widely distributed in natural ecosystems around the world and are very much significant for the recycling of organic matter [25,33].

Bhitarkanika National Park, Odisha is situated at the delta of the Brahmani and Baitarani rivers meeting the Bay of Bengal. Fifty-five out of 58 reported mangrove species of India are found in Bhitarkanika. The soil associated with mangroves as well as the rhizosphere region of mangroves has been reported to harbor huge varieties of actinomycetes species [34].

Actinomycetes are significant due to their diversity and proven ability to produce new compounds. However, a single genus *Streptomyces* has been considered as a chemical factory due to the discovery of novel antibiotic and nonantibiotic lead molecules. *Streptomyces* can be indigenous to the marine environment, and this environment is likely to harbor diverse strains that may yield numerous novel antibiotics and other compounds. Besides, these marine actinomycetes also play important ecological roles, similar to their saprophytic counterparts in soils, perhaps substantially impacting the cycling of complex carbon substrates in benthic ocean habitats [24]. However, a well-defined biodiversity and taxonomic study of *Streptomyces* is important to understand actinomycetes from the marine environment [35,36]. Thus, this chapter aims to isolate actinomycetes from the sediment samples of mangrove and screening of these isolates for antagonistic activities. Taxonomy of the potent isolates were ascertained by morphological and biochemical studies.

4.2 Materials and Methods

4.2.1 Study Area and Sampling

Sediment samples were collected from the mangrove-associated forest in Bhitarkanika National Park, Odisha, India (20°44.33′N and 086°52.06′E), which is well known for its dense crocodile habitats, other faunas, and natural vegetation of huge varieties of mangrove plants, spreading over 267.14 km^2. Samples were collected from top 4 cm soil profile where most of the microbial activity occurs. Approximately, 500 g sediment was collected by using clean, dry, and sterile polythene bags along with sterile spatula. Samples were stored in icebox and transported to the laboratory for isolation of actinomycetes.

4.2.2 Isolation of Actinomycetes

Each sediment sample was divided into two parts: First part was used as wet sample and the second part was put inside the dryer for 1 week to be used as dry sample. The samples were taken for serial dilution up to 10^{-3} and 0.2 ml of each dilution was inoculated in duplicate plates of the ISP2 medium (yeast extract 4.0 g, malt extract 10 g, dextrose 4.0 g, agar 15 g, seawater 1000 ml, and pH 7.3) for the isolation of actinomycetes by spread plate technique. Nystatin (25 μg/ml) and nalidixic acid (10 μg/ml) were used as antifungal and antibacterial agents, respectively, in plates [37] and incubated at 37 °C for 7 days until powdery actinomycetes colonies were formed on the plates. All the colonies were subsequently collected, subcultured, and maintained in ISP2 slant at 4 °C temperature for further studies.

4.2.3 Screening for Antimicrobial Activity

Screening for antimicrobial activity of selected isolates of actinomycetes was studied preliminarily by cross streak method against five pathogenic bacterial strains (*Escherichia coli, Pseudomonas, Klebsiella, Bacillus*, and *Proteus*). Each isolated culture was streaked as straight line on nutrient agar plate and incubated at 28 °C for 5 days. After observing a good ribbon-like growth of the actinomycetes on the petri dish, test pathogens were streaked at right angles to the original streak of actinomycetes and incubated at 37 °C. The growth inhibition was noticed after 24 and 48 h. A control plate was also maintained without inoculating the actinomycetes to assess the normal growth of the test pathogens.

4.2.4 Identification and Systematics

Based on the results of antagonistic activity, the strains were selected for further taxonomic studies following Refs [36,38,39,40].

4.2.4.1 Phenotypic Characterization

For phenotypic characterization, strains were grown on ISP-2 medium for 7–10 days at 37 °C and the aerial mass color, reverse-side pigment, and melanoid pigments were recorded. Spore chain surface morphology was observed directly under light microscope and spore surface morphology was observed under scanning electron microscope.

4.2.4.2 Physiological and Biochemical Characterization

Assimilation of Carbon Source The ability of actinomycetes strains to utilize various carbon compounds, that is, D-glucose, L-arabinose, sucrose, D-fructose, D-xylose, raffinose, D-mannitol, cellulose, rhamnose, and inositol, as source of energy was studied by following the method recommended in International Streptomyces Project (ISP) [41]. Growth of actinomycetes strains were checked by taking 1% carbon source in Basal mineral salts medium as mentioned in ISP. Plates were streaked by inoculation loop by flame sterilization technique and incubated at 37 °C for 7–10 days. Growths were observed by comparing them with positive and negative controls.

Sodium Chloride Tolerance The test was performed using different concentrations of sodium chloride (0, 5, 10, 15, 20, 25, and 35%) on the ISP-2 medium. Isolate was streaked on the agar medium, incubated at 37 °C for 7–15 days, and the presence or absence of growth was recorded on seventh day onward.

Degradation of Cellulose One percent of carboxymethyl cellulose (CMC) was added to the ISP-2 medium. Plates were inoculated and incubated for 7–15 days. Control plate was used as standard to check the growth of actinomycetes after 7–15 days for cellulose degradation activity that may be visually observed.

Hydrogen Sulfide Production Tryptone yeast extract agar slants were used to study H_2S production. Observations on the presence of the characteristic greenish-brown, brown, bluish-black, or black color of the substrate, indicative of H_2S production, were recorded on day 7, 10, and 15. The incubated tubes were compared with uninoculated controls.

Gelatin Liquefaction Gelatin deep tubes were used to demonstrate the hydrolytic activity of gelatinase. The medium consists of nutrient supplemented with 12% gelatin; this high gelatin concentration results in a stiff medium and also serves as substrate for the activity of gelatinase. Gelatin liquefaction was studied by subculturing the strain on gelatin agar medium.

Hydrolysis of Starch The test was performed on ISP-2 medium supplemented with 1% starch. After 7 days of incubation at 37 °C, the plates were flooded with Lugol's iodine solution.

Coagulation of Milk Milk coagulation was studied with skimmed milk (HiMedia, India). The skimmed milk tubes were inoculated and incubated at 37 °C. The extent of coagulation was recorded on day 7 and 10 of incubation.

Ability to Grow in Different pH This test was carried out on ISP-2 media and pH was set at values 5, 6, 7, 8, 9, and 10. Duplicate slants were prepared for each strain of each value. After incubating for 10–12 days, readings were taken for each strain.

Lipolytic Activity This test was done by taking 1% Tween 20 (HiMedia, India) on ISP-2 media, incubated at a temperature of 37 °C for 7–10 days and the reading was taken subsequently.

4.2.4.3 Chemotaxonomical Characterization

Considering the importance of chemotaxonomical characteristics in the identification of actinomycetes, an attempt was made to identify the actinomycetes by analyzing their cell wall components. The whole-cell sugars and amino acids were analyzed according to the method of Becker *et al.* [42]. Galactose, arabinose, xylose, and mannose were used as standards for sugar. LL-DAP and *meso*-DAP (diaminopimelic acid) were used as amino acid standards for thin layer chromatography of cell wall hydrolysate.

Finally, after all these experiments, results were matched with the keys given for 458 species of actinomycetes included in ISP for taxonomic identification.

4.3 Result

From the collected sediment samples, four isolates (B2, CS, C11, and D1) were found, based on the result of antimicrobial activity by cross streak method against five pathogens (*Pseudomonas, E. coli, Klebsiella, Bacillus,* and *Proteus*). Among these, only D1 exhibited good antimicrobial activity against *Klebsiella* and *Proteus*. Growth of *Proteus* was inhibited by all strains. C11 showed antimicrobial activity against all bacterial strains. *Pseudomonas* was the least inhibited pathogenic strain. However, C11 and D1 showed moderate inhibition against it. *E. coli* also showed least resistance to four isolates (Table 4.1).

Table 4.1 Actinomycetes strains showing antimicrobial activity against test pathogens.

Strains	*Pseudomonas*	*Bacillus*	*E. coli*	*Proteus*	*Klebsiella*
B2	−	−	−	+	−
CS	±	+	−	+	−
C11	±	+	+	+	+
D1	±	−	−	++	++

++: good, +: positive, −: negative, ±: moderate.

Table 4.2 Phenotypic features of isolated actinomycetes strain.

Isolates	Aerial mass color	Spore chain morphology	Spore surface morphology	Soluble pigments	Melanoid pigments	Reverse-side pigment
B2	W	S	S	1	0	0
CS	Gy	S	Sp	1	1	0
C11	W	S/RF	S	1	0	0
D1	Y	S	S	1	0	0

W: white, Gy: gray, Y: yellow, 1: distinctive/present, 0: nondistinctive/absent, S: smooth, Sp: spiny, S: spiral spira.

Phenotypic parameters like aerial mass color, reverse-side pigment, melanoid pigments, spore chain and spore surface morphology were recorded (Table 4.2). Spore chain and spore surface morphology of the isolates were found to be spiral, smooth, and spiny (Figure 4.1).

Isolates were also studied for carbon assimilation for different monosaccharides. C11 and CS were able to utilize all monosaccharides, whereas B2 and D1 were specific in terms of carbon source (Table 4.3). In terms of optimum growth at different pH, B2 and CS could grow at pH 5–9, C11 at pH 5–8, and D1 at pH 7 and 8 (Table 4.4).

Biochemical assay and enzymatic activities like cellulase and H_2S production, and gelatinase, amylase, caseinase, and lipolytic activities were carried out in respective media (Table 4.5).

Cell wall sugar and amino acid composition often provide imperative information on actinomycetes classification and identification. B2, CS, and D1 cell wall composition matches with type II cell wall, that is, glycine (+), LL-DAP (−), *meso*-DAP (+), arabinose (−), and galactose (−), whereas C11 matches with type V, that is, ornithine (+), lysin (+), arabinose (−), and galactose (−) cell wall. Cell wall type II and V are particular for genus *Streptomyces* and *Actinomyces*, respectively. After obtaining all the results from the experiments conducted, they were matched with the keys given for 458 species of actinomycetes included in ISP. From the keys given, the strains B2, CS, C11, and D1 were identified as *Streptomyces almquisti*,

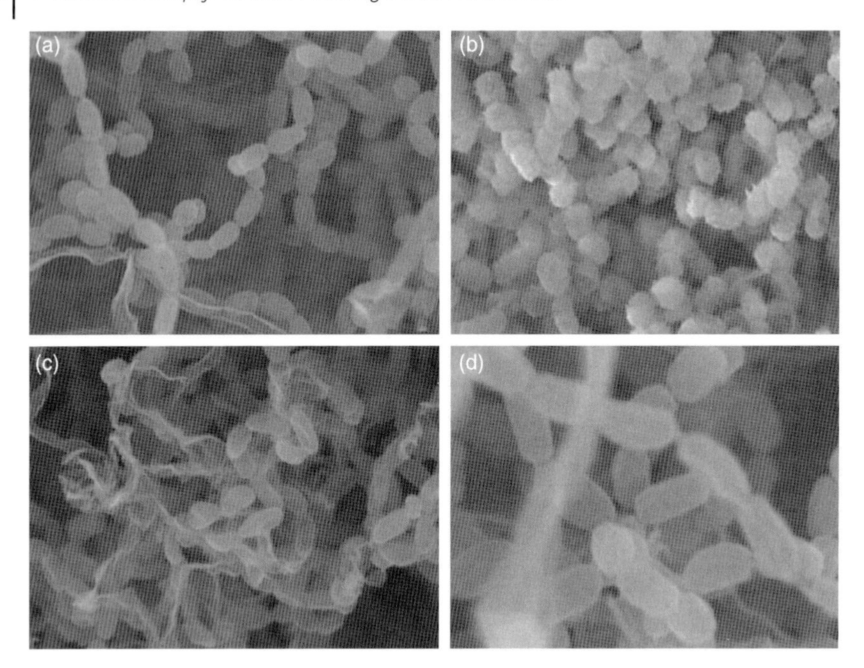

Figure 4.1 Scanning electron microscopy of spore surface morphology of actinomycetes. (a) B2, (b) CS, (c) C11, and (d) D1.

Table 4.3 Assimilation of carbon sources by different actinomycetes strains.

Strains	Glucose	Xylose	Inositol	Sucrose	Raffinose	Fructose	Rhamnose	Mannitol	Arabinose
B2	+	+	+	−	−	±	−	+	−
CS	+	+	+	+	+	±	+	+	±
C11	+	+	+	+	+	+	+	+	±
D1	+	+	−	+	±	+	+	+	±

+: positive, −: negative, ±: moderate.

Table 4.4 Growth of actinomycetes at various pH.

Strains	pH 5	pH 6	pH 7	pH 8	pH 9
B2	+	+	+	+	+
CS	+	+	+	+	+
C11	+	+	+	+	−
D1	−	−	+	+	−

+: positive, −: negative, ±: moderate.

Table 4.5 Biochemical assay of the isolated actinomycetes.

Strains	Cellulase activity	H_2S production	Gelatinase activity	Amylase activity	Caseinase activity	Lipolytic activity
B2	++	+	−	+++	+	−
CS	+	+	−	+	±	−
C11	+	+	−	±	−	−
D1	+	+	−	−	+	−

+: positive, −: negative, ±: moderate.

Streptomyces luteogriseus, *Actinomyces aureocirculatus*, and *Streptomyces spheroides*, respectively. The match was done on the basis of maximum percentage of resemblance of characteristics.

4.4 Discussion

Marine actinomycetes have been least explored, compared to other groups of microorganisms. As they inhabit in very adverse conditions, the chances of isolating novel, potential bioactive compounds are high from these organisms. However, without knowing the evolution of secondary metabolic pathways in marine actinomycetes that have been transferred among the different taxa, the correlation between taxonomic and biosynthetic novelties and hence any firm conclusions cannot be drawn. In this regard, taxonomic studies play an important role in correctly identifying the isolate and further characterization of the same. It helps in correlation of the isolated taxa and their secreted compounds with others from the same or different environmental conditions.

Goodfellow and Haynes [43] reviewed the literature on isolation of actinomycetes from marine sediments and suggested that the marine sediments may be valuable for the isolation of novel actinomycetes. In this chapter, four actinomycetes were isolated from mangrove soil and phonotypical features were studied. For species affiliation, various physiological and biochemical characteristics were analyzed [35]. The aerial mass color of almost all strains was whitish gray and only one strain D1 showed yellow color. Vanajakumar and Natarajan [44] reported that white color actinomycetes were the dominant forms. The isolates were screened for their antimicrobial activity. Isolate D1 showed good zone of inhibition against *Klebsiella* and *Proteus*. Almost all strains have shown inhibition against *Proteus*, whereas *E. coli* has shown substantial sensitivity. Gurung *et al.* [45] have reported 27 different actinomycetes showing antimicrobial activity against the human pathogenic *Salmonella typhi*, *E. coli*, *Shigella* sp., *Proteus mirabilis*, and *Bacillus subtilis*. As per their result, actinomycetes were more active against Gram-positive pathogens than the Gram-negative pathogens, which have been well established in this study. Actinomycetes isolates identified as *Streptomyces* sp., *Micromonospora* sp., and *Nocadia* sp. were reported to have antimicrobial properties by their cell-free extract toward human pathogens such as *S. aureus*, *Proteus vulgaris*, *Proteus aeroginosa*, *E. coli*, *B. subtilis*, *Bacillus megaterium*, *Klebsiella pneumoniae*, *Candida albicans*, *Aspergillus niger*, and *Saccharomyces cerevisiae* [46], and the results of the recent study fall under the same trend. The bioactive compounds isolated from actinomycetes also showed antimicrobial activity against multiantibiotic-resistant *S. aureus* and *E. coli* [47], which has also been well established in this chapter. The new drugs, notably antibiotics from marine actinomycetes, urgently need to cease and reverse the relentless spread of antibiotic-resistant pathogens that cause life-threatening infections [48]. Filamentous bacteria belonging to the order Actinomycetales, especially *Micromonospora* and *Streptomyces* strains, have a unique and proven capacity to produce novel antibiotics [49]; hence, the continued interest in screening such organisms for new bioactive compounds prevails. It is also becoming increasingly clear that unexplored and underexplored habitats, such as desert biomes and marine ecosystems, are very rich sources of novel actinomycetes that have the capacity to produce interesting new bioactive compounds, including antibiotics [50].

Major milestone in the identification is the assimilation of carbon by actinomycetes. Utilization of carbon sources, namely, arabinose, xylose, inositol, mannitol, fructose, rhamnose, sucrose, and raffinose were analyzed for classification. These carbon sources were separately supplemented (1%) on each ISP-2 medium. Test includes 10 carbon sources that are sterilized by membrane filtration method. Almost all the isolates have shown very luxuriant growth. Pandey *et al.* [51] showed that for the optimum production of antibiotics, certain carbon sources are required. In this study, the authors also suggested that pH might play an important role in the production of antibiotics by actinomycetes. pH tolerance test was conducted for all isolates, which showed positive results. As the samples were collected from the mangroves, it is often expected that the isolates can tolerate high range of salinity. Study done by Vasavada *et al.* [52] showed that the use of media, pH, salinity, and carbon and nitrogen affect the growth and antibiotic production by actinomycetes. The morphology of the spore-bearing hyphae with entire spore chain along with substrate and aerial mycelium was examined under light microscope as well as scanning electron microscope. All this screening was done by the cover slip culture technique following Thenmozhi and Kannabiran [53]. Hydrolysis of starch was evaluated by using the media of Gordon *et al.* [54]. Liquefaction of gelatin was evaluated by the method of Waksman [55]. H_2S production test was conducted by preparing the slant culture and all the nine isolates showed positive result. To know the overall activity of all the strains, various enzymatic screening have been performed. These include the cellulase activity, caseinase activity, amylase activity, lipolytic activity, and gelatinase activity. Gelatin hydrolysis was not shown by any strain. But the amylase activity was shown by almost all the strains. Cellulase and caseinase activities were shown almost in the same manner by all strains. Chemotaxonomical characteristics like cell wall sugar and amino acid composition often provide imperative information on actinomycetes identification up to genus level. In addition to the glucosamine and muramic acid of peptidoglycan, actinomycete cells contain some kinds of unambiguous sugars. The sugar pattern plays a key role in the identification of sporulating actinomycetes that have *meso*-DAP in their cell walls. B2, CS, and D1 cell wall composition matches with type II cell wall, while C11 cell wall matches type V cell wall. Cell wall types II and V are particular for genus *Streptomyces* and *Actinomyces*, respectively. Presence of diaminopimelic acid isomers is one of the most important cell wall properties of actinomycetes type II [38]. Finally, all these experimental results matched with the keys given for 458 species of actinomycetes included in ISP, and the species identification was done and it was found that all the isolates have been grouped under *Streptomyces* genus.

All the isolates studied in this chapter were identified to be under *Streptomyces* genus. Anzai *et al.* [56] found *Streptomyces* is the most abundant genus among actinomycetes distributed in the environment, and it produces a variety of bioactive compounds that are active against pathogenic bacteria and hemolytic activities. A potent actinomycetes strain isolated from Vellar Estuary, Tamil Nadu, India, producing antibacterial, antifungal, and anti-yeast compounds was identified to be *Streptomyces bikiniensis* by its cultural and physiological properties and from DNA homology study [57]. Another study from marine environment reveals that *Streptomycetes* spp. are most prevalent in both water and sediment samples showing bioactive potential [58].

4.5 Conclusion

It can be concluded that actinomycetes diversity is least explored in the marine environments and that actinomycetes are the greatest source of bioactive compounds. They are responsible for antibacterial, antifungal, anti-yeast, antitumor, anticancer, and antimalarial activities. The bioactive compounds produced by marine actinomycetes are more active against the Gram-positive pathogens than against the Gram-negative human pathogens. *Streptomyces* are the most abundant genus among the marine actinomycetes present in the mangrove-associated marine environments.

Acknowledgments

Authors would like to acknowledge the authorities of NIT, Rourkela for providing the required facilities. N.M and H.R.D gratefully acknowledge the receipt of research fellowship from the Ministry of Human Resource Development, Government of India. S.D thanks the Department of Biotechnology, Government of India for a research grant.

References

1 Goodfellow, M. and Williams, S.T. (1983) Ecology of actinomycetes. *Ann. Rev. Microbiol*, **37**, 189–126.

2 Kumar, S.V., Sahu, M.K., and Kathiresan, K. (2005) Isolation and characterization of *Streptomycetes* producing antibiotics from a

mangrove environment. *Asian J. Microbial. Biotechnol. Environ. Sci.*, 7 (3), 457–464.

3 Bryskier, A. (2005) *Antimicrobial Agents: Antibacterials and Antifungals*, ASM Press, Washington, DC.

4 Strohl, W.R. (1997) *Biotechnology of Antibiotics*, 2nd edn, Marcel Dekker, Inc., New York, NY.

5 Brotze-Oesterhelt, H. and Brunner, N.A. (2008) How many modes of action should an antibiotic have? *Curr. Opin. Pharmacol.*, **8**, 564–573.

6 Nikolaou, K.C. and Montagnon, T. (2008) *Molecules That Changed the World*, Wiley-VCH Verlag GmbH, Weinheim, Germany.

7 Hodgkin, D.C. (1949) The X-ray analysis of the structure of penicillin. *Adv. Sci.*, **6**, 85–89.

8 Sheehan, J. and Henery-Logan, K.R. (1959) The total synthesis of penicillin V. *J. Am. Chem. Soc*, **81**, 3089–3094.

9 Gale, E.F., Cundliffe, E., Reynolds, P.E., Richmond, M.H., and Waring, M.J. (eds) (1981) *The Molecular Basis of Antibiotic Action*, 2nd edn, John Wiley & Sons, Inc., New York.

10 Walsh, C. (2003) *Antibiotics: Actions, Origins, Resistance*, ASM Press, Washington, DC.

11 Veiga, M., Esparis, A., and Fabregas, J. (1983) Isolation of cellulolytic actinomycetes from marine sediments. *Appl. Environ. Microbiol.*, **46**, 286–287.

12 Baskaran, R., Vijayakumar, R., and Mohan, P.M. (2011) Enrichment method for the isolation of bioactive actinomycetes from mangrove sediments of Andaman Islands, India. *Malaysian J. Microbiol.*, **7** (1), 1–7.

13 Bredholt, H., Fjaervik, E., Jhonsen, G., and Zotechev, S.B. (2008) Actinomycetes from sediments in the Trondhein Fjrod, Norway: diversity and biological activity. *J. Marine Drugs*, **6**, 12–24.

14 Kpehn, F.E. and Carter, G.T. (2005) The evolving role of natural products in drug discovery. *Nat. Rev. Drug Discov.*, **4**, 206–220.

15 Baltz, R.H. (2006) Marcel Faber Roundtable: is our antibiotic pipeline unproductive because of starvation, constitution or lack of inspiration? *J. Ind. Microbiol. Biotechnol.*, **33**, 507–513.

16 Pelaez, F. (2006) The historical derive of antibiotic from microbial natural product: can history repeat? *J. Biochem. Pharmacol.*, **71**, 981–990.

17 Weyland, H. (1969) Actinomycetes in North Sea and Atlantic Ocean sediments. *Nature*, **223**, 858.

18 Jensen, P.R.E., Gontang, C., Mafnas, T., Mincer, J., and Fenical, W. (2005) Culturable marine actinomycetes diversity from tropical Pacific Ocean sediments. *Appl. Environ. Microbiol*, **7**, 1039–1048.

19 Stach, J.E.M., Maldonado, L.A., Ward, A.C., Goodfellow, M., and Bull, A.T. (2003) New primers for the class Actinobacteria: application to marine and terrestrial environments. *Environ. Microbiol.*, **5**, 828–841.

20 Colquhoun, J.A., Mexson, J., Goodfellow, M., Ward, A.C., Horikoshi, K., and Bull, A.T. (1998) Novel rhodococci and other mycolata actinomycetes from the deep sea. *Antonie van Leeuwenhoek*, **74**, 27–40.

21 Takami, H., Inoue, A., Fuji, F., and Horikoshi, K. (1997) Microbial flora in the deepest sea mud of the Mariana Trench. *FEMS Microb. Lett.*, **152**, 279–285.

22 Pathom-aree, W., Stach, J.E.M., Ward, A.C., Horikoshi, K., Bull, A.T., and Goodfellow, M. (2006) Diversity of actinomycetes isolated from Challenger Deep sediment (10, 898m) from the Mariana Trench. *Extremophiles*, **10**, 181–189.

23 Murphy, P. and Hill, R.T. (1998) Marine vision becomes reality: drugs from the sea. *Biofuture*, **179**, 34–37.

24 Mincer, T.J., Jensen, P.R., Kauffman, C.A., and Fenical, W. (2002) Widespread and persistent populations of a major new marine actinomycete taxon in ocean sediments. *Appl. Environ. Microbiol.*, **68**, 5005–5011.

25 Das, S., Ward, L.R., and Burke, C. (2008) Prospects of using marine Actinobacteria as probiotics in aquaculture. *Appl. Microbiol. Biotechnol*, **81**, 419–429.

26 Jeffrey, L.S.H., Sahilah, A.M., Son, R., and Tosiah, S. (2007) Isolation and screening of actinomycetes from Malaysian soil for their enzymatic and antimicrobial activities. *J. Trop. Agric. Food Sci.*, **35**, 159–164.

27 Oskay, M., Same, A., and Azeri, C. (2004) Antibacterial activity of some actinomycetes isolated from farming soils of Turkey. *Afr. J. Biotechnol.*, **3**, 441–446.

28 Jeffrey, L.S.H. (2008) Isolation, characterization and identification of actinomycetes from agriculture soils at Semongok, Sarawak. *Afr. J. Biotechnol.*, **7** (20), 3697–3702.

29 Riedlinger, J., Reicke, A., Zahner, H., Krismer, B., Bull, A.T., Maldonado, L.A., Ward, A.C., Goodfellow, M., Bister, B., Bischoff, D., Süssmuth, R.D., and Fiedler, H.P. (2004) Abyssomicins, inhibitors of the *para*-aminobenzoic acid pathway produced by the marine *Verrucosispora* strain AB-18–032. *J. Antibiot.*, **57**, 271–279.

30 Williams, P.G., Buchanan, G.O., Feling, R.H., Kauffman, C.A., Jensen, P.R., and Fenical, W. (2005) New cytotoxic salinosporamides from marine actinomycete *Salinispora tropica*. *J. Org. Chem.*, **70**, 6196–6203.

31 Boonlarppradab, C., Kauffman, C.A., Jensen, P.R., and Fenical, W. (2008) Marineosins A and B, cytotoxic spiroaminals from a marine-derived actinomycete. *Org. Lett.*, **10**, 5505–5508.

32 Prudhomme, J., McDaniel, E., Ponts, N., Bertani, S., and Fenical, W. (2008) Marine actinomycetes: a new source of compounds against the human malaria parasite. *PLoS ONE*, **3** (6), e2335.

33 Srinivasan, M.C., Laxman, R.S., and Deshpande, M.V. (1991) Physiology and nutrition aspects of actinomycetes: an overview. *World J. Microbiol. Biotechnol.*, **7**, 171–184.

34 Xiao, J., Wang, Y., Luo, Y., Xie, S.J., Ruan, J.S., and Xu, J. (2009) *Streptomyces avicenniae* sp. nov., a novel actinomycete isolated from the rhizosphere of the mangrove plant *Avicennia marina*. *Int. Syst. Evol. Microbiol.*, **59**, 2624–2628.

35 Das, S., Lyla, P.S., and Khan, S.A. (2008) Distribution and generic composition of culturable marine actinomycetes from the sediments of Indian continental slope of Bay of Bengal. *Chin. J. Oceanol. Limnol.*, **26** (2), 166–177.

36 Das, S., Lyla, P.S., and Khan, S.A. (2008) Characterization and identification of marine actinomycetes existing systems, complexities and future directions. *Natl. Acad. Sci. Lett.*, **31** (5), 149–160.

37 Ravel, J., Amoroso, M.J., Colwell, R.R., and Hill, R.T. (1998) Mercury-resistant actinomycetes from the Chesapeake Bay. *FEMS Microbiol. Lett.*, **162**, 177–184.

38 Lechevalier, M.P. and Lechevalier, H.A. (1970) Chamical composition as a criterion in classification of aerobic actinomycetes, *Int. J. Syst. Bacteriol.*, **20**, 435–443.

39 Shirling, E.B. and Gottlieb, D. (1966) Methods for characterization of *Streptomyces* species. *Int. J. Syst. Bacteriol.*, **16**, 313–340.

40 Nonomura, H. (1974) Key for classification and identification of 458 species of Streptomycetes included in ISP. *J. Ferment. Technol.*, **52**, 78–92.

41 Pridham, T.G., Anderson, P., Foley, E., Lindenfelser, L.A., Hesseltine, E.W., and Benedict, R.G. (1957) A selection of media for maintenance and taxonomic study of *Streptomyces*. *Antibiot Annu.*, 947–953.

42 Becker, B. and Lechevalier, M.P. (1965) Chemical composition of cell wall preparations from strains of various form genera of aerobic actinomycetes. *Appl. Environ. Microbiol.*, **13**, 236–243.

43 Goodfellow, M. and Haynes, J.A. (1984) Actinomycetes in marine sediments, in *Biological, Biochemical and Biomedical Aspects of Actinomycetes* (eds L. Ortiz-Ortiz, L.F. Bjalil, and V. Yakoleff), Academic Press, London, pp. 453–472.

44 Vanajakumar, S.N. and Natarajan, R. (1995) Antagonistic properties of actinomycetes isolated from mollusks of the Porto Novo region, South India, pp. 267–274.

45 Gurung, T.V., Sherpa, C., Agrawal, V.P., and Lekhak, B. (2009) Isolation and characterization of antibacterial actinomycetes from soil samples of Kalapatthar, Mount Everest region. *Nepal. J. Sci. Technol.*, **10**, 173–182.

46 Usha, R.J., Hema, S.N., and Devi, D.K. (2011) Antagonistic activity of actinomycetes isolates against human pathogen. *J. Microbiol. Biotechnol. Res.*, **1**, 74–79.

47 Sharma, D., Kaur, T., Chadha, B.S., and Manhas, R.K. (2011) Antimicrobial activity of actinomycetes against multidrug resistant *Staphylococcus aureus*, *E. coli* and

various other pathogens. *Trop. J. Pharma. Res.*, **10**, 801–808.

48 Talbot, G.H., Bradley, J., Edwards, J.E., Jr., Gilbert, D., Scheld, M., and Bartlett, J.G. (2006) Bad bugs need drugs: an update on the development pipeline from the Antimicrobial Availability Task Force of the Infectious Diseases Society of America. *Clin. Infect. Dis.*, **42**, 657–668.

49 Bentley, S.D., Chater, K.F., Cerdeno-Tarraga, A.M., Challis, G.L., Thompson, N.R., James, K.D., Harris, D.E., Quail, M.A., Kieser, H., and Harper, D. (2002) Complete genome sequence of the model actinomycete *Streptomyces coelicolor. Nature*, **417**, 141–147.

50 Hong, K., Gao, A.H., Xie, Q.Y., Gao, H., Zhuang, L., Lin, H.P., Yu, H.P., Li, J., and Yao, X.S. (2009) Actinomycetes for marine drug discovery isolated from mangrove soils and plants in China. *Mar. Drugs*, **7** (1), 24–44.

51 Pandey, A., Shukla, A., and Majumdar, S.K. (2005) Utilization of carbon and nitrogen sources by *Streptomyces kanamyceticus* M27 for the production of an antibacterial antibiotic. *Afr. J. Biotechnol.*, **4**, 909–910.

52 Vasavada, S.H., Thumar, J.T., and Singh, S. P. (2006) Secretion of a potent antibiotic by salt-tolerant and alkaliphilic actinomycete *Streptomyces sannanensis* strain RJT-1. *Curr. Sci.*, **91**, 1393–1397.

53 Thenmozhi, M. and Kannabiran, K. (2010) Studies on isolation, classification and phylogenetic characterization of novel antifungal *Streptomyces* sp. VITSTK7 in India. *Curr. Res. J. Biol. Sci*, **2** (5), 306–312.

54 Gordon, R.E., Barnett, D.A., Handerhan, J.E., and Pang, C.H.N. (1974) *Nocardia coeliaca, Nocardia autotrophica* and the nocardin strain. *Int. J. Syst. Bacteriol.*, **24**, 54–63.

55 Waksman, S.A. (1961) *The Actinomycetes Classification, Identification and Description of Genera and Species*, Williams & Wilkins Company, Baltimore, MD, p. 327.

56 Anzai, K., Ohno, M., Nakashima, T., Kuwahara, N., Suzuki, R., Tamura, T., Komaki, H., Miyadoh, S., Harayama, S., and Ando, K. (2008) Taxonomic distribution of *Streptomyces* species capable of producing bioactive compounds among strains preserved at NITE/NBRC. *Appl. Microbiol. Biotechnol*, **80**, 287–295.

57 Dhanasekaran, D., Selvamani, S., Panneerselvam, A., and Thajuddin, N. (2008) Isolation and characterization of actinomycetes in Vellar Estuary, Annagkoil, Tamil Nadu. *Afr. J. Biotechnol*, **8**, 4159–4162.

58 Ogunmwonyi, I.H., Mazomba, N., Mabinya, L., Ngwenya, E., Green, E., David, A., Akinpelu, D.A., Olaniran, A.O., Bernard, K., and Okoh, A.I. (2010) Studies on the culturable marine actinomycetes isolated from the Nahoon beach in the Eastern Cape Province of South Africa. *Afr. J. Microbiol.*, **4**, 2223–2230.

5
Marine Cyanobacteria: A Prolific Source of Bioactive Natural Products as Drug Leads

Lik Tong Tan

5.1
Introduction

The prokaryotic marine cyanobacteria, especially filamentous strains belonging to the *Lyngbya* and *Symploca* genera, are a cornucopia of novel bioactive natural products. More than 400 compounds have been reported thus far, among which most of them belong to the hybrid polyketide–polypeptide structural class [1,2]. These compounds are characterized by the linking of polyketide-derived fragment, for example, long carbon chain or α-hydroxy/amino acid, with a variety of proteinogenic amino acids. They are further functionalized with unique oxidations, methylations (usually by SAM), and halogenations, resulting in a tremendous structural diversity [3]. The pharmaceutical potential of marine cyanobacteria was realized in the 1970s through the pioneering work of Richard Moore at the University of Hawaii [4]. These earlier studies revealed marine cyanobacterial molecules to be biologically active with potential to be utilized as molecular tools in dissecting cellular processes as well as anticancer agents. The high potency of these natural molecules is due to their specific interactions with cellular targets, including microtubules, actin filaments, and enzymes.

The importance of marine cyanobacteria as source of drug leads is illustrated by the recent approval of an antibody drug conjugate (ADC), brentuximab vedotin, for the treatment of Hodgkin lymphoma and anaplastic large cell lymphoma [5]. The ADC consists of monomethyl auristatin E, which is a synthetic analogue of the cyanobacterial compound dolastatin 10. When the predicted biosynthetic source of marine-derived molecules currently in clinical trials is taken into consideration, marine cyanobacterial compounds constitutes about 20% [6]. Moreover, about 24% (or 29 compounds) of marine natural products commercially available for biomedical research are of cyanobacterial origin [6]. The main aim of this chapter is to provide an overview of the bioactive marine cyanobacterial compounds isolated in the past 3 years with

Marine Microbiology: Bioactive Compounds and Biotechnological Applications, First Edition.
Edited by Se-Kwon Kim

potential use in various disease areas. The various molecules discussed are organized based on their biological properties, such as anticancer agents, neuromodulating agents, anti-infective agents, and anti-inflammatory agents. Updates on the development of important compounds, such as largazole and apratoxins, shown previously to be promising drug leads will also be provided.

5.2 Bioactive Secondary Metabolites from Marine Cyanobacteria

5.2.1 Anticancer Agents

A large proportion of bioactive compounds from marine cyanobacteria are reported to be highly cytotoxic. These compounds were found to target cellular targets, such as proteins associated with cytoskeleton (e.g., microtubules and actin filaments) and enzymes (e.g., histone deacetylase and proteasome). This section will highlight cyanobacterial compounds that interfere with the dynamics of microtubule and actin as well as MDM2–p53 interaction and function of enzymes, including histone deacetylase and proteases. Compounds that modulate the apoptotic process will also be presented.

5.2.1.1 Microtubule-Interfering Compounds

The dolastatins, particularly 10 (**1**) and 15 (**2**), constitute an important class of natural products having remarkable anticancer properties [7]. These compounds were originally isolated and characterized from the Indian Ocean sea hare, *Dolabella auricularia*, in the late 1980s and 1990s. However, it is believed that marine cyanobacteria are the true biogenic sources and the presence of the dolastatins in the sea hare is sequestered via the invertebrate's diet [8]. The potent cytotoxic nature of dolastatins 10 (**1**) and 15 (**2**) is attributed to their specific molecular interference with the dynamics of cellular microtubule. Subsequently, these molecules and their synthetic analogues, including auristatin PE (**3**) and tasidotin (**4**), underwent extensive clinical testing as potential anticancer drugs. One of the auristatin series of compounds, monomethyl auristatin E, formulated as an ADC brentuximab vedotin (**5**), was subsequently approved for the treatment of Hodgkin lymphoma and anaplastic large cell lymphoma [5,9]. Furthermore, three additional dolastatin 10 analogues, formulated as ADCs, are currently in the clinical pipeline as anticancer drugs [6]. A new dolastatin 10 analogue, auristatin TP as sodium salt (**6**), was recently reported from the laboratory of Dr. Pettit [10]. This molecule, a tyramide phosphate modification of dolastatin 10, was demonstrated to possess markedly improved anticancer properties due to the increased bioavailability and the presence of the phosphate group. The ED_{50} values of compound **6** ranged from <1.2 to 54.6 nM when tested against a panel of cancer cell lines, including P388, NCI-H460, and MCF-7 [10].

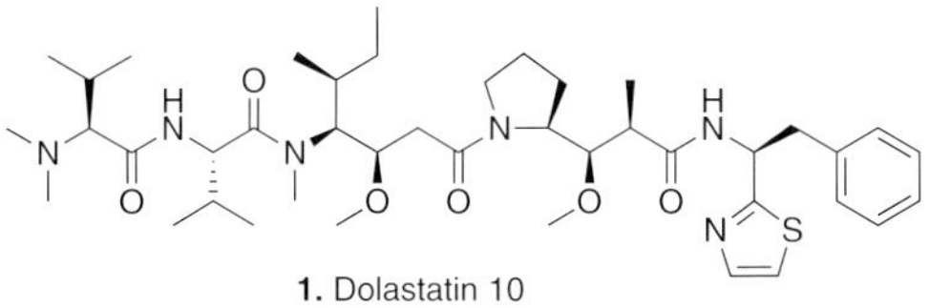

1. Dolastatin 10

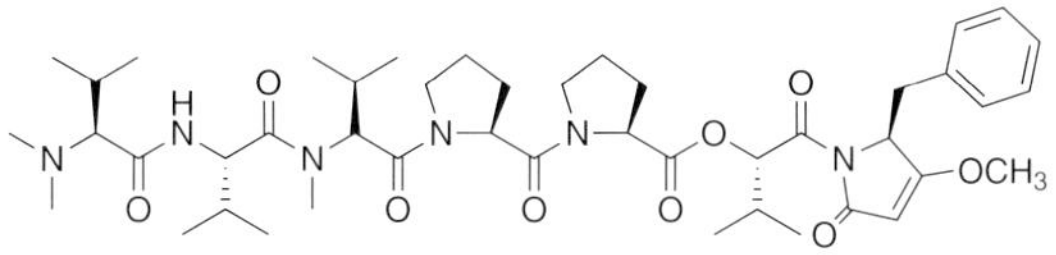

2. Dolastatin 15

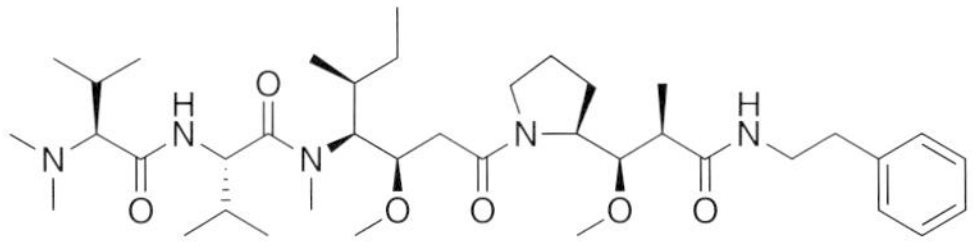

3. Auristatin PE (= TZT-1027, Soblidotin)

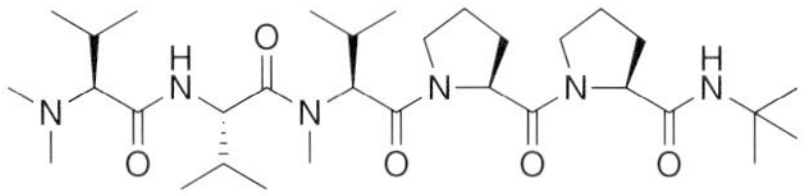

4. Tasidotin (= Synthadotin, ILX651)

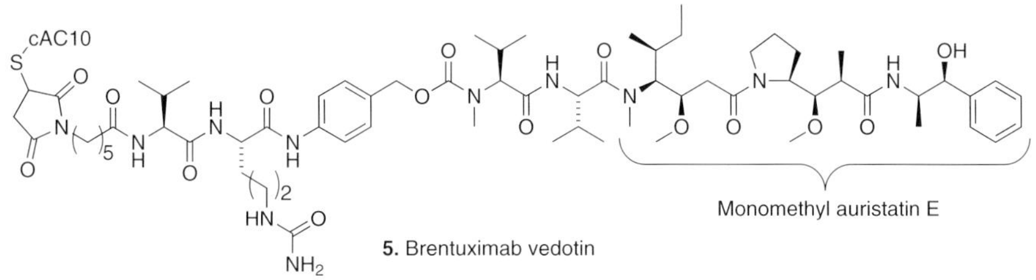

Monomethyl auristatin E

5. Brentuximab vedotin

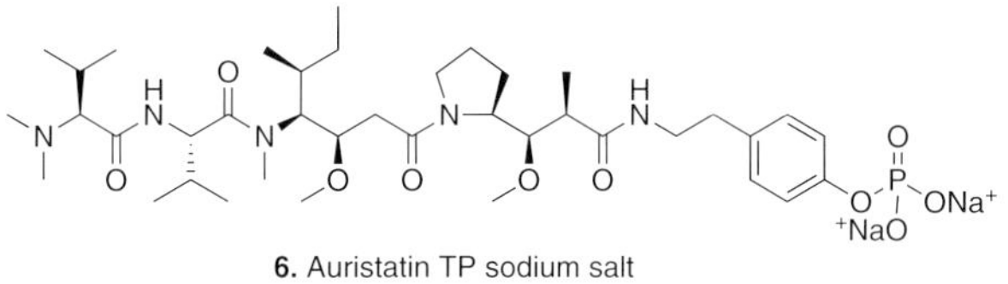

6. Auristatin TP sodium salt

5.2.1.2 **Actin-Stabilizing Agents**

A host of potent actin-disrupting molecules have been previously uncovered from marine cyanobacteria, including cyclic depsipeptides [e.g., dolastatin 11 (**7**)] as well as thiazole-containing cyclic depsipeptides [e.g., lyngbyabellin A (**8**) and hectochlorin (**9**)]. Recently, a linear peptide, bisebromoamide (**10**), having potent cytotoxic properties, was reported from an Okinawan strain of the marine cyanobacterium, *Lyngbya* sp. [11]. This novel compound contained a unique *N*-methyl-3-bromotyrosine, a modified 4-methylproline, a 2-(1-oxo-propyl) pyrrolidine, and a *N*-pivalamide unit. Based on the total synthesis of bisebromoamide, as reported by Gao *et al.* [12], the stereochemistry of the molecule has been revised to compound **10** and subsequently confirmed by Sasaki and coworkers [13]. In addition, a highly convergent method of synthesizing bisebromoamide and other simplified analogues was recently reported by Li *et al.* [14].

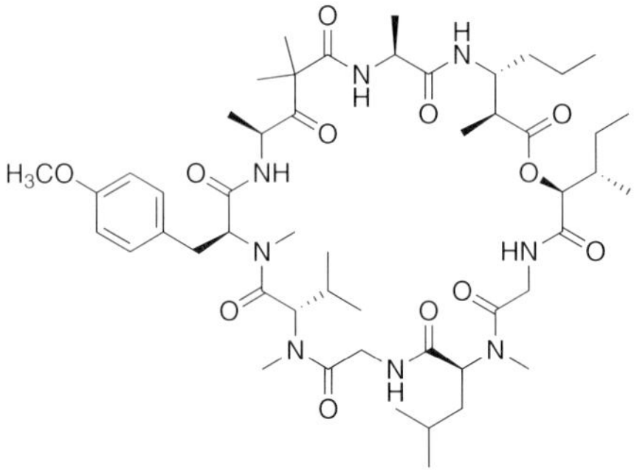

7. Dolastatin 11

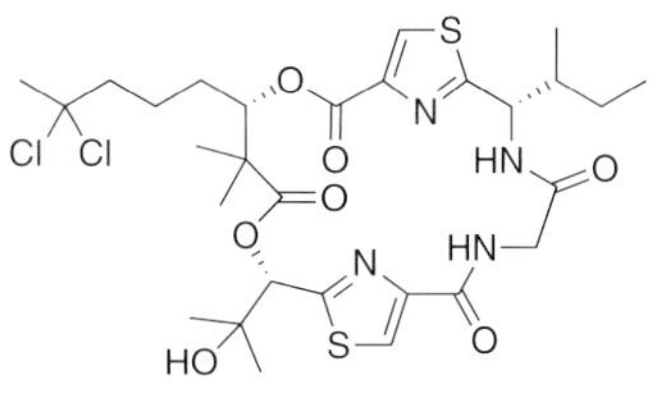

8. Lyngbyabellin A

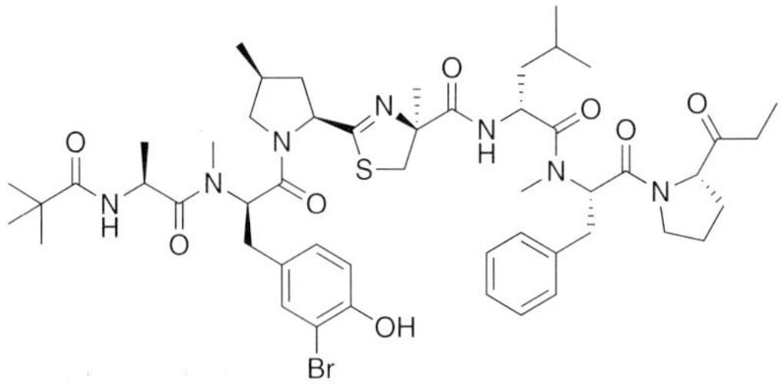

10. Bisebromoamide

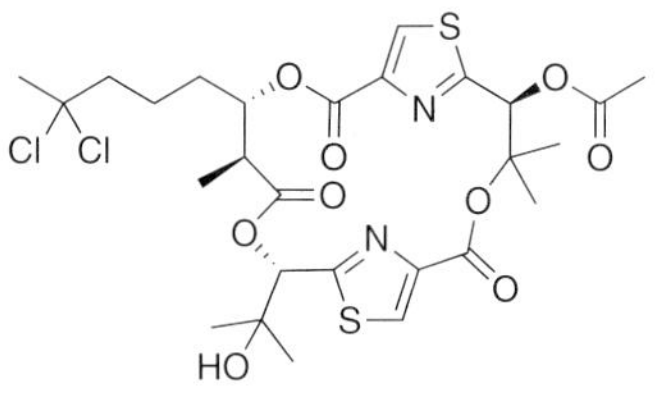

9. Hectochlorin

Bisebromoamide (**10**) possessed exquisite cytotoxic property against HeLa S_3 cells with an IC_{50} value of 0.04 μg/ml. When tested against a panel of 39 human cancer cell lines, bisebromoamide gave an average GI_{50} value of 40 nM. Furthermore, biochemical data suggested that the ERK (extracellular signal regulated protein kinase) signaling pathways could potentially be a target for this compound. SAR studies on bisebromoamide and synthetic analogues revealed that the stereochemistry of the methylthiazoline moiety and methyl group at the 4-methylproline unit did not influence the cytotoxicity activity significantly [14]. Bisebromoamide was subsequently identified as an actin filament stabilizer based on cell morphological profiling analysis [15]. A recent report by Suzuki *et al.* revealed that bisebromoamide inhibits phosphorylation of extracellular signal-regulated kinase and AKT (protein kinase) when tested in renal cell carcinoma [16].

5.2.1.3 **Histone Deacetylase Inhibitors**

Largazole (**11**), a cyclic depsipeptide, is a highly potent class I histone deacetylase (HDAC) inhibitor and is currently being investigated development as a potential anticancer drug. It was originally discovered from the marine cyanobacterium, *Symploca* sp., obtained from Key Largo, Florida [17]. Largazole is a prodrug and upon hydrolysis of the thioester provides the highly active largazole thiol (**12**), which is probably the most potent inhibitor of the HDAC enzymes. Due to its potent activity

Table 5.1 *In vivo* and *in vitro* studies performed on largazole (**11**).

	Biological activities	References
1	Largazole is a selective inhibitor of the ubiquitin-activating enzyme (E1). Destabilization of the ubiquitination process has therapeutic usage for the treatment of cancer	[18]
2	Largazole was shown to induce expression of EBV (Epstein–Barr virus) lytic-phase gene and sensitize lymphoma cells to nucleoside antiviral drugs. Results show potential use of largazole in the treatment of EBV-associated lymphomas	[19]
3	Largazole and dexamethasone shown to cooperate by inducing localization of E-cadherin to the plasma membrane in breast cancers as well as to suppress *in vitro* cellular invasion of cancer cells	[20]
4	Largazole exhibited *in vitro* and *in vivo* osteogenic activity via increased expression of Runx2 (runt-related transcription factor 2) and BMPs (bone morphogenetic proteins). Expression of alkaline phosphatase (ALP) and osteopontin (OPN) was significantly induced by largazole. The molecule further demonstrated *in vivo* bone formation in the mouse calvarical bone formation assay	[21]
5	Largazole is more active against colon cancer cell types. Molecule was shown to regulate transcription of genes involved in the induction of cell cycle arrest and apoptosis	[22,23]

and unique structure, it has attracted tremendous interest of synthetic chemists regarding its total synthesis. To date, 12 total syntheses of largazole have been reported and have been recently reviewed by Li *et al.* [24]. In addition, extensive SAR studies were carried out and the results of such studies are summarized recently by Hong and Luesch [25]. Various *in vivo* and *in vitro* biological studies were performed on this molecule since its discovery, and these data are summarized in Table 5.1.

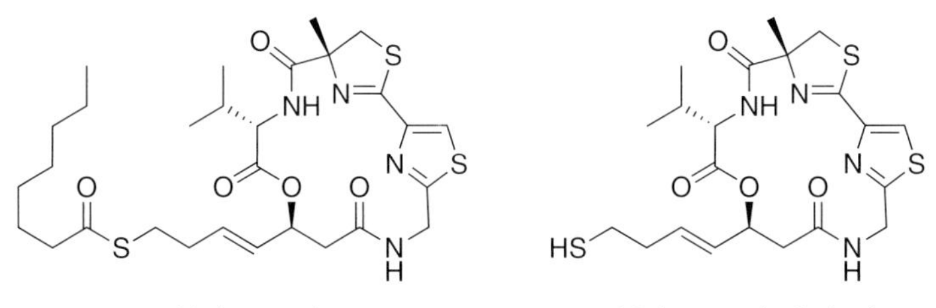

11. Largazole **12.** Largazole derivative

5.2.1.4 p53/MDM2 Inhibitor

The p53 protein is a well-studied tumor suppressor that is encoded by the TP53 gene in human cells. It regulates cell cycle and is involved in prevention of tumor formation. Its function can be regulated either by deletion or mutation of the TP53 gene or by interaction with MDM2, a murine ubiquitin ligase [26]. MDM2 is known to negatively regulate p53 and any disruption to MDM2–p53 interaction would result in the reactivation of p53 function. Thus, potential small-molecule inhibitors of MDM2–p53 interaction present a viable target for anticancer drug development [27]. Hoiamide D was recently isolated as its carboxylic acid (**13**) and conjugate base (**14**) forms from the spongy, purple cyanobacterium *Symploca* sp. collected in the vicinity of Kape Point at Papua New Guinea [28]. Hoiamide D is structurally related to hoiamides A (**50**)–C (**52**), reported to possess significant sodium channel-activating properties. The structure of hoiamide D contains a unique triheterocyclic system comprising two consecutive thiazolines and a thiazole unit as well as an unusual isoleucine moiety. Based on a series of bioassays, it was revealed that the carboxylate anion of hoiamide D exhibited significant inhibition of p53–HDM2 (human homologue) interaction with EC_{50} reported at 4.5 μM. This activity is noteworthy as hoiamide D is one of the most potent peptide-based natural product inhibitors of p53–MDM2 interaction reported till date.

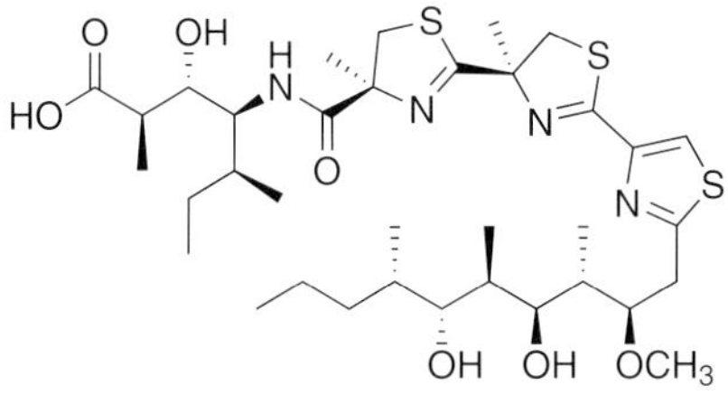

13. Hoiamide D, acid form

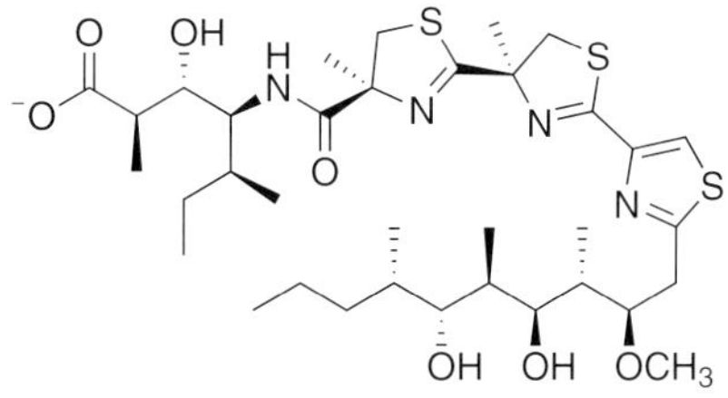

14. Hoiamide D, conjugate base form

5.2.1.5 Proteasome Inhibitors

Potent inhibitors of cellular proteasome have emerged as important therapeutic agents for the treatment of cancer and other types of diseases associated with inflammation and bacterial infection [29]. Carmaphycins A (**15**) and B (**16**) are the first reported marine cyanobacterial molecules having exquisite antiproteasome activity. These compounds were isolated in small amounts from a sample of *Symploca* sp. found attached to an anchor rope south of CARMABI beach, Curacao [30]. These compounds are peptidic in nature and feature a terminal α,β-epoxyketone moiety and either a methionine sulfoxide in **15** or methionine sulfone in **16**. An efficient and scalable convergent method was also employed for the total synthesis of these molecules [30]. Due to the structural similarities with the known proteasome inhibitor epoxomicin (**17**), the inhibitory properties of carmaphycins A and B were evaluated against *Saccharomyces cerevisiae* 20S proteasome and found to possess IC_{50} values of 2.5 and 2.6 nM, respectively. The inhibitory activities of these compounds are comparable with those reported for epoxomicin and the marine-derived salinosporamide A with IC_{50} values of 2.7 and 1.4 nM, respectively. The structure–activity relationships of these cyanobacterial compounds are currently underway.

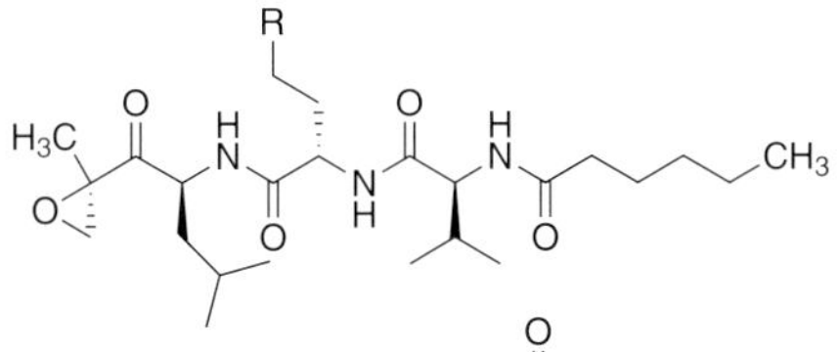

15. Carmaphycin A

16. Carmaphycin B

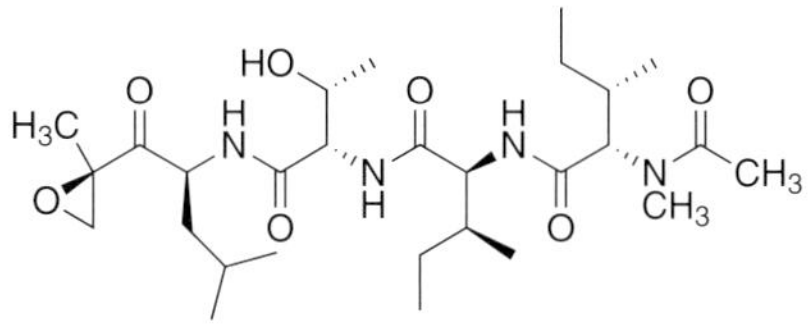

17. Epoxomicin

5.2.1.6 Protease Inhibitors

Proteases are ubiquitous enzymes found in humans and microbial life-forms such as bacteria and viruses. They account for about 2% of the genes in human cells and are involved primarily in proteins activation, synthesis, and turnover [31]. Interference with the function of proteases will therefore lead to potential treatment of diseases, including cancer, inflammation, and microbial infections [31]. It has been reported that certain strains of marine cyanobacteria are sources of potent serine protease inhibitors. These compounds mainly belong to the 3-amino-6-hydroxypiperidone (Ahp)-containing cyclic depsipeptides, for example, lyngbyastatin 4 (**18**), kempopeptin A (**19**), and somamide B (**20**). The chemistry and biology of these molecules have been recently reviewed [32].

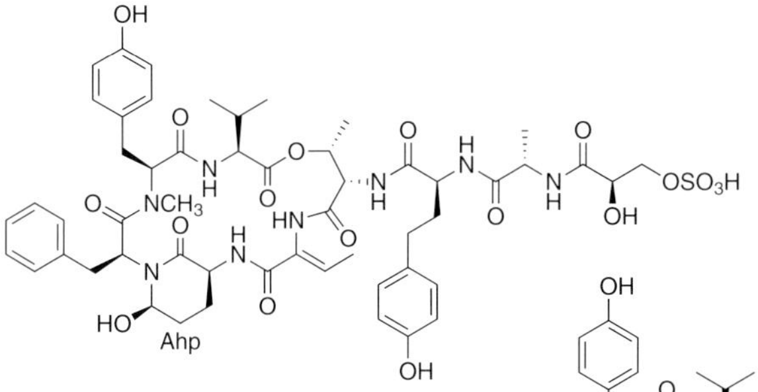

18. Lyngbyastatin 4

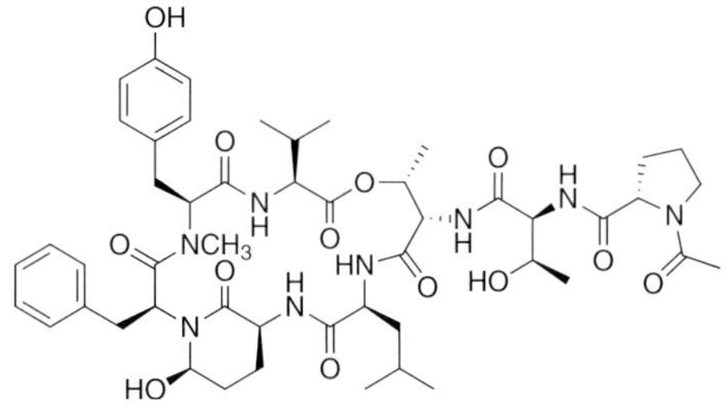

19. Kempopeptin A

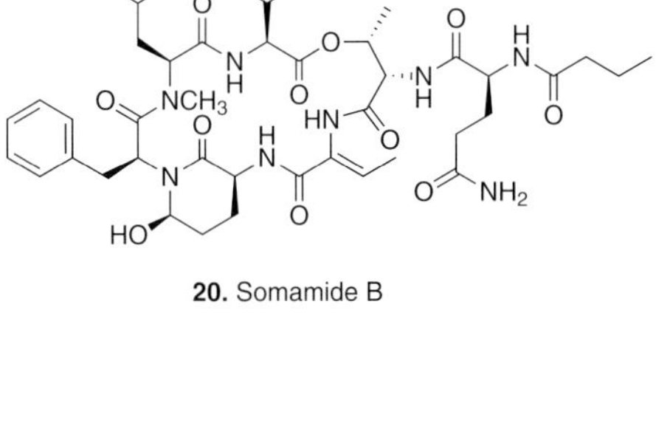

20. Somamide B

In addition to serine protease inhibitors, a number of recently reported statin-containing linear depsipeptides, including grassystatins A (**21**)–C (**23**) and symplocin A (**24**), were found to be potent inhibitors of cathepsins D and E. Cathepsins D and E are classified as aspartate proteases and their over-expression have been observed in various cancer forms, such as pancreatic ductal adenoma, cervical adenocarcinoma, lung carcinoma, and gastric adenocarcinoma [33,34]. The linear decadepsipeptides, grassystatins A (**21**)–C (**23**), were isolated from the marine cyanobacterium, *Lyngbya confervoides*, collected at Grassy Key, Florida [35]. These statin unit-containing molecules were isolated based on a screening program by profiling the inhibitory activities of natural products against 59 proteases. Grassystatins A (**21**) and B (**22**) displayed potent inhibitory activity against cathepsins D and E with IC_{50} values averaging at 16.9 and 0.62 nM, respectively. In addition, grassystatin A was able to reduce antigen presentation by dendritic cells [35].

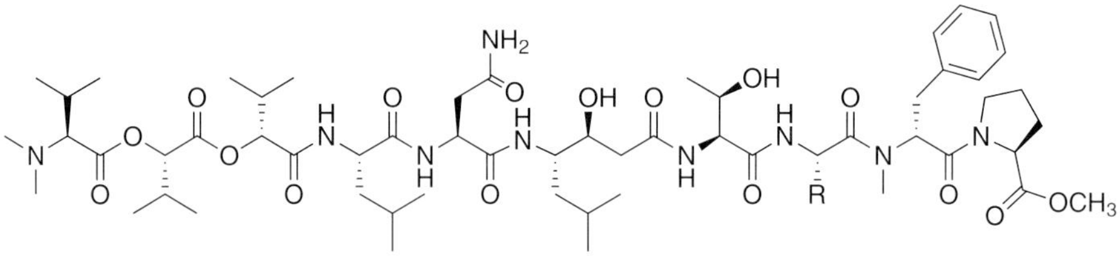

21. Grassystatin A R = Me
22. Grassystatin B R = Et

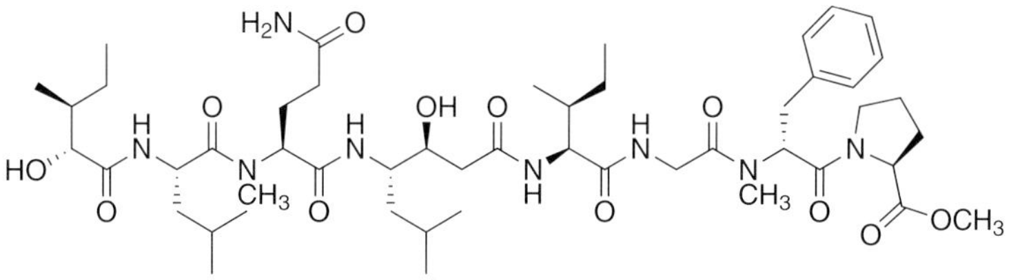

23. Grassystatin C

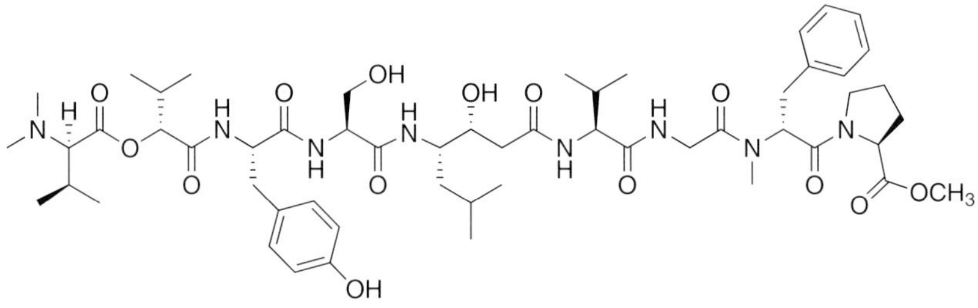

24. Symplocin A

Symplocin A (**24**) is a new cathepsin E inhibitor recently isolated from the Bahamian cyanobacterium *Symploca* sp. [36]. Compound **24** is structurally related to grassypeptins in that both have a statin unit. In addition to applying

Marfey's method, a new strategy using 2-naphthacyl esters of *N,N*-dimethylamino and 2-hydroxy acids was also employed for the absolute stereochemistry determination of this molecule. Symplocin A is a potent inhibitor of cathepsin E with IC_{50} value of 300 pM, which is comparable to that of pepstatin [36].

5.2.1.7 **Apoptosis-Inducing Agents**

The apratoxins [e.g., apratoxin A (**25**)] are a novel class of potent cytotoxic cyclic depsipeptides isolated from various strains of *Lyngbya* sp. procured from Guam, Palau, and Papua New Guinea [37,38]. These molecules belong to the hybrid PKS–NRPS structural class and they are highly cytotoxic when tested against a panel of various cancer cell lines, including HT29, HeLa, and U2OS, with IC_{50} values in the nanomolar range. The chemistry and biology of the apratoxin class of molecules have been recently reviewed [2,32]. Through a series of biochemical studies, it was demonstrated that the apratoxins exert its apoptotic effect by downregulating receptors and associated growth ligands in cancer cells [39]. SAR studies were subsequently conducted and based on the structural features of apratoxins A and E, a hybrid synthetic molecule **26** with improved *in vivo* antitumor properties was synthesized [39]. In addition, the putative gene cluster involved in the biosynthesis of apratoxin A was recently identified using a two-pronged approach of single-cell genomic sequencing and metagenomic library screening [40]. The identification of the biosynthetic gene cluster of this molecule will lead the way for production of apratoxin A and its analogues.

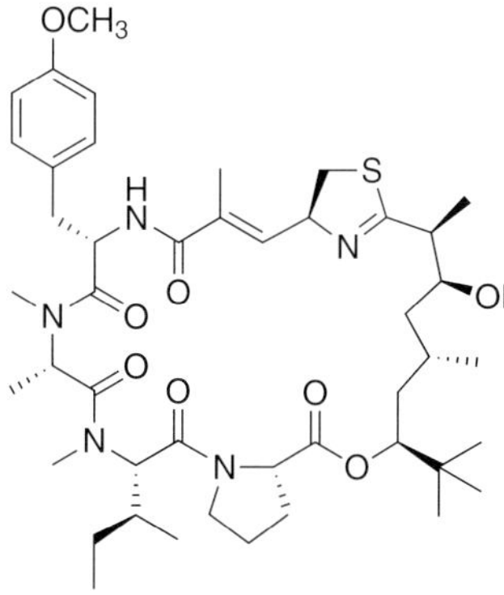

OCH3 H N S N OH N O O O O O N N O

25. Apratoxin A

OCH3 H N S N OH N O O O O O N N O

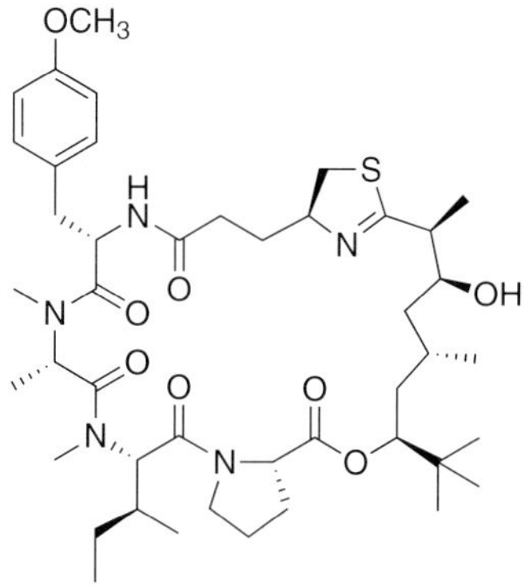

26. Hybrid apratoxin A/E molecule

Aurilide (**27**) and related compounds are another important class of potent cytotoxic molecules with activities in the picomolar–nanomolar range [2,32]. To date, six aurilide-related compounds have been discovered, with lagunamides A (**28**) and B (**29**) being the latest additions [41]. The importance of this series of compounds as potential anticancer agents is attested by the growing number of synthetic efforts geared toward their total synthesis [42–45]. Recent biochemical investigation by Sato *et al.* showed aurilide (**27**) to be a potent inhibitor of mitochondrial prohibitin 1 [46]. The binding of aurilide to prohibitin 1 leads to the activation of the proteolytic process of optic atrophy 1, resulting in mitochondria-induced apoptosis [46].

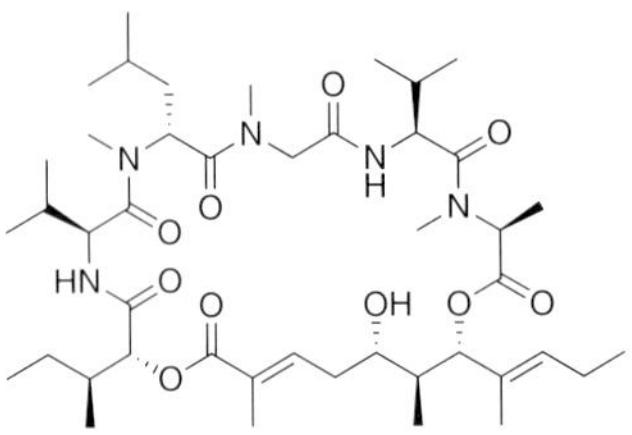

27. Aurilide

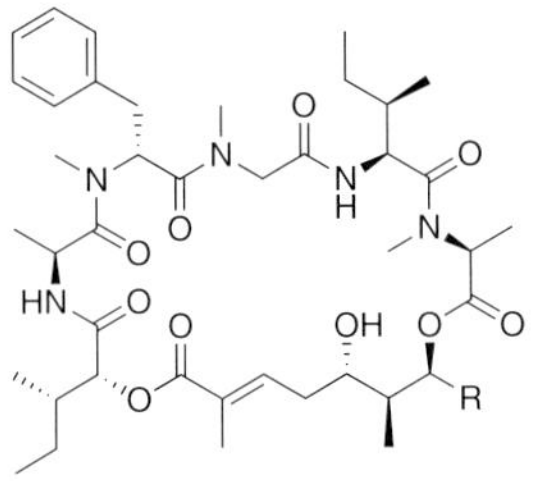

28. Lagunamide A R=

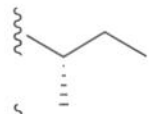

29. Lagunamide B R=

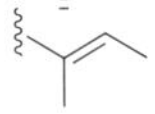

5.2.1.8 **Other Potent Cytotoxic Compounds**

The discovery of coibamide A, grassypeptolides, and veraguamides in recent years adds to the impressive list of cytotoxic marine cyanobacterial molecules already discussed. A potent antiproliferative lariat-type cyclic depsipeptide, coibamide A (**30**), was isolated from the Panamanian marine cyanobacterium, *Leptolyngbya* sp. [47]. The chemical structure of coibamide A (**30**) was deduced by extensive 2D Nuclear magnetic resonance (NMR) spectroscopic experiments, including COSY, TOCSY, multiplicity-edited HSQC, HSQC–TOCSY, HMBC, H2BC, ^{1}H—^{15}H gHMBC, and ROESY, as well as from mass spectroscopic data. This molecule possesses a high degree of *N*-methylation, with 8 out of 11 residues being *N*-methylated. More importantly, coibamide A displayed potent cytotoxicity against NCI-H460 lung cancer cells and mouse Neuro-2a cells, with LC_{50} values less than 23 nM. In addition, the compound was evaluated in the NCI's panel of 60 cancer cell lines, and it exhibited significant activities against MDA-MB-231, LOX IMVI, HL-60(TB), and SNB-75 at 2.8, 7.4, 7.4, and 7.6 nM, respectively. Furthermore, COMPARE analysis indicated that coibamide A might inhibit cancer cell proliferation via a novel mechanism [47]. Total synthesis of this molecule is currently underway by the research group of Dr McPhail at Oregon State University.

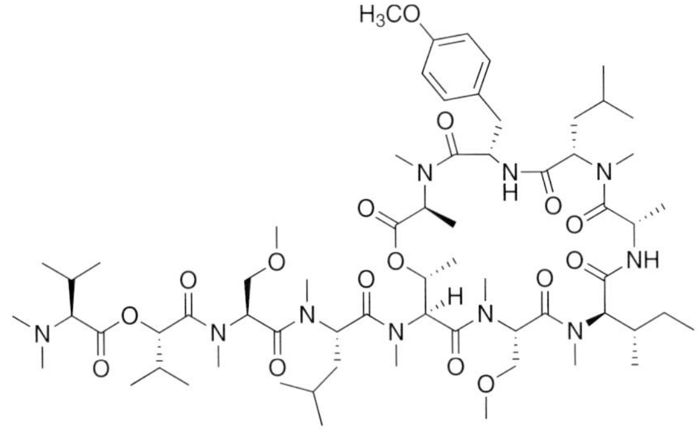

30. Coibamide A

Grassypeptolides A (**31**)–C (**33**) are cytotoxic bis-thiazoline-containing cyclic depsipeptides isolated from the marine cyanobacterium *L. confervoides* from Grassy Key, Florida [48,49]. Their complete structures, including 3D structures, were determined by a combination of NMR, mass spectrometry (MS), X-ray crystallography, chemical degradation, and molecular modeling methods. Grassypeptolide A (**31**) displayed significant anticancer activity against four cell lines – human osteosarcoma (U2OS), cervical carcinoma (HeLa), colorectal adenocarcinoma (HT29), and neuroblastoma (IMR-32) – with IC_{50} values of 2.2, 1.0, 1.5, and 4.2 μM, respectively. Further biological studies showed that at lower concentrations, compounds **31** and **33** arrest cell cycle at G1 phase, while at higher concentrations, they arrest cell cycle at the G2/M phase [48]. Liu *et al.* accomplished successfully the first synthesis of grassypeptolide A [23].

Four additional new analogues, grassypeptolides D (**34**)–G (**37**) were isolated from two marine cyanobacterial species, *Leptolyngbya* sp. and *Lyngbya majuscula*. Grassypeptolides D (**34**) and E (**35**) were purified from field collections of the marine cyanobacterium *Leptolyngbya* sp., obtained from a shipwreck in the Red Sea [50]. Grassypeptolides F (**36**) and G (**37**) were obtained from the marine cyanobacterium *L. majuscula*, collected from Palau [51]. Grassypeptolides D (**34**) and E (**35**) exhibited significant activity against the HeLa and mouse Neuro-2a blastoma cell lines with IC_{50} values of 335 and 192 nM and 599 and 407 nM, respectively.

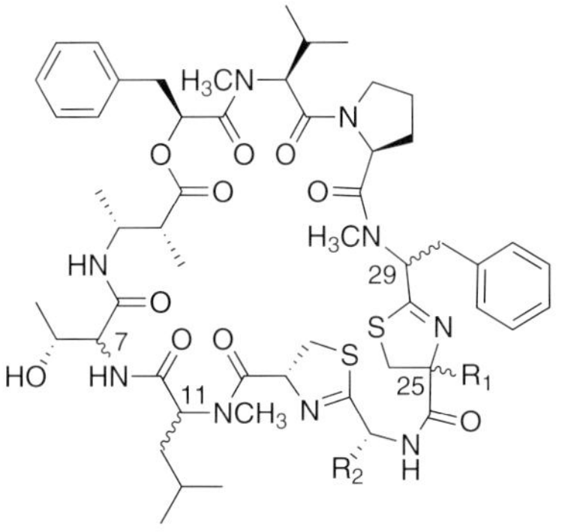

		R_1	R_2
31. Grassypeptolide A	7*R*, 11*R*, 25*R*, *29*R*	H	Et
32. Grassypeptolide B	7*R*, 11*R*, 25*R*, *29*R*	H	Me
33. Grassypeptolide C	7*R*, 11*R*, 25*R*, *29*S*	H	Et
34. Grassypeptolide D	7*R*, 11*R*, 25*S*, 29*S*	Me	Et
35. Grassypeptolide E	7*S*, 11*S*, 25*S*, 29*S*	Me	Et

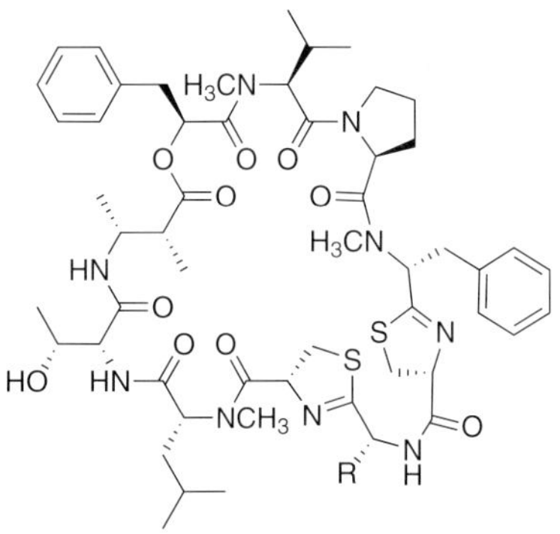

36. Grassypeptolide F R = Et
37. Grassypeptolide G R = Me

From a dolastatin 16-producing marine cyanobacterium *Symploca* cf. *hydnoides*, collected from Cetti Bay, Guam, seven new cyclodepsipeptides – veraguamides A (**38**)–G (**44**) – were isolated and characterized [52]. In addition to α-amino/hydroxy acid units, these compounds contain a C_8-polyketide-derived β-hydroxy acid residue with tail end occurring either as alkynyl bromide, alkyne, or vinyl functional group. Veraguamides D (**41**) and E (**42**) are the most cytotoxic in this series of molecules, with IC_{50} values of 0.54–1.5 μM when tested against the HT29 and HeLa cancer cell lines [52].

Additional natural analogues, veraguamides H (**45**)–L (**49**), were recently isolated from the marine cyanobacterium, cf. *Oscillatoria margaritifera*, from the Coiba National Park, Panama [53]. These new analogues were isolated along with previously reported veraguamides A–C. In this study, only veraguamide A (**38**) showed potent activity with LD_{50} value of 141 nM when tested on H-460 cell line. Due to the structural similarities of veraguamide A with other known compounds, such as kulomo'opunalide-1 and kulomo'opunalide-2, it was suggested that the alkynyl bromide functional group in **38** is essential for the observed biological activity.

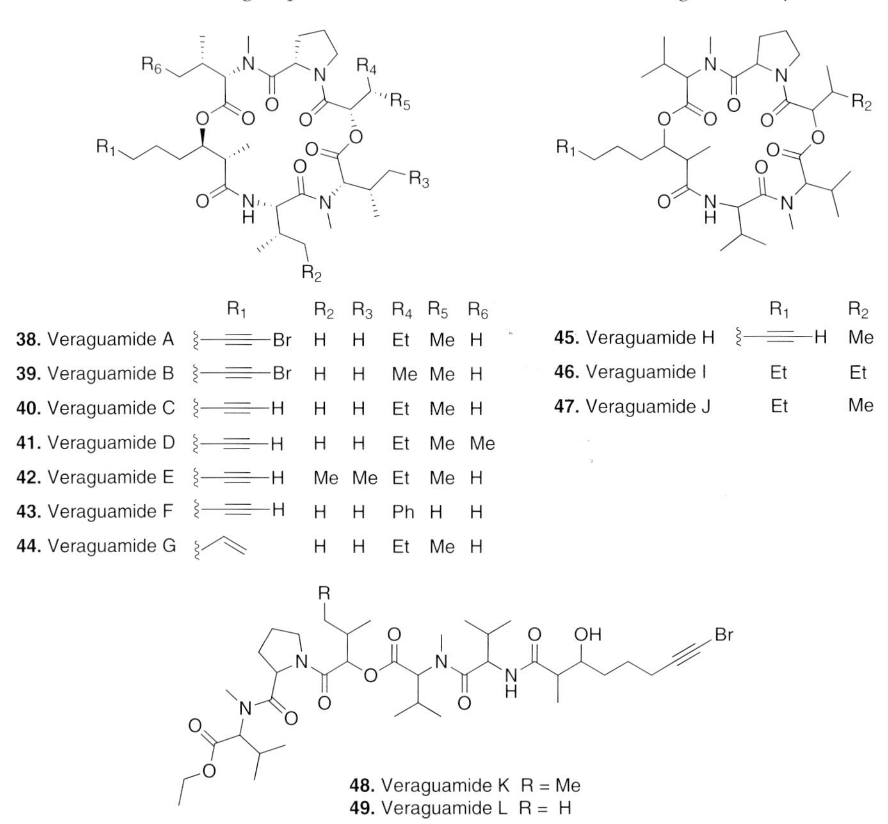

5.2.2
Neuromodulating Agents

In addition to anticancer agents, marine filamentous cyanobacteria are a prolific source of novel neuromodulating molecules that target either the cholinergic synapses or the voltage-gated sodium channels. Natural products that target the voltage-gated sodium channels generally fall into two groups: voltage-gated sodium channel blockers or activators. Recent review by Araoz *et al.* [54] provided a comprehensive coverage on the chemistry and biology of marine cyanobacterial compounds, including molecules that block (e.g., kalkitoxin and jamaicamide) and activate (e.g., antillatoxin) voltage-gated sodium channels. Natural products exhibiting significant neuromodulating properties can be utilized for the treatment of various ailments, including neurodegenerative disorders, stroke and epilepsy, chronic pain, and inflammation. This section will highlight the novel neurotoxic agents reported since the publication of Ref. [49].

5.2.3 Modulators of the Voltage-Gated Sodium Channels and Calcium Oscillation

The hoiamides (**50**–**52**) are a new class of neurotoxic lipopeptides isolated from a consortium of marine cyanobacteria collected from various locations at Papua New Guinea [55,56]. These compounds contain several unusual structural features, including a consecutive linkage of three heterocyclic (units and a highly oxygenated and methylated polyketide chains. Total structural elucidation of these complex molecules was established by extensive 1D and 2D NMR experiments as well as various chemical manipulations, including Mosher's method. The relative stereochemistry of the 4-amino-3-hydroxy-2,5-dimethylheptanoic acid (Ahdhe) residue and 5,7-dihydroxy-3-methoxy-4,6,8-trimethylundecanoyl-derived unit (Dmetua) were determined by analysis of homonuclear and heteronuclear coupling constants and ROE correlations. Biosynthetically, these molecules are assembled from a variety of primary metabolites, including acetyl-CoA, SAM (*S*-adenosyl methionine), cysteine, Ile and/or Val, and Thr (Figure 5.1). Of these series of compounds, the cyclic forms, hoiamides A (**50**) and B (**51**), showed nanomolar range inhibition of calcium oscillations in neocortical neurons with EC_{50} values of 45.6 and 79.8 nM, respectively. In addition, hoiamides A and B showed significant sodium channel-activating properties with EC_{50} values of 1.7 and 3.9 μM, respectively. Furthermore, studies by Pereira *et al.* revealed hoiamide A to be a partial agonist of site 2 on the voltage-gated sodium channel [55]. Batrachotoxin and veratridine are known to interact with site 2 leading to the stabilization of the channel in the open state. The cyclic nature of these compounds is essential for observed biological activity due to the inactivity of the liner form of hoiamide C (**52**). The structure of hoiamide C was recently confirmed by its total synthesis in 16 steps (overall yield of 1.8%) by Wang *et al.* [57].

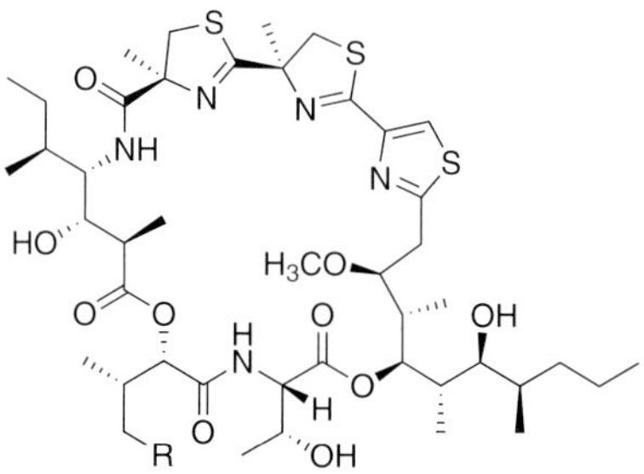

50. Hoiamide A R = H
51. Hoiamide B R = CH_3

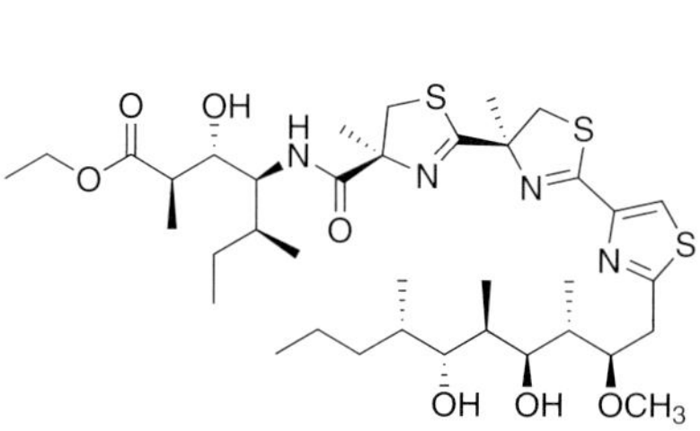

52. Hoiamide C

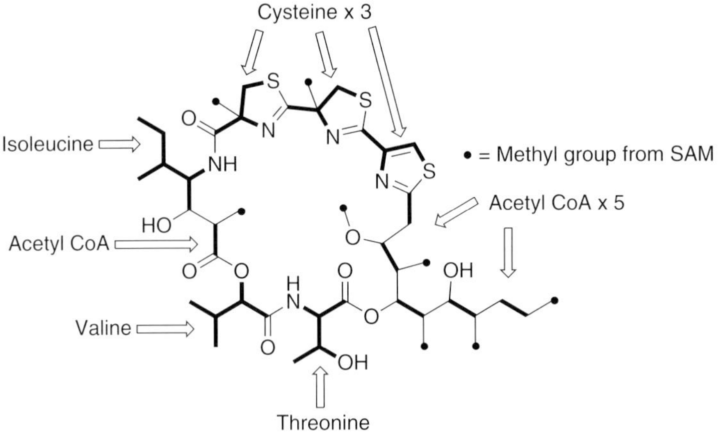

Figure 5.1 Proposed biosynthesis of hoiamide A (**50**).

Credneramides A (**53**) and B (**54**) are phenethylamine and isopentylamine derivatives of a vinyl chloride-containing fatty acid, recently isolated from cf. *Trichodesmium* sp. nov. obtained from Papua New Guinea [58]. These molecules showed significant inhibition of calcium oscillations in cerebrocortical mouse neurons with IC_{50} values of 4.0 and 3.9 μM for compounds **53** and **54**, respectively. Calcium oscillations are tightly regulated by action of pumps and channels and are involved in a variety of cellular physiological processes, including apoptosis and necrosis. The discovery of credneramide A joins a growing class of phenethylamine fatty acid derivatives that are increasingly being reported from cyanobacteria. It has been speculated that these molecules might play a role in cyanobacterial intercellular communication and quorum sensing [58]. These molecules are akin to *N*-acyl homoserine lactones, which are well-studied signaling molecules involved in bacterial quorum sensing. Shortly after the publication on the isolation of credneramides A and B, their total synthesis, based on Alder-ene reactions between terminal alkene and internal alkyne, was achieved by Erver and Hilt [59].

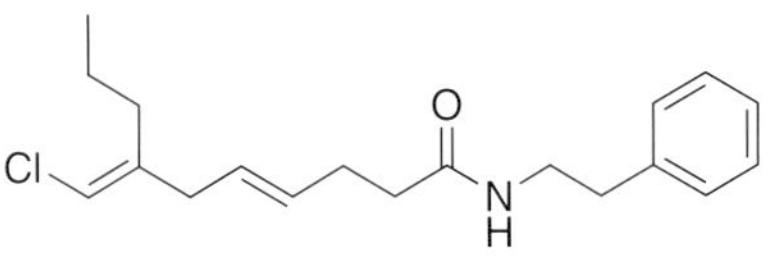

53. Credneramide A

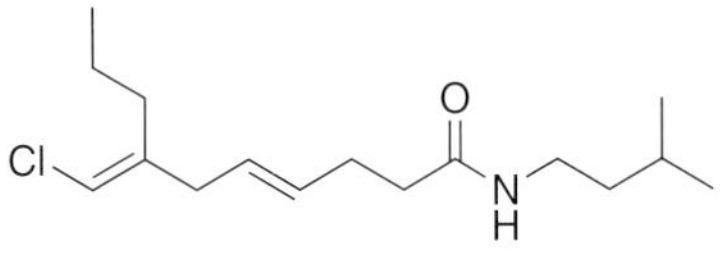

54. Credneramide B

Structurally related to the credneramides are a series of chlorinated lipoamides, janthielamide A (**55**) and kimbeamides A (**56**)–C (**58**), recently characterized from two collections of marine cyanobacteria from Curacao and Papua New Guinea [60]. Their chemical structures were determined by key COSY and HMBC correlations as well as by the analysis of 1D NMR data. The stereochemistry chiral centers in janthielamide A and kimbeamide A were determined via chemical degradation, derivatization with reagents, and comparative analysis with derivatized standards on either chiral or achiral gas chromatography–mass spectrometry (GC–MS)/liquid chromatography–mass spectrometry (LC–MS). Janthielamide A (**55**), a major constituent from the Curacao sample, possessed sodium channel-blocking activity with IC_{50} value of 11.5 μM when tested on murine Neuro-2a cells. It was also found to be an agonist of veratridine-induced sodium influx with an IC_{50} value of 5.2 μM in cerebrocortical neurons [60].

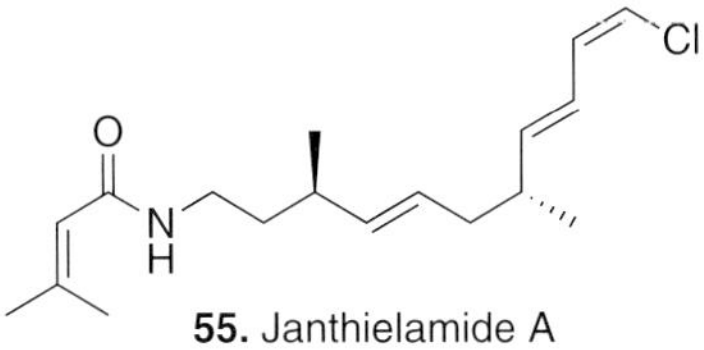

55. Janthielamide A

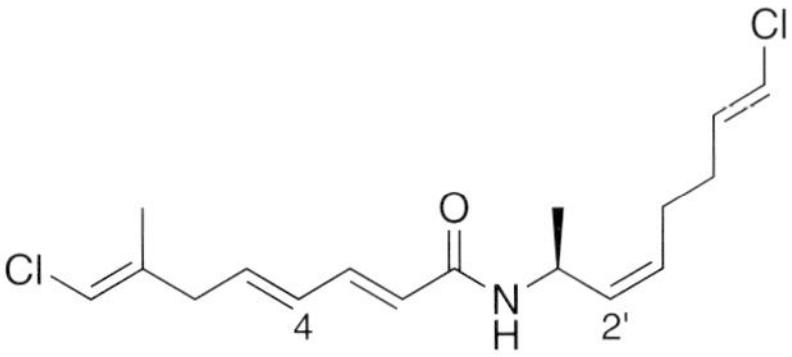

56. Kimbeamide A 4*E*, 2′*Z*
57. Kimbeamide B 4*Z*, 2′*Z*
58. Kimbeamide C 4*Z*, 2′*E*

Palmyramide A (**59**) is a 19-membered cyclodepsipeptide isolated from an assemblage consisting of the marine cyanobacterium *L. majuscula* and the red alga *Centroceras* sp., collected from a Palmyra Atoll [61]. This molecule consists of six residues, including three α-amino acids of Val, *N*Me-Val, and Pro, and three hydroxy acids of 2,2-dimethyl-3-hydroxyhexanoic acid (Dmhha), lactic acid, and 3-phenyllactic acid. Its complete structure was accomplished by NMR, MS spectral data, and chemical manipulation, including Marfey's analysis. The absolute stereochemistry of the β-hydroxy acid, Dmhha, was determined by the stereo-selective synthesis of *S*- and *R*-Dmhha and subjecting them to chiral GC–MS analysis together with the Dmhha unit derived from the acid hydrolysis of palmyramide A. It was only through the use of MALDI-TOF imaging technique that the authors were able to identify the marine cyanobacterium as the true producer of the natural product. When evaluated in the sodium channel-blocking assay, compound **59** inhibited veratridine- and ouabain-induced sodium overload, which resulted in the cytotoxicity of Neuro-2a cells with an IC_{50} of 17.2 μM. In addition, palmyramide A displayed moderate cytotoxicity against the H-460 human lung carcinoma cells with an IC_{50} value of 39.7 μM [61].

Palmyrolide A (**60**) is a new neurotoxin isolated from a consortium of marine cyanobacteria consisting of *Leptolyngbya* cf. and *Oscillatoria* sp. collected from Palmyra Atoll [62]. The structure of this molecule consists of an extensive polyketide fragment linked with peptidic residues. This molecule is structurally related to the laingolides by featuring a *tert*-butyl group possibly derived from malonyl-CoA with the methyl groups contributed by *S*-adenosyl-L-methionine (SAM) [63]. Palmyrolide A was found to suppress calcium influx in cerebrocortical neurons with IC_{50} value of 3.7 μM. In addition, it possesses moderate sodium channel-blocking activity in Neuro-2a cells with IC_{50} value of 5.2 μM [62]. The total synthesis of palmyrolide A has recently been accomplished by the research group of Maio and coworkers [64].

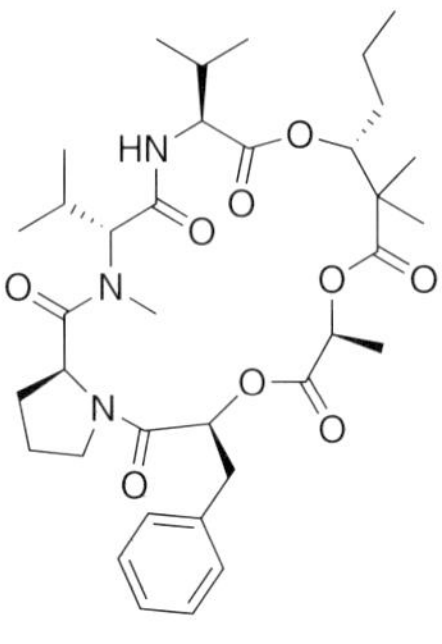

59. Palmyramide A

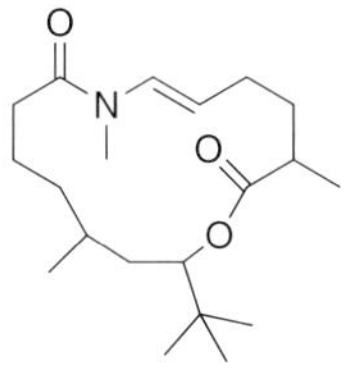

60. Palmyrolide A

5.2.4
Cannabinomimetic Agents

Unusual alkyl amides that are structurally related to anandamide and other endocannabinoids have been isolated from marine organisms, particularly from marine cyanobacteria. From an organic extract of *L. majuscula*, collected from Papua New Guinea and Panama, two new cannabinomimetic lipids, serinolamide A (**61**) and propenediester (**62**), were isolated and characterized [65]. Due to their structural similarities with known endocannabinoids, they were evaluated in a radioligand binding assay using human CB_1 and CB_2 cannabinoid receptors. Only serinolamide A showed moderate selective affinity for CB_1 receptor having a K_i value of 1.3 μM [65].

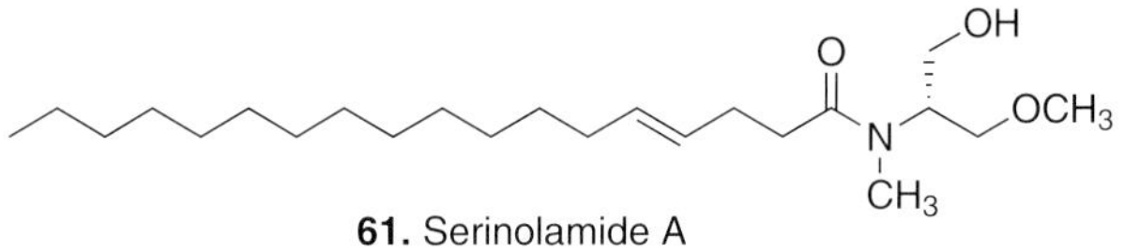

61. Serinolamide A

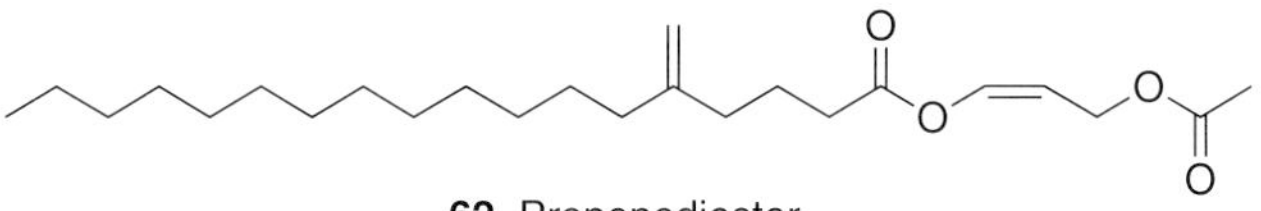

62. Propenediester

5.2.5
Anti-Infective Agents

5.2.5.1 **Antiprotozoal Agents**

Over the years, a number of linear and cyclic depsipeptides from marine cyanobacteria have been reported to possess significant antiprotozoal activities. Most of these discoveries stemmed from the International Cooperative Biodiversity

Groups program, sponsored by the NIH Fogarty Center, Panama. These compounds represent new classes of antiparasitic agents and they are ideal as drug leads for further development as antiprotozoal drugs due to the moderate to low cytotoxic properties reported in a number of molecules. For instance, a new synthetic molecule, **73**, with improved antileishmanial activity was synthesized based on the structures of the almiramides as well as related compounds [66]. Table 5.2 provides a summary of recently discovered marine cyanobacterial compounds with significant antiprotozoal activities.

Table 5.2 IC_{50} values (μM) of antiprotozoal marine cyanobacterial compounds.

Source	Compounds	Bioactivity (μM)			References
		Plasmodium falciparum	*Leishmania donovani*	*Trypanosoma cruzi*	
Linear depsipeptides					
Lyngbya majuscula	Carmabin A (**63**)	4.3	n.t.	n.t.	[67]
	Dragomabin (**64**)	6.0	n.t.	n.t.	
	Dragonamide A (**65**)	7.7	6.5	n.t.	
L. majuscula	Dragonamide E (**66**)	n.t.	5.1	n.t.	[68]
	Herbamide B (**67**)	n.t.	5.9	n.t.	
Schizothrix sp.	Gallinamide A (**68**) (= Symplostatin 4)	8.4	n.t	n.t.	[69]
Oscillatoria nigroviridis	Viridamide A (**69**)	1.5	1.5 (*L. mexicana*)	1.1	[70]
L. majuscula	Almiramides A (**70**)	n.t..	>13.5	n.t.	[66]
	Almiramides B (**71**)	n.t.	2.4	n.t.	
	Almiramides C (**72**)	n.t.	1.9	n.t.	
Cyclic peptides/depsipeptides					
Oscillatoria sp.	Venturamides A (**74**)	8.2	>20	14.6	[71]
	B (**75**)	5.2	>19	15.8	
L. majuscula	Lagunamides A (**28**)	0.19	n.t.	n.t.	[41]
	B (**29**)	0.91	n.t.	n.t.	

n.t.: not tested.

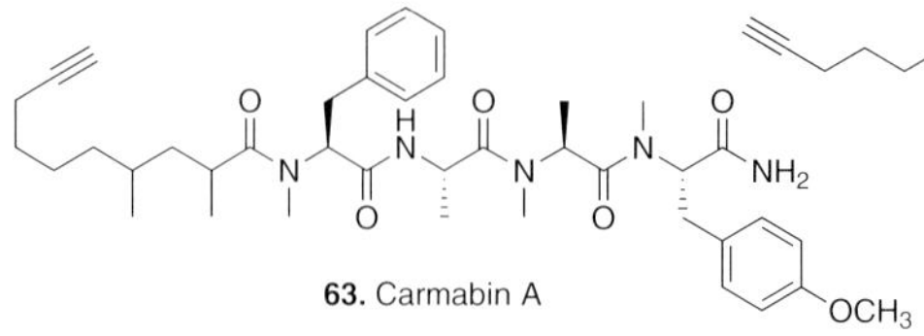

63. Carmabin A

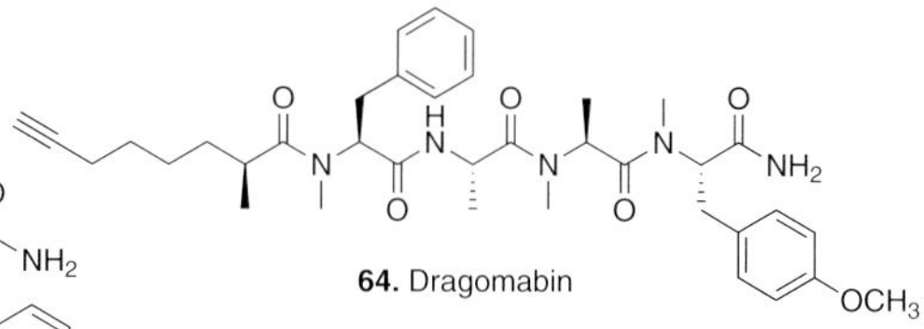

64. Dragomabin

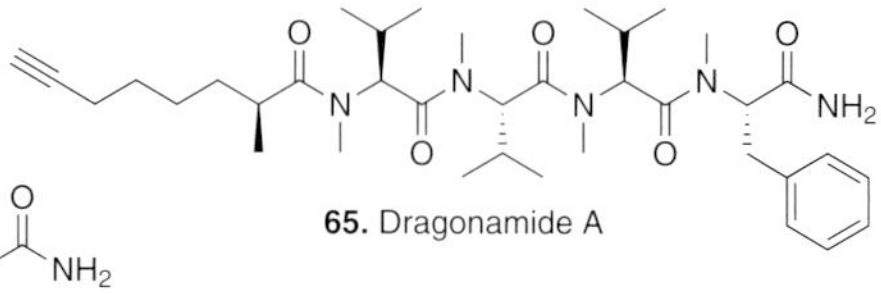

65. Dragonamide A

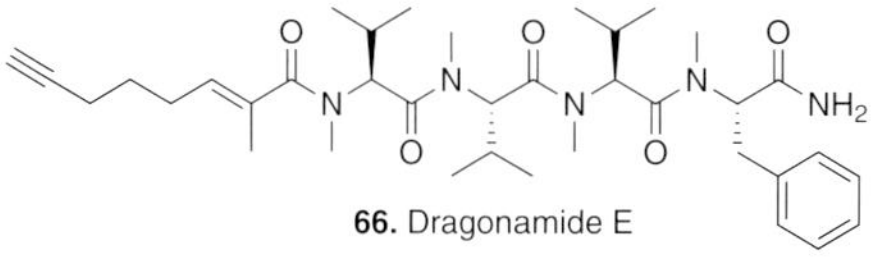

66. Dragonamide E

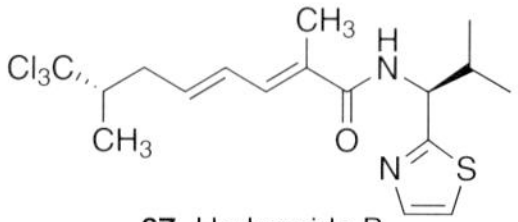

67. Herbamide B

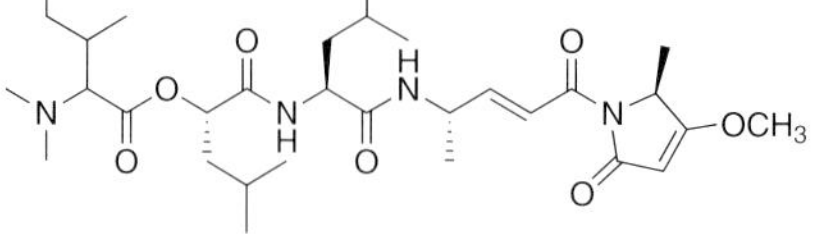

68. Gallinamide A

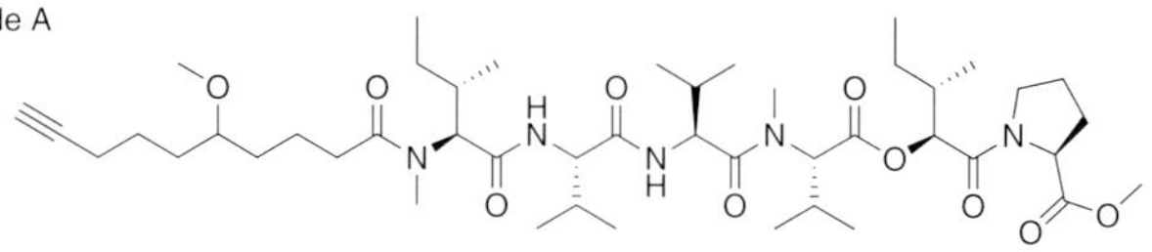

69. Viridamide A

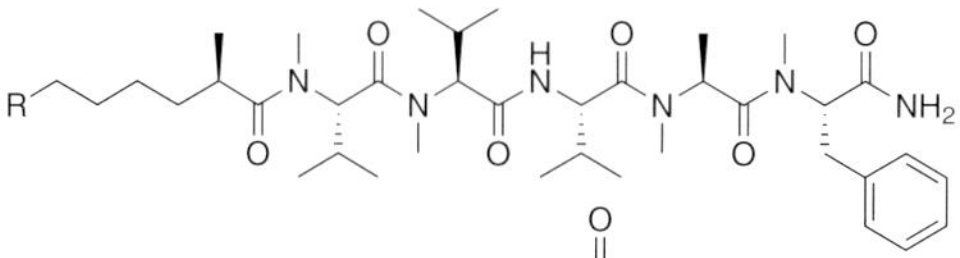

70. Almiramide A R =

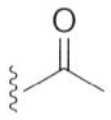

71. Almiramide B R =

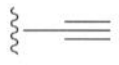

72. Almiramide C R =

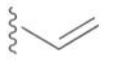

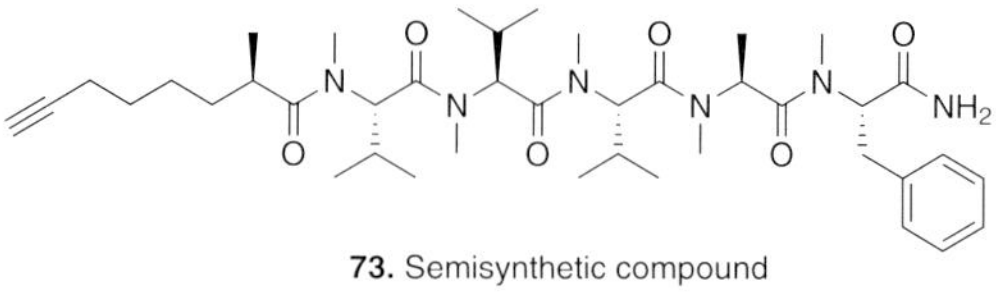

73. Semisynthetic compound

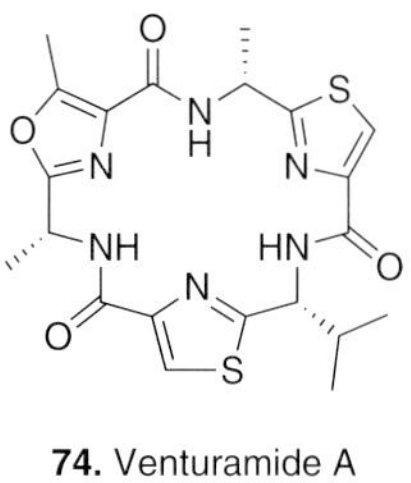

74. Venturamide A

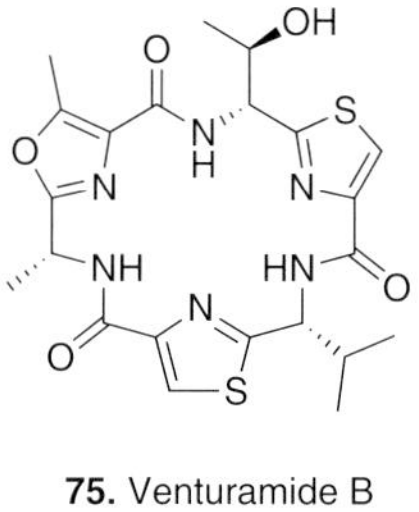

75. Venturamide B

5.2.5.2 Antimycobacterial Agents

The filamentous marine cyanobacteria are prolific producers of β-hydroxy/amino acid-containing cyclic hexadepsipeptides. The pitipeptolides are one such class recently reported from the marine cyanobacterium *L. majuscula* from Guam [72]. These compounds differ mainly in the composition of the amino acids as well as in the degree of unsaturation at the terminal end of the polyketide-derived β-hydroxy acid fragment. These compounds were reported to exhibit moderate cytotoxic activities but significant antimycobacterial properties, with pitipeptolide F (**76**) being the most active natural analogue [72].

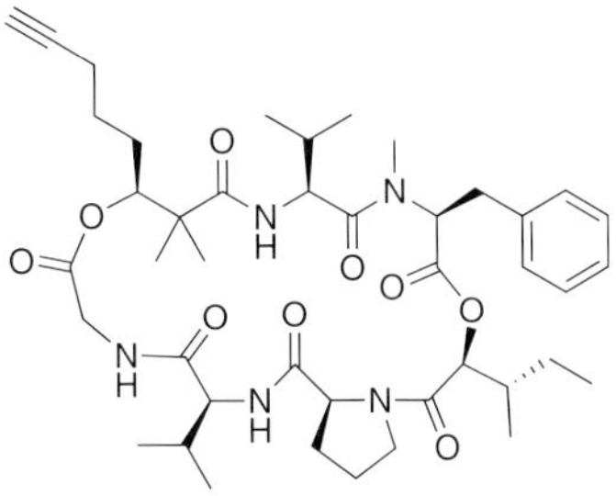

76. Pitipeptolide F

5.2.5.3 Anti-Inflammatory Agents

The malyngamides are marine cyanobacterial lipoamide compounds characterized by the fatty acid-derived portion of different carbon chain lengths coupled via amide bond to various amino acid residues. These compounds have been reported to exhibit a host of biological activities such as antifeedant, ichthyotoxicity, and cytotoxicity of modest potencies. Recent research by Villa *et al.* on the screening of a library of marine natural products revealed malyngamide F acetate (**77**) to possess strong anti-inflammatory property using the NO (nitric oxide) assay with IC_{50} value of 7.1 μM [73]. In addition, this molecule was found to selectively inhibit the MyD88-dependent pathway. Malyngamide 2 (**78**) is another recently reported cyanobacterial molecule having significant anti-inflammatory activity in LPS (lipopolysaccharide)-induced RAW macrophage cells with IC_{50} value of 8.0 μM. Malyngamide 2 (**78**) was isolated along with three known compounds from a collection of cf. *Lyngbya sordida* obtained from Papua New Guinea [74].

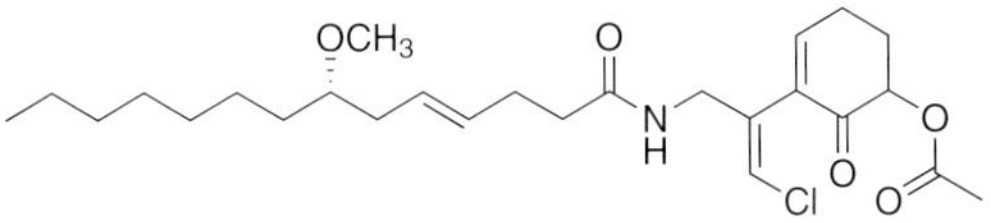

77. Malyngamide F acetate

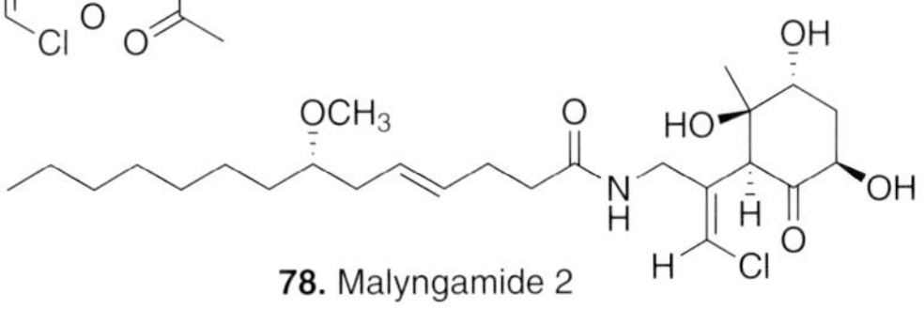

78. Malyngamide 2

5.3
Conclusions

This chapter highlighted the biological activities of more than 70 marine cyanobacterial natural products. These compounds predominantly belong to the hybrid polyketide–polypeptide structural class. A high proportion of these compounds is found to have potent anticancer properties due to their interaction with molecular targets, such as actin, microtubule filaments, protease, proteasome, and histone deacetylase enzymes. Due to their high potencies, a number of these compounds, including apratoxins, largazole, and aurilide-related compounds, are currently in preclinical testing as potential anticancer drugs. Another pharmacological trend observed is the report of marine cyanobacterial compounds with neuromodulating properties. Furthermore, novel marine cyanobacterial compounds having significant anti-infective activities have also been recently reported. In conclusion, marine cyanobacteria, particularly filamentous strains, are exciting sources of structurally novel bioactive compounds. Undoubtedly, a continued research on marine cyanobacterial natural products will be fruitful in yielding potential lead compounds in drug discovery efforts.

Acknowledgment

The author would like to thank the NIE AcRF grant (RI 11/10 TLT) for financial support.

References

1 Tan, L.T. (2007) Bioactive natural products from marine cyanobacteria for drug discovery. *Phytochemistry*, **68**, 954–979.
2 Tan, L.T. (2012) Marine cyanobacteria: a treasure trove of bioactive secondary metabolites for drug discovery. *Studies Nat. Prod. Chem.*, **36**, 67–110.
3 Jones, A.C., Monroe, E.A., Eisman, E.B., Gerwick, L., Sherman, D.H., and Gerwick, W.H. (2010) The unique mechanistic transformations involved in the biosynthesis of modular natural products from marine cyanobacteria. *Nat. Prod. Rep.*, **27**, 1048–1065.
4 Moore, R.E. (1996) Cyclic peptides and depsipeptides from cyanobacteria: a review. *J. Ind. Microbiol.*, **16**, 134–143.
5 Minich, S.S. (2012) Brentuximab vedotin: a new age in the treatment of Hodgkin lymphoma and anaplastic large cell lymphoma. *Ann. Pharmacother.*, **46**, 377–383.
6 Gerwick, W.H. and Moore, B.S. (2012) Lessons from the past and charting the future of marine natural products drug discovery and chemical biology. *Chem. Biol.*, **19**, 85–98.
7 Poncet, J. (1999) The dolastatins, a family of promising antineoplastic agents. *Curr. Pharm. Des.*, **5**, 139–162.
8 Luesch, H., Harrigan, G.G., Goetz, G., and Horgen, F.D. (2002) The cyanobacterial origin of potent anticancer agents originally isolated from sea hares. *Curr. Med. Chem.*, **9**, 1791–1806.
9 Haddley, K. (2012) Brentuximab vedotin: its role in the treatment of anaplastic large cell and Hodgkin's lymphoma. *Drugs Today*, **48**, 259–270.

10 Pettit, G.R., Hogan, F., and Toms, S. (2011) Antineoplastic agents. 592. Highly effective cancer cell growth inhibitory structural modifications of dolastatin 10. *J. Nat. Prod.*, **74**, 962–968.

11 Teruya, T., Sasaki, H., Fukazawa, H., and Suenaga, K. (2009) Bisebromoamide, a potent cytotoxic peptide from the marine cyanobacterium *Lyngbya* sp.: isolation, stereostructure and biological activity. *Org. Lett.*, **11**, 5062–5065.

12 Gao, X., Liu, Y., Kwong, S., Xu, Z., and Ye, T. (2010) Total synthesis and stereochemical reassignment of bisebromoamide. *Org. Lett.*, **12**, 3018–3021.

13 Sasaki, H., Teruya, T., Fukazawa, H., and Suenaga, K. (2011) Revised structure and structure–activity relationship of bisebromoamide and structure of norbisebromoamide from the marine cyanobacterium *Lyngbya* sp. *Tetrahedron*, **67**, 990–994.

14 Li, W., Yu, S., Jin, M., Xia, H., and Ma, D. (2011) Total synthesis and cytotoxicity of bisebromoamide and its analogues. *Tetrahedron Lett.*, **52**, 2124–2127.

15 Sumiya, E., Shimogawa, H., Sasaki, H., Tsutsumi, M., Yoshita, K., Ojika, M., Suenaga, K., and Uesugi, M. (2011) Cell-morphology profiling of a natural product library identifies bisebromoamide and miuraenamide A as actin filament stabilizers. *ACS Chem. Biol.*, **6**, 425–431.

16 Suzuki, K., Mizuno, R., Suennaga, K., Kosaka, T., Tanaka, N., Shinoda, K., Kono, H., Kikuchi, E., Nagata, H., Asanuma, H., Miyajima, A., Nakagawa, K., and Oya, M. (2012) 307: Bisebromoamide, as a novel molecular target drug inhibiting phosphorylation of both extracellular signal-regulated kinase and AKT in renal cell carcinoma. *J. Urol.*, **187**, e124–e125.

17 Taori, K., Paul, V.J., and Luesch, H. (2008) Structure and activity of largazole, a potent antiproliferative agent from the Floridian marine cyanobacterium *Symploca* sp. *J. Am. Chem. Soc.*, **130**, 1806–1807.

18 Li, S., Yao, H., Xu, J., and Jiang, S. (2011) Synthetic routes and biological evaluation of largazole and its analogues as potent histone deacetylase inhibitors. *Molecules*, **16**, 4681–4694.

19 Hong, J. and Luesch, H. (2012) Largazole: from discovery to broad-spectrum therapy. *Nat. Prod. Rep.*, **29**, 449–456.

20 Pei, D., Zhang, Y., and Zheng, J. (2012) Regulation of p53: a collaboration between Mdm2 and Mdmx. *Oncotarget*, **3**, 228–235.

21 Shangary, S. and Wang, S. (2009) Small-molecule inhibitors of the MDM2-p53 protein–protein interaction to reactivate p53 function: a novel approach for cancer therapy. *Annu. Rev. Pharmacol. Toxicol.*, **49**, 223–241.

22 Malloy, K.L., Choi, H., Fiorilla, C., Valeriote, F.A., Matainaho, T., and Gerwick, W.H. (2012) Hoiamide D, a marine cyanobacteria-derived inhibitor of p53/MDM2 interaction. *Bioorg. Med. Chem. Lett.*, **22**, 683–688.

23 Kisselev, A.F., and van derLinden, W.A., and Overkleeft, H.S. (2012) Proteasome inhibitors: an expanding army attacking a unique target. *Chem. Biol.*, **19**, 99–115.

24 Pereira, A., Kale, A.J., Fenley, A.T., Byrum, T., Debonsi, H.M., Gilson, M.K., Valeriote, F.A., Moore, B.S., and Gerwick, W.H. (2012) The carmaphycins: new proteasome inhibitors exhibiting an α,β-epoxyketone warhead from a marine cyanobacterium. *ChemBioChem*, **13**, 810–817.

25 Turk, B. (2006) Targeting proteases: successes, failures and future prospects. *Nat. Rev. Drug Discov.*, **5**, 785–799.

26 Tan, L.T. (2010) Filamentous tropical marine cyanobacteria: a rich source of natural products for anticancer drug discovery. *J. Appl. Phycol.*, **22**, 659–676.

27 Chlabicz, M., Gacko, M., Worowska, A., and Lapinski, R. (2011) Cathepsin E (EC 3.4.23.34): a review. *Folia Histochem. Cytobiol.*, **49**, 547–557.

28 Chai, Y., Wu, W., Zhou, C., and Zhou, J. (2012) The potential prognostic value of cathepsin D protein in serous ovarian cancer. *Arch. Gynecol. Obstet.*, **286**, 465–471.

29 Kwan, J.C., Eksioglu, E.A., Liu, C., Paul, V.J., and Luesch, H. (2009) Grassystatins A–C from marine cyanobacteria, potent cathepsin E inhibitors that reduce antigen presentation. *J. Med. Chem.*, **52**, 5732–5747.

30 Molinski, T.F., Reynolds, K.A., and Morinaka, B.I. (2012) Symplocin A, a linear peptide from the Bahamian cyanobacterium *Symploca* sp.: configurational analysis of *N*,

N-dimethylamino acids by chiral-phase HPLC of naphthacyl esters. *J. Nat. Prod.*, **75**, 425–431.

31 Luesch, H., Yoshida, W.Y., Moore, R.E., Paul, V.J., and Corbett, T.H. (2001) Total structure determination of apratoxin A, a potent novel cytotoxin from the marine cyanobacterium *Lyngbya majuscula*. *J. Am. Chem. Soc.*, **123**, 5418–5423.

32 Luesch, H., Yoshida, W.Y., Moore, R.E., and Paul, V.J. (2002) New apratoxins of marine cyanobacterial origin from Guam and Palau. *Bioorg. Med. Chem.*, **10**, 1973–1978.

33 Chen, Q.-Y., Liu, Y., and Luesch, H. (2011) Systematic chemical mutagenesis identifies a potent novel apratoxin A/E hybrid with improved *in vivo* antitumor activity. *ACS Med. Chem. Lett.*, **2**, 861–865.

34 Grindberg, R.V., Ishoey, T., Brinza, D., Esquenazi, E., Coates, R.C., Liu, W.-T., Gerwick, L., Dorrestein, P.C., Pevzner, P., Lasken, R., and Gerwick, W.H. (2011) Single cell genome amplification accelerates identification of the apratoxin biosynthetic pathway from a complex microbial assemblage. *PLoS One*, **6**, e18565.

35 Tripathi, A., Puddick, J., Prinsep, M.R., Rottmann, M., and Tan, L.T. (2010) Lagunamides A and B: cytotoxic and antimalarial cyclodepsipeptides from the marine cyanobacterium *Lyngbya majuscula*. *J. Nat. Prod.*, **73**, 1810–1814.

36 Suenaga, K., Mutou, T., Shibata, T., Itoh, T., Fujita, T., Takada, N., Hayamizu, K., Takagi, M., Irifune, T., Kigoshi, H., and Yamada, K. (2004) Aurilide, a cytotoxic depsipeptide from the sea hare *Dolabella auricularia*: isolation, structure determination, synthesis, and biological activity. *Tetrahedron*, **60**, 8509–8527.

37 Suenaga, K., Kajiwara, S., Kuribayashi, S., Handa, T., and Kigoshi, H. (2008) Synthesis and cytotoxicity of aurilide analogs. *Bioorg. Med. Chem. Lett.*, **18**, 3902–3905.

38 Takahashi, T., Nagamiya, H., Doi, T., Griffiths, P.G., and Bray, A.M. (2003) Solid phase library synthesis of cyclic depsipeptides: aurilide and aurilide analogues. *J. Comb. Chem.*, **5**, 414–428.

39 Takada, Y., Umehara, M., Katsumata, R., Nakao, Y., and Kimura, J. (2012) The total synthesis and structure–activity relationships of a highly cytotoxic depsipeptide kulokekahilide-2 and its analogs. *Tetrahedron*, **68**, 659–669.

40 Sato, S.-I., Murata, A., Orihara, T., Shirakawa, T., Suenaga, K., Kigoshi, H., and Uesugi, M. (2011) Marine natural product aurilide activates the OPA1-mediated apoptosis by binding to prohibitin. *Chem. Biol.*, **18**, 131–139.

41 Medina, R.A., Goeger, D.E., Hills, P., Mooberry, S.L., Huang, N., Romero, L.I., Ortega-Barria, E., Gerwick, W.H., and McPhail, K.L. (2008) Coibamide A, a potent antiproliferative cyclic depsipeptide from the Panamanian marine cyanobacterium *Leptolyngbya* sp. *J. Am. Chem. Soc.*, **130**, 6324–6325.

42 Kwan, J.C., Rocca, J.R., Abboud, K.A., Paul, V.J., and Luesch, H. (2008) Total structure determination of grassypeptolide, a new marine cyanobacterial cytotoxin. *Org. Lett.*, **10**, 789–792.

43 Kwan, J.C., Ratnayake, R., Abboud, K.A., Paul, V.J., and Luesch, H. (2010) Grassypeptolides A–C, cytotoxic bis-thiazoline containing marine cyclodepsipeptides. *J. Org. Chem.*, **75**, 8012–8023.

44 Liu, H., Liu, Y., Xing, X., Xu, Z., and Ye, T. (2010) Total synthesis of grassypeptolide. *Chem. Commun.*, **46**, 7486–7488.

45 Thornburg, C.C., Thimmaiah, M., Shaala, L.A., Hau, A.M., Malmo, J.M., Ishmael, J.E., Youssef, D.T.A., and McPhail, K.L. (2011) Cyclic depsipeptides, grassypeptolides D and E and ibu-epidemethoxylyngbyastatin 3, from a Red Sea *Leptolyngbya* cyanobacterium. *J. Nat. Prod.*, **74**, 1677–1685.

46 Popplewell, W.L., Ratnayake, R., Wilson, J. A., Beutler, J.A., Colburn, N.H., Henrich, C. J., McMahon, J.B., and McKee, T.C. (2011) Grassypeptolides F and G, cyanobacterial peptides from *Lyngbya majuscula*. *J. Nat. Prod.*, **74**, 1686–1691.

47 Salvador, L.A., Biggs, J.S., Paul, V.J., and Luesch, H. (2011) Veraguamides A–G, cyclic hexadepsipeptides from a dolastatin 16-producing cyanobacterium *Symploca* cf. *hydnoides* from Guam. *J. Nat. Prod.*, **74**, 917–927.

48 Mevers, E., Liu, W.-T., Engene, N., Mohimani, H., Byrum, T., Pevzner, P.A., Dorrestein, P.C., Spafora, C., and Gerwick,

W.H. (2011) Cytotoxic veraguamides, alkynyl bromide-containing cyclic depsipeptides from the marine cyanobacterium cf. *Oscillatoria margaritifera*. *J. Nat. Prod.*, **74**, 928–936.

49 Araoz, R., and Molgo, J., and deMarsac, N.T. (2010) Neurotoxic cyanobacterial toxins. *Toxicon*, **56**, 813–828.

50 Pereira, A., Cao, Z., Murray, T.F., and Gerwick, W.H. (2009) Hoiamide A, a sodium channel activator of unusual architecture from a consortium of two Papua New Guinea cyanobacteria. *Chem. Biol.*, **16**, 893–906.

51 Choi, H., Pereira, A.R., Cao, Z., Shuman, C.F., Engene, N., Byrum, T., Matainaho, T., Murray, T.F., Mangoni, A., and Gerwick, W.H. (2010) The hoiamides, structurally intriguing neurotoxic lipopeptides from Papua New Guinea marine cyanobacteria. *J. Nat. Prod.*, **73**, 1411–1421.

52 Wang, L., Xu, Z., and Ye, T. (2011) Total synthesis of hoiamide C. *Org. Lett.*, **13**, 2506–2509.

53 Malloy, K.L., Suyama, T.L., Engene, N., Debonsi, H., Cao, Z., Matainaho, T., Spadafora, C., Murray, T.F., and Gerwick, W.H. (2012) Credneramides A and B: neuromodulatory phenethylamine and isopentylamine derivatives of a vinyl chloride-containing fatty acid from cf. *Trichodesmium* sp. nov. *J. Nat. Prod.*, **75**, 60–66.

54 Erver, F. and Hilt, G. (2012) Cobalt- versus ruthenium-catalyzed Alder-ene reaction for the synthesis of credneramide A and B. *J. Org. Chem.*, **77**, 5215–5219.

55 Nunnery, J.K., Engene, N., Byrum, T., Cao, Z., Jabba, S.V., Pereira, A.R., Matainaho, T., Murray, T.F., and Gerwick, W.H. (2012) Biosynthetically intriguing chlorinated lipophilic metabolites from geographically distant tropical marine cyanobacteria. *J. Org. Chem.*, **77**, 4198–4208.

56 Taniguchi, M., Nunnery, J.K., Engene, N., Esquenazi, E., Byrum, T., Dorrestein, P.C., and Gerwick, W.H. (2010) Palmyramide A, a cyclic depsipeptide from a Palmyra Atoll collection of the marine cyanobacterium *Lyngbya majuscula*. *J. Nat. Prod.*, **73**, 393–398.

57 Pereira, A.R., Cao, Z., Engene, N., Soria-Mercado, I.E., Murray, T.F., and Gerwick, W.H. (2010) Palmyrolide A, an unusually stabilized neuroactive macrolide from Palmyra Atoll cyanobacteria. *Org. Lett.*, **12**, 4490–4493.

58 Klein, D., Braekman, J.C., and Daloze, D. (1996) Laingolide, a novel 15-membered macrolide from *Lyngbya bouillonii* (Cyanophyceae). *Tetrahedron Lett.*, **37**, 7519–7520.

59 Tello-Aburto, R., Johnson, E.M., Valdez, C.K., and Maio, W.A. (2012) Asymmetric total synthesis and absolute stereochemistry of the neuroactive marine macrolide palmyrolide A. *Org. Lett.*, **14**, 2150–2153.

60 Gutierrez, M., Pereira, A.R., Debonsi, H.M., Ligresti, A., Marzo, V.D., and Gerwick, W.H. (2011) Cannabinomimetic lipid from a marine cyanobacterium. *J. Nat. Prod.*, **74**, 2313–2317.

61 Sanchez, L.M., Lopez, D., Vesely, B.A., Della Togna, G., Gerwick, W.H., Kyle, D.E., and Linington, R.G. (2010) Almiramides A–C: discovery and development of a new class of leishmaniasis lead compounds. *J. Med. Chem.*, **53**, 4187–4197.

62 Montaser, R., Paul, V.J., and Luesch, H. (2011) Pitipeptolides C-F, antimycobacterial cyclodepsipeptides from the marine cyanobacterium *Lyngbya majuscula* from Guam. *Phytochemistry*, **72**, 2068–2074.

63 Villa, F.A., Lieske, K., and Gerwick, L. (2010) Selective MyD88-dependent pathway inhibition by the cyanobacterial natural product malyngamide F acetate. *Eur. J. Pharmacol.*, **629**, 140–146.

64 Malloy, K.L., Villa, F., Engene, N., Matainaho, T., Gerwick, L., and Gerwick, W.H. (2011) Malyngamide 2, an oxidized lipopeptide with nitric oxide inhibiting activity from a Papua New Guinea marine cyanobacterium. *J. Nat. Prod.*, **74**, 95–98.

6
Marine Bacteria Are an Attractive Source to Overcome the Problems of Antibiotic-Resistant *Staphylococcus aureus**

Dae-Sung Lee, Sung-Hwan Eom, Myung-Suk Lee, and Young-Mog Kim

6.1
Introduction

6.1.1
Mechanisms of Bacterial Resistance to Antibiotics

When bacteria are exposed to antibiotics, most of the cells die. However, some of the cells that acquire mechanism of antibiotic resistance will survive and reproduce, and the new population will be drug resistant. Many bacteria now exhibit multidrug resistance, including *Staphylococci*, *Enterococci*, *Gonococci*, *Streptococci*, and others. The following are the four main mechanisms by which microorganisms exhibit resistance to antimicrobials:

1) *Drug inactivation or modification*: For example, enzymatic deactivation of penicillin G in some penicillin-resistant bacteria through the production of β-lactamases.
2) *Alteration of target site*: For example, alteration of PBP – the binding target site of penicillins – in methicillin-resistant *Staphylococcus aureus* (MRSA) and other penicillin-resistant bacteria.
3) *Alteration of metabolic pathway*: For example, some sulfonamide-resistant bacteria do not require *para*-aminobenzoic acid (PABA), an important precursor for the synthesis of folic acid and nucleic acids in bacteria inhibited by sulfonamides. Instead, like mammalian cells, they turn to utilizing preformed folic acid.
4) *Reduced drug accumulation*: By decreasing drug permeability and/or increasing active efflux (pumping out) of the drugs across cell membrane [1].

Antibiotic resistance of *S. aureus* was almost unknown when penicillin was first introduced in 1943; indeed, the original petri dish on which Alexander Fleming

*The PhD thesis by Lee [1] and the article written by Shin *et al.* [2] have largely been referred to while writing this chapter.

Marine Microbiology: Bioactive Compounds and Biotechnological Applications, First Edition.
Edited by Se-Kwon Kim

observed the antibacterial activity of the *Penicillium* mold was growing a culture of *S. aureus*. However, today, *S. aureus* has become resistant to many commonly used antibiotics. The increasing antibiotic resistance of *S. aureus* has become a serious problem, especially in the nosocomial- and community-associated infections [3–5]. By 1950, 40% of hospital *S. aureus* isolates was penicillin resistant; and by 1960, this had risen to 80% [6,7]. By 1960, many hospitals had outbreaks of virulent multiresistant *S. aureus*. These were overcome with penicillinase-stable penicillins (methicillin, oxacillin, cloxacillin, and flucloxacillin), but only 2 years later, the first case of MRSA was reported in England [8]. Recently, MRSA has been the most problematic Gram-positive bacterium in public health because it has become resistant to almost all available antibiotics except vancomycin and teicoplanin [5,9].

The mechanism of resistance to methicillin is mediated via the *mec* operon, part of the staphylococcal cassette chromosome *mec* (SCCmec). Resistance is conferred by the *mec*A gene, which codes for an altered penicillin-binding protein (PBP2a or PBP2′) that has a lower affinity for binding β-lactams (penicillins, cephalosporins, and carbapenems). This allows resistance to all β-lactam antibiotics and obviates their clinical use during MRSA infections. MRSA has become resistant to almost all available antibiotics except vancomycin and teicoplanin. As a result, the glycopeptide, vancomycin, is often deployed against MRSA.

6.1.2 Prevalence of MRSA

MRSA is often referred to as a "superbug." In the last decade or so, the number of MRSA infections in the United States has increased significantly. A 2007 report published in Ref. [10] estimated that the number of MRSA infections treated in hospitals doubled nationwide, from approximately 127 000 in 1999 to 278 000 in

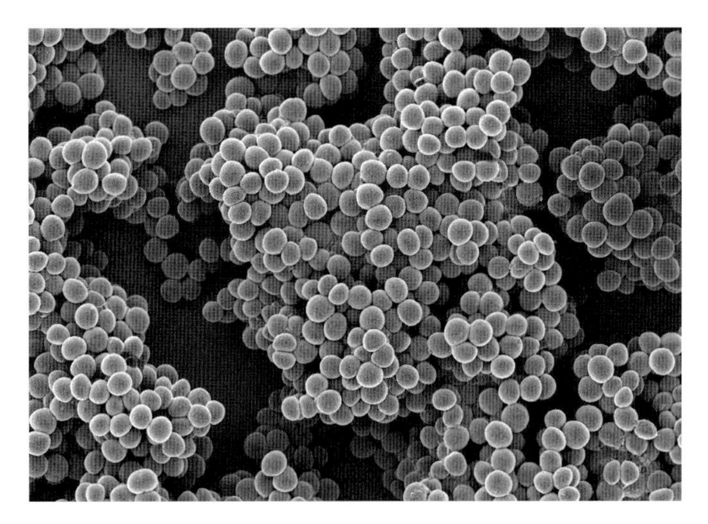

Figure 6.1 Scanning electron micrograph of *S. aureus*.

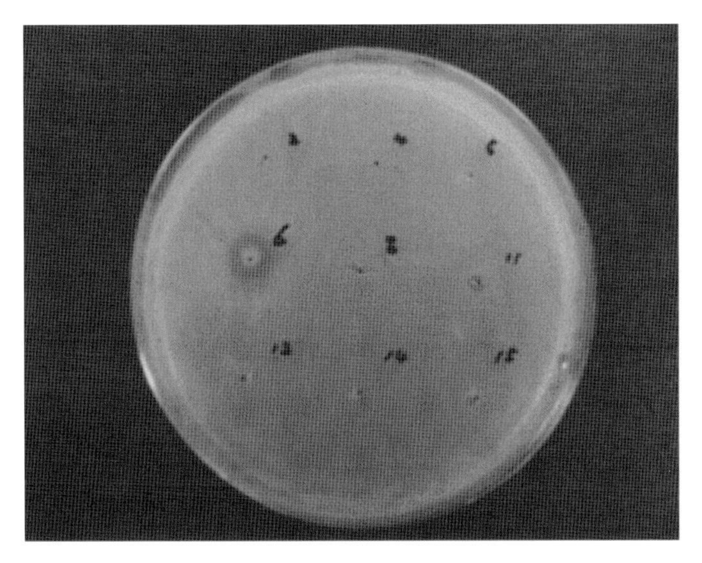

Figure 6.2 Isolation of marine bacteria producing anti-MRSA substance. Arrow bar indicated a bacterial strain producing an anti-MRSA substance. [1].

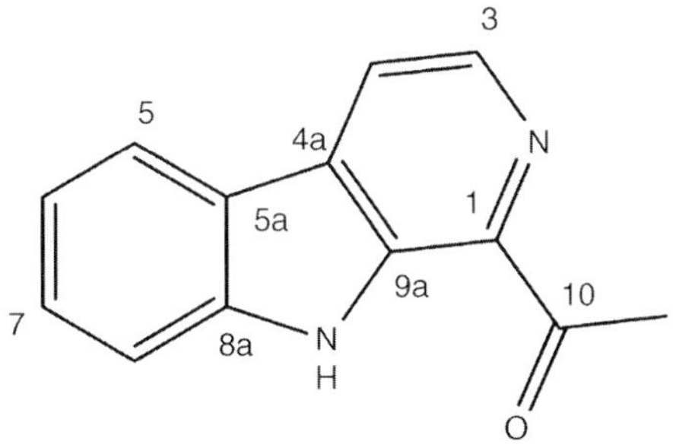

Figure 6.3 Structure of 1-acetyl-β-carboline.

2005, and deaths increased from 11 000 to more than 17 000. Another report published in Ref. [11] estimated that MRSA would have been responsible for 94 360 serious infections and associated with 18 650 hospital stay-related deaths in the United States in 2005. The UK Office for National Statistics reported 1629 MRSA-related deaths in England and Wales during 2005, indicating the MRSA-related mortality rate half the rate of that in the United States for 2005, although the figures from the British source were explained to be high because of the "improved levels of reporting, possibly brought about by the continued high public profile of the disease" during the 2005 UK General Election. MRSA is thought to have caused 1652 deaths in 2006 in the United Kingdom up from 51 in 1993 [12].

It has been argued that the observed mortality increased among MRSA-infected patients may be the result of the increased underlying morbidity of these patients. However, several studies, including Ref. [13], that have agreed to the assumption of underlying disease still found MRSA bacteremia to have a higher attributable mortality than the methicillin-susceptible *S. aureus* (MSSA) bacteremia. While the statistics suggest a national epidemic growing out of control, it has been difficult to quantify the degree of morbidity and mortality attributable to MRSA. A population-based study of the incidence of MRSA infections in San Francisco during 2004–2005 demonstrated that nearly 1 in 300 residents suffered from such an infection in the course of a year and that more than 85% of these infections occurred outside the health care setting [14]. A 2004 study showed that patients in the United States with *S. aureus* infection had, on average, three times the length of hospital stay (14.3 versus 4.5 days), incurred three times the total cost ($48 824 versus $14 141), and experienced five times the risk of in-hospital death (11.2% versus 2.3%) than patients without this infection [15]. In a meta-analysis of 31 studies, Cosgrove *et al.* [16] concluded that MRSA bacterium is associated with increased mortality compared to MSSA bacterium [17]. In addition, Wyllie *et al.* [18] reported a death rate of 34% within 30 days among patients infected with MRSA, compared to the death rate of 27% seen among MSSA-infected patients.

6.2 Strategies for Overcoming Antibiotic Resistance of Bacteria

Since the problem of bacterial resistance has increased, the use of antibacterial drugs in the future is uncertain. Thus, the development of new drugs or alternative therapies is required, which is active against resistant bacteria [19]. In response to bacterial resistance, many researchers have made attempts to develop alternative antibiotics and therapies. In the following sections, several strategies for overcoming antibiotic resistance of bacteria have been discussed.

6.2.1 Synthesis of New Chemical Entities

Many researchers have attempted to develop new antibiotics, either synthetic or natural. They are synthesizing new chemical entities (NCEs) in search of new β-lactams, such as cephalosporin-based antibiotics [20], dithiocarbamate carbapenems, and 2-metabiphenyl carbapenems [21].

The concept of anti-PBP2a β-lactams was pursued. Compassionate use of the older molecules is being employed in cases of difficult-to-treat MRSA or glycopeptide-intermediate *S. aureus* (GISA) infections [22,23]. In parallel, drug chemistry succeeded in generating semisynthetic drugs with greatly improved PBP2a affinity. β-Lactams with anti-MRSA activity were described in all the three classes of penams, cephems, and carbapenems. However, novel anti-MRSA molecules mainly cluster among cephems and carbapenems. This could be related to the difficulty of modifying functional groups at the ring fused to the β-lactam ring in penam molecules [24].

Despite these developmental hindrances, clinically useful anti-MRSA β-lactams are likely to be imminent. Because they mimic an adduct (D-ala-D-ala) unique to the bacterial world, β-lactams can be given in large doses to humans. Thus, new anti-PBP2a β-lactams should provide the most appropriate response to the increase of multiresistant MRSA, including the recent glycopeptide-resistant isolates.

6.2.2 Screening NCEs for New Antibiotics

An alternative strategy to the synthesis of NCEs is their identification by screening natural sources such as bacteria, fungi, and plants. Especially, screening bacteria from marine ecosystems, such as the deep sea floor, could lead to the isolation of new antibiotics [25]. The chemical compounds produced by marine bacteria are less well known than those produced by their terrestrial counterparts. Many such organisms produce marine natural products that possess unique structural features compared to terrestrial metabolites [26]. Among them, a few and novel antibiotics from marine microorganisms have been reported, including loloatins from *Bacillus* [27], agrochelin and sesbanimides from *Agrobacterium* [28,29], pelagiomicins from *Pelagiobacter variabilis* [30], δ-indomycinone from *Streptomyces* sp. [31], and dihydrophencomycin methyl ester from *Streptomyces* sp. [32]. Screening of antibiotics has been traditionally performed with *Actinomyces* such as *Streptomyces* species. Therefore, screening of novel secondary metabolites such as antibiotics from marine bacteria is attracting attention since marine bacteria are believed to have physiological and genetic characteristics different from that of soil bacteria such as *Streptomyces* sp.

6.2.3 Synergistic Effect of Combination of Commercial Antibiotics

During the past 20 years, the number of new drugs has been declining. Concomitantly, resistance is rising to high levels among bacterial pathogens. This means that the rate of loss of efficacy of old antibiotics is outstripping their replacement with new ones for many species of pathogenic bacteria [33]. It is not uncommon for the effect of two chemicals on an organism to be greater than the effect of each chemical individually, or the sum of the individual effects. The presence of one chemical enhances the effects of the second. It is designed to either improve efficacy through synergistic action of the agents or to overcome bacterial resistance.

6.2.4 The Genomics Revolution

The complete sequencing of the genomes of many pathogenic bacteria has led to an explosion in knowledge about these organisms [34]. Genomics is used to select potential antibacterial targets [35,36] and can also be used to provide insights into, for example, pathogenesis [37,38] and antibiotic resistance [39,40]. An antibacterial target may be the receptor of a ligand in a bacterial molecule, a specific function of a bacterial molecule such as an enzyme or a metabolic pathway. Libraries of compounds may be screened at this time to find a lead molecule that, for example, inhibits the enzymatic function of the target. Further development includes searching for antibacterial activity of the molecule. Several lead compounds are chosen and lead optimization then proceeds to complete the preclinical program, which is required before entry into clinical trials. In the case of antibiotic discovery, this approach and many other similar ones have been and are being attempted [41]. However, so far, no marketed antibiotics have reached the marketplace via the genomics route [42]. The reason for this failure is not entirely clear. Peptide deformylase is one target that has been identified with the assistance of genomic information [43], and inhibitors of this enzyme have been developed, which have entered into human clinical trials [44]. Unfortunately, these inhibitors seem to generate mutants at a high rate [45,46] and hence development has not proceeded as expected.

6.2.5 Bacteriophages

Bacteriophages and their fragments [47] kill bacteria. It is estimated that every 2 days, half of the world's bacterial population is destroyed by bacteriophages [48]. Bacteriophages have been used as antibacterials in humans in some countries [49]. Potential advantages of this approach are that the mechanism of action is likely to be completely different from that of current antimicrobials. Its disadvantages are that quality control and standardization are difficult. Also, when used systemically in patients, phages are likely to be immunogenic and may induce neutralizing antibodies [50]. Furthermore, massive bacterial lysis might lead to toxic shock [51]. One approach that has been developed to address this problem is lysis-deficient phages, which can still kill bacteria [51]. The development of phage gene products is another potential route for new antibacterials. Phage lysins that are cell wall hydrolases and are produced late in the viral infection cycle bind to peptidoglycan and disrupt the cell wall of Gram-positive bacteria, which results in hypotonic lysis. Lysins have potential uses as antibacterials for human use. Systemic use of lysins has also been described [52], and may have advantages over whole bacteriophages because, in preclinical studies, they do not induce resistance, neutralizing antibodies, or toxic shock. At present, there is a lack of good human clinical trial results, although animal studies suggest that in certain circumstances, bacteriophage therapy may be useful.

6.2.6 Marine Bacteria Producing an Antibacterial Substance

Since the problem of antibiotic-resistant bacteria has increased, the use of antibacterial drugs in the future is uncertain. Thus, the development of new drugs or alternative therapies is required, which is active against resistant bacteria [19]. In these respects, enormous efforts to developing antimicrobial agents from various natural resources have been made, which have been potential candidates during the last few decades.

One of the preventing antibiotic resistances is to use new compounds that are not based on the existing synthetic antimicrobial agents. It is believed that novel metabolites produced by marine organisms have unique structural features compared to metabolites produced by terrestrial organisms [53–55]. Especially, screening natural sources from marine ecosystems, such as the deep sea floor, could lead to isolation of new antibiotics [56]. Thus, screening of novel secondary metabolites such as antibiotics from marine organisms is attracting attention since the marine organisms are believed to have physiological and genetic characteristics different from that of terrestrial organism. In fact, several novel antibiotics from marine microorganisms have been reported, including loloatins from *Bacillus* [27], agrochelin and sesbanimides from *Agrobacterium* [28,29], pelagiomicins from *P. variabilis* [30], and δ-indomycinone from a *Streptomyces* sp. [31], although the chemical compounds produced by marine microorganisms are less well known than those produced by their terrestrial counterparts.

6.3 Marine Bacteria Are Attractive Natural Sources for Overcoming Antibiotic Resistance of MRSA

MRSA has become a worldwide concern because it is highly prevalent, capable of developing new clones, resistant to almost all currently available antibiotics except vancomycin and teicoplanin, and can potentially cause death [9]. Moreover, glycopeptide-resistant strains have been emerging owing to the increasing use of glycopeptides [57]. Recently, linezolid and daptomycin have also become available for MRSA infections. However, new resistant strains against many drugs besides these have already been reported [55]. Furthermore, a decrease in the susceptibility of MRSA to vancomycin and teicoplanin has been reported in several hospitals worldwide [54]. Therefore, new and potent anti-MRSA agents are urgently required. In Section 6.3.1, the isolation of an anti-MRSA substance from marine bacteria and its antibacterial characterization against MRSA have been described.

6.3.1 Isolation of an Anti-MRSA Substance from Marine Bacteria

As a part of our ongoing screening program of bioactive secondary metabolites, we isolated marine actinomycete strains from marine sediments and marine organisms collected at various sites of South Korea and the western Pacific Ocean. Strain 04DH52 was isolated from shallow water sediment at 2 m depth of Ayajin Bay, on the East Sea of Korea. This strain was identified as *Streptomyces* sp. by 16S rRNA gene sequence analysis. The crude extracts of strain 04DH52 exhibited

Table 6.1 ^{1}H and ^{13}C NMR data of 1-acetyl-β-carboline in CD_3OD-d_4 [2].

No.	δ_H	Mult (J in Hz)	δ_c	COSY	HMBC
1			137.4		
2					
3	8.45	d (5.4)	138.8	H-4	C-1, C-4, C-4a
4	8.30	d (5.4)	120.4	H-3	C-3, C-9a, C-5a
4a			133.4		
5a			121.8		
5	8.21	d (7.8)	122.8	H-6	C-7, C-8a
6	7.31	td (7.8, 1.0)	121.8	H-5, H-7	C-5a, C-8
7	7.59	td (7.8, 1.0)	130.5	H-8	C-5, C-8a
8	7.70	d (7.8)	113.6		C-5a, C-6
8a			143.6		
9					
9a			136.3		
10			203.4		
11	2.82	s	26.2		C-1, C-10

anti-MRSA activity. Bioassay-guided fractionation by solvent partitioning ODS vacuum flash chromatography and purification with a reversed-phase HPLC gave a pure anti-MRSA compound, 1-acetyl-β-carboline [2]. Purification and isolation of 1-acetyl-β-carboline were guided by the anti-MRSA activity. The molecular formula of 1-acetyl-β-carboline was established as $C_{13}H_{10}N_2O$ by mass spectrometry (MS) analysis [m/z 209 $(M-H)-$, $211(M+H)+$] and 13C NMR spectral data (Table 6.1) [2].

6.3.2
Antibacterial Activity of 1-Acetyl-β-Carboline

In order to evaluate the antibacterial activity of 1-acetyl-β-carboline against MRSA strains and other bacteria, the minimum inhibitory concentrations (MICs) were determined by the twofold serial dilution method, as described by the National Committee for Clinical Laboratory Standards [58]. The MICs of 1-acetyl-β-carboline against MSSA and MRSA strains are shown in Table 6.2 . The MRSA strains tested in the present study were highly resistant to the β-lactams. 1-Acetyl-β-carboline

Table 6.2 MIC of 1-acetyl-β-carboline and β-lactams against MSSA and MRSA [1,2].

Strain	*mecA*[a)]	MIC (μg/ml)			
		1-Acetyl-β-carboline	Ampicillin	Penicillin	Oxacillin
MSSA (KCTC 1927)	−	16	<1	<1	<1
MRSA (KCCM 40510)	+	32	512	512	128
MRSA (KCCM 40511)	+	64	512	256	128

a) +, *mec*A positive; -, *mec*A negative.

Table 6.3 Antibacterial activity of 1-acetyl-β-carboline against pathogenic bacteria [2].

	Strain	MIC (μg/ml)	
		1-Acetyl-β-carboline	**Vancomycin**
Gram-positive	*Bacillus subtilis* KCTC 1028	64	0.5
	Enterococcus faecalis KCTC 2011	64	0.5
Gram-negative	*Escherichia coli* KCTC 1682	128	512
	Klebsiella pneumonia KCTC 2242	128	512
	Legionella birminghamensis KCTC 2057	128	256
	Pseudomonas aeruginosa KCTC 1637	128	256
	Salmonella typhimurium KCTC 1925	256	512

showed antibacterial activity against MRSA strains tested with MICs ranging from 16 to 64 μg/ml. The MICs of 1-acetyl-β-carboline were lower than those of β-lactams against MRSA strains (Table 6.2). 1-Acetyl-β-carboline also showed antibacterial activity against MSSA, suggesting the anti-MRSA activity of 1-acetyl-β-carboline may not be related to penicillin-binding protein 2A, which decreases the binding affinity to β-lactams, because the antibacterial activity of 1-acetyl-β87q-carboline is not specific to MRSA. As shown in Table 6.3 , 1-acetyl-β-carboline also exhibited antibacterial activity against Gram-negative bacteria with MICs ranging from 128 to 256 μg/ml, even though the compound was less effective against Gram-negative bacteria than against other Gram-positive bacteria. These results suggest that 1-acetyl-β-carboline shows broad-spectrum antibacterial activities [2].

6.3.3 Combination Effect of 1-Acetyl-β-Carboline and β-Lactams against MRSA

It has been known that one of the effective approaches to overcome bacterial resistance is restoration of antibiotic activity through the synergistic action of antibacterial materials from natural and old agents [59,60]. Therefore, we compared MICs of 1-acetyl-β-carboline used alone and in combination with β-lactams against MRSA strains. The synergistic effects of 1-acetyl-β-carboline and β-lactams, including ampicillin, penicillin, and oxacillin, were investigated by the checker-board method and were evaluated as the fractional inhibitory concentration (FIC) index [61]. As shown in Table 6.4, the FIC indices of 1-acetyl-β-carboline and ampicillin ranged from 0.156 to 0.266 in combination with 32 and 64 μg/ml of 1-acetyl-β-carboline against all tested MRSA strains and a synergistic effect was observed. The FIC indices of penicillin ranged from 0.188 to 0.313 in combination with 1-acetyl-β-carboline (32 and 64 μg/ml) and a synergistic effect was also observed [2]. However, no synergy was observed between 1-acetyl-β-carboline and oxacillin against MRSA. Differences in synergistic effects of anti-MRSA substance–β-lactams combination were also reported in epigallocatechin gallate (EGCg) and dieckol [60,62]. This discrepancy in susceptibilities may be attributable to structural differences between dieckol and EGCg. Dieckol is a hexamer of phloroglucinol;

Table 6.4 MICs and FIC indices of 1-acetyl-β-carboline in combination with β-lactams against MRSA2.

Strain	Ampicillin					Penicillin					Oxacillin				
	MIC (μg/ml)			FIC index		MIC (μg/ml)			FIC index		MIC (μg/ml)			FIC index	
	A	B	C	b	c	A	B	C	b	c	A	B	C	b	c
MRSA40510	512	16	8	0.156	0.266	512	32	16	0.188	0.281	128	128	64	1.125	0.750
MRSA40511	512	16	8	0.156	0.266	256	32	16	0.250	0.313	128	128	128	1.125	1.250

A: without 1-acetyl-β-carboline, B–C and b–c: 1-acetyl-β-carboline at 32 and 64 μg/ml, respectively. The FIC was calculated as the MIC of 1-acetyl-β-carboline or each antibiotic in combination divided by the MIC of 1-acetyl-β-carboline or each antibiotic alone. The FIC index was obtained by the sum of FICs. The FIC index indicated synergy: 0.5, synergic; >0.5–1, additive; >1–2, independent; >2, antagonistic.

thus, it has a rather bulkier structure than EGCg [62]. It appears that dieckol (or EGCg) and penicillin synergistically attack the same site of the cell wall. However, no synergistic binding between dieckol and oxacillin will occur at the target site(s), as has been reported previously between EGCg and β-lactams [60,62].

Some significant infections should not be treated with single antibiotic, because the bacteria can rapidly develop resistance when such a single antibiotic is used. Combinations of two or more compounds are generally superior to the use of a single compound, especially for the treatment of serious infections caused by antibiotics-resistant bacteria. The ability of 1-acetyl-β-carboline to enhance the antibacterial activity of ampicillin for MRSA is a useful property and these combinations could decrease the emergence of MRSA. The mechanism of the synergistic effect of 1-acetyl-β-carboline and β-lactams is unknown at present, but the authors speculate that a possible mechanism for the synergistic effect may be attributed to their different structure and that they attack the same target site of the cell wall. Studies on the synergistic anti-MRSA mechanism and *in vivo* efficacy of 1-acetyl-β-carboline, alone and in combination with ampicillin, deserve further investigation [2].

Indole alkaloids are pharmacologically active natural products that have been shown to possess a wide range of biological activities, including cytotoxic, antiviral, antimicrobial, anti-inflammatory, antiserotonin, and enzyme-inhibitory activities [63]. One important subclass of indole alkaloids is β-carbolines, which possess a common tricyclic pyrido indole ring structure [9,64]. The β-carboline skeleton is an important structure in drug discovery [53], and drugs on the market such as tadalafil possess this indole nucleus. β-Carboline alkaloids are widespread in plants, but are very rare in microorganisms [65].

These results strongly suggest that marine bacteria appear to harbor some potential as promising pharmaceuticals, including both anti-MRSA substances and bioactive nutraceuticals, and marine bacteria will be a good source for the development of an alternative phytotherapeutic agent against MRSA and for the applications of the treatment of MRSA infections.

References

1 Lee, D.S. (2009) Isolation and characterization of anti-MRSA (methicillin-resistant *Staphylococcus aureus*) substances from marine organisms: microbiology. PhD thesis. Pukyong National University, Busan.

2 Shin, H.J., Lee, H.S., and Lee, D.S. (2010) The synergistic antibacterial activity of 1-acetyl-β-carboline and β-lactams against methicillin-resistant *Staphylococcus aureus* (MRSA). *J. Microbiol. Biotechnol.*, **20**, 501–505.

3 Levy, S.B. (2005) Antibiotic resistance: the problem intensifies. *Adv. Drug Deliv. Rev.*, **57**, 1446–1450.

4 Schaberg, D.R., Culver, D.H., and Gaynes, R.P. (1991) Major trends in the microbial etiology of nosocomial infection. *Am. J. Med.*, **91**, 72S–75S.

5 Witte, W. (1999) Antibiotic resistance in Gram-positive bacteria: epidemiological aspects. *J. Antimicrob. Chemother.*, **44**, 1–9.

6 Chambers, H.F. (2001) The changing epidemiology of *Staphylococcus aureus*? *Emerg. Infect. Dis.*, **7**, 178–182.

7 Finland, M. (1979) Emergence of antibiotic resistance in hospitals, 1935–1975. *Rev. Infect. Dis.*, **1**, 4–21.

8 Jevons, M.P. (1961) "Celbenin"-resistant *Staphylococci*. *Br. Med. J.*, **1**, 124–125.
9 Isnansetyo, A. and Kamei, Y. (2009) Anti-methicillin-resistant *Staphylococcus aureus* (MRSA) activity of MC21-B, an antibacterial compound produced by the marine bacterium *Pseudoalteromonas phenolica* O-B30^T. *Int. J. Antimicrob. Agents*, **34**, 131–135.
10 Klein, E., Smith, D.L., and Laxminarayan, R. (2007) Hospitalizations and deaths caused by methicillin-resistant *Staphylococcus aureus*, United States, 1999–2005. *Emerg. Infect. Dis.*, **13**, 1840–1846.
11 Klevens, R.M., Morrison, M.A., Nadle, J., Petit, S., Gershman, K., Ray, S., Harrison, L. H., Lynfield, R., Dumyati, G., Townes, J.M., Craig, A.S., Zell, E.R., Fosheim, G.E., McDougal, L.K., Carey, R.B., and Fridkin, S. K. (2007) Invasive methicillin-resistant *Staphylococcus aureus* infections in the United States. *J. Am. Med. Assoc.*, **298**, 1763–1771.
12 Office for National Statistics (2005) MRSA deaths continue to rise in 2005. Retrieved from http://www.statistics.gov.uk/cci/nugget.asp?id=1067 on October 24, 2007.
13 Blot, S.I., Vandewoude, K.H., Hoste, E.A., and Colardyn, F.A. (2002) Outcome and attributable mortality in critically ill patients with bacteremia involving methicillin-susceptible and methicillin-resistant *Staphylococcus aureus*. *Arch. Intern. Med.*, **162**, 2229–2235.
14 Liu, C., Graber, C.J., Karr, M., Diep, B.A., Basuino, L., Schwartz, B.S., Enright, M.C., O'Hanlon, S.J., Thomas, J.C., Perdreau-Remington, F., Gordon, S., Gunthorpe, H., Jacobs, R., Jensen, P., Leoung, G., Rumack, J.S., and Chambers, H.F. (2008) A population-based study of the incidence and molecular epidemiology of methicillin-resistant *Staphylococcus aureus* disease in San Francisco, 2004–2005. *Clin. Infect. Dis.*, **46**, 1637–1646.
15 Noskin, G.A., Rubin, R.J., Schentag, J.J., Kluytmans, J., Hedblom, E.C., Smulders, M., Lapetina, E., and Gemmen, E. (2005) The burden of *Staphylococcus aureus* infections on hospitals in the United States: an analysis of the 2000 and 2001 nationwide inpatient sample database. *Arch. Intern. Med.*, **165**, 1756–1761.
16 Cosgrove, S.E., Qi, Y., Kaye, K.S., Harbarth, S., Karchmer, A.W., and Carmeli, Y. (2005) The impact of methicillin resistance in *Staphylococcus aureus* bacteremia on patient outcomes: mortality, length of stay, and hospital charges. *Infect. Control Hosp. Epidemiol.*, **26**, 166–174.
17 Hardy, K.J., Hawkey, P.M., Gao, F., and Oppenheim, B.A. (2004) Methicillin resistant *Staphylococcus aureus* in the critically ill. *Br. J. Anaesth.*, **92**, 121–130.
18 Wyllie, D., Crook, D.W., and Peto, T.E.A. (2006) Mortality after *Staphylococcus aureus* bacteraemia in two hospitals in Oxfordshire, 1997–2003: cohort study. *BMJ*, **333**, 269–270.
19 Nascimento, G.G.F., Locatelli, J., Freitas, P. C., and Silva, G.L. (2000) Antibacterial activity of plant extracts and phytochemicals on antibiotic-resistant bacteria. *Braz. J. Microbiol.*, **31**, 247–256.
20 Lloyd, A.W. (1998) Monitor: molecules and profiles. *Drug Discov. Today*, **3**, 89.
21 Lloyd, A.W. (1998) Monitor: molecules and profiles. *Drug Discov. Today*, **3**, 43–46.
22 Hiramatsu, K., Hanaki, H., Ino, T., Yabuta, K., Oguri, T., and Tenover, F.C. (1997) Methicillin-resistant *Staphylococcus aureus* clinical strain with reduced vancomycin susceptibility. *J. Antimicrob. Chemother.*, **40**, 135–136.
23 Nicolas, M.H., Kitzis, M.D., and Karim, A. (1993) Stérilisation par 2 grammes d'Augmentin (R) des urines infectées à *Staphylococcus aureus* résistant à la méticilline. *Med. Mal. Infect.*, **23**, 82–94.
24 Guignard, B., Entenza, J.M., and Moreillon, P. (2005) β-Lactams against methicillin-resistant *Staphylococcus aureus*. *Curr. Opin. Pharmacol.*, **5**, 479–489.
25 Tan, Y.T., Tillett, D.J., and McKay, I.A. (2000) Molecular strategies for overcoming antibiotic resistance in bacteria. *Mol. Med. Today*, **6**, 309–314.
26 Larsen, T.O., Smedsgaard, J., Nielsen, K.F., Hansen, M.E., and Frisvad, J.C. (2005) Phenotypic taxonomy and metabolite profiling in microbial drug discovery. *Nat. Prod. Rep.*, **22**, 672–695.
27 Gerard, J.M., Haden, P., Kelly, M.T., and Andersen, R.J. (1999) Loloatins A–D, cyclic decapeptide antibiotics produced in culture

by a tropical marine bacterium. *J. Nat. Prod.*, **62**, 80–85.

28 Acebal, C., Cañedo, L.M., Puentes, J.L., Baz, J.P., Romero, F., de laCalle, F., Grávalos, M. D., and Rodrigues, P. (1999) Agrochelin, a new cytotoxic antibiotic from a marine *Agrobacterium*: taxonomy, fermentation, isolation, physicochemical properties and biological activity. *J. Antibiot.*, **52**, 983–987.

29 Acebal, C., Alcazar, R., Cañedo, L.M., de laCalle, F., Rodriguez, P., Romero, F., and Fernandez Puentes, J.L. (1998) Two marine agrobacterium producers of sesbanimide antibiotics. *J. Antibiot.*, **51**, 64–67.

30 Imamura, N., Nishijima, M., Takadera, T., Adachi, K., Sakai, M., and Sano, H. (1997) New anticancer antibiotics, pelagiomicins produced by a new marine bacterium *Pelagiobacter variabilis*. *J. Antibiot.*, **50**, 8–12.

31 Biabani, M.A., Laatsch, H., Helmke, E., and Weyland, H. (1997) δ-Indomycinone: a new member of pluramycin class of antibiotics isolated from marine *Streptomyces* sp. *J. Antibiot.*, **50**, 874–877.

32 Pusecker, K., Laatsch, H., Helmke, E., and Weyland, H. (1997) Dihydrophencomycin methyl ester, a new phenazine derivative from a marine *Streptomycete*. *J. Antibiot.*, **50**, 479–483.

33 Hancock, R.E.W. (2007) The end of an era? *Nat. Rev. Drug Discov.*, **6**, 28.

34 Monaghan, R.L. and Barrett, J.F. (2006) Antibacterial drug discovery: then, now and the genomics future. *Biochem. Pharmacol.*, **71**, 901–909.

35 Arcus, V.L., Lott, J.S., Johnston, J.M., and Baker, E.N. (2006) The potential impact of structural genomics on tuberculosis drug discovery. *Drug Discov. Today*, **11**, 28–34.

36 Sakharkar, K.R., Sakharkar, M.K., and Chow, V.T. (2004) A novel genomics approach for the identification of drug targets in pathogens, with special reference to *Pseudomonas aeruginosa*. *In Silico Biol.*, **4**, 355–360.

37 Field, D., Hughes, J., and Moxon, E.R. (2004) Using the genome to understand pathogenicity. *Methods Mol. Biol.*, **266**, 261–287.

38 Polissi, A. and Soria, M.R. (2005) Functional genomics of bacterial pathogens: from post-genomics to therapeutic targets. *Mol. Microbiol.*, **57**, 307–312.

39 Black, M.T. and Hodgson, J. (2005) Novel target sites in bacteria for overcoming antibiotic resistance. *Adv. Drug Deliv. Rev.*, **57**, 1528–1538.

40 Fournier, P.E., Vallenet, D., Barbe, V., Audic, S., Ogata, H., Poirel, L., Richet, H., Robert, C., Mangenot, S., Abergel, C., Nordmann, P., Weissenbach, J., Raoult, D., and Claverie, J.M. (2006) Comparative genomics of multidrug resistance in *Acinetobacter baumannii*. *PLoS Genet.*, **2**, e7.

41 Freiberg, C. and Brötz-Oesterhelt, H. (2005) Functional genomics in antibacterial drug discovery. *Drug Discov. Today*, **10**, 927–935.

42 Mills, S.D. (2006) When will the genomics investment pay off for antibacterial discovery? *Biochem. Pharmacol.*, **71**, 1096–1102.

43 Yuan, Z. and White, R.J. (2006) The evolution of peptide deformylase as a target: contribution of biochemistry, genetics and genomics. *Biochem. Pharmacol.*, **71**, 1042–1047.

44 Clements, J.M., Ayscough, A.P., Keavey, K., and East, S.P. (2002) Peptide deformylase inhibitors, potential for a new class of broad spectrum antibacterials. *Curr. Med. Chem.*, **1**, 239–249.

45 Apfel, C.M., Locher, H., Evers, S., Takács, B., Hubschwerlen, C., Pirson, W., Page, M. G., and Keck, W. (2001) Peptide deformylase as an antibacterial drug target: target validation and resistance development. *Antimicrob. Agents Chemother.*, **45**, 1058–1064.

46 Margolis, P.S., Hackbarth, C.J., Young, D. C., Wang, W., Chen, D., Yuan, Z., White, R., and Trias, J. (2000) Peptide deformylase in *Staphylococcus aureus*: resistance to inhibition is mediated by mutations in the formyltransferase gene. *Antimicrob. Agents Chemother.*, **44**, 1825–1831.

47 Borysowski, J., Weber-Dabrowska, B., and Górski, A. (2006) Bacteriophage endolysins as a novel class of antibacterial agents. *Exp. Biol. Med.*, **231**, 366–377.

48 Hendrix, R.W. (2002) Bacteriophages: evolution of the majority. *Theor. Popul. Biol.*, **61**, 471–480.

49 Sulakvelidze, A., Alavidze, Z., and Morris, J. G., Jr. (2001) Bacteriophage therapy. *Antimicrob. Agents Chemother.*, **45**, 649–659.

50 Dabrowska, K., Switala-Jelen, K., Opolski, A., Weber-Dabrowska, B., and Gorski, A. (2005) Bacteriophage penetration in vertebrates. *J. Appl. Microbiol.*, **98**, 7–13.

51 Matsuda, T., Freeman, T.A., Hilbert, D.W., Duff, M., Fuortes, M., Stapleton, P.P., and Daly, J.M. (2005) Lysis-deficient bacteriophage therapy decreases endotoxin and inflammatory mediator release and improves survival in a murine peritonitis model. *Surgery*, **137**, 639–646.

52 Fischetti, V.A., Nelson, D., and Schuch, R. (2006) Reinventing phage therapy: are the parts greater than the sum? *Nat. Biotechnol.*, **24**, 1508–1511.

53 Trujillo, J.I., Meyers, M.J., Anderson, D.R., Hegde, S., Mahoney, M.W., Vernier, W.F., Buchler, I.P., Wu, K.K., Yang, S., Hartmann, S.J., and Reitz, D.B. (2007) Novel tetrahydro-β-carboline-1-carboxylic acids as inhibitors of mitogen activated protein kinase-activated protein kinase 2 (MK-2). *Bioorg. Med. Chem. Lett.*, **17**, 4657–4663.

54 Vaudaux, P., Francois, P., Berger-Bächi, B., and Lew, D.P. (2001) *In vivo* emergence of subpopulations expressing teicoplanin or vancomycin resistance phenotypes in a glycopeptides-susceptible, methicillin-resistant strain of *Staphylococcus aureus*. *J. Antimicrob. Chemother.*, **47**, 163–170.

55 Woodford, N. (2005) Biological counterstrike: antibiotic resistance mechanisms of Gram-positive cocci. *Clin. Microbiol. Infect.*, **11**, 2–21.

56 Tan, R.X. and Zou, W.X. (2001) Endophytes: a rich source of functional metabolites. *Nat. Prod. Rep.*, **18**, 448–459.

57 Maruyama, T., Yamamoto, Y., Kano, Y., Kurazono, M., Shitara, E., Iwamatsu, K., and Atsumi, K. (2008) Synthesis of novel di- and tricationic carbapenems with potent anti-MRSA activity. *Bioorg. Med. Chem. Lett.*, **19**, 447–450.

58 CLSI (2006) *Methods for Dilution Antimicrobial Susceptibility Tests for Bacteria That Grow Aerobically: Approved Standard*, 7th edn, CLSI M7-A7. Clinical and Laboratory Standards Institute, Wayne, PA.

59 Shiota, S., Shimizu, M., Sugiyama, J., Morita, Y., Mizushima, T., and Tsuchiya, T. (2004) Mechanisms of action of corilagin and tellimagrandin I that remarkably potentiate the activity of β-lactams against methicillin-resistant *Staphylococcus aureus*. *Microbiol. Immunol.*, **48**, 67–73.

60 Zhao, W.H., Hu, Z.Q., Okubo, S., Hara, Y., and Shimamura, T. (2001) Mechanism of synergy between epigallocatechin gallate and β-lactams against methicillin-resistant *Staphylococcus aureus*. *Antimicrob. Agents Chemother.*, **45**, 1737–1742.

61 Norden, C.W., Wentzel, H., and Keleti, E. (1979) Comparison of techniques of measurement of *in vitro* antibiotic synergism. *J. Infect. Dis.*, **140**, 629–633.

62 Lee, D.S., Kang, M.S., Hwang, H.J., Eom, S. H., Yang, J.Y., Lee, M.S., Lee, W.J., Jeon, Y.J., Choi, J.S., and Kim, Y.M. (2008) Synergistic effect between dieckol from *Ecklonia stolonifera* and β-lactams against methicillin-resistant *Staphylococcus aureus*. *Biotechnol. Bioprocess Eng.*, **13**, 758–764.

63 Mansoor, T.A., Ramalhete, C., Molnár, J., Mulhovo, S., and Ferreira, M.J.U. (2009) Tabernines A–C, β-carbolines from the leaves of *Tabernaemontana elegans*. *J. Nat. Prod.*, **72**, 1147–1150.

64 Zhang, H. and Larock, R.C. (2001) Synthesis of beta- and gamma-carbolines by the palladium-catalyzed iminoannulation of internal alkynes. *Org. Lett.*, **3**, 3083–3086.

65 Allen, J.R.F. and Holmstedt, B.R. (1980) The simple β-carboline alkaloids. *Phytochemistry*, **19**, 1573–1582.

7
Marine Bacteria as Probiotics and Their Applications in Aquaculture

Chamilani Nikapitiya

7.1
Introduction

Aquaculture of aquatic organisms including animals and algal plants is one of the fastest growing food producing sectors in the world. Disease outbreaks such as bacterial and viral epizootics and high mortality are considered as major constraints on successful aquaculture production and development in terms of quantity, quality, and regularity and also on trade, affecting the economic development of the aquaculture sector. It is important to produce aquaculture products that are more acceptable to consumers by reducing chemical and drugs use in aquaculture. In community medicine, agriculture, and aquaculture, the usage and abuse of antimicrobial drugs are a growing concern. High usage of antimicrobials for disease control and to increase the growth exerted selective pressure on the microbial environment and paved the way for natural emergence of bacterial resistance and also the resistant bacteria that have ability to transfer the antibacterial resistance genes to other bacteria that have never been exposed to antibiotics, as evident in World Health Organization's (WHO) (facts sheet 194, http://www.who.int/inf-fs/en/fact194.html). The clinically resistant bacterial strains could be hazardous to human health and cause possible failure of antibiotic therapy [1]. Even though few antibiotics have been approved for finfish farming, no antibiotics were available for shrimp farming for controlling diseases. Vibriosis and black shell disease are bacterial diseases that cause outbreaks in fish and shellfish, showing a significant constraint on their production [2–4]. The aquatic environment consists of a number of bacteria with beneficial and nonpathogenic bacterial strains as well as opportunistic and secondary bacterial pathogens. Although disinfection of the hatchery and rearing facilities creates clean and pathogen-free environment in the system, it can cause loss in a stable microbial balance through proliferation of opportunistic bacteria that enter into the system (natural or artificial food sources and inlet water vertical transmission through broodstock) and thereby limit the natural biological control of opportunistic pathogens [5].

Marine Microbiology: Bioactive Compounds and Biotechnological Applications, First Edition.
Edited by Se-Kwon Kim

New strategies on health management and feeding have drawn much attention with the ban on antibiotic and growth promoters in fish aquaculture practices. The worldwide demand for safe food has led to search for more natural alternative growth promoters to be used in aquaculture. In this context, much research has been focused on developing novel dietary supplement strategies to produce health- and growth-promoting compounds in aquaculture as prebiotics, probiotics, symbiotics, phytobiotics, and other functional dietary supplements.

Composition of bacteria in the culture environment as well as in the indigenous protective flora of the cultured organisms can change in the intensive aquaculture cultivation systems. Due to imbalance, microbiota in the intestinal tract could increase the susceptibility of the host animal to opportunistic or secondary pathogens and cause decrease in feed conversion ratio. It has been reported that by manipulation of the gut/intestinal microbial environment, the survival and health of the organism in intensive rearing system could be improved [6].

Microorganisms can play an important role in aquaculture as probiotics and provide greater nonspecific and broad-spectrum disease protection. Gram-positive and Gram-negative bacteria (*Bacillus, Lactobacillus, Lactococcus, Enterococcus, Clostridium, Leuconostoc, Pseudomonas, Vibrio, Aeromonas,* and *Shewanella*) [3,7] yeasts [8], bacteriophages [9], and unicellular algae [10] have been studied for use as probiotics in aquaculture [11]. Most of these probiotic strains have been isolated from indigenous and exogenous microbiota of aquatic animals. Since application of probiotic bacteria in aquaculture to control microbial pathogen is an environment-friendly method, it has gained considerable research interest as alternative management practices for disease control and prevention.

Using a single beneficial bacterial strain as probiotic in culture of many aquatic species has shown promising results, as reported in a number of scientific publications on probiotics [12–22]. As suggested by Noh *et al.* [13] and Bogut *et al.* [14], considering the possibility of using different species is also important. Although many probiotic-related studies have been carried out, the mode of action of probiotic is still not completely understood [23–26]. The generally accepted mechanism of probiotics is antagonism (inhibitory interactions), inhibitory compound production, microbial balance improvement, competition for adhesion sites and chemicals, immune stimulation and modulation, and bioremediation. In most of the articles published, the functional role of probiotics, mode of actions, scope of applications in larviculture and mechanism of actions *in vivo* and *in vitro*, application of molecular techniques to study microbe–microbe and host–microbe interactions, and screening methods have been reported.

This chapter gives an overview of bacteria as probiotics in aquaculture, specially related to mollusks, crustaceans, and finfish, and also possible modes of action of bacteria and recent research findings related to probiotics in aquaculture. It also reports the future prospects and direction for further research related to probiotics in aquaculture.

7.2 Definition of Probiotics in Aquaculture

The term probiotics, meaning "for life," was derived from two Greek words "pro" and "bios" [27]. Lilly and Stillwell [28] were the first to put forward the concept of probiotics and described them as "substances secreted by one microorganism to stimulate other microorganisms" [29]. Parker in 1974 [30] modified the term as "organisms and substances which contribute to intestinal microbial balance," which is closer to the definition used today. Fuller [31] improved the Parker's definition as "A live microbial feed supplement which beneficially affects the host animal by improving its intestinal microbial balance" and this was a revision of the original definition of probiotic by Lilly and Stillwell [28] and referred to them as protozoans producing substances that stimulate the growth of other protozoans. This definition demonstrates the importance of live cells as the main component of a potential probiotic, introduces the aspect of beneficial effect on host, and clears the term "substances" that created confusions with antibiotics. Havenaar *et al.* [32] broadened the definition of probiotic as "A viable mono or mixed culture of microorganisms which applied to animal or man, beneficially affects the host by improving the properties of the indigenous micro biota." This concept was further broadened by Salminen [33] and Schaafsma [34] by no longer limiting the proposed health effects that has an influence on the indigenous microbiota. Tannock [35] proposed the definition of probiotics as "living microbial cells administered as dietary supplement with the aim of improving health." Since the first studies of probiotics were conducted for terrestrial species such as human and farm animals (poultry, ruminants, and pigs), these definitions were originally used for these species. Since probiotic applications are being used in aquaculture industry and it is relatively new, the definitions need to be modified accordingly. The potential pathogens could live by themselves in the external environment of the aquatic animals, proliferate independent of the host, and are taken up constantly by the aquatic animals through the process of feeding and osmoregulation. Especially in bivalve larvae, whose feeding is through filtration, a continuous water flow exists through the organisms and constantly interacts with opportunistic pathogens.

Almost all the aquatic animals, including fish and shellfish larval forms, are released into the external environment at an early ontogenetic stage. As their digestive tract and immune system is still incomplete and not fully developed, the probiotic treatments are required at larval stages [36]. Hence, larvae are highly exposed to gastrointestinal microbiota-associated disorders [37]. It has been reported that Atlantic halibut (*Hippoglossus hippoglossus*) with the first feeding [38] showed transition from prevailing *Flavobacterium* sp. intestinal flora to *Vibrio* sp. and an *Aeromonas* sp. to be dominant. This suggests that external environment and feeding are responsible for microbial status of the fish. Furthermore, larvae could maintain specific intestinal flora different from that of the external tank, suggesting that even though the external environment influences the internal environment of microbial flora in aquatic animals, they could still maintain host specific flora at any given time. This phenomenon is not applicable to bivalve larvae studied by Jorquera *et al.* [39] that the transition time of bacteria in bivalve larvae was too short to establish the microbial population different from surrounding water. Since there are many differences in the level of interactions between the terrestrial and aquatic animals as well as between the intestinal microbiota and surrounding environment, it is essential to modify the definition of probiotic in aquatic environments. Host–microbe interactions are qualitatively and quantitatively different for terrestrial and aquatic species as host and microorganisms share the same ecosystem in aquatic environment. In contrast, water-limited environment or moist habitat is present in the gut of terrestrial systems. Therefore, probiotics were defined by Gatesoupe [36] as "microbial cells that are administered in such a way as to enter the gastrointestinal tract and to keep alive, with the aim of improving health." This definition was modified by removing the restriction to the improvement of the intestine and defined as "a live microbial supplement which beneficially affects the host animal by improving its microbial balance." Aquatic environment microbes have many living associations with potential host intestinal tract, gills, or skin; however, in terrestrial environment, activities may be limited to the gut of host animals.

In aquaculture systems, the immediate ambient environment has larger influence on the health status than the human or terrestrial animals. Bacteria in the aquatic environment influence the composition of the gut microorganisms and vice versa. Especially with filter feeders, bacteria from the culture water could ingest at high rate and cause natural interaction between microorganisms of the aquatic ambient environment and the live food. So the opportunistic pathogens can reach at high densities around the animals. Aquatic animals spawn axenic eggs in the water and bacteria in the ambient environment colonize the egg surface. Freshly hatched larvae or newborn animals do not have fully developed intestine or a microbial community in the intestinal tract, skin, or gills. The probiotics could activate on other tissues and, therefore, the properties of the bacteria in the ambient water in which aquatic larvae were reared are most important. Also, lots of probiotics are developed mainly from the culture environment and not directly from the feed. Concerning all these aspects, definition of probiotics can be modified to be applicable in broader sense as "live microbial adjunct which has a beneficial effect on the host by modifying the host associated or ambient microbial community, by ensuring improved use of the feed or enhancing its nutritional value, by enhancing the host response towards disease, or by improving the quality of its ambient environment" [3]. The Food and Agriculture Organization and World Health Organization Joint Action Committee define the probiotic as "live microorganisms, conferring a healthy benefit on the host when being consumed in adequate amounts" (FAO/WHO 2001) [40] and considered it as an alternative viable therapy that becomes an integral part of the aquaculture practices for improving growth and disease resistance.

Above definitions reflect that probiotics prevent pathogens from proliferating in the intestinal tract and in the culture environment of the culture species they ensure optimal use of the feed by aiding in its digestion, improve water quality, and stimulate the immune system of the host. Also, bacteria produce single-cell proteins that are essential nutrients to host without being active in the host or without interacting with other bacteria, with the environment of the host, or with the host itself. Probiotics are not necessarily a natural enemy of the pathogen, but are able to prevent damage caused by pathogen through competition mainly by producing substances that inhibit the growth or attachment of harmful organisms. Utilization of natural enemies to reduce, control, or regulate the damage caused by pathogens or noxious organisms to tolerable level could be described as biological control [41]. Since probiotic microorganisms do not necessarily attack the pathogen or noxious agents, they should not be classified as a biological control agent. Since probiotic action does not involve ingrowth improvement but are associated with only a general improvement of health, probiotics should not be classified as growth promoters [42].

7.3 Selecting and Developing Probiotics in Aquaculture

To develop a probiotic for commercial use in aquaculture is a multistep and multidisciplinary process. It requires experimental and fundamental research, full-scale trials, and economic assessments of its use. Understanding the mechanism of probiotic action and defining the selection criteria for potential probiotics are essential in probiotic research. Biosafety considerations, methods of production and processing, method of administration of the probiotic, and locations in the body where the microorganisms are expected to be active are the general criteria for selection of probiotics [43]. With the limited scientific evidence, selection of probiotics has been an empirical process. Inappropriate microorganism selection leads to failures in probiotic research. Even though the selection steps have been defined, they need to be adapted for different host species and environments. Mainly, all the microbial probiotic strains need to be nonpathogenic and nontoxic to the host in order to avoid undesirable side effects when administered to aquatic animals. Also, all the microbial probiotics should have beneficial effect on the host and should not cause any harm to the host. Probiotics are mainly used to maintain and reestablish a favorable relationship between pathogenic and friendly microorganisms that constitute the flora of skin, mucus, and intestinal tract of aquatic organisms. Probiotic strains have been isolated from endogenous as well as exogenous microbiota of aquatic animals and majority of marine animals constitute indigenous Gram-negative facultative anaerobic bacteria such as *Vibrio* sp. and *Pseudomonas* sp. The lactic acid-producing bacteria are mainly prevalent in mammals and bird gut, are subdominant in fish, and represent the genus *Carnobacterium* [44,45]. It has been reported that some research and commercial products emphasize the multifactorial effect of the probiotics [46,47] and this would not agree with evidence; this type of overestimation could affect the perspective of real probiotic design for aquaculture industry [48]. For the potential probiotic, it is desirable to have different modes of action or properties such as antagonism to pathogens [44,49], colonization or adhesion properties [50], ability of cells to produce beneficial compounds, enzymes, and other metabolites (e.g., vitamins) [51], and enhance the immune system [52]. Selection criteria for probiotic bacteria in the larviculture of aquatic animals could include collection of background information, acquisition of potential probiotics, evaluation of the ability of potential probiotic to compete with pathogenic strains, assessment of the pathogenicity of potential probiotics, evaluation of the effect of the potential probiotics in larvae, and economic cost-benefit analysis [42]. Certain beneficial characteristics need to be present for a microorganism to be a probiotic in aquaculture: they must isolate from the environment where the probiotic will be applied and avoid risk of introduction of all autochthonous organisms to the system, show beneficial effect, and be nontoxic or nonpathogenic to the host and to other living organisms in the system such as phytoplankton. To find potential probiotics, these properties should be examined stepwise. Many of these properties could be tested via *in vivo* experimentation with the target animal. A probiotic as a novel, effective, and safe product in aquaculture should possess certain properties: It should not be harmful to the host it is desired for, it should be accepted by the host and potential colonization replication within the host, it should reach the location where the effect is required, it should really work *in vivo* as opposed to *in vitro* findings, and it should not contain virulence and antibiotics resistance genes.

With these properties, novel probiotics for aquaculture could be produced for safety evaluation. Probiotics in the field of aquaculture is attracting attention considerably, and the number of commercial products is growing rapidly, particularly in shrimp larviculture [53]. Due to the increasing demand for alternatives to antibiotic products applied in aquaculture, a number of better commercial probiotics could be produced in the field of larviculture. Currently, a number of probiotic preparations are commercially available and have been introduced in molluscan, shellfish, and fish as feed additives and also in culture waters [24,54,55], since according to the producers these products are effective, health supporting, and safe. Safety considerations regarding antimicrobial resistance need to be taken into account for the development and marketing of probiotics [56] because genes can be transferred between microorganisms (between probiotics and gut microflora or pathogen) and for the possibility of multiple resistance genes transfer from probiotic strains to bacterial pathogens or from aquatic commensals to probiotics [57]. New probiotic screening techniques together with initial *in vivo* studies will allow a wide range of bacteria to be identified as probiotics. Successful acquisition of such novel probiotics depends on obtaining a better understanding of microbial ecology of a cultured species as well as on restricting the probiotic screens to the bacterial species that share the immediate environment with the cultured species. Probiotics that had already adapted through natural process to the dynamics of an aquaculture production system should not manipulate the farm management involving environmental manipulation practices, particularly in mollusk production [53]. Recently, culture-independent molecular methods have been developed to evaluate the potential probiotics [58] such as multiplex PCR, random amplified polymorphic DNA (RAPD), arbitrary primed (AP) PCR, pulse field gel electrophoresis (PFGE), denaturing gradient gel electrophoresis (DGGE), terminal restriction fragment length polymorphism (TRFLP), and temperature-gradient gel electrophoresis [59]. For the safety evaluation of probiotics to ensure that healthy benefits are accessible to aquatic animals, molecular biology techniques, other advanced modern techniques, and cell culture techniques can be used. Primarily cultured cell model was established from epithelial cells of tilapia (*Oreochromis niloticus*), it evaluated the cell viability morphologic characters, livability, and permeability of probiotic *Rhodopseudomonas palustris* [60], and cell culture technique is one of the promising approaches for safety evaluation of probiotics.

7.4 Effects of Probiotics on Aquatic Organisms

Protection against pathogens by probiotics is due to different mode of actions or properties of probiotics such as antagonism to pathogens, competitive exclusion for adhesion sites, ability to produce organic acids, hydrogen peroxide, metabolites like vitamins, enzymes, and other compounds such as lysozyme, antimicrobials, siderophores, and modulation of host immunity and physiology [44,49,50–52]. However, the mechanisms through which probiotics effectively modulate the immune system are not well elucidated. It is clear that the data given on probiotic stimulation are fragmentary and probably bacterial strain or aquatic species dependent; hence, much more research is necessary to explain the protection reported after feeding probiotics.

7.4.1 Possible Mode of Action

7.4.1.1 Competitive Exclusion

The established microflora preventing or reducing the colonization of bacteria competing for the same location in the intestine could be explained as competitive exclusion in the gastrointestinal tract. Some probiotic products have been designed in aquaculture for adhesion to mucosal surface by a collection of microorganisms based on competitive exclusion principles, which may help improve the host health status. These factors are important for intestinal homeostasis, digestion, adhesion of microorganisms to intestinal epithelial cells, and activation of immune system [61]. When designing the probiotic products in aquaculture, competitive exclusion is considered. Probiotic products designed under competitive exclusion are obtained from stable, controlled microbiota in culture based on competition for attachment sites on the mucosa and from inhibitory substance produced by microflora that prevents replication and/or destroys the challenging bacteria and reduces its colonization and competition for nutrients [3].

7.4.1.2 Antagonisms

Microorganisms are sources of different bioactive natural metabolites that are of commercial interest due to their inhibitory effects on microbial growth [62]. Bacterial antagonism is a natural phenomenon in Nature and microbial interactions play a role in the equilibrium between beneficial and potentially pathogenic competing microorganisms. The chemical substances produced by microorganisms (bacteria here) are called antagonistic compounds that have inhibitory (bacteriostatic) or toxic (bactericidal) effects on other microorganisms. Bacteria producing antibacterial compounds in the host intestine, on the surface, or in culture water would prevent proliferation of pathogenic bacteria and sometimes could eliminate those.

Antibiotic compounds can directly or indirectly affect pathogens [48]. Bacteriocins produced by lactic acid bacteria are active only against closely related species [63] and most of the pathogens related to aquaculture belong to Gram-negative bacteria. Therefore, bacteriocins may not inhibit fish pathogenic bacteria and it is important to perform selective test before introducing them into the host or water.

Probiotic bacteria can produce proteinaceous as well as nonproteinaceous substrates. The structures of antibacterial compounds and their mode of actions need to be elucidated in most of the compounds isolated from such bacteria. Furthermore, it is important to find out whether these compounds are produced *in vivo* and demonstrate antibacterial properties *in vivo*. If the antibacterial compound production is the only mode of action against pathogen, it is possible that pathogen could eventually develop resistance toward compound, thus making treatment ineffective [48]. Therefore, it is important to prevent the pathogen from developing resistance to the active compound to ensure a stable effect of probiotic bacterium. The origin of probiotic strain needs to be considered in the antagonism test. Microorganisms have different physiologies or biochemical activities during their development based on their environment (fresh or marine). These may create false impression about the ability of probiotics to inhibit in *in vivo* test, for example, the attachment sites [64].

7.4.1.3 Probiotics as Immune Stimulants

Probiotics have beneficial effects on the intestinal ecosystem [65]. The probiotics causing beneficial and protective effects on gut microbiota and several microbial disorders have been reported [27]. Probiotics can stimulate the nonspecific immune system of fish. *Clostridium butyricum* bacteria given orally to rainbow trout have shown increased phagocytic activity of leukocytes and enhanced the resistance of fish to Vibriosis [66]. Addition of *Bacillus* species (strain S11) to tiger shrimp (*Penaeus monodon*) provided disease protection by activating both humoral and cellular immune defenses [67]. Introduction of mixture of *Bacillus* and *Vibrio* spp. to white shrimp juveniles positively influenced its survival and growth and showed protective effect against the pathogens *Vibrio harveyi* and white spot syndrome virus by activating immune system by increasing antibacterial and phagocytosis activities [68]. It was shown that respiratory burst was stimulated in rainbow trout by administration of 10^5 cfu/g of *Lactobacillus rhamnosus* strain

ATCC53103 [69]. Probiotics produce immune modulators that can stimulate the immune system against pathogens. Immune modulators can be extracted from the cell walls of microorganisms such as peptidoglycan from Gram-positive bacteria, lipopolysaccharides from Gram-negative bacteria, and β-1,3-glucan from fungi. Results of experimentally tested small-scale cultures suggest that above immune stimulants could be considered for large-scale cultures of crustaceans as immune stimulants for controlling diseases [70,71]. Various studies have shown that β-1,3-glucan can improve the resistance against various infectious diseases in bivalves, shrimps, and fish [72]. These immune stimulants can be applied by immersion, injection, and, most practically, by integration with the feed. Different mechanisms of probiotic immune stimulation in the fish immune system related to immune cells, lysozymes, antibodies, antimicrobial peptides, and acid phosphates have been reported. Increase in innate immune parameters such as lysozyme and phagocytic activities with the use of *Lactobacillus rhamnosus* or their cell wall components on *Oncorhynchus mykiss* and as prophylactic factor to low stress in *Oreochromis niloticus* has been reported [73,74]. Immune system has been stimulated in *Miichthys miiuy* fed by *Clostridium butyricum* showing an increase in acid phosphatase and lysozyme activities [75]. Disease protection was provided to *P. monodon* by activating immune defenses by using *Bacillus* sp. Resistance of *Litopenaeus vannamei* against *V. harveyi* was promoted by addition of mixture of *Vibrio* sp. and *Bacillus* sp. [76]. Autochthonous intestinal bacteria are also considered to be potential probionts [77]. Recent studies have reported that early administration of autochthonous bacterial strains to gilthead sea bream [78,79] and European sea bass [80] have shown promising effects on gut physiology, intestinal microflora, and immunity of fish larvae. Probiotic formulation containing autochthonous *Lactobacillus fructivorans* and *Lactobacillus plantarum* from human feces increased the intestinal IgM^+ and acidophilic granulocytes [81]. When administered early, this multispecies probiotic formulation was significantly more effective for fry survival and stress responses. Viable form of probiotic *L. rhamnosus* showed better immune stimulatory activity than killed form in rainbow trout [73], suggesting the ability of live bacteria to colonize the host gut. However, diets supplemented with inactivated *Lactobacillus delbrueckii* sp. lactis and *Bacillus subtilis* (individually or combined) fed to juvenile sea bream showed higher number of gut IgM^+ cells, higher total serum IgM, and acidophilic granulocytes than the untreated controls [82]. Autochthonous bacterium *Lactobacillus delbrueckii delbrueckii* added to sea bass showed down-regulation of key inflammatory cytokine (IL-1β, Cox-2, IL-10, and TGF-β) expression, increased survival, increased weight gain, and decreased cortisol levels and improved stress responses were monitored [15,79]. Furthermore, it showed increased number of intestinal T cells with increased total body TcR-β transcripts and acidophilic granulocytes. When rainbow trout head kidney (HK) leukocytes were incubated with probiotics that were isolated previously from the host [83], the significant upregulation of cytokine expression was observed. Upregulation of IL-1β, TNF1 and TNF2, and TGF-β in the spleen and the head kidney was observed in probiotic-fed rainbow trout [84]. Recent proteomic analysis of probiotic stimulated in cod also showed lower level of immune stimulation, but upregulation of growth- and development-related proteins [85]. Immune modulatory capability of potential probiotics has been studied by using intestinal bacteria of the Atlantic cod (*Gadus morhua*) itself and by assessing transcriptional responses in the head kidney leukocytes [77] *in vitro*.

Gut Microbiota and Probiotic Immune-Stimulation in the Fish Although microecology of the fish gut has been relatively less studied over the past years, application of probiotics to the fish has been studied extensively [86]. Dominant microorganisms in the fish intestine are bacteria, belonging mainly to the phylogenetic Proteobacteria [87–89]. Gut microbiota of the fish is highly influenced by the surrounding complex environment. Due to this complex ecosystem, microbes compete for the space and nutrients for their survival. Similar to mammals, interaction between host and commensal bacteria and mechanisms related to their benefits are still not understood properly. Significant role in intestinal immune responses might occur through resident gut microbes, occasional pathogens, and artificially delivered microbes that act as probiotics. Through the continued and active signaling, these organisms coexist in a dynamic equilibrium within host gut and beneficial microbial consortium uses multiple independent signals to interact with the host and coexist with the pathogenic bacteria. With the first feeding at the larval-phase, the symbiotic consortium of microorganisms starts and reaches to stable state at the juvenile stage [90,91]. Microbiota compositional changes are influential at the time of intestinal colonization and immune system development. It has been reported that arrested differentiation and altered functions have occurred in the absence of microbiota during the development stages in zebrafish [92].

7.4.1.4 Antiviral Effects

The laboratory experiments showed that candidate probiotics have antiviral activities. Recently, it has been found that probiotics have beneficial effects against viral and protozoan infections such as iridovirus in *Epinephelus coioides* [93] and Ich in *O. mykiss*, respectively [94], apart from the protection against a variety of bacterial diseases. It has been reported that strains of *Vibrio* sp., *Aeromonas* sp., *Pseudomonas* sp., and groups of coryneforms isolated from freshwater salmon hatcheries showed antiviral activity (more than 50% plaque reduction) against infectious hematopoietic necrosis virus (IHNV) [95]. Two strains of *Vibrio* sp. (NICA 1030 and NICA 1031) isolated from black tiger shrimp hatchery showed antiviral activity against IHNV and *Oncorhynchus masou* virus (OMV) showing plaque reduction between 62 and 99%. The exact mechanism of these antiviral properties is not known and it can occur through chemical and biological substances as extracellular agents of bacteria [96].

7.4.1.5 Digestive Process

Most of the *in vitro* studies in probiotics in aquaculture have used specific bacteria as antagonists of pathogen to find out the effect of probiotics in health management, measuring their survival in the fish gut, immune responses, and disease-resistant fish. Among the important effects of probiotics requiring much attention are the feed efficiency and growth promotion of aquatic animals by probiotic supplements [97,98]. Probiotics consume large amounts of carbohydrates for their growth when they reside in the intestine of the host and produce digestive enzymes such as amylase, protease, and lipase, which enhance the digestibility of the organic matter and proteins and higher growth. These enzymes are also useful for preventing intestinal disorders and stimulating or causing predigestion of secondary compounds in plant protein sources [98,99]. Similarly, microbial flora of adult penaeid shrimp (*Penaeus chinensis*) also produces complement of enzymes for digestion [100]. Usage of lactic acid bacteria and yeast has shown beneficial effects of feed efficiency, growth, and digestibility of organic matter and proteins in fish [13,14,98,101–103]. Fatty acid and vitamin supplementation by *Bacteroides* and *Clostridium* sp. contribute to the fish nutrition [104]. Contribution of *Pseudomonas* sp., *Brevibacterium* sp., *Agrobacterium* sp., *Staphylococcus* sp., and *Microbacterium* sp. for nutritional process in Arctic charr (*Salvelinus alpinus* L.) has also been reported [105]. The diet supplementation with probiotics demonstrated high activity of alkaline phosphatase in Nile tilapia (*O. niloticus*) [106], which helps in the development of brush border membranes of enterocytes. These could be stimulated by probiotics and would reflect as an indicator of carbohydrate and lipid absorption and explain the higher weight and the best feed conversion. Bivalves and shrimp (*Penaeus chinensis*) also produce extracellular enzymes (proteases and lipases) and growth factors [100,107] that might be helpful in the digestion process of bivalves.

7.4.1.6 Adhesion

In aquatic animals such as fish and shrimps, soon after hatching, the colonization of the gastrointestinal tract starts and is completed within few hours to begin modulation of gene expression in the digestive tract creating desirable habitat for the microbiota and prevent and protect against invasion of other bacteria introduced later into the ecosystem [108]. This helps in competitive exclusion mechanisms and improved immune system development and maturation. The main objective of adhesion is to obtain a significant level of bacteria in the host and prevent them from being flushed out by the movement of food through digestive tract. Adhesion and colonization of the mucosal surfaces are protective mechanisms against pathogens via competition for binding sites and nutrients [109] or immune modulations [110]. By attaching, probiotics can stay considerably for a long time within the gut and could influence the gastrointestinal microflora of their host. Based on the ability to attach to the mucous, the growth of the pathogen could be suppressed by the candidate probiotics due to competition for essential nutrient, space, and so on. Probiotics constitute a part of resident microflora in the intestine and contribute to health or well-being of their host. Colonization process includes attraction of bacteria toward the mucosal surface followed by association with the mucous gel or attachment to epithelial cells [48]. Ability of adhesion to mucus, gastrointestinal tract, epithelial cells, and other tissues needs to be considered in the probiotic selection since it is associated with bacterial colonization.

It is important to use probiotics in early larval stage before definitive installation of competitive indigenous microbiota. Addition of high dose of probiotics is possible only for artificial and temporary dominance after installation of indigenous microbiota. Microbes are able to colonize the gastrointestinal tract when multiplication rate is higher than the expulsion rate [111]. It has been reported that there was a sharp decrease in the probiotic population in the gastrointestinal tract during the period when the intake was stopped [112].

By knowing and understanding the mechanisms between gut microbiota and probiotics and also how the immune system of aquatic animals generally responds to gut microbiota would be of great advantage in identifying the molecular targets of probiotics and the biomarkers of their effects. This will offer evidences of their benefits on host in physiological conditions and immune-mediated disorders.

7.5 Probiotics in the Larviculture

7.5.1 Probiotics in the Larviculture of Mollusks

In the last few years, research has been focused on bacteria having antimicrobial activity to control microbiota composition associated with mollusk larvae in hatcheries to avoid pathogens and improve larval survival. Probiotics have been used to control pathogenic bacteria *Vibrio* sp. in bivalve aquaculture such as oyster *Crassostrea gigas* larvae [113], scallops *Pecten maximus* [114], and *A. purpuratus* [115]. Different bacterial strains have been tested for their ability to reduce the heavy losses experienced within oyster production due to disease outbreaks. Different taxa such as *Phaeobacter* or *Pseudoalteromonas* have been considered as probiotics and experiments have been conducted in both the laboratory and the field. Effect of antibiotic-producing marine bacteria on scallop (*Pecten ziczac*) larval survival was published in Ref. [116]. Deeper study was carried out on the probiotic bacteria *Pseudoalteromonas haloplanktis* (formerly *Alteromonas haloplanktis*), scallop *Argopecten purpuratus* larvae [115], and a range of other bacteria, including genus *Vibrio*, and found that the active compound produced was probably an intracellular component that produces a secondary metabolite at the stationary phase of the growth. *In vivo* studies showed protection of larvae against pathogenic bacteria *Vibrio anguillarum* strain. Pretreatment with *Pseudomonas* sp. and *Vibrio* sp. strains protected the Chilean scallop (*A. purpuratus*) larvae against subsequent challenges of different *V. anguillarum* strains [117] with a strong antibacterial activity. The production of inhibitory substances against *V. anguillarum*-related strains was screened from different isolated strains, from hatchery and laboratory sources. Axenic microalgal culture of *Isochrysis galbana* was used to incorporate active bacteria into the culture [118] and studies were conducted for large-scale larvicultures [119], demonstrating that larval phase could complete by avoiding the usage of commercial chemotherapeutics. Ingestion of potential probionts (PPs) by scallop depended on the type of strain employed [120]. Pacific oyster (*C. gigas*) larvae fed with algae and a bacterium (CA2) *Alteromonas* sp. [121,122] enhanced survival (21–22%) and growth (16–21%) compared with those fed algae alone with the optimal of 10^5 cells/ml. Bacteria have supported with essential nutrients that are not present in the algae or provide enzymes that are useful for improving larval digestion and have probiotic effect [122]. Bacteriocin-like inhibitory substances were produced by *Aeromonas media* (A199) that showed antagonistic activity against many potential pathogenic bacteria. A199 was used as a probiont to control infections of *C. gigas* larvae with *Vibrio tubiashii* strain as its survival was ~95% when adding a combination of *V. tubiashii* and A199, whereas all larvae died in the presence of *V. tubiashii* alone [123]. In scallop (*P. maximus*), cell extracts of the *Roseobacter* strain BS107 lowered mortality rates of larvae in a challenge test with *Aeromonas salmonicida*. However, whole cells of the *Roseobacter* strain did not enhance survival of the larvae [124]. It was suggested that antibacterial effect of strain BS107 showed only in the presence of another bacterium that produce proteinaceous molecule that act as an effector. *In vitro* studies of inhibition of pathogenic *Vibrios* by the *Phaeobacter gallaeciensis* formally named *Roseobacter gallaeciensis* confirmed the experiments conducted using phytoplankton cultures with oyster (*Ostrea edulis*) and clam (*R. philippinarum*) larvicultures [125]. It was concluded that it is necessary to introduce probiotic isolates to the system previously or simultaneously to the pathogen suggesting that *P. gallaeciensis* could be used as a control method in mollusk hatcheries, if the probiotic action is allowed before the pathogen reaches high concentration in the system. Probiotic supplement containing both bacteria strains and yeast improved the growth rate of *Haliotis midae* [126]. In abalone challenged with *V. anguillarum*, 62% of the probiotic-fed animals survival compared to 25% for nontreated animals was observed. It has been reported that the biofilms of pigmented marine bacteria stimulate the induction of larval fixation process of *Crassostrea virginica* [127]. The bacteria was identified as *Alteromonas colwelliana* [128] and later reclassified as *Shewanella colwelliana* [129] and produces hydrosoluble exopolysaccharide [130] and polysaccharide adhesive viscous exopolymer (PAVE), which promotes adhesion of microorganisms to the surfaces [131]. However, *S. colwelliana* could also induce fixation of *O. edulis* but not *P. maximus* [132] and in hatcheries of *C. gigas* and *C. virginica*, the bacteria *S. colwelliana* was successfully used to induce the fixation [133].

Probiotic bacteria supplementation with algal feed has shown to enhance the nutritional value of algae to the larvae [53]. This helps in early colonization of microflora in the gut to improve digestion. It has also shown to stimulate or speed the development of the innate immune response to potentially pathogenic bacteria in shellfish [134]. Efficacy of new probiotic bacteria for the shellfish hatcheries was evaluated using 26 probiotic candidate bacteria isolated from oysters, scallops, and mass culture of green algae [135]. Fifteen isolates were shown to inhibit known scallop pathogenic bacteria *Vibrio* sp. strains B183 (*Vibrio coralliilyticus*) and B122 *in vitro*. Survival of oysters larvae exposed to these 15 probiotic candidates for 48 h was more than 90% (similar to control oyster larvae). Seven distinct strains were reisolated from the larvae in a challenge test and 5 day challenge was performed to test larval survival against B183 strain to confirm positive effect of these candidates. Probiotic candidate *Vibrio* sp. strain OY15 significantly improved the survival of the oyster larvae when challenged with known *Vibrio* sp. shellfish larvae pathogen alone in bench scale experiments. These probiotic bacteria were isolated from bay scallop *Argopecten irradians* (Lamarck 1819) and the eastern oyster *C. virginica* (Gmelin 1791). OY15 was not toxic on microalgal feed strain *Isochrysis* sp. (T-ISO) in the range of 10^2–10^4 cfu/ml. Effectiveness of OY15 to improve survival of oyster larvae to metamorphosis with and without pathogen B183 challenge was conducted under pilot-scale culture conditions [136]. The effective dose for the significant larval survival for OY15 was 10^3 cfu/ml and LD_{50} for pathogen B183 was 9.6×10^4 cfu/ml. Addition of probiotic candidate OY15 significantly improved the survival of oyster larvae to metamorphosis when challenged with pathogen B183 in pilot-scale trial, which could be considered as the basis for the development of functional food for shellfish larviculture using naturally occurring probiotic bacterial strains.

7.5.2 Probiotics in the Larviculture of Crustaceans

The addition of probiotics is a common practice in commercial shrimp hatcheries in many countries such as Mexico [137] and Ecuador [3]. However, not much scientific information has been published on the use of bacteria as probiotics in shrimp larviculture. It has been reported that in South American larval rearing systems, in some hatcheries commercial table sugar was added to stimulate the growth of the sucrose-fermenting *Vibrio* spp. which was considered as "good *Vibrios*" [138]. These bacteria were cultured in a manner similar to microalgae and then added to the larval rearing tanks [138,139]. Since late 1992, *Vibrio alginolyticus* has been used as probiotic in several Ecuadorian shrimp hatcheries and overall antibiotic usage was reported to be decreased by 94% during 1991–1994 and production volumes increased by 35% [140]. One brief anonymous report without data stated that several bacterial species are being used in the larviculture of *P. monodon* and *Penaeus penicillatus* in Asia with promising results [141]. Addition of sterile "soil extract" along with diatom to *P. monodon* larval rearing tanks has high survival rate of over 4 days compared to the addition of diatom alone; however, both treatments gave similar results at the end of the experiment [142,143]. Since very limited data were presented, clear interpretation is difficult and perhaps the addition of soil extracts allowed the growth of other microorganisms that might have shown beneficial effect on larval performance. *V. alginolyticus* is a bacterium that is mostly tested as probiotics in shrimp hatcheries with promising results. *V. alginolyticus* bacteria was isolated from the seawater and tested on *L. vannamei* larvae by Garriques and Arevalo [144] in their hatchery. In the bath challenge (2×10^3 cells/ml) pathogenicity test, no mortality was observed for this strain; however, 100% mortality was observed with *V. parahaemolyticus* strain after 96 h. *V. alginolyticus* in healthy rotifer was observed by Gatesoupe [145] who showed a positive correlation between the turbot larvae survival rate and the proportion of *V. alginolyticus* in the rearing environment. *V. alginolyticus* may have some capability of providing protection against disease and could be a probiotic candidate for shrimp larviculture. Since some strains could be pathogenic, so cautions need to be taken before using them as probiotics [146]. *Vibrio* sp. is the most common cause of infections in shrimps industry. Today, there is already a broad application of probiotic bacteria in commercial shrimp hatcheries. Most studies of probiotic bacteria for application in shrimp productions have focused on *Bacillus* strains. One or several strains of *Bacillus* were included in following commercial products such as Liqualife, BaoZyme-Aqua, and Promarine [147]. With the abundance of *Bacillus* strains, shrimps could be cultured for more than 160 days, whereas mass mortality within 80 days occurred in cultures without *Bacillus* strains, in the presence of luminous *Vibrio* strains [24]. The main bacterial pathogen in shrimp *V. harveyi* causes significant production losses in Asian countries [24,148]. Due to a disease outbreak caused by this pathogen during 1995–1997 in Philippines, the shrimp production has dropped to 55%. In Thailand during 1994–1997, due to *V. harveyi* and shrimp virus [24], the shrimp production has dropped to 40%. It has been reported that addition of the *Bacillus* strains to shrimp cultures resulted in a reduced mortality when challenged with *V. harveyi* [149,150]. This disease-reducing effect of the *Bacillus* strains may be due to improvement of the water quality by removal of ammonia [151,152]. *Bacillus* sp. and photosynthetic bacteria *Rhodobacter sphaeroides* were used on shrimp *Penaeus vannamei* to investigate the growth performance and digestive enzyme activity and the effect was related with the supplement concentration of the probiotics and the use of 10 g/kg (wet weight) that stimulated the productive performance [153].

Bacterial strains coded PM-4 were isolated from seawater [154,155] and identified as *Thalassobacter utilis* [156]. They were inoculated into blue crab *Portunus trituberculatus* larval rearing tanks (10^6 cells/ml). Survival of 27.2% in the challenged tank compared with 6.8% in the control tank (no bacteria) was observed. The 33 trials of the experiment showed the average survival rate of 28.3% in PM-4 strain in inoculated tanks compared with 15.6% in control tanks of more than 42 trials. These trials were continued for more than 4 years [156] and were observed that PM-4 could inhibit "supposedly pathogenic" strain of *V. anguillarum in vitro* experiments and suppress the presence of other *Vibrio* sp. Furthermore, *V. anguillarum* could inhibit the growth of the fungus *Haliphthoros* sp. and some pigmented bacteria in the larval tanks. Another finding was that repeated inoculation of strain PM-4 into the tanks did not exceed the bacterial level 10^6 cells/ml [154]. This effect is due to grazing behavior of protozoa [157] and lack of required nutrients. Gomez-Gil *et al.* [42] and Maeda [158] stated that aquaculture ecosystem cannot support bacterial level greater than 10^6 cells/ml. It was reported that when 10^7 cells/ml bacteria was inoculated in sterile shrimp hatchery seawater, count decreased after 72 h [159]. Adding microbial nutrients (glucose, urea, and potassium phosphate) to produce a mixture of bacteria improved the survival of blue crab (*P. trituberculatus*) larvae; however, better conclusion was difficult to obtain due to variable results depending on the composition of the microbial flora

[160] in the tanks. Interestingly, research on sponge-associated marine bacteria as potential source of novel shrimp probiotics is under progress and endosymbiotic marine acitinobacterium *Nocardiopsis alba* MSA10 was isolated from the marine sponge *Fasciospongia cavernosa*. These bacteria showed antagonistic activity against *Vibrio* pathogens of *P. monodon* in *in vitro* and *in vivo* [161], unpublished data cited by Ninawa and Selvin [6].

In addition, symbiotics as a combination of prebiotics and probiotics has been studied for the synergistic effects in aquatic organisms. It has been reported that growth-promoting chemotherapeutics could be replaced in the aquaculture industry by using combined application of probiotics and prebiotics and would be useful tool in the aquatic animal rearing systems. Combined and individual effects of dietary application of commercial probiotic (*Bacillus* sp.) and mannan oligosaccharides (MOS) on the growth performance, feed cost-benefits, and survival of European lobster (*Homarus gammarus*) larvae were studied and the results strongly suggest that the combination of dietary supplements of *Bacillus* sp. and MOS was cost-effective in terms of survival and growth performance than their individual supplementation [162]. Similar results have been observed for the shrimp *L. vannamei*, and their disease resistance was improved by enhancing the immunity and modulating the microflora in the shrimp gut [17].

7.5.3
Probiotics in the Larviculture of Finfish

Microorganisms can cause mortality in marine fish eggs. Mucosal surface of egg and larval skin are good substrates for adhesion of bacteria. Marine fish larvae ingest considerable amount of bacteria and later some get established as gut microbiota. Mass production of cod (*G. morhua*) and halibut (*H. hippoglossus*) egg and larvae are being conducted in intensive rearing systems. Bacterial overgrowth causes adverse conditions during hatching, affecting development and health of the fish. The antibiotics are continuously used during hatching and early larval stage to control bacterial colonization. Unpredictable and variable results occur due to uncontrolled development of microbial communities in hatcheries. Probiotic applications may cause beneficial effect on the cultures in hatcheries. Screening and preselecting of potential probiotics are important and this is based on extensive experimental work conducted *in vivo*. Atlantic cod (*G. morhua*) egg microbiota was manipulated by incubating gnotobiotic eggs in cultures of known bacterial strains; however, colonization of eggs failed due to the strains naturally present in microbiota in the incubator [163]. Members of the genera *Pseudomonas, Aeromonas, Alteromonas, Flavobacterium*, and filamentous bacterium *Leucothrix mucor* were found to dominate on the surface of both cod and halibut eggs. The same bacterial genera were found in the eggs colonized in the gastrointestinal tract of Atlantic cod (*G. morhua*) larvae soon after hatching [163]. Majority of them were *Pseudomonas, Flexibacter*, and *Cytophaga*. Because of the substantial amount of bacteria grazing on egg debris and suspended particles, the egg microbiota would affect the primary colonization of the fish larvae.

Fish digestive tract contains around 10^8 cells/g microorganisms, which is much higher than the surrounding ambient environment [105]. The establishment of primary intestinal microflora persists beyond first feeding, at the time of ingestion of the bacteria at yolk sac stage of cold fish. The bacterial succession will be followed until adult microflora is established [90]. Thus, it is important to add the potential probiotic (PP) as soon as possible after hatching to obtain effective colonization of the larval gut before the introduction of live food [44]. Introduction of artificial dominant groups of fish-associated microbiota at initial feeding time to the rearing water or to the culture medium of the live food will give beneficial effect. The mortality of nonfeeding turbot larvae (*Scophthalmus maximus*) was significantly reduced to 8 from 12 *Roseobacter* strains tested of which all showed antagonistic activity against *Vibrio* sp. [164]. By daily addition of lactic acid bacteria to the live food (enrichment medium of rotifer) for the turbot larvae (*S. maximus*) improved the survival rate. Significant reduction of larval mortality was observed when larvae were challenged with pathogenic *Vibrio* at day 9. Lactic acid bacteria would act as a microbial barrier against pathogens and prevent the invasion of pathogens in the larvae [165]. Similarly, *Lactobacillus bulgaricus* and *Streptococcus lactis* was added to turbot larva feeding (Artemia and Brachionus) [166]. After addition of living lactic acid bacteria, 55% survival was observed at d17 and 66% survival was found with disabled ones compared to control group (34%), suggesting that the bacterial cells, live or disabled, showed improved survival of turbot larvae. Turbot larvae was fed with culture medium of rotifer containing spores of *Bacillus* strain IP5832 and was found that Vibrionaceae in the rotifer decreased and there was a significant increase in turbot larvae mean weight at day 10 with the spore fed rotifer. The mean survival of spore-fed rotifers (31%) on day 10 was significantly higher than the control (10%), suggesting that production of antibiotics may be the mode of action; however, improvement of nutritional status of the larvae may also be a reason for the improvement of resistance to infection is not clear [12]. In turbot larvae receiving rotifers with both *Roseobacter* 27-4 and *V. anguillarum*, the accumulated survival was significantly reduced compared to treatment with *V. anguillarum* alone [167,168], suggesting that even though no correlation between larval survival rates and the number of bacteria in the gut was observed, the gut microbial flora plays an important role in determining the survival of larval turbot. Positive effects on the accumulated survival of turbot larvae have also been reported by addition of *Vibrio mediterranei* [169] and *Vibrio pelagius* [44]. Positive effect of a *Roseobacter* strain on survival rate of gilthead sea bream larvae (*Sparus aurata*) was also found, but no significant difference was found between treatment with or without *Roseobacter* [170]. Another lactic acid bacterium, *Lactococcus lactis*, enhanced the growth of *B. plicatilis* and had an inhibitory effect against *V. anguillarum* [171]. *Bacillus* sp. spores inhibited the growth of *Vibrio* sp. in the culture of rotifers, with an improvement in the mean weight of turbot [12]. Moreover, *Bacillus* probiotic mixture (*B. subtilis*, *B. pumilus*, and *B. licheniformis*) added into gilthead sea bream (*S. aurata*) larviculture showed beneficial effect against growth and stress responses [172], and that scientific and technical information was useful for implementation of sustainable development of sea bream aquaculture [57]. Olive flounder (*Paralichthys olivaceus*) supplemented with individual as well as mixture of *Lactobacillus acidophilus, L. plantarum, L. sakei, B. subtilis*, and *Saccharomyces cerevisiae* also showed the similar results [173]. Better immune response and resistance against lymphocystis disease virus infected olive flounder were observed in groups supplemented with mixed probiotics with *Lactobacil* probiotics added individually or mixed with Sporolac-enriched diet [174]. Different results were shown in survival rates in the feeding experiment of 600 *O. niloticus* using the diet containing single or mixed isolated probiotic bacteria [175]. However, the beneficial effect of probiotics fed with aquatic animals depends on probiotic strain, isolation species, culture animals, and the water quality.

7.5.3.1 Probiotics in Fish Juvenile and Adults

Diet supplement containing *L. plantarum* later classified as *Carnobacterium divergens* was challenged with cohabitant Atlantic salmon (*Salmo salar*) infected with *A. salmonicida* through intraperitoneal injection [176]. Even though the lactic bacteria given as diet supplement could colonize the intestine, no protection was observed against *A. salmonicida* showing highest mortality for the fish given the diet containing lactic acid bacteria over the 4 week of observation. However, when Atlantic cod fry that was fed with *C. divergens* isolated from Atlantic cod (*G. morhua*) intestine supplemented with dry feed was exposed to *V. anguillarum* virulent strain, it showed improved disease resistance. It was shown that 3 weeks after the challenge, the lactic acid bacteria dominated the intestinal microbiota of the surviving fish [177]. The two strains of *C. divergens* were isolated from the intestine of Atlantic salmon and Atlantic cod and added to commercial dry feed and given to Atlantic cod for 3 weeks. It showed reduced cumulative mortality after 12 days of infection with same virulent *V. anguillarum* [178]. However, after 4 weeks of infection, cumulative mortality reached to same values in all groups, suggesting that this infection trial only delayed the cod fry mortality due to probiotic treatment. Ability of *Carnobacterium* strain K1 to colonize the intestinal tract of rainbow trout (*O. mykiss*) to inhibit *V. anguillarum* and *A. salmonicida* in mucus and fecal extracts has been reported [179]. *In vitro* studies have shown the production of growth inhibitors in both extracts of mucus and fecal against *V. anguillarum* and *A. salmonicida*. Furthermore, it was found that *V. anguillarum* growth in fecal extracts of turbot juveniles was inhibited by *Carnobacterium* cells [180] suggesting that *V. anguillarum* enrichment sites are mainly an intestinal tract and feces of turbot and inhibitory activity having intestinal bacteria could be used to decrease the pathogenic *Vibrio* sp. in turbot hatcheries. The adult marine flat fish, turbot (*S. maximus*) and dab (*Limanda limanda*) harbor the bacteria that can suppress the growth of *V. anguillarum* [50] and could act as probiotics against such pathogens. *V. alginolyticus* strain (isolated from a shrimp hatchery in Ecuador) was effective in reducing disease caused by two pathogenic *Vibrio* species *V. anguillarum* (from 10 to 26% survival rates) and *Vibrio ordalii* (0–26%) and *A. salmonicida* (0–82%) in Atlantic salmon; however, no beneficial effect was found when challenged with *Yersinia ruckeri* [181]. Siderophore-producing *Pseudomonas fluorescens* have been successfully used as biological control agents. *P. fluorescens* AH2 was able to exclude *A. salmonicida* strain from the Atlantic salmon presmolts with stress-inducible furunculosis infection [182]. *P. fluorescens* strain (AH2) reduced the mortality of rainbow trout *O. mykiss* (40 g) infected with a pathogenic *V. anguillarum* (90-11-287, serotype 01) strain [23]. After 7 days, treated fish showed cumulative mortality of 32%, whereas controls with pathogenic bacterium inoculated fish showed cumulative mortality of 47%. Both show good correlation between production of siderophore and protective action of *P. fluorescens* suggesting that completion of free irons is involved in the mode of action. A recent research on combined probiotics and herbs supplemented in the diet gives promising results on aquatic organisms to their health. This could be a promising alternative tool to supplement and supplant antibiotics, chemicals, or vaccines [86,183]. Harikrishnan *et al.* [184] reported that the administration of probiotic (*Lactobacillus sakei* BK19) and herb (*Scutellaria baicalensis*) can effectively restore the altered hematological parameters, enhance the innate immunity, and minimize the mortality in *O. fasciatus* against *Edwardsiella tarda* indicating promising application of mixture of herbs and probiotics to prevent disease outbreaks in aquaculture. Similar results have been shown by the same research group that reported enhanced growth, blood biochemical constituents, and nonspecific immunity in *P. olivaceus* against *Streptococcus parauberis* when administered with probiotic and herbal mixture supplementation diets [185]. Further investigation of other functional additives and probiotics and their interactions at molecular level would be beneficial to the aquaculture industry.

7.6 Problems Associated with Probiotics Development

At present, the probiotic production is mainly carried out using conventional batch fermentation and suspended cultures [186]. Therefore, development of fermentation technologies requires new approaches. The probiotic strains developed through novel cultivation techniques require an intensive process optimization. Experimental evidences are accumulating about prophylactic use of probiotics in aquatic species. Nevertheless, there is still lack of knowledge about the exact mode of action involved in probiotic effects. Generally, correlations are made between *in vitro* observations and *in vivo* probiotic effect of the same bacterial strain. Thus, when the pathogen is inhibited by the production of bacteriostatic agent or antimicrobial or by competition for nutrients (e.g., production of a siderophore), the same effect was considered for *in vivo* studies, where it may not be the exact effect of the probiotic *in vivo* [3]. It is important to study when the pathogen is available and how many opportunities are available for putative pathogen to grow and become a threat. Careful monitoring of the shift of overall microbial communities provides an insight into how a community of microbial species is evolved, its instability, and potential evolution of unwanted microbial associations. Use of immunoassay procedure, 16S rDNA probes, and green fluorescent protein are some of the reliable techniques useful for distinguishing exogenous probiotics strains in a mixed microbial community to locate and quantify the probiotic strain [3]. Widely accepted method of 16S rDNA sequencing analysis is the best tool for taxonomic positioning of probiotic cultures. However, it has limitations in discriminating closely related lactic acid bacterial species use in probiotic production [147]. Furthermore, Huys *et al.* [187] reported that misidentifications in species levels could happen in taxonomical complex genera such as *Lactobacillus* and *Bifidobacterium*. It has been reported that more than 28% of commercial cultures intended for probiotics were misidentified at the genus or species levels [6]. Furthermore, terrestrial probiotic strains are seldom successful in aquatic species due to their lack of competency in marine aquatic environment. It has been reported that marine endosymbiotic antagonistic microbes might have potential probiotic efficiency, however, these bacteria are uncultivable with available media [161]. Hence, exploring the marine bacterial endosymbionts would be a potential approach for finding novel marine probiotics.

7.7 Further Work and Conclusions

In the larviculture of aquatic animals, it is important to provide healthy environment with beneficial microbial community. Addition of antimicrobials or elimination of bacteria from rearing tanks to control bacterial population may cause problems as an unfavorable alteration of the microbiota. Probiotic has good potential in this aspect; however, its usage in aquaculture is still in infant stage with many questions remaining unanswered. Gomez *et al.* [47] raises few questions in dealing with probiotic studies research: whether the probiotics are effective, if yes then how they affect, its mode of action and whether they mainly act as a food or compete with potentially harmful bacteria, how they perform when the larvae are weakened and in stressful situations, can probiotics become pathogenic and how the probiotic strain will be differentiated from potentially pathogenic strain, and so on. Some of these unanswered queries still crop up in bacteria associated with any aquatic organisms under culture conditions. In the larviculture, many bacteria have been considered as probiotics depending on the empirical observations rather than on scientific data hard to extract definite conclusions from these information. In future, one should consider identifying the criteria or characteristics of bacteria for easy identification of probiotic strains since so far the identifications are done on trial and error basis. Although most of the probiotics are generally safe, especially when newly introduced candidate species is available, safety evaluation consideration is much important. Even though, yet unreported in aquaculture practices, as nonpathogens, probiotics and pathogens coexist in the intestinal tract of aquatic species, there is a possibility of acquisition of virulence genes and antimicrobial drug-resistant traits from pathogens to probiotic through horizontal gene transfer. Therefore, more fundamental research is needed to evaluate their safety aspects [86]. Oral administration of different probiotics demonstrated strong health beneficial effects; however, immune mechanism behind these effects has been less exploited, not properly understood, and need to be studied in depth. Understanding of such responses in the host will be useful for understanding underlying mechanisms of their immune functions. More attention has been paid for the studies on immunomodulatory effects of probiotics in aquaculture species recently. However, a complete understanding of interactions between gut immune system, intestinal epithelium, and gut microbes is important. This will be useful for developing proper strategy to stimulate the local as well as systemic immunity through manipulation of gut microbiota with suitable probiotic/symbiotics/prebiotics without changing intestinal homeostasis [86]. Gnotobiotic fish models that were developed recently would be useful to study the immune effects of microbial community and probiotics in aquatic teleost, which could be beneficial for other aquatic species as well. Future probiotic research should mainly be directed toward the aquatic species where vaccines are not available to protect against infectious diseases such as mollusks and crustaceans larvae, and/or diseases caused by bacteria where no vaccines are available (e.g., *Flavobacterium psychrophilum*). Multifaceted studies are required to understand prospects and challenge of probiotic development. Therefore, much concern is necessary to understand avenue and challenges of probiotics in aquaculture, particularly innovative methods in probiotic development, regulatory aspects, potential strains, mechanisms of actions *in vitro* and *in vivo*, field realities, and considerations for future research are revived and commented whenever necessary.

References

1 Witte, W., Klare, I., and Werner, G. (1999) Selective pressure by antibiotics as feed additives. *Infection*, **27** (Suppl. 2), 35–38.
2 Bachere, E., Mialhe, E., Noel, D. *et al.* (1995) Knowledge and research prospects in marine mollusk and crustacean immunology. *Aquaculture*, **132**, 17–32.
3 Verschuere, L., Rombaut, G., Sorgeloos, P., and Veerstraete, W. (2000) Probiotic bacteria as biological control agents in aquaculture. *Microbiol. Mol. Biol. Rev.*, **64**, 655–671.
4 Selvin, J., Huxley, A.J., and Lipton, A.P. (2005) Pathogenicity, antibiogram and biochemical characteristics of luminescent *Vibrio harveyi*, associated with "black shell disease" of *Penaeus monodon*. *Fish Technol.*, **42**, 191–196.
5 Olafsen, J.A. (2001) Interactions between fish larvae and bacteria in marine aquaculture. *Aquaculture*, **200**, 223–247.
6 Ninawe, A.S. and Selvin, J. (2009) Probiotics in shrimp aquaculture: avenues and challenges. *Crit. Rev. Microbiol.*, **35** (1), 43–66.
7 Gram, L. and Ringo, E. (2005) Prospects of fish probiotics, in *Microbial Ecology in Growing Animals* (eds W.H. Holzapfel and P.J. Naughton), Elsevier, pp. 379–417.
8 Tovar, D., Zambonino, J., Cahu, C. *et al.* (2002) Effect of live yeast incorporation in compound diet on digestive enzyme activity in sea bass (*Dicentrarchus labrax*) larvae. *Aquaculture*, **204**, 113–123.
9 Nakai, T. and Park, S.C. (2002) Bacteriophage therapy of infectious diseases in aquaculture. *Res. Microbiol.*, **153**, 13–18.
10 Austin, B., Baudet, E., and Stobie, M. (1992) Inhibition of bacterial fish pathogens by *Tetraselmis suecica*. *J. Fish Dis.*, **15**, 55–61.

11 Irianto, A. and Austin, B. (2002) Use of probiotics to control furunculosis in rainbow trout, *Oncorhynchus mykiss* (Walbaum). *J. Fish Dis.*, **25** (6), 333–342.

12 Gatesoupe, F.J. (1991) The effect of three strains of lactic bacteria on the production rate of rotifers, *Brachionus plicatilis*, and their dietary value for larval turbot, *Scophthalmus maximus*. *Aquaculture*, **96**, 335–342.

13 Noh, S.H., Han, K., Won, T.H., and Choi, Y.J. (1994) Effect of antibiotics, enzyme, yeast culture and probiotics on the growth performance of Israeli carp. *Korean J. Anim. Sci.*, **36**, 480–486.

14 Bogut, I., Milakovic, Z., Bukvic, Z. *et al.* (1998) Influence of probiotic *Streptococcus faecium* M74 on growth and content of intestinal microflora in carp *Cyprinus carpio*. *Czech J. Anim. Sci.*, **43**, 231–235.

15 Carnevali, O., Vivo, L., Sulpizio, R. *et al.* (2006) Growth improvement by probiotic in European sea bass juveniles (*Dicentrarchus labrax*, L.), with particular attention to IGF-1, myostatin and cortisol gene expression. *Aquaculture*, **258**, 430–438.

16 Diaz-Rosales, P., Arijo, S., Chabrillon, M. *et al.* (2009) Effects of two closely related probiotics on respiratory burst activity of Senegalese sole (*Solea senegalensis*, Kaup) phagocytes, and protection against *Photobacterium damselae* subsp. *piscicida*. *Aquaculture*, **293**, 16–21.

17 Li, J., Tan, B., and Mai, K. (2009) Dietary probiotic *Bacillus* OJ and isomaltooligosaccharides influence the intestine microbial populations, immune responses and resistance to white spot syndrome virus in shrimp (*Litopenaeus vannamei*). *Aquaculture*, **291**, 35–40.

18 Zhou, X., Wang, Y., and Li, W. (2009) Effect of probiotic on larvae shrimp (*Penaeus vannamei*) based on water quality, survival rate and digestive enzyme activities. *Aquaculture*, **287**, 349–353.

19 Tovar-Ramirez, D., Mazurais, D., Gatesoupe, J.F. *et al.* (2010) Dietary probiotic live yeast modulates antioxidant enzyme activities and gene expression of sea bass (*Dicentrarchus labrax*) larvae. *Aquaculture*, **300**, 142–147.

20 Wang, Y. and Gu, Q. (2010) Effect of probiotics on white shrimp (*Penaeus vannamei*) growth performance and immune response. *Marine Biol. Res.*, **6**, 327–332.

21 Zhou, X., Tian, Z., Wang, Y., and Li, W. (2010) Effect of treatment with probiotics as water additives on tilapia (*Oreochromis niloticus*) growth performance and immune response. *Fish Physiol. Biochem.*, **36**, 501–509.

22 Wang, Y. (2011) Use of probiotics *Bacillus coagulans*, *Rhodopseudomonas palustris* and *Lactobacillus acidophilus* as growth promoters in grass carp (*Ctenopharyngodon idella*) fingerlings. *Aqua. Nutr.*, **17**, 372–378.

23 Gram, L., Melchiorsen, J., Sapnggaard, B. *et al.* (1999) Inhibition of *Vibrio anguillarum* by *Pseudomonas fluroscens* AH2, a possible probiotic treatment for fish. *Appl. Environ. Microbiol.*, **65**, 969–973.

24 Moriarty, D.J. (1998) Control of luminous *Vibrio* species in penaeid aquaculture ponds. *Aquaculture*, **164**, 351–358.

25 Panigrahi, A., Kiron, V., Kobayashi, T. *et al.* (2004) Immune responses in rainbow trout *Oncorhynchus mykiss* induced by a potential probiotics bacteria *Lactobacillus rhamnosus* JCM 1136. *Vet. Immunol. Immunopathol.*, **102**, 379–388.

26 Salinas, I., Cuesta, A., Esteban, M.A., and Meseguer, J. (2005) Dietary administration of *Lactobacillus delbrueckii* and *Bacillus subtilis*, single or combined, on gilthead seabream cellular innate immune responses. *Fish Shellfish Immunol.*, **19**, 67–77.

27 Gismondo, M.R., Drago, L., and Lombardi, A. (1999) Review of probiotics available to modify gastrointestinal flora. *Int. J. Antimicrob. Agents*, **12**, 287–292.

28 Lilly, D.M. and Stillwell, R.H. (1965) Probiotics: growth promoting factors produced by microorganisms. *Science*, **147**, 747–748.

29 Chukeatirote, E. (2003) Potential use of probiotics. *Songklanakarin J. Sci. Technol.*, **25**, 275–282.

30 Parker, R.B. (1974) Probiotics, the other half of the antibiotic story. *Anim. Nutr. Health*, **29**, 4–8.

31 Fuller, R. (1989) Probiotics in man and animals. *J. Appl. Bacteriol.*, **66**, 365–378.

32 Havenaar, R. and HuisIn't Veld, M.J.H. (1992) Probiotics: a general view, in *Lactic Acid Bacteria in Health and Disease*, vol. **1**, (ed. B.J.B. Wood) pp 151–170. Elsevier Applied Science Publishers, Amsterdam.

33 Salminen, S. (1996) Uniqueness of probiotic strains. *IDF Nutr. News Lett.*, **5**, 16–18.

34 Schaafsma, G. (1996) State of art concerning probiotic strains in milk products. *IDF Nutr. News Lett.*, **5**, 23–24.

35 Tannock, G.W. (1997) Modification of the normal microbiota by diet, stress, antimicrobial agents, and probiotics, in *Gastrointestinal Microbiology*, vol. 2 (eds R.I. Mackie, B.A. With, and R.E. Isaacsson), Chapman and Hall, New York, pp. 1219–1228.

36 Gatesoupe, F.J. (1999) The use of probiotics in aquaculture. *Aquaculture*, **180**, 147–165.

37 Timmermans, L.P.M. (1987) Early development and differentiation in fish. *Sarsia*, **72**, 331–339.

38 Bergh, O., Naas, K., and Harboe, T. (1994) Shift in the intestinal microflora of Atlantic halibut *Hippoglossus hippoglossus* L. during first feeding. *Can. J. Fish. Aquat. Sci.*, **51**, 1899–1903.

39 Jorquera, M.A., Silva, F.R., and Riquelme, C.E. (2001) Bacteria in the culture of the scallop *Argopecten purpuratus* (Lamarck, 1819). *Aqua. Int.*, **9**, 285–303.

40 FAO/WHO (2001) Joint FAO/WHO Expert Consultation Report on Evaluation of Health and Nutritional Properties of Probiotics in Food Including Powder Milk with Live Lactic Acid Bacteria. Córdoba, Argentina.

41 Debach, P. and Rosen, D. (1991) *Biological Control by Natural Enemies*, Cambridge University Press, Cambridge, p. 440.

42 Gomez-Gil, B., Roque, A., and Turnbull, J.F. (2000) The use and selection of probiotic bacteria for use in the culture of larval aquatic organisms. *Aquaculture*, **191**, 259–270.

43 Huis in't Veld, J.H.J., Havenaar, R., and Marteau, P.H. (1994) Establishing a scientific basis for probiotic R&D. *Trends Biotechnol.*, **12**, 6–8.

44 Ringo, E. and Vadstein, O. (1998) Colonization of *Vibrio pelagius* and *Aeromonas caviae* in early developing turbot (*Scophtalmus maximus* L.) larvae. *J. Appl. Microbiol.*, **84**, 227–233.

45 Hagi, T., Tanaka, D., Iwamura, Y., and Hoshino, T. (2004) Diversity and seasonal changes in lactic acid bacteria in the intestinal tract of cultured freshwater fish. *Aquaculture*, **234**, 335–346.

46 Tuohy, K.M., Probert, H.M., Smejkal, C.W., and Gibson, G.R. (2003) Using probiotics and prebiotics to improve gut health. *Drug Discov. Today*, **8**, 693–700.

47 Gomez, R., Geovanny, D., Balcazar, J.L., and Shen, M.A. (2007) Probiotics as control agents in aquaculture. *J. Ocean Univ. China*, **6** (1), 76–79.

48 Lara-Flores, M. (2011) The use of probiotic in aquaculture: an overview. *Int. Res. J. Microbiol.*, **2** (12), 471–478.

49 Gram, L. and Melchiorsen, J. (1996) Interaction between fish spoilage bacteria *Pseudomonas* sp. and *Shewanella putrefaciens* in fish extracts and on fish tissue. *J. Appl. Bacteriol.*, **80**, 589–595.

50 Olsson, J.C., Westerdahk, A., Conway, P.L., and Kjelleberg, S. (1992) Intestinal colonization potential of turbot (*Scophthalmus maximus*) and dab (*Limanda limanda*) associated bacteria with inhibitory effects against *Vibrio anguillarum*. *Appl. Environ. Microbiol.*, **58**, 551–556.

51 Ali, A. (2000) *Probiotic in Fish Farming: Evaluation of a Candidate Bacterial Mixture*, Sveriges Lantbruks Universitet, Umea, Senegal.

52 Perdigon, G., Alvarez, S., Rachid, M. *et al.* (1995) Probiotic bacteria for humans: clinical systems for evaluation of effectiveness – immune system stimulation by probiotics. *J. Dairy Sci.*, **78**, 1597–1606.

53 Kesarcodi-Watson, A., Kaspar, H., Lategan, M.J., and Gibson, L. (2008) Probiotics in aquaculture: the need, principles and mechanisms of action and screening processes. *Aquaculture*, **274**, 1–14.

54 Wang, Y., Xu, Z., and Xia, M. (2005) The effectiveness of commercial probiotics in northern white shrimp (*Penaeus vannamei* L.) ponds. *Fish. Sci.*, **71**, 1034–1039.

55 Prado, S., Romalde, J.L., and Barja, J.L. (2010) Review of probiotics for use in bivalve hatcheries. *Vet. Microbiol.*, **145**, 187–197.

56 Courvalin, P. (2006) Antibiotic resistance: the pros and cons of probiotics. *Dig. Liver Dis.*, **38** (Suppl. 2), S261–S265.

57 Zhou, X. and Wang, Y. (2012) Probiotics in Aquaculture: Benefits to the Health, Technological Applications and Safety. In *Health and Environment in Aquaculture*, (ed. E.D. Carvalho, G.S. David and R.J. Silva), pp. 1–13. InTech, Croatia.

58 Vergin, K.L., Rappe, M.S., and Giovannoni, S.J. (2001) Streamlined method to analyze 16S rRNA gene clone libraries. *Biotechniques*, **30**, 938–944.

59 Pond, M.J., Stone, D.M., and Alderman, D.J. (2006) Comparison of conventional and molecular techniques to investigate the intestinal microflora of rainbow trout (*Oncorhynchus mykiss*). *Aquaculture*, **261**, 194–203.

60 Wang, Y. and Xu, Z. (2007) Safety evaluation of probiotic PSB0201 (*Rhodopseudomonas palustris*) using primary culture epithelial cells isolated from tilapia (*Oreochromis nilotica*) intestine. *J. Agric. Biotechnol.*, **15**, 233–236.

61 Aguirre-Guzman, G. (1992) Uso de probióticos en acuacultura, in *Avances en Nutrición Acuǒcola. 2do Simposio Internacional sobre Nutrición y Tecnologǒa de Alimentos para Acuacultura* (eds L.E. Cruz-Suárez, D. Ricque, and R. Mendoza), Facultad de Ciencia Biológicas de la Universidad Autónoma de Nuevo León, Monterrey, Nuevo León, México, pp. 332–337.

62 Das, S., Lyla, P.S., and Khan, S.A. (2006) Application of *Streptomyces* as a probiotic in the laboratory culture of *Penaeus monodon* (Fabricius). *Isr. J. Aquac.*, **58**, 198–204.

63 Klaenhammer, T.R. (1993) Genetics of bacteriocins produced by lactic acid bacteria. *FEMS Microbiol. Rev.*, **12**, 39–86.

64 Vanbelle, M., Teller, E., and Focant, M. (1990) Probiotics in animal nutrition: a review (Berlin). *Arch. Tierrenahr.*, **40**, 542–567.

65 Julio, A. and Marie-Josee, B. (2011) Proteomics, human gut microbiota and probiotics. *Expert Rev. Proteom.*, **8**, 279–288.

66 Sakai, M., Yoshida, T., Astuta, S., and Kobayashi, M. (1995) Enhancement of resistance to vibriosis in rainbow trout, *Oncorhynchus mykiss* (Walbaum), by oral administration of *Clostridium butyricum* bacteria. *J. Fish Dis.*, **18**, 187–190.

67 Rengpipat, S., Rukpratanporn, S., Piyatiratitivorakul, S., and Menasaveta, P. (2000) Immunity enhancement in black tiger shrimp (*Penaeus monodon*) by a probiont bacterium (*Bacillus* S11). *Aquaculture*, **191**, 271–288.

68 Balcazar, J.L. (2003) Evaluation of probiotic bacterial strains in *Litopenaeus vannamei*. Final Report, National Center for Marine and Aquaculture Research, Guayaquil, Ecuador.

69 Nikoskelainen, S., Ouwehand, A., Bylund, G. *et al.* (2003) Immune enhancement in rainbow trout (*Oncorhynchus mykiss*) by potential probiotic bacteria (*Lactobacillus rhamnosus*). *Fish Shellfish Immunol.*, **15**, 443–452.

70 Gullian, M., Thompson, F., and Rodrıguez, J. (2004) Selection of probiotic bacteria and study of their immunostimulatory effect in *Penaeus vannamei*. *Aquaculture*, **233**, 1–14.

71 Pais, R., Khushiramani, R., Karunasagar, I., and Karunasagar, I. (2008) Effect of immunostimulants on the haemolymph haemagglutinins of tiger shrimp *Penaeus monodon*. *Aquac. Res.*, **39** (12), 1339–1345.

72 Rodriguez, J., Espinosa, Y., Echeverria, F. *et al.* (2007) Exposure to probiotics and β-1,3/1,6-glucans in larviculture modifies the immune response of *Penaeus vannamei* juveniles and both the survival to White Spot Syndrome Virus challenge and pond culture. *Aquaculture*, **273** (4), 405–415.

73 Panigrahi, A., Kiron, V., Puangkaew, J. *et al.* (2005) The viability of probiotic bacteria as a factor influencing the immune response in rainbow trout *Oncorhynchus mykiss*. *Aquaculture*, **243** (1–4), 241–254.

74 Goncalves, A.T., Maita, M., Futami, K. *et al.* (2011) Effects of a probiotic bacterial

Lactobacillus rhamnosus dietary supplement on the crowding stress response of juvenile Nile tilapia *Oreochromis niloticus*. *Fish. Sci.*, **77** (4), 633–642.

75 Song, Z., Wu, T., Cai, L. *et al.* (2006) Effects of dietary supplementation with *Clostridium butyricum* on the growth performance and humoral immune response in *Miichthys miiuy*. *J. Zhejiang Univ. Sci. B*, **7** (7), 596–602.

76 Balcazar, J.L., Vendrell, D., Ruiz-Zarzuela, I., and Muzquiz, J.L. (2004) Probiotics: a tool for the future of fish and shellfish health management. *J. Aquac. Trop.*, **19** (4), 239–242.

77 Lazado, C.C., Caipang, C.M.A., Rajan, B. *et al.* (2010) Characterisation of GP21 and GP12-two potential probiotic bacteria isolated from the gastrointestinal tract of Atlantic cod. *Probiotics Antimicrob. Proteins*, **2**, 126–134.

78 Picchietti, S., Mazzini, M., Taddei, A.R. *et al.* (2007) Effects of administration of probiotic strains on GALT of larval gilthead seabream: immunohistochemical and ultrastructural studies. *Fish Shellfish Immunol.*, **22**, 57–67.

79 Abelli, L., Randelli, E., Carnevali, O., and Picchietti, S. (2009) Stimulation of gut immune system by early administration of probiotic strains in *Dicentrarchus labrax* and *Sparus aurata*, *Ann. N Y Acad. Sci.*, **1163**: 340–342.

80 Picchietti, S., Fausto, A.M., Randelli, E. *et al.* (2009) Early treatment with *Lactobacillus delbrueckii* strain induces rise in intestinal T cells and granulocytes and modulates immune related genes of larval *Dicentrarchus labrax* (L.). *Fish Shellfish Immunol.*, **26**, 368–376.

81 Sepulcre, M.P., Pelegrȍn, P., Mulero, V., and Meseguer, J. (2002) Characterization of gilthead seabream acidophilic granulocytes by a monoclonal antibody unequivocally points to their involvement in fish phagocytic response. *Cell Tissue Res.*, **308**, 97–102.

82 Salinas, I., Abelli, L., Bertoni, F. *et al.* (2008) Monospecies and multispecies probiotic formulations produce different systemic and local immunostimulatory effects in the gilthead seabream (*Sparus aurata* L.). *Fish Shellfish Immunol.*, **25**, 114–123.

83 Kim, D.H. and Austin, B. (2006) Innate immune responses in rainbow trout (*Oncorhynchus mykiss*, Walbaum) induced by probiotics. *Fish Shellfish Immunol.*, **21**, 513–524.

84 Panigrahi, A., Kiron, V., Satoh, S. *et al.* (2007) Immune modulation and expression of cytokine genes in rainbow trout *Oncorhynchus mykiss* upon probiotic feeding. *Dev. Comp. Immunol.*, **31** (4), 372–382.

85 Sveinsdottir, H., Steinarsson, A., and Gudmundsdottir, A. (2009) Differential protein expression in early Atlantic cod larvae (*Gadus morhua*) in response to treatment with probiotic bacteria. *Comp. Biochem. Physiol. D*, **4**, 249–254.

86 Nayak, S.K. (2010) Probiotics and immunity: a fish perspective. *Fish Shellfish Immunol.*, **29**, 2–14.

87 Cahill, M.M. (1990) Bacterial flora of fishes: a review. *Microb. Ecol.*, **19**, 21–41.

88 Huber, I., Spanggaard, B., Appel, K.F. *et al.* (2004) Phylogenetic analysis and *in situ* identification of the intestinal microbial community of rainbow trout (*Oncorhynchus mykiss*, Walbaum). *J. Appl. Microbiol.*, **96**, 117–132.

89 Romero, J. and Navarrete, P. (2006) 16S rDNA-based analysis of dominant bacterial populations associated with early life stages of coho salmon (*Oncorhynchus kisutch*). *Microb. Ecol.*, **51**, 422–430.

90 Hansen, G.H. and Olafsen, J.A. (1999) Bacterial interactions in early life stages of marine cold water fish. *Microb. Ecol.*, **38**, 1–26.

91 Raz, E. (2010) Mucosal immunity: aliment and ailments. *Mucosal Immunol.*, **3**, 4–7.

92 Bates, J.M., Mittge, E., Kuhlman, J. *et al.* (2006) Distinct signals from the microbiota promote different aspects of zebrafish gut differentiation. *Dev. Biol.*, **297**, 374–386.

93 Son, V.M., Changa, C.C., Wu, M.C. *et al.* (2009) Dietary administration of the probiotic, *Lactobacillus plantarum*, enhanced the growth, innate immune responses, and disease resistance of the grouper *Epinephelus coioides*. *Fish Shellfish Immunol.*, **26**, 691–698.

94 Pieters, N., Brunt, J., Austin, B., and Lyndon, A.R. (2008) Efficacy of in-feed probiotics against *Aeromonas bestiarum* and *Ichthyophthirius multifiliis* skin infections in rainbow trout (*Oncorhynchus mykiss*, Walbaum). *J. Appl. Microbiol.*, **105**, 723–732.

95 Kamei, Y., Yoshimizu, M., Ezura, Y., and Kimura, T. (1988) Screening of bacteria with antiviral activity from fresh water salmonid hatcheries. *Microbiol. Immunol.*, **32**, 67–73.

96 Direkbusarakom, S., Yoshimizu, M., Ezura, Y. *et al.* (1998) *Vibrio* spp., the dominant flora in shrimp hatchery against some fish pathogenic viruses. *J. Mar. Biotechnol.*, **6**, 266–267.

97 Gatesoupe, F.J. (2002) Probiotic and formaldehyde treatments of *Artemia* nauplii as food for larval pollack, *Pollachius pollachius*. *Aquaculture*, **212**, 347–360.

98 Lara-Flores, M., Olvera-Novoa, M.A., Guzmán-Mendez, B.E., and Lopez-Madrid, W.G. (2003) Use of the bacteria *Streptococcus faecium* and *Lactobacillus acidophilus*, and the yeast *Saccharomyces cerevisiae* as growth promoters in Nile tilapia (*Oreochromis niloticus*). *Aquaculture*, **216**, 193–201.

99 El-Haroun, E.R., Goda, A.M., and Chowdhury, M.A.K. (2006) Effect of dietary probiotic Biogen® supplementation as a growth promoter on growth performance and feed utilization of Nile tilapia *Oreochromis niloticus* (L.). *Aquac. Res.*, **37**, 1473–1480.

100 Wang, X., Li, H., Zhang, X. *et al.* (2000) Microbial flora in the digestive tract of adult penaeid shrimp (*Penaeus chinensis*). *J. Ocean. Univ. Quingdao.*, **30**, 493–498.

101 Vazquez-Juarez, R., Ascencio, F., and Andlid, T. (1993) The expression of potential probiotic colonization factors of yeast isolated from fish during different growth conditions. *Can. J. Microbiol.*, **39**, 1135–1141.

102 Ringo, E. and Gatesoupe, F.J. (1998) Lactic acid bacteria in fish: a review. *Aquaculture*, **160**, 177–203.

103 DeSchrijver, R. and Ollevier, F. (2000) Protein digestion in juvenile turbot (*Scophthalmus maximus*) and effects of dietary administration of *Vibrio proteolyticus*. *Aquaculture*, **186**, 107–116.

104 Sakata, T. (1990) Microflora in the digestive tract of fish and shellfish, (ed. R. Lesel) In *Microbiology in Poecilotherms. Proceedings of the International Symposium on Microbiology in Poecilotherms*. pp 171–176. Elsevier, France.

105 Ringo, E., Strom, E., and Tabacheck, J. (1995) Intestinal microflora of salmonids: a review. *Aquac. Res.*, **26**, 773–789.

106 Lara-Flores, M., Olivera-Castillo, L., and Olvera-Novoa, M.A. (2010) Effect of the inclusion of a bacterial mix (*Streptococcus faecium* and *Lactobacillus acidophilus*), and the yeast (*Saccharomyces cerevisiae*) on growth, feed utilization and intestinal enzymatic activity of Nile tilapia (*Oreochromis niloticus*). *Int. J. Fish. Aquac.*, **2**, 93–101.

107 Prieur, G., Nicolas, J.L., Plusquellec, A., and Vigneulle, M. (1990) Interactions between bivalves mollusks and bacteria in the marine environment. *Oceanogr. Mar. Biol. Annu. Rev.*, **28**, 227–252.

108 Balcazar, J.L., Vendrell, D., DeBlas, I. *et al.* (2006) The role of probiotic in aquaculture. *Vet. Microbiol.*, **114**, 173–186.

109 Westerdahl, A., Olsson, J., Kjelleberg, S., and Conway, P. (1991) Isolation and characterization of turbot (*Scophthalmus maximus*) associated bacteria with inhibitory effects against *Vibrio anguillarum*. *Appl. Environ. Microbiol.*, **57**, 2223–2228.

110 Salminen, S., Ouwehan, A., Benno, Y., and Lee, Y.K. (1999) Probiotics: how they be defined? *Trends Food Sci. Technol.*, **10**, 107–110.

111 Conway, P.L. (2006) Development of intestinal microbiota, in *Gastrointestinal Microbiology* (eds R.I. Mackie, B.A. White, and R.E. Isaacson), Chapman and Hall., New York, pp. 3–38.

112 Fuller, R. (1992) History and development of probiotics, in *Probiotics: The Scientific Basis* (ed. R. Fuller), Chapman and Hall, London, pp. 1–45.

113 Elston, R., Leibovitz, L., Relyea, D., and Zatila, J. (1981) Diagnosis of vibriosis in a commercial oyster hatchery epizootic: diagnostic tools and management features. *Aquaculture*, **24**, 53–62.

114 Nicolas, J.L., Corre, S., Gauthier, G. *et al.* (1996) Bacterial problems associated with scallop *Pecten maximus* larval culture. *Dis. Aquat. Organ.*, **27**, 67–76.

115 Riquelme, C., Hayashida, G., Araya, R. *et al.* (1996) Isolation of a native bacterial strain from scallop *Argopecten purpuratus* with inhibitory effects against pathogenic vibrios. *J. Shellfish Res.*, **15**, 369–374.

116 Lodeiros, C., Freites, L., Fernandez, E. *et al.* (1989) Efecto antibiotico de tres bacterias marinas en la supervivencia de larvas de la vieira *Pecten ziczac* infectadas con el germen *Vibrio anguillarum*. *Bol. Inst. Oceanogr. Venezuela Univ. Oriente.*, **28**, 165–169.

117 Riquelme, C., Araya, R., Vergara, N. *et al.* (1997) Potential probiotic strains in the culture of the Chilean scallop *Argopecten purpuratus* (Lamarck, 1819). *Aquaculture*, **154**, 17–26.

118 Avendano, R.E. and Riquelme, C. (1999) Establishment of mixed probiotics and microalgae as food for bivalve larvae. *Aquac. Res.*, **30**, 893–900.

119 Riquelme, C., Jorquera, M.A., Rojas, A.I. *et al.* (2001) Addition of inhibitor-producing bacteria to mass cultures of *Argopecten purpuratus* larvae (Lamarck, 1819). *Aquaculture*, **192**, 111–119.

120 Riquelme, C., Araya, R., and Escribano, R. (2000) Selective incorporation of bacteria by *Argopecten purpuratus* larvae: implications for the use of probiotics in culturing systems of the Chilean scallop. *Aquaculture*, **181**, 25–36.

121 Douillet, P. and Langdon, C.J. (1993) Effect of marine bacteria on the culture of axenic oyster *Crassostrea gigas* (Thunberg) larvae. *Biol. Bull.*, **184**, 36–51.

122 Douillet, P. and Langdon, C.J. (1994) Use of probiotic for the culture of larvae of the Pacific oyster (*Crassostrea gigas* Thunberg). *Aquaculture*, **119**, 25–40.

123 Gibson, L.F., Woodworth, J., and George, A.M. (1998) Probiotic activity of *Aeromonas media* on the Pacific oyster, *Crassostrea gigas*, when challenged with *Vibrio tubiashii*. *Aquaculture*, **169**, 111–120.

124 Ruiz-Ponte, C., Samain, J.F., Sanchez, J.L., and Nicolas, J.L. (1999) The benefit of a *Roseobacter* species on the survival of scallop larvae. *Mar. Biotechnol.*, **1**, 52–59.

125 Prado, S., Montes, J., Romalde, J.L., and Barja, J.L. (2009) Inhibitory activity of *Phaeobacter* strains against aquaculture pathogenic bacteria. *Int. Microbiol.*, **12**, 107–114.

126 Macey, B.M. and Coyne, V.E. (2005) Improved growth rate and disease resistance in farmed *Haliotis midae* through probiotic treatment. *Aquaculture*, **245** (1–4), 249–261.

127 Weiner, R.M. and Colwell, R.R. (1982) Induction of settlement and metamorphosis in *Crassostrea virginica* by a melanin-synthesizing bacterium. Technical Report Maryland Sea Grant Program, UM-SG-TS-82-05, pp. 1–44.

128 Weiner, R.M., Walch, M., Labare, M.P. *et al.* (1989) Effect of biofilms of the marine bacterium *Alteromonas colwelliana* on set of the oysters *Crassostrea gigas* and *Crassostrea virginica*. *J. Shellfish Res.*, **8**, 117–123.

129 Coyne, V.E., Pillidge, C.J., Sledjeski, D.D. *et al.* (1989) Reclassification of *Alteromonas colwelliana* to the genus *Shewanella* by DNA–DNA hybridization, serology and 5S ribosomal RNA sequence data. *Syst. Appl. Microbiol.*, **12**, 275–279.

130 Abu, G.O., Weiner, R.M., Bonar, D.B., and Colwell, R.R. (1986) Extracellular polysaccharide production by a marine bacterium, in *Proceedings of the Sixth International Biodeterioration Symposium* (eds S. Barry and D.R. Houghton), Cambrian News Ltd., pp. 543–549.

131 Zobell, C.E. (1943) The effect of solid surfaces upon bacterial activity. *J. Bacteriol.*, **46**, 39–56.

132 Tritar, S., Prieur, D., and Weiner, R. (1992) Effects of bacterial films on the settlement of the oysters, *Crassostrea gigas* and *Ostrea edulis*, and the scallop *Pecten maximus*. *J. Shellfish Res.*, **11**, 325–330.

133 Walch, M., Weiner, R.M., Colwell, R.R., and Coon, S.L. (1999) Use of L-DOPA and soluble bacterial products to improve set of *Crassostrea virginica* (Gmelin, 1791) and *C. gigas* (Thunberg, 1793). *J. Shellfish Res.*, **18**, 133–138.

134 Vaughan, E.E., deVries, M.C., Zoetendal, E.G. *et al.* (2002) *The intestinal LABs*. *Antonie Van Leeuwenhoek*, **82**, 341–352.

135 Lim, H.J., Kapareiko, D., Schott, E.J. *et al.* (2011) Isolation and evaluation of new probiotic bacteria for use in shellfish hatcheries: I. Isolation and screening for bioactivity. *J. Shellfish Res.*, **30** (3), 609–615.

136 Kapareiko, D., Lim, H.J., Schott, E.J. *et al.* (2011) Isolation and evaluation of new probiotic bacteria for use in shellfish hatcheries: II. Effects of a *Vibrio* sp. probiotic candidate upon survival of oyster larvae (*Crassostrea virginica*) in pilot scale trials. *J. Shellfish Res.*, **30** (3), 617–625.

137 Rico-Mora, R., Voltolina, D., and Villaescusa-Celaya, J.A. (1998) Biological control of *Vibrio alginolyticus* in *Skeletonema costatum* (Bacillariophyceae) cultures. *Aquac. Eng.*, **19**, 1–6.

138 Garriques, D. and Wyban, J. (1993) Up to date advances on *Penaeus vannamei* maturation, nauplii and postlarvae production. Associacao Brasileira de Aquiculture. IV Simposio Brasileiro Sobre Cultivo de Camarao, November 22–27, Brasil, pp. 217–235.

139 Daniels, H.V. (1993) Disease control in shrimp ponds and hatcheries in Ecuador. Associacao Brasileira de Aquicultura. IV Simposio Brasileiro Sobre Cultivo de Camarao, November 22–27, Brasil, pp. 175–184.

140 Griffith, D.R.W. (1995) Microbiology and the role of probiotics in Ecuadorian shrimp hatcheries, in *Larvi' 95: Fish and Shellfish Larviculture Symposium*, vol. 24 (eds P. Lavens, E. Jaspers, and I. Roelants), European Aquaculture Society, Gent, Belgium, p. 478.

141 Anonymous (1991) Disease control in the hatchery by microbiological techniques. Asian Shrimp News 8, 4th quarter.

142 Maeda, M. (1992) Effect of bacterial population on the growth of a prawn larva, *Penaeus monodon. Bull. Natl. Res. Inst. Aquac.*, **21**, 25–29.

143 Maeda, M. (1988) Microorganisms and protozoa as feed in mariculture. *Prog. Oceanogr.*, **21**, 201–206.

144 Garriques, D. and Arevalo, G. (1995) An evaluation of the production and use of a live bacterial isolate to manipulate the microbial flora in the commercial production of *Penaeus vannamei* postlarvae in Ecuador, in *Swimming through Troubled Water: Proceedings of the Special Session on Shrimp Farming, Aquaculture'95* (eds C.L. Browdy and J.S. Hopkins), World Aquaculture Society, Baton Rouge, LA, pp. 53–59.

145 Gatesoupe, F.J. (1990) The continuous feeding of turbot larvae, *Scophthalmus maximus*, and control of the bacterial environment of rotifers. *Aquaculture*, **89**, 139–148.

146 Lightner, D.V. (1993) Diseases of cultured penaeid shrimps, in *CRC Hand book of Mariculture* (ed. J.P. McVey), CRC press, Boca Raton, FL, pp. 393–486.

147 Hong, H.A., Duc, L.H., and Cutting, S.M. (2005) The use of bacterial spore formers as probiotics. *FEMS Microbiol. Rev.*, **29**, 813–835.

148 Karunasagar, I., Pai, R., Malathi, G.R., and Karunasagar, I. (1994) Mass mortality of *Penaeus monodon* larvae due to antibiotic-resistant *Vibrio harveyi* infection. *Aquaculture*, **128**, 203–209.

149 Rengpipar, S., Tunyannun, A., Fast, A.W. *et al.* (2003) Enhanced growth and resistance to *Vibrio* challenge in pond reared black tiger shrimp *Penaeus monodon* fed a *Bacillus* probiotic. *Dis. Aquat. Organ.*, **55**, 169–173.

150 Vaseeharan, B. and Ramasamy, P. (2003) Control of pathogenic *Vibrio* spp. by *Bacillus subtilis* BT23, a possible probiotic treatment for black tiger shrimp *Penaeus monodon*. *Lett. Appl. Microbiol.*, **36**, 83–87.

151 Chen, C.C. and Chen, S.N. (2001) Water quality management with *Bacillus* sp. in the high-density culture of red-parrot fish *Cichlasoma citrinellum* × C-synspilum. *North Am. J. Aquac.*, **63**, 66–73.

152 Farzanfar, A. (2006) The use of probiotics in shrimp aquaculture. *FEMS Immunol. Med. Microbiol.*, **48**, 149–158.

153 Wang, Y. (2007) Effect of probiotics on growth performance and digestive enzyme activity of the shrimp *Penaeus vannamei*. *Aquaculture*, **269**, 259–264.

154 Nogami, K. and Maeda, M. (1992) Bacteria as biocontrol agents for rearing larvae of the crab *Portunus triruberculatus*. *Can. J. Fish Aquac. Sci.*, **49**, 2373–2376.

155 Maeda, M. (1992) Fry production with biocontrol. *Isr. J. Aquac.*, **44**, 142–143.

156 Nogami, K., Hamasaki, K., Maeda, M., and Hirayama, K. (1997) Biocontrol method in

aquaculture for rearing the swimming crab larvae *Portunus trituberculatus*. *Hydrobiologia*, **358**, 291–295.
157 Maeda, M., Nogami, K., Kanematsu, M., and Hirayama, K. (1997) The concept of biological control methods in aquaculture. *Hydrobiologia*, **358**, 285–190.
158 Maeda, M. (1994) Biocontrol of the larval rearing biotope in aquaculture. *Bull. Natl. Res. Inst. Aquac.*, **1**, 71–74.
159 Gomez-Gil, B. (1998) *Evaluation of Potential Probionts for Use in Penaeid shrimp Larval Culture*. University of Stirling, p. 269
160 Maeda, M., Nogami, K., and Ishibashi, N. (1992) Utility of microbial food assemblages for culturing crab, *Portunus trituberculatus*. *Bull. Natl. Res. Inst. Aquac.*, **21**, 31–38.
161 Selvin, J., Huxley, A.J., and Lipton, A.P. (2004) Immunomodulatory potential of marine secondary metabolites against bacterial diseases of shrimp. *Aquaculture*, **230**, 241–248.
162 Daniels, C.L., Merrifield, D.L., Boothroyd, D.P. *et al.* (2010) Effect of dietary *Bacillus* sp. and mannan oligosaccharides (MOS) on European lobster (*Homarus gammarus* L.) larvae growth performance, gut morphology and gut microbiota. *Aquaculture*, **304**, 49–57.
163 Hansen, G.H. and Olafsen, J.A. (1989) Bacterial colonization of cod (*Gadus morhua* L.) and halibut (*Hippoglosus hippoglossus*) eggs in marine aquaculture. *Appl. Environ. Microbiol.*, **55**, 1435–1446.
164 Hjelm, M., Bergh, O., Riaza, A. *et al.* (2004) Selection and identification of autochthonous potential probiotic bacteria from turbot larvae (*Scophthalmus maximus*) rearing units. *Syst. Appl. Microbiol.*, **27**, 360–371.
165 Gatesoupe, F.J. (1994) Lactic acid bacteria increase the resistance of turbot larvae, *Scophthalmus maximus*, against pathogenic *Vibrio*. *Aquat. Living Resour.*, **7**, 277–282.
166 de laBanda, I., Chereguini, O., and Rasines, I. (1992) Influencia de la adición de bacteria lácticas en el cultivo larvario delrodaballo (*Scophthalmus maximus* L.). *Bol. Inst. Esp. Oceanogr.*, **8**, 247–254.
167 Planas, M., Perez-Lorenzo, M., Hjelm, M. *et al.* (2006) Probiotic effect *in vivo* of *Roseobacter* strain 27-4 against *Vibrio* (*Listonella*) *anguillarum* infections in turbot (*Scophthalmus maximus* L.) larvae. *Aquaculture*, **255**, 323–333.
168 Munro, P.D., Barbour, A., and Birkbeck, T.H. (1995) Comparison of the growth and survival of larval Turbot in the absence of culturable bacteria with those in the presence of *Vibrio anguillarum*, *Vibrio alginolyticus*, or a marine *Aeromonas* sp. *Appl. Environ. Microbiol.*, **61**, 4425–4428.
169 Huys, L., Dhert, P., Robles, R. *et al.* (2001) Search for beneficial bacterial strains for turbot (*Scophthalmus maximus* L.) larviculture. *Aquaculture*, **193**, 25–37.
170 Makridis, P., Martins, S., Vercauteren, T. *et al.* (2005) Evaluation of candidate probiotic strains for gilthead sea bream larvae (*Sparus aurata*) using an *in vivo* approach. *Lett. Appl. Microbiol.*, **40**, 274–277.
171 Harzevili, A.R.S., VanDuffel, H., Dhert, P. *et al.* (1998) Use of a potential probiotic *Lactococcus lactis* AR21 strain for the enhancement of growth in the rotifer *Brachionus plicatilis* (Muller). *Aquac. Res.*, **29**, 411–417.
172 Avella, M.A., Gioacchini, G., Decamp, O. *et al.* (2010) Application of multi-species of *Bacillus* in sea bream larviculture. *Aquaculture*, **305**, 12–19.
173 Harikrishnan, R., Kim, M.C., Kim, J.S. *et al.* (2011) Immunomodulatory effect of probiotics enriched diets on *Uronema marinum* infected olive flounder. *Fish Shellfish Immunol.*, **30**, 964–971.
174 Harikrishnan, R., Balasundaramb, C., and Heo, M.S. (2010) Effect of probiotics enriched diet on *Paralichthys olivaceus* infected with lymphocystis disease virus (LCDV). *Fish Shellfish Immunol.*, **29**, 868–874.
175 Aly, S.M., Abd-El-Rahman, A.M., John, G., and Mohamed, M.F. (2008) Characterization of some bacteria isolated from *Oreochromis niloticus* and their potential use as probiotics. *Aquaculture*, **277**, 1–6.
176 Gildberg, A., Johansen, A., and Bogwald, J. (1995) Growth and survival of Atlantic salmon (*Salmo salar*) fry given diets supplemented with fish protein hydrolysate and lactic acid bacteria during

a challenge trial with *Aeromonas salmonicida*. *Aquaculture*, **138**, 23–34.

177 Gildberg, A., Mikkelsen, H., Sandaker, E., and Ringo, E. (1997) Probiotic effect of lactic acid bacteria in the feed on growth and survival of fry of Atlantic cod (*Gadus morhua*). *Hydrobiologia*, **352**, 279–285.

178 Gildberg, A. and Mikkelsen, H. (1998) Effect of supplementing the feed of Atlantic cod (*Gadus morhua*) fry with lactic acid bacteria and immunostimulating peptides during a challenge trial with *Vibrio anguillarum*. *Aquaculture*, **167**, 103–113.

179 Jobon, A., Olsson, J.C., Westerdahl, A. *et al.* (1997) Colonization in the fish intestinal tract and production of inhibitory substances in intestinal mucus and fecal extracts by *Carnobacterium* sp. strain K1. *J. Fish Dis.*, **20**, 383–392.

180 Olsson, J.C., Joborn, A., Westerdahl, A. *et al.* (1998) Survival, persistence and proliferation of *Vibrio anguillarum* in juvenile turbot, *Scophthalmus maximus* (L.), intestine and feces. *J. Fish Dis.*, **21**, 1–9.

181 Austin, B., Stuckey, L.F., Robertson, P.A.W. *et al.* (1995) A probiotic strain of *Vibrio alginolyticus* effective in reducing diseases caused by *Aeromonas salmonicida*, *Vibrio anguillarum* and *Vibrio ordalii*. *J. Fish Dis.*, **18**, 93–96.

182 Smith, P. and Davey, S. (1993) Evidence for the competitive exclusion of *Aeromonas salmonicida* from fish with stress inducible furunculosis by a fluorescent Pseudomonad. *J. Fish Dis.*, **16**, 521–524.

183 Sahu, M.K., Swarnakumar, N.S., Sivakumar, K. *et al.* (2008) Probiotics in aquaculture: importance and future perspectives. *Indian J. Microbiol.*, **48**, 299–308.

184 Harikrishnan, R., Kim, M.C., Kim, J.S. *et al.* (2011) Protective effect of herbal and probiotics enriched diet on haematological and immunity status of *Oplegnathus fasciatus* (Temminck & Schlegel) against *Edwardsiella tarda*. *Fish Shellfish Immunol.*, **30**, 886–893.

185 Harikrishnan, R., Kim, M.C., Kim, J.S. *et al.* (2011) Probiotics and herbal mixtures enhance the growth, blood constituents, and nonspecific immune response in *Paralichthys olivaceus* against *Streptococcus parauberis*. *Fish Shellfish Immunol.*, **31**, 310–317.

186 Lacroix, C. and Yildirim, S. (2007) Fermentation technologies for the production of probiotics with high viability and functionality. *Curr. Opin. Biotechnol.*, **18**, 1–8.

187 Huys, G., Vancanneyt, M., Haene, K.D. *et al.* (2006) Accuracy of species identity of commercial bacterial cultures intended for probiotic or nutritional use. *Res. Microbiol.*, **157**, 803–810.

8
Small-Molecule Antibiotics from Marine Bacteria and Strategies to Prevent Rediscovery of Known Compounds

Matthias Wietz, Maria Månsson, Nikolaj G. Vynne, and Lone Gram

8.1
Antibiotic Activity of Marine Bacteria

Many antibiotics are becoming ineffective due to the development of antibiotic resistance in pathogenic bacteria [1]. The need for novel antibacterial compounds and lead structures has stimulated the exploration of marine environments for bioactive secondary metabolites. Marine eukaryotes such as macroalgae and invertebrates were the first sources of marine natural products [2] and have provided numerous bioactive compounds [3]. Also, marine bacteria, totaling approximately 10^{29} cells in the world ocean [4], have a significant potential for bioprospecting, with annual increases in reported natural products of up to 38% [5]. In 2010, almost one-fifth of new entries in the microbial natural products database, AntiBase [6], were from marine bacteria.

This chapter will focus on small-molecule (<3 kDa) antibacterial compounds from marine bacteria. To date, studies of marine antibiotic-producing bacteria have largely focused on the euphotic zone, leading to the isolation of antagonistic bacteria from surface waters [7,8] as well as from immersed biotic and abiotic surfaces [7,9,10]. Antibiotic-producing bacteria have also been found in the deep sea [11,12], sediments [13], aquaculture systems [14,15], polar oceans [16,17], and sea ice [18,19]. Surface-associated microbial communities attached to organic particles [20], abiotic surfaces, or eukaryotic hosts [7,9,21] are particularly a prolific resource of antagonistic bacterial strains [8,22]. Prokaryotic and prokaryote–eukaryote interactions in nutrient-rich and complex surface niches [23,24] constitute a competitive environment that can select for bioactive microorganisms [25]. Accordingly, many antibiotic-producing strains originate from eukaryotic hosts, including sponges [26–28], zooplankton [18], macroalgae [29–31], fish [7,32], corals [33–35], mollusks [36], worms [37–39], crustaceans [14,40], and bryozoans [41]. Several bioactive compounds originally believed to originate from eukaryotic organisms have been traced to bacterial symbionts, as demonstrated for bacteria associated with bryozoans [42], ascidians [43], and sponges [27,44]. Also, the

Marine Microbiology: Bioactive Compounds and Biotechnological Applications, First Edition.
Edited by Se-Kwon Kim

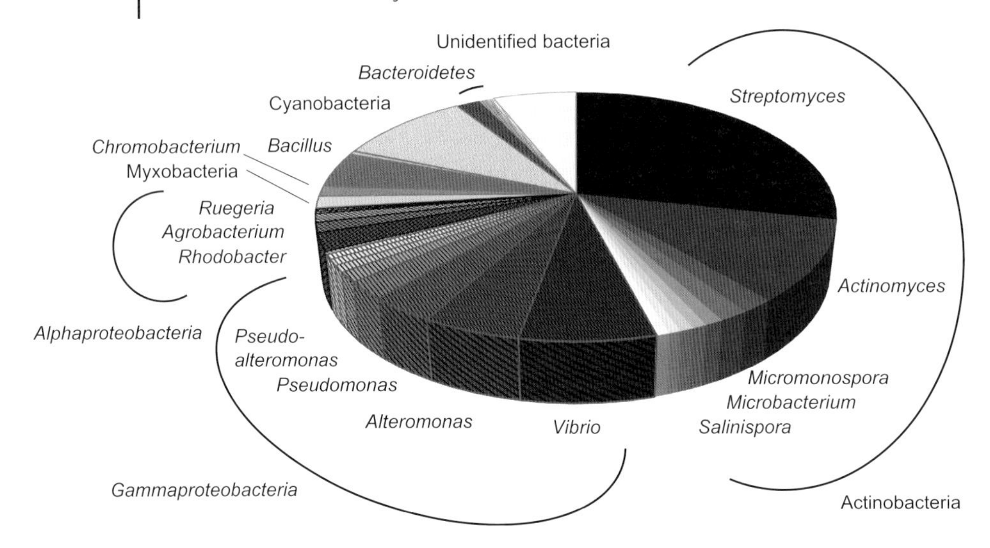

Figure 8.1 Distribution of reported marine bacterial metabolites in AntiBase 2010 according to their taxonomic origin [6]. Figure adapted and updated with permission from Laatsch, 2006 [50]. For the taxa including both marine and non-marine species, only compounds reported from a 'marine' source were included.

FDA-approved alkaloid trabectedin (Yondelis®) is likely produced by "*Candidatus* Endoecteinascidia frumentensis" and not by the tunicate host [45].

Antibiotic-producing marine bacteria are found in many taxonomic groups, with the majority of antibiotics originating from actinomycetes, in particular, *Streptomyces* spp. (Figure 8.1). Marine bacterial antibiotics are structurally diverse and include unique chemical scaffolds not yet found from terrestrial sources [46], often featuring a high degree of halogenation [47]. In addition to the small-molecule antibiotics covered in this chapter, antibiotic macromolecules, including amino acid oxidases [48] and glycoproteins [49], have also been isolated from marine bacteria.

8.2 Structurally Elucidated Marine Bacterial Antibiotics

8.2.1 Actinobacteria

Terrestrial Actinobacteria are prolific producers of antibiotic secondary metabolites [51,52], with an estimated total repertoire of 10^5 compounds for the genus *Streptomyces* alone [53]. More recently, marine Actinobacteria have become a focus of natural product research [47]. The biosynthetic potential among marine Actinobacteria even surpasses that of their terrestrial counterparts [54], since oceanic assemblages harbor a greater diversity of bioactive strains [55]. Antibiotic-producing Actinobacteria have been isolated from sediment [13,55,56], pelagic waters [18,57], sponges [58,59], macroalgae [29,30], and zooplankton [18]. In one study, more than 40% of eukaryote-associated actinomycetes inhibited other bacteria [60]. Metabolites from sponge-associated Actinobacteria were suggested to maintain the balance of the associated microbiota, thereby conferring a beneficial effect on the host [61].

There are several excellent reviews on the biosynthetic capacities of marine Actinobacteria [47,62,63]. Here, we highlight a few examples of recently discovered compounds, the majority from *Streptomyces* spp. Some of the structurally most unique compounds are the marinopyrroles (**1**–**4**) from deep-sea sediment *Streptomyces* sp., which possess an unusual N,C-linked bipyrrole with several chlorine and bromine substituents [64,65]. The marinopyrroles are strong inhibitors of methicillin-resistant *Staphylococcus aureus* (MRSA) with minimum inhibitory concentration (MIC) of less than 2 μM [64]. A80915A (**5**) and A80915B (**6**), two marine-derived streptomycetal antibiotics of the napyradiomycin class [66], are unique meroterpenoids with unusual levels of halogenations and potent and rapid bactericidal anti-MRSA activity [67]. Comparable activity was observed from two bisanthraquinones BE-43472A (**7**) and BE-43472B (**8**) [68] from *Streptomyces* sp. isolated from a tunicate-associated cyanobacterium [69]. Despite structural similarity to tetracyclines, the compounds were inhibitory towards a number of tetracycline-resistant *S. aureus* strains (Figure 8.2) [69].

Other actinobacterial genera, such as *Salinispora* spp., also produce antibacterial compounds [13,70]. A strain of *Salinispora arenicola* isolated from the tunicate

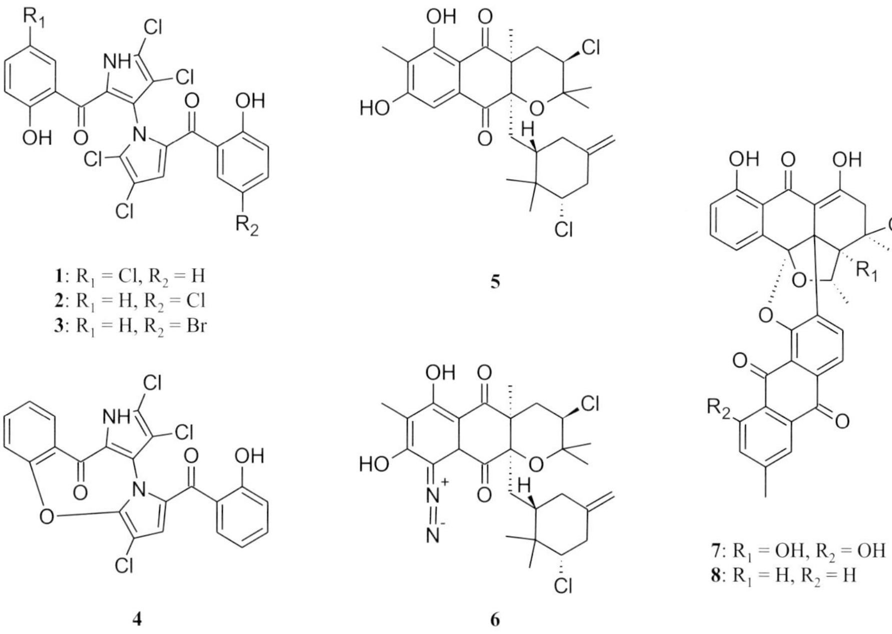

1: R_1 = Cl, R_2 = H
2: R_1 = H, R_2 = Cl
3: R_1 = H, R_2 = Br

7: R_1 = OH, R_2 = OH
8: R_1 = H, R_2 = H

Figure 8.2 Selected antibiotics from marine *Streptomyces* spp.

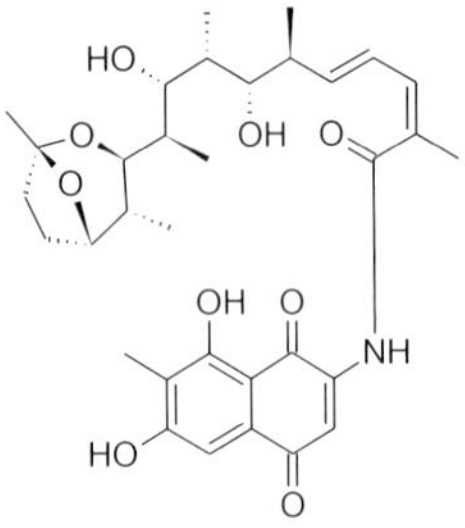

Figure 8.3 Selected antibiotics from marine Actinobacteria.

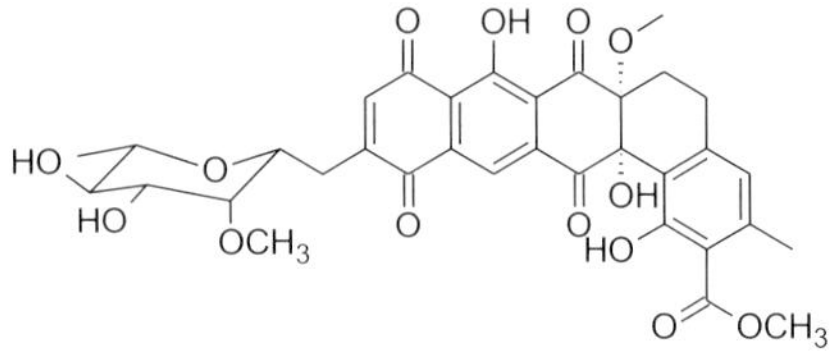

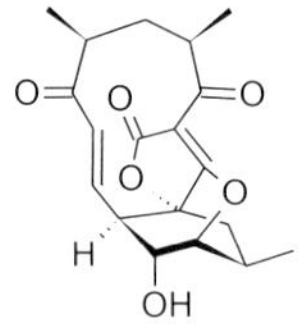

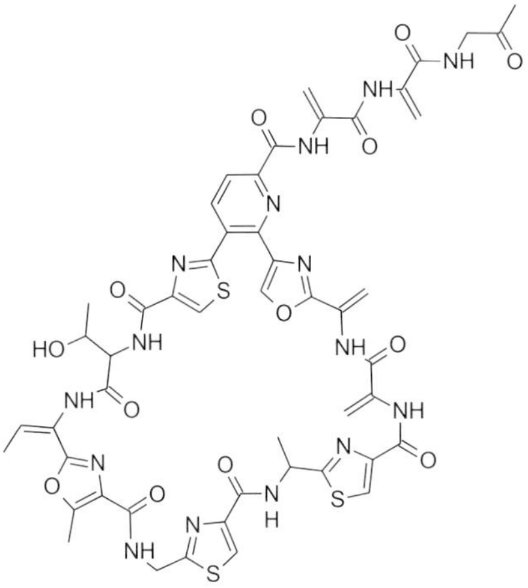

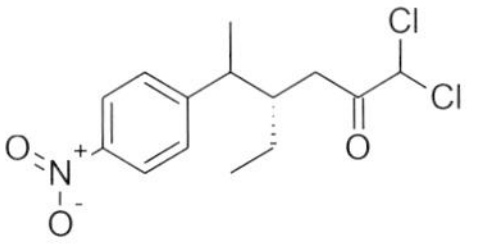

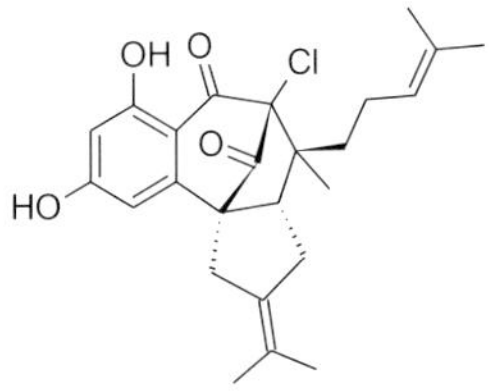

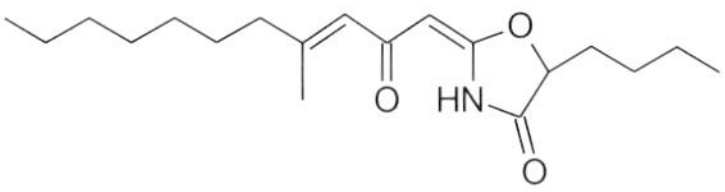

Ecteinascidia turbinata produces the rifamycin antibiotic salinisporamycin (**9**) [71] as well as the benzo[α]naphthacene quinine, arenimycin (**10**), with marked activity against multiresistant *S. aureus* (Figure 8.3) [72]. *Verrucosispora* sp. was the source of three aminofuran antibiotics [73] as well as of abyssomicin C (**11**), a polycyclic polyketide being the first natural inhibitor of *para*-aminobenzoic acid biosynthesis [74]. The thiopeptide antibiotic TP-1161 (**12**) with a rare aminoacetone moiety was isolated from sponge-associated *Nocardiopsis* sp. [75], while a strain of *Nocardia* was the source of a chlorinated alamycin analogue (**13**) containing a rarely observed nitro group [76]. A tropical strain of the genus *Marinispora* produced a series of 2-alkylidene-5-alkyl-4-oxazolidinones, exemplified by lipoxazolidinone A (**14**), with a novel pharmacophore and broad-spectrum activity similar to the commercial antibiotic linezolid [77]. Recently, the novel meroterpenoid merochlorin A (**15**) was shown to be highly bactericidal against *Clostridium difficile* and MRSA, and susceptibility to the antibiotic was unaffected by linezolid or daptomycin resistance. Antibacterial activity was, however, markedly reduced in human serum, which would counteract medicinal applications [78].

8.2.2
Pseudoalteromonas spp.

Antibiotic-producing strains from the Gram-negative, obligate marine genus *Pseudoalteromonas* (Gammaproteobacteria) are found in a variety of niches [7,79–81] and are a considerable source of bioactive natural products [82,83]. Pigmentation of pseudoalteromonads is often correlated with production of bioactive compounds [21,82,84], mirroring observations in terrestrial bacteria [85,86]. On a global scale, antagonistic *Pseudoalteromonas* spp. are more likely to originate from biotic or abiotic surfaces than from the bacterioplankton [21]. In addition, some strains produce antibiotics only when attached to organic particles [87].

Pentabromopseudilin (**16**), a highly brominated compound with strong antibacterial activity against a variety of human pathogens [88,89], has been isolated from *Pseudoalteromonas luteoviolacea* [90] and *Pseudoalteromonas phenolica* (Figure 8.4) [21]. Production of pentabromopseudilin in *P. luteoviolacea* strains determined one of multiple specific chemotypes with unique antibiotic profiles [91]. While all strains produced the purple pigment and antibiotic violacein, they could be grouped according to their production of pentabromopseudilin and another antibiotic, indolmycin (**17**) [91,92]. Several *Pseudoalteromonas* species produce violacein (**18**) [93–95], and it has been suggested that violacein plays a role in grazing resistance [96,97]. The antibiotic korormycin (**19**) selectively inhibits the Na^+-translocating NADH:quinone in marine halophilic Gram-negative bacteria, thus displaying a very narrow spectrum of activity [98]. Other antibiotics from *Pseudoalteromonas* spp. include prodigiosin analogues (**20**) [21,99,100] and multiple brominated metabolites [101,102]. The antibiotic tetrabromopyrrole (**21**) not only inhibits pathogenic bacteria and is autotoxic to the producer [90] but also induces metamorphosis of coral larvae [103]. 2,4-Dibromo-6-chlorophenol (**22**) has antibacterial activity against MRSA and the cystic fibrosis pathogen *Burkholderia cepacia*, and production is induced by two diketopiperazines [104]. Two highly brominated anti-MRSA antibiotics, including 2,3,5,7-tetrabromobenzofuro[3,2-*b*]pyrrole (**23**) with a tricyclic benzofuropyrrole core, have been isolated from *Pseudoalteromonas* sp. collected from a Hawaiian nudibranch [105]. This compound is related to bromophene (**24**) from *P. phenolica*, a bactericidal phenyldiol that may be bound to the cell membrane. The level of activity against *Enterococcus* spp. and MRSA is comparable to that of vancomycin and apparently relates to permeabilization of the target cell membrane [106]. Another series of antibiotics, thiomarinols A with analogues (**25**), in which a pyrrothine moiety is linked to close analogues of the medically relevant antibiotic mupirocin, were isolated from *Pseudoalteromonas* sp. SANK73390 [107–109]. The thiomarinols are strongly inhibitory toward MRSA ($MIC \leq 0.01\,\mu g/ml$) by apparently inhibiting isoleucyl–transfer RNA synthetase [110] and may be able to overcome emerging resistances against mupirocin [111]. Genetically engineered strains of SANK73390 produced additional anti-MRSA compounds with considerable potential for clinical applications [112]. Besides their pharmaceutical potential, antibiotic compounds from *Pseudoalteromonas* may play a role in natural community dynamics by altering bacterial community composition

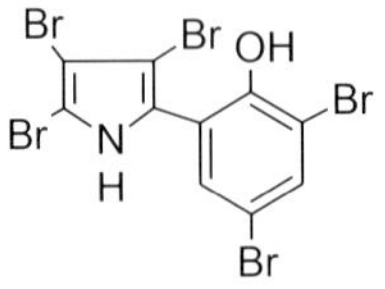

16 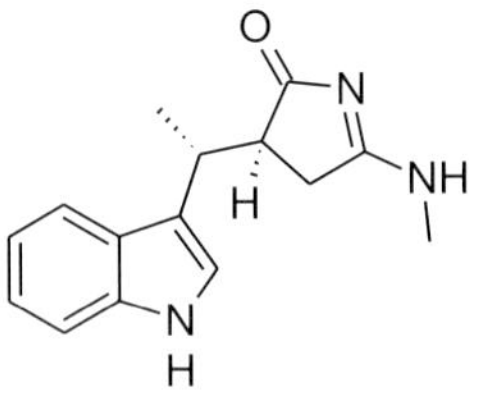 17 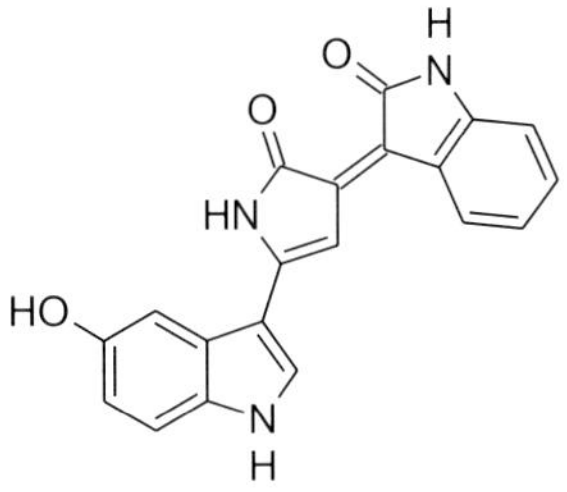 18

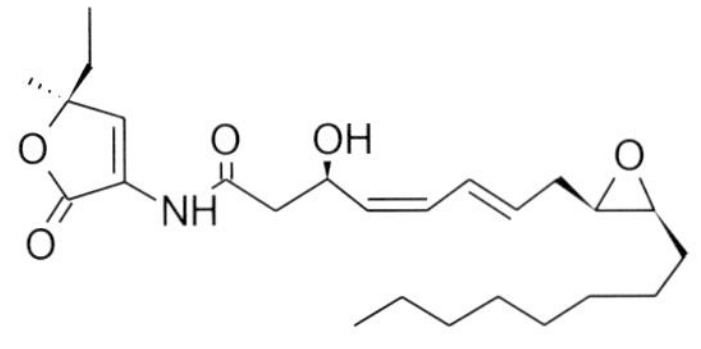

19 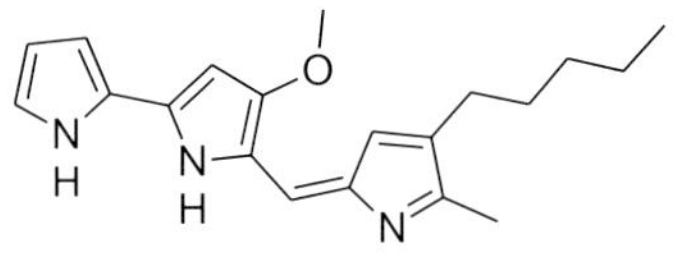 20

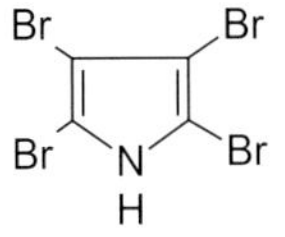

21 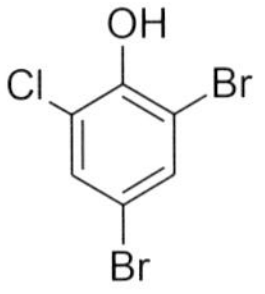 22 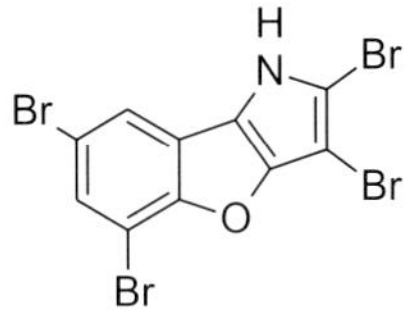 23 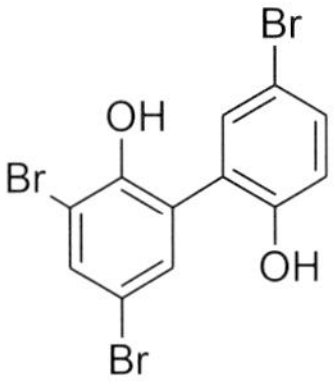 24

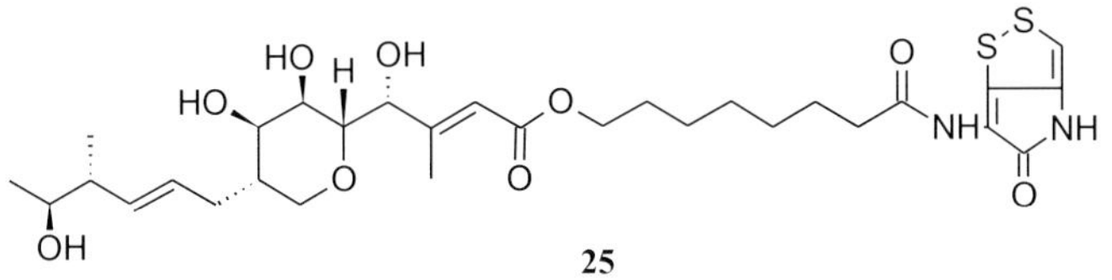

25

Figure 8.4 Selected antibiotics from *Pseudoalteromonas* spp.

on organic particles [87], shaping biofilm communities [95,113,114], and driving mutualistic relationships with macroalgae [115,116].

8.2.3
Vibrionaceae

The *Vibrionaceae* (Gammaproteobacteria) are ubiquitous in marine and brackish environments and have mainly been investigated regarding their symbiotic and pathogenic characteristics [117], but several species also produce antibacterial compounds [8,118]. The worldwide isolation of antagonistic strains during a global marine expedition indicated a widespread occurrence of antibiosis in vibrios [7] that may contribute to their cosmopolitan distribution. Thus, vibrios may represent an untapped resource of bioactive compounds [119].

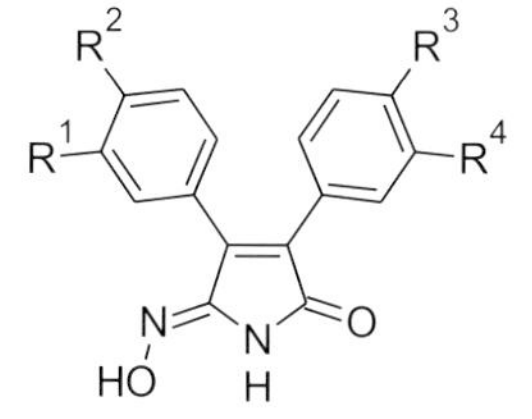

26: $R_1 = NO_2$, $R_2 = OH$, $R_3 = H$, $R_4 = H$
27: $R_1 = NO_2$, $R_2 = OH$, $R_3 = OH$, $R_4 = NO_2$

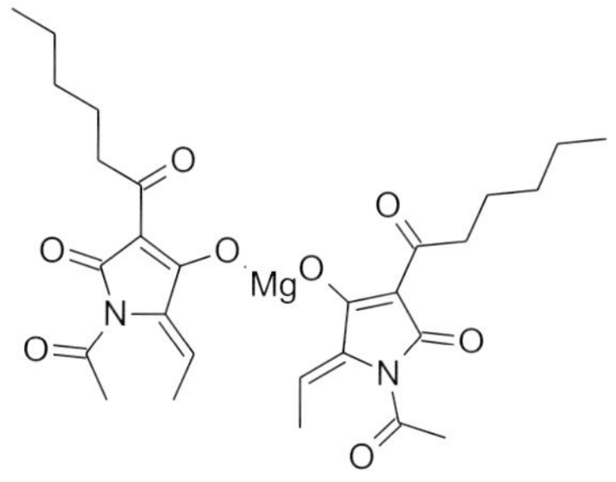

28

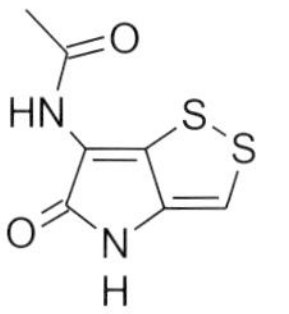

29

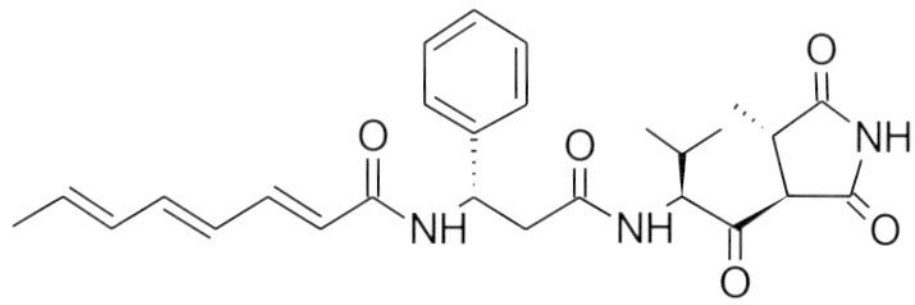

30

Figure 8.5 Selected antibiotics from *Vibrionaceae*.

Antimicrobials from *Vibrio* spp. can reduce the number of other community members and influence microscale variations in competing bacterial populations [8]. However, the frequency of antagonism among vibrios is still largely unknown.

Antibacterial activities have been described for several strains [15,34,118,120,121] and have been recently reviewed [119]. Some structurally intriguing examples are the aqabamycins [122], a series of nitrosubstituted maleimide analogues isolated from a soft coral-associated *Vibrio* sp. The aqabamycins represent a unique structural group due to both their high degree of nitrosubstitution that is rare in nature [123] and the maleimide monoxime present in aqabamycins E (**26**) and F (**27**) (Figure 8.5). An estuarine strain of *Vibrio gazogenes* antagonized an indigenous *Bacillus* sp. by production of prodigiosin (**20**), suggesting that the compound may provide a competitive advantage in the environment [124]. Interestingly, *V. gazogenes* also synthesizes the unique magnesium-containing antibiotic magnesidin A (**28**) [125]. Holomycin (**29**) from a mussel-associated tropical *Photobacterium* sp. [118] is a pyrrothine antibiotic with a characteristic pyrrolinonodithiole core [126], and its strong bacteriostatic effect against a wide range of bacteria [118] is likely due to interference with RNA synthesis [127]. The antibiotic andrimid (**30**) has been isolated from two strains of *Vibrio coralliilyticus* collected from sediment in the Indian Ocean and seaweed in the Eastern Pacific [118], and has previously been detected in a *Vibrio* sp. where it was shown to impede the proliferation of *Vibrio cholerae* on suspended particles [129]. Andrimid is a hybrid peptide–polyketide antibiotic [130] inhibitory toward a wide range of Gram-negative and Gram-positive bacteria [131] by acetyl-CoA carboxylase inhibition targeting fatty acid synthesis [132].

The pyrrolidinedione head was shown as being essential for activity, while variations in the fatty acid tail were more tolerable [133].

8.2.4 Antibiotic Compounds from Other Phylogenetic Groups

8.2.4.1 Firmicutes

Production of antibacterial compounds by marine Firmicutes is mainly seen in *Bacillus* spp. [30,134,135]. The genus is widespread in marine settings and has been isolated from invertebrates [136,137], algae [30], and seawater [138]. Marine bacilli have provided several peptide antibiotics, for instance, bogorol A (**31**) from *Brevibacillus laterosporus* with both C-terminal aminol and N-terminal *R*-hydroxy acid modifications, which represents a new structural template for cationic peptide antibiotics with selective and potent activity against MRSA and vancomycin-resistant enterococci (VRE) (Figure 8.6) [139,140]. Different novel macrolactin antibiotics, exemplified by macrolactin T (**32**) from *Bacillus marinus* [141], have

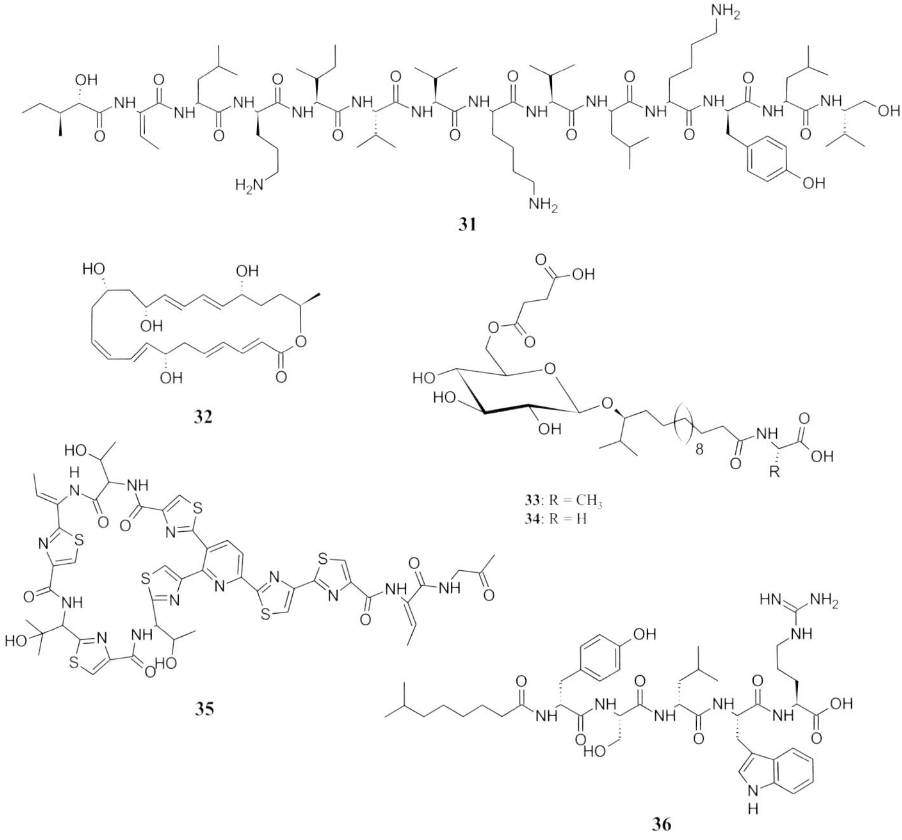

31

32

33: R = CH_3
34: R = H

35

36

Figure 8.6 Selected antibiotics from marine Firmicutes.

been isolated from strains associated with gorgonians [142], macroalgae [143], and sediment [144]. A strain of *Bacillus licheniformis* produced ieodoglucomides A and B (**33** and **34**), unique glycolipopeptides comprising an amino acid, a new fatty acid, a succinic acid, and a sugar [145]. The thiopeptide YM-266183 (**35**) from sponge-associated *B. cereus* [146,147] is closely related to the terrestrial thiocillins, illustrated by analogous biosynthetic gene clusters [148]. The potent activity of YM-266183 against MRSA (MIC = 0.68 μg/ml) and VRE (MIC = 0.025 μg/ml) is based on inhibition of ribosomal protein synthesis [146]. From a tropical strain of *Brevibacillus laterosporus*, the lipopeptide tauramamide (**36**) with potent and selective inhibition of pathogenic *Enterococcus* has been isolated [149].

8.2.4.2 *Roseobacter* clade

The *Roseobacter* clade (Alphaproteobacteria) is a diverse group of obligate marine bacteria involved in global biogeochemical cycles [150]. Roseobacters occur in pelagic and coastal systems worldwide [151,152] and are abundant in association with phytoplankton [150]. Antibiotic traits have been described in different species [7,153,154] and may in part explain the global occurrence of these bacteria [155]. Particle-associated roseobacters were demonstrated to be 10 times more likely to produce antimicrobial compounds than the planktonic representatives [8]. The most prominent antibiotic from roseobacters is the disulfide tropolone tropodithietic acid (TDA) (**37**) produced by strains of *Silicibacter, Ruegeria, Phaeobacter,* and *Pseudovibrio* (Figure 8.7) [15,153,156–159]. Production of TDA in worldwide collected *Ruegeria* spp. [7] as well as TDA-mediated inhibition of indigenous bacterial and algal isolates [160] indicated that the compound may serve an ecological function. TDA is able to enhance the survival of fish larvae [15,161,162], with tolerance hardly arising among pathogenic strains [163]. Five strongly antibiotic diketopiperazines, cyclo(D)—Pro—(D)—Phe, cyclo(D)—Pro—(D)—Leu, cyclo(D)—Pro—(D)—Val, and cyclo(D)—Pro—(D)—Ile and cyclo-*trans*-4-OH-(D)—Pro—(D)—Phe were isolated from *Roseobacter* spp. collected from bivalve aquaculture [164,165] and inhibited *V. anguillarum* with MICs as low as 0.03 μg/ml, being 10-fold superior to antibiotics presently used in aquaculture [164]. The genetic repertoire of roseobacters for natural product biosynthesis [155] indicates an array of novel antibiotic compounds to be unveiled, potentially once appropriate cultivation conditions have been determined [166]. On the other hand, a recent study on 32 whole-genome sequences revealed that polyketide synthase (PKS)–

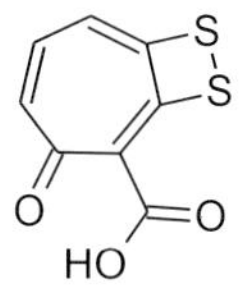

Figure 8.7 Tropodithietic acid.

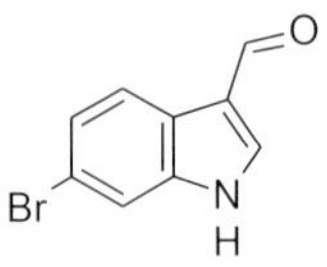

38

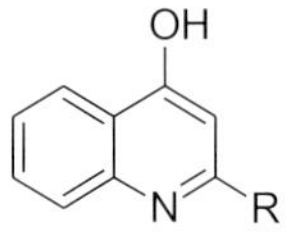

39: R = C_5H_{11}
40: R = C_7H_{15}

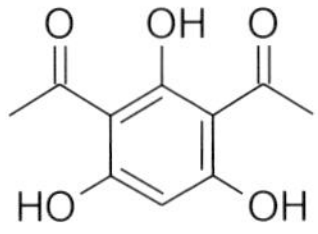

41

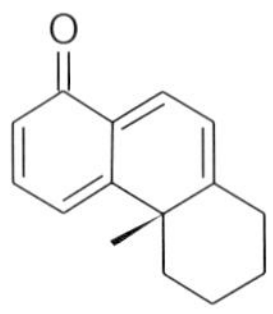

42

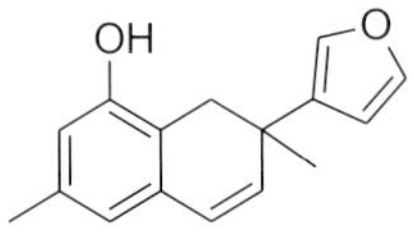

43

Figure 8.8 Selected antibiotics from marine *Pseudomonas* spp.

nonribosomal peptide synthetase (NRPS) and the TDA gene clusters were present only in half of the strains [167].

8.2.4.3 *Pseudomonas* spp.

Marine *Pseudomonas* spp. (Gammaproteobacteria) may possess similar biosynthetic capacities as their terrestrial counterparts, sometimes sharing identical metabolites [168]. A *Pseudomonas* from a coastal tide pool produced an antibiotic mixture of 6-bromoindole carboxaldehyde (**38**), 2-*n*-pentyl- (**39**), and 2-*n*-heptylquinolinol (**40**), the latter being a known antibiotic from terrestrial *Pseudomonas aeruginosa* (Figure 8.8) [169]. An algae-associated *Pseudomonas* sp. produces the antibiotic 2,4-diacetylphloroglucinol (DAPG) (**41**) [170]. DAPG is also produced by terrestrial *Pseudomonas fluorescens* [171] and causes lysis of MRSA at approximately 1 μg/ml, thus being stronger than many commercial antibiotics [170]. The antibiotic was also highly inhibitory toward multiple vancomycin-resistant *S. aureus* [172]. *Pseudomonas stutzeri* was the source of two cell wall-targeting antibiotics, zafrin (**42**) [173] and bushrin (**43**) [174]. The substantial number of marine *Pseudomonas* spp. that lack structural elucidation of their produced antimicrobials [168] suggests that future discoveries are likely.

8.2.4.4 Cyanobacteria

Autotrophic marine cyanobacteria are a rich source of secondary metabolites with activity against eukaryotic cells [175]. In contrast, little is known about marine cyanobacterial antibiotics, although oceanic habitats harbor antagonistic strains [176] with PKS and NRPS genes [177]. Some planktonic cyanobacteria potentially match secondary metabolite production of the streptomycetes, but with much smaller genomes [178]. A shallow water strain of *Lyngbya majuscula* provided the δ-lactone antibiotic malyngolide (**44**) inhibitory toward *Mycobacterium* and *Streptococcus* (Figure 8.9) [179]. In addition, a highly lipophilic cyclic peptide, the cytotoxic

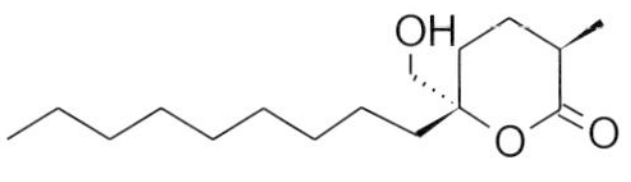

Figure 8.9 Selected cyanobacterial antibiotics.

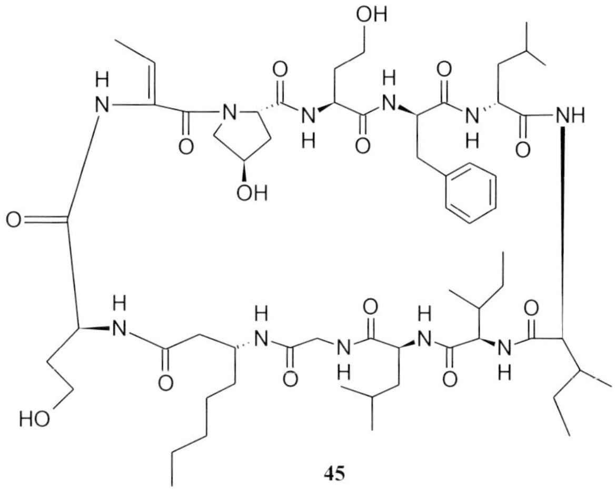

antibiotic hormothamnin A (**45**), has been isolated from the tropical species *Hormothamnion enteromorphoides* [180]. Since terrestrial and freshwater cyanobacteria are known to produce a variety of antibiotic metabolites (Ref. [181] and references therein), future research may yield further antibacterial compounds also from marine strains.

Also other phylogenetic groups may emerge as antibiotic producers. For instance, the Gram-negative phylum Bacteroidetes, often abundant on particulate organic matter [182] and mainly known for polymer degradation [183], also harbors antagonistic strains [29,30,184]. A *Salegentibacter* isolate from Arctic sea ice was shown to produce several aromatic nitrogenous antibiotics [19], including 2-nitro-4-(2-nitroethenyl)-phenol (**46**), while the polyketide–peptides ariakemicins A (**47**) and B with potent antistaphylococcal activity were isolated from *Rapidithrix* sp. (Figure 8.10) [185]. Considering the isolation of several antibiotics from terrestrial strains [186–189], marine Bacteroidetes may be a promising target for natural product research. This may also be valid for, e.g., the Deltaproteobacteria, as

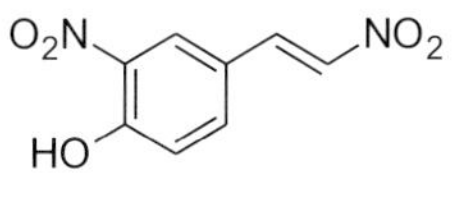

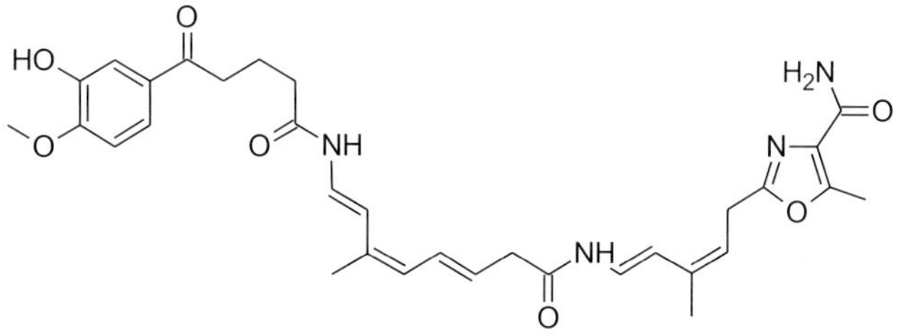

Figure 8.10 Antibiotics from other phylogenetic groups.

illustrated by a marine strain of *Desulfovibrio* whose broad-spectrum antibiotic activity is based on at least eight (so far unidentified) secondary metabolites with different polarities and likely varying modes of action [190].

8.3 Cosmopolitan Antibiotics: the Rediscovery Problem

As of February 2013, seven marine-derived natural products have been approved as therapeutics, with a further 11 being in clinical trials (http://marinepharmacology.midwestern.edu/clinPipeline.htm). However, only two of the 18 compounds – none of which are antibiotics – have a confirmed microbial origin. Despite the substantial number of antibiotic-producing marine bacterial strains and reported natural products [191], research into microbial natural products has overall declined [192].

One major obstacle is the rediscovery of known compounds, as illustrated by the considerable fraction of antibiotics from terrestrial actinomycetes that were later also found in marine strains. For instance, etamycin from terrestrial *Streptomyces griseus* was, 45 years later, also isolated from a marine actinomycete from the coast of Fiji [193]. Rediscoveries of known antibiotics are common among Actinobacteria, including *Streptomyces* [194,195], *Micromonospora* [196], and *Salinispora* [197], but also occur in evolutionary distant microbes with distinct niches [198]. Examples include violacein from marine *Pseudoalteromonas* and terrestrial *Chromobacterium*, indolmycin from marine *Pseudoalteromonas* and terrestrial *Streptomyces*, holomycin from marine *Photobacterium* and soil actinomycetes, magnesidin from *Pseudomonas* and *Vibrio*, and prodigiosin from *Vibrio*, *Pseudoalteromonas*, *Serratia*, *Hahella*, and *Streptomyces* (Table 8.1) [199]. Also, five of eight isolated compounds from *Salinispora* are produced by unrelated taxa [200]. Antibiotic cosmopolitanism can even cross the distinct border between prokaryotes and eukaryotes, as the compound cephalosporin and some β-lactams occur in both bacteria and fungi [201]. However, the full extent of this phenomenon remains unexplored as literature is biased toward reporting novel chemistry.

While convergent evolution of biosynthetic pathways is one possible explanation, the phenomenon of cosmopolitan antibiotics likely results from horizontal gene transfer (HGT) [198]. A high frequency of HGT has been demonstrated in the oceans, illustrated by a common occurrence of gene transfer agents with broad host range, providing interspecific gene transfer under ecologically relevant conditions [202]. The occurrence of the antibiotic andrimid in marine vibrios [118,129,203] and pseudomonads [204], an insect endosymbiont [205], and an enterobacterium [130] is likely facilitated by the presence of transposase pseudogenes adjacent to the andrimid gene cluster [130]. Close contact of bacterial cells in pelagic habitats [206], wide dispersal through ocean circulation [207], vertical transfer by "hitch-hiking" on larger organisms [208,209] and sinking organic particles [210], and alternating

Table 8.1 Examples of cosmopolitan antibiotics in marine bacteria. Compiled and modified from Ref. [6]. Where a species has been re-named and/or reclassified, this is changed in the table compared to AntiBase.

Antibiotic	Examples of producer strains
2-n-heptyl-(1H)-quinolin-4-one	*Pseudomonas methanica, Pseudomonas aeruginosa, Chromobacterium* sp.
2-n-nonyl-(1H)-quinolin-4-one	*P. aeruginosa, Chromobacterium* sp.
Andrimid	*Enterobacter* sp., *Pseudomonas fluorescens, Vibrio* sp.
Bistribromopyrrole	*Chromobacterium* sp., *Pseudoalteromonas luteoviolacea*
Griseoluteic acid	*Streptomyces griseus, Vibrio* sp.
Holomycin	*Streptomyces* sp., *Photobacterium* sp.
Hydroxy-3-nitrophenyl-propionic acid	*Streptomyces hygroscopicus, Salegentibacter* sp.
Hydroxypseudomonic acid C amide	*Alteromonas* sp., *P. fluorescens*
Indolmycin	*Streptomyces* sp., *P. luteoviolacea*
Leupeptin C	*Streptomyces roseus, Alteromonas* sp.
Magnesidin A	*Pseudomonas magnesiorubra, Vibrio gazogenes*
Pentabromopseudilin	*Chromobacterium* sp., *P. luteoviolaceae*
Prodigiosin	*Pseudoalteromonas rubra, Serratia marcescens, Actinomadura pellereti, Nocardia* sp., *P. magnesiorubra, Vibrio psychroerythreus*
Pseudomonic acid C	*P. fluorescens, Alteromonas* sp.
Tropodithietic acid	*Phaeobacter* sp., *Ruegeria* sp.
Violacein	*Chromobacterium* sp., *P. luteoviolacea, Pseudoalteromonas phenolica, Pseudoalteromonas tunicata*

free-living and surface-associated life stages [211] further facilitate the exchange of genetic information. Bacterial groups harboring many mobile genetic elements [212] may have acquired a variety of biosynthetic genes during evolution and are candidates for intensified natural product research.

8.4 Future Strategies for the Discovery of Marine Bacterial Antibiotics

The phenomenon of cosmopolitan antibiotics affects natural product research, since even extensive microbiological and chemical studies of yet uncharacterized microbes can result in rediscoveries. Efficient dereplication is necessary to access the full potential of these microbes, and new chemical tools such as explorative solid-phase extraction (E-SPE) [92] and high-throughput profiling [213,214] linking antibiotic activities directly to compounds produced have decreased the number of futile screening efforts. However, analysis of microbial genome sequences indicates that only about 1% of the natural product diversity is known and that 90% of the metabolites of recognized producer species remain uncharacterized [198]. To counteract the declining discovery rate and minimize the risk of further rediscoveries, methodological improvements are imperative.

8.4.1
Accessing Novel Marine Bacterial Natural Products through Improved Cultivation and Sampling Approaches

The majority of marine bacteria cannot be cultivated under standard laboratory conditions due to a combination of inadequate substrate, oxidative stress, formation of viable but nonculturable cells, induction of lysogenic phages, and lack of cellular interactions [215]. However, many uncultivated phyla may possess interesting physiological and chemical traits [216]. Adjusting isolation procedures to mimick the physicochemical characteristics of the original sample may increase the cultivation efficacy. The incubation of diluted sediment samples in aquarium diffusion chambers increased the number of colony-forming units to up to 40% of inoculated cells [217], being considerably higher than commonly recorded culturable counts [7,218,219]. The recovery of bacteria intimately associated with marine invertebrates has been improved by cultivation in presence of eukaryotic tissue [220], addition of sponge extracts [58,221], and supplementing catalase or sodium pyruvate [222]. For selective isolation of marine actinomycetes, different sample pretreatments exist [223], while incubation with amino acid analogues aids the selective isolation of antibiotic-producing *Bacillus* spp. [224]. An alternative approach is the long-term encapsulation of individual cells in gel microdroplets at a constant flow of very low nutrient concentrations, allowing the exchange of metabolites and/or signaling molecules. Flow cytometric separation of gel beads containing cells that had grown into microcolonies resulted in the isolation of previously uncultured groups [225]. Also, low-nutrient media in combination with dilution-to-extinction [226,227] and virus-depleted incubation conditions [54] can assist in the recovery of greater species richness. Low-nutrient regimes have facilitated the isolation of slow-growing marine Myxobacteria, which could represent an equally prolific resource of antimicrobials as their terrestrial counterparts [228]. The first efforts in this direction allowed the isolation of marine antagonistic Myxobacteria [229] and an antifungal compound [230].

It has been proposed that sampling underexplored biological niches will result in the discovery of new natural product frameworks [231]. Recent examples include a deep sea *Streptomyces* sp., which was the source of the benzoxazole caboxamycin with antibacterial and antibiofilm activities [12]. The deep sea was also the source of antibiotic-producing *Pseudonocardia* [232] and *Marinactinospora* spp. [233]. The polar oceans contain a distinct microbiota compared to warmer waters [151,234], while their comparatively low species diversity [234] may reduce the discovery rate. However, many Arctic and Antarctic bacteria are antibacterial [16,17,19,235], and functional analyses have revealed a great diversity of PKS and NRPS genes in

Antarctic sediment microbiota [236]. Chemical analyses pointed to the existence of novel antimicrobials [237], including alkaloids from *Pseudomonas* [238], aromatic nitro compounds from *Salegentibacter* [19], a naphthalene compound from the cyanobacterium *Nostoc* [239], and a novel angucyclinone from *Streptomyces* [240]. In contrast, a recent survey of marine antagonistic bacteria in the central Arctic Ocean only yielded the rediscovery of three known arthrobacilin analogues from *Arthrobacter* spp. [18].

8.4.2
Eliciting Production of Antibiotics by Activating "Silent" Biosynthetic Pathways

The activation of silent biosynthetic pathways may require the presence of certain elicitors [241]. The source of carbon [242,243] and nutrient availability [244] can be crucial to trigger production of secondary metabolites [50,245–247], and the adjustment of physicochemical factors can influence biosynthesis *in vitro* [223,248,249]. The one strain-many compounds (OSMAC) approach, which aims to stimulate multiple secondary metabolite profiles by small variations in, for example, aeration rate, type of culture vessel, and addition of enzyme inhibitors [250] yielded enhanced antibiotic production in *Pseudoalteromonas* by determining a hydrophilic vessel surface and abundant air supply, but low shear stress as important elicitors [251]. In the marine filamentous cyanobacterium *Geitlerinema*, addition of NaBr to the medium, cell immobilization in vegetable sponge pieces, and temperature positively affected the production of bioactive compounds [252].

Considering the ecological functions of secondary metabolites [253], it is reasonable that antibiotic production can also be stimulated by culturing under "natural conditions." This comprises the use of "niche–mimic" bioreactors [254], roller bottle cultivation that mimic oceanic movements [255], and bioreactor fermentations [256]. Growth substrate formulations must also be considered, as the nutrient regimes generally characteristic for laboratory experiments – such as an excess of nutrients typically not encountered in the oligotrophic marine environment – can shift the phenotype toward an uncharacteristic metabolic state [257] and suppress antibiotic production [258]. The use of standard marine growth substrates such as Marine Broth 2216 therefore possibly limits the discovery of novel natural products; in addition, the complexity of the medium results in artefacts that might eclipse bacterial metabolites during chemical analysis or even cause false positive/negative results in antibacterial assays. In contrast, "marine nutrients" can act as stimulant of secondary metabolism [259]. For instance, antibiosis of a marine actinomycetal strain depends on macroalgal extracts [260], and phytoplankton-derived metabolites can trigger bacterial secondary metabolism [261]. In *V. coralliilyticus*, cultivation with chitin – one of the most abundant marine nutrients [262] – resulted in a twofold higher yield of andrimid per cell, while the biosynthesis of other metabolites was largely abolished [128]. The production of fewer (but interesting) compounds when cultured on natural substrates also facilitates subsequent dereplication and fractionation.

Stimulation of antibiotic production can also result from cocultures and mixed fermentation as synergetic effects [263,264] and a more competitive environment may induce expression of silent biosynthetic genes [265]. This also allows the exchange of signaling molecules, which may elicit antibiosis [10,157,266,267]. Such "competitive induction" was, for instance, observed in a marine *Bacillus* strain, which produced an antibiotic diketopiperazine only upon challenging with another *Bacillus* [268]. In a *Streptomyces* sp., istamycin production was induced by 23% of cocultured bacterial species [269]. Increasing antimicrobial activity as a response to competitors was also seen among marine epibionts [270,271]. Methodological advances within imaging mass spectrometry to detect natural products on surfaces in live cultures, such as desorption electrospray ionization mass spectrometry (DESI-MS) [272,273] and matrix-assisted laser desorption ionization time-of-flight mass spectrometry (MALDI-TOF) imaging [274,275], could further advance such ecology-driven natural product research as in the case of directional release of *Bacillus* antibiotics when cocultured with *S. aureus* [276].

8.4.3 Genome-Based Natural Product Research

The low culturability of marine bacteria, even if improved by the above-mentioned approaches, implies that only a limited amount of microbial diversity can be obtained by culture-based methods. From the Global Ocean Sampling project [277], it has become apparent that there is still a wealth of novel protein folds and other natural products to be discovered among marine bacteria [278]. As a result, natural product research increasingly incorporates molecular tools. The growing amount of bacterial sequence data and whole-genome sequences, in combination with enhanced bioinformatic capacities and automated data analysis, make genome-based approaches a major future direction in natural product research. This includes the screening of metagenomic clone libraries for functional activities [279] and single-cell genomics to study the biosynthetic repertoire of selected environmental microbes [280,281]. In addition, whole-genome sequences can be "mined" for biosynthetic gene clusters, mainly PKS and NRPS [282,283]. The presence and sequence diversity of biosynthetic genes are indicative of a broad secondary metabolite spectrum, which aids in the selection of potentially prolific producers [61]. PCR targeting sequences that encode catalytic domains, including ketosynthases that are informative in predicting the associated biosynthetic pathway [284], are a complementary approach. Analysis of PKS sequences from marine sponge-associated Actinobacteria resulted in the prediction of novel rifamycin antibiotics [197]. This is especially valuable for chemically poorly studied bacterial groups. For instance, genomic analyses enabled to isolate a small cyclic peptide from the ubiquitous marine cyanobacterium *Trichodesmium erythraeum* [285]. Genomic analyses of a terrestrial Bacteroidetes strain allowed the targeted isolation of two elansolid-type antibiotics [286], and comparable studies on marine isolates might yield similar results. Genomics on a marine Planctomycetes revealed several ORFs potentially coding for polyketide and nonribosomal peptide antibiotics, with unusually long coding regions that would result in proteins of up to 3665 amino acids [287]. While the biosynthetic potential of marine Planctomycetes remains uncertain [288], future efforts integrating directed cultivation and genome mining may result in the discovery of novel natural products.

8.5
Conclusions and Perspectives

Marine bacteria are a promising resource of novel antibacterial compounds to overcome evolving resistances toward traditional antibiotics. However, the discovery of marine bacterial antibiotics will benefit from more sophisticated isolation, cultivation, and screening techniques in combination with genomic analyses. Since even the investigation of largely unexplored bacterial groups or ecological niches can yield the rediscovery of known compounds, the scale of bacterial gene transfer resulting in "cosmopolitan antibiotics" may be greater than expected and limit natural product discovery. Nevertheless, the vast and still largely uncharacterized bacterial diversity in the oceans remains a treasure for future scientific discoveries, with both an academic value and a benefit for society.

References

1 Taubes, G. (2008) The bacteria fight back. *Science*, **321**, 356–361.
2 Molinski, T. (2009) Marine natural products. *Clin. Adv. Hematol. Oncol.*, **7**, 383–385.
3 Mayer, A.M.S., Rodríguez, A.D., Berlinck, R.G.S., and Fusetani, N. (2011) Marine pharmacology in 2007–8: marine compounds with antibacterial, anticoagulant, antifungal, anti-inflammatory, antimalarial, antiprotozoal, antituberculosis, and antiviral activities; affecting the immune and nervous system, and other miscellaneous mec. *Comp. Biochem. Physiol. C Toxicol. Pharmacol.*, **153**, 191–222.
4 Whitman, W.B. (1998) Prokaryotes: the unseen majority. *Proc. Natl. Acad. Sci. USA*, **95**, 6578–6583.
5 Blunt, J.W., Copp, B.R., Hu, W.P., Munro, M.H.G., Northcote, P.T., and Prinsep, M.R. (2009) Marine natural products. *Nat. Prod. Rep.*, **26**, 170–244.
6 Laatsch, H. (2010) AntiBase, 2010.
7 Gram, L., Melchiorsen, J., and Bruhn, J.B. (2010) Antibacterial activity of marine culturable bacteria collected from a global sampling of ocean surface waters and surface swabs of marine organisms. *Mar. Biotechnol.*, **12**, 439–451.
8 Long, R.A. and Azam, F. (2001) Antagonistic interactions among marine pelagic bacteria. *Appl. Environ. Microbiol.*, **67**, 4975–4983.
9 Penesyan, A., Kjelleberg, S., and Egan, S. (2010) Development of novel drugs from marine surface associated microorganisms. *Mar. Drugs*, **8**, 438–459.
10 Burgess, J.G., Jordan, E.M., Bregu, M., Mearns-Spragg, A., and Boyd, K.G. (1999) Microbial antagonism: a neglected avenue of natural products research. *J. Biotechnol.*, **70**, 27–32.
11 Fusetani, N., Ejima, D., Matsunaga, S., Hashimoto, K., Itagaki, K., Akagi, Y., Taga, N., and Suzuki, K. (1987) 3-Amino-3-deoxy-D-glucose: an antibiotic produced by a deep-sea bacterium. *Experientia*, **43**, 464–465.
12 Hohmann, C., Schneider, K., Bruntner, C., Irran, E., Nicholson, G., Bull, A.T., Jones, A.L., Brown, R., Stach, J.E.M., Goodfellow, M., Beil, W., Kraemer, M., Imhoff, J.F., Sussmuth, R.D., and Fiedler, H.-P. (2009) Caboxamycin, a new

antibiotic of the benzoxazole family produced by the deep-sea strain *Streptomyces* sp. NTK 937. *J. Antibiot.*, **62**, 99–104.

13 Jensen, P., Gontang, E., Mafnas, C., Mincer, T., and Fenical, W. (2005) Culturable marine actinomycete diversity from tropical Pacific Ocean sediments. *Environ. Microbiol.*, **7**, 1039–1048.

14 Jorquera, M., Riquelme, C., Loyola, L., and Munoz, L. (2000) Production of bactericidal substances by a marine *Vibrio* isolated from cultures of the scallop *Argopecten purpuratus*. *Aquac. Int.*, **7**, 433–448.

15 Hjelm, M., Bergh, Ø., Riaza, A., Nielsen, J., Melchiorsen, J., Jensen, S., Duncan, H., Ahrens, P., Birkbeck, H., and Gram, L. (2004) Selection and identification of autochthonous potential probiotic bacteria from turbot larvae (*Scophthalmus maximus*) rearing units. *Syst. Appl. Microbiol.*, **27**, 360–371.

16 O'Brien, A., Sharp, R., Russell, N., and Roller, S. (2004) Antarctic bacteria inhibit growth of food-borne microorganisms at low temperatures. *FEMS Microbiol. Ecol.*, **48**, 157–167.

17 Lo Giudice, A., Brun, V., and Michaud, L. (2007) Characterization of Antarctic psychrotrophic bacteria with antibacterial activities against terrestrial microorganisms. *J. Basic Microbiol.*, **47**, 496.

18 Wietz, M., Månsson, M., Bowman, J.S., Blom, N., Ng, Y., and Gram, L. (2012) Wide distribution of closely related, antibiotic-producing *Arthrobacter* strains throughout the Arctic Ocean. *Appl. Environ. Microbiol.*, **78**, 2039–2042.

19 Al-Zereini, W., Schuhmann, I., Laatsch, H., Helmke, E., and Anke, H. (2007) New aromatic nitro compounds from *Salegentibacter* sp. T436, an Arctic Sea ice bacterium: taxonomy, fermentation, isolation and biological activities. *J. Antibiot.*, **60**, 301–308.

20 Grossart, H., Levold, F., Allgaier, M., Simon, M., and Brinkhoff, T. (2005) Marine diatom species harbour distinct bacterial communities. *Environ. Microbiol.*, **7**, 860–873.

21 Vynne, N.G., Månsson, M., Nielsen, K.F., and Gram, L. (2011) Bioactivity, chemical profiling and 16S rRNA based phylogeny of *Pseudoalteromonas* strains collected on a global research cruise. *Mar. Biotechnol.*, **13**, 1062–1073.

22 Nair, S. and Simidu, U. (1987) Distribution and significance of heterotrophic marine bacteria with antibacterial activity. *Appl. Environ. Microbiol.*, **53**, 2957–2962.

23 Steinberg, P., DeNys, R., and Kjelleberg, S. (2002) Chemical cues for surface colonization. *J. Chem. Ecol.*, **28**, 1935–1951.

24 DeNys, R., Steinberg, P.D., Willemsen, P., Dworjanyn, S.A., Gabelish, C.L., and King, R.J. (1995) Broad spectrum effects of secondary metabolites from the red alga *Delisea pulchra* in antifouling assays. *Biofouling*, **8**, 259–271.

25 Egan, S., Thomas, T., and Kjelleberg, S. (2008) Unlocking the diversity and biotechnological potential of marine surface associated microbial communities. *Curr. Opin. Microbiol.*, **11**, 219–225.

26 Taylor, M.W., Hill, R.T., Piel, J., Thacker, R.W., and Hentschel, U. (2007) Soaking it up: the complex lives of marine sponges and their microbial associates. *ISME J.*, **1**, 187–190.

27 Taylor, M.W., Radax, R., Steger, D., and Wagner, M. (2007) Sponge-associated microorganisms: evolution, ecology, and biotechnological potential. *Microbiol. Mol. Biol. Rev.*, **71**, 295.

28 Thomas, T., Rusch, D., DeMaere, M.Z., Yung, P.Y., Lewis, M., Halpern, A., Heidelberg, K.B., Egan, S., Steinberg, P.D., and Kjelleberg, S. (2010) Functional genomic signatures of sponge bacteria reveal unique and shared features of symbiosis. *ISME J.*, **4**, 1557–1567.

29 Penesyan, A., Marshall-Jones, Z., Holmström, C., Kjelleberg, S., and Egan, S. (2009) Antimicrobial activity observed among cultured marine epiphytic bacteria reflects their potential as a source of new drugs. *FEMS Microbiol. Ecol.*, **69**, 113–124.

30 Wiese, J., Thiel, V., Nagel, K., Staufenberger, T., and Imhoff, J.F. (2009) Diversity of antibiotic-active bacteria associated with the brown alga *Laminaria saccharina* from the Baltic Sea. *Mar. Biotechnol.*, **11**, 287–300.

31 Boyd, K.G., Adams, D.R., and Burgess, J.G. (1999) Antibacterial and repellent activities of marine bacteria associated with algal surfaces. *Biofouling*, **14**, 227–236.

32 Makridis, P., Martins, S., Tsalavouta, M., Dionisio, L., Kotoulas, G., Magoulas, A., and Dinis, M. (2005) Antimicrobial activity in bacteria isolated from Senegalese sole, *Solea senegalensis*, fed with natural prey. *Aquac. Res.*, **36**, 1619–1627.

33 Nissimov, J., Rosenberg, E., and Munn, C.B. (2009) Antimicrobial properties of resident coral mucus bacteria of *Oculina patagonica*. *FEMS Microbiol. Lett.*, **292**, 210–215.

34 Rypien, K.L., Ward, J.R., and Azam, F. (2010) Antagonistic interactions among coral-associated bacteria. *Environ. Microbiol.*, **12**, 28–39.

35 Ritchie, K.B. (2006) Regulation of microbial populations by coral surface mucus and mucus-associated bacteria. *Mar. Ecol. Prog. Ser.*, **322**, 1–14.

36 Romanenko, L.A., Uchino, M., Kalinovskaya, N.I., and Mikhailov, V.V. (2008) Isolation, phylogenetic analysis and screening of marine mollusc-associated bacteria for antimicrobial, hemolytic and surface activities. *Microbiol. Res.*, **163**, 633–644.

37 Gerard, J., Haden, P., Kelly, M., and Andersen, R. (1996) Loloatin B, a cyclic decapeptide antibiotic produced in culture by a tropical marine bacterium. *Tetrahedron Lett.*, **37**, 7201–7204.

38 Trindade-Silva, A.E., Machado-Ferreira, E., Senra, M.V.X., Vizzoni, V.F., Yparraguirre, L.A., Leoncini, O., and Soares, C.A.G. (2009) Physiological traits of the symbiotic bacterium *Teredinibacter turnerae* isolated from the mangrove shipworm *Neoteredo reynei*. *Genet. Mol. Biol.*, **32**, 572–581.

39 Shankar, S., Punitha, M.J., and Hepziba, J.M.A. (2012) Antimicrobial activity of marine bacteria associated with Polychaetes. *Biores. Bull.*, **1**, 24–28.

40 Gil-Turnes, M.S., Hay, M.E., and Fenical, W. (1989) Symbiotic marine bacteria chemically defend crustacean embryos from a pathogenic fungus. *Science*, **246**, 116.

41 Heindl, H., Wiese, J., Thiel, V., and Imhoff, J.F. (2010) Phylogenetic diversity and antimicrobial activities of bryozoan-associated bacteria isolated from Mediterranean and Baltic Sea habitats. *Syst. Appl. Microbiol.*, **33**, 94–104.

42 Davidson, S., Allen, S., Lim, G., Anderson, C., and Haygood, M. (2001) Evidence for the biosynthesis of bryostatins by the bacterial symbiont "*Candidatus* Endobugula sertula" of the bryozoan *Bugula neritina*. *Appl. Environ. Microbiol.*, **67**, 4531–4537.

43 Schmidt, E.W. and Donia, M.S. (2010) Life in cellulose houses: symbiotic bacterial biosynthesis of ascidian drugs and drug leads. *Curr. Opin. Biotechnol.*, **21**, 827–833.

44 Unson, M. and Faulkner, D. (1993) Cyanobacterial symbiont biosynthesis of chlorinated metabolites from *Dysidea herbacea* (Porifera). *Experientia*, **49**, 349–353.

45 Rath, C.M., Janto, B., Earl, J., Ahmed, A., Hu, F.Z., Hiller, L., Dahlgren, M., Kreft, R., Yu, F., Wolff, J.J., Kweon, H.K., Christiansen, M.A., Hakansson, K., Williams, R.M., Ehrlich, G.D., and Sherman, D.H. (2011) Meta-omic characterization of the marine invertebrate microbial consortium that produces the chemotherapeutic natural product ET-743. *ACS Chem. Biol.*, **6**, 1244–1256.

46 Fenical, W. (1993) Chemical studies of marine bacteria: developing a new resource. *Chem. Rev.*, **93**, 1673–1683.

47 Fenical, W. and Jensen, P.R. (2006) Developing a new resource for drug discovery: marine actinomycete bacteria. *Nat. Chem. Biol.*, **2**, 666–673.

48 Gomez, D., Espinosa, E., Bertazzo, M., Lucas-Elio, P., Solano, F., and Sanchez-Amat, A. (2008) The macromolecule with antimicrobial activity synthesized by *Pseudoalteromonas luteoviolacea* strains is an L-amino acid oxidase. *Appl. Environ. Microbiol.*, **79**, 925–930.

49 Barja, J.L., Lemos, M.L., and Toranzo, A.E. (1989) Purification and characterization of an antibacterial substance produced by a marine *Alteromonas* species. *Antimicrob. Agents Chemother.*, **33**, 1674.

50 Laatsch, H. (2006) Marine bacterial metabolites. In: *Front. Mar. Biotechnol.* (eds Proksch, P. and Müller, W.E.G.), p. 225–288. Horizon Bioscience, Norfolk, UK. ISBN 1-904933-18-1
51 Demain, A.L. and Sanchez, S. (2009) Microbial drug discovery: 80 years of progress. *J. Antibiot.*, **62**, 5–16.
52 Bérdy, J. (2005) Bioactive microbial metabolites. *J. Antibiot.*, **58**, 1–26.
53 Watve, M., Tickoo, R., Jog, M., and Bhole, B. (2001) How many antibiotics are produced by the genus *Streptomyces*? *Arch. Microbiol.*, **176**, 386–390.
54 Bull, A.T. and Stach, J.E.M. (2007) Marine Actinobacteria: new opportunities for natural product search and discovery. *Trends Microbiol.*, **15**, 491–499.
55 Bredholdt, H., Galatenko, O.A., Engelhardt, K., Fjaervik, E., Terekhova, L. P., and Zotchev, S.B. (2007) Rare actinomycete bacteria from the shallow water sediments of the Trondheim fjord, Norway: isolation, diversity and biological activity. *Environ. Microbiol.*, **9**, 2756–2764.
56 Magarvey, N., Keller, J., Bernan, V., Dworkin, M., and Sherman, D. (2004) Isolation and characterization of novel marine-derived actinomycete taxa rich in bioactive metabolites. *Appl. Environ. Microbiol.*, **70**, 7520–7529.
57 Hakvåg, S., Fjaervik, E., Josefsen, K.D., Ian, E., Ellingsen, T.E., and Zotchev, S.B. (2008) Characterization of *Streptomyces* spp. isolated from the sea surface microlayer in the Trondheim Fjord, Norway. *Mar. Drugs*, **6**, 620–635.
58 Li, Z.-Y. and Liu, Y. (2006) Marine sponge *Craniella austrialiensis*-associated bacterial diversity revelation based on 16S rDNA library and biologically active actinomycetes screening, phylogenetic analysis. *Lett. Appl. Microbiol.*, **43**, 410–416.
59 Thomas, T.R.A., Kavlekar, D.P., and LokaBharathi, P.A. (2010) Marine drugs from sponge–microbe association: a review. *Mar. Drugs*, **8**, 1417–1468.
60 Zheng, Z.H., Zeng, W., Huang, Y.J., Yang, Z.Y., Li, J., Cai, H.R., and Su, W. (2000) Detection of antitumor and antimicrobial activities in marine organism associated actinomycetes isolated from the Taiwan Strait, China. *FEMS Microbiol. Lett.*, **188**, 87–91.
61 Schneemann, I., Nagel, K., Kajahn, I., Labes, A., Wiese, J., and Imhoff, J.F. (2010) Comprehensive investigation of marine Actinobacteria associated with the sponge *Halichondria panicea*. *Appl. Environ. Microbiol.*, **76**, 3702–3714.
62 Lam, K. (2006) Discovery of novel metabolites from marine actinomycetes. *Curr. Opin. Microbiol.*, **9**, 245–251.
63 Dharmaraj, S. (2010) Marine *Streptomyces* as a novel source of bioactive substances. *World J. Microbiol. Biotechnol.*, **26**, 2123–2139.
64 Hughes, C.C., Prieto-Davo, A., Jensen, P.R., and Fenical, W. (2008) The marinopyrroles, antibiotics of an unprecedented structure class from a marine *Streptomyces* sp. *Org. Lett.*, **10**, 629–631.
65 Hughes, C.C., Kauffman, C.A., Jensen, P.R., and Fenical, W. (2010) Structures, reactivities, and antibiotic properties of the marinopyrroles A–F. *J. Org. Chem.*, **75**, 3240–3250.
66 Soria-Mercado, I.E., Prieto-Davo, A., Jensen, P.R., and Fenical, W. (2005) Antibiotic terpenoid chloro-dihydroquinones from a new marine actinomycete. *J. Nat. Prod.*, **68**, 904.
67 Haste, N.M., Farnaes, L., Perera, V.R., Fenical, W., Nizet, V., and Hensler, M.E. (2011) Bactericidal kinetics of marine-derived napyradiomycins against contemporary methicillin-resistant *Staphylococcus aureus*. *Mar. Drugs*, **9**, 680–689.
68 Kushida, S., Nakajima, S., Koyama, T., Suzuki, H., Ojiri, K., and Suda, H. (1996) JP 08143569. U.S. Patent No. JP 08143569.
69 Socha, A.M., LaPlante, K.L., and Rowley, D.C. (2006) New bisanthraquinone antibiotics and semi-synthetic derivatives with potent activity against clinical *Staphylococcus aureus* and *Enterococcus faecium* isolates. *Bioorg. Med. Chem.*, **14**, 8446–8454.
70 Jensen, P.R., Williams, P.G., Oh, D.C., Zeigler, L., and Fenical, W. (2007) Species-specific secondary metabolite production in marine actinomycetes of the genus *Salinispora*. *Appl. Environ. Microbiol.*, **73**, 1146–1152.

71 Matsuda, S., Adachi, K., Matsuo, Y., Nukina, M., and Shizuri, Y. (2009) Salinisporamycin, a novel metabolite from *Salinispora arenicola*. *J. Antibiot.*, **62**, 519–526.

72 Asolkar, R.N., Kirkland, T.N., Jensen, P.R., and Fenical, W. (2010) Arenimycin, an antibiotic effective against rifampin- and methicillin-resistant *Staphylococcus aureus* from the marine actinomycete *Salinispora arenicola*. *J. Antibiot.*, **63**, 37–39.

73 Fiedler, H.P., Bruntner, C., Riedlinger, J., Bull, A.T., Knutsen, G., Goodfellow, M., Jones, A., Maldonado, L., Wasu, P., Winfried, B., Kathrin, S., Simone, K., and Sussmuth, R.D. (2008) Proximicin A, B and C, novel aminofuran antibiotic and anticancer compounds isolated from marine strains of the actinomycete *Verrucosispora*. *J. Antibiot.*, **61**, 158–163.

74 Bister, B., Bischoff, D., Stroebele, M., Riedlinger, J., Reicke, A., Wolter, F., Bull, A.T., Zaehner, H., Fiedler, H.P., and Sussmuth, R.D. (2004) Abyssomicin C: a polycyclic antibiotic from a marine *Verrucosispora* strain as an inhibitor of the *p*-aminobenzoic acid/tetrahydrofolate biosynthesis pathway. *Angew. Chem., Int. Ed.*, **43**, 2574–2576.

75 Engelhardt, K., Degnes, K.F., Kemmler, M., Bredholt, H., Fjaervik, E., Klinkenberg, G., Sletta, H., Ellingsen, T., and Zotchev, S.B. (2010) Production of a new thiopeptide antibiotic, TP-1161, by a marine *Nocardiopsis* species. *Appl. Environ. Microbiol.*, **76**, 4969–4976.

76 El-Gendy, M.M.A., Hawas, U.W., and Jaspars, M. (2008) Novel bioactive metabolites from a marine derived bacterium *Nocardia* sp ALAA 2000. *J. Antibiot.*, **61**, 379–386.

77 Macherla, V.R., Liu, J., Sunga, M., White, D.J., Grodberg, J., Teisan, S., Lam, K.S., and Potts, B.C.M. (2007) Lipoxazolidinones A, B, and C: antibacterial 4-oxazolidinones from a marine actinomycete isolated from a Guam marine sediment. *J. Nat. Prod.*, **70**, 1454–1457.

78 Sakoulas, G., Nam, S.-J., Loesgen, S., Fenical, W., Jensen, P.R., Nizet, V., and Hensler, M. (2012) Novel bacterial metabolite merochlorin A demonstrates *in vitro* activity against multi-drug resistant methicillin-resistant *Staphylococcus aureus*. *PLoS One*, **7**, e29439.

79 Radjasa, O.K., Martens, T., Grossart, H., Brinkhoff, T., Sabdono, A., and Simon, M. (2007) Antagonistic activity of a marine bacterium *Pseudoalteromonas luteoviolacea* TAB4.2 associated with coral *Acropora* sp. *J. Biol. Sci.*, **7**, 239–246.

80 Longeon, A., Peduzzi, J., Barthelemy, M., Corre, S., Nicolas, J.L., and Guyot, M. (2004) Purification and partial identification of novel antimicrobial protein from marine bacterium *Pseudoalteromonas* species strain X153. *Mar. Biotechnol.*, **6**, 633–641.

81 Hentschel, U., Schmid, M., Wagner, M., Fieseler, L., Gernert, C., and Hacker, J. (2001) Isolation and phylogenetic analysis of bacteria with antimicrobial activities from the Mediterranean sponges *Aplysina aerophoba* and *Aplysina cavernicola*. *FEMS Microbiol. Ecol.*, **35**, 305–312.

82 Bowman, J.P. (2007) Bioactive compound synthetic capacity and ecological significance of marine bacterial genus *Pseudoalteromonas*. *Mar. Drugs*, **5**, 220–241.

83 Holmström, C. and Kjelleberg, S. (1999) Marine *Pseudoalteromonas* species are associated with higher organisms and produce biologically active extracellular agents. *FEMS Microbiol. Ecol.*, **30**, 285–293.

84 Egan, S., James, S., Holmström, C., and Kjelleberg, S. (2002) Correlation between pigmentation and antifouling compounds produced by *Pseudoalteromonas tunicata*. *Environ. Microbiol.*, **4**, 433–442.

85 Rudd, B.A. and Hopwood, D.A. (1980) A pigmented mycelial antibiotic in *Streptomyces coelicolor*: control by a chromosomal gene cluster. *J. Gen. Microbiol.*, **119**, 333–340.

86 Durán, N. and Menck, C.F. (2001) *Chromobacterium violaceum*: a review of pharmacological and industrial perspectives. *Crit. Rev. Microbiol.*, **27**, 201–222.

87 Long, R.A., Qureshi, A., Faulkner, D.J., and Azam, F. (2003) 2-*n*-Pentyl-4-quinolinol produced by a marine *Alteromonas* sp. and its potential ecological and biogeochemical roles. *Appl. Environ. Microbiol.*, **69**, 568–576.

88 Burkholder, P.R., Pfister, R.M., and Leitz, F.H. (1966) Production of a pyrrole antibiotic by a marine bacterium. *Appl. Environ. Microbiol.*, **14**, 649–653.

89 Fehér, D., Barlow, R., McAtee, J., and Hemscheidt, T.K. (2010) Highly brominated antimicrobial metabolites from a marine *Pseudoalteromonas* sp. *J. Nat. Prod.*, **73**, 1963–1966.

90 Andersen, R.J., Wolfe, M.S., and Faulkner, D.J. (1974) Autotoxic antibiotic production by a marine *Chromobacterium*. *Mar. Biol.*, **27**, 281–285.

91 Vynne, N.G., Månsson, M., and Gram, L. (2012) Gene sequence based clustering assists in dereplication of *Pseudoalteromonas luteoviolacea* strains with identical inhibitory activity and antibiotic production. *Mar. Drugs*, **10**, 1729–1740.

92 Månsson, M., Phipps, R.K., Gram, L., Munro, M.H., Larsen, T.O., and Nielsen, K.F. (2010) Explorative solid-phase extraction (E-SPE) for accelerated microbial natural product discovery. *J. Nat. Prod.*, **73**, 1126–1132.

93 McCarthy, S.A., Johnson, R.M., Kakimoto, D., and Sakata, T. (1985) Effects of various agents on the pigment (violacein) and antibiotic production of *Alteromonas luteoviolacea*. *Nippon Suisan Gakkai. Shi.*, **51**, 1115–1121.

94 Zhang, X. and Enomoto, K. (2011) Characterization of a gene cluster and its putative promoter region for violacein biosynthesis in *Pseudoalteromonas* sp. 520P1. *Appl. Microbiol. Biotechnol.*, **90**, 1963–1971.

95 Thomas, T., Evans, F.F., Schleheck, D., Mai-Prochnow, A., Burke, C., Penesyan, A., Dalisay, D.S., Stelzer-Braid, S., Saunders, N., Johnson, J., Ferriera, S., Kjelleberg, S., and Egan, S. (2008) Analysis of the *Pseudoalteromonas tunicata* genome reveals properties of a surface-associated life style in the marine environment. *PLoS One*, **3**, e3252.

96 Matz, C., Deines, P., Boenigk, J., Arndt, H., Eberl, L., Kjelleberg, S., and Jurgens, K. (2004) Impact of violacein-producing bacteria on survival and feeding of bacterivorous nanoflagellates. *Appl. Environ. Microbiol.*, **70**, 1590–1593.

97 Matz, C., Webb, J.S., Schupp, P.J., Phang, S.Y., Penesyan, A., Egan, S., Steinberg, P., and Kjelleberg, S. (2008) Marine biofilm bacteria evade eukaryotic predation by targeted chemical defense. *PLoS One*, **3**, e2744.

98 Yoshikawa, K., Nakayama, Y., Hayashi, M., Unemoto, T., and Mochida, K. (1999) Korormicin, an antibiotic specific for Gram-negative marine bacteria, strongly inhibits the respiratory chain-linked Na^+-translocating NADH: quinone reductase from the marine *Vibrio alginolyticus*. *J. Antibiot.*, **52**, 182–185.

99 Gerber, N.N. and Gauthier, M.J. (1979) New prodigiosin-like pigment from *Alteromonas rubra*. *Appl. Environ. Microbiol.*, **37**, 1176–1179.

100 Fehér, D., Barlow, R.S., Lorenzo, P.S., and Hemscheidt, T.K. (2008) A 2-substituted prodiginine, 2-(*p*-hydroxybenzyl) prodigiosin, from *Pseudoalteromonas rubra*. *J. Nat. Prod.*, **71**, 1970–1972.

101 Gauthier, M.J. and Flatau, G.N. (1976) Antibacterial activity of marine violet-pigmented *Alteromonas* with special reference to the production of brominated compounds. *Can. J. Microbiol.*, **22**, 1612–1619.

102 Yu, M., Wang, J., Tang, K., Shi, X., Wang, S., Zhu, W.-M., and Zhang, X.-H. (2012) Purification and characterization of antibacterial compounds of *Pseudoalteromonas flavipulchra* JG1. *Microbiology*, **158**, 835–842.

103 Tebben, J., Tapiolas, D.M., Motti, C.A., Abrego, D., Negri, A.P., Blackall, L.L., Steinberg, P.D., and Harder, T. (2011) Induction of larval metamorphosis of the coral *Acropora millepora* by tetrabromopyrrole isolated from a *Pseudoalteromonas* bacterium. *PLoS One*, **6**, e19082.

104 Jiang, Z., Boyd, K.G., Mearns-Spragg, A., Adams, D.R., Wright, P.C., and Burgess, J.G. (2000) Two diketopiperazines and one halogenated phenol from cultures of the marine bacterium, *Pseudoalteromonas luteoviolacea*. *Nat. Prod. Lett.*, **14**, 435–440.

105 Fehér, D., Barlow, R., McAtee, J., and Hemscheidt, T.K. (2010) Highly brominated antimicrobial metabolites

from a marine *Pseudoalteromonas* sp. *J. Nat. Prod.*, **73**, 1963–1966.

106 Isnansetyo, A. and Kamei, Y. (2003) *Pseudoalteromonas phenolica* sp. nov., a novel marine bacterium that produces phenolic anti-methicillin-resistant *Staphylococcus aureus* substances. *Int. J. Syst. Evol. Microbiol.*, **53**, 583–588.

107 Shiozawa, H., Kagasaki, T., Kinoshita, T., Haruyama, H., Domon, H., Utsui, Y., Kodama, K., and Takahashi, S. (1993) Thiomarinol, a new hybrid antimicrobial antibiotic produced by a marine bacterium: fermentation, isolation, structure, and antimicrobial activity. *J. Antibiot.*, **46**, 1834–1842.

108 Shiozawa, H., Kagasaki, T., Torikata, A., Tanaka, N., Fujimoto, K., Hata, T., Furukawa, Y., and Takahashi, S. (1995) Thiomarinol-B and thiomarinol-C, new antimicrobial antibiotics produced by a marine bacterium. *J. Antibiot.*, **48**, 907–909.

109 Shiozawa, H., Shimada, A., and Takahashi, S. (1997) Thiomarinols D, E, F and G, new hybrid antimicrobial antibiotics produced by a marine bacterium: isolation, structure, and antimicrobial activity. *J. Antibiot.*, **50**, 449–452.

110 Thomas, C.M., Hothersall, J., Willis, C.L., and Simpson, T.J. (2010) Resistance to and synthesis of the antibiotic mupirocin. *Nat. Rev. Microbiol.*, **8**, 281–289.

111 Gurney, R. and Thomas, C.M. (2011) Mupirocin: biosynthesis, special features and applications of an antibiotic from a Gram-negative bacterium. *Appl. Microbiol. Biotechnol.*, **90**, 11–21.

112 Murphy, A.C., Fukuda, D., Song, Z., Hothersall, J., Cox, R.J., Willis, C.L., Thomas, C.M., and Simpson, T.J. (2011) Engineered thiomarinol antibiotics active against MRSA are generated by mutagenesis and mutasynthesis of *Pseudoalteromonas* SANK73390. *Angew. Chem., Int. Ed.*, **50**, 3271–3274.

113 Mai-Prochnow, A., Evans, F., Isay-Saludes, D., Stelzer, S., Egan, S., James, S., Webb, J.S., and Kjelleberg, S. (2004) Biofilm development and cell death in the marine bacterium *Pseudoalteromonas tunicata*. *Appl. Environ. Microbiol.*, **70**, 3232–3238.

114 Rao, D., Skovhus, T., Tujula, N., Holmström, C., Dahllöf, I., Webb, J.S., and Kjelleberg, S. (2010) Ability of *Pseudoalteromonas tunicata* to colonize natural biofilms and its effect on microbial community structure. *FEMS Microbiol. Ecol.*, **73**, 450–457.

115 Egan, S., Thomas, T., Holmström, C., and Kjelleberg, S. (2000) Phylogenetic relationship and antifouling activity of bacterial epiphytes from the marine alga *Ulva lactuca*. *Environ. Microbiol.*, **2**, 343–347.

116 Holmström, C., Egan, S., Franks, A., McCloy, S., and Kjelleberg, S. (2002) Antifouling activities expressed by marine surface associated *Pseudoalteromonas* species. *FEMS Microbiol. Ecol.*, **41**, 47–58.

117 Thompson, F.L., Iida, T., and Swings, J. (2004) Biodiversity of vibrios. *Microbiol. Mol. Biol. Rev.*, **68**, 403–431.

118 Wietz, M., Månsson, M., Gotfredsen, C.H., Larsen, T.O., and Gram, L. (2010) Antibacterial compounds from marine *Vibrionaceae* isolated on a global expedition. *Mar. Drugs*, **8**, 2946–2960.

119 Månsson, M., Gram, L., and Larsen, T.O. (2011) Production of bioactive secondary metabolites by marine *Vibrionaceae*. *Mar. Drugs*, **9**, 1440–1468.

120 Austin, B., Stuckey, L.F., Robertson, P.A. W., Effendi, I., and Griffith, D.R.W. (1995) A probiotic strain of *Vibrio alginolyticus* effective in reducing diseases caused by *Aeromonas salmonicida*, *Vibrio anguillarum* and *Vibrio ordalii*. *J. Fish Dis.*, **18**, 93–96.

121 Castro, D., Pujalte, M.J., Lopez-Cortes, L., Garay, E., and Borrego, J.J. (2002) Vibrios isolated from the cultured manila clam (*Ruditapes philippinarum*): numerical taxonomy and antibacterial activities. *J. Appl. Microbiol.*, **93**, 438–447.

122 Al-Zereini, W., Fotso Fondja Yao, C.B., Laatsch, H., and Anke, H. (2010) Aqabamycins A–G: novel nitro maleimides from a marine *Vibrio* species: I. Taxonomy, fermentation, isolation and biological activities. *J. Antibiot.*, **63**, 297–301.

123 Winkler, R. and Hertweck, C. (2007) Biosynthesis of nitro compounds. *ChemBioChem*, **8**, 973–977.

124 Starič, N., Danevčič, T., and Stopar, D. (2010) *Vibrio* sp. DSM 14379 pigment production: a competitive advantage in the environment? *Microb. Ecol.*, **60**, 592–598.

125 Imamura, N., Adachi, K., and Sano, H. (1994) Magnesidin A, a component of marine antibiotic magnesidin, produced by *Vibrio gazogenes* ATCC29988. *J. Antibiot.*, **47**, 257–261.

126 Celmer, W.D. and Solomons, I.A. (1955) The structures of thiolutin and aureothricin, antibiotics containing a unique pyrrolinonodithiole nucleus. *J. Am. Chem. Soc.*, **77**, 2861–2865.

127 Oliva, B., O'Neill, A., Wilson, J.M., O'Hanlon, P.J., and Chopra, I. (2001) Antimicrobial properties and mode of action of the pyrrothine holomycin. *Antimicrob. Agents Chemother.*, **45**, 532–539.

128 Wietz, M., Månsson, M., and Gram, L. (2011) Chitin stimulates production of the antibiotic andrimid in a *Vibrio coralliilyticus* strain. *Environ. Microbiol. Rep.*, **3**, 559–564.

129 Long, R.A., Rowley, D.C., Zamora, E., Liu, J., Bartlett, D.H., and Azam, F. (2005) Antagonistic interactions among marine bacteria impede the proliferation of *Vibrio cholerae*. *Appl. Environ. Microbiol.*, **71**, 8531–8536.

130 Jin, M., Fischbach, M.A., and Clardy, J. (2006) A biosynthetic gene cluster for the acetyl-CoA carboxylase inhibitor andrimid. *J. Am. Chem. Soc.*, **128**, 10660–10661.

131 Singh, M.P., Mroczenski-Wildey, M.J., Steinberg, D.A., Andersen, R.J., Maiese, W.M., and Greenstein, M. (1997) Biological activity and mechanistic studies of andrimid. *J. Antibiot.*, **50**, 270–273.

132 Freiberg, C., Brunner, N.A., Schiffer, G., Lampe, T., Pohlmann, J., Brands, M., Raabe, M., Häbich, D., and Ziegelbauer, K. (2004) Identification and characterization of the first class of potent bacterial acetyl-CoA carboxylase inhibitors with antibacterial activity. *J. Biol. Chem.*, **279**, 26066–26073.

133 Pohlmann, J., Lampe, T., Shimada, M., Nell, P.G., Pernerstorfer, J., Svenstrup, N., Brunner, N.A., Schiffer, G., and Freiberg, C. (2005) Pyrrolidinedione derivatives as antibacterial agents with a novel mode of action. *Bioorg. Med. Chem. Lett.*, **15**, 1189–1192.

134 Webster, N.S., Wilson, K.J., Blackall, L.L., and Hill, R.T. (2001) Phylogenetic diversity of bacteria associated with the marine sponge *Rhopaloeides odorabile*. *Appl. Environ. Microbiol.*, **67**, 434–444.

135 Leyton, Y. and Riquelme, C. (2010) Marine *Bacillus* spp. associated with the egg capsule of *Concholepas concholepas* (common name "loco") have an inhibitory activity toward the pathogen *Vibrio parahaemolyticus*. *Microb. Ecol.*, **60**, 599–605.

136 Santos, O.C.S., Pontes, P.V.M.L., Santos, J.F.M., Muricy, G., Giambiagi-deMarval, M., and Laport, M.S. (2010) Isolation, characterization and phylogeny of sponge-associated bacteria with antimicrobial activities from Brazil. *Res. Microbiol.*, **161**, 604–612.

137 Beleneva, I.A. (2008) Distribution and characteristics of *Bacillus* bacteria associated with hydrobionts and the waters of the Peter the Great Bay, Sea of Japan. *Mikrobiologiia*, **77**, 558–565.

138 Ivanova, E.P., Vysotskii, M.V., Svetashev, V.I., Nedashkovskaya, O.I., Gorshkova, N.M., Mikhailov, V.V., Yumoto, N., Shigeri, Y., Taguchi, T., and Yoshikawa, S. (1999) Characterization of *Bacillus* strains of marine origin. *Int. Microbiol.*, **2**, 267–271.

139 Barsby, T., Kelly, M.T., Gagné, S.M., and Andersen, R.J. (2001) Bogorol A produced in culture by a marine *Bacillus* sp. reveals a novel template for cationic peptide antibiotics. *Org. Lett.*, **3**, 437–440.

140 Barsby, T., Warabi, K., Sørensen, D., Zimmerman, W.T., Kelly, M.T., and Andersen, R.J. (2006) The Bogorol family of antibiotics: template-based structure elucidation and a new approach to positioning enantiomeric pairs of amino acids. *J. Org. Chem.*, **71**, 6031–6037.

141 Xue, C., Tian, L., Xu, M., Deng, Z., and Lin, W. (2008) A new 24-membered lactone and a new polyene delta-lactone from the marine bacterium *Bacillus marinus*. *J. Antibiot.*, **61**, 668–674.

142 Gao, C.-H., Tian, X.-P., Qi, S.-H., Luo, X.-M., Wang, P., and Zhang, S. (2010) Antibacterial and antilarval compounds from marine gorgonian-associated

bacterium *Bacillus amyloliquefaciens* SCSIO 00856. *J. Antibiot.*, **63**, 191–193.

143 Nagao, T., Adachi, K., Sakai, M., Nishijima, M., and Sano, H. (2001) Novel macrolactins as antibiotic lactones from a marine bacterium. *J. Antibiot.*, **54**, 333–339.

144 Lu, X.L., Xu, Q.Z., Shen, Y.H., Liu, X.Y., Jiao, B.H., Zhang, W.D., and Ni, K.Y. (2008) Macrolactin S, a novel macrolactin antibiotic from marine *Bacillus* sp. *Nat. Prod. Res.*, **22**, 342–347.

145 Tareq, F.S., Kim, J.H., Lee, M.A., Lee, H.-S., Lee, Y.-J., Lee, J.S., and Shin, H.J. (2012) Ieodoglucomides A and B from a marine-derived bacterium *Bacillus licheniformis*. *Org. Lett.*, **14**, 1464–1467.

146 Nagai, K., Kamigiri, K., Arao, N., Suzumura, K., Kawano, Y., Yamaoka, M., Zhang, H., Watanabe, M., and Suzuki, K. (2003) YM-266183 and YM-266184, novel thiopeptide antibiotics produced by *Bacillus cereus* isolated from a marine sponge. I. Taxonomy, fermentation, isolation, physico-chemical properties and biological properties. *J. Antibiot.*, **56**, 123–128.

147 Suzumura, K., Yokoi, T., Funatsu, M., Nagai, K., Tanaka, K., Zhang, H., and Suzuki, K. (2003) YM-266183 and YM-266184, novel thiopeptide antibiotics produced by *Bacillus cereus* isolated from a marine sponge. II. structure elucidation. *J. Antibiot.*, **56**, 129–134.

148 Shoji, J., Kato, T., Yoshimura, Y., and Tori, K. (1981) Structural studies on thiocillins I, II and III (studies on antibiotics from the genus *Bacillus* XXIX). *J. Antibiot.*, **34**, 1126–1136.

149 Desjardine, K., Pereira, A., Wright, H., Matainaho, T., Kelly, M., and Andersen, R. J. (2007) Tauramamide, a lipopeptide antibiotic produced in culture by *Brevibacillus laterosporus* isolated from a marine habitat: structure elucidation and synthesis. *J. Nat. Prod.*, **70**, 1850–1853.

150 Wagner-Döbler, I. and Biebl, H. (2006) Environmental biology of the marine *Roseobacter* lineage. *Annu. Rev. Microbiol.*, **60**, 255–280.

151 Wietz, M., Gram, L., Jørgensen, B., and Schramm, A. (2010) Latitudinal patterns in the abundance of major marine bacterioplankton groups. *Aquat. Microb. Ecol.*, **61**, 179–189.

152 Pommier, T., Pinhassi, J., and Hagström, Å. (2005) Biogeographic analysis of ribosomal RNA clusters from marine bacterioplankton. *Aquat. Microb. Ecol.*, **41**, 79–89.

153 Porsby, C.H., Nielsen, K.F., and Gram, L. (2008) *Phaeobacter* and *Ruegeria* species of the *Roseobacter* clade colonize separate niches in a Danish turbot (*Scophthalmus maximus*)-rearing farm and antagonize *Vibrio anguillarum* under different growth conditions. *Appl. Environ. Microbiol.*, **74**, 7356–7364.

154 Wagner-Döbler, I., Rheims, H., Felske, A., El-Ghezal, A., Flade-Schröder, D., Laatsch, H., Lang, S., Pukall, R., and Tindall, B.J. (2004) *Oceanibulbus indolifex* gen. nov., sp. nov., a North Sea alphaproteobacterium that produces bioactive metabolites. *Int. J. Syst. Evol. Microbiol.*, **54**, 1177–1184.

155 Martens, T., Gram, L., Grossart, H.P., Kessler, D., Müller, R., Simon, M., Wenzel, S.C., and Brinkhoff, T. (2007) Bacteria of the *Roseobacter* clade show potential for secondary metabolite production. *Microb. Ecol.*, **54**, 31–42.

156 Ruiz-Ponte, C., Cilia, V., Lambert, C., and Nicolas, J.L. (1998) *Roseobacter gallaeciensis* sp. nov., a new marine bacterium isolated from rearings and collectors of the scallop *Pecten maximus*. *Int. J. Syst. Bacteriol.*, **48**, 537–542.

157 Rao, D., Webb, J.S., and Kjelleberg, S. (2005) Competitive interactions in mixed-species biofilms containing the marine bacterium *Pseudoalteromonas tunicata*. *Appl. Environ. Microbiol.*, **71**, 1729–1736.

158 Geng, H., Bruhn, J.B., Nielsen, K.F., Gram, L., and Belas, R. (2008) Genetic dissection of tropodithietic acid biosynthesis by marine roseobacters. *Appl. Environ. Microbiol.*, **74**, 1535–1545.

159 Penesyan, A., Tebben, J., Lee, M., Thomas, T., Kjelleberg, S., Harder, T., and Egan, S. (2011) Identification of the antibacterial compound produced by the marine epiphytic bacterium *Pseudovibrio* sp. D323 and related sponge-associated bacteria. *Mar. Drugs*, **9**, 1391–1402.

160 Brinkhoff, T., Bach, G., Heidorn, T., Liang, L., Schlingloff, A., and Simon, M. (2004) Antibiotic production by a *Roseobacter* clade-affiliated species from the German Wadden Sea and its antagonistic effects on indigenous isolates. *Appl. Environ. Microbiol.*, **70**, 2560.

161 Planas, M., Pérez-Lorenzo, M., Hjelm, M., Gram, L., Uglenes Fiksdal, I., Bergh, Ø., and Pintado, J. (2006) Probiotic effect *in vivo* of *Roseobacter* strain 27-4 against *Vibrio* (*Listonella*) *anguillarum* infections in turbot (*Scophthalmus maximus* L.) larvae. *Aquaculture*, **255**, 323–333.

162 Fjellheim, A.J., Klinkenberg, G., Skjermo, J., Aasen, I.M., and Vadstein, O. (2010) Selection of candidate probionts by two different screening strategies from Atlantic cod (*Gadus morhua* L.) larvae. *Vet. Microbiol.*, **144**, 153–159.

163 Porsby, C.H., Webber, M.A., Nielsen, K.F., Piddock, L.J.V., and Gram, L. (2011) Resistance and tolerance to tropodithietic acid, an antimicrobial in aquaculture, is hard to select. *Antimicrob. Agents Chemother.*, **55**, 1332–1337.

164 Fdhila, F., Vázquez, V., Sánchez, J.L., and Riguera, R. (2003) dd-diketopiperazines: antibiotics active against *Vibrio anguillarum* isolated from marine bacteria associated with cultures of *Pecten maximus*. *J. Nat. Prod.*, **66**, 1299–1301.

165 Riguera Vega, R., Sanchez Lopez, J.L., and Ben Mohamed Fdhila, F. (2004) Novel antibiotics against *Vibrio anguillarum* and the applications thereof in cultures of fish, crustaceans, molluscs and other aquaculture activities. U.S. Patent No. PCT/ES2003/000325.

166 Slightom, R.N. and Buchan, A. (2009) Surface colonization by marine roseobacters: integrating genotype and phenotype. *Appl. Environ. Microbiol.*, **75**, 6027–6037.

167 Newton, R.J., Griffin, L.E., Bowles, K.M., Meile, C., Gifford, S., Givens, C.E., Howard, E.C., King, E., Oakley, C.A., Reisch, C.R., Rinta-Kanto, J.M., Sharma, S., Sun, S., Varaljay, V., Vila-Costa, M., Westrich, J.R., and Moran, M.A. (2010) Genome characteristics of a generalist marine bacterial lineage. *ISME J.*, **4**, 784–798.

168 Isnansetyo, A. and Kamei, Y. (2009) Bioactive substances produced by marine isolates of *Pseudomonas*. *J. Ind. Microbiol. Biotechnol.*, **36**, 1239–1248.

169 Wratten, S.J., Wolfe, M.S., Andersen, R.J., and Faulkner, D.J. (1977) Antibiotic metabolites from a marine pseudomonad. *Antimicrob. Agents Chemother.*, **11**, 411–414.

170 Isnansetyo, A., Horikawa, M., and Kamei, Y. (2001) *In vitro* anti-methicillin-resistant *Staphylococcus aureus* activity of 2,4-diacetylphloroglucinol produced by *Pseudomonas* sp. AMSN isolated from a marine alga. *J. Antimicrob. Chemother.*, **47**, 724–725.

171 Shanahan, P., O'Sullivan, D.J., Simpson, P., Glennon, J.D., and O'Gara, F. (1992) Isolation of 2,4-diacetylphloroglucinol from a fluorescent pseudomonad and investigation of physiological parameters influencing its production. *Appl. Environ. Microbiol.*, **58**, 353–358.

172 Isnansetyo, A., Cui, L., Hiramatsu, K., and Kamei, Y. (2003) Antibacterial activity of 2,4-diacetylphloroglucinol produced by *Pseudomonas* sp. AMSN isolated from a marine alga, against vancomycin-resistant *Staphylococcus aureus*. *Int. J. Antimicrob. Agents*, **22**, 545–547.

173 Uzair, B., Ahmed, N., Ahmad, V.U., Mohammad, F.V., and Edwards, D.H. (2008) The isolation, purification and biological activity of a novel antibacterial compound produced by *Pseudomonas stutzeri*. *FEMS Microbiol. Lett.*, **279**, 243–250.

174 Ahmed, N., Uzair, B., Ahmad, V.U., and Kousar, F. (2008) Antibiotic Bushrin. U.S. Patent No. 2008/0090900 A1.

175 Nunnery, J.K., Mevers, E., and Gerwick, W.H. (2010) Biologically active secondary metabolites from marine cyanobacteria. *Curr. Opin. Biotechnol.*, **21**, 787–793.

176 Martins, R.F., Ramos, M.F., Herfindal, L., Sousa, J.A., Skaerven, K., and Vasconcelos, V.M. (2008) Antimicrobial and cytotoxic assessment of marine cyanobacteria: *Synechocystis* and *Synechococcus*. *Mar. Drugs*, **6**, 1–11.

177 Ehrenreich, I.M., Waterbury, J.B., and Webb, E.A. (2005) Distribution and diversity of natural product genes in marine and freshwater cyanobacterial cultures and genomes. *Appl. Environ. Microbiol.*, **71**, 7401–7413.

178 Li, B., Sher, D., Kelly, L., Shi, Y., Huang, K., Knerr, P.J., Joewono, I., Rusch, D., Chisholm, S.W., and van derDonk, W.A. (2010) Catalytic promiscuity in the biosynthesis of cyclic peptide secondary metabolites in planktonic marine cyanobacteria. *Proc. Natl. Acad. Sci. USA*, **107**, 10430–10435.

179 Cardellina, J.H., Moore, R.E., Arnold, E.V., and Clardy, J. (1979) Structure and absolute configuration of malyngolide, an antibiotic from the marine blue-green alga *Lyngbya majuscula* Gomont. *J. Org. Chem.*, **44**, 4039–4042.

180 Gerwick, W.H., Jiang, Z.D., Agarwal, S.K., and Farmer, B.T. (1992) Total structure of hormothamnin A, a toxic cyclic undecapeptide from the tropical marine cyanobacterium *Hormothamnion enteromorphoides*. *Tetrahedron*, **48**, 2313–2324.

181 Singh, R.K., Tiwari, S.P., Rai, A.K., and Mohapatra, T.M. (2011) Cyanobacteria: an emerging source for drug discovery. *J. Antibiot.*, **64**, 401–412.

182 Kirchman, D.L. (2002) The ecology of *Cytophaga-Flavobacteria* in aquatic environments. *FEMS Microbiol. Ecol.*, **39**, 91–100.

183 Bauer, M., Kube, M., Teeling, H., Richter, M., Lombardot, T., Allers, E., Würdemann, C.A., Quast, C., Kuhl, H., Knaust, F., Woebken, D., Bischof, K., Mussmann, M., Choudhuri, J.V., Meyer, F., Reinhardt, R., Amann, R.I., and Glöckner, F.O. (2006) Whole genome analysis of the marine Bacteroidetes '*Gramella forsetii*' reveals adaptations to degradation of polymeric organic matter. *Environ. Microbiol.*, **8**, 2201–2213.

184 Wagner-Döbler, I., Beil, W., Lang, S., Meiners, M., and Laatsch, H. (2002) Integrated approach to explore the potential of marine microorganisms for the production of bioactive metabolites. *Adv. Biochem. Eng. Biotechnol.*, **74**, 207–238.

185 Oku, N., Adachi, K., Matsuda, S., Kasai, H., Takatsuki, A., and Shizuri, Y. (2008) Ariakemicins A and B, novel polyketide–peptide antibiotics from a marine gliding bacterium of the genus *Rapidithrix*. *Org. Lett.*, **10**, 2481–2484.

186 Katayama, N., Nozaki, Y., Okonogi, K., Ono, H., Harada, S., and Okazaki, H. (1985) Formadicins, new monocyclic beta-lactam antibiotics of bacterial origin: I. Taxonomy, fermentation and biological activities. *J. Antibiot.*, **38**, 1117–1127.

187 Shoji, J., Hinoo, H., Matsumoto, K., Hattori, T., Yoshida, T., Matsuura, S., and Kondo, E. (1988) Isolation and characterization of katanosins A and B. *J. Antibiot.*, **41**, 713–718.

188 Katayama, N., Fukusumi, S., Funabashi, Y., Iwahi, T., and Ono, H. (1993) TAN-1057 A–D, new antibiotics with potent antibacterial activity against methicillin-resistant *Staphylococcus aureus*: taxonomy, fermentation and biological activity. *J. Antibiot.*, **46**, 606–613.

189 Kamigiri, K., Tokunaga, T., Sugawara, T., Nagai, K., Shibazaki, M., Setiawan, B., Rantiatmodjo, R.M., Morioka, M., and Suzuki, K. (1997) YM-32890 A and B, new types of macrolide antibiotics produced by *Cytophaga* sp. *J. Antibiot.*, **50**, 556–561.

190 Zhang, Y., Mu, J., Gu, X., Zhao, C., Wang, X., and Xie, Z. (2009) A marine sulfate-reducing bacterium producing multiple antibiotics: biological and chemical investigation. *Mar. Drugs*, **7**, 341–354.

191 Blunt, J.W., Copp, B.R., Keyzers, R.A., Munro, M.H.G., and Prinsep, M.R. (2012) Marine natural products. *Nat. Prod. Rep.*, **29**, 144–222.

192 Li, J.W.-H. and Vederas, J.C. (2009) Drug discovery and natural products: end of an era or an endless frontier? *Science*, **325**, 161–165.

193 Haste, N.M., Perera, V.R., Maloney, K.N., Tran, D.N., Jensen, P., Fenical, W., Nizet, V., and Hensler, M.E. (2010) Activity of the streptogramin antibiotic etamycin against methicillin-resistant *Staphylococcus aureus*. *J. Antibiot.*, **63**, 219–224.

194 Kock, I., Maskey, R.P., Biabani, M.A.F., Helmke, E., and Laatsch, H. (2005) 1-Hydroxy-1-norresistomycin and resistoflavin methyl ether: new antibiotics

from marine-derived streptomycetes. *J. Antibiot.*, **58**, 530–534.

195 Kang, H., Jensen, P.R., and Fenical, W. (1996) Isolation of microbial antibiotics from a marine ascidian of the genus *Didemnum*. *J.Org. Chem.*, **61**, 1543–1546.

196 Fiedler, H.P., Bruntner, C., Bull, A.T., Ward, A.C., Goodfellow, M., Potterat, O., Puder, C., and Mihm, G. (2005) Marine actinomycetes as a source of novel secondary metabolites. *Antonie Van Leeuwenhoek*, **87**, 37–42.

197 Kim, T.K., Hewavitharana, A.K., Shaw, P.N., and Fuerst, J.A. (2006) Discovery of a new source of rifamycin antibiotics in marine sponge Actinobacteria by phylogenetic prediction. *Appl. Environ. Microbiol.*, **72**, 2118–2125.

198 Fischbach, M.A. (2009) Antibiotics from microbes: converging to kill. *Curr. Opin. Microbiol.*, **12**, 520–527.

199 Månsson, M. (2011) Discovery of bioactive natural products from marine bacteria, PhD Thesis. Technical University of Denmark, Kgs. Lyngby, Denmark.

200 Penn, K., Jenkins, C., Nett, M., Udwary, D.W., Gontang, E.A., McGlinchey, R.P., Foster, B., Lapidus, A., Podell, S., Allen, E.E., Moore, B.S., and Jensen, P.R. (2009) Genomic islands link secondary metabolism to functional adaptation in marine Actinobacteria. *ISME J.*, **3**, 1193–1203.

201 Liras, P. and Martín, J.F. (2006) Gene clusters for beta-lactam antibiotics and control of their expression: why have clusters evolved, and from where did they originate? *Int. Microbiol.*, **9**, 9–19.

202 McDaniel, L.D., Young, E., Delaney, J., Ruhnau, F., Ritchie, K.B., and Paul, J.H. (2010) High frequency of horizontal gene transfer in the oceans. *Science*, **330**, 50.

203 Oclarit, J.M., Okada, H., Ohta, S., Kaminura, K., Yamaoka, Y., Iizuka, T., Miyashiro, S., and Ikegami, S. (1994) Anti-bacillus substance in the marine sponge, *Hyatella* species, produced by an associated *Vibrio* species bacterium. *Microbios*, **78**, 7–16.

204 Needham, J., Kelly, M.T., Ishige, M., and Andersen, R.J. (1994) Andrimid and moiramides A–C, metabolites produced in culture by a marine isolate of the bacterium *Pseudomonas fluorescens*: structure elucidation and biosynthesis. *J. Org. Chem.*, **59**, 2058–2063.

205 Fredenhagen, A., Tamura, S.Y., Kenny, P.T. M., Komura, H., Naya, Y., Nakanishi, K., Nishiyama, K., Sugiura, M., and Kita, H. (1987) Andrimid, a new peptide antibiotic produced by an intracellular bacterial symbiont isolated from a brown planthopper. *J. Am. Chem. Soc.*, **109**, 4409–4411.

206 Malfatti, F. and Azam, F. (2009) Atomic force microscopy reveals microscale networks and possible symbioses among pelagic marine bacteria. *Aquat. Microb. Ecol.*, **58**, 1–14.

207 Colling, A. (2001) Ocean circulation, in *Ocean Circulation* (ed. G. Bearman), 2nd edn, Butterworth-Heinemann, p. 208–209.

208 Grossart, H.-P., Dziallas, C., Leunert, F., and Tang, K.W. (2010) Bacteria dispersal by hitchhiking on zooplankton. *Proc. Natl. Acad. Sci. USA*, **107**, 11959–11964.

209 Zarubin, M., Belkin, S., Ionescu, M., and Genin, A. (2012) Bacterial bioluminescence as a lure for marine zooplankton and fish. *Proc. Natl. Acad. Sci. USA*, **109**, 853–857.

210 Tang, K.W., Turk, V., and Grossart, H.-P. (2010) Linkage between crustacean zooplankton and aquatic bacteria. *Aquat. Microb. Ecol.*, **61**, 261–277.

211 Grossart, H.-P. (2010) Ecological consequences of bacterioplankton lifestyles: changes in concepts are needed. *Environ. Microbiol. Rep.*, **2**, 706–714.

212 Hazen, T.H., Pan, L., Gu, J.-D., and Sobecky, P.A. (2010) The contribution of mobile genetic elements to the evolution and ecology of Vibrios. *FEMS Microbiol. Ecol.*, **74**, 485–499.

213 Ito, T., Odake, T., Katoh, H., Yamaguchi, Y., and Aoki, M. (2011) High-throughput profiling of microbial extracts. *J. Nat. Prod.*, **74**, 983–988.

214 Bugni, T.S., Harper, M.K., McCulloch, M.W.B., Reppart, J., and Ireland, C.M. (2008) Fractionated marine invertebrate extract libraries for drug discovery. *Molecules*, **13**, 1372–1383.

215 Bruns, A., Cypionka, H., and Overmann, J. (2002) Cyclic AMP and acyl homoserine lactones increase the cultivation efficiency

of heterotrophic bacteria from the central Baltic Sea. *Appl. Environ. Microbiol.*, **68**, 3978–3987.

216 Rappé, M.S. and Giovannoni, S.J. (2003) The uncultured microbial majority. *Annu. Rev. Microbiol.*, **57**, 369–394.

217 Kaeberlein, T., Lewis, K., and Epstein, S.S. (2002) Isolating "uncultivable" microorganisms in pure culture in a simulated natural environment. *Science*, **296**, 1127–1129.

218 Staley, J.T. and Konopka, A. (1985) Measurement of *in situ* activities of nonphotosynthetic microorganisms in aquatic and terrestrial habitats. *Annu. Rev. Microbiol.*, **39**, 321–346.

219 Amann, R.I., Ludwig, W., and Schleifer, K.H. (1995) Phylogenetic identification and *in situ* detection of individual microbial cells without cultivation. *Microbiol. Rev.*, **59**, 143–169.

220 Pernice, M., Pichon, D., Domart-Coulon, I., Favet, J., and Boucher-Rodoni, R. (2006) Primary co-culture as a complementary approach to explore the diversity of bacterial associations in marine invertebrates: the example of *Nautilus macromphalus* (Cephalopoda: Nautiloidea). *Mar. Biol.*, **150**, 749–757.

221 Selvin, J., Joseph, S., Asha, K.R.T., Manjusha, W.A., Sangeetha, V.S., Jayaseema, D.M., Antony, M.C., and Denslin Vinitha, A.J. (2004) Antibacterial potential of antagonistic *Streptomyces* sp. isolated from marine sponge *Dendrilla nigra*. *FEMS Microbiol. Ecol.*, **50**, 117–122.

222 Olson, J., Lord, C., and McCarthy, P. (2000) Improved recoverability of microbial colonies from marine sponge samples. *Microb. Ecol.*, **40**, 139–147.

223 Hameş-Kocabaş, E.E. and Uzel, A. (2012) Isolation strategies of marine-derived actinomycetes from sponge and sediment samples. *J. Microbiol. Methods*, **88**, 342–347.

224 Imada, C., Hotta, K., and Okami, Y. (1998) A novel marine *Bacillus* with multiple amino acid analog resistance and selenomethionine-dependent antibiotic productivity. *J. Mar. Biotechnol.*, **6**, 189–192.

225 Zengler, K., Toledo, G., Rappe, M., Elkins, J., Mathur, E.J., Short, J.M., and Keller, M. (2002) Cultivating the uncultured. *Proc. Natl. Acad. Sci. USA*, **99**, 15681–15686.

226 Connon, S.A. and Giovannoni, S.J. (2002) High-throughput methods for culturing microorganisms in very-low-nutrient media yield diverse new marine isolates. *Appl. Environ. Microbiol.*, **68**, 3878–3885.

227 Bruns, A., Hoffelner, H., and Overmann, J. (2003) A novel approach for high throughput cultivation assays and the isolation of planktonic bacteria. *FEMS Microbiol. Ecol.*, **45**, 161–171.

228 Wenzel, S.C. and Müller, R. (2009) The biosynthetic potential of myxobacteria and their impact in drug discovery. *Curr. Opin. Drug Discov. Devel.*, **12**, 220–230.

229 Schäberle, T.F., Goralski, E., Neu, E., Erol, O., Hölzl, G., Dörmann, P., Bierbaum, G., and König, G.M. (2010) Marine myxobacteria as a source of antibiotics: comparison of physiology, polyketide-type genes and antibiotic production of three new isolates of *Enhygromyxa salina*. *Mar. Drugs*, **8**, 2466–2479.

230 Fudou, R., Iizuka, T., and Yamanaka, S. (2001) Haliangicin, a novel antifungal metabolite produced by a marine myxobacterium: 1. Fermentation and biological characteristics. *J. Antibiot.*, **54**, 149–152.

231 Clardy, J. and Walsh, C. (2004) Lessons from natural molecules. *Nature*, **432**, 829–837.

232 Li, S., Tian, X., Niu, S., Zhang, W., Chen, Y., Zhang, H., Yang, X., Zhang, W., Li, W., Zhang, S., Ju, J., and Zhang, C. (2011) Pseudonocardians A–C, new diazaanthraquinone derivatives from a deep-sea actinomycete *Pseudonocardia* sp. SCSIO 01299. *Mar. Drugs*, **9**, 1428–1439.

233 Zhu, Q., Li, J., Ma, J., Luo, M., Wang, B., Huang, H., Tian, X., Li, W., Zhang, S., Zhang, C., and Ju, J. (2012) Discovery and engineered overproduction of antimicrobial nucleoside antibiotic A201A from the deep-sea marine actinomycete *Marinactinospora thermotolerans* SCSIO 00652. *Antimicrob. Agents Chemother.*, **56**, 110–114.

234 Fuhrman, J.A., Steele, J.A., Hewson, I., Schwalbach, M.S., Brown, M.V., Green, J.

L., and Brown, J.H. (2008) A latitudinal diversity gradient in planktonic marine bacteria. *Proc. Natl. Acad. Sci. USA*, **105**, 7774–7778.

235 Mangano, S., Michaud, L., Caruso, C., Brilli, M., Bruni, V., Fani, R., and Lo Giudice, A. (2009) Antagonistic interactions between psychrotrophic cultivable bacteria isolated from Antarctic sponges: a preliminary analysis. *Res. Microbiol.*, **160**, 27–37.

236 Zhao, J., Yang, N., and Zeng, R. (2008) Phylogenetic analysis of type I polyketide synthase and nonribosomal peptide synthetase genes in Antarctic sediment. *Extremophiles*, **12**, 97–105.

237 Biondi, N., Tredici, M.R., Taton, A., Wilmotte, A., Hodgson, D.A., Losi, D., and Marinelli, F. (2008) Cyanobacteria from benthic mats of Antarctic lakes as a source of new bioactivities. *J. Appl. Microbiol.*, **105**, 105–115.

238 Jayatilake, G.S., Thornton, M.P., Leonard, A.C., Grimwade, J.E., and Baker, B.J. (1996) Metabolites from an Antarctic sponge-associated bacterium, *Pseudomonas aeruginosa*. *J. Nat. Prod.*, **59**, 293–296.

239 Asthana, R.K., Tripathi, M.K., Srivastava, A., Singh, A.P., Singh, S.P., Nath, G., Srivastava, R., and Srivastava, B.S. (2008) Isolation and identification of a new antibacterial entity from the Antarctic cyanobacterium *Nostoc* CCC 537. *J. Appl. Phycol.*, **21**, 81–88.

240 Bruntner, C., Binder, T., Pathom-aree, W., Goodfellow, M., Bull, A.T., Potterat, O., Puder, C., Hörer, S., Schmid, A., Bolek, W., Wagner, K., Mihm, G., and Fiedler, H.-P. (2005) Frigocyclinone, a novel angucyclinone antibiotic produced by a *Streptomyces griseus* strain from Antarctica. *J. Antibiot.*, **58**, 346–349.

241 Knight, V., Sanglier, J.J., DiTullio, D., Braccili, S., Bonner, P., Waters, J., Hughes, D., and Zhang, L. (2003) Diversifying microbial natural products for drug discovery. *Appl. Microbiol. Biotechnol.*, **62**, 446–458.

242 Das, P., Mukherjee, S., and Sen, R. (2009) Substrate dependent production of extracellular biosurfactant by a marine bacterium. *Bioresour. Technol.*, **100**, 1015–1019.

243 Sanchez, S., Chavez, A., Forero, A., Garcia-Huante, Y., Romero, A., Sanchez, M., Rocha, D., Sanchez, B., Avalos, M., Guzman-Trampe, S., Rodriguez-Sanoja, R., Langley, E., and Ruiz, B. (2010) Carbon source regulation of antibiotic production. *J. Antibiot.*, **63**, 442–459.

244 Demain, A.L., Aharonowitz, Y., and Martin, J.-F. (1983) Metabolic control of secondary biosynthetic pathways. In *Biochemistry and Genetic Regulation of Commercially Important Antibiotics*. Vining, L.C. (ed.). Reading, MA, USA: Addison-Wesley, pp. 49–72.

245 Sujatha, P., Bapi Raju, K.V.V.S.N., and Ramana, T. (2005) Studies on a new marine streptomycete BT-408 producing polyketide antibiotic SBR-22 effective against methicillin resistant *Staphylococcus aureus*. *Microbiol. Res.*, **160**, 119–126.

246 Seghal Kiran, G., Anto Thomas, T., Selvin, J., Sabarathnam, B., and Lipton, A.P. (2010) Optimization and characterization of a new lipopeptide biosurfactant produced by marine *Brevibacterium aureum* MSA13 in solid state culture. *Bioresour. Technol.*, **101**, 2389–2396.

247 Saurav, K. and Kannabiran, K. (2010) Diversity and optimization of process parameters for the growth of *Streptomyces* VITSVK9 spp isolated from Bay of Bengal, India. *J. Nat. Environ. Sci.*, **1**, 56–65.

248 Newman, D.J. and Cragg, G.M. (2007) Natural products as sources of new drugs over the last 25 years. *J. Nat. Prod.*, **70**, 461–477.

249 Bull, A.T. (2010) The renaissance of continuous culture in the post-genomics age. *J. Ind. Microbiol. Biotechnol.*, **37**, 993–1021.

250 Bode, H.B., Bethe, B., Höfs, R., and Zeeck, A. (2002) Big effects from small changes: possible ways to explore nature's chemical diversity. *ChemBioChem*, **3**, 619–627.

251 Mitra, S., Sarkar, S., Gachhui, R., and Mukherjee, J. (2011) A novel conico-cylindrical flask aids easy identification of critical process parameters for cultivation of marine bacteria. *Appl. Microbiol. Biotechnol.*, **90**, 321–330.

252 Caicedo, N.H., Heyduck-Söller, B., Fischer, U., and Thöming, J. (2010) Bioproduction of antimicrobial compounds by using

marine filamentous cyanobacterium cultivation. *J. Appl. Phycol.*, **23**, 811–818.
253 Hibbing, M.E., Fuqua, C., Parsek, M.R., and Peterson, S.B. (2009) Bacterial competition: surviving and thriving in the microbial jungle. *Nat. Rev. Microbiol.*, **8**, 15–25.
254 Sarkar, S., Saha, M., Roy, D., Jaisankar, P., Das, S., Roy, L.G., Gachhui, R., Sen, T., and Mukherjee, J. (2008) Enhanced production of antimicrobial compounds by three salt-tolerant actinobacterial strains isolated from the Sundarbans in a niche-mimic bioreactor. *Mar. Biotechnol.*, **10**, 518–526.
255 Yan, L., Boyd, K.G., and Burgess, J. (2002) Surface attachment induced production of antimicrobial compounds by marine epiphytic bacteria using modified roller bottle cultivation. *Mar. Biotechnol.*, **4**, 356–366.
256 Marwick, J., Wright, P., and Burgess, J. (1999) Bioprocess intensification for production of novel marine bacterial antibiotics through bioreactor operation and design. *Mar. Biotechnol.*, **1**, 495–507.
257 Palková, Z. (2004) Multicellular microorganisms: laboratory versus nature. *EMBO Rep.*, **5**, 470–476.
258 Doull, J.L. and Vining, L.C. (1990) Nutritional control of actinorhodin production by *Streptomyces coelicolor* A3(2): suppressive effects of nitrogen and phosphate. *Appl. Microbiol. Biotechnol.*, **32**, 449–454.
259 deCarvalho, C.C.C.R. and Fernandes, P. (2010) Production of metabolites as bacterial responses to the marine environment. *Mar. Drugs*, **8**, 705–727.
260 Okazaki, T., Kitahara, T., and Okami, Y. (1975) Studies on marine microorganisms: IV. A new antibiotic SS-228 Y produced by *Chainia* isolated from shallow sea mud. *J. Antibiot. (Tokyo)*, **28**, 176–184.
261 Sharifah, E.N. and Eguchi, M. (2011) The phytoplankton *Nannochloropsis oculata* enhances the ability of *Roseobacter* clade bacteria to inhibit the growth of fish pathogen *Vibrio anguillarum*. *PLoS One*, **6**, e26756.
262 Gooday, G.W. (1990) The ecology of chitin degradation. *Adv. Microb. Ecol.*, **11**, 387–430.
263 Angell, S., Bench, B.J., Williams, H., and Watanabe, C.M.H. (2006) Pyocyanin isolated from a marine microbial population: synergistic production between two distinct bacterial species and mode of action. *Chem. Biol.*, **13**, 1349–1359.
264 Sher, D., Thompson, J.W., Kashtan, N., Croal, L., and Chisholm, S.W. (2011) Response of *Prochlorococcus* ecotypes to co-culture with diverse marine bacteria. *ISME J.*, **5**, 1125–1132.
265 Pettit, G.R., Herald, C.L., Doubek, D.L., Herald, D.L., Arnold, E., and Clardy, J. (1982) Isolation and structure of bryostatin 1. *J. Am. Chem. Soc.*, **104**, 6846–6848.
266 Mitova, M.I., Lang, G., Wiese, J., and Imhoff, J.F. (2008) Subinhibitory concentrations of antibiotics induce phenazine production in a marine *Streptomyces* sp. *J. Nat. Prod.*, **71**, 824–827.
267 Guo, X.C., Zheng, L., Zhou, W.H., Cui, Z.S., Han, P., Tian, L., and Wang, X.R. (2011) A case study on chemical defense based on quorum sensing: antibacterial activity of sponge-associated bacterium *Pseudoalteromonas* sp NJ6-3-1 induced by quorum sensing mechanisms. *Ann. Microbiol.*, **61**, 247–255.
268 Trischman, J.A., Oeffner, R.E., deLuna, M.G., and Kazaoka, M. (2004) Competitive induction and enhancement of indole and a diketopiperazine in marine bacteria. *Mar. Biotechnol.*, **6**, 215–220.
269 Slattery, M., Rajbhandari, I., and Wesson, K. (2001) Competition-mediated antibiotic induction in the marine bacterium Streptomyces tenjimariensis. *Microb. Ecol.*, **41**, 90–96.
270 Mearns-Spragg, A., Bregu, M., Boyd, K.G., and Burgess, J.G. (1998) Cross-species induction and enhancement of antimicrobial activity produced by epibiotic bacteria from marine algae and invertebrates, after exposure to terrestrial bacteria. *Lett. Appl. Microbiol.*, **27**, 142–146.
271 Dusane, D.H., Matkar, P., Venugopalan, V. P., Kumar, A.R., and Zinjarde, S.S. (2011) Cross-species induction of antimicrobial compounds, biosurfactants and quorum-sensing inhibitors in tropical marine epibiotic bacteria by pathogens and biofouling microorganisms. *Curr. Microbiol.*, **62**, 974–980.
272 Nyadong, L., Hohenstein, E.G., Galhena, A., Lane, A.L., Kubanek, J.,

Sherrill, C.D., and Fernández, F.M. (2009) Reactive desorption electrospray ionization mass spectrometry (DESI-MS) of natural products of a marine alga. *Anal. Bioanal. Chem.*, **394**, 245–254.

273 Nyadong, L., Galhena, A.S., and Fernández, F.M. (2009) Desorption electrospray/metastable-induced ionization: a flexible multimode ambient ion generation technique. *Anal. Chem.*, **81**, 7788–7794.

274 Esquenazi, E., Coates, C., Simmons, L., Gonzalez, D., Gerwick, W.H., and Dorrestein, P.C. (2008) Visualizing the spatial distribution of secondary metabolites produced by marine cyanobacteria and sponges via MALDI-TOF imaging. *Mol. Biosyst.*, **4**, 562–570.

275 Lane, A.L., Nyadong, L., Galhena, A.S., Shearer, T.L., Stout, E.P., Parry, R.M., Kwasnik, M., Wang, M.D., Hay, M.E., Fernandez, F.M., and Kubanek, J. (2009) Desorption electrospray ionization mass spectrometry reveals surface-mediated antifungal chemical defense of a tropical seaweed. *Proc. Natl. Acad. Sci. USA*, **106**, 7314–7319.

276 Gonzalez, D.J., Haste, N.M., Hollands, A., Fleming, T.C., Hamby, M., Pogliano, K., Nizet, V., and Dorrestein, P.C. (2011) Microbial competition between *Bacillus subtilis* and *Staphylococcus aureus* monitored by imaging mass spectrometry. *Microbiology*, **157**, 2485–2492.

277 Rusch, D.B., Halpern, A.L., Sutton, G., Heidelberg, K.B., Williamson, S., Yooseph, S., Wu, D., Eisen, J.A., Hoffman, J.M., Remington, K., Beeson, K., Tran, B., Smith, H., Baden-Tillson, H., Stewart, C., Thorpe, J., Freeman, J., Andrews-Pfannkoch, C., Venter, J.E., Li, K., Kravitz, S., Heidelberg, J.F., Utterback, T., Rogers, Y.-H., Falcón, L.I., Souza, V., Bonilla-Rosso, G., Eguiarte, L.E., Karl, D.M., Sathyendranath, S., Platt, T., Bermingham, E., Gallardo, V., Tamayo-Castillo, G., Ferrari, M.R., Strausberg, R.L., Nealson, K., Friedman, R., Frazier, M., and Venter, J.C. (2007) The *Sorcerer II* Global Ocean Sampling expedition: northwest Atlantic through eastern tropical Pacific. *PLoS Biol.*, **5**, e77.

278 Yooseph, S., Sutton, G., Rusch, D.B., Halpern, A.L., Williamson, S.J., Remington, K., Eisen, J.A., Heidelberg, K.B., Manning, G., and Li, W. (2007) The *Sorcerer II* Global Ocean Sampling expedition: expanding the universe of protein families. *PLoS Biol.*, **5**, e16.

279 Lorenz, P. and Eck, J. (2005) Metagenomics and industrial applications. *Nat. Rev. Microbiol.*, **3**, 510–516.

280 Woyke, T., Xie, G., Copeland, A., González, J.M., Han, C., Kiss, H., Saw, J.H., Senin, P., Yang, C., Chatterji, S., Cheng, J.-F., Eisen, J.A., Sieracki, M.E., and Stepanauskas, R. (2009) Assembling the marine metagenome, one cell at a time. *PLoS One*, **4**, e5299.

281 Siegl, A. and Hentschel, U. (2009) PKS and NRPS gene clusters from microbial symbiont cells of marine sponges by whole genome amplification. *Environ. Microbiol. Rep.*, **2**, 507–513.

282 Foerstner, K.U., Doerks, T., Creevey, C.J., Doerks, A., and Bork, P. (2008) A computational screen for type I polyketide synthases in metagenomics shotgun data. *PLoS One*, **3**, e3515.

283 Miao, V. and Davies, J. (2008) Metagenomics and antibiotic discovery from uncultivated bacteria, in *Uncultivated Microorganisms* (ed. S. Epstein), Springer, p. 217–236.

284 Ginolhac, A., Jarrin, C., Gillet, B., Robe, P., Pujic, P., Tuphile, K., Bertrand, H., Vogel, T.M., Perrière, G., Simonet, P., and Nalin, R. (2004) Phylogenetic analysis of polyketide synthase I domains from soil metagenomic libraries allows selection of promising clones. *Appl. Environ. Microbiol.*, **70**, 5522–5527.

285 Sudek, S., Haygood, M.G., Youssef, D.T.A., and Schmidt, E.W. (2006) Structure of trichamide, a cyclic peptide from the bloom-forming cyanobacterium *Trichodesmium erythraeum*, predicted from the genome sequence. *Appl. Environ. Microbiol.*, **72**, 4382–4387.

286 Teta, R., Gurgui, M., Helfrich, E.J.N., Künne, S., Schneider, A., VanEchten-Deckert, G., Mangoni, A., and Piel, J. (2010) Genome mining reveals *trans*-AT polyketide synthase directed antibiotic biosynthesis in the bacterial phylum

Bacteroidetes. *ChemBioChem*, **11**, 2506–2512.

287 Glöckner, F.O., Kube, M., Bauer, M., Teeling, H., Lombardot, T., Ludwig, W., Gade, D., Beck, A., Borzym, K., Heitmann, K., Rabus, R., Schlesner, H., Amann, R., and Reinhardt, R. (2003) Complete genome sequence of the marine planctomycete *Pirellula* sp. strain 1. *Proc. Natl. Acad. Sci. USA*, **100**, 8298–8303.

288 Labutti, K., Sikorski, J., Schneider, S., Nolan, M., Lucas, S., Glavina Del Rio, T., Tice, H., Cheng, J.-F., Goodwin, L., Pitluck, S., Liolios, K., Ivanova, N., Mavromatis, K., Mikhailova, N., Pati, A., Chen, A., Palaniappan, K., Land, M., Hauser, L., Chang, Y.-J., Jeffries, C.D., Tindall, B.J., Rohde, M., Göker, M., Woyke, T., Bristow, J., Eisen, J.A., Markowitz, V., Hugenholtz, P., Kyrpides, N.C., Klenk, H.-P., and Lapidus, A. (2010) Complete genome sequence of *Planctomyces limnophilus* type strain (Mü 290). *Stand. Genomic Sci.*, **3**, 47–56.

9
Marine Bacteriophages for the Biocontrol of Fish and Shellfish Diseases

Mahanama De Zoysa

9.1
Introduction

Global aquaculture of marine fish and shellfish has intensified in recent decades together with the various types of infectious diseases. Particularly, marine species under natural and commercial aquaculture are challenged by serious bacterial diseases such as vibriosis, edwardsiellosis, streptococcosis, lactococcosis, photobacteriosis, furunculosis, and flexibacteriosis, resulting in huge economic losses [1,2]. Farmers are continuously applying different strategies including vaccines, probiotics, antibiotics, immune stimulants, and other methods (hormones, vitamins, etc.) for preventing (prophylactic) and treating (therapeutic) these bacterial diseases [3]. Among them application of antibiotics is the most popular and widely used therapeutic method in aquaculture. Selection of antibiotics is mainly due to their higher therapeutic power against a broad spectrum of bacteria. A limited number of approved antibiotics such as oxytetracycline, sulfonamides, fluoroquinolones, and florfenicol are available in aquaculture [4]. However, several scientific studies highlighted that antibiotics have not been applied in a responsible manner in aquaculture practices. The misuse of antibiotics in aquaculture has led to several critical issues such as development of new drug-resistant bacteria, contamination of antibiotic-resistant genes in natural environment, degradation of marine ecosystem, residual accumulation in edible seafood, and thereby significant risk to human health [5]. One example of this situation is emergence of new drug-resistant strains of *Lactococcus garvieae* in yellowtail (*Seriola quinqueradiata*) fish farms where antibiotics have been applied frequently [6]. Also, many pathogenic *Streptococcus* strains for yellowtail are resistant to antibiotics such as lincomycin, tetracycline, and chloramphenicol [7]. Due to adverse effects, many countries have implemented strict regulation on the use of antibiotics such as chloramphenicol, furazolidone, and sulfonamide in hatcheries and aquaculture farms [3]. At present, people have understood that the paramount goal for fish and shellfish aquaculture should be based on safe production in a sustainable manner. Antimicrobial

Marine Microbiology: Bioactive Compounds and Biotechnological Applications, First Edition.
Edited by Se-Kwon Kim

peptides (AMPs) and vaccines are popular and promising alternatives for antibiotics. However, there are some limitations of those methods such as the difficulty in administering vaccines at the larval stage of fish. Administration of vaccines by injection is stressful, time-consuming, and requires more work force [2]. Also, shellfish species are invertebrate animals and do not have adaptive immune components to produce antibodies to develop disease resistance after vaccination. Therefore, it is important to implement biocontrol-based approach for treating bacterial diseases in fish and shellfish.

Bacterial viruses that infect bacteria are called as bacteriophages or "phages." The practice of phage therapy dates back almost a century. In 1915, British bacteriologist Frederick Twort discovered a small organism that can infect and kill the bacteria [8]. Later in 1919, the first therapeutic use of phage was conducted in human subjects [9]. Then, phages have been successfully used in various fields including food industry, medicine, agriculture, and so on [10,11]. Since 2006, Food and Drug Administration (FDA) and United States Department of Agriculture (USDA) have approved several phage products as food additives such as ListShield™, EcoShield™, and SalmoFresh™ [12,13]. In human medicine, it has been attempted to control skin, urinary tract, intestinal infections, and dysentery [14]. After scientific verification, it has now been considered that phage treatment is a safe and effective method to control pathogenic bacteria on a large-scale basis without serious drawbacks like in antibiotics. Hence, application of phage therapy strategy has a great potential for biocontrolling pathogens of bacterial infections in fish and shellfish as well as in farm operations such as cleaning of sewage. Several studies have already confirmed that bacterial infections in fish and shellfish can be successfully controlled by phages [15,16]. Even though phages are available largely in marine ecosystems, a limited number of marine phages have been identified and utilized under biocontrol applications in marine fish. Also, there are very few studies related to phage therapy on bacterial diseases of shellfish such as mollusk and crustaceans, which are very important groups under marine aquaculture systems. The aim of this chapter is to summarize the recent phage therapy applied to control or reduce bacterial diseases in marine fish and shellfish. Furthermore, it discusses the feasibility, future potentials, and limitations of phage treatments as biocontrol approach with integration of biotechnology and genetic engineering applications.

9.2 Mode of Action of Phages

Similar to other viruses, phages lack their own metabolism and carry their genetic information in the form of either DNA or RNA. Phages synthesize their nucleic acid and protein coats by sharing the energy and resources from bacterial host [17]. There are two types of phages, namely, lytic (virulent) and lysogenic (temperate), based on their life cycle in the bacterial host. Both types of phages first attach (adsorption) to the outer membranes of host bacteria, and then insert the nucleic acid. Only the lytic phage can replicate inside the host cell. Newly synthesized progeny phages are involved in the lysis of the host cell to cause bactericidal effect. This mode of action of lytic phage is beneficial for killing bacteria and thereby controlling bacterial diseases. On the other hand, nucleic acid of the lysogenic phages is integrated at the specific section of bacterial chromosome and thereby duplicates during every normal cell division. This lysogenic phage is not involved in the infection of host bacterial cells and is usually not important as an antibacterial agent [18]. The potential advantages of lytic phages result from their unique replication strategy in bacterial cells. Lytic phages replicate until the target bacteria is controlled and it eliminates from the body when the host is absent [19]. The burst size is defined as the number of new phages produced during one infection cycle; it varies and is generally between 50 and 200 [19]. Phage titer may increase during bacterial infection, and therefore determination of initial dose is not very important [16]. Also, a single phage is sufficient to destroy a specific host bacterium [20]. Selection of specific lytic phages is important to develop phage therapy application. Phage adsorption to a receptor on bacteria cell surface is the first stage in virus–host interaction and is the key factor for phage–host specificity. Rakhuba *et al.* [21] have described the factors associated with phage–host interactions and mechanisms of phage adsorption and penetration into host bacteria. Most phages consist of tails with specific receptors that bind carbohydrate, protein, and lipopolysaccharide molecules on the surface of bacterial host. The nature of phage receptors such as chemical composition, structure spatial configuration and structure of the receptor binding proteins play an important role in phage-host interactions. Loss of or changes in receptor molecules of host bacteria can lead to the development of phage-resistant bacteria. Adsorption, replication, lytic activity, and survival of the phage depend on the metabolic status of the host. It has been observed that maximum yield of phages is obtained under optimum growth conditions of the host. Also, low nutrient availability often results in increased latent period and reduced burst size. Phage therapy can be very effective under certain conditions and shows unique advantages against antibiotics, and hence phages can be considered as natural antibiotics (Table 9.1) [22–24]. Based on the listed advantages of phages over antibiotics, phage therapy has received attention in the recent past as a biocontrol method of bacterial diseases in fish and shellfish.

9.3 Diversity of Marine Phages

Widely spread phages occur in every corner of the biosphere, their main habitats are being ocean and topsoil. Sir Macfarlane Burnet was the first person to classify phages in 1937 [25]. In 1971, the International Committee on Taxonomy of Viruses (ICTV) issued the first report with six phage genera [26]; now it has been extended to 6 orders, 87 families, 19 subfamilies, and 348 genera [12]. Marine phages are the most diverse form of biological entities in the oceans due to the largest oceanic

Table 9.1 Comparison between phage therapy and antibiotic treatment.

Phages	Antibiotics	Comments	References
Less side and toxic effects	Multiple side effects	Only minor side effects have been reported from bacteria lysed by the phage	[22]
Bacteria that resist to a phage may have susceptible to another phage	Resistance to specific antibiotic is not limited to target bacteria	Antibiotics are broad spectrum and develop many resistant bacteria species	[20]
If target bacteria become phage resistant, other bacteria may not develop resistant Rate of phage resistance development is low	Increase the frequency of antibiotic resistance	Rate of resistance development to phage is one-tenth of the resistance to antibiotics	[23]
Selection and development of a new phage is a rapid process that can be completed in several months	Developing antibiotics especially for antibiotic-resistant bacteria may takes several years	Active phages can be selected against antibiotic-resistant or phage-resistant bacterium by the process of natural selection	[4]
Narrow host range (only affect the target bacteria)	Target both pathogenic and nonpathogenic bacteria species and damaging the natural balance of microorganisms (broad host range)	High specificity of phages in disease outbreaks needs correct identification of the bacteria before treating the phage therapy	[22]
Replicate at the site of infection	Not concentrated at the site of infection	Phage administration is not required frequently	[20]
Phages can undergo some mutations and overcome the bacterial mutation to resistance	No inherent ability of chemicals to overcome the problems of resistance	Phages can respond to bacteria resistance and modify themselves to overcome the resistance	[24]

reservoirs [27]. Approximately there are 10 million viruses per milliliter on the surface of seawater. According to Sulakvelidze [28] around 70% of marine bacteria may be infected by phages. Although it has been largely neglected during the last few decades, marine phages were first introduced and described in 1955 [29]. They have been isolated from various sources including natural seawater, disease-infected fish or shellfish tissues, water collected from hatcheries, and so on [30]. Upon realizing their importance, researchers have emphasized further studies on marine phages in basic (carbon cycling, gene transfer, etc.) and applied research (phage therapy) areas of aquaculture. Analysis of phage genomes is essential to gain insight into genetic composition and molecular structures of phages, which ensures the safety of phage therapy in applications. The first marine phage genome that was completely sequenced was *Pseudoalteromonas espejiana* BAL-31 phi PM2 [31]. Also, complete genome sequences from several other marine phages such as *Vibrio parahaemolyticus* and *Vibrio harveyi* are available and nearly one-third of sequenced marine phages are classified under vibriophages [18]. The genome size of the marine phages ranges from 39 to 243 kb, while family Podoviridae is the most abundant group among them (genome size is 39–60 kb). Myoviridae phages often have a broader host range than other tailed phages such as podoviruses. Moreover, it has been reported that the large genome-sized viruses are the least abundant group (280–500 kb) in marine ecosystem [32]. However, genomic information is significantly less compared with the predicted number and diversity of phages from marine resources. According to Paul and Sullivan [18], there are three main areas that should be considered in marine phage studies: (i) genome sequencing of important phage families such as Leviviridae or Microviridae; (ii) proper understanding about lysogeny; and (iii) experiments on phage isolation, expression, and protein analysis. Recent studies on metagenomics of uncultured viral population open novel insights into the ecology of environmental bacteriophages, which will be important while screening potential phage candidates for phage therapy [33]. At present genome sequencing of phages is much easier, rapid, and low cost with next-generation sequencing approach due to small genome size of phages. In future, more marine phage genome sequencing will supply wealth of useful information on potential phages for biocontrol of pathogenic bacteria related to marine animals. Also, screening of phage genome can identify the putative antibacterial genes like lysine and holin that are lethal to bacteria and can be used as templates for producing recombinant protein or synthetic analogues. Pharmaceutical industry can utilize such genetic information to develop antibacterial products. Also, phage genome analysis has led to the discovery of toxic genes, and avoiding the use of such phages in therapy can minimize the side effects. Finally, it indicates that marine resources have huge potential in isolation of phages but it is essential to select safe phage candidates for applying biocontrol of bacterial infections in fish and shellfish.

9.4 Application of Marine Phages to Control Fish and Shellfish Diseases

The use of phages for controlling fish- and shellfish-associated bacteria pathogens has been investigated by several research groups. However, based on literature review, we evidenced that a considerable number of studies (in fish and shellfish) are mainly focused on phage classification, characterization, and survival rate of host in laboratory-level or small-scale trials in farms [34,35]. More studies are conducted on phage therapy in fish [16,30] than that in shrimps [36] and mollusks such as abalone [37] and oyster [38]. Also, the availability of commercial phage-based products for marine animals is very limited. Oliveira *et al.* [30] described the efficiency of phage treatment that varies with many factors such as purity of phage isolate, administration (method or routes, dose, time), host specificity, host defense responses (immune response), environmental conditions (salinity, temperature, acidity of water), development of phage-resistant bacteria, bacterial residuals (debris) after treatment, and safety (availability of toxic genes). These factors should be well investigated and optimized before the establishment of therapy in aquaculture. This section summarizes the recent phage treatments that have been applied to control selected diseases such as vibriosis, edwardsiellosis, streptococcosis, and lactococcosis in marine fish and shellfish species.

Vibriosis infections occur through Gram-negative bacteria in the family of Vibrionaceae, genus *Vibrio*. They are typically marine and brackish water organisms and one of the most common bacteria in surface waters. Some *Vibrio* species have freshwater habitats under association with aquatic animals; however, vibriosis is occasionally reported in freshwater fish. Also, some *Vibrio* species (*V. cholera*, *V. parahaemolyticus*, and *V. vulnificus*) are known to cause diseases such as diarrhea, gastroenteritis, and wound infections in human following the consumption of contaminated shellfish [39]. Several species of *Vibrio* such as *V. anguillarum*, *V. alginolyticus*, *V. parahaemolyticus*, *V. ordalli*, *V. damsela*, *V. carchariae*, *V. vulnificus*, *V. salmonicid*, and *V. harveyi* are associated with disease infections in fish and shellfish [40,41]. *Vibrio* species-related phages have been isolated from marine culturing systems such as infected fish, farm water, and natural seawater. More than 10 phages with lytic activity against *V. harveyi* (caused luminous vibriosis) have been isolated and characterized [35,36,42–48]. Most of the phages that used the *V. harveyi* control are classified under Siphoviridae and Myoviridae. Vinod *et al.* [42] described higher survival of *Penaeus monodon* larvae by phage treatments in both laboratory and hatchery trials compared with untreated control and antibiotic-treated groups. Five phages against *V. parahaemolyticus* have been isolated from shrimp tissues and ponds [49]. One of the phages (Myoviridae) named as Vp1 has shown lysis of eight strains of *V. parahaemolyticus* (V1, V3–V6, V9, V11, and V12). Application of phage cocktail can increase the efficiency of phage therapy by controlling a wide range of host bacteria and eight phages have been used to develop phage cocktail for controlling phyllosoma disease caused by *V. harveyi* in rock lobster *Panulirus ornatus* [48]. Based on the current trials, we can assume that further isolation of *Vibrio*-based phages has excellent potential to control pathogenic *Vibrio* species in fish and shellfish aquaculture systems. Moreover, application of phage cocktails for biosanitation of aquaculture farms can minimize the spreading of *Vibrio* infections into human and other animals.

Edwardsiellosis caused by *Edwardsiella tarda* is septicemic disease affecting host species of fish, reptiles, and mammals including humans [50]. Several studies have reported the successful isolation of *E. tarda* phages aiming to control edwardsiellosis. One of the earliest attempts was made by Wu and Chao in 1982 and the results showed that 25 of 27 *E. tarda* strains were controlled by *E. tarda* phage (φET-1) *in vitro* [51]. Application of phage φET-1 has been tested (*in vivo*) by exposing the loach (*Misgurnus anguillicaudatus*) to phage φET-1 inoculated with *E. tarda* at varying time intervals. Higher survival rate (90%) has been observed when the fish was challenged to pathogenic *E. tarda* after 8 h of φET-1 treatment, suggesting that protection is greater when there is a longer gap between phage treatment and initial infection. Kwon *et al.* [52] have generated the *E. tarda* ghost using phage (PhiX174) lysis gene E to improve the efficiency of vaccination against edwardsiellosis. The killing of both *E. tarda* and *Escherichia coli* strains has occurred at 2 h and 30 min, respectively, after controlled induction of E gene indicating that *E. tarda* has more resistance for lysis protein made by gene E. However, the results showed the same killing activity for both bacteria by PhiX174 lysis gene E. These data indicate that phage lysis genes have the potential to control pathogenic bacteria. *E. tada* phage typing has been done from diseased eel and collected water [53]. Despite the origin from freshwater environment, it may have effective control on other marine *E. tarda* strains and ultimately result in a broader host range.

Streptococcosis has become a major disease in cultured marine and freshwater fish worldwide. It is caused by different strains of *Streptococcus* such as *Streptococcus iniae* and *Streptococcus agalactiae* [54]. Several *S. iniae* phages classified under Siphoviridae have been investigated for their therapeutic power against *S. iniae* strains in Japanese flounder *Paralichthys olivaceus* [55]. Intraperitoneal injection (10^8 PFU/fish) of *S. iniae*-based phages has shown significantly lower mortalities in *S. iniae*-injected (10^5–10^7 CFU/ml) fish than in control fish; however, phage-resistant *S. iniae* strains were noticed in dead animals in treated group. In order to control the phage-resistant strains, screening and development of highly multivalent phages or testing with different phage cocktail are required.

L. garvieae (previously *S. garvieae*) is the etiological agent of lactococcosis, a major disease that has been reported in marine and freshwater aquaculture. It causes higher mortality in yellowtail *S. quinqueradiata*; application of frequent chemotherapy has developed the resistant strains of *L. garvieae* [6]. Several phages, namely, PLgY [56], PLgY-16, PLgY-30, and PLgW-1 [6], have been isolated from *L. garvieae*-infected yellowtail cultures. Higher survival rate in yellowtail has resulted after phage administration by intraperitoneal as well as phage-impregnated feed. In marine aquaculture, phage persistence under environment fluctuations is important to maintain long-term protection against the pathogenic bacteria. In this regard, *L. garvieae* phages have shown strong tolerance to water temperature (5–37 °C) at high and low salinity. Although most of the phage treatment has shown effective control of pathogenic bacteria, in some phages it is ineffective particularly in the frunculosis disease caused by *Aeromonas salmonicida* in Atlantic salmon [57] and brook trout [58]. Among the shellfish species, most of the shrimp vibriosis has been effectively controlled by phage therapy [36]. In mollusk, phage therapy has not been applied widely indicating that there is a great potential to apply this environment-friendly biocontrol disease control concept.

9.5
Potentials and Limitations of Phage Therapy in Marine Fish and Shellfish

Successful phage therapy data collected from different studies have opened a new window for applying next-generation phage therapy concepts in aquaculture, particularly in fish and shellfish. Phage administration into fish and shellfish is possible with feed (oral), injection or immersion into water. In general, susceptibility of fish and shellfish for bacterial infection can occur at any stage of their life cycle (egg, larval, juvenile, and adult), and potential of combined administration of phage is a great advantage. However, each administration method has shown specific advantages and disadvantages in previous studies [5,15,16].

Virolysins are phage-encoded lytic enzymes (lysins or endolysins) produced at the late stage of lytic cycle, which are critical for bacterial lysis [59]. Courchesne *et al.* [60] summarized the production and potential application of phage-encoded lysins for veterinary and human medicine. Several patents have been obtained for virolysin products that can be used to treat human and other animal bacterial infections caused by *Staphylococcus aureus* [61] and *Salmonella typhimurium* [62]. Application of phage-based virolysin is alternative to whole-phage therapy and it can prevent the emergence of phage-resistant bacteria as well as antiphage responses from host immune system. Moreover, use of lysine can act immediately on bacterial peptidoglycan for direct degradation after administration. The research on marine phage-based virolysins has great potential for the development of antibacterial agents at a commercial level. Also, a recent report described that filamentous phages can be used as a delivery vector for immunogenetic peptides for development of vaccines with multiple benefits such as high immunogenicity, low production cost, and stable phage preparations [63]. Since there are many fish vaccines available for preventing bacterial diseases, identified marine phages can be tested to increase the effectiveness of those vaccines.

Phages have been applied with antibiotics as cotherapy to prevent the emergence of bacterial resistance to antibiotics as well as phages [20]. Therefore, phage–antibiotic cotherapy is a new area in marine aquaculture where research should be focused. The application of phages as stand-alone therapy and cotherapeutic agent (adjuvant) shows the unlimited potential areas to promote the minimum usage of antibiotics and prevent the emergence of resistant species in marine system.

Phage therapy also has several drawbacks. Hence there is a need to overcome those by manipulating treatment conditions and modern biotechnological applications. One of the main limitations of phage therapy is that it has relatively narrow host range for lysis of bacteria. In general, marine environment has a complex of pathogenic bacteria communities such as *Vibrios*, and the narrow host range of phages is a critical factor for effective control of such opportunistic *Vibrio* species. Possible ways to overcome this issue is by developing multivalent phages or phage cocktails as suggested by Imbeault *et al.* [58]. Establishment of marine phage library from different sources can be a sound foundation that ensures one of the phages will have a higher chance to lyse targeted bacteria. On the other hand, narrow host specificity gives advantage of minimum effect on the natural host.

Intracellular bacteria show the capacity to survive and replicate inside the phagocyte and sometimes invade epithelial cells [64]. *E. tarda* is considered as such intracellular bacteria with type III recreation system that is essential for their intracellular replication [65]. Several studies revealed that phages are unable to reach the intracellular bacteria such as *E. tarda* that localize in phagocytes [24]. This limitation can be overcome to some extent by administration of phage before bacterial infection, as described by Nakai [16]. However, further research is required to develop new strategies with phage application for controlling intracellular bacteria infecting marine animals. Furthermore, bacteria residue after phage lysis is accumulated in body fluid and tissues. Also, proper phage preparation can include endotoxin and debris into phage solution. As a result of these conditions, some inflammatory reactions can cause fatality to host animals. However, it can be minimized to some extent by using advance centrifugation and purification steps during the phage preparation.

Before application on a large scale, it is essential to investigate the lytic cycle of virulent phages extensively for safety measures. Most of the identified phages from marine resources have not been investigated for their genomes and transcriptional profiles, and therefore direct use of such phages may involve a high risk while transferring toxic genes to other organisms. Due to technical (high specificity, resistance from host immune system, etc.) and nontechnical reasons (regulatory approvals, patent protection, market acceptance), phage products are not widely developed on a commercial scale or applied in marine aquaculture. However, with the safety assurance and approval given by FDA to certain phage products, next-generation phage therapy may accelerate the development of commercial phage formulations that can be used in marine sectors. This approach can be easier due to the availability of large numbers of marine phages resources.

Acknowledgment

This study was financially supported by a grant from the National Research Foundation (NRF) program of Ministry of Education, Science and Technology, Republic of Korea (NRF-2011-0022671).

References

1 Toranzo, A.E., Magarinos, B., and Romalde, J.L. (2005) A review of the main bacterial fish diseases in mariculture systems. *Aquaculture*, **246**, 37–61.

2 Pereira, C., Silva, Y.J., Santos, A.L., Cunha, A., Gomes, N.C., and Almeida, A. (2011) Bacteriophages with potential for inactivation of fish pathogenic bacteria: survival, host specificity and effect on bacterial community structure. *Mar. Drugs*, **9**, 2236–2255.

3 Chanu, T.I., Sharma, A., Roy, S.D., Chaudhuri, A.K., and Biswas, P. (2012) Herbal biomedicine – an alternative to synthetic chemicals in aquaculture feed in Asia. *World Aquacult. Mag.*, **43**, 3.

4 Morrison, S. and Rainnie, D.J. (2004) Bacteriophage therapy: an alternative to antibiotic therapy in aquaculture? Canadian Technical Report of Fisheries and Aquatic Sciences, 2532.

5 Cabello, F.C. (2006) Heavy use of prophylactic antibiotics in aquaculture: a growing problem for human and animal health and for the environment. *Environ. Microbiol.*, **8**, 1137–1144.

6 Nakai, T., Sugimoto, R., Park, K.H., Matsuoka, S., Mori, K., Nishioka, T. *et al.* (1999) Protective effects of bacteriophage on experimental *Lactococcus garvieae* infection in yellowtail. *Dis. Aquat. Organ.*, **37**, 33–41.

7 Aoki, T., Takami, K., and Kitao, T. (1990) Drug resistance in a non-hemolytic *Streptococcus* sp. isolated from cultured yellowtail *Seriola quinqueradiata*. *Dis. Aquat. Organ.*, **8**, 171–177.

8 D'Herelle, F. (1917) Sur un microbe invisible antagoniste des bac. Dysenteriques. *Crit. Rev. Acad. Sci. Paris*, **165**, 373.

9 Topley, W. and Wilson, G.S. (1936) *The Principles of Bacteriology and Immunity*, 2nd edn, Edward Arnold.

10 McGrath, S., Fitzgerald, G.F., and vanSinderen, D. (2004) The impact of bacteriophage genomics. *Curr. Opin. Biotechnol.*, **15**, 94–99.

11 Lu, T.K. and Koeris, M.S. (2011) The next generation of bacteriophage therapy. *Curr. Opin. Microbiol.*, **14**, 524–531.

12 http://en.wikipedia.org/wiki/Bacteriophage (accessed July 24, 2012).

13 http://www.intralytix.com/ (accessed July 24, 2012).

14 Sulakvelidze, A. and Kutter, E. (2005) Bacteriophage therapy in humans, in *Bacteriophages: Biology and Applications* (eds E. Kutter and A. Sulakvelidze), CRC Press, Boca Raton, FL, pp. 381–436.

15 Park, S.C., Shimamura, I., Fukunaga, M., Mori, K.I., and Nakai, T. (2000) Isolation of bacteriophages specific to a fish pathogen, *Pseudomonas plecoglossicida*, as a candidate for disease control. *Appl. Environ. Microbiol.*, **66**, 1416–1422.

16 Nakai, T. (2010) Application of bacteriophages for control of infectious diseases in aquaculture, in *Bacteriophages in the Control of Food and Waterborne Pathogens* (eds P.M. Sabour and M.W. Griffiths), American Society for Microbiology Press, Washington, pp. 257–272.

17 Ackermann, H.W. and Dubow, M.S. (1987) Viruses of prokaryotes I: general properties of bacteriophages (Chapter 7), in *Practical Applications of Bacteriophages*, CRC Press, Boca Raton, FL.

18 Paul, J.H. and Sullivan, M.B. (2005) Marine phage genomics: what have we learned? *Curr. Opin. Biotechnol.*, **16**, 299–307.

19 Marks, T. and Sharp, R. (2000) Bacteriophages and biotechnology: a review. *J. Chem. Technol. Biotechnol.*, **75**, 6–17.

20 Carlton, R.M. (1999) Phage therapy: past history and future prospects. *Arch. Immunol. Ther. Exp.*, **47**, 267–274.

21 Rakhuba, D.V., Kolomiest, E.I., Dey, E.S., and Novik, G.I. (2010) Bacteriophage receptors, mechanisms of phage adsorption and penetration into host cells. *Pol. J. Microbiol.*, **59**, 145–155.

22 Barrow, P.A. and Soothill, J.S. (1997) Bacteriophage therapy and prophylaxis: rediscovery and renewed assessment of potential. *Trends Microbiol.*, **5**, 268–271.

23 Salyers, A.A. and Amabile-Cuevas, C.F. (1997) Why are antibiotic resistance genes so resistant to elimination? *Antimicrob. Agents Chemother.*, **41**, 2321–2325.

24 Sulakvelidze, A., Alavidze, Z., and Morris, J.G.Jr. (2001) Bacteriophage therapy. *Antimicrob. Agents Chemother.*, **45**, 649–659.

25 Ackermann, H.W. (2011) Bacteriophage taxonomy. *Microbiol. Australia*, **32**, 90–94.

26 Wildy, P. (1971) Classification and nomenclature of viruses. First report of the international committee on nomenclature of viruses. *Monogr. Virol.*, **5**, 81.

27 Angly, F.E., Felts, B., Breitbart, M., Salamon, P., Edwards, R.A., Carlson, C. *et al.* (2006) The marine viromes of four oceanic regions. *PLoS Biol.*, **4**, 2121–2131.

28 Sulakvelidze, A. (2011) The challenges of bacteriophage therapy. *Eur. Indust. Pharm.*, **10**, 14–18.

29 Spencer, R. (1955) A marine bacteriophage. *Nature*, **175**, 690–691.

30 Oliveira, J., Castilho, F., Cunha, A., and Pereira, M.J. (2012) Bacteriophage therapy as a bacterial control strategy in aquaculture. *Aquac. Int.*, **20**, 879–910.

31 Paul, J.H., Sullivan, M.S., Segall, A.M., and Rohwer, F. (2002) Marine phage genomics. *Comp. Biochem. Physiol. B*, **133**, 463–476.

32 Sandaa, R.A. (2008) Burden or benefit? Virus–host interactions in the marine environment. *Res. Microbiol.*, **159**, 374–381.

33 Schmitz, J.E., Schuch, R., and Fischetti, V.A. (2010) Identify active phage lysins through functional viral metagenomics. *Appl. Environ. Microbiol.*, **76**, 7181–7187.

34 Matsuzaki, S., Inoue, T., Tanaka, S., Koga, T., Kuroda, M., Kimura, S. *et al.* (2000) Characterization of a novel *Vibrio parahaemolyticus* phage KVP241, and its

relatives frequently isolated from seawater. *Microbial. Immunol.*, **44**, 953–956.

35 Yuksel, S.A., Thompson, K.D., Ellis, A.E., and Adams, A. (2001) Purification of *Piscirickettsia salmonis* and associated phage particles. *Dis. Aquat. Org.*, **44**, 231–235.

36 Karunasagar, I., Shivu, M.M., Girisha, S.K., Krohne, G., and Karunasagar, I. (2007) Biocontrol of pathogens in shrimp hatcheries using bacteriophages. *Aquaculture*, **268**, 288–292.

37 Tai-Wu, L., Xiang, J., and Liu, R. (2000) Studies on phage control of pustule disease in abalone *Haliotis discus hannai*. *J. Shellfish. Res.*, **19**, 535.

38 Pelon, W., Luftig, R.B., and Johnston, K.H. (2005) *Vibrio vulnificus* load reduction in oysters after combined exposure to *Vibrio vulnificus* specific bacteriophage and to an oyster extract component. *J. Food Prot.*, **68**, 1188–1191.

39 Daniels, N.A. and Shafaie, A. (2000) A review of pathogenic *Vibrio* infections for Clinicians. *Infect. Med.*, **17**, 665–685.

40 Hanna, P.J., Altmann, K., Chen, D., Smith, A., Cosic, S., Moon, P., and Hammond, L.S. (1992) Development of monoclonal antibodies for the rapid identification of epizootic *Vibrio* species. *J. Fish Dis.*, **15**, 63–69.

41 Sung, H.H., Li, H.C., Tsai, F.M., Ting, Y.Y., and Chao, W.L. (1999) Changes in the composition of *Vibrio* communities in pond water during tiger shrimp (*Penaeus monodon*) cultivation and in the hepatopancreas of healthy and diseased shrimp. *J. Exp. Mar. Biol. Ecol.*, **239**, 261–271.

42 Vinod, M.G., Shivu, M.M., Umesha, K.R., Rajeeva, B.C., Krohne, G., Karunasagar, I. *et al.* (2006) Isolation of *Vibrio harveyi* bacteriophage with a potential for biocontrol of luminous vibriosis in hatchery environments. *Aquaculture*, **255**, 117–124.

43 Oakey, H.J. and Owens, L. (2000) A new bacteriophage, VHML, isolated from a toxin-producing strain of *Vibrio harveyi* in tropical Australia. *J. Appl. Microbiol.*, **89**, 702–709.

44 Oakey, H.J., Cullen, B.R., and Owens, L. (2002) The complete nucleotide sequence of the *Vibrio harveyi* bacteriophage VHML. *J. Appl. Microbiol.*, **93**, 1089–1098.

45 Shivu, M.M., Rajeeva, B.C., Girisha, S.K., Karunasagar, I., Krohne, G., and Karunasagar, I. (2007) Molecular characterization of *Vibrio harveyi* bacteriophages isolated from aquaculture environments along the coast of India. *Environ. Microbiol.*, **9**, 322–331.

46 Phumkhachorn, P. and Rattanachaikunsopon, P. (2010) Isolation and partial characterization of a bacteriophage infecting the shrimp pathogen *Vibrio harveyi*. *Afr. J. Microbiol. Res.*, **4**, 1794–1800.

47 Srinivasan, P., Ramasamy, P., Brennan, G.P., and Hanna, R.E.B. (2007) Inhibitory effects of bacteriophages on the growth of *Vibrio* sp., pathogens of shrimp in the Indian aquaculture environment. *Asian J. Anim. Vet. Adv.*, **2**, 166–183.

48 Crothers-Stomps, C., Høj, L., Bourne, D.G., Hall, M.R., and Owens, L. (2010) Isolation of lytic bacteriophage against *Vibrio harveyi*. *J. Appl. Microbiol.*, **108**, 1744–1750.

49 Alagappan, K.M., Deivasigamani, B., Somasundaram, S.T., and Kumaran, S. (2010) Occurrence of *Vibrio parahaemolyticus* and its specific phages from shrimp ponds in East coast of India. *Curr. Microbiol.*, **61**, 235–240.

50 Park, S.B., Aoki, T., and Jung, T.S. (2012) Pathogenesis of and strategies for preventing *Edwardsiella tarda* infection in fish. *Vet. Res.*, **43**, 67–78.

51 Wu, J.L. and Chao, W.J. (1982) Isolation and application of a new bacteriophage, φET-1, which infect *Edwardsiella tarda*, the pathogen of edwardsiellosis. *Rep. Fish. Dis. Res.*, **8**, 8–17.

52 Kwon, S.R., Nam, Y.K., Kim, S.D., and Kim, K.H. (2005) Generation of *Edwardsiella tarda* ghosts by bacteriophage PhiX174 lysis gene E. *Aquaculture*, **250**, 16–21.

53 Yamamoto, A. and Maegawa, T. (2008) Phage typing of *Edwardsiella tarda* from eel farm and diseased eel. *Aquac. Sci.*, **56**, 611–612.

54 Hernández, E., Figueroa, J., and Iregui, C. (2009) Streptococcosis on a red tilapia, *Oreochromis* sp., farm: a case study. *J. Fish Dis.*, **32**, 247–252.

55 Matsuoka, S., Hashizume, T., Kanzaki, H., Iwamoto, E., Park, S.C., Yoshida, T. *et al.* (2007) Phage therapy against beta-hemolytic streptococcosis of Japanese flounder *Paralichthys olivaceus*. *Fish Pathol.*, **42**, 181–190.

56 Park, K.H., Matsuoka, S., Nakai, T., and Muroga, K. (1997) A virulent bacteriophage of *Lactococcus garvieae* (formerly *Enterococcus seriolicida*) isolated from yellowtail *Seriola quinqueradiata*. *Dis. Aquat. Org.*, **29**, 145–149.

57 Verner-Jeffreys, D.W., Algot, M., Pond, M.J. *et al.* (2007) Furunculosis in Atlantic salmon (*Salmon salar* L.) is not readily controllable by bacteriophage therapy. *Aquaculture*, **270**, 475–484.

58 Imbeault, S., Parent, S., Lagace, M. *et al.* (2006) Using bacteriophage to prevent furunculosis caused by *Aeromonas salmonicida* in farmed Brook Trout. *J. Aquat. Anim. Health.*, **18**, 203–214.

59 Fischetti, V.A. (2005) Bacteriophage lytic enzymes: novel anti-infectives. *Trends Microbiol*, **13**, 491–496.

60 Courchesne, N.M.D., Parisien, A., and Lan, C.Q. (2009) Production and application of bacteriophages and bacteriophage encoded lysins. *Recent Pat. Biotechnol.*, **3**, 37–45.

61 Yoon, S., Kang, S., Kyoung, S., Choi, Y., and Son, J. (2008) Patent No. WO2008016240.

62 Harris, D.L. and Lee, N. (2003) Patent No. WO03103578.

63 Samoylova, T.I., Noris, M.D., Samoylov, M.A., Cochran, A.M. *et al.* (2012) Infective and inactivated filamentous phage as carriers for immunogenetic peptides. *J. Virol. Methods*, **183**, 63–68.

64 Janda, J.M. and Abbott, S.L. (1993) Infections associated with the genus *Edwardsiella: the role of Edwardsiella tarda in human disease. Clin. Infect. Dis.*, **17**, 742–748.

65 Okuda, J., Kiriyama, M., Suzaki, E., Kataoka, K., Nishibuchi, M., and Nakai, T. (2009) Characterization of proteins secreted from a Type II secretion system of *Edwardsiella tarda* and their roles in macrophage infection. *Dis. Aquat. Org.*, **84**, 115–121.

10
Marine Actinomycetes as Source of Pharmaceutically Important Compounds

M.L. Arvinda swamy

10.1
Introduction

The search for novel pharmaceutical compounds from different natural resources is practiced from ancient times. The therapeutic compounds mostly obtained from plant sources are used for treating various types of diseases. However, these drugs had one or other side effects on the human body. Moreover, microorganisms gained resistance against these diseases that resulted in the search for the diverse natural products from oceans.

More than 70% of our earth comprises oceans, which have a higher number of diverse fauna and flora compared with the terrestrial life. Marine microbes, plants, and invertebrates have gained much importance in the recent decades due to their capability of producing biologically active pharmaceutical compounds for treating various human diseases.

Various researchers across the world are focused on the rich biodiversity potential of seas using advanced technologies to discover the untapped natural products in the deep seas.

This chapter is focused on marine actinomycetes having potential therapeutic value in treating various human and plant diseases. More emphasis is given on diverse secondary metabolites or bioactive compounds that can be used as antibacterials, antiproliferatives/antitumorogenic, antifungals, and antimalarials.

10.2
Marine Actinomycetes as Source of Therapeutics

Actinomycetes are a group of bacteria found at varying depths with diverse temperatures, pH, and salinity. They have the ability to produce a range of secondary metabolites having potential therapeutic values. Marine actinomycetes are Gram-positive bacteria that are filamentous in nature, and produce diverse biologically active metabolic compounds that can serve as enzymes, or for treating cancer, malaria, fungal, and bacterial diseases.

Marine Microbiology: Bioactive Compounds and Biotechnological Applications, First Edition.
Edited by Se-Kwon Kim

10.3
Marine Actinomycete Compounds as Antibacterials

Products derived from natural resources have a number of advantages over manmade synthetic compounds. Novel antibacterial compounds derived from various marine actinomycetes can be utilized for treating drug-resistant strains or multidrug-resistant bacteria in humans. *Streptomyces* species are the largest group known to produce secondary metabolites of immense importance.

Different classes of compounds have been isolated from *Streptomyces* species. Bonactin was isolated from *Streptomyces* species BD21-2 that has shown action against both Gram-positive and Gram-negative bacteria. This strain was isolated from the Kailua beach in Hawaii [1]. Essramycin, a triazolopyrimidine antibiotic isolated from the *Streptomyces* species Merv8102, has shown antibiotic action against both Gram-positive and Gram-negative bacteria at minimum inhibitory concentration of 2–8 μg/mL [2].

Lajollamycin is a yellow compound derived from *Streptomyces nodosus* (NPS007994) that has shown a broad range of antibiotic activity against both drug-sensitive and drug-resistant Gram-positive and Gram-negative bacteria. *S. nodosus* was collected from Scripps Canyon, La Jolla, California [3]. *Streptomyces* strain CNQ-525 produces napyradiomycin and its derivatives. A80915A and A80915B have shown potent antibacterial action against methicillin-resistant *Staphylococcus aureus* (MRSA). These compounds are also produced by the terrestrial *Streptomyces* species, and exhibit antimicrobial action against the MRSA [4].

Etamycin isolated from strain CNS-575 from the coast of Fiji has shown action against MRSA and many Gram-negative and Gram-positive bacteria; this compound is nontoxic at concentrations of 20-fold. This is also evident from the significant protection provided by etamycin against MRSA infection in the mouse model [5]. Novel bioactive compounds are produced from the protoplast fusion of two different *Streptomyces* strains, Merv1996 and Merv7409. The compounds isolated from these fusions are benzopyrone derivatives (**1–3**). 7-Methylcoumarin (**1**) and two flavonoids, rhamnazin and cirsimaritin, can be used as antibacterials after further testing on various strains of bacteria [6].

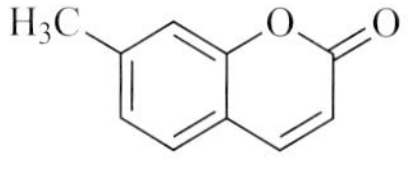

1 (7-methylcoumarin)

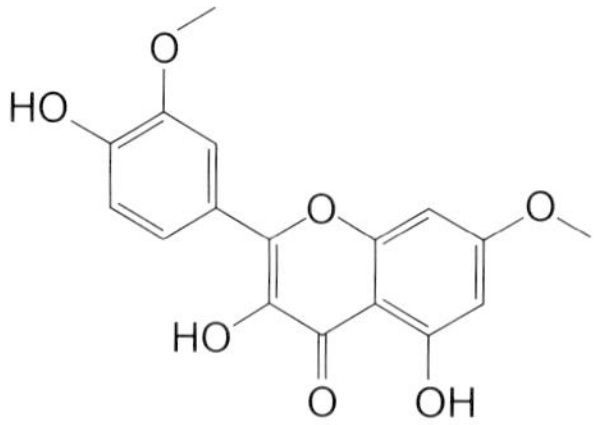

2 (rhamnazim)

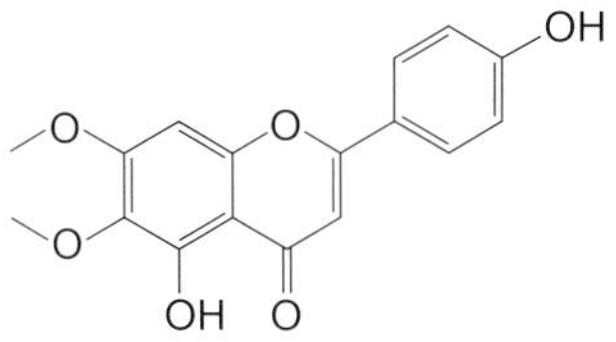

3 (cirsimaritin)

Anthraquinone class of compounds, 1,8-dihydroxy-2-ethyl-3-methylanthraquinone (**4**), octadecanoic acid (**5**), and cholest-4-en-3-one (**6**), are isolated from *Streptomyces* species FX-58 [7]. Other anthraquinone compounds isolated from this species are 1,6-dihydroxy-8-propylanthraquinone and anthraquinone, along with 3,8-dihydroxy-1-propylanthraquinone-2-carboxylic acid [8]. Bioactive compound resistoflavine has shown weak antibacterial action against both Gram-positive and Gram-negative bacteria [9]. Marinomycins (**7**) isolated from CNQ-140 possess antibiotic action against MRSA and vancomycin-resistant *Enterococcus faecium* at minimum inhibitory concentrations ranging from 0.1 to 0.6 μM [10].

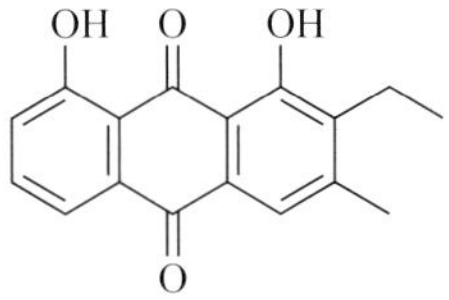

4 (1,8-dihydroxy-2 ethyl-3-methylanthraquinone)

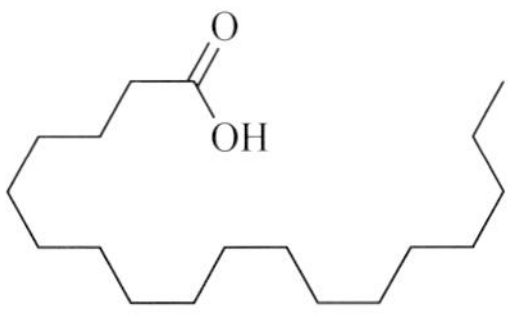

5 (octadecanoic acid)

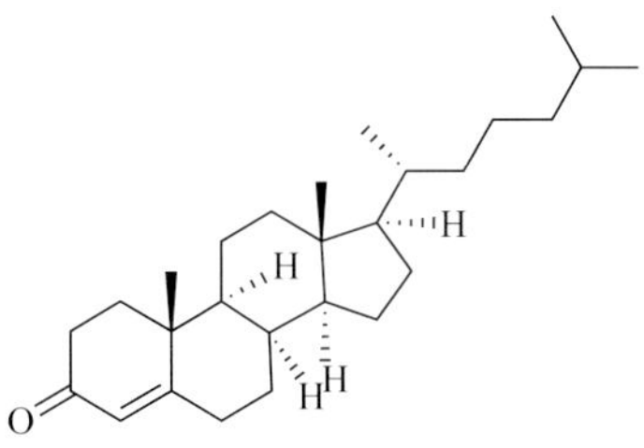

6 (cholest-4-en-3-one)

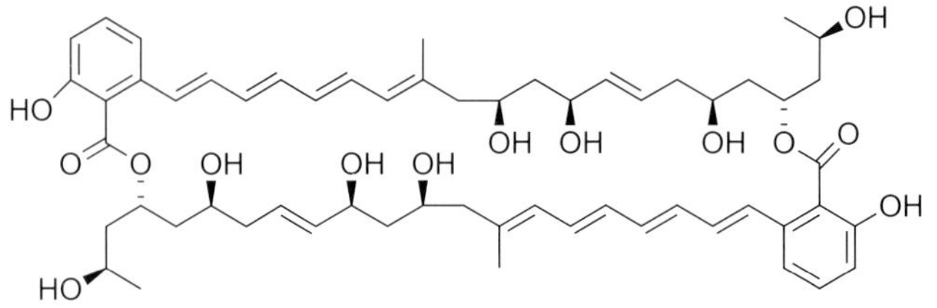

7 (marinomycin A)

Mansouramycin A (7-methylamino-3,4-dimethylisoquinoline-5,8-dione), a red compound, was tested on various bacteria (*S. aureus, Bacillus subtilis,* and *Escherichia coli*) and found to exhibit moderate antibacterial activity; the zone of inhibition varied from 10 to 12 mm diameter [11].

Apart from *Streptomyces,* species belonging to *Micromonospora, Nocardia,* and *Actinomadura* also produce antibacterial compounds. Studies conducted for screening *Actinomycetes* species possessing antimicrobial compounds from the Norwegian fjord have shown antibacterial and antifungal action against tested organisms, and among all the strains, *Micromonospora* strains have a high abundance [12].

Novel antibiotic chandrananimycin A, B, and C were isolated from the *Actinomycetes* species *Actinomadura* isolate M045 [13]. *Nocardiopsis* species produces a novel thiopeptide antibiotic that has an aminoacetone moiety with potential antibiotic activity. This thiopeptide was isolated from the isolate TFS65-07 and designated as TP-1161 [14]. Nocapyrones E–G, diketopiperazine derivatives nocazines A and C (**8** and **9**), and oxazoline compound nocazoline A were isolated from the *Actinomycetes* species *Nocardiopsis dassonvillei* HR10-5 and exhibited antibiotic action against *B. subtilis* (Table 10.1) [15].

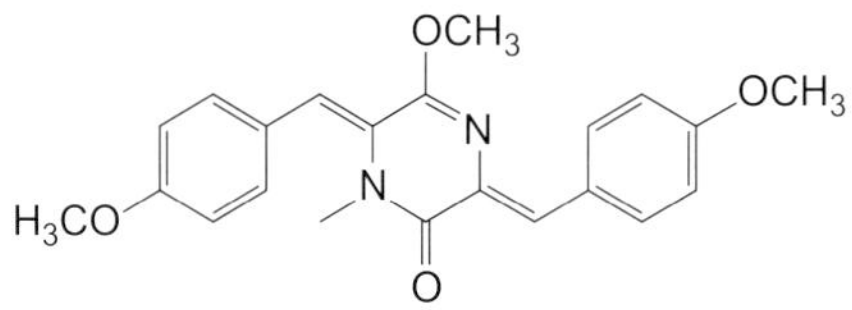

8 (nocazine A)

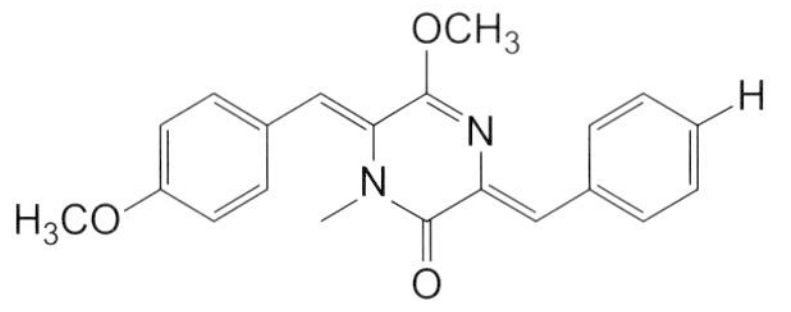

9 (Nocazine C)

Table 10.1 Marine actinomycete compounds as antibacterials.

Source	Compound	Activity
Streptomyces sp. BD21-2	Bonactin	Antibacterial
Streptomyces sp. Merv8102	Triazolopyrimidine	Antibacterial
Streptomyces sp. CNQ-525	Napyradiomycin	Antibacterial
Streptomyces sp. CNS-575	Etamycin	Antibacterial
Streptomyces sp. FX-58	1,8-Dihydroxy-2-ethyl-3-methylanthra-quinone	Antibacterial
Streptomyces sp. M045	Chinkomycins	Antibacterial
Streptomyces sp. CNQ-140	Macrolides and marinomycins	Antibacterial
Streptomyces strains Merv1996 and Merv7409	Rhamnazin and cirsimaritin	Antibacterial
Streptomyces strain Mei37	Mansouramycin A	Antibacterial
Streptomyces sp. CNQ-525	Daryamides	Antibacterial
S. nodosus NPS007994	Lajollamycin	Antibacterial
Streptomyces chinaensis AUBN1/7	Resistoflavine	Antibacterial
Nocardiopsis sp. TFS65-07	Thiopeptide	Antibacterial
N. dassonvillei HR10-5	Nocapyrones	Antibacterial
Actinomadura sp. M045	Chandrananimycin	Antibacterial

10.4
Marine Actinomycete Compounds as Antitumors/Antiproliferative

The widespread occurrence of different types of cancers throughout the world, side effects of drugs, and high treatment costs urge the need for novel secondary compounds that can cure different types of cancers. Different *Actinomycetes* species produce bioactive compounds that have antitumor or antiproliferative activity.

Four daryamides (daryamide A, daryamide B, daryamide C, and (2*E*,4*E*)-7-methylocta-2,4-dienoic acid amide) belonging to polyketide family are isolated from the marine *Streptomyces* species CNQ-085. This strain is isolated from a depth of 50 m. The compounds daryamide A and (2*E*,4*E*)-7-methylocta-2,4-dienoic acid amide showed higher action against the colon cancer cell line (HCT-116) at (IC_{50}) 3.15 μM/mL compared with other compounds [16]. Streptocarbazoles A and B are ethyl acetate-extracted compounds from the *Streptomyces* species FMA. This strain was collected from the mangrove soil in Sanya Hainan province, China. Streptocarbazoles that belong to the indolocarbazole family have shown cytotoxic activity against different cell lines. Streptocarbazole A showed the effect on HL-60, A-549, and P388 cell lines at concentrations ranging from 1.4 to 34.5 μM, and compound **10** concentration of 12.8 μM in HL-60 and 22.5 μM in P388 cell lines. These compounds are known to affect the cell cycle in G2/M phase; compound **11** arrests cell cycle at a concentration of 10 μM in HL-60 cell line [17].

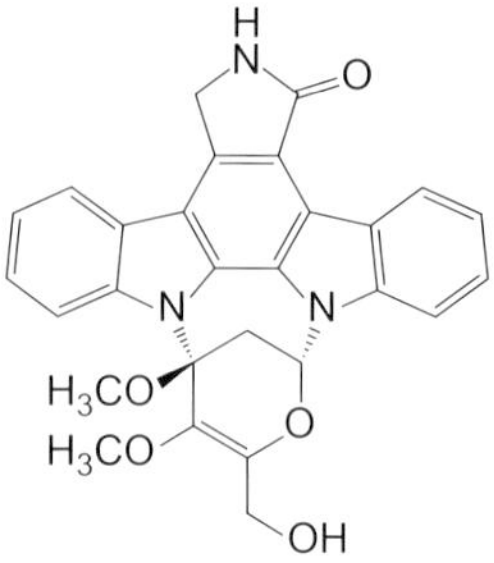

10 (streptocarbazole A)

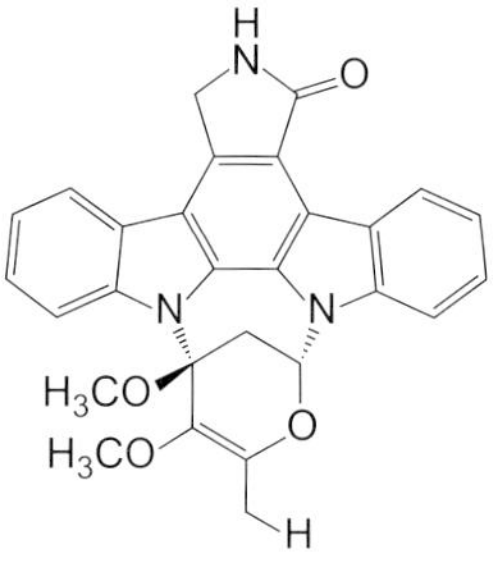

11 (streptocarbazole B)

The antitumorogenic properties of four different types of daryamides (A, B, C, and 2*E*,4*E*-7-methylocta-2,4-dienoic acid amide) isolated from *Streptomyces* strain CNQ-085 were tested on HCT-116 cell lines (human colon carcinoma cells), showed weak to moderate cytotoxic effect on these cell lines [9]. Antitumorogenic and antibacterial compound resistoflavine isolated from *Streptomyces chibaensis* AUBN1/7 has shown cytotoxic activity on tumors of gastric adenocarcinoma (HMO2) and hepatic carcinoma (HepG2) cell lines. This compound is related to quione antibiotic and has shown weak antibacterial action against both Gram-positive and Gram-negative bacteria. Strain AUBN1/7 was isolated from a marine sediment sample collected at a depth of 30 m at a distance of 8 km from Machilipatnam coast of Bay of Bengal [18].

Actinomycetes (ACT 01, ACT 02, ACT 03, ACT 04, and ACT 05) belonging to *Streptomyces* species were isolated from the mangrove sediments from Kanyakumari district, Tamil Nadu, India. Compounds that were extracted using ethyl acetate from these actinomycetes were tested against the breast cancer cell (MCF-7 and MDA-MB-231). The crude extracts from all the samples exhibited inhibitory action against breast cancer cell lines at different time points (24 and 48 h). ACT 05 shown IC value ranging from 85 to >100 μg/mL in MCF-7 and MDA-MB-231 cell lines, whereas ACT 04 action was observed at IC 58 to >100 μg/mL in respective cell lines [19]. *Streptomyces* species 04DH10 was isolated from marine sediments at a depth of 1 m in Ayajin Bay, east sea of Korea. Methanol extracts of 04DH10 produce a yellowish amorphous solid streptochlorine that showed slight action on human leukemia cell line K-562, at inhibitory concentration (IC_{50}) of 1.05 and 10.9 μg/mL in hepatocytes [20]. Barmumycin that was isolated from the *Streptomyces* species BOSC-022A has wide action on different human cancer cell lines. Apart from Barmumycin, other two compounds pretomaymycin and oxotomaymycins are also extracted from BOSC-022A [21].

A new class of compounds (*R*)-10-methlyl-6-undecanolide and (6*R*,10*S*)-10-methyl-6-dodecanolide) belonging to caprolactones are extracted by ethyl acetate from the marine actinomycete *Streptomyces* sp. B6007. This species is isolated from the mangrove sediment from Papua New Guinea. The compounds (6*R*,10*S*)-10-methyl-6-dodecanolide have shown action against gastric adenocarcinoma (HM02), hepatocarcinoma cell line (HepG2), and breast cancer cell line (MCF-7) [22]. *Streptomyces aureoverticillatus* produces a white compound aureoverticillactum that belongs to the macrocyclic lactum family. This compound was tested against different cancer cell lines (colorectal adenocarcinoma, Jurkat cells, leukemia) where it showed moderate activity. The effective concentration (EC_{50}) was 3.3 μM in colorectal adenocarcinoma (HT-29), 2.2 μM in acute T-cell leukemia (Jurkat cells), and 2.3 μM in melanoma cell line (B16-F10) [23].

Marmycin A (red compound) and Marmycin B (pink compound) were extracted from marine actinomycete belonging to *Streptomyces* strain CNH990. This strain was isolated from the sea sediments of Cortez, Baja California Sur, Mexico. Marmycin A exhibited action against colon cancer cell line HCT-116 at an inhibitory concentration of 60.5 nM, whereas marmycin B showed action at 1.09 μM. When these compounds were tested against drug-resistant bacteria (MRSA and VREF), they did not show any significant activity [24]. *Streptomyces* strain BL-49–58-005 was isolated from the marine invertebrates in Mexico. This strain produced compounds belonging to indole family. Bioactive compounds 3,6-disubstituted indoles (**12–14**) were tested on 14 different cancer cell lines. Compound **12**, 6-prenyltryptophol with molecular formula $C_{15}H_{19}ON$, showed highest activity against leukemia cell line (K-562) at GI_{50} of 8.46 μM and 6-prenyltryptophol analogue (Compound **13**) with molecular formula $C_{15}H_{18}ON_2$ exhibited action on a wide range of cancerous cell lines such as prostate cancer (LN-caP), pancreatic cancer (PANC1), endothelial cancer (HMEC1), colon cancer (LOVO-DOX and LOVO), and leukemia (K-562) [25].

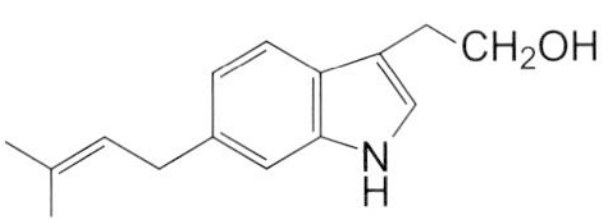

CH_2OH N H

12 (6-prenyltryptohol)

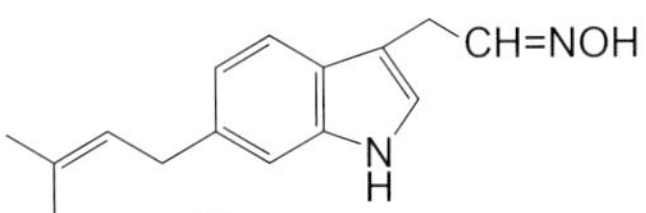

CH=NOH N H

13

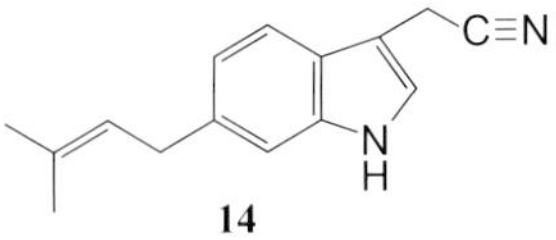

14

Mansouramycin are a new class of compounds belonging to isoquinoline quinones. Mansouramycin A was isolated from *Streptomyces* strain Mei37 from North Sea coast, Germany. Various mansouramycins extracted from this species are mansouramycin A (7-methylamino-3,4-dimethylisoquinoline-5,8-dione), mansouramycin B (6-chloro-3-methylisoquinoline-5,8-dione), mansouramycin C (3-carbomethoxy-7-methylaminoisoquinoline-5,8-dione), and Mansouramycin D (3-(1*H*-indole-3-yl)-7-methylaminoisoquinoline-5,8-dione). Different mansouramycins are tested against the 36 cancer cell lines. Mansouramycin C (3-carbomethoxy-7-methylaminoisoquinoline-5,8-dione), dark red in color, showed highest cytotoxic action compared with other mansouramycins. When tested on 36 different cancer cell lines it exhibited action against 10 cell lines (T-24, SF-268, LXFA629L, MEXF 520L, OVCAR-3, RXF944L, and UXF1138L), with inhibition concentrations ranging from 0.008 to 0.02 μM [11].

Marinispora CNQ-140 strain was isolated from the sediments at a depth of 56 m offshore of La Jolla, CA. The biologically active compounds produced by this strain are active on cancerous cell lines and bacteria. Compound extracted by using ethyl acetate showed action against the human colon cancer cell line (HCT-116) at inhibitory concentration of (IC_{50}) 1.2 μg/mL. Marinomycins, yellowish in color, showed strong action against NCI 60 cancerous cell lines; its efficient action (LC_{50}) was observed at average concentrations of 0.2–2.7 μM [10].

Five bipyridine alkaloids named as caerulomycin F, caerulomycin G, caerulomycin H, caerulomycin I, caerulomycin J, and one phenylpyrideine alkaloid named caerulomycin K (**15**–**20**) were isolated from *Actinomycetes* species *Actinoalloteichus cyanogriseus* WH1–2216-6 which showed cytotoxic effect on cancerous cell lines HL-60, K562, KB, and A549 cell lines. Analogues of caerulomycins showed antibiotic action against *E. coli, Aerobacter aerogenes,* and *Candida albicans* at minimum inhibitory concentrations ranging from 9.7 to 38.6 μM [26].

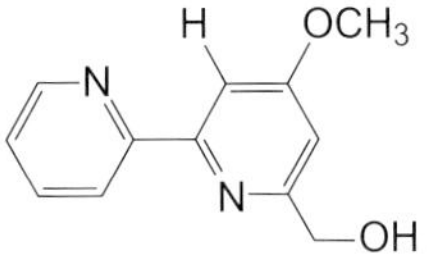

15 (caerulomycin F)

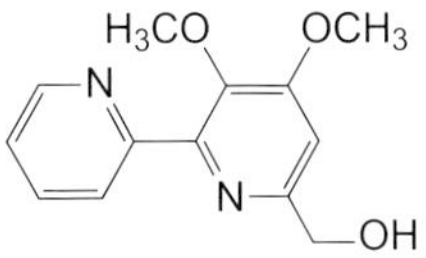

16 (caerulomycin G)

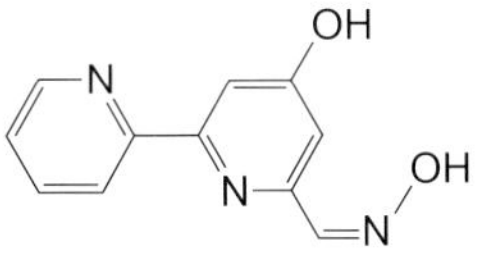

17 (caerulomycin H)

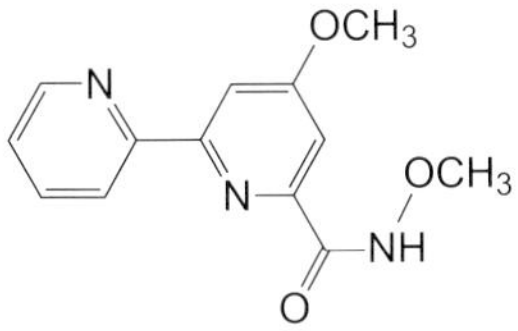

18 (caerulomycin I)

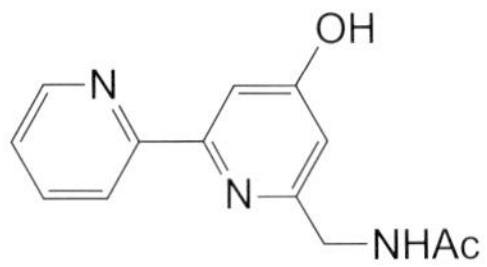

19 (caerulomycin J)

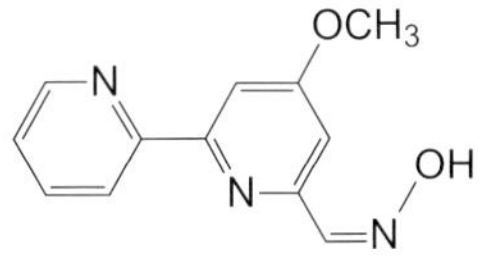

20 (caerulomycin K)

Antitumor compound tartrolon D was isolated from the marine actinomycete *Streptomyces* species MDG-04–17–069 from the east coast of Madagaskar at a depth of 30 m. This compound had a whitish color and exhibited potent effect on cancer cell lines A549 (lung carcinoma), HT-29 (colorectal carcinoma), and MDA-MB-231 (breast adenocarcinoma) (Table 10.2) [27].

10.5 Marine Actinomycete Enzymes as Antiproliferatives

The enzymes inhibitors isolated from various *Actinomycetes* species from different marine environments have shown high pharmaceutical activity. They can be used for the treatment of various human cancers.

L-Asparaginase enzyme extracted from PDK7 and PDK2 actinomycetes isolated from the coastal areas of Cochin have shown strong cytostatic action against antilymphoblastic leukemia and chronic myelogenous leukemia cell lines (JURKAT and K562). The enzyme is stable at 80 °C and has shown strong action at 60 °C in

Table 10.2 Marine actinomycete compounds as antitumors/antiproliferatives.

Source	Compound/family	Activity
Streptomyces strain CNQ-085	Daryamides	Antitumor
Streptomyces sp. FMA	Streptocarbazole A	Antitumor
S. chibaensis AUBN1/7T	Resistoflavine	Antitumor/antibacterial
Streptomyces sp. 04DH10	Streptochlorine	Antitumor
Streptomyces sp. BOSC-022A	Barmumycin	Antitumor
Streptomyces sp. B6007	Caprolactone	Antitumor
S. aureoverticillatus NPS001583	Aureoverticillactum	Antitumor
Streptomyces strain CNH990	Marmycin A	Antitumor
Streptomyces strain BL-49–58-005	Indoles	Antitumor
Streptomyces strain Mei37	Mansouramycin	Antitumor
Marinispora sp. CNQ-140	Marinomycins	Antitumor/antibacterial
A. cyanogriseus WH1–2216-6	Caerulomycins	Antitumor
Streptomyces sp. MDG-04–17–069	Tartrolon	Antitumor

the pH range of 8–9. The enzyme showed its effect on these cancerous cell lines within 48 h [28].

Thirty-four strains belonging to actinomycetes were isolated from the marine sediments of Parangipettai, east coast of India. Only seven strains have shown L-asparaginase activity ranging from 0.09 to 2.49 μM ammonia/(ml/h). *Streptomyces* species strain EPD 27 has shown maximum asparaginase activity; this strain can grow at temperatures ranging from 38 to 40 °C, pH from 5 to 10, and can tolerate high salt concentrations. 16S rRNA analysis revealed that it is very close to *Streptomyces* species (FJ799173 and 799170) [29].

Partially purified L-asparaginase from actinomycete strain LA-29 isolated from gut of *Mugil cephalus* inhibited the growth of leukemia in Wistar rats; the enzyme was effective at 100 units (13.57 IU/mg) [30].

Streptomyces sp. DA11 is associated with the marine sponge from China (*Craniella australiensis*), and chitinase isolated from this showed action against fungus (*Aspergillus niger* and *C. albicans*). The chitinase activity of this species was influenced by temperature, pH, and salinity [31].

10.6 Marine Actinomycete Compounds as Antimalarials

Malaria is an arthropod-borne disease prevalent in many countries. Antimalarial drugs are produced through chemical synthesis and plant extracts are widely used to cure malarial diseases but the emergence of drug/multidrug-resistant strains poses a great challenge. The search for novel compounds from marine sources yielded potent antimalarial compounds from the *Actinomycetes* species.

The compounds SCSIO 00652 isolated from the marine *Actinomycetes* species *Marinactinospora thermotolerans*, belonging to Nocardiopsaceae, produced bioactive compounds belonging to alkaloids. These alkaloids are β-carboline alkaloids (marinacarbolines A–D (**21–24**)) and indolactam alkaloids (13-*N*-demethyl-methylpendolmycin and methylpendolmycin-14-*O*-α-glucoside). These compounds have shown action against *Plasmodium falciparum* strains (3D7 and Dd2) at varying inhibitory concentrations from 1.9 to 36.03 μM [32].

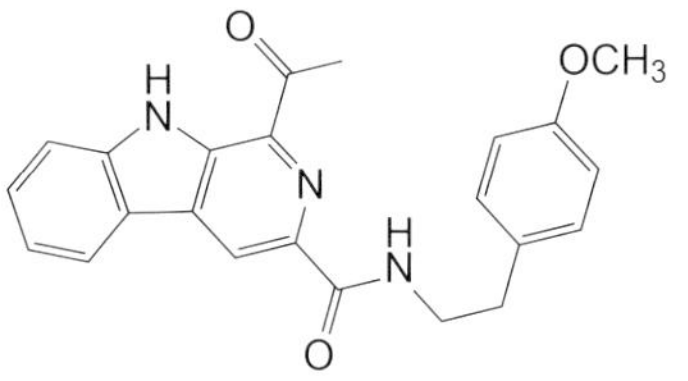

21 (marinacarboline A)

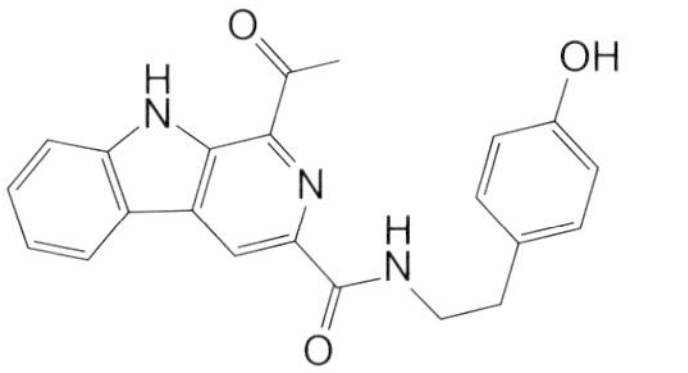

22 (marinacarboline B)

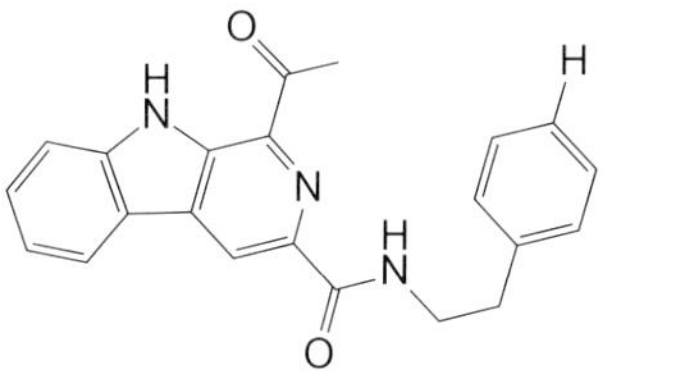

23 (marinacarboline C)

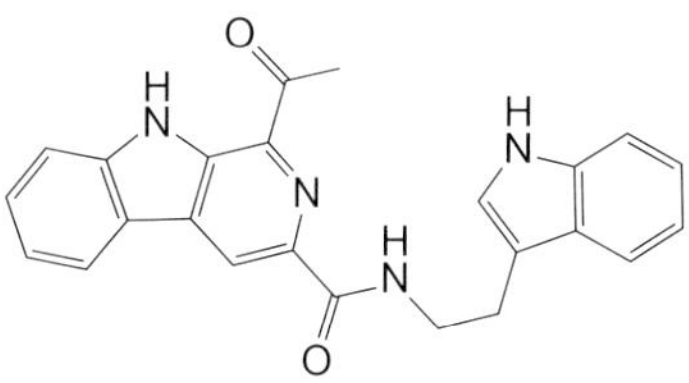

24 (marinacarboline D)

Secondary metabolite salinosporamide A isolated from *Salinospora tropica* has shown protective action against the malaria parasite in the erythrocytic stage by inhibiting the 20S proteosome. Studies showed that introduction of low doses of salinosporamide A to the *Plasmodium*-infected mice protected them from severe malarial parasite infection (Table 10.3) [33].

Table 10.3 Marine actinomycete compounds as antimalarials.

Source	Bioactive compound	Activity
M. thermotolerans SCSIO 00652	Marina carbolines	Antimalarial
S. tropica	Salinosporamide A	Antimalarial

10.7 Marine Actinomycete Compounds as Antifungals

Different fungal species causes different diseases in both humans and plants. The search for novel antifungal pharmaceutical compounds of interest from various marine habitat actinomycetes is going on across the world.

Antimycotic compounds saadamycin and 5,7-dimethoxy-4-*p*-methoxylphenylcoumarin were isolated from endophytic *Streptomyces* species Hedaya 48 that is associated with the Egyptian sponge (*Aplysina fistularis*) from Red Sea. The yield of saadamycin was greatly improved by UV irradiation, and the mutant strain Ah22 developed by this process produced yields greater than 10-fold from the wild-type strain. These compounds have shown action against various fungi that cause skin infections [34]. Daryamide A and daryamide B isolated from marine *Streptomyces* species CNQ-085 exhibited weak fungal activity against *C. albicans* [13].

Fungal disease affects commercial crops causing huge loss to the farmers and impacting the society and government. The fungi that infect rice crops and cause disease are *Rhizoctonia solani* (sheath blight), *Pyricularia oryzae* (blast of rice), and *Helminthosporium oryzae* (leaf spot disease). In sugarcane, fungus *Colletotrichum falcatum* causes red rot diseases. One hundred and sixty marine actinomycetes were isolated from different coastal areas of India and tested against these disease-causing fungal strains. Majority of the isolates exhibited antifungal activity on one or two tested fungal varieties. Ten actinomycete isolates exhibited antifungal activity on all the tested fungi. *Streptomyces fradiae* (RHI1), *Streptomyces roseolilacinus* (MP 1), *Streptomyces auriomonopodiales* (MI 1), and *Streptomyces albidoflavus* (SEA 5) have shown highest antifungal activity on all the tested fungi and their zone of inhibition ranged from highest 20 mm to lowest 10 mm in different species. All other isolates *Streptomyces orientalis* (BC 3), *Streptomyces roseiscleroticus* (BC 1), *Streptomyces sclerotialus* (SD 3), and *Streptomyces galtieri* (AN 1C) exhibited inhibition zones ranging from 9 to 15.5 mm against different fungi (Table 10.4) [35].

10.8 Bioactive Compounds from Sponge-Associated Actinomycetes

The association of microbes with various other higher organisms in the marine environment is due to various factors such as symbiosis, food, protection, and so on. The association of *Actinomycetes* species with the sponges is observed in various

Table 10.4 Marine actinomycete compounds as antifungals.

Source	Bioactive compound	Activity
Streptomyces sp. Hedaya 48	Saadamycin	Antifungal
Streptomyces sp. CNQ-085	Daryamide A and daryamide B	Antifungal
S. fradiae	Unidentified	Antifungal
S. roseolilacinus	Unidentified	Antifungal
S. auriomonopodiales	Unidentified	Antifungal
S. albidoflavus	Unidentified	Antifungal
S. orientalis	Unidentified	Antifungal
S. roseiscleroticus	Unidentified	Antifungal
S. sclerotialus	Unidentified	Antifungal
S. galtieri	Unidentified	Antifungal

actinomycete species that provide secondary metabolic compounds that can be used as antibacterial or antiproliferative/anticancer compounds.

Sixty-three marine actinomycetes isolated from Arabian Sea, southwest coast of India, have shown antibacterial action on at least one human pathogen. The number of *Actinomycetes* species isolated from different seasons varied and the maximum number was found during the southwest monsoon season. Sponge-associated *Actinomycetes* species have shown highest antibacterial action when compared with other isolated *Actinomycetes* species. *Streptomyces albogroseolus, Streptomyces aureocirculatus, Streptomyces achromogenes, Streptomyces raceochromogenes,* and *Streptomyces furlongus* have shown action on various human eye pathogens (10 strains), antibiotic-resistant bacteria (5 strains), and sensitive bacteria (10 strains). The zone of inhibition for most of the actinomycetes ranged from 7 to 12 mm diameter [36]. Actinomycete *N. dassonvillei* MAD08 collected from the southwest coast of India, associated with *Dendrilla nigra*, has shown action against the multidrug-resistant bacteria. This actinomycete has also produced enzymes such as cellulase, amylase, lipases, and proteases [37].

Actinomycetes associated with the sponge *Callyspongia diffusa* were isolated from the Bay of Bengal at a depth of 10–15 m. Out of 26 endosymbiotic strains isolated, 4 strains (CPI3, CPI9, CPI12, and CPI13) showed the maximum antibacterial activity, and their zone of inhibition ranged from 31 to 66 mm diameter [38].

Antibacterial compound mayamycin was extracted from the marine actinomycete *Streptomyces* strain HB202. This strain was isolated from the marine sponge (*Halichondria panicea*) in Baltic Sea. Mayamycin has potent antibacterial activity on different bacteria including the drug-resistant bacteria [39]. Mayamycin, when tested on different cancer cell lines, exhibited strong cytotoxic activity compared with the standard drug adriamycin on hepatocellular carcinoma and colon adenocarcinoma cancer cell line (HepG2, HT-29). The inhibition concentrations of mayamycin ranged from 0.13 to 0.33 μM in the tested cell lines (HepG2, HT-29, GXF251L (gastric cancer), LXF529L (nonsmall-cell lung cancer), MAXF401NL (mammary cancer), MEXF462NL (melanoma cancer), PAXF1657L (pancreatic cancer), and RXF486L (renal cancer)) [40]. Three compounds 3,5,6-trisubstituted 2(1*H*)pyrazinone, JBIR-56, and JBIR-57 are isolated from the *Streptomyces* species SpD081030SC-03. This isolate is isolated from the unidentified marine sponge [41].

Novel tetrapeptide compounds JBIR-34 and JBIR-35 are isolated from the marine sponge (*Haliclona* species) associated *Streptomyces* species Sp080513GE-26 from Chiba Prefecture, Japan. These two compounds have shown DPPH radical scavenging action at inhibitory concentration of 1.0 and 2.5 mM, respectively. These compounds did not exhibit any action against bacteria and cancer cell lines [42].

Three novel histone deacylases (HDACs) compounds JBIR-109, JBIR-110, and JBIR-111 (trichostatin analogues) are isolated from the marine sponge associated *Streptomyces* sp. RM72. This strain was isolated from Takara Island, Kagoshima Prefecture, Japan, at a depth of 195 m. The HDACs play an important role in gene expression, differentiation, cell cycle, and death of the tumor cells; these compounds can serve as potent antitumor compounds [43].

Three compounds belonging to isoprenoids (JBIR-46, JBIR-47, and JBIR-48) were isolated from *Streptomyces* species SpC080624SC-11, which has association with the marine sponge belonging to *Cinachyra* species. *H. panicea* (marine sponge) associated strain HB383 was isolated from the Baltic Sea. This strain belongs to the *Nocardiopsis* species and the compounds extracted from these are nocapyrones. Similar compounds isolated from different strains exhibited antibacterial and cytotoxicity against different cancer cell lines [44].

Bendigoles D–F and 3-ketosterols are isolated from *Actinomadura* species SBMs009 that has an association with marine sponge. The compound bendigole 1 has shown cytotoxic action against the mouse fibroblast cell line L929 at a concentration of 30 μM. This compound can inhibit glucocorticoid receptor (GR) activity while bendigole 3 has shown action against GR at a concentration of 71 μM (Tables 10.5 and 10.6) [45].

Table 10.5 Bioactive compounds from sponge-associated actinomycetes.

Source	Sponge	Activity
S. albogroseolus (MSUKR-29)	Unidentified	Broad-spectrum antibiotic
S. aureocirculatus (MSUKR-39)	Unidentified	Broad-spectrum antibiotic
S. raceochromogenes (MSUKR-44)	Unidentified	Broad-spectrum antibiotic
S. achromogenes (MSUKR-46)	Unidentified	Broad-spectrum antibiotic
S. furlongus (MSUKR-57)	Unidentified	Broad-spectrum antibiotic
Streptomyces sp. (CPI3, CPI9, CPI12, and CPI13	*C. diffusa*	Broad-spectrum antibiotic
N. dassonvillei MAD08	*D. nigra*	Multidrug-resistant bacteria
Streptomyces sp. HB202	*H. panicea*	Drug-resistant bacteria

Table 10.6 Bioactive compounds from sponge-associated actinomycetes.

Source	Sponge	Compound	Activity
Streptomyces sp. SpD081030SC-03	Unidentified	JBIR-56 and JBIR-57 (tetrapeptide compounds)	Unidentified
Streptomyces sp. HB202	*H. panicea*	Mayamycin	Antitumor
Streptomyces. sp. Sp080513GE-26	*Haliclona* sp.	JBIR-34 and JBIR-35	Radical scavenging
Streptomyces sp. RM72.	Unidentified	JBIR-109, JBIR-110, and JBIR-111 (trichostatin analogues)	Antitumor
Streptomyces sp. SpC080624SC-11	*Cinachyra* sp.	JBIR-46, JBIR-47, and JBIR-48 (isoprenoids)	Antitumor
Nocardiopsis sp. HB383	*H. panicea*	Nocapyrones	Antitumor and antibacterial
Actinomadura sp. SBMs009	Unidentified	Bendigoles	Antitumor

10.9 Conclusion

Marine actinomycetes can survive at different temperatures, pH, and depths of the sea. They provide a diverse group of bioactive compounds that have novel mechanisms of action in treating various diseases. Various isolates belonging to *Streptomyces* provide a large variety of bioactive secondary metabolic compounds that play a potential role as antitumors, antibacterials, antimalarials, and antifungals compared with *Marinispora*, *Actinomadura*, and *Nocardiopsis* species.

Acknowledgment

The author thanks Mrs. Ira Bhatnagar for drawing the structures and providing required information for improving the manuscript.

References

1 Schumacher, R.W., Talmage, S.C., Miller, S.A., Sarris, K.E., Davidson, B.S., and Goldberg, A. (2003) Isolation and structure determination of an antimicrobial ester from a marine sediment-derived bacterium. *J. Nat. Prod.*, **66**, 1291–1293.

2 El-Gendy, M.M., Shaaban, M., Shaaban, K.A., El-Bondkly, A.M., and Laatsch, H. (2008) Essramycin: a first triazolopyrimidine antibiotic isolated from nature. *J. Antibiot. (Tokyo)*, **61**, 149–157.

3 Manam, R.R., Teisan, S., White, D.J., Nicholson, B., Grodberg, J., Neuteboom,

S.T., Lam, K.S., Mosca, D.A., Lloyd, G.K., and Potts, B.C. (2005) Lajollamycin, a nitro-tetraene spiro-beta-lactone-gamma-lactam antibiotic from the marine actinomycete *Streptomyces nodosus*. *J. Nat. Prod.*, **68**, 240–243.

4 Haste, N.M., Farnaes, L., Perera, V.R., Fenical, W., Nizet, V., and Hensler, M.E. (2011) Bactericidal kinetics of marine-derived napyradiomycins against contemporary methicillin-resistant *Staphylococcus aureus*. *Mar. Drugs*, **9**, 680–689.

5 Haste, N.M., Perera, V.R., Maloney, K.N., Tran, D.N., Jensen, P., Fenical, W., Nizet, V., and Hensler, M.E. (2010) Activity of the streptogramin antibiotic etamycin against methicillin-resistant *Staphylococcus aureus*. *J. Antibiot. (Tokyo)*, **63**, 219–224.

6 El-Gendy, M.M., Shaaban, M., El-Bondkly, A.M., and Shaaban, K.A. (2008) Bioactive benzopyrone derivatives from new recombinant fusant of marine *Streptomyces*. *Appl. Biochem. Biotechnol.*, **150**, 85–96.

7 Huang, Y.F., Tian, L., Fu, H.W., Hua, H.M., and Pei, Y.H. (2006) One new anthraquinone from marine *Streptomyces* sp. FX-58. *Nat. Prod. Res.*, **20**, 1207–1210.

8 Huang, Y.F., Tian, L., Sun, Y., and Pei, Y.H. (2006) Two new compounds from marine *Streptomyces* sp. FX-58. *J. Asian Nat. Prod. Res.*, **8**, 495–498.

9 Asolkar, R.N., Jensen, P.R., Kauffman, C.A., and Fenical, W. (2006) Daryamides A–C, weakly cytotoxic polyketides from a marine-derived actinomycete of the genus *Streptomyces* strain CNQ-085. *J. Nat. Prod.*, **69**, 1756–1759.

10 Kwon, H.C., Kauffman, C.A., Jensen, P.R., and Fenical, W. (2006) Marinomycins A–D, antitumor–antibiotics of a new structure class from a marine actinomycete of the recently discovered genus "*Marinispora*". *J. Am. Chem. Soc.*, **128**, 1622–1632.

11 Hawas, U.W., Shaaban, M., Shaaban, K.A., Speitling, M., Maier, A., Kelter, G., Fiebig, H.H., Meiners, M., Helmke, E., and Laatsch, H. (2009) Mansouramycins A–D, cytotoxic isoquinoline quinones from a marine streptomycete. *J. Nat. Prod.*, **72**, 2120–2124.

12 Bredholt, H., Fjaervik, E., Johnsen, G., and Zotchev, S.B. (2008) Actinomycetes from sediments in the Trondheim fjord, Norway: diversity and biological activity. *Mar. Drugs*, **6**, 12–24.

13 Maskey, R.P., Li, F., Qin, S., Fiebig, H.H., and Laatsch, H. (2003) Chandrananimycins A approximately C: production of novel anticancer antibiotics from a marine *Actinomadura* sp. isolate M048 by variation of medium composition and growth conditions. *J. Antibiot. (Tokyo)*, **56**, 622–629.

14 Engelhardt, K., Degnes, K.F., Kemmler, M., Bredholt, H., Fjaervik, E., Klinkenberg, G., Sletta, H., Ellingsen, T.E., and Zotchev, S.B. (2010) Production of a new thiopeptide antibiotic, TP-1161, by a marine *Nocardiopsis* species. *Appl. Environ. Microbiol.*, **76**, 4969–4976.

15 Fu, P., Liu, P., Qu, H., Wang, Y., Chen, D., Wang, H., Li, J., and Zhu, W. (2011) A-pyrones and diketopiperazine derivatives from the marine-derived actinomycete *Nocardiopsis dassonvillei* HR10-5. *J. Nat. Prod.*, **74**, 2219–2223.

16 Macherla, V.R., Liu, J., Bellows, C., Teisan, S., Nicholson, B., Lam, K.S., and Potts, B.C. (2005) Glaciapyrroles A, B, and C, pyrrolosesquiterpenes from a *Streptomyces* sp. isolated from an Alaskan marine sediment. *J. Nat. Prod.*, **68**, 780–783.

17 Fu, P., Yang, C., Wang, Y., Liu, P., Ma, Y., Xu, L., Su, M., Hong, K., and Zhu, W. (2012) Streptocarbazoles A and B, two novel indolocarbazoles from the marine-derived actinomycete strain *Streptomyces* sp. FMA. *Org. Lett.*, **14**, 2422–2425.

18 Gorajana, A., Venkatesan, M., Vinjamuri, S., Kurada, B.V., Peela, S., Jangam, P., Poluri, E., and Zeeck, A. (2007) Resistoflavine, cytotoxic compound from a marine actinomycete, *Streptomyces chibaensis* AUBN1/7. *Microbiol. Res.*, **162**, 322–327.

19 Ravikumar, S., Fredimoses, M., and Gnanadesigan, M. (2012) Anticancer property of sediment actinomycetes against MCF-7 and MDA-MB-231 cell lines. *Asia Pac. J. Trop. Biomed.*, **2**, 92–96.

20 Shin, H.J., Jeong, H.S., Lee, H.S., Park, S.K., Kim, H.M., and Kwon, H.J. (2007) Isolation and structure determination of streptochlorine, an antiproliferative agent from a marine-derived *Streptomyces* sp.

04DH110. *J. Microbiol. Biotechnol.*, **17**, 1403–1406.

21 Lorente, A., Pla, D., Cañedo, L.M., Albericio, F., and Alvarez, M. (2010) Isolation, structural assignment, and total synthesis of barmumycin. *J. Org. Chem.*, **75**, 8508–8515.

22 Stritzke, K., Schulz, S., Laatsch, H., Helmke, E., and Beil, W. (2004) Novel caprolactones from a marine streptomycete. *J. Nat. Prod.*, **67**, 395–401.

23 Mitchell, S.S., Nicholson, B., Teisan, S., Lam, K.S., and Potts, B.C. (2004) Aureoverticillactam, a novel 22-atom macrocyclic lactam from the marine actinomycete *Streptomyces aureoverticillatus*. *J. Nat. Prod.*, **67**, 1400–1402.

24 Martin, G.D., Tan, L.T., Jensen, P.R., Dimayuga, R.E., Fairchild, C.R., Raventos-Suarez, C., and Fenical, W. (2007) Marmycins A and B, cytotoxic pentacyclic C-glycosides from a marine sediment-derived actinomycete related to the genus *Streptomyces*. *J. Nat. Prod.*, **70**, 1406–1409.

25 Sánchez López, J.M., Martínez Insua, M., Pérez Baz, J., Fernández Puentes, J.L., and Cañedo Hernández, L.M. (2003) New cytotoxic indolic metabolites from a marine *Streptomyces*. *J. Nat. Prod.*, **66**, 863–864.

26 Fu, P., Wang, S., Hong, K., Li, X., Liu, P., Wang, Y., and Zhu, W. (2011) Cytotoxic bipyridines from the marine-derived actinomycete *Actinoalloteichus cyanogriseus* WH1–2216-6. *J. Nat. Prod.*, **74**, 1751–1756.

27 Pérez, M., Crespo, C., Schleissner, C., Rodríguez, P., Zúñiga, P., and Reyes, F. (2009) Tartrolon D, a cytotoxic macrodiolide from the marine-derived actinomycete *Streptomyces* sp. MDG-04–17–069. *J. Nat. Prod.*, **72**, 2192–2194.

28 Dhevagi, P. and Poorani, E. (2006) Isolation and characterization of L-asparaginase from marine actinomycetes. *Indian J. Biotechnol.*, **5**, 514–520.

29 Poorani, E., Saseetharan, M.K., and Dhevagi, P. (2009) L-Asparaginase production by newly isolated marine *Streptomyces* sp. strain EPD 27: molecular identification. *Int. J. Integr. Biol.*, **7**, 150–155.

30 Sahu, M.K., Poorani, E., Sivakumar, K., Thangaradjou, T., and Kannan, L. (2007) Partial purification and anti-leukemic activity of L-asparaginase enzyme of the actinomycete strain LA-29 isolated from the estuarine fish, *Mugil cephalus* (Linn.). *J. Environ. Biol.*, **28**, 645–650.

31 Han, Y., Yang, B., Zhang, F., Miao, X., and Li, Z. (2009) Characterization of antifungal chitinase from marine *Streptomyces* sp. DA11 associated with South China Sea sponge *Craniella australiensis*. *Mar. Biotechnol.*, **11**, 132–140.

32 Huang, H., Yao, Y., He, Z., Yang, T., Ma, J., Tian, X., Li, Y., Huang, C., Chen, X., Li, W., Zhang, S., Zhang, C., and Ju, J. (2011) Antimalarial β-carboline and indolactam alkaloids from *Marinactinospora thermotolerans*, a deep sea isolate. *J. Nat. Prod.*, **74**, 2122–2127.

33 Prudhomme, J., McDaniel, E., Ponts, N., Bertani, S., Fenical, W., Jensen, P., and LeRoch, K. (2008) Marine actinomycetes: a new source of compounds against the human malaria parasite. *PLoS One*, **3**, e2335.

34 El-Gendy, M.M. and El-Bondkly, A.M. (2010) Production and genetic improvement of a novel antimycotic agent, saadamycin, against dermatophytes and other clinical fungi from endophytic *Streptomyces* sp. Hedaya 48. *J. Ind. Microbiol. Biotechnol.*, **37**, 831–841.

35 Kathiresan, K., Balamuruganathan, R., and Masilamani Selvam, M. (2005) Fungicidal activity of marine actinomycetes against phytopathogenic fungi. *Indian J. Biotechnol.*, **4**, 271–276.

36 Ravikumar, S., Krishnakumar, S., Jacob Inbaneson, S., and Gnanadesigan, M. (2010) Antagonistic activity of marine actinomycetes from Arabian Sea coast. *Arch. Appl. Sci. Res.*, **2**, 273–280.

37 Selvin, J., Shanmughapriya, S., Gandhimathi, R., Seghal Kiran, G., Rajeetha Ravji, T., Natarajaseenivasan, K., and Hema, T.A. (2009) Optimization and production of novel antimicrobial agents from sponge associated marine actinomycetes *Nocardiopsis dassonvillei* MAD08. *Appl. Microbiol. Biotechnol.*, **83**, 435–845.

38 Gandhimathi, R., Arunkumar, M., Selvin, J., Thangavelu, T., Sivaramakrishnan, S., Kiran, G.S., Shanmughapriya, S., and

Natarajaseenivasan, K. (2008) Antimicrobial potential of sponge associated marine actinomycetes. *J. Med. Mycol.*, **18**, 16–22.

39 Schneemann, I., Kajahn, I., Ohlendorf, B., Zinecker, H., Erhard, A., Nagel, K., Wiese, J., and Imhoff, J.F. (2010) Mayamycin, a cytotoxic polyketide from a *Streptomyces* strain isolated from the marine sponge *Halichondria panicea*. *J. Nat. Prod.*, **73**, 1309–1312.

40 Motohashi, K., Inaba, K., Fuse, S., Doi, T., Izumikawa, M., Khan, S.T., Takagi, M., Takahashi, T., and Shin-ya, K. (2011) JBIR-56 and JBIR-57, 2(1*H*)-pyrazinones from a marine sponge-derived *Streptomyces* sp. SpD081030SC-03. *J. Nat. Prod.*, **74**, 1630–1635.

41 Motohashi, K., Takagi, M., and Shin-Ya, K. (2010) Tetrapeptides possessing a unique skeleton, JBIR-34 and JBIR-35, isolated from a sponge-derived actinomycete, *Streptomyces* sp. Sp080513GE-23. *J. Nat. Prod.*, **73**, 226–228.

42 Hosoya, T., Hirokawa, T., Takagi, M., and Shin-ya, K. (2012) Trichostatin analogues JBIR-109, JBIR-110, and JBIR-111 from the marine sponge-derived *Streptomyces* sp. RM72. *J. Nat. Prod.*, **75**, 285–289.

43 Izumikawa, M., Khan, S.T., Takagi, M., and Shin-ya, K. (2010) Sponge-derived *Streptomyces* producing isoprenoids via the mevalonate pathway. *J. Nat. Prod.*, **73**, 208–212.

44 Schneemann, I., Ohlendorf, B., Zinecker, H., Nagel, K., Wiese, J., and Imhoff, J.F. (2010) Nocapyrones A–D, gamma-pyrones from a *Nocardiopsis* strain isolated from the marine sponge *Halichondria panicea*. *J. Nat. Prod.*, **73**, 1444–1447.

45 Simmons, L., Kaufmann, K., Garcia, R., Schwär, G., Huch, V., and Müller, R. (2011) Bendigoles D–F, bioactive sterols from the marine sponge-derived *Actinomadura* sp. SBMs009. *Bioorg. Med. Chem.*, **19**, 6570–6575.

11
Antimicrobial Agents from Marine Cyanobacteria and Actinomycetes

Arnab Pramanik, Malay Saha, and Barindra Sana

11.1
Introduction

Continuous discovery of new antibiotics is essential due to the ever-growing resistance of microorganisms to existing drugs and the emergence of new pathogens. Many antibiotics in today's clinical use are secondary metabolites of actinomycetes or their semisynthetic derivatives. Antibiotic from actinomycetes is an established yet interesting area for researchers, while secondary metabolite of cyanobacteria is still an emerging research field. Most antibiotics from actinomycetes were discovered with traditional whole-cell screening method by testing crude actinomycetes culture/extract against pathogens; examples include erythromycin, tetracycline, rifampicin, and daptomycin. Later many successful modifications were made to improve their activity – the basic idea was to retain the core scaffold of a chemical class and develop derivatives by changing chemical groups attached to the main scaffold. Often the modified antibiotics encounter the preexisting resistance mechanisms and questions arise regarding how many modified molecules can be developed using the known scaffolds. More recently, target-based discovery approach using bacterial genomics, combinatorial chemistry, and high-throughput screening was expected to change the scenario dramatically. Hundreds of microbial genes were sequenced and thousands of molecules were screened but that resulted in very few lead compounds, and none of them were approved for clinical use. These difficulties have led to a return to whole-cell screening as a meaningful way for searching new antimicrobial agents.

Like other bioactive compounds, most of the antimicrobial agents have been derived from terrestrial organisms. Actinomycetes and fungi showed extreme productivity for novel antimicrobial agents, but the number of reports of novel antimicrobial scaffold from this group of microorganisms is gradually declining, which gives rise to the urge for widening the research area. As a result, bioprospecting has focused on various unexplored environments. Marine ecosystem is unique with diverse environmental parameters such as high salt concentration, extreme temperature and pH, high

Marine Microbiology: Bioactive Compounds and Biotechnological Applications, First Edition.
Edited by Se-Kwon Kim

pressure, and low nutrient availability. Marine microorganisms developed unique metabolic pathways to adopt in this environment, which make them potential target as a source of unique secondary metabolites. Within the last two decades, marine natural product research expanded significantly and actinomycetes emerged as a lead source of marine antimicrobial agents. Chemically diverse antimicrobial agents showed their potential to combat new multiple drug-resistant (MDR) pathogens. Recent research on antimicrobial agents has expanded beyond actinomycetes and fungi; other microbial genera are included in the study and marine cyanobacteria emerge as a promising source of novel antimicrobial agents.

Historically, exploitation of cyanobacteria was neglected due to their slow growth rate and a number of issues related to their handling. Most of these problems are now resolved and study of natural products from cyanobacteria got a new dimension in the last decade. Exploration of this traditionally neglected microorganism proved huge diversity in chemical composition and biological activity of their metabolites. Investigations into the biochemical reactions responsible for synthesis of these compounds have revealed unique metabolic pathways that are rarely observed in other microorganisms [1]. Several antimicrobial agents were isolated from marine cyanobacteria, which include antibiotics, antifungal, antiviral, and antimycobacterial compounds [2–5]. A recent review reports the number of striking structural features of many marine cyanobacterial metabolites, which confer their unique biological activities and high chemical stability [6]. However, most of these studies deal with isolation and primary characterization of antimicrobial moieties from marine cyanobacteria. Intensive advanced research is essential to justify their potential for clinical application – future research in this direction may place marine cyanobacteria in the first row of commercial antibiotic producers.

11.2 Antimicrobials from Marine Actinomycetes

Antibiotics from marine actinomycetes are of particular interest as very few have been reported as compared to their terrestrial counterpart. Several bioactive compounds from marine sources were reported to be 100 times more potent than similar terrestrial products. Antibacterials and other bioactive compounds from actinomycetes constitute more than 70% of antimicrobial agents reported since today; out of them about 68% is from genus *Streptomyces* and the rest (32%) is from *Actinomycetes*. This observation suggests that actinomycetes should be the most important target in antimicrobial screening programs from any natural resources [7]. Table 11.1 summarizes various antimicrobial agents reported from marine Actinobacteria.

11.2.1 Antibacterial Activity

Actinomycetes are one of the oldest reported lives on earth. In billion years of evolution they have developed suitable ways to inhibit other bacterial growth, and

Table 11.1 Antimicrobial agents from marine Actinobacteria.

Compound	Source microorganism	Activity	Reference
Bithiazole-type antibiotics	*Myxococcus* sp.	Antibacterial	[8]
Bonactin	*Streptomyces* sp. BD21-2	Antibacterial, antifungal	[9]
Chandrananimycins A, B, and C	*Actinomadura* sp.	Antibacterial	[10]
Abyssomicin	*Verrucosispora* strain	Antibacterial	[11]
Essramycin (triazolopyrimidine)	*Streptomyces* sp.	Antibacterial	[12]
Lynamicins (bisindole pyrrole)	*Marinispora* sp.	Antibacterial	[13]
Marinopyrroles (bispyrrole)	*Streptomyces* sp.	Antibacterial	[14]
TP-1161 (thiazolyl peptide)	*Nocardiopsis* sp.	Antibacterial	[15]
2-Allyloxyphenol	*Streptomyces* sp.	Antimicrobial, antioxidant	[16]

thus are a prominent source of antibiotics [17]. Several antibiotics were reported from marine actinomycetes, mostly by last two decades. Macherla *et al.* [18] reported a group of antibacterial compounds Lipoxazolidinones A–C from the novel marine actinomycetes strain NPS008920, isolated from marine sediments. All the isolated compounds showed broad-spectrum antimicrobial activity similar to that of the commercial antibiotic linezolid.

Exploration of intertidal zone of Bay of Bengal established its microbial diversity. Antimicrobial activity was observed in several actinomycetes isolated from marine ediments of the Sunderbans – world's largest mangrove forest [16,19–21]. A broad-spectrum antimicrobial agent was isolated from the spore-forming Actinobacteria MS3/20 [19]. The compound was active against Gram-positive and Gram-negative bacteria, molds, yeast, and several MDR pathogens, including methicillin-resistant *Staphylococcus aureus* (MRSA). Production of this compound was optimized in a special type of fermenter [21]. Arumugam *et al.* [16] reported a previously known synthetic 2-allyloxyphenol of molecular weight 150 and empirical molecular formula $C_9H_{10}O_2$ from another isolate Actinobacteria MS1/7. This compound is inhibitory to 21 bacteria and 3 fungi and also possesses strong antioxidant property. Hydroxyl and allyloxy groups in 2-allyloxyphenol were shown to be responsible for antimicrobial and antioxidant activities.

Another screening program of marine streptomycetes reported several known and unknown bioactive compounds [22]. Two novel antimicrobial agents designated as himalomycins A and B were isolated from the ethyl acetate extract of *Streptomyces* B6921. Both the compounds were highly active against *S. aureus*, *Escherichia coli*, *Bacillus subtilis*, and *Streptomyces viridochromogenes*. The same researchers reported three novel antibiotics designated chandrananimycins A–C from the culture broth of the marine *Actinomadura* sp. isolate M045 [10]. They also isolated a new macrolide antibiotic designated as chalcomycin B from the culture broth of a marine *Streptomyces* B7064 [23]. Structures of all new antibiotics were determined by the application of mass spectroscopy, and 1D and 2D NMR as well by comparing the data with similar known compounds.

A novel strategy was used to induce antibiotic production by a marine *Streptomyces* [24]. *Streptomyces tenjimariensis* produces the antibiotic istamycin A & B only when cocultured with other bacterial species. Twelve different microbial species were reported to induce istamycin production. Limiting natural resources may cause competition within a microbial community that triggers biosynthesis of antimicrobial compounds to counter the competitive species. The isolated compound was reported to inhibit the growth of competitor colonies. This technique may be useful for induction of antimicrobial compounds against drug-resistant pathogens, and if successful, it would open a new research area for stress-induced drug development using marine actinomycetes.

Streptomyces sp. NS 13239 is a producer of novel antibiotics isolated from the east coast of Korea [25]. It can efficiently inhibit methicillin-resistant *S. aureus* and several other Gram-positive bacteria but has no activity against Gram-negative bacteria. The optimum growth conditions and antimicrobial activity of the *Streptomyces* were optimized in various media. A recent report described submerged fermentation of extracellular antimicrobial compounds by *Amycolatopsis alba* var. nov. DVR D4 that showed broad antibacterial spectra against Gram-positive and Gram-negative bacteria. Several parameters were optimized for maximum antibacterial activity [26].

11.2.2
Antifungal Activity

Fungi are no more under control of traditional antifungal agents and even some newer azole drugs are being challenged by alarming emergence of resistant strains. Confrontation of fungal infection becomes a challenging job due to emerging numbers of drug-resistant fungi. Reports on actinomycetes-derived antifungal agents are really scanty in comparison to antibacterial compounds. Walters *et al.* [27] reported the structure of a member of the polyene macrolide class of antifungal agents that is a natural product obtained by fermentation of *Streptomyces nodosus* M4575 isolated from a Venezuelan soil sample. A huge number of marine actinomycetes were screened and studied for antifungal activity [28,29]). Zheng *et al.* [29] studied 17 antimicrobial isolates and reported inhibition of fungal growth by two marine-derived actinomycetes. In another study 160 marine actinomycetes were studied, out of which 10 strains were active against several fungal strains [28]. Media compositions were optimized for production of fungicidal compounds by all 10 actinomycetes. A new compound, assigned the trivial name bonactin, was isolated from the liquid culture of a *Streptomyces* sp. BD21-2 obtained from a shallow-water sediment sample collected at Kailua Beach, Oahu, Hawaii [9]. The structure of the compound was elucidated by 1D and 2D NMR, HRFABMS, and IR. This broad-range antimicrobial compound displayed activity against several fungus species as well as Gram-positive and Gram-negative bacteria.

11.2.3
Antiviral Activity

Viral infections became a major concern in recent years when they broke out as epidemics and took a toll of several lives. Some of the pathogens were controlled by existing antiviral agents while others were resistant to them. Antiviral compounds are yet to be developed for several viral diseases, including AIDS. Several antiviral agents were reported from terrestrial actinomycetes but their marine counterparts are yet to be explored. Kumar *et al.* [30] screened marine actinomycetic compounds for treatment of white spot syndrome in a marine shrimp and reported decreased mortality of the postlarval shrimp by the action of concentrated fermentation broth of six marine actinomycetes [30]. This observation suggests antiviral potential of marine actinomycetic metabolites but further exploration is essential for isolation and identification of those compounds.

Figure 11.1 shows chemical structure of various bioactive compounds isolated from marine actinomycetes.

Diazepinomicin

Resistoflavine

Abyssomicin C

Salinosporamide A

Proximicin C

Mansouramycin C

Lynamicin B

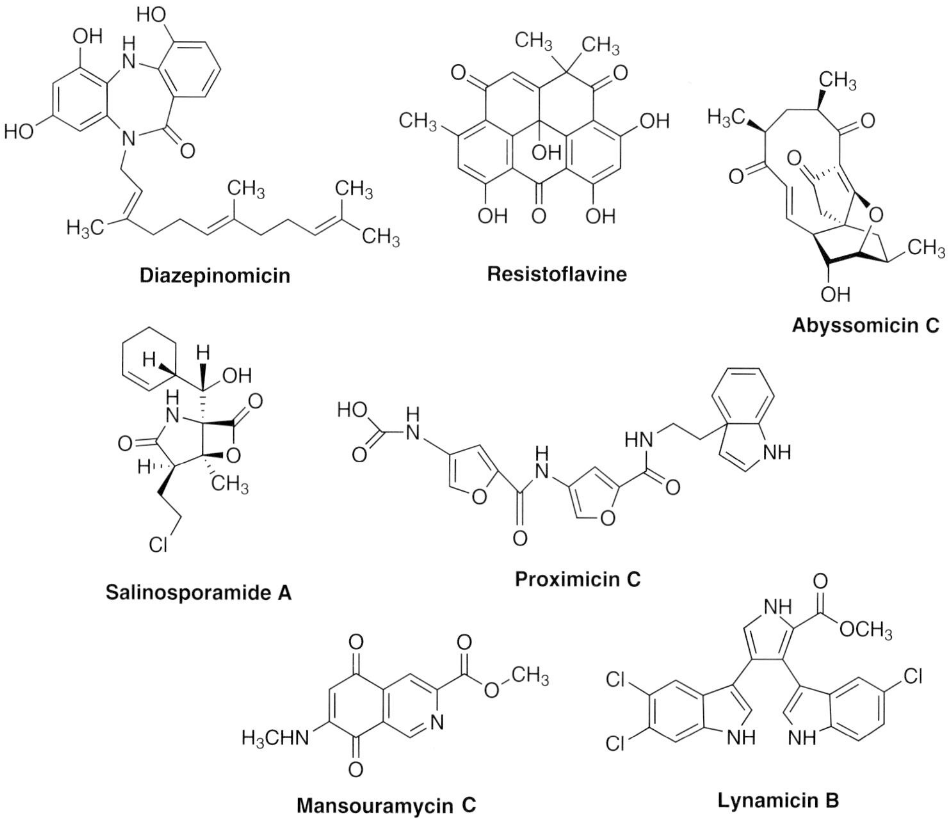

Figure 11.1 Chemical structures of the bioactive compounds from marine actinomycetes.

11.3 Antimicrobials from Marine Cyanobacteria

Cyanobacteria have been shown to be a source of antibiotic compounds [31]. Most of the antibiotic metabolites isolated until now were accumulated in the cyanobacterial cell but cyanobacteria are also known to produce extracellular antibiotic compounds [32]. In spite of the studies carried out so far, many marine cyanobacterial compounds are still largely unexplored and the chemicals involved are mostly unidentified, thus providing an opportunity for discovery of new bioactive compounds including antimicrobial agents.

11.3.1 Antibacterial Activity

Filamentous tropical marine cyanobacteria are prokaryotic microorganisms usually found in the littoral zones as conspicuous intertidal and infralittoral mats. Out of the 800 reported secondary marine cyanobacterial metabolites, 49% are derived from *Oscillatoriales* followed by *Nostocales* and *Chroococcales* with 26 and 15%, respectively [33]. To date, more than 300 nitrogen-containing metabolites have been isolated mainly from two marine cyanobacterial genera *Lyngbya* and *Symploca* species [6].

Antibacterial substances identified also include fatty acids, bromophenols, terpenoids, *N*-glycosides, cyclic depsipeptides, lipopeptides, cyclic peptides, cylicundecapeptides, and isonitrile-containing indole. *Nodularia harveyana*, a dinitrogen-fixing cyanobacteria isolate from the Mediterranean Sea, was assayed for its bioactive compounds. The active substances were lipophilic and showed strong activity against Gram-positive pathogenic bacteria [34]. Bioassay guided fractionation of the lipophilic extract of the marine cyanobacterium *Phormidium ectocarpi* yielded a new natural product hierridin B and 2,4-dimethoxy-6-heptadecyl-phenol [35]. The alkaloids ambiguine isonitriles H and I were isolated from a marine cyanobacterium *Fischerella* sp. These compounds possessed antibacterial activity [36]. Antimicrobial activity was present in the Portuguese marine cyanobacteria *Synechocystis* sp. and *Synechococcus* sp., toward Gram-positive bacteria [2]. Besarhanamides A and B are fatty acid amides purified from the marine cyanobacterium, *Lyngbya majuscula*, collected from Pulau Hantu, Singapore [37]. The marine cyanobacterium *Leptolyngbya crosbyana* forms extensive blooms on Hawaiian coral reefs and results in significant damage to the subtending corals. This *L. crosbyana* produced four heptabrominated polyphenolic ethers, crossbyanols A–D. Crossbyanol B showed antibiotic activity with an MIC value of 2.0–3.9 μg/ml against MRSA [38]. Extracts of several cyanobacterial species collected from different marine locations in Florida (USA) were screened for their ability to disrupt quorum sensing in the reporter strain *Chromobacterium violaceum* CV017 [39]. Eight obligately halophilic, euryhaline cyanobacteria were isolated from Sundarbans. These isolates showed antimicrobial activity against *S. aureus, E. coli, B. subtilis, Pseudomonas aeruginosa*, and multiple drug-resistant clinical isolates [40].

Relatively little has been known about the antimicrobial activity of cyanobacterial compounds from Chroococcales group (e.g., *Synechocystis* and *Synechococcus*). However, Martins *et al.* emphasized potential of these genera as a source of antibiotic compounds that produce substances with inhibitory effects on prokaryotic cells and with apoptotic activity in eukaryotic cell lines, which highlights the importance of these organisms as potential pharmacological agents [2]. Unfortunately, almost all of the studies have used only *in vitro* assays. It is likely that most of the compounds responsible for antibiotic activity will have little or no application in medicine as they are either too toxic or inactive *in vivo* [41]. They may, however, serve as useful leads to create new synthetic antibiotics that might represent in the development of new pharmaceutical agents.

11.3.2
Antifungal Activity

Several extracts of cyanobacteria belonging to Stigonematales, Nostocales, and Oscillatoriales have shown antifungal activity in *in vitro* test systems. Antifungal compounds isolated from these extracts include hapalindoles, tolytoxin, scytophycins, toyocamycin, tjipanazoles, hassallidin A, nostocyclamide, and nostodione [42].

A new antifungal lactone compound, tanikolide, has been isolated from the lipid extract of a Madagascan collection of the marine cyanobacterium *L. majuscula* [43]. Lyngbyabellin B was isolated from a marine cyanobacterium, *L. majuscula*, collected near the Dry Tortugas National Park, Florida. This new cyclic depsipeptide displayed potent toxicity toward the fungus *Candida albicans* [3]. The novel antifungal compound majusculoic acid was isolated from a marine cyanobacterial mat microbial community. Majusculoic acid exhibits antifungal activity against *C. albicans* ATCC 14503 (MIC 8 μM) [44]. Antifungal lipopeptides, lobocyclamides A–C, were isolated from marine cyanobacterium *Lyngbya confervoides* collected from Cay Lobos, Bahamas. Of them, peptides B and C have an unusual amino acid known as 4-hydroxythreonine in addition to rare long b-amino acids, 3-aminooctanoic acid, and 3-aminodecanoic acid. Moreover, peptides A and B exhibited synergism in antifungal activity, but showed less activity when tested individually [45]. Thus, natural products fractionated from marine cyanobacteria seemed to possess strong growth-inhibitory potential against severe disease-causing prokaryotic pathogens. Synthesis of highly complex bioactive metabolites in laboratories is a mammoth task. It can be reduced by synthesizing specifically active unusual moieties of bioproducts with required modifications. Interestingly, some compounds, like venturamides, strongly inhibit specific pathogens. The mode of action of such metabolites lies in interrupting pathways specifically pertaining to the target organism. Identification of such metabolites may not elicit any side effects to the host organism. Cyanobacterial metabolites with antimicrobial properties are listed in Table 11.2.

Table 11.2 Antimicrobial agents from marine cyanobacteria.

Compound	Source microorganism	Activity	Reference
Tanikolide	*L. majuscula*	Antifungal	[43]
Lyngbyabellin B	*L. majuscula*	Antifungal	[3]
Pitipeptolides A and B	*L. majuscula*	Antimycobacterial	[5]
Hectochlorin	*L. majuscula*	Antifungal	[46]
Majusculoic acid	Cyanobacterial mat assemblage	Antifungal	[44]
Carbamidocyclophanes A–C	*Nostoc* sp.	Antibacterial	[47]
Scytoscalarol	*Scytonema* sp.	Antibacterial, antifungal and antimycobacterial	[48]
Lyngbyazothrins A–D	*Lyngbya* sp.	Antibacterial	[49]
Crossbyanol B	*L. crosbyana*	Antibacterial	[38]

11.3.3
Antiviral Activity

Marine cyanobacteria also appear to be a rich source of new antiviral compounds. The initial screening program in Ref. [50] indicated that a large percentage of extracts of field-collected cyanophytes exhibited antiviral activity when assayed against herpes simplex virus, type II (HSV-2). Screening programs of the University of Hawaii and the U.S. National Cancer Institute have demonstrated antiviral activity in approximately 10% of extracts tested (some 600 cyanophyte strains) using live virus test systems for inhibition of HSV-2 and human immunodeficiency virus, type 1 (HIV-1). However, a smaller percentage (2.5%) of the extracts were active against respiratory syncytial virus [51]. Lau *et al.* [52] have also screened extracts of over 900 strains of cultured cyanobacteria for inhibition of the reverse transcriptases of avian myeloblastosis virus (AMV) and HIV-1 and they found that over 2% of extracts showed activity against AMV and HIV reverse transcriptases. Few of the antiviral compounds were isolated from marine cyanobacteria so far. Gustafson *et al.* [53] used a tetrazolium-based microculture to screen extracts of cultured marine cyanobacteria, *Lyngbya lagerheimii* and *Phormidium tenue* (strain CN-2-1), for inhibition of HIV-1. This led to the discovery of sulfonic acid-containing glycolipids as a new class of HIV-1-inhibitory compounds. A cytotoxic and antiviral compound was also produced by the marine cyanobacterium *Lyngbya* sp. Pearl strain in large laboratory-scale batch cultures. Antiviral activity against influenza virus PR8 was found in extracts prepared from early exponential growth phase cells but antiviral activity was not detected in extracts of mid-log or late-log growth phase cells [54]. The potential of these drug candidates to achieve clinical success holds great strategic promise for exploiting the structural complexity of marine cyanobacterial metabolites across diverse therapeutic areas in future drug discovery.

Figure 11.2 shows chemical structure of various bioactive compounds isolated from marine cyanobacteria.

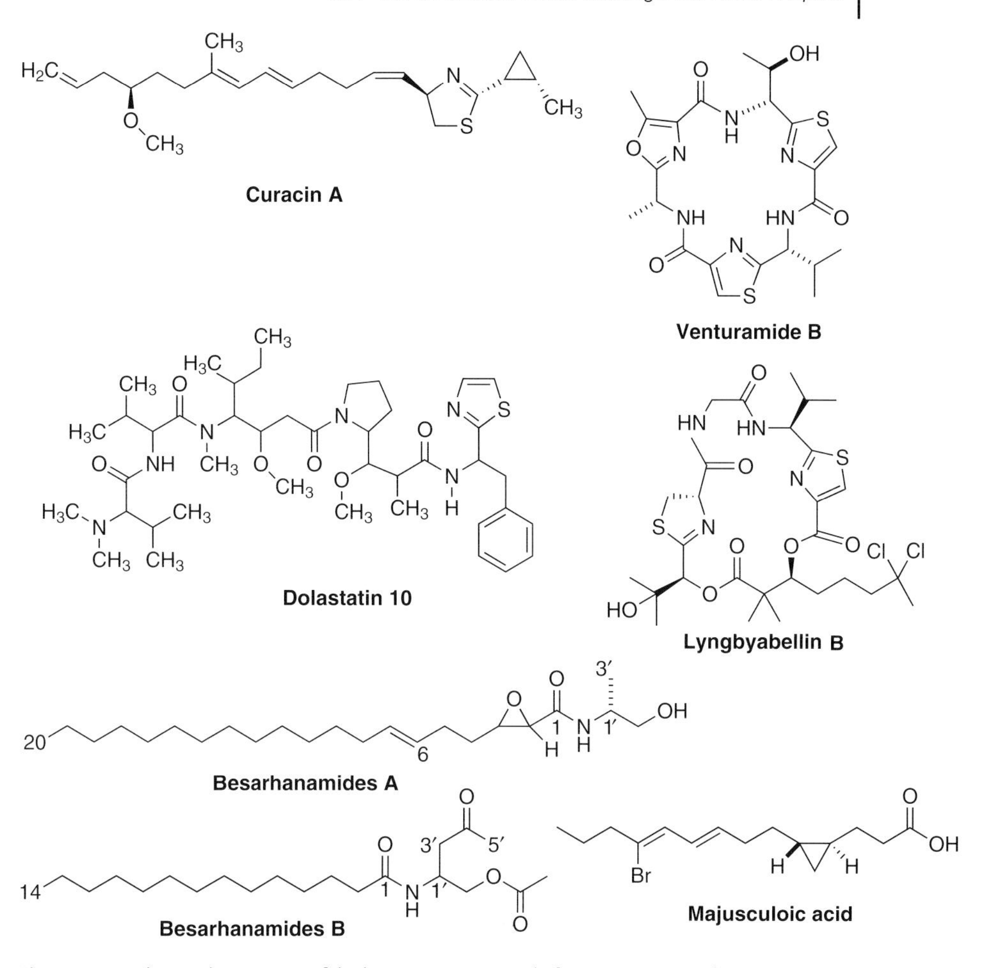

Figure 11.2 Chemical structures of the bioactive compounds from marine cyanobacteria.

11.4
Current Research Status: Challenges and Future Prospects

Thousands of actinomycetes are being screened every year from several fields that cover only a tiny part of the earth's surface and present only a small fraction of actinomycete taxons [55,56]. Screening major microbial resources of the earth is a mammoth task that needs the use of modern tools and technologies. Alternatively, uncommon actinomycetes can be screened from special ecological niches such as marine environments. For a long time, insufficient knowledge about cultivation parameters restricted marine microbial research within some relatively easily cultivable strains from easily accessible near-shore areas. However, better understanding of marine microbiology and advancement of marine technology allowed cultivation of rare marine species from previously inaccessible environments.

In recent years, diverse actinomycetes and cyanobacteria have been reported from diverse marine environment ranging from marine sediments of intertidal zone, shallow-water sediments to the deep sea sediments from hot equatorial region to frozen polar sea [6,19,40,57–61]. On the other hand, several genome-based techniques have been adopted for boosting antibiotic search from actinomycetes. Bioinformatics and chemoinformatics tools along with several databases are being used for antimicrobial research – a countable number of databases and online resources are generated for this purpose, and the number is increasing every day [62]. Phylogenetic approaches were developed to assess the biosynthetic potential and novelty of individual strains that successfully reported several antimicrobial compounds from marine actinomycetes [63,64]. Analysis of 5.2 MB genome of the marine actinomycetes *Salinispora tropica* identified 17 potential secondary metabolites including some antimicrobial compounds [65]. A high-throughput genome scanning technique was used to detect secondary metabolite gene clusters, which reported two novel antimicrobial molecules from two different actinomycetes after predicting the structure and biological activity of the compounds [66,67]. Identification of antibiotic biosynthetic pathways and sequencing of secondary metabolic gene clusters would be helpful for the development of metabolic engineering techniques essential to divert microbial metabolism toward biosynthesis of targeted antimicrobial compounds [68]. Integration between metabolic engineering, molecular modeling, and postgenomic technologies may be extremely effective for biological synthesis of hypothetically active small molecules. There is a recent trend of engineering cyanobacteria for biofuel production [69,70]; similar strategy may be useful for production of other small molecules including antimicrobial agents. Application of latest biotechnological tools is a must for advancement of marine natural product research but their economic feasibility and practicability at a large scale would be extremely important factors at a later stage.

Most reported studies are performed at a small scale, but large-scale production and downstream processing of microbial products are quite different from laboratory-scale studies. The success of screening, purification, characterization, or molecular tailoring would be of no use without successful scale-up of fermentation and downstream processing. Large-scale production of the antimicrobial lead compounds is essential to meet its actual demand where economic viability is a very big issue. For many compounds, large-scale production is essential for detailed studies during several rounds of clinical trial. Clinical trials of some marine metabolites are being performed after their production by aquaculture, chemical synthesis, or fermentation [71]. In recent years, fermentation processes gained considerable importance for antibiotic production from actinomycetes. Antibiotic fermentations by marine actinomycetes have optimized in standard stirred-tank reactors and new fermenters are being designed for this purpose [19,21,72,73]). However, the scenario for cyanobacteria is quite different. Mandatory requirement of suitable light source is a challenge for their large-scale production. Their cultivation is not possible in very large fermenters due to uneven light intensity in different parts of the vessel. Most of the traditional commercial fermenters are not suitable for this purpose but recently developed photobioreactors may be perfect for this job [74,75]. Fueled by successful cyanobacterial biofuel production, several types of photobioreactors are being designed for easier cultivation of different species of cyanobacteria. A recent report even describes development of a novel benthic photobioreactor system for surface-growing/mat-forming cyanobacteria [76]. However, minimal nutrient requirement and less susceptibility to bacterial/fungal contamination are advantageous for large-scale cultures of cyanobacteria. A much cheaper and easy-to-operate technique is being used for rooftop cultivation of cyanobacteria in transparent cylindrical photobioreactors, mostly for biofuel production [74,77]. This method utilizes natural light (sunlight) and partially solves the problem associated with large floor-space requirement, which ensures economic feasibility and sustainability for commercial-scale cyanobacteria cultivation by this approach.

11.5 Conclusions

Immense environmental and biological diversity of the marine world logically argue in favor of vast chemical diversity of marine microbial metabolites. Marine actinomycetes and cyanobacteria have developed the greatest genomic and metabolic diversity in their long journey of evolution. Study of the microorganisms for searching novel antimicrobial agents is still in its infancy, but within a short period they are reported to produce a significant number of novel antimicrobial compounds. More significantly, some of these compounds contain novel core scaffold that can be used for synthesis of new antibiotics. Unique metabolic pathways of marine actinomycetes and cyanobacteria proved their potential as a source of novel antimicrobial agents essential to fight ever-growing number of drug-resistant pathogens. Further improvement of microbial screening and product isolation strategies is of utmost importance for ensuring success in this area. Continuous exploration of untapped marine fields may screen new molecular entities with better antimicrobial prospects but research should also be directed toward engineering of newly screened (or even existing) microorganisms and tailoring of isolated molecules. Exploration of untapped fields, genomic analysis, proper metabolic engineering, and successful modification of isolated molecules would enhance potential of marine cyanobacteria and actinomycetes for confronting the emergence of new drug-resistant pathogens in near future.

References

1 Kehr, J.-C., Picchi, D.G., and Dittmann, E. (2011) Natural product biosyntheses in cyanobacteria: a treasure trove of unique enzymes. *Beilstein J. Org. Chem.*, **7**, 1622–1635.

2 Martins, R.F., Ramos, M.F., Herfindal, L., Sousa, J.A., Skaerven, K., and Vasconcelos, V.M. (2008) Antimicrobial and cytotoxic assessment of marine cyanobacteria – *Synechocystis* and *Synechococcus*. *Mar. Drugs*, **6** (1), 1–11.

3 Milligan, K.E., Marquez, B.L., Williamson, R.T., and Gerwick, W.H. (2000) Lyngbyabellin B, a toxic and antifungal

secondary metabolite from the marine cyanobacterium *Lyngbya majuscula*. *J. Nat. Prod.*, **63** (10), 1440–1443.

4 Uzair, B., Tabassum, S., Rasheed, M., and Rehman, S.F. (2012) Exploring marine cyanobacteria for lead compounds of pharmaceutical importance. *Scientific World J.* doi: 10.1100/2012/179782.

5 Luesch, H., Pangilinan, R., Yoshida, W.Y., Moore, R.E., and Paul, V.J. (2001) Pitipeptolides A and B, new cyclodepsipeptides from the marine cyanobacterium *Lyngbya majuscula*. *J. Nat. Prod.*, **64** (3), 304–307.

6 Tan, L.T. (2007) Bioactive natural products from marine cyanobacteria for drug discovery. *Phytochemistry*, **68** (7), 954–979.

7 Dairi, T., Hamano, Y., Furumai, T., and Oki, T. (1999) Development of a self-cloning system for *Actinomadura verrucosospora* and identification of polyketide synthase genes essential for production of the angucyclic antibiotic pradimicin. *Appl. Environ. Microbiol.*, **65** (6), 2703–2709.

8 Ahn, J.-W. and Kim, B.-S. (2002) Isolation and *in vivo* activities of antifungal compounds from *Myxococcus* sp. JW154 (Myxobacteria). *Korean J. Appl. Microbiol. Biotechnol.*, **30** (2), 162–166.

9 Schumacher, R.W., Talmage, S.C., Miller, S.A., Sarris, K.E., Davidson, B.S., and Goldberg, A. (2003) Isolation and structure determination of an antimicrobial ester from a marine sediment-derived bacterium. *J. Nat. Prod.*, **66** (9), 1291–1293.

10 Maskey, R.P., Li, F.C., Qin, S., Fiebig, H.H., and Laatsch, H. (2003) Chandrananimycins A–C: production of novel anticancer antibiotics from a marine *Actinomadura* sp. isolate M048 by variation of medium composition and growth conditions. *J. Antibiot. (Tokyo)*, **56** (7), 622–629.

11 Bister, B., Bischoff, D., Ströbele, M., Riedlinger, J., Reicke, A., Wolter, F., Bull, A.T., Zähner, H., Fiedler, H.P., and Süssmuth, R.D. (2004) Abyssomicin C – a polycyclic antibiotic from a marine Verrucosispora strain as an inhibitor of the *p*-aminobenzoic acid/tetrahydrofolate biosynthesis pathway. *Angew. Chem., Int. Ed.*, **43** (19), 2574–2576.

12 El-Gendy, M.M.A., Shaaban, M., Shaaban, K.A., El-Bondkly, A.M., and Laatsch, H. (2008) Essramycin: a first triazolopyrimidine antibiotic isolated from nature. *J. Antibiot. (Tokyo)*, **61** (3), 149–157.

13 McArthur, K.A., Mitchell, S.S., Tsueng, G., Rheingold, A., White, D.J., Grodberg, J., Lam, K.S., and Potts, B.C.M. (2008) Lynamicins A–E, chlorinated bisindole pyrrole antibiotics from a novel marine actinomycete. *J. Nat. Prod.*, **71** (10), 1732–1737.

14 Hughes, C.C., Prieto-Davo, A., Jensen, P.R., and Fenical, W. (2008) The marinopyrroles, antibiotics of an unprecedented structure class from a marine *Streptomyces* sp. *Org. Lett.*, **10** (4), 629–631.

15 Engelhardt, K., Degnes, K.F., Kemmler, M., Bredholt, H., Fjaervik, E., Klinkenberg, G., Sletta, H., Ellingsen, T.E., and Zotchev, S.B. (2010) Production of a new thiopeptide antibiotic, TP-1161, by a marine *Nocardiopsis* species. *Appl. Environ. Microbiol.*, **76** (15), 4969–4976.

16 Arumugam, M., Mitra, A., Jaisankar, P., Dasgupta, S., Sen, T., Gachhui, R., Kumar Mukhopadhyay, U., and Mukherjee, J. (2010) Isolation of an unusual metabolite 2 allyloxyphenol from a marine actinobacterium, its biological activities and applications. *Appl. Microbiol. Biotechnol.*, **86** (1), 109–117.

17 Baltz, R.H. (2008) Renaissance in antibacterial discovery from actinomycetes. *Curr. Opin. Pharmacol.*, **8** (5), 557–563.

18 Macherla, V.R., Liu, J., Sunga, M., White, D. J., Grodberg, J., Teisan, S., Lam, K.S., and Potts, B.C.M. (2007) Lipoxazolidinones A, B, and C: antibacterial 4-oxazolidinones from a marine actinomycete isolated from a Guam marine sediment. *J. Nat. Prod.*, **70** (9), 1454–1457.

19 Saha, M., Ghosh, D.Jr., Ghosh, D., Garai, D., Jaisankar, P., Sarkar, K.K., Dutta, P.K., Das, S., Jha, T., and Mukherjee, J. (2005) Studies on the production and purification of an antimicrobial compound and taxonomy of the producer isolated from the marine environment of the Sundarbans. *Appl. Microbiol. Biotechnol.*, **66** (5), 497–505.

20 Saha, M., Jaisankar, P., Das, S., Sarkar, K.K., Roy, S., Besra, S.E., Vedasiromani, J.R.,

Ghosh, D., Sana, B., and Mukherjee, J. (2006) Production and purification of a bioactive substance inhibiting multiple drug resistant bacteria and human leukemia cells from a salt-tolerant marine *Actinobacterium* sp. isolated from the Bay of Bengal. *Biotechnol. Lett.*, **28** (14), 1083–1088.

21 Sarkar, S., Saha, M., Roy, D., Jaisankar, P., Das, S., Gauri Roy, L., Gachhui, R., Sen, T., and Mukherjee, J. (2008) Enhanced production of antimicrobial compounds by three salt-tolerant actinobacterial strains isolated from the Sundarbans in a niche-mimic bioreactor. *Mar. Biotechnol.*, **10** (5), 518–526.

22 Maskey, R.P., Helmke, E., and Laatsch, H. (2003) Himalomycin A and B: isolation and structure elucidation of new fridamycin type antibiotics from a marine Streptomyces isolate. *J. Antibiot. (Tokyo)*, **56** (11), 942–949.

23 Asolkar, R.N., Maskey, R.P., Helmke, E., and Laatsch, H. (2002) Chalcomycin B, a new macrolide antibiotic from the marine isolate *Streptomyces* sp. B7064. *J. Antibiot. (Tokyo)*, **55** (10), 893–898.

24 Slattery, M., Rajbhandari, I., and Wesson, K. (2001) Competition-mediated antibiotic induction in the marine bacterium *Streptomyces tenjimariensis*. *Microbial. Ecol.*, **41** (2), 90–96.

25 Shin, I.S., Lee, J.M., and Park, U.Y. (2000) Optimum condition of marine actinomycetes, *Streptomyces* sp. NS 13239 for growth and producing antibiotics. *J. Fish. Sci. Technol.*, **3** (3–4), 217–221.

26 Dasari, V.R.R.K., Nikku, M.Y., and Donthireddy, S.R.R. (2011) Screening of antagonistic marine actinomycetes: optimization of process parameters for the production of novel antibiotic by *Amycolatopsis alba* var. nov. DVR D4. *J. Microbial Biochem. Technol.*, **3** (5), 92–98.

27 Walters, D.R., Dutcher, J.D., and Wintersteiner, O. (1957) The structure of mycosamine. *J. Am. Chem. Soc.*, **79**, 5076–5077.

28 Kathiresan, K., Balagurunathan, R., and Selvam, M.M. (2005) Fungicidal activity of marine actinomycetes against phytopathogenic fungi. *Indian J. Biotechnol.*, **4** (2), 271–276.

29 Zheng, Z., Zeng, W., Huang, Y., Yang, Z., Li, J., Cai, H., and Su, W. (2000) Detection of antitumor and antimicrobial activities in marine organism associated actinomycetes isolated from the Taiwan Strait, China. *FEMS Microbiol. Lett.*, **188** (1), 87–91.

30 Kumar, S.S., Philip, R., and Achuthankutty, C.T. (2006) Antiviral property of marine actinomycetes against White Spot Syndrome Virus in penaeid shrimps. *Curr. Sci.*, **91** (6), 807–811.

31 Jaki, B., Heilmann, J., and Sticher, O. (2000) New antibacterial metabolites from the cyanobacterium *Nostoc commune* (EAWAG 122b). *J. Nat. Prod.*, **63** (9), 1283–1285.

32 Jaiswal, P., Singh, P.K., and Prasanna, R. (2008) Cyanobacterial bioactive molecules – an overview of their toxic properties. *Can. J. Microbiol.*, **54** (9), 701–717.

33 Gerwick, WH., Coates, R.C., Engene, N., Gerwick, L., Grindberg, R.V., Jones, A.C., and Sorrels, C.M. (2008) Giant marine cyanobacteria produce exciting potential pharmaceuticals. *Microbe*, **3** (6), 277–284.

34 Pushparaj, B., Pelosi, E., and Jüttner, F. (1998) Toxicological analysis of the marine cyanobacterium *Nodularia harveyana*. *J. Appl. Phycol.*, **10** (6), 527–530.

35 Papendorf, O., König, G.M., and Wright, A.D. (1998) Hierridin B and 2,4-dimethoxy-6 heptadecyl-phenol, secondary metabolites from the cyanobacterium *Phormidium ectocarpi* with antiplasmodial activity. *Phytochemistry*, **49** (8), 2383–2386.

36 Raveh, A. and Carmeli, S. (2007) Antimicrobial ambiguines from the cyanobacterium *Fischerella* sp. collected in Israel. *J. Nat. Prod.*, **70** (2), 196–201.

37 Tan, L.T., Chang, Y.Y., and Ashootosh, T. (2008) Besarhanamides A and B from the marine cyanobacterium *Lyngbya majuscula*. *Phytochemistry*, **69** (10), 2067–2069.

38 Choi, H., Engene, N., Smith, J.E., Preskitt, L.B., and Gerwick, W.H. (2010) Crossbyanols A–D, toxic brominated polyphenyl ethers from the Hawaiian bloomforming cyanobacterium *Leptolyngbya crosbyana*. *J. Nat. Prod.*, **73** (4), 517–522.

39 Dobretsov, S., Teplitski, M., Alagely, A., Gunasekera, S.P., and Paul, V.J. (2010) Malyngolide from the cyanobacterium *Lyngbya majuscula* interferes with quorum

sensing circuitry. *Environ. Microbiol. Rep.*, **2** (6), 739–744.

40 Pramanik, A., Sundararaman, M., Das, S., Ghosh, U., and Mukherjee, J. (2011) Isolation and characterization of cyanobacteria possessing antimicrobial activity from the Sundarbans, the worlds largest tidal mangrove forest. *J. Phycol.*, **47** (4), 731–743.

41 Borowitzka, M.A. (1995) Microalgae as sources of pharmaceuticals and other biologically active compounds. *J. Appl. Phycol.*, **7** (1), 3–15.

42 Abed, R.M., Dobretsov, S., and Sudesh, K. (2009) Applications of cyanobacteria in biotechnology. *J. Appl. Microbiol.*, **106** (1), 1–12.

43 Singh, I.P., Milligan, K.E., and Gerwick, W.H. (1999) Tanikolide, a toxic and antifungal lactone from the marine cyanobacterium *Lyngbya majuscula*. *J. Nat. Prod.*, **62** (9), 1333–1335.

44 MacMillan, J.B. and Molinski, T.F. (2005) Majusculoic acid, a brominated cyclopropyl fatty acid from a marine cyanobacterial mat assemblage. *J. Nat. Prod.*, **68** (4), 604–606.

45 MacMillan, J.B., Ernst-Russell, M.A., DeRopp, J.S., and Molinski, T.F. (2002) Lobocyclamides A–C, lipopeptides from a cryptic cyanobacterial mat containing *Lyngbya confervoides*. *J. Org. Chem.*, **67** (23), 8210–8215.

46 Marquez, B.L., Watts, K.S., Yokochi, A., Roberts, M.A., Verdier-Pinard, P., Jimenez, J.I., Hamel, E., Scheuer, P.J., and Gerwick, W.H. (2002) Structure and absolute stereochemistry of hectochlorin, a potent stimulator of actin assembly. *J. Nat. Prod.*, **65** (6), 866–871.

47 Bui, H.T.N., Jansen, R., Pham, H.T.L., and Mundt, S. (2007) Carbamidocyclophanes A–E, chlorinated paracyclophanes with cytotoxic and antibiotic activity from the Vietnamese cyanobacterium *Nostoc* sp. *J. Nat. Prod.*, **70** (4), 499–503.

48 Mo, S., Krunic, A., Pegan, S.D., Franzblau, S.G., and Orjala, J. (2009) An antimicrobial guanidine bearing sesterterpene from the cultured cyanobacterium *Scytonema* sp. *J. Nat. Prod.*, **72** (11), 2043–2045.

49 Zainuddin, E.N., Jansen, R., Nimtz, M., Wray, V., Preisitsch, M., Lalk, M., and Mundt, S. (2009) Lyngbyazothrins A–D, antimicrobial cyclic undecapeptides from the cultured cyanobacterium *Lyngbya* sp. *J. Nat. Prod.*, **72** (8), 1373–1378.

50 Rinehart, K.L.Jr., Shaw, P.D., and Shield, L.S. (1981) Marine natural products as sources of antiviral, antimicrobial, and antineoplastic agents. *Pure Appl. Chem.*, **53** (4), 795–817.

51 Patterson, G.M.L., Baker, K.K., Baldwin, C.L., Bolis, C.M., Caplan, F.R., Larsen, L.K. *et al.* (1993) Antiviral activity of cultured blue-green algae (Cyanophyta). *J. Phycol.*, **29** (1), 125–130.

52 Lau, A.F., Siedlecki, J., Anleitner, J., Patterson, G.M.L., Caplan, F.R., and Moore, R.E. (1993) Inhibition of reverse transcriptase activity by extracts of cultured blue-green algae (Cyanophyta). *Planta Med.*, **59** (2), 148–151.

53 Gustafson, K.R., Cardellina, J.H.II, Fuller, R.W., Weislow, O.S., Kiser, R.F., Snader, K.M., Patterson, G.M.L., and Boyd, M.R. (1989) AIDS-antiviral sulfolipids from cyanobacteria (blue-green algae). *J. Natl. Cancer Inst.*, **81** (16), 1254–1258.

54 Armstrong, J.E., Janda, K.E., Alvarado, B., and Wright, A.E. (1991) Cytotoxin production by a marine *Lyngbya* strain (cyanobacterium) in a large-scale laboratory bioreactor. *J. Appl. Phycol.*, **3** (3), 277–282.

55 Baltz, R.H. (2005) Antibiotic discovery from actinomycetes: will a renaissance follow the decline and fall? *SIM News*, **55**, 186–196.

56 Baltz, R.H. (2007) Antimicrobials from actinomycetes: back to the future. *Microbe*, **2** (3), 125–131.

57 Bredholdt, H., Galatenko, O.A., Engelhardt, K., Fjñrvik, E., Terekhova, L.P., and Zotchev, S.B. (2007) Rare actinomycete bacteria from the shallow water sediments of the Trondheim fjord, Norway: isolation, diversity and biological activity. *Environ. Microbiol.*, **9** (11), 2756–2764.

58 Bull, A.T. and Stachm, J.E.M. (2007) Marine actinobacteria: new opportunities for natural product search and discovery. *Trends Microbiol.*, **15** (11), 491–499.

59 Fenical, W. and Jensen, P.R. (2006) Developing a new resource for drug discovery: marine actinomycete bacteria. *Nat. Chem. Biol.*, **2** (12), 666–673.

60 Fiedler, H.-P., Bruntner, C., Bull, A.T., Ward, A.C., Goodfellow, M., Potterat, O.,

Puder, C., and Mihm, G. (2005) Marine actinomycetes as a source of novel secondary metabolites. *Antonie Van Leeuwenhoek*, **87** (1), 37–42.

61 Panthom-aree, W., Stach, J.E.M., Ward, A.C., Horikoshi, K., Bull, A.T., and Goodfellow, M. (2006) Diversity of actinomycetes isolated from challenger deep sediment (10 898m) from the Mariana Trench. *Extremophiles*, **10** (3), 181–189.

62 Hammami, R. and Fliss, I. (2010) Current trends in antimicrobial agent research: chemo- and bioinformatics approaches. *Drug Discov. Today*, **15** (13–14), 540–546.

63 Jensen, P.R., Mincer, T.J., Williams, P.G., and Fenical, W. (2005) Marine actinomycete diversity and natural product discovery. *Antonie Van Leeuwenhoek*, **87** (1), 43–48.

64 Jensen, P.R., Williams, P.G., Oh, D.-C., Zeigler, L., and Fenical, W. (2007) Species specific secondary metabolite production in marine actinomycetes of the genus *Salinispora*. *Appl. Environ. Microbiol.*, **73** (4), 1146–1152.

65 Udwary, D.W., Zeigler, L., Asolkar, R.N., Singan, V., Lapidus, A., Fenical, W., Jensen, P.R., and Moore, B.S. (2007) Genome sequencing reveals complex secondary metabolome in the marine actinomycetes *Salinispora tropica*. *Proc. Natl. Acad. Sci. USA*, **104** (25), 10376–10381.

66 McAlpine, J.B., Bachmann, B.O., Piraee, M., Tremblay, S., Alarco, A.M., Zazopoulos, E., and Farnet, C.M. (2005) Microbial genomics as a guide to drug discovery and structure elucidation: ECO-02301, a novel antifungal agent, as an example. *J. Nat. Prod.*, **68** (4), 493–496.

67 Banskota, A.H., McAlpine, J.B., Sorensen, D., Ibrahim, A., Aouidate, M., Piraee, M., Alarco, A.-M., Farnet, C.M., and Zazopoulos, E. (2006) Genomic analyses lead to novel secondary metabolites. Part 3. ECO-0501, a novel antibacterial of a new class. *J. Antibiot. (Tokyo)*, **59** (9), 533–542.

68 Williams, G.J., Zhang, C., and Thorson, J.S. (2007) Expanding the promiscuity of a natural-product glycosyltransferase by directed evolution. *Nat. Chem. Biol.*, **3** (10), 657–662.

69 Lu, X. (2010) A perspective: photosynthetic production of fatty acid-based biofuels in genetically engineered cyanobacteria. *Biotechnol. Adv.*, **28** (6), 742–746.

70 Ruffing, A.M. (2011) Engineering cyanobacteria: teaching an old bug new tricks. *Bioeng. Bugs*, **2** (3), 136–149.

71 Munro, M.H.G., Blunt, J.W., Dumdei, E.J., Hickford, S.J.H., Lill, R.E., Li, S.X., Battershill, C.N., and Duckworth, A.R. (1999) The discovery and development of marine compounds with pharmaceutical potential. *J. Biotechnol.*, **70** (1–3), 15–25.

72 Sarkar, S., Mukherjee, J., and Roy, D. (2009) Antibiotic production by a marine isolate (MS 310) in an ultra-low-speed rotating disk bioreactor. *Biotechnol. Bioprocess Eng.*, **14** (6), 775–780.

73 Sarkar, S., Roy, D., and Mukherjee, J. (2010) Production of a potentially novel antimicrobial compound by a biofilm-forming marine *Streptomyces* sp. in a niche-mimic rotating disk bioreactor. *Bioprocess Biosyst. Eng.*, **33** (2), 207–217.

74 Kim, H.W., Vannela, R., Zhou, C., Harto, C., and Rittmann., B.E. (2010) Photoautotrophic nutrient utilization and limitation during semi-continuous growth of *Synechocystis* sp. PCC6803. *Biotechnol. Bioeng.*, **106** (4), 553–563.

75 Sheng, J., Kim, H.W., and Badalamenti, J.P. (2011) Effects of temperature shifts on growth rate and lipid characteristics of *Synechocystis* sp. PCC 6803 in a bench-top photobioreactor. *Bioresour. Technol.*, **102** (24), 11218–11225.

76 Esson, D., Wood, S.A., and Packer, M.A. (2011) Harnessing the self-harvesting capability of benthic cyanobacteria for use in benthic photobioreactors. *AMB Express*, **1** (1), 19.

77 Tsygankov, A.A., Fedorov, A.S., Kosourov, S.N., and Rao, K.K. (2002) Hydrogen production by cyanobacteria in an automated outdoor photobioreactor under aerobic conditions. *Biotechnol. Bioeng.*, **80**, 777–783.

12
Bioactive Compounds from Marine Actinomycetes

Ana M. Íñiguez-Martínez, Graciela Guerra-Rivas, Nahara E. Ayala-Sánchez, and Irma E. Soria-Mercado

12.1
Introduction

The oceans cover 70% of the Earth's surface, including diverse habitats such as shallow waters, tropical coral reefs, and deep sea trenches. Since environmental variations are extreme in terms of pressure, temperature, and salinity, the organisms develop physiological and metabolic adaptations in order to survive, resulting in the production of new chemical compounds.

Over the past 20 years, an increased interest has been put into marine microorganisms, including bacteria, fungi, and algae. Since the "golden age" of antibiotics, which started in 1928 with the discovery of penicillin by Alexander Fleming, studies in bacteria and fungi, especially those from marine sediments, have received a special interest for being a rich source of bioactive secondary metabolites [1].

The use of novel culturing methods and sampling techniques in the 1990s enabled both access to the unknown deep sea environments by the scientists and the discovery of novel bioactive secondary metabolites from the microbial sources endemic to these regions [2]. Secondary metabolites are defined as nonessential metabolic products that can have significant positive effects on the fitness and ecology of the bacterial populations that produce them and are products of large gene collectives that are subject to horizontal gene transfer (HGT) [3,4] and whose distributions among closely related bacterial populations remain largely unknown [5]. Even though the world's oceans comprise most of the Earth's surface, the extensive drug discovery efforts involving soil bacteria have not been extended to this ecosystem [1]. So, special effort should be put into seawater and sediments, since the typical microbial abundance is of $10^6\ ml^{-1}$ in the water column and $10^9\ ml^{-1}$ in the ocean bottom sediments.

12.2
Actinomycetes

Bacteria belonging to the order Actinomycetales, commonly called actinomycetes, are the most prolific source of structurally diverse secondary metabolites known to man

Marine Microbiology: Bioactive Compounds and Biotechnological Applications, First Edition.
Edited by Se-Kwon Kim

[6,7]. The actinomycetes are Gram-positive bacteria whose DNA base composition is comprised of 63–78% guanine plus cytosine (G + C), a percentage higher than that found in any other bacteria. Most of these organisms are aerobic (oxidative), although some of them are facultative or forced anaerobes (fermentation) [8]. These filamentous microorganisms grow as networks called mycelium. If the structure remains on the surface, it is called aerial mycelium; and if it attaches to the substrate surface, it is called substrate mycelium. The individual filaments of the mycelium or hyphae are divided into units as a result of growth of the cell wall into the hyphae at regular intervals along this structure. This process is called septation and each of the resulting septa contains one DNA molecule. The mycelium bacterium is analogous to the mycelium forming filamentous fungi [9]. A particular feature is that the reproduction of Actinobacteria leads to the formation of spores that are produced in specialized hyphae, many of which are developed on the aerial filament. Generally, these structures are unable to move, although some genera produce flagellated spores. Actinobacteria not only inhabit the soil but can also be found in aquatic environments. These bacteria are capable of degrading many complex substances such as chitin and, consequently, play an important role in soil chemistry. In fact, the characteristic odor of soil is due to special metabolites that are known as Geominas [6]. For the past 55 years, academic and pharmaceutical researchers have focused their work mainly on soil-derived actinomycetes and more than 15 000 bioactive molecules have been produced from these microbes, with many of them being used as drugs today. More than 50% of the microbial antibiotics discovered to date are obtained from the soil-derived actinomycete bacteria *Streptomyces* and *Micromonospora*. The belief that actinomycetes isolated from marine sources were largely of terrestrial origin and had been washed to shore and existed in the ocean as metabolically inactive spores [10] has been discarded, since phylogenetic analyses of their 16S rRNA genes indicate that many of these strains belong to a new taxa widely distributed in ocean sediments [11], including some that appear to be unique and obligate marine actinomycete bacteria [12]. These strains represent the most significant source of naturally occurring microbial antibiotics [6,13–15] and antitumor compounds [12,14,15] with specific metabolic and physiological capabilities that had not been observed before in terrestrial microorganisms [16,17]. Actinomycetes are the most economically and biotechnologically valuable prokaryotes [18]. Members of this group are producers of clinically useful antitumor drugs such as anthracyclines (aclarubicin, daunomycin, and doxorubicin), glycopeptides (bleomycin and actinomycin D), aureolic acids (mithramycin), enediynes (neocarzinostatin), antimetabolites (pentostatin), carzinophilin, and mitomycins [19]. A large number of actinomycete bacterial populations have been examined from waters all around the world such as San Diego Bay, Gulf of California, Bahamas, Fiji, and Guam Islands.

12.3 Diversity and Distribution of Marine Actinobacteria

The class Actinobacteria is one of the major phyla in the bacterial domain [20]. The current hierarchical classification comprises 5 subclasses (Acidimicrobidae,

Actinobacteridae, Coriobacteridae, Rubrobacteridae, Nitriliruptoridae), 9 orders (Acidimicrobiales, Actinomycetales, Bifidobacteriales, Coriobacteriales, Rubrobacterales, Solirubrobacterales, Thermoleophilales, Nitriliruptorales, Euzebyales), 55 families, 240 genera, and 3000 species.

Actinobacteria comprise the major components of the microbial communities in marine environments such as marine organic aggregates (marine invertebrates and marine snows) and deep sea gas hydrate reservoirs (deep sea floor) [21]. About 30–40% of clone libraries were obtained from hydrate-bearing sediments from the Gulf of Mexico [22] and the Nankai Trough around Japan [23]. The adaptation of actinomycetes to the marine environment makes this unique ecosystem a prolific source for the discovery of novel secondary metabolites. Originally, these bacteria were found as symbiotic or commensal microorganisms of marine invertebrates, with no specific reason for the production of secondary metabolites.

Scientists, in the search of novel bioactive natural products from actinomycetes, have focused on the extreme and neglected environments of marine sediments [24,25]. Earlier reports of marine habitats were restricted to the genera *Micromonospora*, *Rhodococcus*, and *Streptomyces* [10,26], but now the immense actinobacterial diversity that marine habitats hold [1,27–29] is known and have become important sources of new bioactive compounds with an evolutionary pedigree [25,27,30,31].

The biogeographical distribution of microbial populations appears to be limited by dispersion process [32] and the limitations do not appear to be a dominant force for structuring microbial communities [33]. Also, the accumulation of neutral mutations due to geographical isolation was found in *Bacillus* species, with high cross-continent migration [34].

The population distribution is limited within specific habitat types; biogeographical patterning varies among taxa and is closely linked to the ecology and physiology of the organisms [27].

Recently, the biogeographical patterns have been assessed using genes involved in secondary metabolism [35,36]. These genes develop small molecules within important processes. The HGT participates in the evolution of polyketide synthase type I (PKSI) and polyketide synthase genes and is used to make predictions about the structures of the secondary metabolites produced by complex biosynthetic pathways [35,36]. In addition, clustering of subpopulations of bacteria depends on the collection site [37], providing further evidence for endemism associated with secondary metabolism through terminal restriction fragment length polymorphism (T-RFLP) [38].

Despite the diversity of marine actinomycetes, novel research about their marine sediment diversity has been reported in the early 1990s [39]. In this report, they found the first group of new unusual marine-derived actinomycetes with specific adaptations, like the requirement of seawater for growth and a pan tropical distribution, since they were isolated from marine sediments collected in the Caribbean Sea, the Sea of Cortez, the Red Sea, and the Guam Island. These strains were later named *Salinospora* [11]. To date, three species of *Salinispora* with several phylotypes have been isolated. Two of them (*Salinispora tropica* and *Salinispora arenicola*) have been formally described [40] and the third one (*Salinispora pacifica*) is the newest proposal made [41]. The diversity, distribution, and phylogeny of the latter were investigated employing a sequential screening method for antibiotic activity, RFLP patterns, 16S rRNA, and internal transcribed spacer (ITS) sequence analyses in the sediment samples from Fijian Island [27]. They found little evidence of geographical isolation based on 16S, ITS, or secondary metabolite biosynthetic gene fingerprinting.

12.4 Bioactive Compounds

Members of the order Actinomycetales, mainly the *Streptomyces* genus, are the richest source of bioactive metabolites as antibiotics and antitumor agents [19,42–44]. Forty-five percent of all microbial bioactive secondary metabolites have been isolated from actinomycetes, with 80% of them from *Streptomyces* genus [42]. During 1985–2008, about 100 novel compounds were discovered from actinomycetes [45]. However, this amount represents about 10% of the total synthesized compounds by these bacteria [46]. The procedures for selective isolation of previously unknown bacteria, including actinomycetes that are undergoing pharmacological screening programs, present a small fraction of the vast uncultivated genetic diversity of prokaryotes present in natural habitats [6,20,47–49].

Jensen *et al.* [50] reported a seminested PCR method that incorporates signature nucleotides diagnostic for *Salinospora*, using a specific amplification step in general bacteria or even Actinobacteria, followed by second amplifications with a specific forward primer genus. This Clone library displays RFLP cutting patterns characteristic of the two species: *S. tropica* and *S. arenicola*. The results of cultured strains from sediment samples suggest that in most samples, *Salinospora* occurs largely as spores. The chemical studies of *Salinospora* strains led to the discovery of an unusual bicyclic β-lactone γ-lactam called salinosporamide A (**1**) (Figure 12.1)

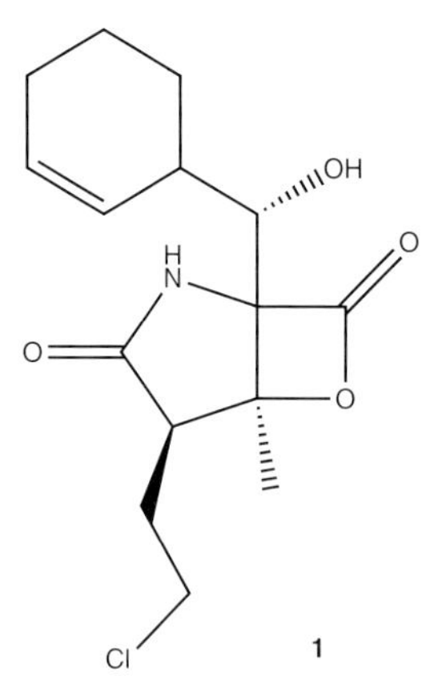

Figure 12.1 Structure of salinosporamide A (**1**) [12].

[12] that displays highly cytotoxic activity in the 60 cell line cancer panel performed by the National Cancer Institute and is also a potent inhibitor of activity of the chymotrypsin-like proteolytic mammalian 20S proteasome (mean $IC_{50} < 10$ nM).

Second-generation total synthesis of salinosporamide A has been described by Kaiya *et al.* [51], establishing three contiguous stereocenters in the γ-lactam structure of the natural product and stereoselective functionalization of a D-arabinose scaffold using a Lewis acid-mediated skeletal rearrangement of a pyranose structure, which enabled the practical conversion of the carbohydrate scaffold to the γ-lactam structure.

Soria-Mercado *et al.* [14], interested in exploring the chemistry of Actinobacteria adapted for survival in ocean sediments, found three new chlorinated dihydroquinones (**2**–**4**) and two previously reported analogues called napyradiomycins (**5** and **6**) (Figure 12.2) [52,53] from the new strain CNQ-525 isolated from sediments collected to 152 m near La Jolla, CA. These new strains belong to the Streptomycetaceae family. Compounds **2**–**6** possess significant antibiotic properties against methicillin-resistant *Staphylococcus aureus* (MRSA) and vancomycin-resistant *Enterococcus faecium* (VREF) and also cytotoxic activity against human colon adenocarcinoma HCT-116.

Similar compounds with napyradiomycin-like structures were discovered by Motohashi *et al.* [54]: napyradiomycin SR, 16-dechloro-16-hydroxynapyradiomycin C2, 18-hydroxynapyradiomycin A1, 18-oxonapyradiomycin A1, 16-oxonapyradiomycin A2, 7-demethyl SF2415A3, 7-demethyl A80915B, and (*R*)-3-chloro-6-hydroxy-8-methoxy-α-lapachone. The last two compounds showed significant antibacterial activity against *S. aureus, Bacillus subtilis, Enterococcus faecalis, E. faecium*, and *Streptococcus pyogenes*.

A novel meroterpenoid called azamerone (**7**) (Figure 12.3) was isolated from the saline culture of a new marine-derived bacterium related to the genus *Streptomyces* strain CNQ-766, belonging to the group "MAR4" [50]. Azamerone is an unprecedented unusual meroterpenoid phthalazinone [55], and this is the first time that a unique phthalazinone ring system has been found as a natural product. The compound displays weak *in vitro* cytotoxicity against mouse splenocyte populations of T cells and macrophages with an IC_{50} value of 40 μM, though it is not clear whether this activity is due to inhibition of a topoisomerase.

The new genus *Marinispora* sp. produced four novel compounds named marinomycines A (**8**), B (**9**), C (**10**), and D, which are a novel class of polyketides, isolated from the CNQ-140 sediment sample collected offshore of La Jolla, CA (Figure 12.4). These compounds inhibit cancer cell proliferation with an average LC_{50} value of 0.2–2.7 μM against 60 cell line cancer panels tested in the National Cancer Institute. Marinomycin A showed significant activity against human melanoma cell lines LOX IMVI, M14, SK-MEL-2, SK-MEL-5, UACC-257, and UACC-62. Marinomycins B and C also showed potent antitumor activities with LC_{50} values of 0.9 and 0.2 μM, respectively [56]. The majority of the polyketide compounds isolated from marine actinomycetes have diverse structures that show antitumor activity. They have been synthesized by multifunctional enzymes called polyketides synthases [57]. Some of the polyketide compounds as well as other terpenoid derivatives isolated from different regions of the world are shown in Table 12.1.

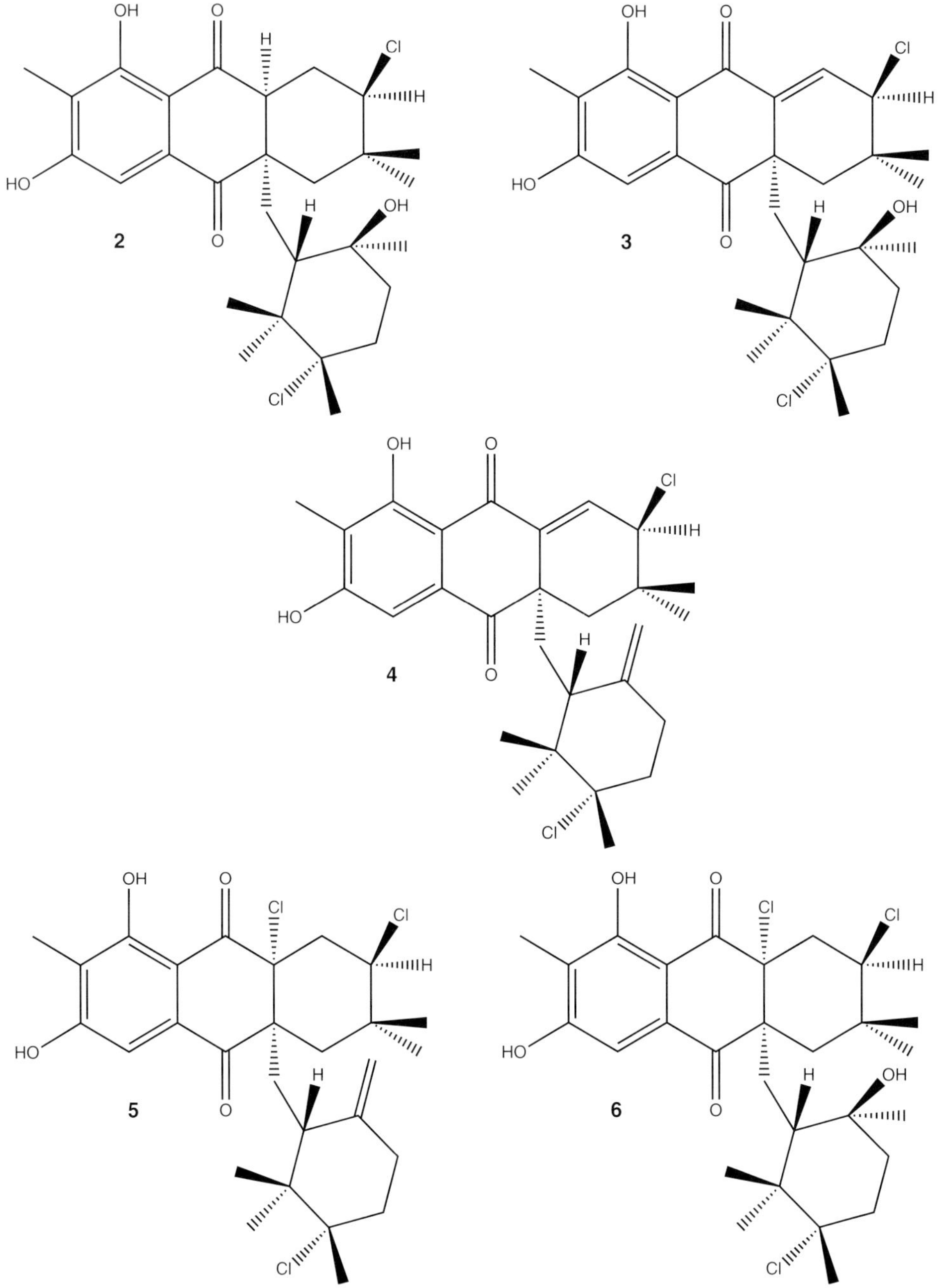

Figure 12.2 New chlorinated dihydroquinones (**2**–**4**) [14] from the new strain CNQ-525, and two previously reported analogues (**5** and **6**) [52,53].

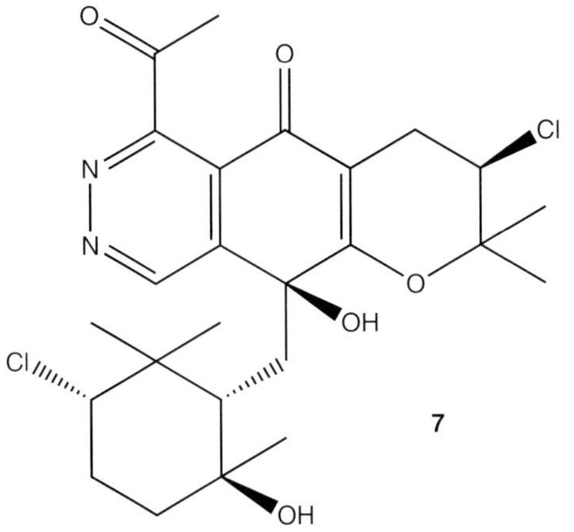

Figure 12.3 Structure of azamerone (**7**), meroterpenoid phthalazinone, isolated from the actinomycete strain CNQ-766 [55].

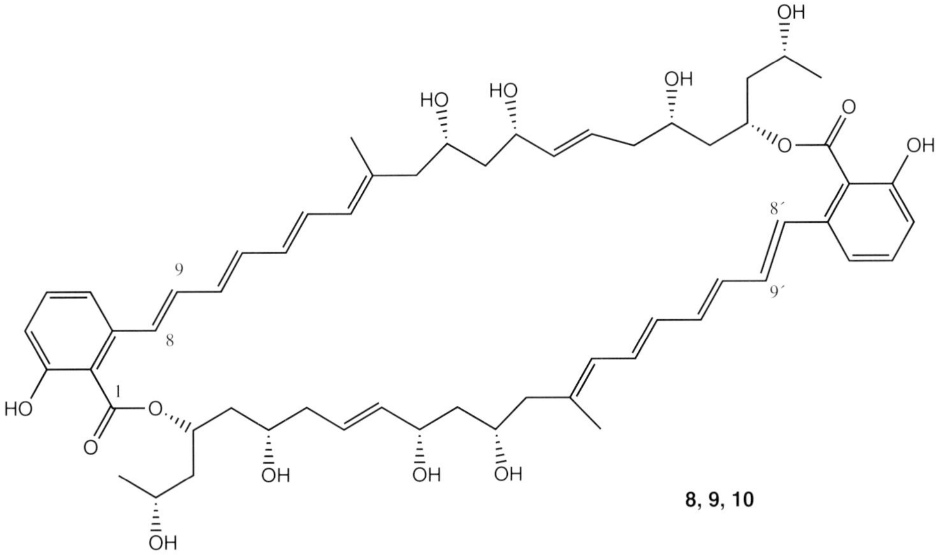

Figure 12.4 Marinomycins A (**8**: *trans*-Δ-8,9-*trans*-Δ-8′9′), B (**9**: *cis*-Δ-8,9-*cis*-Δ-8′9′), and C (**10**: *trans*-Δ-8,9-*cis*-Δ-8′9′), isolated from the strain CNQ-140 [56].

Based on DNA–DNA comparisons, a new *Streptomyces* species was found (actinomycete strain CNQ-418). This new species shared 98.1% 16S rRNA gene sequence identity with its closest neighbor (*Streptomyces sannurensis*). Since 98% of 16S rRNA gene sequence identity is highly conservative in terms of delineating species-level relationship, it was concluded that the CNQ-418 strain must be a new species.

The culture extracts represent the first strain of a new marine lineage to be examined for its ability to produce biologically active natural products; both

Table 12.1 Some bioactive compounds isolated from marine actinomycetes.

Compounds	Organism	Biological activity	Reference
Actinofuranones	*Streptomyces* sp.	Mouse splenocyte T cells	[58]
Ammosamides	*Streptomyces* sp.	Human colon cancer HCT-116 cell line	[59]
Arenicolides	*S. arenicola*	Human colon cancer HCT-116 cell line	[60]
Bohemamines	*Streptomyces* sp.	Human colon cancer HCT-116 cell line	[61]
Carboxamycin	*Streptomyces* sp.	Breast carcinoma MCF7 cell line	[62]
Chartreusin	*Streptomyces* sp.	Murine P388 and L1210 leukemia cells lines	[63]
N-Carboxamido-staurosporine	*Streptomyces* sp.	Human tumor cell lines I	[63]
Chalcomycin	*Streptomyces* sp.	HeLa human cervix carcinoma cell line	[64]
Chinikomycins A and B	*Streptomyces* sp.	Mammary melanoma MEXF 462NL cell line	[65]
Chlorinated dihydroquinones isoprenoid	*Streptomyces* sp.	Human colon cancer HCT-116 cell line	[14]
Cyanosporasides	*S. pacifica*	Human colon cancer HCT-116 cell line	[66]
Daryamides	*Streptomyces* sp.	Human colon cancer HCT-116 cell line	[67]
Echinosporins	*Streptomyces albogriseolus*	Human myelogenous leukemia K562 cell line	[68]
1-Hydroxy-1-norresistomycin	*Streptomyces chinensis*	HepG2 hepatic carcinoma cell line	[69]
Resitoflavine	*S. chinensis*	Human colon cancer HT-29, Jurkat leukemia and mouse melanoma B16F10 cells lines	[69]
Marinomycins A–D	*Marinispora* sp.	Melanoma SK-MEL-5 cell line	[56]
T-muurolol sesquiterpenes	*Streptomyces* sp.	Human tumor cell lines	[70]
Nonactin	*Streptomyces* sp.	Human erythroleukemia K-562 cell line	[71]
Streptokordin	*Streptomyces* sp.	Human cancer cell lines	[71]
Piericidins C7 and C8	*Streptomyces* sp.	YM14–060 Neuro-2a cells	[72]
Saliniketals	*S. arenicola*	Human bladder carcinoma T24 cell line	[73]
Salinipyrones A and B	*S. pacifica*	Mouse splenocyte inhibition of interleukin-5 cell line	[74]
Salinosporamide A	*S. tropica*	Human colon cancer HCT-116 cell line	[12]
Staurosporine analogue indocarbazoles	*Actinomadura* sp.	Human lung cancer A549 cell line	[75]
Undecylprodigiosin	*Saccharopolyspora* sp.	Hepatic carcinoma BEL-7402 cell line	[76]
K252c arcyriaflavin A	*Actinomycete*	Myelogenous leukemia K562 cell line	[77]
Streptochlorin	*Streptomyces* sp.	Human leukemia cells	[78]

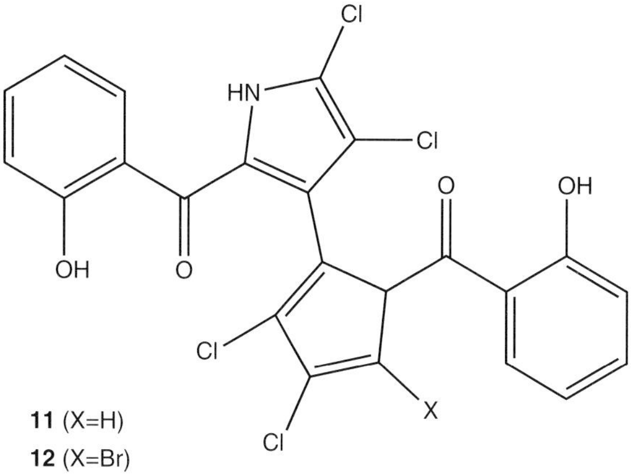

11 (X=H)
12 (X=Br)

Figure 12.5 Marinopyrroles A (**11**) and B (**12**), isolated from a new *Streptomyces* species (actinomycete strain CNQ-418) [79].

marinopyrroles A (**11**) and B (**12**) (Figure 12.5) [79] displayed significant antibiotic activity against MRSA, as well as cytotoxicity against human cancer cell line HCT-116.

From the saline strain CNQ-509 belonging to the "MAR4" group of marine actinomycetes, which have been demonstrated to be a rich source of hybrid isoprenoid secondary metabolites with farnesyl-α-nitropyrroles structures, five new nitropyrrolins have been isolated – nitropyrrolins A–E (**13**–**17**) (Figure 12.6) [80]. The structures of the nitropyrrolins are composed of *R*-nitropyrroles with functionalized farnesyl groups at the C-4 position. These compounds are the first examples of naturally occurring terpenyl-α-nitropyrroles. Nitropyrrolins A, B, and D showed cytotoxic activity against HCT-116 human colon adenocarcinoma cells, and nitropyrrolin D also showed weak antibacterial activity against MRSA.

Nowadays scientists are applying new methods for the selective isolation of uncommon actinomycetes and their novel secondary metabolites, with a few of these methods making use of innovative screening and fermentation techniques, metagenomic screening of DNA [81], and biosynthesis [24]. The new strategies are based on both the bioprospection and the hypothesis that taxonomic diversity is a surrogate for chemical diversity [24,30,82] as well as on the thought that metabolites may act in the bacterial evolution [82,83], followed by expression and detection techniques, and concluding with the full taxonomic characterization.

The hypothesis that the associated biosynthetic pathway is inherited from a common ancestor and, subsequently, evolved independently into two species of *Salinispora* was reported with the recent discovery of salinosporamide K from *S. pacifica* strain CNT-133 [84]. This compound was previously found exclusively in *S. tropica* [85]. Considering that *S. tropica* and *S. pacifica* are closely related taxa, this observation provided the opportunity to test the hypothesis that the associated biosynthetic pathway was inherited from a common ancestor and subsequently evolved independently into two species [5].

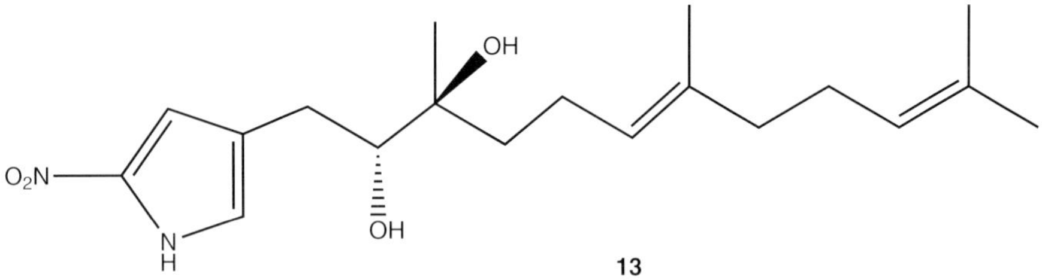

13

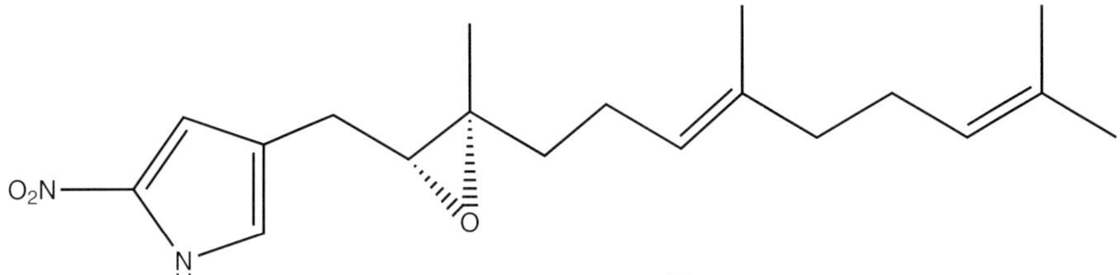

14

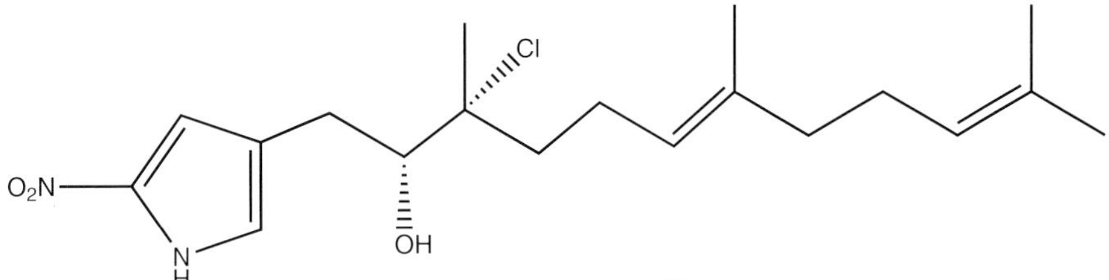

15

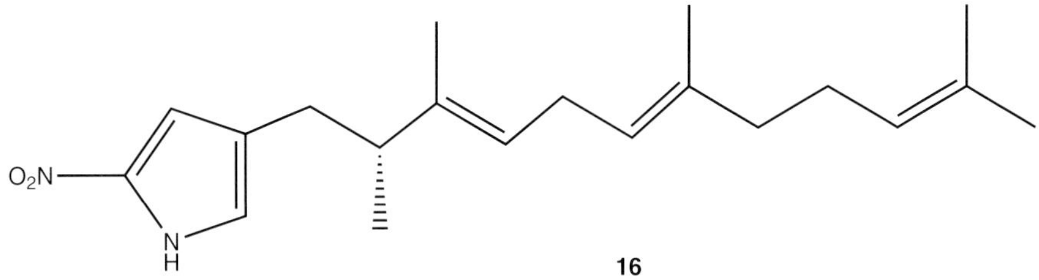

16

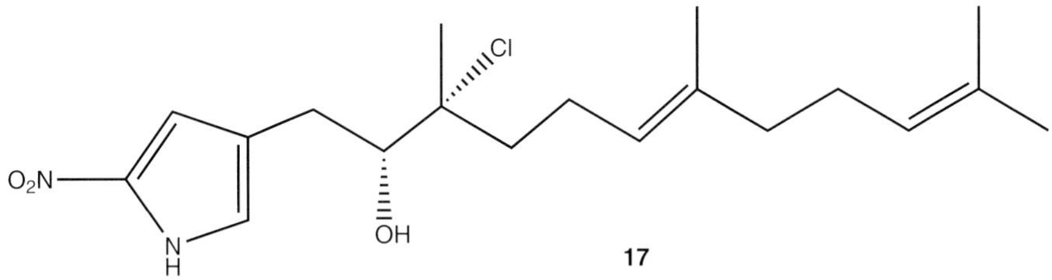

17

Figure 12.6 New nitropyrrolins A (**13**), B (**14**), C (**15**), D (**16**), and E (**17**), isolated from strain CNQ-509 belong to the "MAR4" group of marine actinomycete [80].

New web tools that provide automated methods to assess the secondary metabolite gene diversity have been recently published [86]. The Natural Product Domain Seeker (NaPDoS) analysis is based on the phylogenetic relationships of sequence tags derived from the polyketide synthase and nonribosomal peptide synthetase (NRPS) genes and then compared with an internal database of experimentally characterized biosynthetic genes. NaPDoS also provides a rapid mechanism to infer the generalized structures of secondary metabolite biosynthetic gene richness and diversity within a genome or environmental sample by extraction and classification of ketosynthase and condensation domains from PCR products, genomes, and metagenomic data sets. With this, an exponential progress increase in the field of science has begun, rendering outstanding benefits in the field of drug discovery.

12.5 Conclusions

A large number of secondary metabolites with biological activities in actinomycetes have been found through mean innovative screening programs around the world, using new fermentative techniques and metagenomic analysis of DNA, by selective isolation of unknown actinomycetes species and novel bioactive secondary metabolites produced by them, as well as biosynthesis methods. All this have led to the hypothesis that taxonomic diversity is a surrogate for chemical diversity and the metabolites may act in the bacterial evolution, followed by expression and detection techniques, and concluding with the full taxonomic characterization.

Acknowledgment

The authors would like to thank the Autonomous University of Baja California for the financial support in the 16th Internal Conv. (Grant No. 369).

References

1 Fenical, W. and Jensen, P.R. (2006) Developing a new resource for drug discovery: marine actinomycete bacteria. *Nat. Chem. Biol.*, **2**, 666–673.

2 Clardy, J. and Walsh, C. (2004) Lessons from natural molecules. *Nature*, **432**, 829–837.

3 Fischbach, M.A., Walsh, C.T., and Clardy, J. (2008) The evolution of gene collectives: how natural selection drives chemical innovation. *Proc. Natl. Acad. Sci. USA*, **105**, 4601–4608.

4 Jenke-Kodama, H., Sandmann, A., Muller, R., and Dittmann, E. (2005) Evolutionary implications of bacterial polyketide synthases. *Mol. Biol. Evol.*, **22**, 2027–2039.

5 Freel, K.C., Nam, S.J., Fenical, W., and Jensen, P.R. (2011) Evolution of secondary metabolite genes in three closely related marine actinomycete species. *Appl. Environ. Microbiol.*, **77**, 7261–7270.

6 Bull, A.T., Ward, A.C., and Goodfellow, M. (2000) Search and discovery strategies for

biotechnology: the paradigm shift. *Microbiol. Mol. Biol. Rev.*, **64**, 573–604.

7 Okami, Y. and Hotta, K. (1988) Search and discovery of new antibiotics, in *Actinomycetes in Biotechnology* (eds M. Goodfellow, S.T. Williams, and M. Mordarski), Academic Press, New York, pp. 33–67.

8 Logan, N. (1994) *Bacterial Systematic*, Blackwell Scientific Publications, Oxford, p. 272.

9 Nathan, A., Magarvey, K.J.M., Bernan, V., and Dworkin, M. (2004) Isolation and characterization of novel marine-derived *Actinomycete* taxa rich in bioactive metabolites. *Appl. Environ. Microbiol.*, **70**, 7520–7529.

10 Goodfellow, M. and Haynes, J.A. (1984) Actinomycetes in marine sediments, in *Biological, Biochemical, and Biomedical Aspects of Actinomycetes* (eds L. Ortíz-Ortíz, L.F. Bojalil, and V. Yakoleff), Academic Press, Orlando, pp. 453–472.

11 Mincer, T.J., Jensen, P.R., Kauffman, C.A., and Fenical, W. (2002) Widespread and persistent populations of a major new marine actinomycetes taxon in ocean sediments. *Appl. Environ. Microbiol.*, **68**, 5005–5011.

12 Feling, R.H., Buchanan, G.O., Mincer, T.J., Kauffman, C.A., Jensen, P.R., and Fenical, W. (2003) Salinosporamide A: a highly cytotoxic proteasome inhibitor from a novel microbial source, a marine bacterium of the new genus *Salinospora*. *Angew. Chem., Int. Ed.*, **42**, 355–357.

13 Bhatnagar, I. (2010) Immense essence of excellence: marine microbial compounds. *Mar. Drugs*, **8**, 2673–2701.

14 Soria-Mercado, I.E., Prieto-Davó, A., Jensen, P.R., and Fenical, W. (2005) Antibiotic terpenoid chloro-dihydroquinones from a new marine actinomycete. *J. Nat. Prod.*, **68**, 904–910.

15 Strohl, W.R. (2004) Antimicrobials, in *Microbial Diversity and Bioprospecting* (ed. A.T. Bull), ASM Press, pp. 336–355.

16 Blunt, J.W., Copp, B.R., Munro, M.H., Northcote, P.T., and Prinsep, M.R. (2006) Marine natural products. *Nat. Prod. Rep.*, **23**, 26–78.

17 Magarvey, N.A., Keller, J.M., Bernan, V., Dworkin, M., and Sherman, D.H. (2004) Isolation and characterization of novel marine-derived actinomycete taxa rich in bioactive metabolites. *Appl. Environ. Microbiol.*, **70**, 7520–7529.

18 Jemimah, N.S., Srinivasan, V.M., and Devi, C.S. (2011) Novel anticancer compounds from marine actinomycetes: a review. *J. Pharm. Res.*, **4**, 1285–1287.

19 Olano, C., Méndez, C., and Salas, J.A. (2009) Antitumor compounds from actinomycetes from gene clusters to new derivatives by combinatorial synthesis. *Nat. Prod. Rep.*, **26**, 628–660.

20 Goodfellow, M. and Fiedler, H.P. (2010) A guide to successful bioprospecting: informed by actinobacterial systematics. *Antonie Van Leeuwenhoek*, **98**, 119–142.

21 Lam, K.S. (2006) Discovery of novel metabolites from marine actinomycetes. *Curr. Opin. Microbiol.*, **9**, 245–251.

22 Lanoll, B.D., Sassen, R., LaDuc, M.T., Sweet, S.T., and Nealson, K.H. (2001) Bacteria and archae physically associated with Gulf of Mexico gas hydrates. *Appl. Environ. Microbiol.*, **69**, 7224–7235.

23 Colweel, F., Matsumoto, R., and Reed, D. (2004) A review of gas hydrates, geology, and biology of Nankai Trough. *Chem. Geol.*, **205**, 391–404.

24 Bull, A.T. and Stach, J.E.M. (2007) Marine actinobacteria: new opportunities for natural product search and discovery. *Trends Microbiol.*, **15**, 491–499.

25 Maldonado, L.A., Stach, J.E.M., Pathom-Aree, W., Ward, A.C., Bull, A.T., and Goodfellow, M. (2005) Diversity of cultivable actinobacteria in geographically widespread marine sediments. *Antonie Van Leeuwenhoek*, **87**, 11–18.

26 Colquhoun, J.A., Mexson, J., Goodfellow, M., Ward, A.C., Horikoshi, K., and Bull, A. T. (1998) Novel rhodococci and other mycolate actinomycetes from the deep sea. *Antonie Van Leeuwenhoek*, **74**, 27–40.

27 Freel, K.C., Edlund, A., and Jensen, P.R. (2012) Microdiversity and evidence for high dispersal rates in the marine actinomycete "*Salinispora pacifica*". *Environ. Microbiol.*, **14**, 480–493.

28 Gontang, E.A., Fenical, W., and Jensen, P.R. (2007) Phylogenetic diversity of Gram-positive bacteria cultured from marine sediments. *Appl. Environ. Microbiol.*, **73**, 3272–3282.

29 Jensen, P.R. and Lauro, F.M. (2008) An assessment of actinobacterial diversity in the marine environment. *Antonie Van Leeuwenhoek*, **94**, 51–62.

30 Ward, A.C. and Goodfellow, M. (2004) Phylogeny and functionality: taxonomy as a roadmap to genes, in *Microbial Diversity and Bioprospecting* (ed. A.T. Bull), ASM Press, Washington, DC, pp. 288–313.

31 Maldonado, L.A., Fragoso-Yañez, D., Pérez-García, A., Rosellón-Druker, J., and Quintana, E.T. (2009) Actinobacterial diversity from marine sediments collected in Mexico. *Antonie Van Leeuwenhoek*, **95**, 111–120.

32 Cermeno, P. and Falkowski, P.G. (2009) Controls on diatom biogeography in the ocean. *Science*, **325**, 1539–1541.

33 Chu, H., Fierer, N., Lauber, C.L., Caporaso, J.G., Knight, R., and Grogan, P. (2010) Soil bacterial diversity in the Arctic is not fundamentally different from that found in other biomes. *Environ. Microbiol.*, **12**, 2998–3006.

34 Roberts, M.S. and Cohan, F.M. (1995) Recombination and migration rates in natural populations of *Bacillus subtilis* and *Bacillus mojavensis*. *Evolution*, **49**, 1081–1094.

35 Ginolhac, A., Jarrin, C., Robe, P., Perrière, G., Vogel, T., Simonet, P., and Nalin, R. (2005) Type I polyketide synthases may have evolved through horizontal gene transfer. *J. Mol. Evol.*, **60**, 716–725.

36 Gontang, E.A., Gaudíncio, S., Fenical, W., and Jensen, P.R. (2010) Sequence-based analysis of secondary-metabolite biosynthesis in marine actinobacteria. *Appl. Environ. Microbiol.*, **76**, 2487–2499.

37 Edlund, A., Loesgen, S., Fenical, W., and Jensen, P.R. (2011) Geographic distribution of secondary metabolite genes in the marine actinomycete *Salinispora arenicola*. *Appl. Environ. Microbiol.*, **77**, 5916–5925.

38 Wawrik, B., Kutliev, D., Abdivasievna, U.A., Kukor, J.J., Zylstra, G.J., and Kerkhof, L. (2007) Biogeography of actinomycete communities and type II polyketide synthase genes in soils collected in New Jersey and central Asia. *Appl. Environ. Microbiol.*, **73**, 2982.

39 Jensen, P.R., Dwight, R., and Fenical, W. (1991) Distribution of actinomycetes in near-shore tropical marine sediments. *Appl. Environ. Microbiol.*, **57**, 1102–1108.

40 Maldonado, L.A., Fenical, W., Jensen, P.R., Kauffman, C.A., Mincer, T.J., Ward, A.C., Bull, A.T., and Goodfellow, M. (2005) *Salinispora arenicola* gen. nov., sp. nov. and *Salinispora tropica* sp. nov., obligate marine actinomycetes belonging to the family Micromonosporaceae. *Int. J. Syst. Evol. Microbiol.*, **55**, 1759–1766.

41 Jensen, P.R. and Mafnas, C. (2006) Biogeography of the marine actinomycete *Salinispora*. *Appl. Environ. Microbiol.*, **8**, 1881–1888.

42 Bérdy, J. (2005) Bioactive microbial metabolites. *J. Antibiot. (Tokyo)*, **58**, 1–26.

43 Newman, D.J. and Cragg, G.M. (2007) Natural products as sources of new drugs over the last 25 years. *J. Nat. Prod.*, **70**, 461–477.

44 Olano, C., Méndez, C., and Salas, J.A. (2009) Antitumor compounds from marine actinomycetes. *Mar. Drugs*, **7**, 210–248.

45 Hu, G.P., Yuan, J., Sun, L., She, Z.G., Wu, J.H., Lan, X.J., Zhu, X., Lin, Y.C.H., and Chen, S.P. (2011) Statistical research on marine natural products based on data obtained between 1985 and 2008. *Mar. Drugs*, **9**, 514–525.

46 Watve, M.G., Tickoo, R., Jog, M.M., and Bhole, B.D. (2001) How many antibiotics are produced by the genus *Streptomyces*. *Arch. Microbiol.*, **176**, 386–390.

47 Bull, A.T. (2004) Biotechnology, the art of exploiting biology, in *Microbial Diversity and Bioprospecting* (ed. A.T. Bull), ASM Press, Washington, DC, pp. 3–12.

48 Bull, A.T. (2004) The paradigm shift in microbial prospecting, in *Microbial Diversity and Bioprospecting* (ed. A.T. Bull), ASM Press, Washington, DC, pp. 241–249.

49 Sogin, M.L., Morrison, H.G., Huber, J.A., Welch, D.M., Huse, S.M., Neal, P.R., Arrieta, J.M., and Herndl, G.J. (2006) Microbial diversity in the deep sea and the unexplored "rare biosphere". *Proc. Natl. Acad. Sci. USA*, **103**, 12115–12120.

50 Jensen, P.R., Mincer, T.J., Williams, P.G., and Fenical, W. (2005) Marine actinomycete diversity and natural product discovery. *Antonie Van Leeuwenhoek*, **87**, 43–48.

51 Kaiya, Y., Hasegawa, J., Momose, T., Sato, T., and Chida, N. (2011) Total synthesis of (−)-salinosporamide A. *Chemistry*, **6**, 209–219.

52 Fukuda, D.S., Mynderse, J.S., Baker, P.J., Berry, D.M., Boeck, L.D., Yao, R.C., Mertz, F.P., Nakatsukasa, W.M., Mabe, J., Ott, J., Counter, J.T., Ensminger, P.W., Allen, N.E., Alborn, W.E., and Hobbs, J.N. (1990) A80915, a new antibiotic complex produced by *Streptomyces aculeolatus*. Discovery, taxonomy, fermentation, isolation, characterization, and antibacterial evaluation. *J. Antibiot. (Tokyo)*, **43**, 623–633.

53 Fukuda, D.S., Mynderse, J.S., Baker, P.J., Berry, D.M., Boeck, L.D., Yao, R.C., Mertz, F.P., Nakatsukasa, W.M., Mabe, J., and Ott, J. (1990) Manufacture of antibiotics A 80915 with *Streptomyces* and preparation of dehydrochlorinated derivates. U.S. Patent No. 4,904,590 (Cl. 435-147; C12P7/24), February 27; Appl. 290,724, December 27, 1988.

54 Motohashi, K., Masayuki, S., Furihata, K., Ito, S., and Seto, H. (2008) Terpenoids produced by actinomycetes: napyradiomycins from *Streptomyces antimycoticus* NT17. *J. Nat. Prod.*, **71**, 595–601.

55 Cho, J.Y., Kwon, H.C., Williams, P.G., Jensen, P.R., and Fenical, W. (2006) Azamerone, a terpenoid phthalazinone from a marine-derived bacterium related to the genus *Streptomyces* (Actinomycetales). *Org. Lett.*, **8**, 2471–2474.

56 Kwon, H.C., Kauffman, C.A., Jensen, P.R., and Fenical, W. (2006) Marinomycins A–D, antitumor–antibiotics of a new structure class from a marine actinomycete of the recently discovered genus *Marinispora*. *J. Am. Chem. Soc.*, **128**, 1622–1632.

57 Staunton, J. and Weissman, K.J. (2001) Polyketide biosynthesis: a millennium review. *Nat. Prod. Rep.*, **18**, 380–416.

58 Cho, J.Y., Kwon, H.C., Williams, P.G., Kauffman, C.A., Jensen, P.R., and Fenical, W. (2006) Actinofuranones A and B, polyketides from a marine-derived bacterium related to the genus *Streptomyces* (Actinomycetales). *J. Nat. Prod.*, **69**, 425–428.

59 Hughes, C.C., MacMillan, J.B., Gaudêncio, S.P., Jensen, P.R., and Fenical, W. (2009) The ammosamides: structures of cell cycle modulators from a marine-derived *Streptomyces* species. *Angew. Chem., Int. Ed.*, **48**, 725–727.

60 Williams, P.G., Miller, E.D., Asolkar, R.N., Jensen, P.R., and Fenical, W. (2007) Arenicolides A–C, 26-membered ring macrolides from the marine actinomycete *Salinispora arenicola*. *J. Org. Chem.*, **72**, 5025–5034.

61 Bugni, T.S., Woolery, M., Kauffman, C.A., Jensen, P.R., and Fenical, W. (2006) Bohemamines from a marine-derived *Streptomyces* sp. *J. Nat. Prod.*, **69**, 1626–1628.

62 Hohmann, C., Schneider, K., Bruntner, C., Irran, E., Nicholson, G., Bull, A.T., Jones, A.L., Brown, R., Stach, J.E., Goodfellow, M., Beil, W., Krämer, M., Imhoff, J.F., Süssmuth, R.D., and Fiedler, H.P. (2009) Caboxamycin, a new antibiotic of the benzoxazole family produced by the deep-sea strain *Streptomyces* sp. NTK 937. *J. Antibiot. (Tokyo)*, **62**, 99–104.

63 Wu, S.J., Fotso, S., Li, F., Qin, S., Kelter, G., Fiebig, H.H., and Laatsch, H. (2006) *N*-Carboxamidostaurosporine and selina-4(14), 7(11)-diene-8,9-diol, new metabolites from a marine *Streptomyces* sp. *J. Antibiot. (Tokyo)*, **59**, 331–337.

64 Wu, S.J., Fotso, S., Li, F., Qin, S., and Laatsch, H. (2007) Amorphane sesquiterpenes from a marine *Streptomyces* sp. *J. Nat. Prod.*, **70**, 304–306.

65 Li, F., Maskey, R.P., Qin, S., Sattler, I., Fiebig, H.H., Maier, A., Zeeck, A., and Laatsch, H. (2005) Chinikomycins A and B: isolation, structure elucidation, and biological activity of novel antibiotics from a marine *Streptomyces* sp. Isolate M045. *J. Nat. Prod.*, **68**, 349–353.

66 Oh, D.C., Williams, P.G., Kauffman, C.A., Jensen, P.R., and Fenical, W. (2006) Cyanosporasides A and B, chloro- and cyano-cyclopenta[*a*]indene glycosides from the marine actinomycete "*Salinispora pacifica*". *Org. Lett.*, **8**, 1021–1024.

67 Asolkar, R.N., Jensen, P.A., Kauffman, C.A., and Fenical, W. (2006) Daryamides A–C, weakly cytotoxic polyketides from a marine-derived actinomycete of the genus

Streptomyces strain CNQ-085. *J. Nat. Prod.*, **69**, 1756–1759.

68 Cui, C.B., Liu, H.B., Gu, J.Y., Gu, Q.Q., Cai, B., Zhang, D.Y., and Zhu, T.J. (2007) Echinosporins as new cell cycle inhibitors and apoptosis inducers from marine-derived *Streptomyces albogriseolus*. *Fitoterapia*, **78**, 238–240.

69 Gorajana, A., Kurada, B.V., Peela, S., Jangam, P., Vinjamuri, S., Poluri, E., and Zeeck, A. (2005) 1-Hydroxy-1-norresistomycin, a new cytotoxic compound from a marine actinomycete, *Streptomyces chibaensis* AUBN1/7. *J. Antibiot. (Tokyo)*, **58**, 526–529.

70 Ding, L., Pfoh, R., Rühl, S., Qin, S., and Laatsch, H. (2009) T-muurolol sesquiterpenes from the marine *Streptomyces* sp. M491 and revision of the configuration of previously reported amorphanes. *J. Nat. Prod.*, **72**, 99–101.

71 Jeong, S.Y., Shin, H.J., Kim, T.S., Lee, H.S., Park, S.K., and Kim, H.K. (2006) Streptokordin, a new cytotoxic compound of the methylpyridine class from a marine-derived *Streptomyces* sp. KORDI-3238. *J. Antibiot. (Tokyo)*, **59**, 234–240.

72 Hayakawa, Y., Shirasaki, S., Shiba, S., Kawasaki, T., Matsuo, Y., Adachi, K., and Shizuri, Y. (2007) Piericidins C7 and C8, new cytotoxic antibiotics produced by a marine *Streptomyces* sp. *J. Antibiot. (Tokyo)*, **60**, 196–200.

73 Williams, P.G., Asolkar, R.N., Kondratyuk, T., Pezzuto, J.M., Jensen, P.R., and Fenical, W. (2007) Saliniketals A and B, bicyclic polyketides from the marine actinomycete *Salinispora arenicola*. *J. Nat. Prod.*, **70**, 83–88.

74 Oh, D.C., Gontang, E.S., Kauffman, C.A., Jensen, P.R., and Fenical, W. (2008) Salinipyrones and pacificanones, mixed-precursor polyketides from the marine actinomycete *Salinispora pacifica*. *J. Nat. Prod.*, **71**, 570–575.

75 Han, X., Cui, C., Gu, Q., Zhu, W., Liu, H., Gu, J. and Osada, H. (2005) ZHD-0501, a novel naturally occurring staurosporine analog from *Actinomadura* sp. 007. *Tetrahedron Lett.*, **46**, 6137–6140.

76 Liu, R., Cui, C.B., Duan, L., Gu, Q., and Zhu, W.M. (2005) Potent *in vitro* anticancer activity of metacycloprodigiosin and undecylprodigiosin from a sponge-derived actinomycete *Saccharopolyspora* sp. *Arch. Pharm. Res.*, **28**, 1341–1344.

77 Liu, R., Zhu, T., Li, D., Gu, J., Xia, W., Fang, Y., Liu, H., Zhu, W., and Gu, Q. (2007) Two indolocarbazole alkaloids with apoptosis activity from a marine-derived actinomycete Z2039-2. *Arch. Pharm. Res.*, **30**, 70–274.

78 Shin, H.J., Jeong, H.S., Lee, H.S., Park, S.K., Kim, H.M., and Kwon, H.J. (2007) Isolation and structure determination of streptochlorin, an antiproliferative agent from a marine-derived *Streptomyces* sp. 04DH110. *J. Microbiol. Biotechnol.*, **17**, 1403–1406.

79 Hughes, C.C., Prieto-Davo, A., Jensen, P.R., and Fenical, W. (2008) The marinopyrroles, antibiotics of an unprecedented structure class from a marine *Streptomyces* sp. *Org. Lett.*, **10**, 629–631.

80 Kwon, H.C., Espindola, A.P., Park, J.S., Prieto-Davó, A., Mickea, R., Jensen, P.R., and Fenical, W. (2010) Nitropyrrolins A–E, cytotoxic farnesyl-α-nitropyrroles from a marine-derived bacterium within the actinomycete family *Streptomycetaceae*. *J. Nat. Prod.*, **7**, 2047–2052.

81 Handelsman, J. (2004) Soils: the metagenomics approach, in *Microbial Diversity and Bioprospecting* (ed. A.T. Bull), ASM Press, Washington, DC, pp. 109–119.

82 Jensen, P.R. (2010) Linking species concepts to natural product discovery in the post-genomic era. *J. Ind. Microbiol. Biotechnol.*, **37**, 219–224.

83 Czaran, T.L., Hoekstra, R.E., and Page, L. (2002) Chemical warfare between microbes promotes biodiversity. *Proc. Natl. Acad. Sci. USA*, **99**, 786–790.

84 Eustáquio, A.S., Nam, S.J., Penn, K., Lechner, A., Wilson, M.C., Fenical, W., Jensen, P.R., and Moore, B.S. (2011) The discovery of salinosporamide K from the marine bacterium "*Salinispora pacifica*" by genome mining gives insight into pathway evolution. *ChemBioChem*, **12**, 61–64.

85 Jensen, P.R., Williams, P.G., Oh, D.C., Zeigler, L., and Fenical, W. (2007) Species-specific secondary metabolite production in marine actinomycetes of the genus *Salinispora*. *Appl. Environ. Microbiol.*, **73**, 1146–1152.

86 Ziemert, N., Podell, S., Penn, K., Badger, J.H., Allen, E., and Jensen, P.R. (2012) The natural product domain seeker NaPDoS: a phylogeny based bioinformatic tool to classify secondary metabolite gene diversity. *PLoS One*, **7** (3), e34064.

13
Fungal Bioactive Gene Clusters: A Molecular Insight

Ira Bhatnagar and Se-Kwon Kim

13.1
Introduction to Fungal Secondary Metabolites

Fleming discovered the first successful chemotherapeutic produced by microbes that initiated the golden age of antibiotics. This discovery opened the way for the development of many other antibiotics, and till date, penicillin has remained the most active and the least toxic compound among many others [1]. Penicillin, together with cephalosporins, belongs to the group of β-lactam compounds. There are several types of penicillins, for example, F, G, K, N, and V. Penicillins V and G are active against most aerobic Gram-positive organisms. Penicillin G is one of the most widely used antibiotic agents today, and is used against streptococcal, staphylococcal, and meningococcal infections. On the industrial scale, penicillin G is produced by fermentation of *Penicillium chrysogenum* [2]. Zearalenone is another example of a mycotoxin with pharmacologically useful properties. It is a polyketide that is synthesized entirely from acetate–malonate units [3]. Zearalenone is produced by several *Fusarium* species [4]. It resembles 17β-estradiol, a principal hormone produced by the human ovary, to allow binding to the estrogen receptors in mammalian target cells. The reduced form of zearalenone, that is, α-zearalenol, has been found to increase estrogenic activity [5]. A synthetic commercial formulation called zeranol has been successfully marketed for use as an anabolic agent for both sheep and cattle [6]. Zearalenone has also been used for the treatment of postmenopausal symptoms in women [7], and both zearalenol and zearalenone have been patented as oral contraceptives [8]. In summary, the zearalenone family of metabolites is both an example of potentially harmful metabolites and a promising pharmaceutical candidate.

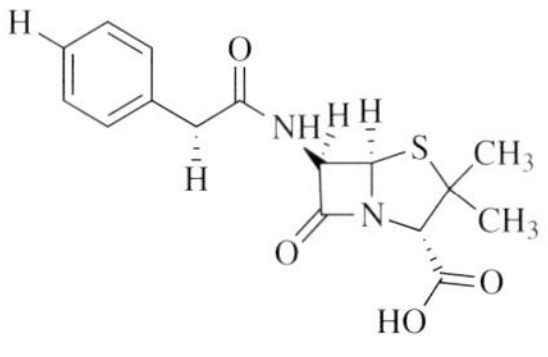

Penicillin G 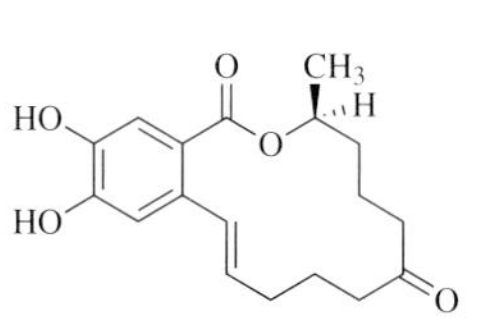 **Zearalenone**

Marine Microbiology: Bioactive Compounds and Biotechnological Applications, First Edition.
Edited by Se-Kwon Kim

Another fungal secondary metabolite, cyclosporin A, was originally discovered as a narrow-spectrum antifungal metabolite produced by the fungus *Tolypocladium inflatum* [9]. This fungal secondary metabolite is associated with the reduction of cytokine formation, and inhibition of activation and/or maturation of various cell types. This includes those cells as well that are involved in cell-mediated immunity, thus cyclosporin A is used as an immunosuppressant in human transplantation surgery and the treatment of autoimmune diseases [10].

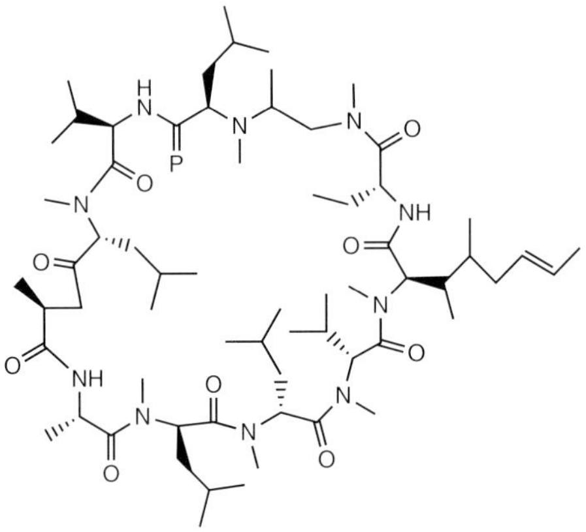

Cyclosporin A

A very old broad-spectrum compound, mycophenolic acid, first discovered in 1896 and never commercialized as an antibiotic, has recently been developed as a new immunosuppressant [11]. Before being developed for an approved immunosuppressant, this organic acid was used to treat psoriasis [12]. 5-Methylorsellinic acid, but not orsellinic acid, is a precursor of mycophenolic acid in *Penicillium brevicompactum* [3]. The members of the statin family of secondary metabolites are potent inhibitors of 3-hydroxy-3-methylglutaryl-coenzyme A (HMG-CoA) reductase, the key enzyme in cholesterol biosynthesis in humans [13]. Besides their main cholesterol-lowering effect, members of the statin family also have strong antifungal activities, especially against yeasts [12]. Brown *et al.* discovered in 1976 the first member of this group – compactin (i.e., ML-236B), as an antibiotic product of *P. brevicompactum*.

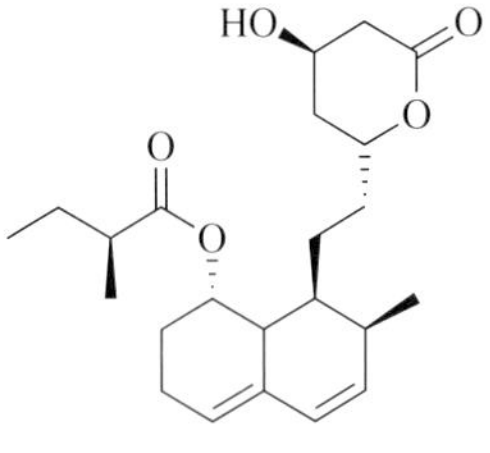

Compactin

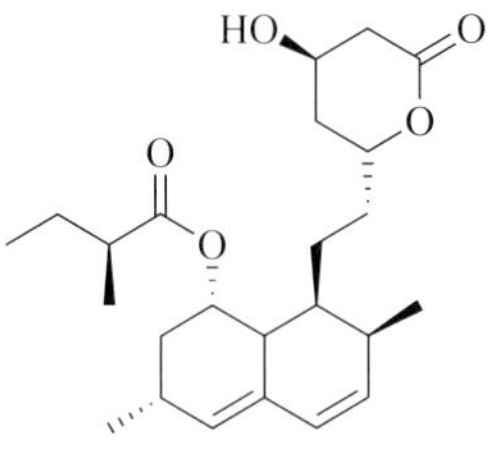

Lovastatin

Independently, in the same year, Endo *et al.* discovered compactin in broths of *Penicillium citrinum* as an inhibitor of HMG-CoA. Few years later the more active methylated form of compactin known as lovastatin (Monacolin K or Mevinolin) was discovered in broths of *Monascus ruber* and *Aspergillus terreus*, respectively. It is important to emphasize that natural statins and their derivatives are examples of multibillion-dollar drugs arising from fungal secondary metabolites [14]. One such derivative is pravastatin, which is produced by bioconversion of compactin [15]. Yet another compound of pharmaceutical importance is a diketopiperazine known as plinabulin (NPI-2358), isolated from a marine alga-associated *Aspergillus* sp. CNC-139. This compound also inhibits tubulin assembly and acts as a vasculature-disrupting agent that destabilizes the tumor vascular endothelial architecture and leads to cell damage. This compound is presently under phase II clinical trials [16].

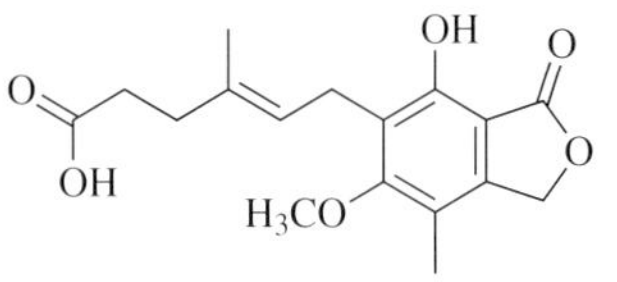

Mycophenolic acid 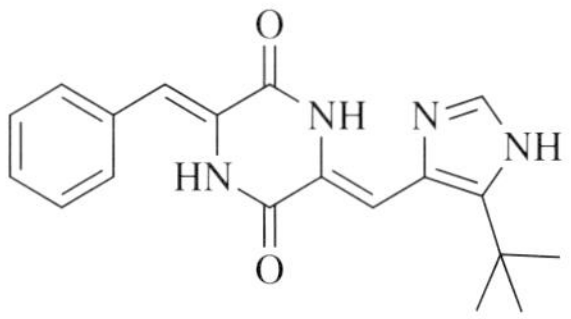 **Plinabulin**

13.2 Polyketide Synthase

More than 100 years ago, Collie coined the term "polyketide" for natural products derived from simple two-carbon acetate building blocks. This proposal was later proved experimentally by Birch who used isotopically labeled acetate in the study of 6-methylsalicylic acid (6-MSAS) biosynthesis in fungi and showed that it was formed from four acetate units. Then, Lynen and coworkers succeeded in detecting MSAS activity in a cell-free extract of *Penicillium patulum*, the first demonstration of polyketide synthase (PKS) function *in vitro*. These chemical and biochemical experiments with fungi established the concepts of "polyketide biosynthesis" and "polyketide synthase" [17]. Nowadays it is obvious that polyketides represent the largest family of structurally diverse secondary metabolites synthesized in both prokaryotic and eukaryotic organisms [18]. The biological activities associated with polyketides encompass antibacterial, antiviral, antitumor, and antihypertensive activities, as well as immunosuppressant and mycotoxin compounds.

Independent of their structural diversity, all polyketides have a common biosynthetic origin. They are derived from highly functionalized carbon chains whose assembly mechanism has close resemblance to the fatty acid biosynthetic pathway [19]. The assembly process is controlled by a multifunctional enzyme complex called PKS [20]. The core of the PKS function is the synthesis of long chains of carbon atoms through repetitive Claisen condensation reactions of small organic acids (such as acetic and malonic acids) via a ketosynthase (KS) enzyme activity. The building units, acetate, propionate, malonate, or methylmalonate, are activated units in the form of coenzyme A (CoA) esters, such as acetyl-CoA and malonyl-CoA, before involvement in the assembly of the polyketide chain. The most common starter-unit acetyl-CoA with two carbon atoms is condensed with a malonyl-CoA, with three carbons, to give a chain of four carbon atoms with loss of one carbon dioxide. Only two carbons are included into the chain in each round of condensation with malonyl-CoA. If the extender unit is methylmalonyl-CoA, the "extra" carbon forms a methyl side branch to keep the original extension speed in the main chain [21]. Each condensation is followed by a cycle of optional modifying reactions that involve the enzymes ketoreductase (KR), dehydratase (DH), and enoylreductase (ER) in the subsequent reduction steps. At this stage, a major difference between fatty acid and polyketide biosynthesis becomes apparent. Fatty acid synthases (FAS) catalyze the full reduction of each β-keto moiety prior to further chain extension in every cycle. The polyketide biosynthesis, however, shows a higher degree of complexity due to full or partial omission of reduction steps following condensation and thus affecting the following functions: β-keto (no reduction), β-hydroxy (keto reduction), enoyl (keto reduction and dehydration), and alkyl (keto reduction, dehydration, and enoyl reduction). This control of β-keto reduction is the key feature of the reducing PKS (R-PKS) that differentiates these enzymes from FAS and leads to a great structural diversity among polyketide compounds [21,22]. The ability to use different chain starter-units (such as acetate, benzoate, cinnamate, and/or amino acids) and alternate extender units (malonate, methylmalonate, and ethylmalonate) by bacterial synthases gives rise to further structural diversity among the polyketides. The assembled polyketide chain can also undergo further modifications such as cyclization, reduction or oxidation, alkylation, and rearrangements after release from PKS.

13.2.1 Classification of Fungal PKS

The iterative type I PKSs are responsible for the biosynthesis of fungal metabolites such as 6-methylsalicylic acid [23] and lovastatin (Figure 13.1) [24]. The iterative type I PKS has only one multidomain protein, in which all the enzyme activities are covalently bound together. The single multifunctional protein is used iteratively to catalyze multiple rounds of chain elongation and appropriate β-keto processing [18].

In contrast to the types I and II PKSs that are composed of ketosynthases and accessory enzymes, the type III PKSs are dimers of KS-like enzymes (more precisely homodimers) that accomplish a complex set of reactions, such as priming of a starter-unit, decarboxylative condensation of extender units, ring closure, and aromatization of the polyketide chain, in a multifunctional active site pocket [25]. Chalcone synthases, the most well-known representatives of this family, are ubiquitous in higher plants, and provide the starting material for a diverse set of biologically important phenylpropanoid metabolites [26]. Type III PKSs were traditionally associated with plants but recently discovered in a number of bacteria [27] as well as in fungi (Figure 13.2) [25,28].

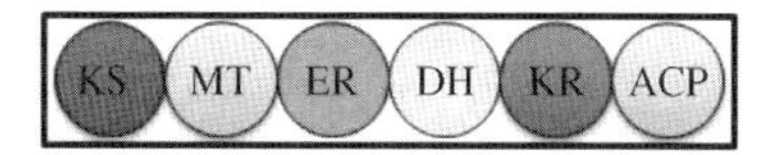

Figure 13.1 Iterative type I PKS found in fungi. KS: β-ketoacyl synthase, MT: methyltransferase, ER: enoylreductase, DH: dehydratase, KR: ketoreductase, ACP: acyl carrier protein.

According to their architecture and the presence or absence of additional β-keto-processing domains, fungal PKSs are grouped into the nonreducing PKS (NR-PKS), partially reducing PKS (PR-PKS), and highly reducing PKS (HR-PKS) [29]. Recent phylogenetic studies on the basis of KS amino acid sequences have provided valuable insights into the evolutionary relationship between different types of fungal PKS [30,31]. Kroken *et al.* showed that amino acid sequences of fungal KS domains cluster according to the degree of reduction of their products into reducing (β-keto-reductive domains: KR, ER, and DH) and nonreducing PKSs (no β-keto reduction), each type being further divided into four subclades [32].

13.2.1.1 Nonreducing PKSs

NR-PKS are shown to be responsible for the biosynthesis of nonreduced polyketides such as 1,3,6,8-tetrahydroxynaphthalene, norsolorinic acid (NA), and naphthopyrone (YWA1) that require no β-keto-reductive steps during their biosynthesis. In all cases, known genes for these synthases encode type I iterative PKS proteins [33]. The main characteristic of NR-PKSs is that they do not contain β-keto-processing domains in their multidomain organization. At the N-terminus, a domain is present that appears to mediate the loading of a starter-unit and is thus named starter unit-ACP transacylase (SAT) component. It is assumed that the starter unit is derived from corresponding FAS, another PKS or an acyl-CoA. The SAT domain is followed by typical KS and AT domains responsible for chain extension and malonate loading. Beyond the AT, there is a conserved domain designated as a product template (PT) with a not yet proven function. Nevertheless, a sequence analysis of this domain suggested its involvement in the control of chain length. The PT domain is followed by one or more ACP domains. Some NR-PKSs appear to terminate after the ACP, but many feature a diverse range of different domains including Claisen cyclase–thioesterases (CLC–TE), MT, and reductases (R). Although not described in the literature, a sequence analysis of the *Monascus purpureus pksCT* sequence [34] showed that it has a C-terminal thioester reductase domain. Similar domains were found in the nonribosomal peptide synthetase (NRPS) systems with reductase domains as chain release mechanisms

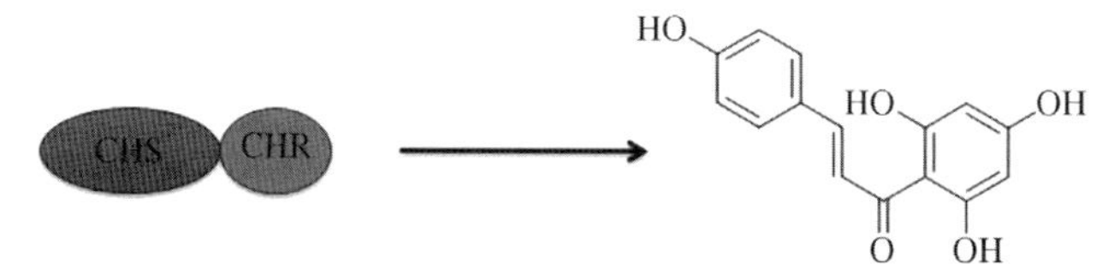

Figure 13.2 Chalcone synthase type III PKS found in plants, bacteria, and some fungi. CHS: chalcone synthase, a homodimer of identical KS monomeric domains; CHR: chalcone reductase.

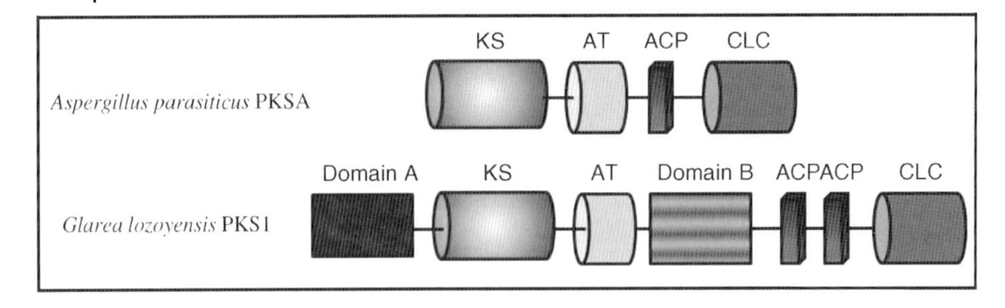

Figure 13.3 The general architecture of NR PKS genes in fungi: domain A that may be starter unit; KS: β-ketoacyl synthase; AT: acyltransferase; domain B (PT): product template; ACP: acyl carrier protein; CLC: Claisen cyclase. There is a possibility that other domains can be included as well at the C-terminus after CLC–TE (e.g., C-MeT).

resulting in an aldehyde or primary alcohol. Very recently, by joint efforts, Cox and coworkers demonstrated the role of the terminal reductase domain in product release via heterologous expression of 3-methylorcinaldehyde synthase (MOS) in *Aspergillus oryzea* [35]. In sum, it appears that these synthases are equipped with an N-terminal loading component, a central chain extension component consisting of KS, AT, and ACP domains with a possible control over a number of extensions, and a C-terminal processing component (Figure 13.3).

13.2.1.2 **Partially Reducing PKSs**

Less is known about the enzymology of the PR-PKS [33]. The domain structure is much closer to mammalian FASs, with an N-terminal KS followed by AT, and DH domains. A so-called "core" domain follows the DH, and this is followed by a KR domain. A typical PR-PKS terminates with an ACP domain as, for example, for MSAS (Figure 13.4). The domain structure differs considerably from the NR-PKS in such a way that there is no SAT or PT domain, and the PKS terminates after the ACP with obviously no requirement for a CLC–TE domain responsible for offloading of the product. Although a number of PR-PKS genes are known from genome sequencing projects, only three genes have been matched to their chemical products – in all cases the tetraketide 6MSAS (e.g., a single round of KR and DH). The first MSAS to be discovered was from *P. patulum* [23]. Ebizuka and coworkers have worked with the *atX* gene from *A. terreus* [36], and most recently an MSAS gene (*pks2*) has been isolated from *Glarea lozoyensis* [36]. Both *P. patulum* and *A. terreus* MSAS form homotetramers [33]. A short region of the core domain was identified by Fujii *et al.*; the presence of this region proved to be essential for

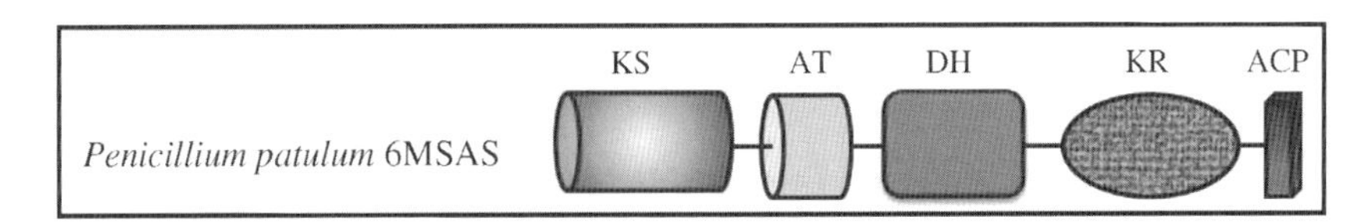

Figure 13.4 Domain architecture of MSAS encoded by *P. patulum* 6MSAS. KS: β-ketoacyl synthase; AT: acyltransferase; DH: dehydratase; KR: ketoreductase; ACP: acyl carrier protein.

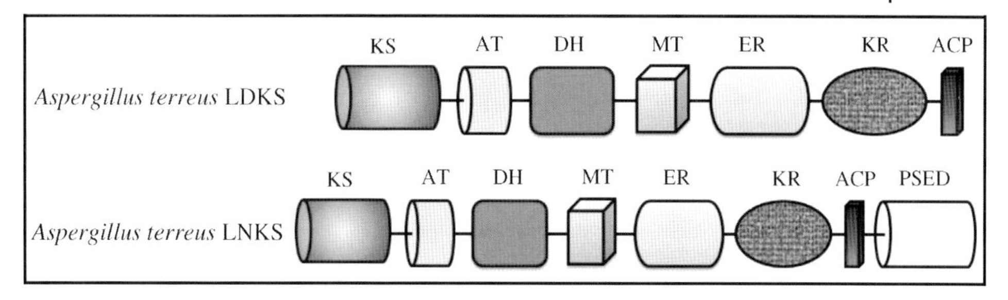

Figure 13.5 General domain organization of HR PKSs. General domain architecture of HR PKSs. KS: β-ketoacyl synthase; AT: acyltransferase; DH: dehydratase; KR: ketoreductase, ER: enoylreductase as optional; ACP: acyl carrier protein; MT: methyltransferase.

successful complementation among diverse deletion mutants of *atX* gene. It was hypothesized that this region of 122 amino acids probably forms a motif required for subunit–subunit interaction. Interestingly, this core sequence is present in other fungal PR-PKS and in the bacterial PKS such as *CalO5* from calicheamicin biosynthesis [37,38].

13.2.1.3 **Highly Reducing PKSs**

The HR-PKSs is the third class of fungal PKSs that produce complex, highly reduced compounds such as lovastatin, T-toxin, fumonisin B1, and squalestatin. These PKSs have an N-terminal KS domain, followed by AT and DH domains. In many cases, the DH is followed by a MT domain. Some HR-PKSs possess an ER domain; in others there is a roughly equivalent length of sequence without known function. An ER domain is succeeded by a KR domain, and finally the PKS often terminates with an ACP. The *lovB*, a gene that encodes for LNKS involved in lovastatin biosynthesis, appears to encode one part of an NRPS condensation (C) domain, immediately downstream of the ACP. It was proposed that this domain plays a role in product release. In general, in HR-PKS there seem to be no domains that are similar to the PT or SAT domains of the NR-PKS, as well as no "core domain" of the PR-PKS (Figure 13.5) [33].

13.3
Nonribosomal Peptide Synthetase

Nonribosomal peptides (NRPs) form a large pool of biologically active natural compounds. The spectrum of the clinical applications of NRPs is broad; for example, they are used as last resort antibiotics (daptomycin), antitumor (bleomycin), or antifungal drugs or as immunosuppressants (cyclosporin) [39]. This diverse bioactivity can be explained by the way how nature synthesizes these molecules. NRPs are produced in the secondary metabolism of bacteria and fungi by the consecutive condensation of amino acids, which is achieved by large multimodular enzymes, NRPSs [39,40]. Notably, this process is not limited to the 20 proteinogenic

Figure 13.6 Predicted module arrangement of the *A. fumigatus* NRP synthetase gene (pesM) (12 kb). The gene is predicted to encode three adenylation (A), three thiolation (T), and three condensation (C) domains, in addition to one epimerase (E) and one thioesterase (TE) domain. Module 1 has a molecular mass of approximately 90 kDa.

amino acids. Some 500 different monomers, including nonproteinogenic amino acids, fatty acids, and α-hydroxy acids, have been identified as building blocks for NRPs [41]. The nonproteinogenic building blocks contribute to structural versatility of NRPs and are likely to contribute substantially to the observed biological activity. In brief, NRPSs are composed of an array of distinct modular sections, each of which is responsible for the incorporation of one defined monomer into the final peptide product. The identity and order of a module in an assembly line specify the following: first, the sequence of monomer units activated and incorporated; second, the chemistry that occurs at each way station in the assembly line; and third, the length and functionality of the product released from the distal end of the assembly line [42]. The modules can be further divided into catalytic domains. Three domains are ubiquitous in NRP synthesis and essential for peptide elongation. The domains are responsible for the activation of the amino acid (adenylation (A) domains), the propagation of the growing peptide chain (thiolation or peptidyl carrier protein (PCP) domains), and the condensation of the amino acids (condensation (C) domains). A fourth essential NRPS catalytic unit associated with product release is the thioesterase (TE) domain. The TE domain is located in the termination module and catalyzes peptide release by either hydrolysis or macrocyclization (Figure 13.6) [43].

The logic and machinery of T domain function in NRPS assembly lines is equivalent to the role of T domains in type I PKS assembly lines. Each 8–10 kDa apo-T domain must be primed by posttranslational modification of a serine side chain with phosphopantetheine, catalyzed by dedicated PPTases. During NRP assembly, aminoacyl and peptidyl chains are tethered in thioester linkage to the terminal thiolate of the phosphopantetheine prosthetic group. The two catalytic domains in a minimal NRPS elongation module have functions similar to the AT and KS domains of PKS modules. The A domain, in analogy to the AT domain, selects the amino acid, activating the carboxylate with ATP to make the aminoacyl-AMP and then installing the aminoacyl group on the thiolate of the adjacent T domain. Like the KS domain, the C domain is the chain-elongating catalyst, joining an upstream peptidyl-*S*-Tn-1 to the downstream aminoacyl-*S*-Tn. Chain elongation by one aminoacyl residue occurs concomitantly with chain translocation to Tn. The initiation module in NRPS assembly lines is often a two-domain A-T module that selects the first amino acid and installs it covalently on T1. The peptidyl chain then elongates, preserving the free amino group of the N-terminal residue.

13.4
PKS and NRPS Products

Among the characteristics of filamentous fungi, the biosynthesis of natural products renders them of great interest to the research community. Fungal metabolites possess a wide range of activities: at times useful for pharmaceutical purposes (e.g., penicillin), yet also possessing potent toxic and carcinogenic properties threatening human, animal, and plant health (e.g., aflatoxins). Research on fungal metabolites dates back to the 1870s, when pigments synthesized in conspicuous mushroom fruiting bodies attracted the attention of organic chemists. The twentieth century marked the discovery, isolation, and chemical characterization of a vast diversity of natural products from filamentous fungi, driven by the discovery of penicillin and its development into a precious antibiotic [44].

13.4.1
Fungal PKS Products: Aflatoxin and Fusarin

Due to their immense toxicological relevance, a huge body of literature on aflatoxin/sterigmatocystin (stc) biosynthesis has accumulated since the early 1990s, and been reviewed [45,46]. *Aspergillus flavus* and *Aspergillus parasiticus* are the most prominent producers of the aflatoxin group of mycotoxins, for example, aflatoxin B1, while sterigmatocystin (the penultimate intermediate of aflatoxin) is a metabolic product of *Aspergillus nidulans*. The biosynthetic loci have been cloned from all three species [47,48]. They were found to include 25 genes arranged in about 70-kb clusters in *A. parasiticus, A. flavus,* and *A. nidulans*. These genes are redesignated as aflA–Y in *A. parasiticus* and *A. flavus,* but as stcA–X in *A. nidulans,* with the exception of aflR and aflJ [44]. As expected, the stc cluster lacks aflP and aflQ homologues, as they code for the specific steps to convert sterigmatocystin into aflatoxins.

Aflatoxin biosynthetic gene cluster is currently the best-described gene cluster, with respect to regulation (Figure 13.7). Aflatoxin is basically a mutagenic, teratogenic, hepatocarcinogenic, and immunosuppressive agent known from *A. parasiticus, Aspergillus oryzae, Aspergillus sojae, Aspergillus nominus, Aspergillus bombycis, Aspergillus pseudotamarii,* and *Aspergillus ochraceoroseus*. Research in the field of aflatoxin production started in 1960, provoked by the death of over 100 000 turkeys in England that were fed with *Aspergillus*-contaminated cottonseeds.

Fusarin C is the most prominent member of the fusarin family that is known as a secondary metabolite from *Fusarium graminearum* and *Fusarium verticillioides* [49]. By probing a cDNA library from *F. graminearum* with a fungal C-methyltransferase

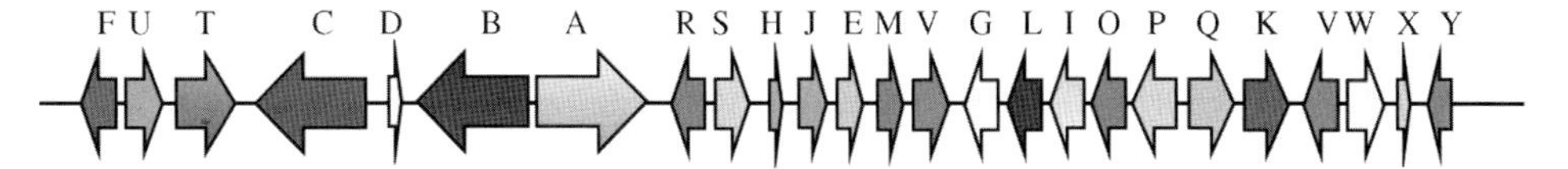

Figure 13.7 Aflatoxin gene cluster.

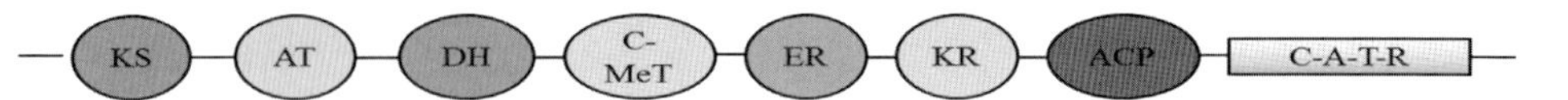

Figure 13.8 Domain structure of fusarin synthase from fusarin biosynthetic gene cluster. KS: β-ketoacyl synthase; AT: acyltransferase; DH: dehydratase; ER: enoylreductase; KR: β-ketoreductase; ACP: acyl carrier protein; and C-A-T-R (NRPS): nonribosomal peptide synthase domain.

PKS-domain, chromosomal clones were identified that covered 26 kb of genomic DNA and included orf1 and orf4 (unknown function), orf2 (a putative hydrolase gene), orf5 and orf6 (a monocarboxylate transporter and an ABC-transporter gene, respectively) [50]. The central gene (fusA) codes for fusarin synthase (FusS), a 3951 aa PKS–NRPS hybrid (KS-AT-DH-MT-ER-KR-ACP-C-A-PCP-R), the first hybrid identified in filamentous fungi and, possibly, the model for a class of tetramic acid natural products. The current biosynthesis model involves heptaketide formation by the PKS domains, and homoserine activation, loading onto the PCP and, eventually, condensation of these two partners by the FusS NRPS domains. In *F. verticillioides* fusA, an unusual intron of length 546 bp was detected that included the genetic information for a partial putative enoylreductase gene (Figure 13.8).

13.4.2
Fungal NRPS Products: Penicillin/Cephalosporin and Diketopiperazine

The second extremely diverse group of fungal natural products comprises small peptides. Nonribosomally assembled peptidic fungal natural products are often generated by means of single multimodular enzymes, the nonribosomal peptide synthetases. Among the most prominent fungal metabolites, the β-lactams such as penicillins (from *A. nidulans* and *P. chrysogenum*) and cephalosporins (from *Cephalosporium acremonium*) are the best-studied candidates. The biosynthesis of β-lactams has been extensively studied and comprehensively reviewed [51,52].

The first enzyme of the penicillin/cephalosporin biosynthetic pathway is ACV-synthetase, a three-module 426 kDa NRPS of 3792 amino acid, encoded by the pcbAB gene [51]. The labile β-lactam moiety is created when IpnA, the isopenicillin N-synthase encoded by pcbC, cyclizes ACV to isopenicillin N. This enzyme's mode of action is a two-step cyclization with the β-lactam ring closure preceding the thiazolidine ring formation [53]. Pathways to penicillin G and cephalosporins, for example, cephalosporin C, diverge after isopenicillin N formation. Penicillin is formed when L-a-aminoadipic acid is substituted for phenylacetic or phenoxyacetic acid by isopenicillin *N*-acyltransferase (IAT), encoded by the *penDE* gene. The genetic region adjacent to the three penicillin genes has been studied and harbors genes that seem to be involved in – but not strictly indispensable for – penicillin production (Figure 13.9) [54].

Figure 13.9 Penicillin biosynthetic gene cluster.

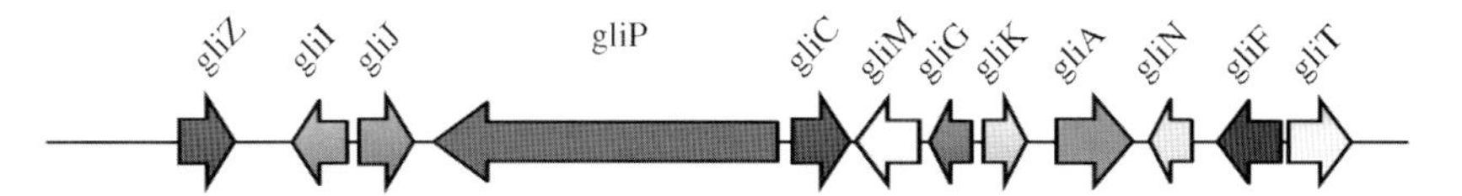

Figure 13.10 Gliotoxin biosynthetic gene clusters.

The diketopiperazines are a widespread and diverse class of natural products found in both prokaryotic and fungal species. Gliotoxin, the most prominent member of this class, is an immunosuppressive mycotoxin from *Aspergillus fumigatus*, the causative agent of human aspergillosis. Other diketopiperazines that causes the canola blackleg disease is a phytotoxin from *Leptosphaeria maculans*, sirodesmin, and the tremorgenic fumitremorgins from *A. fumigatus* and other *Aspergillus* and *Penicillium* species. The basic biosynthetic mechanism underlying the synthesis of diketopiperazines is similar in the sense that two amino acids (for gliotoxin: L-serine and L-phenylalanine; for sirodesmin: L-serine and L-tyrosine) are cyclized by formation of two peptide bonds, to set up the characteristic diketopiperazine core. Due to their decoration with an internal disulfide bridge, gliotoxin and sirodesmin are categorized as epipolythiodioxopiperazines (ETPs). Comprehensive reviews have recently been published on the biosynthesis and toxicity of ETPs [55].

With the sirodesmin cluster as a blueprint and increase in the number of fungal genomes being sequenced, more putative clusters for ETP metabolites have been found. The putative biosynthetic gene cluster of gliotoxin, spanning about 28 kb in the genome of *A. fumigatus*, has recently been worked out, based on the sirodesmin gene cluster as template [56]. Proof of involvement of gli genes in gliotoxin biosynthesis came from the disruption of the NRPS gene gliP, leading to a gliotoxin null chemotype demonstrating the proof of the involvement of gli gene cluster in the synthesis of gliotoxin [57]. The genes in the gli cluster seem to be coregulated as the disruption of gliZ, encoding a Zn_2Cys_6 transcription factor, also resulted in loss of gliotoxin production as well as loss of expression of other gli genes (Figure 13.10) [58].

13.5 Conclusions

The biology of fungal bioactive gene clusters is diverse and holds great promise to understand the basis of complexity of the fungal secondary metabolites. Gene libraries based on the identified fungal bioactive gene clusters could pave way for the effective and efficient isolation of bioactive compounds from fungi leading to a clear vision to natural product chemists and mycologists.

Acknowledgments

The authors are thankful to Marine Bioprocess Research Center of the Marine Bio 21 center funded by the "Ministry of Land, Transport and Maritime," Republic of

Korea for providing the lab space and grant. I.B. is thankful to CSIR, India for her sabbaticals.

References

1 Demain, A.L. (2006) From natural products discovery to commercialization: a success story. *J. Ind. Microbiol. Biotechnol.*, **33** (7), 486–495.

2 Laich, F., Fierro, F., and Martín, J.F. (2002) Production of penicillin by fungi growing on food products: identification of a complete penicillin gene cluster in *Penicillium griseofulvum* and a truncated cluster in *Penicillium verrucosum*. *Appl. Environ. Microbiol.*, **68** (3), 1211–1219.

3 Dewick, P. (2001) The acetate pathway: fatty acids and polyketides, in *Medicinal Natural Products: A Biosynthetic Approach*, 2nd edn, John Wiley & Sons, Ltd, Chichester, pp. 35–117.

4 Lysøe, E., Klemsdal, S.S., Bone, K.R., Frandsen, R.J.N., Johansen, T., Thrane, U., and Giese, H. (2006) The PKS4 gene of *Fusarium graminearum* is essential for zearalenone production. *Appl. Environ. Microbiol.*, **72** (6), 3924–3932.

5 Shier, W., Shier, A., Xie, W., and Mirocha, C. (2001) Structure–activity relationships for human estrogenic activity in zearalenone mycotoxins. *Toxicon*, **39** (9), 1435–1438.

6 Hodge, E.B. (1966) Estrogenic compounds and animal growth promoters. U.S. Patent No. 3,239,345.

7 Utian, W.H. (1973) Comparative trial of P1496, a new non-steroidal oestrogen analogue. *Br. Med. J.*, **1** (5853), 579–581.

8 Hidy, P., Baldwin, R., Greasham, R., Keith, C., and McMullen, J. (1977) Zearalenone and some derivatives: production and biological activities. *Adv. Appl. Microbiol.*, **22**, 59–82.

9 Borel, J.F., Feurer, C., Gubler, H., and Stähelin, H. (1994) Biological effects of cyclosporin A: a new antilymphocytic agent. *Inflamm. Res.*, **43** (3), 179–186.

10 Faulds, D., Goa, K., and Benfield, P. (1993) Cyclosporin: a review of its pharmacodynamic and pharmacokinetic properties, and therapeutic use in immunoregulatory disorders. *Drugs*, **45** (6), 953.

11 Bookstein, R., Lai, C.C., Hoang, T., and Lee, W.H. (1990) PCR-based detection of a polymorphic BamHI site in intron 1 of the human retinoblastoma (RB) gene. *Nucleic Acids Res.*, **18** (6), 1666–11666.

12 Demain, A. (1999) Pharmaceutically active secondary metabolites of microorganisms. *Appl. Microbiol. Biotechnol.*, **52** (4), 455–463.

13 Royer, J.C., Madden, K.T., Norman, T.C., and LoBuglio, K.F. (2004) *Penicillium* genomics. *Appl. Mycol. Biotechnol.*, **4**, 285–293.

14 McAlpine, J. (1998) Unnatural natural products by genetic manipulation, in *Natural Products II: New Technologies to Increase Efficiency and Speed*, (eds D.M. Sapienza and L.M. Savage), International Business Communications, Southborough, MA, pp. 251–278.

15 Peng, Y. and Demain, A. (1998) A new hydroxylase system in *Actinomadura* sp. cells converting compactin to pravastatin. *J. Ind. Microbiol. Biotechnol.*, **20** (6), 373–375.

16 Bhatnagar, I. and Kim, S.K. (2010) Marine antitumor drugs: status, shortfalls and strategies. *Mar. Drugs*, **8** (10), 2702–2720.

17 Fujii, I., Watanabe, A., and Ebizuka, Y. (2004) More functions for multifunctional polyketide synthases, in *Advances in Fungal Biotechnology for Industry, Agriculture, Medicine*, (eds J.S. Tkacz and L. Lange), Kluwer Academic/Plenum Publishers, New York, pp. 97–125.

18 Staunton, J. and Weissman, K.J. (2001) Polyketide biosynthesis: a millennium review. *Nat. Prod. Rep.*, **18** (4), 380–416.

19 O'Hagan, D. and Collie, J.N. (1991) *The Polyketide Metabolites*, Ellis Horwood, Chichester, UK.

20 O'Hagan, D. (1992) Biosynthesis of polyketide metabolites. *Nat. Prod. Rep.*, **9** (5), 447–479.

21 Hopwood, D.A. (2004) Cracking the polyketide code. *PLoS Biol.*, **2** (2), e35.

22 Hutchinson, C.R. and Fujii, I. (1995) Polyketide synthase gene manipulation: a structure–function approach in engineering novel antibiotics. *Annu. Rev. Microbiol.*, **49** (1), 201–238.

23 Beck, J., Ripka, S., Siegner, A., Schiltz, E., and Schweizer, E. (2005) The multifunctional 6-methylsalicylic acid synthase gene of *Penicillium patulum*. *Eur. J. Biochem.*, **192** (2), 487–498.

24 Kennedy, J., Auclair, K., Kendrew, S.G., Park, C., Vederas, J.C., and Hutchinson, C.R. (1999) Modulation of polyketide synthase activity by accessory proteins during lovastatin biosynthesis. *Science*, **284** (5418), 1368–1372.

25 Funa, N., Awakawa, T., and Horinouchi, S. (2007) Pentaketide resorcylic acid synthesis by type III polyketide synthase from *Neurospora crassa*. *J. Biol. Chem.*, **282** (19), 14476–14481.

26 Schröder, J. (1999) Probing plant polyketide biosynthesis. *Nat. Struct. Biol.*, **6** (8), 714.

27 Austin, M.B. and Noel, J.P. (2003) The chalcone synthase superfamily of type III polyketide synthases. *Nat. Prod. Rep.*, **20** (1), 79–110.

28 Seshime, Y., Juvvadi, P.R., Fujii, I., and Kitamoto, K. (2005) Discovery of a novel superfamily of type III polyketide synthases in *Aspergillus oryzae*. *Biochem. Biophys. Res. Commun.*, **331** (1), 253–260.

29 Schümann, J. and Hertweck, C. (2006) Advances in cloning, functional analysis and heterologous expression of fungal polyketide synthase genes. *J. Biotechnol.*, **124** (4), 690–703.

30 Nicholson, T.P., Rudd, B.A.M., Dawson, M., Lazarus, C.M., Simpson, T.J., and Cox, R.J. (2001) Design and utility of oligonucleotide gene probes for fungal polyketide synthases. *Chem. Biol.*, **8** (2), 157–178.

31 Bingle, L.E.H., Simpson, T.J., and Lazarus, C.M. (1999) Ketosynthase domain probes identify two subclasses of fungal polyketide synthase genes. *Fungal Genet. Biol.*, **26** (3), 209–223.

32 Kroken, S., Glass, N.L., Taylor, J.W., Yoder, O., and Turgeon, B.G. (2003) Phylogenomic analysis of type I polyketide synthase genes in pathogenic and saprobic ascomycetes. *Proc. Natl. Acad. Sci. USA*, **100** (26), 15670–15675.

33 Cox, R.J. (2007) Polyketides, proteins and genes in fungi: programmed nano-machines begin to reveal their secrets. *Org. Biomol. Chem.*, **5** (13), 2010–2026.

34 Shimizu, T., Kinoshita, H., Ishihara, S., Sakai, K., Nagai, S., and Nihira, T. (2005) Polyketide synthase gene responsible for citrinin biosynthesis in *Monascus purpureus*. *Appl. Environ. Microbiol.*, **71** (7), 3453–3457.

35 Bailey, A.M., Cox, R.J., Harley, K., Lazarus, C.M., Simpson, T.J., and Skellam, E. (2007) Characterisation of 3-methylorcinaldehyde synthase (MOS) in *Acremonium strictum*: first observation of a reductive release mechanism during polyketide biosynthesis. *Chem. Commun.*, (39), 4053–4055.

36 Fujii, I., Ono, Y., Tada, H., Gomi, K., Ebizuka, Y., and Sankawa, U. (1996) Cloning of the polyketide synthase gene atX from *Aspergillus terreus* and its identification as the 6-methylsalicylic acid synthase gene by heterologous expression. *Mol. Gen. Genet.*, **253** (1), 1–10.

37 Moriguchi, T., Ebizuka, Y., and Fujii, I. (2006) Analysis of subunit interactions in the iterative type I polyketide synthase ATX from *Aspergillus terreus*. *ChemBioChem*, **7** (12), 1869–1874.

38 Moriguchi, T., Ebizuka, Y., and Fujii, I. (2008) Domain–domain interactions in the iterative type I polyketide synthase ATX from *Aspergillus terreus*. *ChemBioChem*, **9** (8), 1207–1212.

39 Walsh, C.T. (2007) The chemical versatility of natural-product assembly lines. *Acc. Chem. Res.*, **41** (1), 4–10.

40 Finking, R. and Marahiel, M.A. (2004) Biosynthesis of nonribosomal peptides 1. *Annu. Rev. Microbiol.*, **58**, 453–488.

41 Caboche, S., Pupin, M., Leclère, V., Fontaine, A., Jacques, P., and Kucherov, G. (2008) NORINE: a database of nonribosomal peptides. *Nucleic Acids Res.*, **36** (Suppl. 1), D326–D331.

42 Fischbach, M.A. and Walsh, C.T. (2006) Assembly-line enzymology for polyketide and nonribosomal peptide antibiotics: logic, machinery, and mechanisms. *Chem. Rev.*, **106** (8), 3468.

43 Kopp, F. and Marahiel, M.A. (2007) Macrocyclization strategies in polyketide and nonribosomal peptide biosynthesis. *Nat. Prod. Rep.*, **24** (4), 735–749.
44 Hoffmeister, D. and Keller, N.P. (2007) Natural products of filamentous fungi: enzymes, genes, and their regulation. *Nat. Prod. Rep.*, **24** (2), 393–416.
45 Yu, J., Chang, P.K., Ehrlich, K.C., Cary, J.W., Bhatnagar, D., Cleveland, T.E., Payne, G.A., Linz, J.E., Woloshuk, C.P., and Bennett, J.W. (2004) Clustered pathway genes in aflatoxin biosynthesis. *Appl. Environ. Microbiol.*, **70** (3), 1253–1262.
46 Yabe, K. and Nakajima, H. (2004) Enzyme reactions and genes in aflatoxin biosynthesis. *Appl. Microbiol. Biotechnol.*, **64** (6), 745–755.
47 Brown, D., Yu, J., Kelkar, H., Fernandes, M., Nesbitt, T., Keller, N., Adams, T., and Leonard, T. (1996) Twenty-five coregulated transcripts define a sterigmatocystin gene cluster in *Aspergillus nidulans*. *Proc. Natl. Acad. Sci. USA*, **93** (4), 1418–1422.
48 Chang, P.K., Ehrlich, K.C., Yu, J., Bhatnagar, D., and Cleveland, T.E. (1995) Increased expression of *Aspergillus parasiticus* aflR, encoding a sequence-specific DNA-binding protein, relieves nitrate inhibition of aflatoxin biosynthesis. *Appl. Environ. Microbiol.*, **61** (6), 2372–2377.
49 Gaffoor, I., Brown, D.W., Plattner, R., Proctor, R.H., Qi, W., and Trail, F. (2005) Functional analysis of the polyketide synthase genes in the filamentous fungus *Gibberella zeae* (anamorph *Fusarium graminearum*). *Eukaryot. Cell*, **4** (11), 1926–1933.
50 Song, Z., Cox, R.J., Lazarus, C.M., and Simpson, T.J. (2004) Fusarin C biosynthesis in *Fusarium moniliforme* and *Fusarium venenatum*. *ChemBioChem*, **5** (9), 1196–1203.
51 Aharonowitz, Y., Cohen, G., and Martin, J. (1992) Penicillin and cephalosporin biosynthetic genes: structure, organization, regulation, and evolution. *Annu. Rev. Microbiol.*, **46** (1), 461–495.
52 Martin, J. (1998) New aspects of genes and enzymes for β-lactam antibiotic biosynthesis. *Appl. Microbiol. Biotechnol.*, **50** (1), 1–15.
53 Burzlaff, N.I., Rutledge, P.J., Clifton, I.J., Hensgens, C.M.H., Pickford, M., Adlington, R.M., Roach, P.L., and Baldwin, J.E. (1999) The reaction cycle of isopenicillin N synthase observed by X-ray diffraction. *Nature*, **401** (6754), 721–724.
54 Fierro, F., García-Estrada, C., Castillo, N.I., Rodríguez, R., Velasco-Conde, T., and Martín, J.F. (2006) Transcriptional and bioinformatic analysis of the 56.8kb DNA region amplified in tandem repeats containing the penicillin gene cluster in *Penicillium chrysogenum*. *Fungal Genet. Biol.*, **43** (9), 618.
55 Gardiner, D.M., Waring, P., and Howlett, B.J. (2005) The epipolythiodioxopiperazine (ETP) class of fungal toxins: distribution, mode of action, functions and biosynthesis. *Microbiology*, **151** (4), 1021–1032.
56 Gardiner, D.M. and Howlett, B.J. (2005) Bioinformatic and expression analysis of the putative gliotoxin biosynthetic gene cluster of *Aspergillus fumigatus*. *FEMS Microbiol. Lett.*, **248** (2), 241–248.
57 Cramer, R.A., Gamcsik, M.P., Brooking, R.M., Najvar, L.K., Kirkpatrick, W.R., Patterson, T.F., Balibar, C.J., Graybill, J.R., Perfect, J.R., and Abraham, S.N. (2006) Disruption of a nonribosomal peptide synthetase in *Aspergillus fumigatus* eliminates gliotoxin production. *Eukaryot. Cell*, **5** (6), 972–980.
58 Bok, J.W., Chung, D.W., Balajee, S.A., Marr, K.A., Andes, D., Nielsen, K.F., Frisvad, J.C., Kirby, K.A., and Keller, N.P. (2006) GliZ, a transcriptional regulator of gliotoxin biosynthesis, contributes to *Aspergillus fumigatus* virulence. *Infect. Immun.*, **74** (12), 6761–6768.

14
Anticancer Potentials of Marine-Derived Fungal Metabolites

Se-Kwon Kim and Pradeep Dewapriya

14.1
Introduction

Cancer causes an estimated one in eight deaths worldwide [1]. Changes in lifestyle, nutrition, and environment factors have steepened the detrimental effect of cancer. Even though cancer research has generated handful of knowledge on initiation, progression, and treatment of cancer, to date there are no potent medicines for some cancers. There has been a great interest in identifying potent anticancer compounds with less number of side effects. Since most of the synthetic compounds have side effects, natural products derived from medicinal plants and animals have gained significance in the treatment of cancer. In fact, natural products and their derivatives represent more than 50% of the drugs clinically used in the world [2]. The progression of drug discovery in natural products led to exploration of marine flora and fauna for novel cures. It is realized within a short period of time that the marine ecosystem is ideal for novel lead compound explorations since this "silent world" represents a much richer biodiversity than that of the terrestrial environment. As a result, every year hundreds of new metabolites with promising pharmaceutical values are derived from marine sources. Published data indicate that several potent chemotherapeutic and chemopreventive agents have been discovered from marine sources [3]. But there are no approved drugs derived from marine sources except Ara-C and ω-conotoxin; therapeutic agents that were synthesized based on the lead compounds originally isolated from marine environment. The development of marine-derived compounds as drug candidates is mainly hampered by the inadequate supply of active ingredients. Potent anticancer compounds such as halichondrin B and bryostatin 1 isolated from marine sources are still sought-after sustainable source to proceed as pharmaceutical agents. To overcome this problem, the direction of marine natural product research was changed to explore sustainable metabolites as drug leads [4]. In this regard, marine microorganisms are in limelight because of their ability to produce unique, novel metabolites [5].

Microbial fermentation is an easier way of achieving the sustainable and economically viable supply of active pharmaceutical ingredients. Thus, various kinds of marine

Marine Microbiology: Bioactive Compounds and Biotechnological Applications, First Edition.
Edited by Se-Kwon Kim

prokaryotic as well as eukaryotic microorganisms have been studied for biologically active compounds. Advance developments of isolation, identification, genomic information, and fermentation techniques have made possible to culture many microorganisms from extreme marine environments [6]. Among them, marine fungi received much attention due to their tremendous ability to produce a wider array of biologically active compounds. These diverse metabolites possess various biological activities including antibiotics and anticancer. In this chapter, anticancer potential of marine-derived fungal metabolites is discussed in a detailed manner.

14.2 Marine Fungi

Since long, fungi have influenced many aspects of human life such as food, medicine, and pathogen. The discovery of penicillin from *Penicillium notatum* in 1929 by Sir Alexander Fleming gave a new insight into the human and fungi association. Furthermore, importance of fungi in exploration of lead compounds for drug discovery was resurged with identification of cyclosporin A from *Tolypocladium inflatum* in 1976. However, interest in marine fungi and in the chemistry of their metabolites was negligible till the 1990s. With advanced technological developments, researchers try to explore unique habitats such as marine environment for new biosynthetic diversity from fungi [7].

Indeed, marine fungus is an ecological group, not a taxonomic group. This group may be obligate, the fungi may grow and sporulate exclusively in seawater, and their spores are capable of germinating in seawater, or facultative, those obtained from freshwater or terrestrial sources that have undergone physiological adaptations for the marine environment. To date over 800 obligate marine fungal strains have been identified. Majority of identified marine fungi belong to the group of ascomycetes. As most of the fungi are heterotrophic eukaryotes, they are common in decomposing plant and animal tissues [8]. Thus, various kinds of marine sources such as algae, mangrove, sediment, and invertebrates have been identified as typical sources for isolating marine fungi. Among them, fungi isolated from marine sponges and algae have contributed to more than 50% of novel compounds that have been isolated from marine fungi. Majority belong to the group of polyketides or their isoprene hybrids. Specifically, fused-ring polyketides are potential candidates of antioxidant, antibiotic, and anticancer lead compounds. In addition, several terpenoids that accounts for around 10% of the identified compounds possess potent antibiotic and cytotoxic properties [9].

14.3 Cancer: Initiation, Progression, and Therapeutics

Generally, normal cells are monitored by various cell-signaling mechanisms to control the cell number while working in cooperation with surrounding cells. This coordination is vital for organ formation during development and cellular hemostasis in adulthood. In a situation that cells violate this natural control mechanism of cell proliferation, cancers arise. Cancer is a complex process that can occur in almost all tissues in the body. Over 100 forms of cancers have been identified and most of them follow unique violated behavior of cell proliferation. Furthermore, uncontrolled growth, lack of response to stop signals, immortality, and random migration are common features of cancers [10]. Several factors that are known to increase the risk of cancer have been identified. Among them, tobacco use, certain infections, radiation, lack of physical activity, obesity, and environmental pollutants are prominent causative agents of cancers. A cell that has undergone mutations due to a risk factor takes a long time to behave like a malignant cell. Thus, it is quite hard to discover the initiation of a cancer [11]. In general, cancers are treated with chemotherapy, radio therapy, and surgery. Even though the prevention has been highlighted as the best way to handle cancers, chemotherapy has proven its ability to control established cancers. Nitrogen mustard became the first systemic chemotherapy agent, and since then, hundreds of compounds were discovered to treat cancers [12]. Common chemotherapy agents act by killing cells that divide rapidly, and thus these compounds also harm normal cells that divide rapidly. This has been identified as the major drawback of chemotherapy since complex side effects are generated as a result of nonselective destruction of cells. Therefore, there is a growing interest in anticancer compounds that act directly against abnormalities of cancer cells, which is termed as targeted therapy [13,14]. Moreover, cancer preventive compounds such as anti-inflammatory and antioxidative have gained increased attention, as inflammation and oxidative stress act as initiators for most cancers [15]. From this point forward, the ability of marine fungi-derived metabolites to prevent and cure cancers is discussed based on the biological effects of the compounds on cancer cells.

14.4 Anticancer Metabolites of Marine Fungal Origin

14.4.1 Cytotoxic and Antitumor Compounds

Traditional cancer chemotherapy relied on killing rapidly dividing cancer cells with little or no specificity, which leads to severe side effects. After few decades of cancer chemotherapy researches, it was realized that cancer cells overexpress many receptors and biomarkers that can be used as targets to deliver cancer drugs. Thus, current exploration for cytotoxic and antitumor compounds as lead drug candidates mainly targets specific receptors or signaling molecules expressed in cancer cells [16].

Protein kinases are responsible for phosphorylation of adapter and intracellular signaling proteins, resulting in the activation of multiple signaling pathways. Inhibitor of protein kinases specifically blocks the action of one or more protein kinases, which subsequently block the downstream signaling. Especially, tyrosine kinase inhibition has been identified as promising target to stop the survival of cancer cells. Several protein kinase inhibitors have been isolated from marine fungi. Lateff [17] isolated a *Chaetomium* sp., which produces chaetominedione as a major secondary metabolite. The screening of biological activity revealed that the compound significantly inhibits the activity of $p56^{lck}$ tyrosine kinase (18.7 and 93.6% enzyme inhibition at 200 μg/ml, respectively). Ulocladol is another $p56^{lck}$ tyrosine kinase inhibitor that was isolated from a culture of the *Ulocladium botrytis*. The compound possesses an ability to reduce the enzyme activity up to 7% at 0.02 μg/ml [18]. Several studies have proved that $p56^{lck}$ tyrosine kinase is involved in the regulation of apoptosis [19]. 1403C is a novel anthraquinone derivative isolated from cultures of the marine-derived mangrove endophytic filamentous fungus *Halorosellinia* sp. Due to its potent inhibition of protein kinase B (PKB), the compound has been identified as an anticancer drug candidate. A recent study showed that marine fungal strain *Halorosellinia* sp. pulse fed with glucose solutions yielded 4.5 g/l of 1403C compound [20]. This strategy is valuable for fermentation scale-up of *Halorosellinia* sp. (No. 1403) for production of 1403C, which can solve the supply problem of 1403C for clinical studies.

Topoisomerases are a group of enzymes that control the three-dimensional structure of DNA; these enzymes are crucial for cellular genetic processes such as DNA replication, transcription, recombination, and chromosome segregation at mitosis. It has been observed that topoisomerase levels have been elevated in several tumor types. Thus, topoisomerase-I (Top I) and topoisomerase-II are key subcellular target for anticancer therapy. Secalonic acid D (SAD) is an excellent Top I inhibitor isolated from marine-derived fungi. This compound displays a considerable inhibition on Top I in a dose-dependent manner with the minimum inhibitory concentration (MIC) of 0.4 μM. The unique property of the compound is to inhibit the binding of Top I to DNA without inducing the formation of Top I–DNA covalent complexes [21]. Yanagihara and his research team isolated several leptosin derivatives from marine fungus *Leptosphaeria* sp. *In vitro* cytotoxic assays showed that compounds are catalytic inhibitors of Top I. Furthermore, Leps F and C inhibited the Akt signaling pathway by dose-dependent and time-dependent dephosphorylation of Akt [22].

In addition, many novel as well as known cytotoxic compounds have been isolated from marine-derived fungi. Coriolin B and three new chlorinated cyclic sesquiterpenes were isolated from marine fungi species belongs to the class *Hyphomycetes*. Biological studies revealed that these metabolites strongly inhibited human breast and central nervous system cancer cell lines with IC50 values of 0.7 (breast) and 0.5 μg (neuroblastoma) [23]. Ethanol extract of a static culture of *Aspergillus niger* isolated from the Mediterranean sponge *Axinella damicornis* yielded a new secondary metabolite bicoumanigrin A that has *in vitro* antiproliferative activities on human cancer cell lines [24]. Furthermore, *Aspergillus versicolor* isolated from a marine sponge yielded a cytotoxic lipopeptide that has been tested against five human tumor cell lines (A549, human lung cancer; SK-OV-3, human ovarian cancer; SK-MEL-2, human skin cancer; XF498, human CNS cancer; and HCT15, human colon cancer). The lipopeptide significantly acts against XF498 and

HCT15 cell lines as compared to commercial drug doxorubicin [25]. Another two new cyclodepsipeptides that have strong antiproliferative activity against pancreatic tumor cells were isolated from a marine sponge-derived fungus *Scopulariopsis brevicaulis*. These compounds have been patented as potential antitumor drug [26]. However, detailed studies have to be conducted to confirm molecular mechanism by which these compounds exhibited their cytotoxicity.

14.4.2 Apoptosis Inducing Metabolites

Apoptosis or programmed cell death is a natural event of any tissue in the body that is essential for homeostasis and proper functioning. Several signaling initiators and regulators are involved in apoptosis, and these signaling molecules are promising targets for pharmacological modulation of cancer cell death [27]. Especially, initiator (caspase 2, −8, −9, and −10) and effector caspases (caspase-3, −6, and −7) are mainly responsible for cell death. The apoptosis has been recognized as a defense strategy against tumor genesis. Furthermore, loss of caspase-8 expression often correlates with amplification of MYCN oncogene and elevated levels of tumor-specific proteins [28]. Recently, it was found that apoptosis signaling triggered by many stimuli and dysfunction of mitochondria plays a pivotal role in coordinating caspase activation through the release of cytochrome *c* [29].

Screening of secondary metabolites of the mangrove endophytic fungi *Halorosellinia* sp. and *Guignardia* sp. led to separate anthracenedione derivatives that are potent inhibitors of mitochondrial function of cancer cell KB and thereby lead to apoptosis [30]. Protuboxepin A, isolated from *Aspergillus* sp. SF-5044, showed potent anticancer activity by inducing the apoptosis of cancer cells. The compound directly binds to α,β-tubulin and stabilizes tubulin polymerization while disrupting microtubule dynamics. This disruption leads to chromosome misalignment and metaphase arrest, which induces apoptosis in cancer cells [31]. Himeic acid is a new ubiquitin-activating enzyme inhibitor isolated from a marine-derived fungus, *Aspergillus* sp. Ubiquitin-activating enzyme plays an important role in cell cycle progression. The compound at a concentration of 50 μM inhibits the formation of ubiquitin-activating enzyme up to 65% and arrests progression of cell cycle leading to apoptosis [32].

14.4.3 Antimetastasis Compounds

In general, metastasis is the spread of a disease from one organ or part to another nonadjacent organ or part. Cancer cells possess this detrimental ability which detach themselves from the primary tumor and invade nonadjacent, healthy tissue. This process allows primary tumor to spread in the body and form *secondary* or *metastatic* tumor, which has characteristics similar to those of the original tumor. The reason for most cancer deaths is secondary tumors and metastases, not primary neoplasias [33].

Marine fungal metabolite 1386A is a newly isolated compound from the mangrove fungus 1386A in the South China Sea. 1386A showed significant anticancer effect against MCF-7 breast cancer cells (the IC50 value at 48 h was 17.1 μmol/l). The global miRNA expression profile of the MCF-7 cells was significantly altered by the compound. These miRNAs are potential targets of many oncogenes and tumor-suppressor genes associated with cancer metastasis [34]. Moreover, chrysophanol, physcion, and emodin, secondary metabolites of marine fungal species of *Microsporum*, have been identified as potent anticancer agents. All three compounds significantly inhibit the expression of matrix metalloproteinase (MMP)-2 and MMP-9 in a dose-dependent manner. The inhibition of the MMP-2 and MMP-9 expression has been achieved via suppression of the JNK and ERK signaling pathways [35].

14.4.4 Anti-inflammatory and Antioxidant Compounds

Chronic inflammation and oxidative stress have been identified as major causes of cancers. Even though these two are normal biological processes of natural defense mechanism of the body, their prolonged existence causes detrimental effect on the biological functions of tissues and organs. Current estimates suggest that over 25% of cancers are associated with chronic inflammation due to several reasons [36]. Infections and irritation of tissues lead to the production of inflammatory cytokines and receptor molecules, which mediates the inflammatory response. The primary role of inflammatory mediators is recovery from injury and healing; however, if targeted destruction and assisted repair are not properly phased, inflammation can lead to persistent tissue damage by leukocytes, lymphocytes, or collagen. Furthermore, increased recruitment of mast cells and leukocytes to the damage site triggers respiratory burst and subsequently releases free radicals. Inflammatory mediators and free radicals can drive carcinogenesis by triggering damages and mutations in DNA, RNA, and cellular macromolecules [37]. Thus, anti-inflammatory and antioxidative compounds serve as preventive agents of this deadly disease.

Microbial extracellular polysaccharides have been recently identified as potent antioxidative metabolites. Interestingly, marine fungal species are well-known producers of exopolysaccharides with potent antioxidant activity. Several recent investigations have showed that exopolysaccharides produced by marine fungi significantly scavenge superoxide and hydroxyl radicals in *in vitro* experiments [38–40]. Moreover, marine fungus is an ideal source for novel antioxidative compounds. Recent researches proved that fungal species in marine environment are capable of producing novel compounds, and their antioxidative properties are comparatively higher than those of some commercially available synthetic products [41]. In addition, several potent anti-inflammatory metabolites have been isolated from marine-derived fungal species. Asperlin isolated from marine-derived fungus *Aspergillus* sp. significantly suppressed the iNOS expression, iNOS-derived NO production, COX-2 expression, and PGE_2 production in lipopolysaccharide (LPS)-stimulated RAW264.7 cells. Molecular investigation revealed that asperlin induced heme oxygenase (HO)-1 expression through nuclear translocation of E2-related factor 2 and increased HO activity in RAW264.7 macrophages [42]. Moreover, many more potent anti-inflammatory compounds with unique inhibitory mechanism of inflammatory signals have been identified from marine fungi. These metabolites represent a new class of anti-inflammatory compounds that do not follow the mechanisms that exert side effects [7].

14.5 Future Prospects and Concluding Remarks

Available literatures are sufficient enough to prove that marine-derived fungi are prolific producers of metabolites that have potent anticancer activity. Here, we revealed that cancer initiation, development, and curing process could be effectively mediated with marine-derived fungal metabolites. Although these metabolites show potent anticancer activity, biological mechanism of some potent anticancer compounds is still not clear. In the process of discovering lead compounds for prevention and treatment of cancer, mechanism of action of a metabolite and its biological consequences are necessary to validate it as a lead compound. Since the fungal metabolites are reproducible, they have clear future in drug development. Thus, comprehensive *in vitro* as well as *in vivo* studies have to be conducted to reveal the mechanism of action of isolated compounds. Furthermore, studies on structure–activity relationship and biosynthetic pathways of these metabolites will pave the way to manipulate the microbial biosynthetic pathways through genetic engineering, which will lead to the production of interesting fungal metabolites with higher potency and productivity.

References

1 Siegel, R., Ward, E., Brawley, O., and Jemal, A. (2011) Cancer statistics. *CA Cancer J. Clin.*, **61** (12), 212–236.

2 Boopathy, N.S. and Kathiresan, K. (2010) Anticancer drugs from marine flora: an overview. *J. Oncol.* doi: 10.1155/2010/214186.

3 Schumacher, M., Kelkel, M., Dicato, M., and Diederich, M. (2011) Gold from the sea: marine compounds as inhibitors of the hallmarks of cancer. *Biotechnol. Adv.*, **29**, 531–547.

4 Waters, A.L., Hill, R.T., Place, R.A., and Hamann, M.T. (2010) The expanding role of marine microbes in pharmaceutical development. *Curr. Opin. Biotechnol.*, **21**, 780–786.

5 Mayer, A.M.S., Glaser, K.B., Cuevas, C., Jacobs, R.S., Kern, W., Little, R.D., McIntosh, J.M., Newman, D.J., Potts, B.C., and Shuster, D.E. (2010) The odyssey of marine pharmaceuticals: a current pipeline perspective. *Trends Pharmacol. Sci.*, **31**, 255–265.

6 Immhoff, J.F., Labes, A., and Wiese, J. (2011) Bio-mining the microbial treasures of the ocean: new natural products. *Biotechnol. Adv.*, **29**, 468–482.

7 Bugni, T.S. and Ireland, C.M. (2004) Marine-derived fungi: a chemically and biologically diverse group of microorganisms. *Nat. Prod. Rep.*, **21**, 143–163.

8 Raghukumar, C. (2008) Marine fungal biotechnology: an ecological perspective. *Fungal Divers.*, **31**, 19–35.

9 Saleem, M., Ali, M.S., Hussain, M., Jabbar, A., Ashraf, M., and Lee, Y.S. (2007) Marine

natural products of fungal origin. *Nat. Prod. Rep.*, **24**, 1142–1152.

10 Weinberg, R.A. (1996) How cancer arises. *Sci. Am.*, **275**, 62–70.

11 Anand, P., Kunnumakara, A.B., Sundaram, C., Harikumar, K.B., Tharakan, S.T., Lai, O.S., Sung, B., and Aggarwal, B.B. (2008) Cancer is a preventable disease that requires major lifestyle changes. *Pharm. Res.*, **25**, 2097–2116.

12 Joensuu, H. (2008) Systemic chemotherapy for cancer: from weapon to treatment. *Lancet Oncol.*, **9**, 304.

13 Zhukov, N.V. and Tjulandin, S.A. (2008) Targeted therapy in the treatment of solid tumors: practice contradicts theory. *Biochemistry*, **73**, 605–618.

14 Katzel, J.A., Fanucchi, M.P., and Li, Z. (2009) Recent advances of novel targeted therapy in non-small cell lung cancer. *J. Hematol. Oncol.*, **2**, 1–18.

15 Nathan, C. (2002) Points of control in inflammation. *Nature*, **240**, 846–852.

16 Jaracz, S., Chen, J., Kuznetsova, L.V., and Ojima, I. (2005) Recent advances in tumor-targeting anticancer drug conjugates. *Bioorg. Med. Chem.*, **13**, 5043–5054.

17 Lateff, A.A. (2008) Chaetominedione, a new tyrosine kinase inhibitor isolated from the algicolous marine fungus *Chaetomium* sp. *Tetrahedron Lett.*, **49**, 6398–6400.

18 Goldberg, D.R., Butz, T., Cardozo, M.G., Eckner, R.J., Hammach, A., Huang, J., Jakes, S., Kapadia, S., Kashem, M., Lukas, S., Morwick, T.M., Panzenbeck, M., Patel, U., Pav, S., Peet, G.W., Peterson, J.D., Prokopowicz, A.S.III, Snow, R.J., Sellati, R., Takahashi, H., Tan, J., Tschantz, M.A., Wang, X.J., Wang, Y., Wolak, J., Xiong, P., and Moss, N. (2003) Optimization of 2-phenylaminoimidazo[4,5-*h*]isoquinolin-9-ones: orally active inhibitors of lck kinase. *J. Med. Chem.*, **46**, 1337–1349.

19 Gruber, C., Henkel, M., Budach, W., Belka, C., and Jendrossek, V. (2004) Involvement of tyrosine kinase p56/Lck in apoptosis induction by anticancer drugs. *Biochem. Pharmacol.*, **67**, 1859–1872.

20 Kang, L., Cai, M., Yu, C., Zhang, Y., and Zhou, X. (2011) Improved production of the anticancer compound 1403C by glucose pulse feeding of marine *Halorosellinia* sp. (No. 1403) in submerged culture. *Bioresour. Technol.*, **102**, 10750–10753.

21 Hong, R. (2011) Secalonic acid D as a novel DNA topoisomerase I inhibitor from marine lichen-derived fungus *Gliocladium* sp. T31. *Pharm. Biol.*, **49**, 796–799.

22 Yanagihara, M., Takahashi, N.S., Sugahara, T., Yamamoto, S., Shinomi, M., Yamashita, L., Hayashida, M., Yamanoha, B., Numata, A., Yamori, T., and Andoh, T. (2005) Leptosins isolated from marine fungus *Leptosphaeria* species inhibit DNA topoisomerases I and/or II and induce apoptosis by inactivation of Akt/protein kinase B. *Cancer Sci.*, **96**, 816–826.

23 Thomas, T.R.A., Kavlekar, D.P., and LokaBharathi, P.A. (2010) Marine drugs from sponge-microbe association – a review. *Mar. Drugs*, **8**, 1417–1468.

24 Hiort, J., Maksimenka, K., Reichert, M., Ottstadt, S.P., Lin, W.H., Wray, V., Steube, K., Schaumann, K., Weber, H., Proksch, P., Ebel, R., Muller, W.E.G., and Bringmann, G. (2004) New natural products from the sponge-derived fungus *Aspergillus niger*. *J. Nat. Prod.*, **67**, 1532–1543.

25 Lee, Y.M., Dang, H.T., Hong, J., Lee, C., Bae, K.S., Kim, D.K., and Jung, J.H. (2010) A cytotoxic lipopeptide from the sponge-derived fungus *Aspergillus versicolor*. *Bull. Korean Chem. Soc.*, **31**, 205–208.

26 Yu, Z., Lang, G., Kajahn, I., Schmaljohann, R., and Imhoff, J.F. (2008) Scopularides A and B, cyclodepsipeptides from a marine sponge-derived fungus, *Scopulariopsis brevicaulis*. *J. Nat. Prod.*, **71**, 1052–1054.

27 Los, M., Burek, C.J., Stroh, C., Benedyk, K., Hug, H., and Mackiewicz, A. (2003) Anticancer drugs of tomorrow: apoptotic pathways as targets for drug design. *Drug Discov. Today*, **8**, 67–77.

28 Olssonand, M. and Zhivotovsky, B. (2011) Caspases and cancer. *Cell Death Differ.*, **18**, 1441–1449.

29 Desagher, S. and Martinou, J.C. (2000) Mitochondria as the central control point of apoptosis. *Trends Cell Biol.*, **10**, 369–377.

30 Zhang, J.Y., Tao, L.Y., Liang, Y.J., Chen, L.M., Mi, Y.J., Zheng, L.S., Wang, F., She, Z. G., Lin, Y.C., To, K.K.W., and Fu, L.W. (2010) Anthracenedione derivatives as anticancer

agents isolated from secondary metabolites of the mangrove endophytic fungi. *Mar. Drugs*, **8**, 1469–1481.

31 Asami, Y., Jang, J.H., Soung, N.K., He, L., Moon, D.O., Kim, J.W., Oh, H., Muroi, M., Osada, H., Kim, B.Y., and Ahn, J.S. (2012) Protuboxepin A, a marine fungal metabolite, inducing metaphase arrest and chromosomal misalignment in tumor cells. *Bioorg. Med. Chem.*, **20**, 3799–3806.

32 Tsukamoto, S., Hirota, H., Imachi, M., Fujimuro, M., Onuki, H., Ohta, T., and Yokosaw, H. (2005) Himeic acid A: a new ubiquitin-activating enzyme inhibitor isolated from a marine-derived fungus, *Aspergillus* sp. *Bioorg. Med. Chem. Lett.*, **15**, 191–194.

33 Leber, M.F. and Efferth, T. (2009) Molecular principles of cancer invasion and metastasis (review). *Int. J. Oncol.*, **34**, 881–895.

34 Tang, B., He, W.L., Zheng, C., Cheang, T.Y., Zhang, X.F., Wu, H., and Yang, H.L. (2012) Marine fungal metabolite 1386A alters the microRNA profile in MCF-7 breast cancer cells. *Mol. Med. Rep.*, **5**, 610–618.

35 Zhang, C. and Kim, S.K. (2012) Antimetastasis effect of anthraquinones from marine fungus, *Microsporum* sp. *Adv. Food Nutr. Res.*, **65**, 415–421.

36 Balkwill, F.R. and Mantovani, A. (2012) Cancer related inflammation: common themes and therapeutic opportunities. *Semin. Cancer Biol.*, **22**, 33–40.

37 Hussain, S.P., Hofseth, L.J., and Harris, C.C. (2003) Radical causes of cancer. *Nat. Rev. Cancer*, **3**, 276–285.

38 Chen, Y., Mao, W., Yang, Y., Teng, X., Zhu, W., Qi, X., Chen, Y., Zhao, C., Hou, Y., Wang, C., and Li, N. (2012) Structure and antioxidant activity of an extracellular polysaccharide from coral associated fungus, *Aspergillus versicolor* LCJ-5-4. *Carbohydr. Polym.*, **87**, 218–226.

39 Sun, C., Wang, J.W., Fang, L., Gao, X.D., and Tan, R.X. (2004) Free radical scavenging and antioxidant activities of EPS2, an exopolysaccharide produced by a marine filamentous fungus *Keissleriella* sp. YS 4108. *Life Sci.*, **75**, 1063–1073.

40 Sun, H.H., Mao, W.J., Chen, Y., Guo, S.D., Li, H.Y., Qi, X.H., Chen, Y.L., and Xu, J. (2009) Isolation, chemical characteristics and antioxidant properties of the polysaccharides from marine fungus *Penicillium* sp. F23-2. *Carbohydr. Polym.*, **78**, 117–124.

41 Li, J.L., Lee, Y.M., Hong, J., Bae, K.S., Choi, J.S., and Jung, J.H. (2011) A new antioxidant from the marine sponge-derived fungus *Aspergillus versicolor*. *Nat. Prod. Sci.*, **17**, 14–18.

42 Lee, D.S., Jeong, G.S., Li, B., Lee, S.U., Oh, H., and Kim, Y.C. (2011) Asperlin from the marine-derived fungus *Aspergillus* sp. SF-5044 exerts anti-inflammatory effects through heme oxygenase-1 expression in murine macrophages. *J. Pharmacol. Sci.*, **116**, 283–295.

15
Antifungal and Antimycotoxin Activities of Marine Actinomycetes and Their Compounds

Pei-Sheng Yan, Li-Xin Cao, and Jing Ren

15.1
Introduction

Since the discovery of actinomycete-derived antibiotics in the 1940s, the actinomycetes have received a great deal of attention. Up to now, about 10 000 of bioactive metabolites are produced by actinomycetes, which accounted for about the half of the known bioactive metabolites [1]. These compounds showed a wide range of biological activities, including antibacterial, antihelminthic, antileukemic, antiparasitic, antiprotozoal, antioxidant, antiviral, antifungal, anticancer, insecticidal, herbicidal, immunosuppressive, enzyme inhibitory, and cell growth inhibitory [2,3].

In the last few decades, most of the bioactive metabolites are derived from terrestrial actinomycetes. But recently, the rate of discovery of new metabolites from terrestrial actinomycetes has decreased, whereas the rate of reisolation of known compounds has increased [4,5]. More important is that the number of resistant and super-resistant pathogens that are no longer susceptible to the currently used drugs has increased [6]. Therefore, the search for new actinomycetes and their novel bioactive metabolites is becoming a global interest from unexplored habitats. As about 70% of the surface of our planet is covered by oceans and typical microbial abundances of $10^6\ ml^{-1}$ in seawater and $10^9\ ml^{-1}$ in ocean-bottom sediments are recognized [5], researchers are looking for novel actinomycetes and their natural metabolites from the marine-derived actinomycete strains [7,8]. In this chapter, the author focuses on the diversity of marine actinomycetes and their bioactive metabolites particularly with antifungal and antimycotoxin activities.

15.2
Diversity of Actinomycetes in the Sea

15.2.1
Free-Living Marine Actinomycetes

Since 1926, it has been recognized that actinomycetes can be recovered not only from the seawater, littoral sediments, and materials suspended in seawater, but also

Marine Microbiology: Bioactive Compounds and Biotechnological Applications, First Edition.
Edited by Se-Kwon Kim

from the deepest known ocean trenches [9–12]. However, it has been argued whether the indigenous marine actinomycetes existed in the sea really. As the terrestrial actinomycetes can produce spores that could be washed from terrestrial soils into the marine environments and remained available but dormant for many years, actinomycetes isolated from marine samples, especially from the littoral zone and inshore localities, are thought to be of terrestrial origin [13]. Until 1984, when the first marine actinomycete *Rhodococcus marinonascens* was described [14], the existence of indigenous marine actinomycetes in the sea was clear [15,16]. Nowadays, at least three marine actinomycete genera have been described, including the genus *Salinibacterium, Serinicoccus,* and *Salinispora* [17–19].

Previous research showed that *Micromonospora, Rhodococcus,* and *Streptomyces* species have been the dominant genera of the marine actinomycetes, and *Streptomyces* are the predominant species and followed by *Micromonospora* [20–22]. Up to now, we know that marine *Actinomycetes* are not restricted to the *Micromonospora–Rhodococcus–Streptomyces* grouping. The low numbers of genera reported in marine environments are unlikely due to low diversity but actually due to under-sampling and the use of inappropriate isolation procedures [23].

Guiding by the cultivation-independent methods, Maldonado *et al.* [23] isolated some novel taxa from deep sea sediments of *Actinomadura, Dietzia, Gordonia, Microbacterium, Mycobacterium, Nocardiopsis, Nonomuraea, Pseudonocardia, Saccharopolyspora, Streptosporangium, Verrucosispora,* and *Williamsia*. Their research was the first report that high diversity of marine actinomycetes was isolated from marine environments and many of the isolates represent novel species as revealed by 16S phylogenetic trees. Bredholdt *et al.* [21] have investigated the diversity of actinomycetes in the shallow water sediments of the Trondheim fjord of Norway by using different selective isolation methods. Although the predominant genera were clearly *Streptomyces* and *Micromonospora,* an unexpected variety of actinomycete genera were isolated, including *Actinocorallia, Actinomadura, Knoellia, Glycomyces, Nocardia, Nocardiopsis, Nonomuraea, Pseudonocardia,* and *Streptosporangium* genera. Among the genera, *Knoellia* and *Glycomyces* species were the first reported isolates from the marine environment. Maldonado *et al.* [24] used 17 different media for the recovery of actinomycetes from marine sediments samples collected in the Gulf of California and the Gulf of Mexico, and nearly 300 actinomycetes were recovered. Full 16S rRNA gene sequencing revealed that the isolates belonged to several actinobacterial taxa, notably to the genera *Actinomadura, Dietzia, Gordonia, Micromonospora, Nonomuraea, Rhodococcus, Saccharomonospora, Saccharopolyspora, Salinispora, Streptomyces, "Solwaraspora,"* and *Verrucosispora*. This is the first systematic study that shows the rich diversity of actinomycetes found in marine sediments collected in Mexico and, probably, worldwide.

MAR1 members are filamentous actinomycetes that failed to grow on an agar medium when seawater was replaced with deionized water. Over 2000 MAR1 isolates have been isolated by Mincer *et al.* [25] from sediment samples collected from the U.S. Virgin Islands and Guam, the Atlantic Ocean, the Red Sea, and the Sea of Cortez. Therefore, MAR1 strains have been recovered from all five tropical and/or subtropical locations, suggesting that they are widely distributed in deep as 600 m as well as shallow sediments. The generic epithet "*Salinospora*" was proposed for MAR1 members. *Salinospora* was the first seawater-obligate marine actinomycete that requires seawater for growth [23]. *Salinospora* strains have been cultivated from marine sediments collected around the world including the Garibbean Sea, the Sea of the Cortez, the Red Sea, and the Tropical Pacific Ocean off Guam supporting the Pan-Pacific distribution. To date, no strains have been recovered from samples collected off San Diego or in the Bering Sea off the coast of Alaska suggesting latitudinal distribution barriers. Only three species of *Salinispora tropica, Salinispora pacifica*, and *Salinispora arenicola* existed in culture and culture-independent method, even more than 2000 strains have been obtained in culture. *Marinophilus* is another genus-level taxon that appears to reside exclusively in the sea [8]. Although only seven strains obtained in culture, the species diversity of the genus *Marinophilus* is significant and at least three species could be recognized by phylogenetic analysis based on almost the complete SSU rRNA gene sequences. A better understanding of the *Marinophilus* diversity might be achieved when large numbers of *Marinophilus* strains are recovered by optimizing the selective isolation methods. Most of the other marine actinomycetes genera, even from deep sea sediments, can grow without seawater, but grow better with seawater [23,24].

Based on incomplete statistics, up to now, more than 40 actinomycete genera have been identified by cultural and molecular techniques from different marine ecological niches, including *Actinomadura, Actinomyces, Actinoplanes, Actinopolyspora, Actinosynnema, Amycolatopsis, Arthrobacter, Blastococcus, Brachybacterium, Corynebacterium, Dietzia, Frankia, Frigoribacterium, Geodermatophilus, Gordonia, Kineococcus, Kitasatospora, Marinophilus, Marinispora, Microbispora, Micromonospora, Micrococcus, Microbacterium, Mycobacterium, Nocardioides, Nocardiopsis, Nonomuraea, Psuedonocardia, Rhodococcus, Saccharomonospora, Saccharopolyspora, Salinispora, Serinicoccus, Solwaraspora, Streptomyces, Streptosporangium, Tsukamurella, Turicella, Verrucosispora*, and *Williamsia* [8, 22–24, 26–32].

15.2.2
Organisms-Associated Marine Actinomycetes

Marine actinomycetes are also widely distributed in biological sources such as fishes, mollusks, sponges, seaweeds, and mangroves [10,33]. Recently, marine actinomycetes from sponges have received particular attention due to rich bioactive compounds frequently obtained from them. Recent studies using both culture-independent and culture-dependent methods demonstrated that novel, abundant actinomycetes are associated with sponges.

A single actinomycete strain NW001 was cultured from the marine sponge *Rhopaloeides odorabile*, which is a member of the α-subgroup of the class Proteobacteria by 16S rRNA gene sequence. This strain was located within the sponge mesohyl visualized by *in situ* hybridization and not isolated from the corresponding seawater samples, indicating that it has a specific, intimate relationship with *R. odorabile* and is not being utilized as a food source [34]. Ten strains related to the "*Salinospora*" group previously reported only from marine sediments were also isolated from the Great Barrier Reef marine sponge *Pseudoceratina clavata* [35]. Three actinomycetes genera obtained from two sponge species in the genus *Xestospongia* were *Gordonia, Micrococcus,* and *Brachybacterium* spp. [36]. A total of 106 Actinobacteria associated with the marine sponge *Hymeniacidon perleve* were isolated using eight different media. These isolates belonged to seven genera of culturable Actinobacteria including *Actinoalloteichus, Micromonospora, Nocardia, Nocardiopsis, Pseudonocardia, Rhodococcus, Streptomyces*; *Streptomyces* was the dominant genus, which represented 74% of the isolates. Among 106 Actinobacteria strains, three candidates are new species [37]. Twenty-four isolates were obtained from a marine sponge *Haliclona* sp. The phylogenetic analysis based on 16S rRNA gene sequencing showed that the isolates belonged to four known genera *Streptomyces, Nocardiopsis, Micromonospora,* and *Verrucosispora*; *Verrucosispora* is the first genus that has been isolated from a marine sponge *Haliclona* sp. The majority of the strains tested belong to the genus *Streptomyces* and three isolates may be new species [38]. In another research, 30 Actinobacteria strains were isolated from the marine sponge *Iotrochota* sp. The phylogenetic analysis based on 16S rRNA gene sequencing showed that the isolates belonged to genera *Streptomyces, Cellulosimicrobium,* and *Nocardiopsis*; *Cellulosimicrobium* is the first reported genus that has been isolated from a marine sponge. The majority of the strains tested also belonged to the genus *Streptomyces* and one of them may be a new species [39]. A total of 162 strains of Actinobacteria were isolated from sponge *Haliclona* sp. using six different culture media by Khan *et al.* [40]. They belonged to seven different genera, namely *Streptomyces* (131 strains), *Cellulomonas* (1 strain), *Micromonospora* (5 strains), *Nocardia* (17 strains), *Nocardiopsis* (6 strains), *Rhodococcus* (1 strain), and *Williamsia* (1 strain). Six strains represent new species in the genus *Streptomyces* for which the names *Streptomyces marinus, Streptomyces tateyamensis, Streptomyces haliclonae,* and *Streptomyces rubrum* have been proposed [40].

Like free-living actinomycetes in the ocean, most of the sponge-associated actinomycetes can grow without seawater. Jiang *et al.* [38] reported that all of the isolates recovered from a marine sponge *Haliclona* could grow on media prepared with both seawater and freshwater, although some of them grew slower and with altered morphology in media without seawater. Khan *et al.* [40] also reported that among 162 strains isolated from sponge *Haliclona*, only seven strains belonging to the genus *Streptomyces* required salt for their growth.

15.3
Diversity of Natural Compounds from Marine Actinomycetes

Marine actinomycetes are a prolific source of secondary metabolites, which show a range of biological activities such as antitumor, antibacterial, antifungal, immunosuppressive, insecticidal, and enzyme inhibition [3,12]. Antifungal metabolites produced by marine actinomycetes can be classified on the basis of their chemical structure as follows.

15.3.1 Aminoglycosides

Kasugamycin, an aminoglycoside antibiotic, can be produced by *Streptomyces rutgersensis* subsp. *gulangyunensis* isolated from the sea sediment at Xiamen, China. The MICs for *Saccharomyces* and *Aspergillus niger* are 6.25 and 50 μg/ml, respectively [41].

15.3.2 Macrolides

Urauchimycins A and B could be produced by *Streptomyces* sp. Ni-80, a strain of the gray series from unidentified sponge. These compounds are antimycin-type antibiotics that are the first reported antimycin with odd number of carbons and branch in the side chain [42].

Four unusual macrolides, namely marinomycins A–D, were isolated from the saline culture of a new group of marine actinomycetes, the genus *Marinispora*. The marinomycins were composed of dimeric 2-hydroxy-6-alkenyl-benzoic acid lactones with conjugated tetraene-pentahydroxy polyketide chains. In room light, marinomycin A slowly isomerizes to its geometrical isomers marinomycins B and C [32]. Two new polyene macrolides, marinisporolides A and B, were also isolated from the saline culture of the marine actinomycete, *Marinispora* strain CNQ-140, by Kwon *et al.* [32]. The marinisporolides are 34-membered macrolides composed of a conjugated pentaene and several pairs of 1,3-dihydroxyl functionalities. Marinisporolide A contains a bicyclic spiro-bis-tetrahydropyran ketal functionality, while marinisporolide B is the corresponding hemiketal [43].

The macrolide of neorustmicin, an inhibitor of inositol phosphoceramide synthase in fungal cells, was isolated from culture broth of a *Micromonospora* sp. FIM03–1149 [44]. Wei *et al.* [45] reported that three macrolides were isolated from the mycelia of marine actinomycete Y12–26, which were bafilomycin D, bafilomycin A1, and bafilomycin K with a novel structure.

15.3.3 Polyketides

A polyketide antibiotic SBR-22 was isolated from a novel strain of *Streptomyces psammoticus* BT-408, which shows inhibitory activity against several fungi [46]. Three polyketides of daryamides A, B, and C were isolated from culture broth of a *Streptomyces* strain, CNQ-085. But these bioactive compounds showed very weak antifungal activities against *Candida albicans* [47].

15.3.4 Enzymes and Proteins

Marine *Streptomyces* sp. DA11 isolated from sponge *Craniella australiensis* produced the enzyme chitinase and showed antifungal activities against *A. niger* and *C. albicans* [48].

A novel antifungal protein (SAP) was found in the culture supernatant of a marine *Streptomyces* sp. strain AP77. The molecular mass of SAP was estimated to be 160 kDa. This was the first report showing that a protein has activity against *Pythium* species [49]. An anticandidal bioactive protein of 87.12 kDa was obtained from the cell-free supernatant of one isolate of *Nocardiopsis dassonvillei* MAD08 isolated from the sponge *Dendrilla nigra* at 80% saturation of ammonium sulfate [50].

15.3.5 Other Substances

An antimicrobial ester, bonactin, was isolated from the *Streptomyces* sp. BD21-2 obtained from marine sediment sample collected at Hawaii displayed antimicrobial activity against both Gram-positive and Gram-negative bacteria as well as antifungal activity [51].

An antifungal acetamide, 4′-phenyl-1-napthyl phenyl acetamide, was characterized from *Streptomyces* sp. DPTB16 isolated from Cuddalore coastal soil, Tamil Nadu, India [52].

An antifungal acid, dehydroabietic acid, was isolated from the culture broth of a marine actinomycete JWH-09 [53].

An antifungal alkaloid, caerulomycin C, was isolated from the marine-derived actinomycete *Actinoalloteichus cyanogriseus* WH1–2216-6 [54].

15.4 Biological Activities

15.4.1 Antifungal Activities Against Clinical and Human Pathogens

The macrolides of urauchimycins A and B produced by a sponge-associated *Streptomyces* sp. Ni-80 could inhibit morphological differentiation of *C. albicans* at 10 μg/ml [42]. Another macrolide of neorustmicin isolated from a *Micromonospora* FIM03–1149 showed strong inhibitory activity against *Cryptococcus* and *C. albicans* [44]. The polyketide SBR-22 isolated from *S. psammoticus* BT-408 could inhibit *C. albicans* with the MIC of 128 μg/ml [46]. Two alkaloids of caerulomycins A and C isolated from the marine actinomycete *A. cyanogriseus* WH1–2216-6 showed antifungal activity against *C. albicans* with MICs of 21.8 and 19.3 μM, respectively [54].

The enzyme chitinase isolated from a sponge-associated *Streptomyces* sp. DA11 could inhibit *C. albicans* with the average diameter of inhibition zone of 10.48 ± 0.45 mm at 20 μl of purified enzyme, indicating that the obtained chitinase has the potential to be an antifungal agent [48]. An anticandidal bioactive protein with a molecular mass of 87.12 kDa purified from a sponge-associated actinomycetes *N. dassonvillei* MAD08 was found to be active against 26 *Candida* strains among 30 *Candida* species tested. The protein showed highest activity against *C. albicans* FC1, a secondary pathogen of HIV infection, and *C. albicans* MTCC 227, and moderate activity against 24 *Candida* strains that were obtained from the oral cavity of tuberculosis patients and were resistant against ceftriaxone, vancomycin, and amphotericin [50].

Some unpurified metabolites also showed antifungal activities. In an early survey, over 100 *Salinispora* spp. strains were cultivated and the pharmacological activities of their secondary metabolites were studied, with 30% of the crude extracts yielding MICs of 19.5 μg/ml or less for amphotericin-resistant *C. albicans* (ARCA) [25]. Out of 26 sponge-associated marine endosymbiotic actinomycete strains isolated from the Bay of Bengal, seven showed antifungal activity against clinical and human pathogens of *Candida tropicalis, Aspergillus fumigatus,* and *Aspergillus flavus.* The isolate CPI 13 showed the highest activity against *C. tropicalis* with the MIC and MFC of 10 and 12.5 μg protein/ml, and *A. fumigatus* with the MIC and MFC of 15.6 and 19.5 μg protein/ml. The isolate CPI 26 showed the highest activity against *A. flavus* with the MIC and MFC of 12.5 and 15.6 μg protein/ml [55]. Vimal *et al.* [56] reported that the chloroform extract obtained from the marine actinomycete *Nocardiopsis* sp. VITSVK5 was very effective against yeasts, *Candida krusei* with the inhibition zone of 18 mm, *C. tropicalis* (15 mm), and *C. albicans* (14 mm) when compared to streptomycin. *Streptomyces* VITSDK 40 and *Micromonospora* VITSDK 45, isolated from the marine sediments, can only inhibit *C. albicans,* but cannot inhibit *A. niger* and *A. fumigatus.* Their ethyl acetate extracts show potent inhibitory activity against *C. albicans* with the MICs of 0.54 and 0.64 μg/ml, respectively, which are equivalent to amphotericin activity [22].

Streptomyces afghaniensis VPTS3-1 was isolated from the region of Palk Strait of Bay of Bengal, Tamil Nadu. The strain VPTS3-1 showed more antifungal activity against *C. albicans* than antibacterial activity. The antimicrobial efficacy of the solvent extracts of the strain VPTS3-1 revealed that the ethyl acetate extract was highly active against *C. albicans* and produced a maximum inhibitory zone of 21 mm at the concentration of 10 μl. The compound was identified as highly oxygenated and derivative of carbohydrates [57]. A total of 68 morphologically different actinomycetes were tested for their antimicrobial activity by the cross streak method against *C. albicans.* Among the strains tested, *Streptomyces* VPTSA18 showed strong antimicrobial activity. Its culture filtrate was extracted with five different solvents of alcohol, chloroform, distilled water, ethyl acetate, and methanol, and tested against two fungal pathogens of *Cryptococcus neoformans* and *C. albicans* using the well-diffusion method. Of the solvent extracts assessed, the ethyl acetate extract showed high antifungal activity with an inhibition zone of 18 and 17 mm, respectively. The other solvent extracts had a moderate to minimum inhibitory effect against all of the pathogens tested. The petroleum ether extract did not show any activity against the tested pathogens [58].

15.4.2
Antifungal Activities Against Plant Fungal Pathogens

A novel protein (SAP) with a molecular mass of 160 kDa purified from marine *Streptomyces* sp. showed a specific inhibitory activity against *Pythium porphyrae,* a causative agent of red rot disease in *Porphyra* sp. with a MIC of 1.6 μg/disk. However, this protein showed no inhibitory effect even up to 100 μg/disk against most of the other fungi including *Alternaria alternata, A. fumigatus, Aspergillus versicolor, Botrytis cinerea, C. albicans, Fusarium graminearum, Mucor circinelloides, Penicillium citrinum, Phytophthora nicotianae, Pythium aphanidermatum, Pythium oligandrum, Trichophyton rubrum, and Ustilago maydis* [49].

The polyketide antibiotic SBR-22 could inhibit the *A. niger* and *Aspergillus wentii* with the MIC of 42 μg/ml [46]. The dehydroabietic acid isolated from a marine actinomycete could control the plant disease caused by *Rhizoctonia solani, Sphaerotheca fuliginea*, and *B. cinerea*, but not for *Colletotrichum lagenarium, Helminthosporium maydis, Gibberella zeae*, and *Pyricularia oryzae* in pot assay. In contrast to pot assay, it could inhibit *C. lagenarium, H. maydis*, and *P. oryzae* in *in vitro* assay, but not for *R. solani* and *B. cinerea* [53]. Three macrolides of bafilomycin D, bafilomycin A1, and bafilomycin K isolated from marine actinomycete were active against *P. oryzae* with the MICs of 125, 16, and 31 μg/ml, respectively. The bafilomycin D also showed weak inhibitory activity against *B. cinerea, R. solani, C. lagenarium, H. maydis*, and *G. zeae*. The MIC of bafilomycin A1 against *B. cinerea* and *C. lagenarium* was 125 μg/ml, and the MIC of bafilomycin K against *C. lagenarium* was 62 μg/ml [45].

Several researchers have also reported that some marine actinomycetes and their metabolites, though unpurified and unidentified, show strong inhibitory activities against some plant fungal pathogens. *Actinopolyspora* sp. AH1 isolated from west coast of India showed good inhibitory activity by agar diffusion method against *A. niger* (the diameter of inhibition zone, 12 mm), *A. fumigatus* (11.5 mm), *A. flavus* (15.8 mm), *Fusarium oxysporum* (16.2 mm), *Trichoderma* sp. (13.5 mm), and *Penicillium* sp. (17.0 mm), but showed no activity against *Cryptococcus* sp. and *Candida albicans*. The antifungal activity was medium dependent. AH1 grown on maltose-yeast extract agar showed good activity against *Trichoderma* sp. compared with starch casein agar (SCA) and tyrosine agar, but no activity if grown on glycerol glycine agar, glycerol asparagine agar, and glucose asparagine agar [59]. Of 160 marine actinomycetes isolates, 10 *Streptomyces* strains showed potent activity against all four fungi tested, *R. solani, P. oryzae, Helminthosporium oryzae*, and *Colletotrichum falcatum*. These strains may prove to be the potent source for isolation of agro-based fungicides. *Streptomyces fradiae* RHI 1 was potent against *H. oryzae, C. falcatum, R. solani, P. oryzae* by paper disk assay impregnated with 0.1 ml of culture supernatant with the diameter of inhibition zone of 20, 27.5, 20, and 25 mm, respectively. *S. fradiae* RHI 1, *Streptomyces roseolilacinus* MP1, *Streptomyces helvaticus* MIII 1, and *Streptomyces albidoflavus* SEA5 all showed inhibition zone greater than 20 mm against both *H. oryzae* (20–22 mm) *and P. oryzae* (20–25 mm). Ethyl acetate extracts from supernatant of *S. fradiae* RHI 1 also showed effective activity against four tested fungi [60]. *Streptomyces* strain B5, isolated from coast of Weihai, China, showed high inhibitory activities against several plant fungal pathogens. Its culture broth could inhibit the mycelia growth by more than 90% against *F. oxysporum* f.sp. *niveum, F. oxysporum* f.sp. *pumkins, F. graminearum,*

Exserohilum turcicum, Bipolaris sorokiniana, Fusarium oxysporium, Curvularia lunata, and *Fulvia fulva,* by more than 80% against *Colletotrichum gloeosporioides, Fusarium coeruleum, Cytospora* sp., *Macrophoma kawatsukai,* and *Aspergillus oryzae,* and by more than 70% against *Alternaria solani, A. alternata,* when incubated on a plate made of 1 ml of B5 culture broth and 9 ml of PDA. The bioactive metabolites were water soluble, heat resistant, and acid stable, but alkaline unstable [61]. The culture broth of a marine actinomycete strain BM-2 could inhibit the mycelia growth of *F. oxysporum* f.sp. *lycopersici, B. cinerea, Ralstonia solanacarum,* and *A. solani* with an inhibition zone of 23.5, 21.0, 18.5, and 15.5 mm, respectively [62]. In another research, the authors demonstrated that the culture broth of a marine actinomycete strain BM-2 could inhibit the mycelia growth of *F. graminearum, F. oxysporium, Helminthosporium carbonum, B. sorokiniana,* and *Phytophthora capsici* with an inhibition zone of 25.1, 20.0, 14, 6.5, and 6.5 mm, respectively [63].

Vimal *et al.* [56] reported that the ethyl acetate extract obtained from the marine actinomycete *Nocardiopsis* sp. VITSVK5 showed a high antifungal activity against *A. fumigatus* (the inhibition zone 23 mm), *A. flavus* (15 mm), and *A. niger* (12 mm) when compared with petroleum ether and chloroform extracts and with amphotericin B. The ethyl acetate extracts of *Actinopolyspora* VITSDK 43 and *Streptomyces* VITSDK 37 showed the strongest activity against *A. niger, A. fumigatus,* and *C. albicans* with the MICs of 0.56, 0.74, and 0.87 μg/ml, respectively [22].

15.4.3 Antimycotoxin Activities

Research on antimycotoxin activity by marine actinomycete is seldom reported. Mycotoxins are toxic secondary metabolites produced by some species and/or strains of filamentous fungi of *Aspergillus, Fusarium,* and *Penicillium* growing on grains before harvest and in storage [64]. Mycotoxin contamination of food and feed is a serious concern not only for great economic losses but also for human and animal health. Aflatoxins, deoxynivalenol (DON), and ochratoxins are the common mycotoxins posing the most serious threats to human health worldwide.

In our efforts to search for effective marine actinomycetes that could inhibit the mycelia growth and mycotoxin production of DON, 10 marine actinomycete strains were obtained with high inhibitory activity against DON-producing *F. graminearum* both on Gause's Synthetic medium and on PDA medium by coculture screening, with an exception of one strain that could not grow on PDA plate. Their inhibition rate reached 58.3–77.1% on Gause's synthetic plate and 52.5–70.0% on PDA plate, respectively. Two organic solvents of chloroform and ethyl acetate were used to extract the bioactive metabolites from the culture supernatants of *Streptomyces* strain MA01. The results showed that the inhibitory activity of crude chloroform extract was significantly higher than that of crude ethyl acetate extract. At 0.25 mg/ml, the inhibition rates on mycelia growth were 55.6 and 1.9% for chloroform extract and ethyl acetate extract, respectively. At 1.0 mg/ml, the chloroform extract could completely inhibit the growth of *F. graminearum* as well as the DON production, whereas it was only 18.9% on mycelia growth for the ethyl acetate extract. In another research, crude chloroform extracts from culture supernatants of marine *Streptomyces labedae* MA08-2 could inhibit the mycelia growth by 97.64%, and completely inhibit the DON production at 0.5 mg/ml, respectively [65].

Aflatoxin production could also be effectively inhibited both by chloroform extract and by ethyl acetate extract from culture supernatants of *Streptomyces* strain MA01. At 0.25 mg/ml, chloroform extract and ethyl acetate extract could inhibit the aflatoxin production by 61.7 and 83.7%, respectively, whereas they showed a weak inhibitory activity against mycelia growth of *Aspergillus parasiticus* by 3.5 and 17.6%, respectively [66]. The active components and their molecular structure are being elucidated. This will be beneficial to the development of biological pesticides for mycotoxin control.

15.5 Conclusions

To date, there is not enough research data related to antifungal and particularly to antimycotoxin activities by marine actinomycetes; however, fungal pathogens are a serious concern not only for humans but also for plants worldwide. Therefore, in this chapter, the author tried to provide information available on marine actinomycetes and their inhibitory ability against fungal pathogens. Based on the available reports, it can be concluded that marine actinomycetes show potential for finding novel compounds with potent inhibitory ability against fungal pathogens and mycotoxin-producing fungi. Further intensive isolation and screening of rare and/or deep sea marine actinomycetes strains, and elucidating the structures of bioactive compounds, will lead to the discovery of novel useful strains and compounds.

Acknowledgments

This work was supported by the Program of Excellent Team in Harbin Institute of Technology and China Ocean Mineral Resources R & D Association (COMRA) Project (No. DY125-15-R-01).

References

1 Lazzarini, A., Cavaletti, L., Toppo, G., and Marinelli, F. (2000) Rare genera of actinomycetes as potential producers of new antibiotics. *Antonie Van Leeuwenhoek*, **78**, 399–405.

2 Berdy, J. (2005) Bioactive microbial metabolites. *J. Antibiot. (Tokyo)*, **58**, 1–26.

3 Solanki, R., Khanna, M., and Lal, R. (2008) Bioactive compounds from marine

actinomycetes. *Indian J. Microbiol.*, **48**, 410–431.

4 Fenical, W., Baden, D., Burg, M., Goyet, C.V., Grimes, J.D., Katz, M., Marcus, N.H., Pomponi, S., Rhines, P., Tester, P., and Vena, J. (1999) Marine-derived pharmaceuticals and related bioactive compounds, in *From Monsoons to Microbes: Understanding the Ocean's Role in Human Health* (ed. W. Fenical), National Academies Press, Washington, pp. 71–86.

5 Fenical, W. and Jensen, P.R. (2006) Developing a new resource for drug discovery: marine actinomycete bacteria. *Nat. Chem. Biol.*, **2**, 666–673.

6 Ekwenye, U.N. and Kazi, E. (2007) Investigation of plasmid DNA and antibiotic resistance in some pathogenic organism. *Afr. J. Biotechnol.*, **6**, 877–880.

7 Bernan, S., Greenstein, M., and Maiese, W.M. (1997) Marine microorganisms as a source of new natural products. *Adv. Appl. Microbiol.*, **43**, 57–90.

8 Jensen, P.R., Mincer, T.J., Williams, P.G., and Fenical, W. (2005) Marine actinomycete diversity and natural product discovery. *Antonie Van Leeuwenhoek*, **87**, 43–48.

9 Aronson, J.D. (1926) Spontaneous tuberculosis in salt water fish. *J. Infect. Dis.*, **39**, 315–320.

10 Grein, A. and Meyers, S.P. (1958) Growth characteristics and antibiotic production of actinomycetes isolated from littoral sediments and materials suspended in sea water. *J. Bacteriol.*, **76** (5), 457–463.

11 Weyland, H. (1969) Actinomycetes in north sea and Atlantic ocean sediments. *Nature*, **223**, 858.

12 Pathom-aree, W. *et al.* (2006) Diversity of actinomycetes isolated from the Challenger Deep sediment (10,898m) from the Mariana Trench. *Extremophiles*, **10**, 181–189.

13 Dharmaraj, S. (2010) Marine *Streptomyces* as a novel source of bioactive substances. *World J. Microbiol. Biotechnol.*, **26** (12), 2123–2139.

14 Helmke, E. and Weyland, H. (1984) *Rhodococcus marinonascens* sp. nov., an actinomycete from the sea. *Int. J. Syst. Bacteriol.*, **34**, 127–138.

15 Jensen, P.R., Gontang, E., Mafnas, C., Mincer, T.J., and Fenical, W. (2005) Culturable marine actinomycete diversity from tropical Pacific Ocean sediments. *Environ. Microbiol.*, **7**, 1039–1048.

16 Lam, K.S. (2006) Discovery of novel metabolites from marine actinomycetes. *Curr. Opin. Microbiol.*, **9**, 245–251.

17 Han, S.K., Nedashkovzkaya, O.I., Mikhailov, V.V., Kim, S.B., and Bae, K.S. (2003) *Salinibacterium amurkyense* gen. nov., sp. nov., a novel genus of the family Microbacteriaceae from the marine environment. *Int. J. Syst. Evol. Microbiol.*, **53**, 2061–2066.

18 Yi, H., Schumann, P., Sohn, K., and Chun, J. (2004) *Serinicoccus marinus* gen. nov., sp. nov., a novel actinomycete with L-ornithine and L-serine in the pedtidoglycan. *Int. J. Syst. Evol. Microbiol.*, **54**, 1585–1589.

19 Maldonado, L. *et al.* (2005) *Salinispora* gen nov., a home for obligate marine actinomycetes belonging to the family Micromonosporaceae. *Int. J. Syst. Evol. Microbiol.*, **55**, 1759–1766.

20 Jensen, P.R., Dwight, R., and Fenical, W. (1991) Distribution of actinomycetes in near-shore tropical marine sediments. *Appl. Environ. Microbiol.*, **57**, 1102–1108.

21 Bredholdt, H., Galatenko, O.A., Engelhardt, K., Fjñrvik, E., Terekhova, L.P., and Zotchev, S.B. (2007) Rare actinomycete bacteria from the shallow water sediments of the Trondheim fjord, Norway: isolation, diversity and biological activity. *Environ. Microbiol.*, **9** (11), 2756–2764.

22 Krish, S. and Krishnan, K. (2010) Diversity and exploration of bioactive marine actinomycetes in the Bay of Bengal of the Puducherry coast of India. *Indian J. Microbiol.*, **50** (1), 76–82.

23 Maldonado, L.A., Stach, J.E., Pathom-aree, W., Ward, A.C., Bull, A.T., and Goodfellow, M. (2005) Diversity of cultivable actinobacteria in geographically widespread marine sediments. *Antonie Van Leeuwenhoek*, **87** (1), 11–18.

24 Maldonado, L.A., Fragoso-Yáñez, D., Pérez-García, A., Rosellón-Druker, J., and Quintana, E.T. (2009) Actinobacterial diversity from marine sediments collected in Mexico. *Antonie Van Leeuwenhoek*, **95** (2), 111–120.

25 Mincer, T.J., Jensen, P.R., Kauffman, C.A., and Fenical, W. (2002) Widespread and persistent populations of a major new marine actinomycete taxon in ocean sediments. *Appl. Environ. Microbiol.*, **68**, 5005–5011.

26 Riedlinger, J., Reicke, A., Zähner, H., Krismer, B., Bull, A.T., Maldonado, L.A., Ward, A.C., Goodfellow, M., Bister, B., Bischo, D., Süssmuth, R.D., and Fiedler, H.P. (2004) Abyssomicins, inhibitors of the *para*-aminobenzoic acid pathway produced by the marine *Verrucosispora* strain AB-18–032. *J. Antibiot. (Tokyo)*, **57**, 271–279.

27 Stach, J.E.M., Maldonado, L.A., Masson, D. G., Ward, A.C., Goodfellow, M., and Bull, A. T. (2003) Statistical approaches to estimating bacterial diversity in marine sediments. *Appl. Environ. Microbiol.*, **69**, 6189–6200.

28 Stach, J.E.M., Maldonado, L.A., Ward, A.C., Bull, A.T., and Goodfellow, M. (2004) *Williamsia maris* sp. nov., a novel actinomycete isolated from the Sea of Japan. *Int. J. Syst. Evol. Microbiol.*, **54**, 191–194.

29 Ward, A.C. and Bora, N. (2006) Diversity and biogeography of marine actinobacteria. *Curr. Opin. Microbiol.*, **9**, 279–286.

30 Das, S., Lyla, P.S., and Ajmal Khan, S. (2006) Marine microbial diversity and ecology: importance and future perspectives. *Curr. Sci.*, **25**, 1325–1335.

31 Magarvey, N.A., Keller, J.M., Bernan, V., Dworkin, M., and Sherman., D.H. (2004) Isolation and characterization of novel marine-derived actinomycete taxa rich in bioactive metabolites. *Appl. Environ. Microbiol.*, **70**, 7520–7529.

32 Kwon, H.C., Kauffman, C.A., Jensen, P.R., and Fenical., W. (2006) Marinomycins A–D, antitumour antibiotics of a new structure class from a marine actinomycete of the recently discovered genus "*Marinispora*". *J. Am. Chem. Soc.*, **128**, 1622–1632.

33 Santavy, D.L., Willenz, P., and Colwell, R.R. (1990) Phenotypic study of bacteria associated with the Caribbean sclerosponge, *Ceratoporella nicholsoni*. *Appl. Environ. Microbiol.*, **56** (6), 1750–1762.

34 Webster, N.S. and Hill, R.T. (2001) The culturable microbial community of the Great Barrier Reef sponge *Rhopaloeides odorabile* is dominated by α-proteobacterium. *Mar. Biol.*, **138**, 843–851.

35 Kim, T.K., Garson, M.J., and Fuerst, J.A. (2005) Marine actinomycetes related to the '*Salinospora*' group from the Great Barrier Reef sponge *Pseudoceratina clavata*. *Environ. Microbiol.*, **7**, 509–518.

36 Montalvo, N.F., Mohamed, N.M., Enticknap, J.J., and Hill, R.T. (2005) Novel actinobacteria from marine sponges. *Antonie Van Leeuwenhoek*, **87**, 29–36.

37 Zhang, H.T., Lee, Y.K., Zhang, W., and Lee, H.K. (2006) Culturable actinobacteria from the marine sponge *Hymeniacidon perleve*: isolation and phylogenetic diversity by 16S rRNA gene-RFLP analysis. *Antonie Van Leeuwenhoek*, **90**, 159–169.

38 Jiang, S., Wei, S., Chen, M., and Dai, S. (2007) Diversity of culturable actinobacteria isolated from marine sponge *Haliclona* sp. *Antonie Van Leeuwenhoek*, **92** (4), 405–416.

39 Jiang, S.M., Li, X., Zhang, L., Sun, W., Dai, S.K., Xie, L.W., Liu, Y.H., and Lee, K.J. (2008) Culturable actinobacteria isolated from marine sponge *Iotrochota* sp. *Mar. Biol.*, **153**, 945–952.

40 Khan, S.T., Komaki, H., Motohashi, K., Kozone, I., Mukai, A., Takagi, M., and Shin-ya, K. (2011) Streptomyces associated with a marine sponge *Haliclona* sp.; biosynthetic genes for secondary metabolites and products: novel compounds from new members of the genus *Streptomyces*. *Environ. Microbiol.*, **13** (2), 391–403.

41 Huang, W.-Z., Su, G.-C., Liu, T.-C., and Fang, J.-R. (1988) Studies on antibiotics 8510. *Chin. J. Antibiot.*, **13** (4), 277–280.

42 Imamura, N., Nishijima, M., Adachi, K., and Sano, H. (1993) Novel antimycin antibiotics, urauchimycins A and B, produced by marine actinomycete. *J. Antibiot.*, **46** (2), 241–246.

43 Kwon, H.C., Kauffman, C.A., Jensen, P.R., and Fenical, W. (2009) Marinisporolides, polyene-polyol macrolides from a marine actinomycete of the new genus *Marinispora*. *J. Org. Chem.*, **74** (2), 675–684.

44 Lin, R., Nie, Y.L., Zhang, H., and Jiang, H. (2010) Isolation and identification on

neorustmicin producing strain FIM03–1149. *J. Microbiol.*, **30** (1), 38–42.

45 Wei, G., Su, C., Zhang, D.-J., Zong, Z.-Y., Si, C.-C., and Tao, L.-M. (2011) Isolation, purification and structure identification of secondary metabolites produced by marine actinomycete Y12–26. *Chin. J. Antibiot.*, **36** (8), 571–575.

46 Sujatha, E., Bapi Raju, K.V.V.S.N., and Ramana, T. (2005) Studies on a new marine streptomycete BT408 producing polyketide antibiotic SBR-22 effective against methicillin resistant *Staphylococcus aureus. Microbiol. Res.*, **160** (2), 119–126.

47 Asolkar, R.N., Jensen, P.R., Kauffman, C.A., and Fenical, W. (2006) Daryamides A–C weakly cytotoxic polyketides from a marine derived actinomycete of the genus *Streptomyces* strain CNQ-085. *J. Nat. Prod.*, **69**, 1756–1759.

48 Han, Y., Yang, B., Zhang, F., Miao, X., and Li, Z. (2009) Characterization of antifungal chitinase from marine *Streptomyces* sp. DA11 associated with South China Sea sponge *Craniella australiensis. Mar. Biotechnol.*, **11**, 132–140.

49 Woo, J.H., Kitamura, E., Myouga, H., and Kamei, Y. (2002) An antifungal protein from the marine bacterium *Streptomyces* sp. strain AP77 is specific for *Pythium porphyrae*, a causative agent of red rot disease in *Porphyra* spp. *Appl. Environ. Microbiol.*, **68**, 2666–2675.

50 Selvin, J., Shanmughapriya, S., Gandhimathi, R., Seghal Kiran, G., Rajeetha Ravji, T., Natarajaseenivasan, K., and Hema, T.A. (2009) Optimization and production of novel antimicrobial agents from sponge associated marine actinomycetes *Nocardiopsis dassonvillei* MAD08. *Appl. Microbiol. Biotechnol.*, **83** (3), 435–445.

51 Schumacher, R.W., Talmage, S.C., Miller, S.A., Sarris, K.E., Davidson, B.S., and Goldberg, A. (2003) Isolation and structure determination of an antimicrobial ester from a marine sediment derived bacterium. *J. Nat. Prod.*, **66**, 1291–1293.

52 Dhanasekaran, D., Thajuddin, N., and Panneerselvam, A. (2008) An antifungal compound: 4′-phenyl-1-napthyl-phenyl acetamide from *Streptomyces* sp. DPTB16. *Med. Biol.*, **15**, 7–12.

53 Wei, G., Xue, S., Zhou, Y., Zong, Z.-Y., and Tao, L.-M. (2010) Isolation, purification and identification of antifungal compound from the metabolites of marine actinomycete JWH-09. *Agrochemicals*, **49** (12), 868–870.

54 Fu, P., Wang, S., Hong, K., Li, X., Liu, P., Wang, Y., and Zhu, W. (2011) Cytotoxic bipyridines from the marine-derived actinomycete *Actinoalloteichus cyanogriseus* WH1-2216-6. *J. Nat. Prod.*, **74** (8), 1751–1756.

55 Gandhimathi, R., Arunkumar, M., Selvin, J., Thangavelu, T., Sivaramakrishnan, S., Kiran, G.S., Shanmughapriya, S., and Natarajaseenivasan, K. (2008) Antimicrobial potential of sponge associated marine actinomycetes. *J. Med. Mycol.*, **18**, 16–22.

56 Vimal, V., Benita, M.R., and Kannabiran, K. (2009) Antimicrobial activity of marine actinomycete, *Nocardiopsis* sp. VITSVK 5 (FJ973467). *Asian J. Med. Sci.*, **1** (2), 57–63.

57 Vijayakumar, R., Panneerselvam, K., Muthukumar, C., Thajuddin, N., and Panneerselvam, A. (2012) Optimization of antimicrobial production by a marine actinomycete *Streptomyces afghaniensis* VPTS3-1 isolated from Palk Strait, East Coast of India. *Indian J. Microbiol.*, **52**, 230–239.

58 Vijayakumar, R., Panneer Selvam, K., Muthukumar, C., Thajuddin, N., Panneerselvam, A., and Saravanamuthu, R. (2012) Antimicrobial potentiality of a halophilic strain of *Streptomyces* sp. VPTSA18 isolated from the saltpan environment of Vedaranyam, India. *Ann. Microbiol.*, **62**, 1039–1047.

59 Kokare, C.R., Mahadik, K.R., Kadam, S.S., and Chopade, B.A. (2004) Isolation, characterization and antimicrobial activity of marine halophilic *Actinopolyspora* species AH1 from the west coast of India. *Curr. Sci.*, **86** (4), 539–598.

60 Kathiresan, K., Balagurunathan, R., and Masilamani Selvam, M. (2005) Fungicidal activity of marine actinomycetes against phytopathogenic fungi. *Indian J. Biotechnol.*, **4**, 271–276.

61 Shao, Y.-P., Fang, L.-P., Wei, S.-P., Li, H.-L., and Ji, Z.-Q. (2007) Studies on antagonistic spectrum and stability of actinomycete B5-strain from marine. *Acta Agr. Boreali-Occidentalis Sin.*, **16** (3), 248–251.

62 Ma, G., Bao, Z., Xia, Z., and Wu, S. (2009) Inhibiting effect of seven marine actinomycete strains against vegetable pathogenic microorganisms. *Crops*, **5**, 3–9.

63 Bao, Z.-H., Ma, G.-Z., Wu, S.-J., and Xia, Z.-Q. (2009) Antibacterial characteristics of actinomycete BM-2 from marine. *Agrochemicals*, **48** (9), 640–643.

64 Reddy, K.R.N., Salleh, B., Saad, B., Abbas, H.K., Abel, C.A., and Sheir, W.T. (2010) An overview of mycotoxin contamination in foods and its implications for human health. *Toxin Rev.*, **29** (1), 3–26.

65 Yan, P.-S., Shi, C.-J., Hou, C.-C., and Kan, G.-F. (2011) Inhibition of vomitoxin-producing Fusarium graminearum by marine actinomycetes and the extracellular metabolites. Proceedings of 2011 IEEE International Conference on Human Health and Biomedical Engineering, HHBE, pp. 454–456.

66 Yan, P.-S., Shi, C.-J., Gao, X.-J., Hou, C.-C., and Li, Q.-W. (2010) Screening of marine actinomycetes and its bioactive metabolites against aflatoxin production. Proceedings of the 8th National Symposium on Marine Biotechnology and Drug Innovation, p. 73.

16
Antituberculosis Materials from Marine Microbes

Quang Van Ta and Se-Kwon Kim

16.1
Introduction

Tuberculosis is the second most common cause of death worldwide. Thirty-two percent of the world's populations are infected with *Mycobacterium tuberculosis* (TB), the main cause of tuberculosis. It is a disease of poverty affecting mostly young adults in their most productive years. Ninety-five percent of TB deaths occur in the developing world. The number of people who suffered from TB dropped to 8.8 million in 2010, including 1.1 million cases among people with HIV. The number has been consistently decreasing since 2005. The estimated global incidence rate fell to 128 cases per 100 000 populations in 2010, after increasing in 2002 at 141 cases per 100 000 [1]. Most forms of active tuberculosis can be treated with 6 months of medication. Unfortunately, outbreaks of multidrug-resistant (MDR) tuberculosis have been occurring since the 1990s [2]. The current first-line TB drug regimen is more than 40 years old and consists primarily of rifampicin and isoniazid. These antibiotics are effective in active, drug-susceptible TB, provided patients complete the course. There is, however, poor patient compliance due to the cost of drugs, adverse effects, the long time required for full treatment (6–12 months), and the required number of drug doses. Noncompliance has contributed to the appearance of multidrug-resistant and extensively drug-resistant (XDR) TB strains. MDR-TB is resistant to, at least, isoniazid and rifampicin, often taking two more years to treat with second-line drugs [3]. However, XDR-TB also exhibits resistance to second-line drugs, including fluoroquinolones and one of capreomycin, kanamycin, or amikacin, and is also virtually incurable [4]. All the above-mentioned reasons make a compelling case for the urgent need for new anti-TB drugs. In particular, shorter but more effective treatments would improve patient compliance and slow down the emergence of drug-resistant strains.

Recently, a great deal of interest has been paid by the consumers toward natural bioactive compounds as functional ingredients in nutraceuticals, cosmeceuticals, and pharmaceuticals. Especially, bioactive compounds derived from marine organisms have been a rich source of health-promoting components. As more than 70% of the world's surface is covered by oceans, the wide diversity of marine

Marine Microbiology: Bioactive Compounds and Biotechnological Applications, First Edition.
Edited by Se-Kwon Kim

organisms offers a rich source of natural products. Marine environment contains a source of functional materials, including polyunsaturated fatty acids (PUFAs), polysaccharides, essential minerals and vitamins, phenolic phlorotannins such as antioxidants, and enzymes and bioactive peptides [5–7]. Among marine organisms, marine microbes are a rich source of structurally diverse bioactive secondary metabolites with unprecedented skeletons and have shown various health beneficial biological activities [8,9]. A number of metabolites from marine-derived microbe possess antioxidant [10,11], antimicrobial [12], antityrosinase [13], cytotoxic or antitumor [14–16], quinone reductase induction [17], and antiplasmodial [18] activities. It has been approved that the marine-derived microbes may produce novel chemical structures and diverse biological activities. This chapter gives an overview on the antituberculosis agents from marine microbes.

16.2
Marine Microbe-Derived Antituberculosis Agents

16.2.1
Alkaloids

Alkaloid is the largest group of antitubercular natural products and includes alkaloids with complex structures and potent inhibitory activity against susceptible *M. tuberculosis* isolates and, in some cases, toward resistant strains. Recent studies on marine microbial alkaloids have shown potential benefits against *M. tuberculosis*. In search for alkaloids from Indo-Pacific sponges, three manzamine alkaloids with a novel ring system, that is, *ent*-12,34-oxamanzamine E, *ent*-12,34-oxamanzamine F, *ent*-8-hydroxymanzamine A, and 12,34-oxamanzamine A, have been isolated. The result of microplate Alamar Blue assay (MABA) showed inhibitory activity against *M. tuberculosis* of these alkaloids. *ent*-12,34-Oxamanzamine E, *ent*-12,34-oxamanzamine F, and *ent*-8-hydroxymanzamine A were active against *M. tuberculosis* with minimum inhibitory concentrations (MICs) of 128, 12.5, and 3.9 μg/ml, respectively. However, (+)-8-hydroxymanzamine A had an MIC 0.91 μg/ml, indicating improved activity for the (+) over the enantiomer. The difference in the activity of these alkaloids against *M. tuberculosis* was suggested to be associated with the changes in the molecule that result during the formation of the new C-12, C-34 oxygen bridge [19]. Furthermore, pseudopteroxazole and *seco*-pseudopteroxazole are two alkaloids from the hexane extracts of the West Indian gorgonian coral *Pseudopterogorgia elisabethae* (Bayer) collected near San Andre's Island, Colombia. Pseudopteroxazole was found to effect potent inhibitory activity (97%) against *M. tuberculosis* H37Rv at a concentration of 12.5 μg/ml, whereas *seco*-pseudopteroxazole inhibited 66% of mycobacterial growth [20]. In addition, it was also reported that the alkaloid isolated from Caribbean Sea sponge *Neopetrosia proxima* showed antituberculosis activities. Neopetrosiamine A was also tested *in vitro* against a pathogenic strain of *M. tuberculosis* (H37Rv) in a MABA exhibiting an MIC value of

7.5 μg/ml [21]. Furthermore, the known (+)-fistularin-3 and 11-deoxyfistularin-3, and the new compound 2-(3-amino-2,4-dibromo-6-hydroxyphenyl)acetic acid were isolated from the sponge *Aplysina cauliformis*. These compounds displayed activity against *M. tuberculosis* H37Rv with MICs of 7.1, 7.3, and 49 μM, respectively. (+)-Fistularin-3 and 11-deoxyfistularin-3 also exhibited low cytotoxicity against J744 macrophages [22]. These reports suggested that marine microbial alkaloids have potential in antituberculosis study.

16.2.2 Lipids

Recent examples of short-chain analogues (C14–C18) of these α-methoxylated fatty acids, in particular those arising from a *Callyspongia* sp. sponge, have been postulated to originate from bacteria in symbiosis with the sponge. All methoxylated fatty acids displayed some degree of inhibition (between 2 and 99%) of *M. tuberculosis* H37Rv at 6.25 μg/ml. The fatty acid exhibiting the most inhibitory effect was 2-methoxydecanoic acid with a minimum inhibitory concentration of 200–239 μM against *M. tuberculosis* H37Rv, as determined by both the microplate Alamar Blue assay and the green fluorescent protein microplate assay. These results are discussed in terms of the possible role of the 2-methoxylated fatty acid as antimicrobial lipids produced either by marine sponges or by associated marine symbiotic bacteria, as a defense mechanism in a highly competitive environment [23]. In addition, three crude extracts of *Aplysina caissara*, a marine sponge endemic to Brazil, were tested against *M. tuberculosis*. The results demonstrate that all the extracts are toxic and capable of inhibiting cellular growth. The results also suggest that reactive oxygen species (ROS) production is not involved in the cytotoxic processes levied by the extracts employed in this study, and that the active metabolites are likely to be present in the polar fractions of the crude extracts. Finally, the results indicate that all three extracts exhibit a moderate antituberculosis activity, and that the removal of an extract's lipid fraction appears to diminish this activity [24]. Moreover, parguesterols A and B from Caribbean Sea sponge *Svenzea zeai* have been reported to show potent antituberculosis activity. Antituberculosis assay on *M. tuberculosis* H37Rv showed that minimum inhibition concentration for antitubercular activity of parguesterols A and B were determined as 7.8 and 11.2 μg/ml, respectively. Moreover, they showed low toxicity against Vero cells ($IC_{50} = 52$ μg/ml). Therefore, they have potency in the development of new antituberculosis agents [25].

16.2.3 Peptides

Marine-derived antimicrobial peptides are well described in the hemolymph of the many marine invertebrates including spider crab, oyster, American lobster, shrimp, and green sea urchin [5]. Interestingly, marine microbial peptides have shown antimicrobial activity, especially antituberculosis activity. Natural products of this group include three novel aminolipopeptides, trichoderin A, A1, and B, isolated from the marine sponge-derived fungus of *Trichoderma* sp. 05FI48 strain, which showed potent inhibitory activity toward *M. tuberculosis* H37Rv under aerobic and hypoxic conditions, with MICs of 0.12, 2.0, and 0.13 μg/ml, respectively [26]. Moreover, the proline-rich peptide pitiprolamide, isolated from the cyanobacteria *Lyngbya majuscula*, was reported to possess weak antituberculosis activity in the disk diffusion assay with the zone of inhibition having a diameter of 13 mm at 50 μg/ml and 26 mm at 100 μg/ml [27].

16.2.4 Terpenes

Three naturally new C-glycosylated benz[*a*]anthraquinone derivatives, urdamycinone E, urdamycinone G, and dehydroxyaquayamycin have been isolated from the marine *Streptomycetes* sp. BCC45596. Urdamycin E has also been identified after a recultivation of the strain. These compounds exhibited potent anti-*M. tuberculosis* with MICs in the range of 3.13–12.50 μg/ml [28]. Moreover, two new antimycobacterial serrulatane diterpenes, erogorgiaene and 7-hydroxyerogorgiaene, and a novel C40 bisditerpene have been isolated from the West Indian gorgonian octocoral *P. elisabethae*. Erogorgiaene induced 96% growth inhibition for *M. tuberculosis* H37Rv at a concentration of 12.5 μg/ml. On the other hand, 7-hydroxyerogorgiaene inhibited 77% of mycobacterial growth at a concentration of 6.25 μg/ml, which indicates that C-7 hydroxylation apparently does not reduce the activity [29]. Furthermore, it was reported that 27 diterpenes, cyanthiwigins, were isolated from the Jamaican sponge *Myrmekioderma styx*. However, only cyanthiwigin C exhibited marginal activity against Mtb with an MIC at 6.25 μg/ml. Cyanthiwigin C has only one hydroxyl group at C-1 and showed the strongest activity against *M. tuberculosis*. Oxidation of the hydroxyl group to ketone at C-1 or introduction of a hydroxyl group at C-8 decreases the activity against tuberculosis [30].

16.3 Conclusions

The marine microbes clearly hold enormous potential for providing new leads for the development of antituberculous agents. The identification of new structural classes active against *M. tuberculosis* will provide undescribed mechanisms of action and better treatments for resistant strains. As a result, the eventual establishment of culture facilities is necessary for the production of bioactive materials from marine microbes. The reasonably high yields of many sophisticated marine microbes products provide an opportunity for the utilization of endemic resources to combat this disease. Collectively, the natural products, derived from marine microbes, have potential to expand their health beneficial value not only in the food industry but also in the pharmaceutical industries.

References

1 World Health Organization (2011) *Global Tuberculosis Control*, WHO Press, ISBN 978 924 156438 0.

2 Gupta, R. and Espinal, M. (2003) Stop TB Working Group on DOTSPlus for MDR-TB. A prioritized research agenda for DOTS-Plus for multidrug-resistant tuberculosis (MDRTB). *Int. J. Tuberc. Lung Dis.*, **7**, 410–414.

3 Johnson, R., Streicher, E.M., Louw, G.E., Warren, R.M., vanHelden, P.D., and Victor, T.C. (2006) Drug resistance in *Mycobacterium tuberculosis*. *Curr. Issues Mol. Biol.*, **8**, 97–112.

4 Burman, W.J., Gallicano, K., and Peloquin, C. (1999) Therapeutic implications of drug interactions in the treatment of human immunodeficiency virus-related tuberculosis. *Clin. Infect. Dis.*, **28**, 419–430.

5 Kim, S.K. and Wijesekara, I. (2010) Development and biological activities of marine-derived bioactive peptides: a review. *J. Funct. Foods*, **2**, 1–9.

6 Ngo, D.H., Wijesekara, I., Vo, T.S., VanTa, Q., and Kim, S.K. (2011) Marine food-derived functional ingredients as potential antioxidants in the food industry: an overview. *Food Res. Int.*, **44**, 523–529.

7 Wijesekara, I., Yoon, N.Y., and Kim, S.K. (2010) Phlorotannins from *Ecklonia cava* (Phaeophyceae): biological activities and potential health benefits. *Biofactors*, **36**, 408–414.

8 Blunt, J.W., Copp, B.R., Munro, M.H.G., Northcote, P.T., and Princep, M.R. (2006) Marine natural products. *Nat. Prod. Rep.*, **23**, 26–78.

9 Smetania, O.F., Kalinovsky, A.I., Khudyakova, Y.V., Pivkin, M.V., Dmitrenok, P.S., Fedorov, S.N., Ji, H., Kwak, J.Y., and Kuznetsova, T.A. (2007) Indole alkaloids produced by a marine fungus isolate of *Penicillium janthinellum* Biourge. *J. Nat. Prod.*, **70**, 906–909.

10 Son, B.H., Kim, J.C., Choi, H.D., and Kang, J.S. (2002) A radical scavenging farnesylhydroquinone from a marine-derived fungus *Penicillium* sp. *Arch. Pharm. Res.*, **25**, 77–79.

11 Chen, L., Fang, Y., Zhu, T., Gu, Q., and Zhu, W. (2008) Gentisyl alcohol derivatives from the marine-derived fungus *Penicillium terrestre*. *J. Nat. Prod.*, **71**, 66–70.

12 Li, X., Kim, S.K., Nam, K.W., Kang, J.S., Choi, H.D., and Son, B.W. (2006) A new antibacterial dioxopiperazine alkaloid related to gliotoxin from a marine isolate of the fungus *Pseudallescheria*. *J. Antibiot. (Tokyo)*, **59**, 248–250.

13 Park, S.H., Kim, D.S., Kim, W.G., Ryoo, I.J., Lee, D.H., Huh, C.H., Youn, S.W., Yoo, I.D., and Park, K.C. (2004) Terrein: a new melanogenesis inhibitor and its mechanism. *Cell. Mol. Life Sci.*, **61**, 2878–2885.

14 Lu, Z., Zhu, H., Fu, P., Wang, Y., Zhang, Z., Lin, H., Liu, P., Zhuang, Y., Hong, K., and Zhu, W. (2010) Cytotoxic polypenols from the marine-derived fungus *Penicillium expansum*. *J. Nat. Prod.*, **73**, 911–914.

15 Neumann, K., Kehraus, S., Gutschow, M., and Konig, G.M. (2009) Cytotoxic and HLE-inhibitory tetramic acid derivatives from marine-derived fungi. *Nat. Prod. Commun.*, **4**, 347–354.

16 Sun, Y., Tian, L., Huang, J., Li, W., and Pei, Y.H. (2006) Cytotoxic sterols from marine-derived fungus *Penicillium* sp. *Nat. Prod. Res.*, **20**, 381–384.

17 Gamal-Eldeen, A.M., Abdel-Lateff, A., and Okino, T. (2009) Modulation of carcinogen metabolizing enzymes by chromanone A; a new chromone derivative from algicolous marine fungus *Penicillium* sp. *Environ. Toxicol. Pharmacol.*, **28**, 317–322.

18 Kasettrathat, C., Ngamrojanavanich, N., Wiyakrutta, S., Mahidol, C., Ruchirawat, S., and Kittakoop, P. (2008) Cytotoxic and antiplasmodial substances from marine-derived fungi, *Nodulisporium* sp. and CRI247-01. *Phytochemistry*, **69**, 2621–2626.

19 Muhammad Yousaf, M., El Sayed, K.A., Rao, K.V., Lim, C.W., Hu, J.F., Kelly, M., Franzblau, S.G., Zhang, F.Q., Peraud, O., Hilld, R.T., and Hamanna, M.T. (2002) 12,34-Oxamanzamines, novel biocatalytic and natural products from manzamine producing Indo-Pacific sponges. *Tetrahedron*, **58**, 7397–7402.

20 Rodríguez, A.D., Ramírez, C., Rodríguez, I.I., and González, E. (1999) novel

antimycobacterial benzoxazole alkaloids, from the West Indian sea whip *Pseudopterogorgia elisabethae*. *Org. Lett.*, **1** (3), 527–530.

21 Wei, X., Nieves, K., and Rodríguez, A.D. (2010) Neopetrosiamine A, biologically active bis-piperidine alkaloid from the Caribbean sea sponge *Neopetrosia proxima*. *Bioorg. Med. Chem. Lett.*, **20**, 5905–5908.

22 De Oliveira, M.F., De Oliveira, J.H.H.L., Galetti, F.C.S., De Souza, A.O., Silva, C.L., Hajdu, E., Peixinho, S., and Berlinck, R.G.S. (2006) Antimycobacterial brominated metabolites from two species of marine sponges. *Planta Med.*, **72** (5), 437–441.

23 Carballeira, N.M., Cruz, H., Kwong, C.D., Wan, B., and Franzblau, S. (2004) 2-Methoxylated fatty acids in marine sponges: defense mechanism against mycobacteria? *Lipids*, **39**, 675–680.

24 Azevedo, L.G., Muccillo-Baisch, A.L., Filgueira, D.D.M.V.B., Boyle, R.T., Ramos, D.F., Soares, A.D., Lerner, C., Silva, P.A., and Trindade, G.S. (2008) Comparative cytotoxic and anti-tuberculosis activity of *Aplysina caissara* marine sponge crude extracts. *Comp. Biochem. Physiol. C Toxicol. Pharmacol.*, **147**, 36–42.

25 Wei, X.M., Rodriguez, A.D., Wang, Y.H., and Franzblau, S.G. (2007) Novel ring B abeo-sterols as growth inhibitors of *Mycobacterium tuberculosis* isolated from a Caribbean Sea sponge, *Svenzea zeai*. *Tetrahedron Lett.*, **48**, 8851–8854.

26 Pruksakorn, P., Arai, M., Kotoku, N., Vilchèze, C., Baughn, A.D., Moodley, P., Jacobs, W.R.Jr., and Kobayashi, M. (2010) Trichoderins, novel aminolipopeptides from a marine sponge-derived *Trichoderma* sp., are active against dormant mycobacteria. *Bioorg. Med. Chem. Lett.*, **20**, 3658–3663.

27 Montaser, R., Abboud, K.A., Paul, V.J., and Luesch, H. (2011) Pitiprolamide, a proline-rich dolastatin 16 analogue from the marine cyanobacterium *Lyngbya majuscule* from Guam. *J. Nat. Prod.*, **74**, 109–112.

28 Supong, K., Thawai, C., Suwanborirux, K., Choowong, W., Supothina, S., and Pittayakhajonwut, P. (2012) Antimalarial and antitubercular C-glycosylated benz[*a*] anthraquinones from the marine-derived *Streptomyces* sp. BCC45596. *Phytochem. Lett.*, **5**, 651–656.

29 Rodríguez, A.D. and Ramírez, C. (2001) Serrulatane diterpenes with antimycobacterial activity isolated from the West Indian sea whip *Pseudopterogorgia elisabethae*. *J. Nat. Prod.*, **64**, 100–102.

30 Peng, J., Walsh, K., Weedman, V., Bergthold, J.D., Lynch, J., Lieu, K.L., Braude, I.A., Kellyc, M., and Hamanna, M. T. (2002) The new bioactive diterpenes cyanthiwigins E–AA from the Jamaican sponge *Myrmekioderma styx*. *Tetrahedron*, **58**, 7809–7819.

17
Harnessing the Chemical and Genetic Diversities of Marine Microorganisms for Medical Applications

Xinqing Zhao, Wence Jiao, and Xiaona Xu

17.1
Introduction

Marine environment has become an exciting source for a wide range of natural products, many of which harbor novel chemical structures and/or potent activities for medical treatments. Although early exploration of marine natural products (MNPs) focused on the studies of macroalgae, sponges, and soft corals, it was later realized that many compounds previously isolated from these macroorganisms actually originate from their associated microbes [1,2]. Marine microorganisms inhabiting various environmental niches have been isolated and identified, which include the symbiotic microorganisms of marine invertebrates (such as marine sponges, tunicates, bryozoans, and mollusks). The freely dwelling microorganisms from marine sediment and seawater, and many new species and taxa of marine microorganisms were discovered, suggesting the great potential of marine microorganisms for the discovery of new chemical structures and novel genes.

Marine microorganisms, including marine bacteria, marine fungi, and marine cyanobacteria, have been extensively explored in the last few years for the chemical diversities for drug discovery in the treatment of cancer, microbial infection, malaria, and other diseases. Of the major producers of useful natural products, marine actinobacteria and marine cyanobacteria are especially notable for their capability to produce diverse, useful natural products [3–6]. Due to the relatively smaller genome of these marine microorganisms compared to those of marine invertebrates, it is easy to decode the genome of these microbial producers and explore the biosynthetic origin of useful natural products. The rapid development of high-throughput genomic sequencing and other omics technologies (e.g., transcriptome, proteome, metabolome), thus, can greatly speed up the discovery of novel MNPs in the marine microbial producers.

Many genes that encode the biosynthetic pathway of small-molecule natural products are clustered in the genome of the microbial producers, some of which are composed of larger than 100 kb of DNA sequence [7]. The identification and characterization of these gene clusters by genome mining greatly facilitate the

Marine Microbiology: Bioactive Compounds and Biotechnological Applications, First Edition.
Edited by Se-Kwon Kim

creation of better drug molecules with improved medical properties by combinational biosynthesis. In the meantime, harnessing the genetic diversity of marine microorganisms can also be achieved by fishing the biosynthetic genes of MNPs out of the metagenomic libraries before the genes were heterologously expressed in *Escherichia coli* or other microbial hosts [8,9].

In this chapter, the most recently reported novel chemical structures and activity studies of marine microorganisms are presented, and progress on genome-based studies for MNPs discovery including gene-based screening, metagenomic studies, and genome mining is summarized. Our objective is to highlight the modern technologies in the exploration of marine natural products for medical treatments. The impact of genetic and enzymatic studies on the identification and creation of chemical diversity of MNPs is emphasized.

17.2
Novel MNPs

MNPs from different marine microorganisms with various novel structures and promising activities have been isolated in recent years. Due to space limitation, major focus in this chapter would be on novel natural products from marine actinobacteria, marine fungi, and marine cyanobacteria. Marine actinobacteria are a diverse group of high G + C Gram-positive bacteria that encompass many potent antibiotic producers, including the microorganisms from *Streptomyces, Micromonospora, Nocardia, Nocardiopsis,* and so on, of which *Streptomyces* are well-known producers of a vast variety of antibiotics [4,9–12]. In the last few years, many novel chemical structures have been discovered from the above-mentioned marine microorganisms; few representative examples are given below, and selected compounds with interesting activities were listed on Table 17.1.

1) Padanamides A and B (compounds **1** and **2**) were isolated from marine-derived *Streptomyces* sp. RJA2928 obtained from a marine sediment collected near the passage Padana Nahua in Papua New Guinea. **1** and **2** are highly modified liner tetrapeptides in which the (2*S*,3*S*,4*S*)-Ahmpp residue has not been previously discovered in any natural product. Both **1** and **2** showed cytotoxic activity to Jurkat T lymphocyte cells (ATCC TIB-152) *in vitro* with an IC_{50} of 20 and 60 μg/ml, respectively [13].

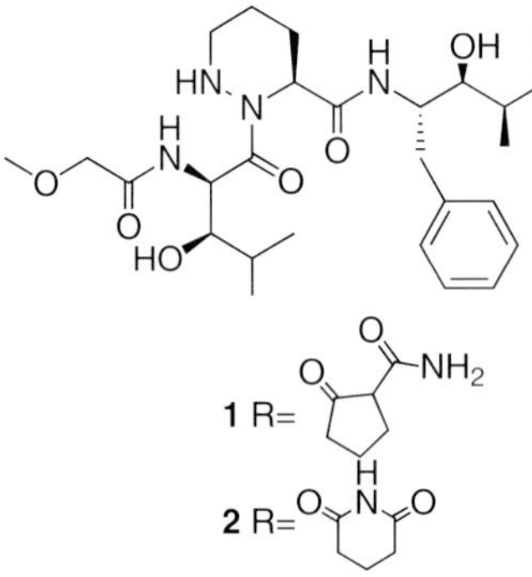

1 R=

2 R=

Table 17.1 Selected MNPs with various biological activities.

Compound (chemical class)	Activity (mode of action)	Isolate	Source	Reference
12-Membered macrolide (3*R*,11*R*),(4*E*,8*E*)-3-hydroxy-11-methyloxacyclododeca-4,8-diene-1,7-dione (**5**)	Anti-HSV-1	*Ascomycetous* fungus strain 222	Baltic Sea, Germany	[15]
Benzopyran (**7**)	Antifibrotic diseases	*S. xiamenensis* sp. nov. DSM 41903^T	Mangrove sediments, China	[2]
Veraguamides A (**16**)	Cytotoxic activity (inhibition of H460 human lung cancer cell)	Cyanobacterium cf. *O. margaritifera*	Coiba National Park, Panama	[22]
Bahamaolides A and B (**17** and **18**)	Antifungal activity	*Streptomyces* sp. CNQ343	North Cat Cay, Bahamas	[23]
Trichostain A	Excellent HDAC inhibitory activity	*Streptomyces* sp. RM72	Marine sponge, Kagoshima Prefecture, Japan	[28]
Bis(dethio)-10*a*-methylthio-3*a*-deoxy-3,3*a*-didehydrogliotoxin and 6-deoxy-5*a*, 6-didehydrogliotoxin (**44** and **45**)	Cytotoxic activity (inhibition of P388 murine leukemia cells)	*Penicillium* sp. strain JMF034	Suruga Bay, Japan	[30]
Marinacarbolines A–D (**50–53**), 13-*N*-demethyl-methylpendolmycin (**54**) and methylpendolmycin-14-*O*-α-glucoside (**55**)	Antiplasmodial activity (inhibition of *P. falciparum* lines 3D7 and Dd2)	*M. thermotolerans* SCSIO 00652	South China Sea, China	[36]
Palmyrolide A	Sodium channel blocking activity in neuro-2a cells	Cyanobacterial assemblage composed of *Leptolyngbya* cf. and *Oscillatoria* spp.	Northwest end of Paradise Island, Palmyra Atoll	[38]

2) Two prenylated diketopiperazines DKPs, nocardioazines A and B (compounds **3** and **4**), accompanied by cyclo-(L-Trp-L-Trp) and cyclo-(L-Trp-D-Trp) were obtained from *Nocardiopsis* sp. CMB-M0232 isolated from a sediment sample from South Molle Island, Australia. Although the nocardioazines were not cytotoxic against Gram-positive and -negative bacteria and human cancer cell lines tested, the bridged DKP scaffold belongs to a new class of noncytotoxic P-glycoprotein (P-gp)

inhibitor that is capable of reversing doxorubicin resistance in a P-gp overexpressing drug-resistant colon cancer cell line [14].

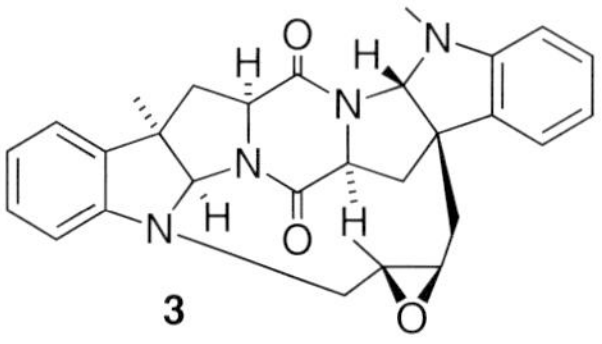

3) A marine *Ascomycetous* fungus strain isolated from driftwood collected from the coast of Greifswalder Bodden, Baltic Sea, Germany produced a new 12-membered macrolide (3*R*,11*R*), (4*E*,8*E*)-3-hydroxy-11-methyloxacyclododeca-4,8-diene-1,7-dione (**5**). Compound **5** displayed anti-HSV-1 activity with an IC_{50} value of 0.45 μM [15].

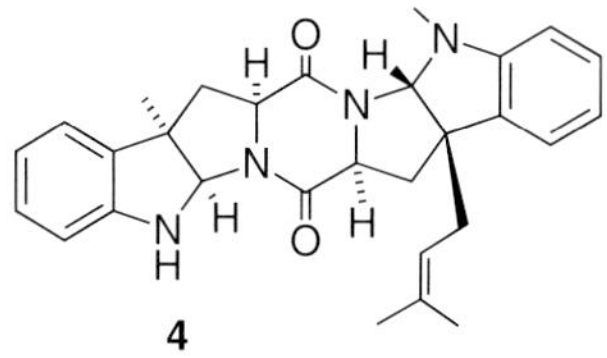

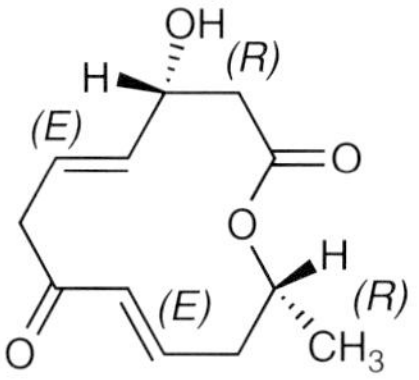

4) Kiamycin (compound **6**), which was isolated from the marine *Streptomyces* sp. strain M268, is a new angucyclinone derivative possessing a 1,12-epoxybenz anthracene ring system. This compound appears to have potential as an anticancer agent with inhibitory activity against the human cell lines HL-60 (leukemia), A549 (lung adenocarcinoma), and BEL-7402 (hepatoma), respectively [16].

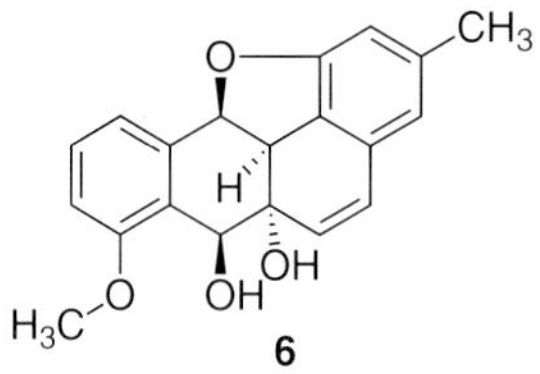

5) *S. xiamenensis* strain 318 producing an antifibrotic benzopyran compound (compound **7**) was isolated from mangrove sediments. Compound **7** was investigated for possible therapeutic effects against fibrosis; it inhibits the proliferation of human lung fibroblasts (WI26), blocks adhesion of human acute monocytic

leukemia cells (THP-1) to a monolayer of WI26 cells, and reduces the contractile capacity of WI26 cells in three-dimensional free-floating collagen gels [17].

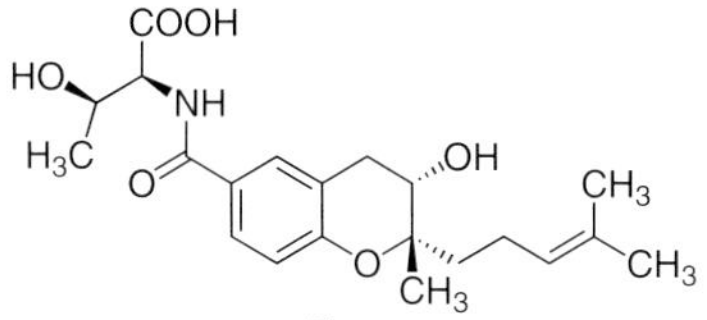

6) Two new antimycin A analogues, Antimycin B1 and B2 (compounds **8** and **9**), were isolated from a spent broth of a marine-derived bacterium, *S. lusitanus* XM52. Compounds **8** and **9** are the first naturally occurring ring-opened antimycin A analogues of marine origin. A bioassay test revealed that compound **8** was inactive against the bacteria while compound **9** showed antibacterial activities against *Staphylococcus aureus* and *Loktanella hongkongensis* with MIC values of 32.0 and 8.0 μg/ml, respectively [18].

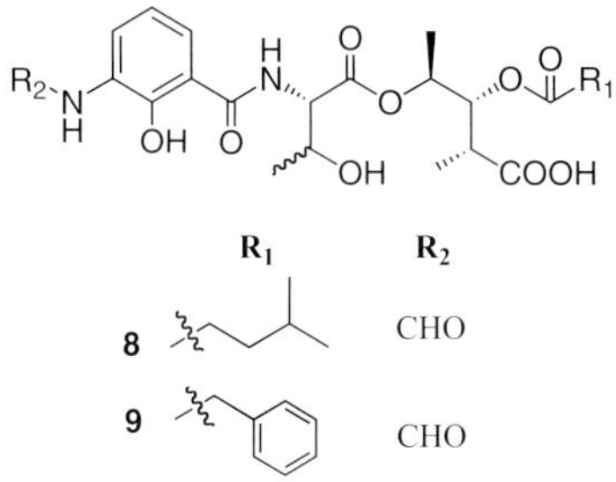

7) Three new members of the rare pyrroloterpene structure class Heronapyrroles A–C (**10–12**) were isolated from a marine-derived *Streptomyces* sp. (CMB-M0423) collected from the beach sand off Heron Island, Australia. Heronapyrroles A–C displayed promising biological activities against the Gram-positive bacteria *S. aureus* ATCC 9144 (IC_{50} 0.6–1.1 μM) and *Bacillus subtilis* ATCC 6633 (IC_{50} 1.1–6.5 μM), but were inactive against Gram-positive bacteria and cytotoxic toward mammalian cell lines [19].

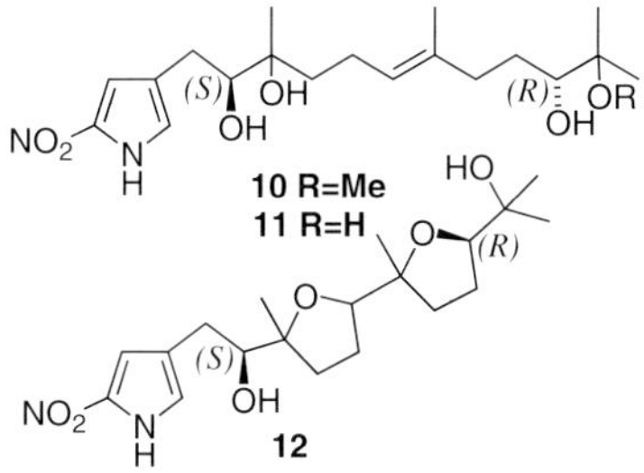

8) Streptocarbazoles A and B (**13a** and **13b**), two novel indolocarbazoles featuring unprecedented cyclic *N*-glycosidic linkages between 1,3-carbon atoms of the glycosyl moiety and two indole nitrogen atoms of the indolocarbazole core, were

isolated from the marine-derived actinomycetes strain *Streptomyces* sp. FMA. Compound **13a** was cytotoxic to HL-60 and A-549 cell lines and could arrest the cell cycle of Hela cells at the G2/M phase [20].

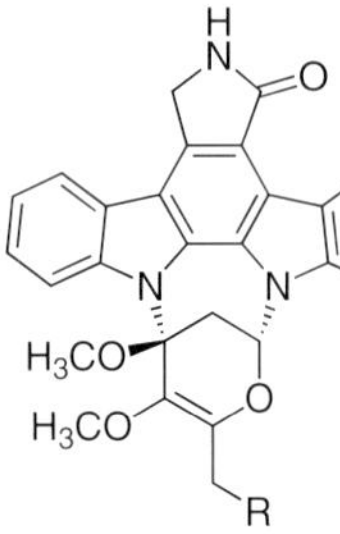

13a R = OH
13b R = H

9) *Streptomyces* sp. LMA-545, isolated from a soil sample collected from La Réunion Island, gave the two novel α,β-unsaturated γ-lactono-hydrazides, geralcin A and B (**14** and **15**). None of the compounds showed antibacterial activity, but compound **15** inhibited the growth of MDA231 breast cancer cells with an IC_{50} of 5 μM [21].

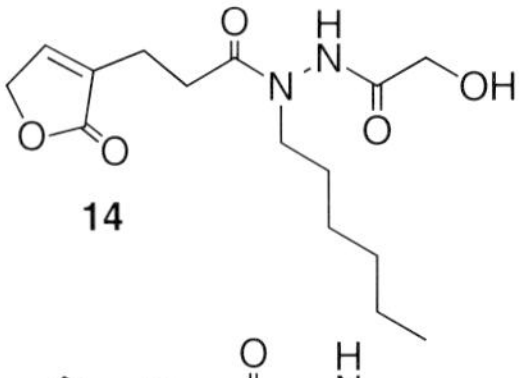

14

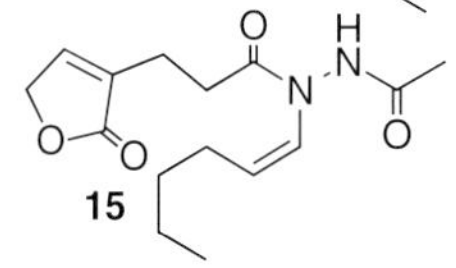

15

10) A cancer cell cytotoxic cyclodepsipeptides, veraguamides A (**16**) was isolated from marine cyanobacterium cf. *Oscillatoria margaritifera* obtained from the Coiba National Park, Panama. Veraguamide A showed potent cytotoxicity to the H460 human lung cancer cell line ($LD_{50} = 141$ nM) [22].

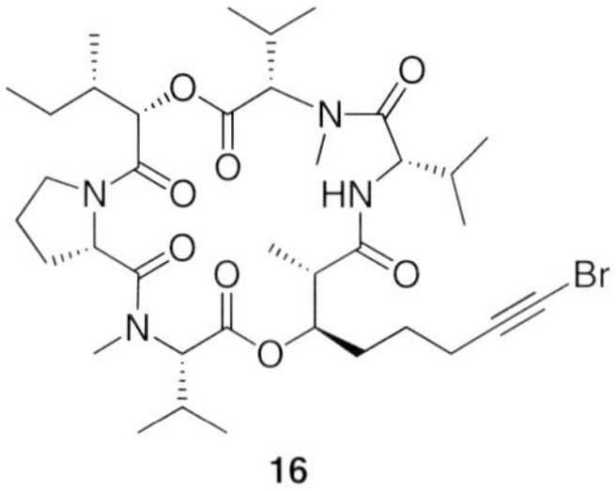

16

11) Cultivation of the marine actinomycete *Streptomyces* sp. CNQ343 gave two polyene polyol macrolides bahamaolides A and B (**17** and **18**), which are

structurally new 36-membered macrocyclic lactones belonging to the hexaene macrolide class, and compound **17** displayed significant inhibitory activity against *Candida albicans* isocitrate lyase and antifungal activity against various pathogenic fungi. However, these two compounds showed no significant activity in antibacterial assays and cytotoxicity [23].

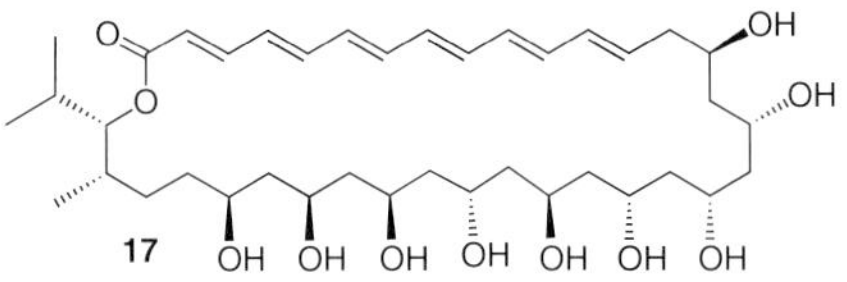

12) Two new 20-membered macrolides, levantilide A and B (**19** and **20**), were isolated from the *Micromonospora* strain M71-A77. Strain M71-A77 was recovered from an Eastern Mediterranean deep sea sediment sample. Both **19** and **20** exhibited moderate antiproliferative activity against several tumor cell lines [24].

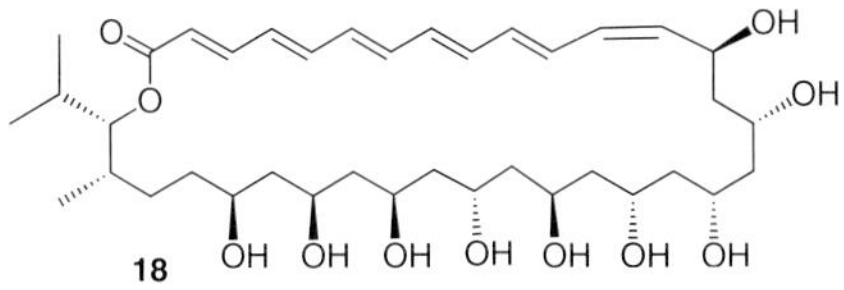

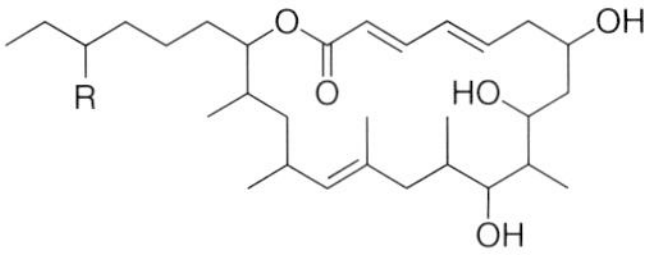

19 R= -OH
20 R= =O

13) A marine *Nocardia* sp., producing five new lipopeptides, peptidolipins B–F (**21–25**), was isolated from the ascidian *Trididemnum orbiculatum* collected from the Florida Keys. Compounds **21** and **24** demonstrated moderate antibacterial activity against methicillin-resistant *S. aureus* and methicillin-sensitive *S. aureus* with an MIC of 64 μg/ml [25].

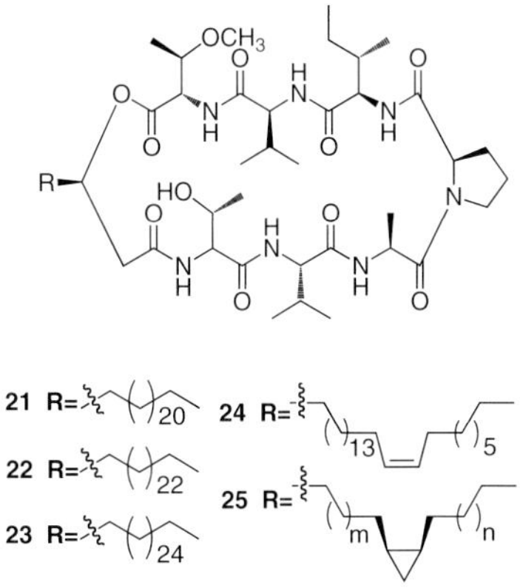

14) Four new anthracyclinones A–D (**26–29**) were isolated from a strain of *Micromonospora* sp. associated with the tunicate *Eudistoma vannamei*. Compounds **26** and **29** were cytotoxic against the HCT-8 human colon adenocarcinoma cell line, with IC_{50} values of 12.7 and 6.2 μM, respectively, while compounds **27** and **28** were inactive [26].

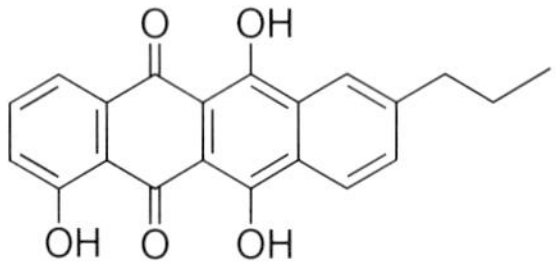

26 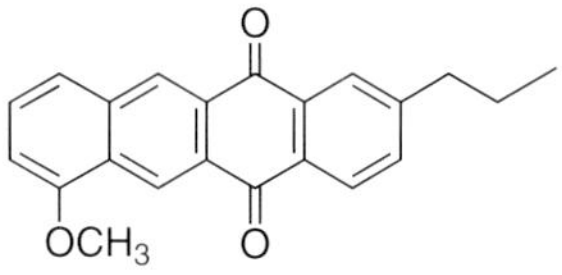 **27** 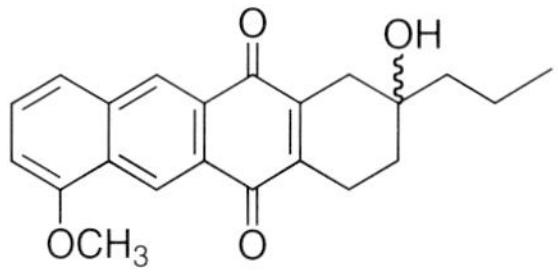 **28** 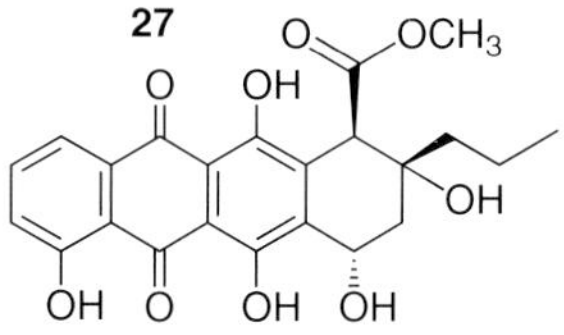 **29**

15) Grincamycins A–F (**30–35**) are new C-glycoside angucyclines antibiotic isolated from *S. lusitanus* SCSIO LR32 of deep sea origin. All compounds except **35** exhibited *in vitro* cytotoxicities against series of cell lines [27].

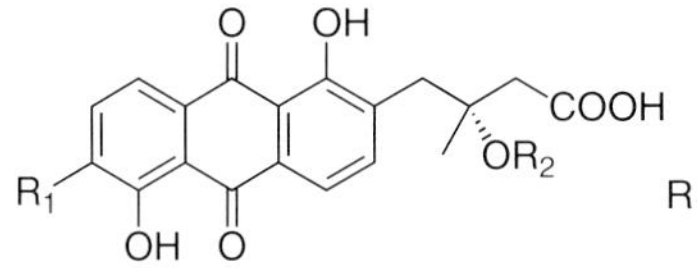

30 R_1 = a R_2 = b
31 R_1 = a R_2 = H
32 R_1 = c R_2 = b

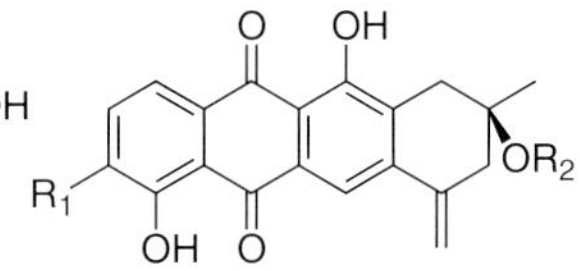

33 R_1 = a R_2 = b

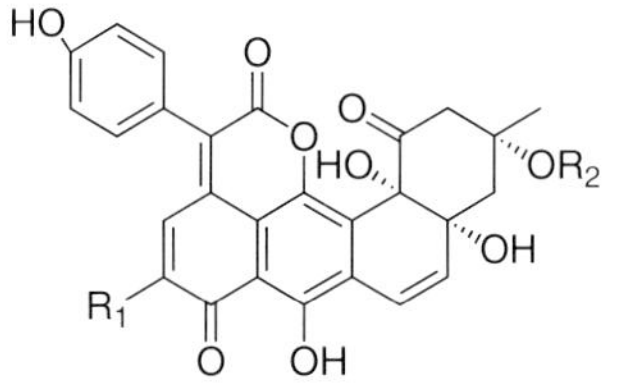

34 R_1 = d R_2 = b

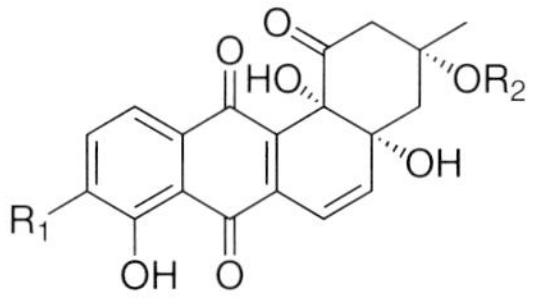

35 R_1 = a R_2 = b

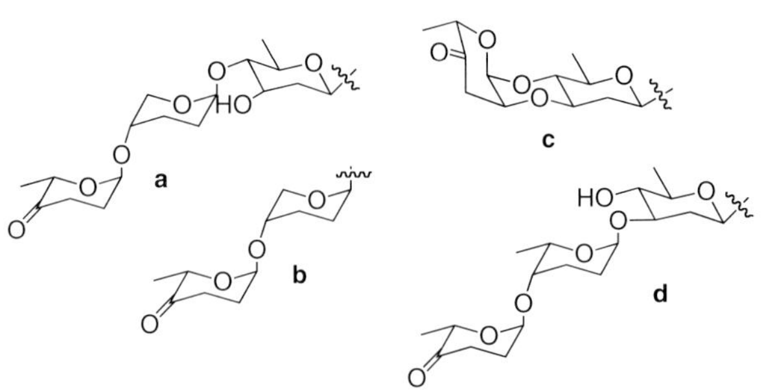

a b c d

16) JBIR-109, 110, 111, trichostain A and trichostatic acid (**36–40**) are trichostatin analogues obtained from *Streptomyces* sp. RM72 isolated from an unidentified marine sponge collected at a depth of 195 m near Takara Island, Kagoshima Prefecture, Japan. Of the five compounds, compound **39** displayed excellent HDAC inhibitory activity [28].

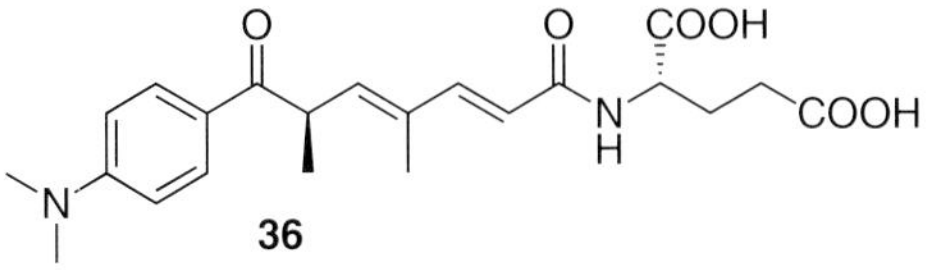

36

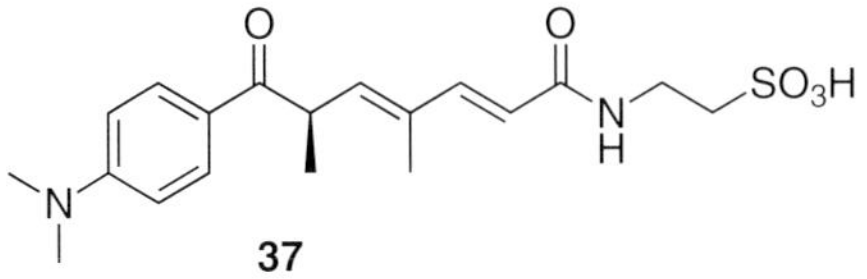

37

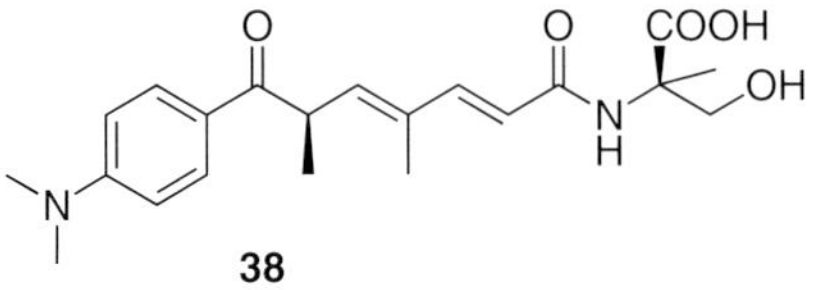

38

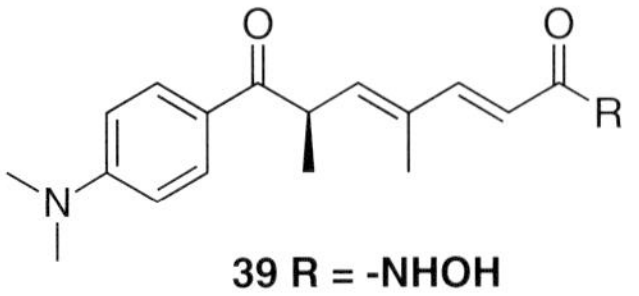

39 R = -NHOH
40 R = -OH

17) A deep sea actinomycete *Pseudonocardia* sp. SCSIO 01299 produced three new diazaanthraquinone derivatives pseudonocardians A–C (**41–43**). Compounds **42** and **43** showed potent cytotoxic activities against three tumor cell lines of SF-268, MCF-7, and NCI-H460, and also exhibited antibacterial activities on *S. aureus* ATCC 29213 and *Enterococcus faecalis* ATCC 29212 [29].

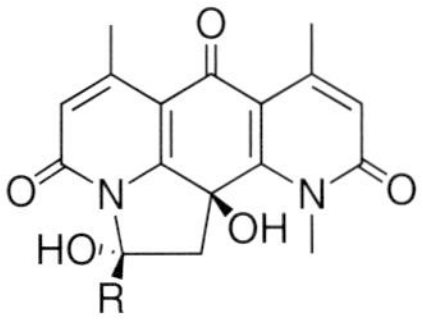

41 R = CH_3

42 R = CH_2CH_3

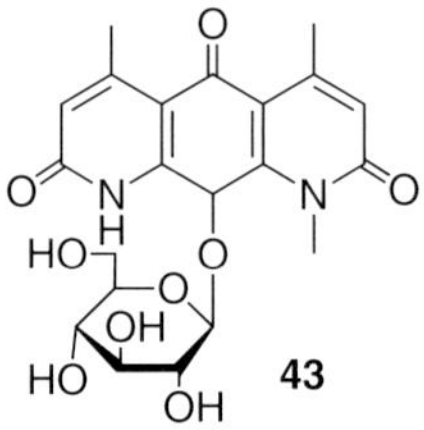

43

18) Two gliotoxin-related compounds bis(dethio)-10*a*-methylthio-3*a*-deoxy-3,3*a*-didehydrogliotoxin and 6-deoxy-5*a*,6-didehydrogliotoxin (compounds **44** and **45**) were isolated from the fungus *Penicillium* sp. strain JMF034, obtained from deep sea sediments of Suruga Bay, Japan. Compounds **44**

and **45** exhibited potential cytotoxic activity against P388 murine leukemia cells [30].

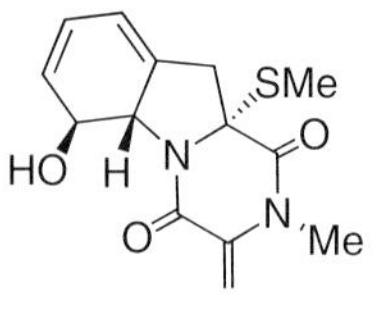

44 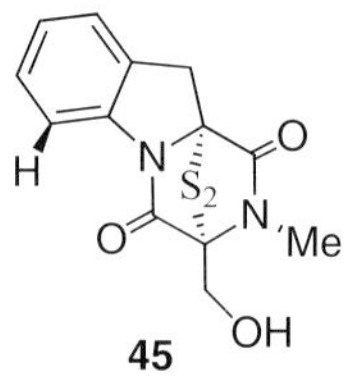 **45**

19) A marine-derived *Streptomyces* sp. strain CNS284 gave two new phenazine derivative compounds (**46** and **47**). In the bioassay test, compounds **46** and **47** showed potential cancer chemopreventive and anti-inflammatory activities [31].

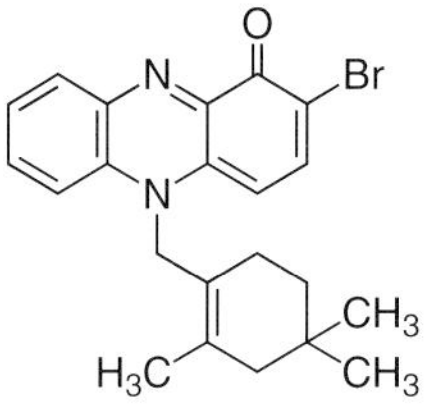

46 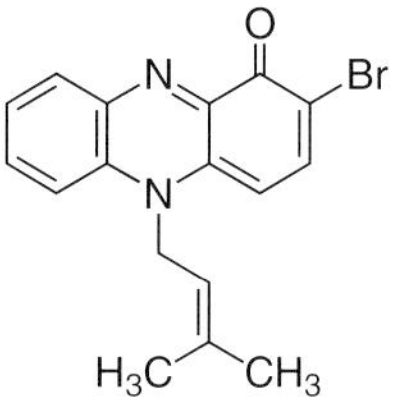 **47**

20) α-Proteobacterium *Tistrella mobilis* gave the depsipeptide didemnin B (**48**), which showed significant *in vitro* cytotoxicity and *in vivo* antitumor activity. Compound **48** was the first marine natural product to enter clinical trials as an anticancer agent. Phase II clinical trials revealed the anticancer activity of **48** against various types of cancer [32].

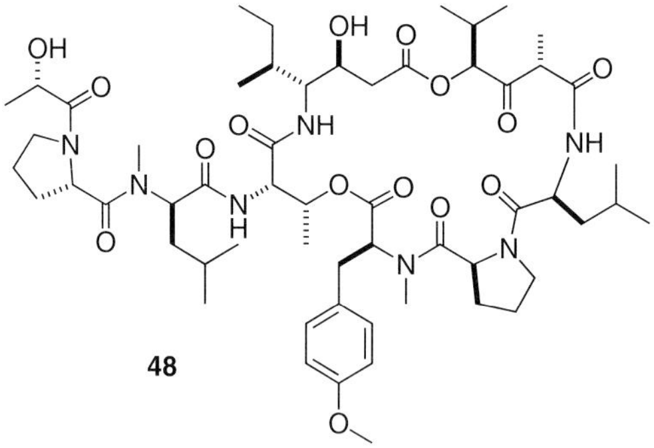

48

21) Ammosamide D (compound **49**) with an oxidized analogue resulting in a 5,6-dioxo-5,6-dihydroquinoline ring system of this class was isolated from a marine-derived *Streptomyces variabilis*. Compound **49** has modest cytotoxicity to the MIA PaCa-2 pancreatic cancer cell line [33].

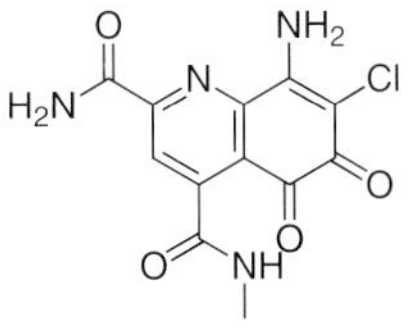

49

22) *Marinactinospora thermotolerans* SCSIO 00652, a newly characterized actinomycete belonging to the family Nocardiopsaceae, produced four new β-carboline alkaloids [34,35], designated marinacarbolines A–D (compounds **50–53**), and two new indolactam alkaloids, 13-*N*-demethyl-methylpendolmycin (**54**) and methylpendolmycin-14-*O*-α-glucoside (**55**). Compounds **50–55** showed no activities against a series of eight tumor cell lines (IC50 > 50 μM) but displayed antiplasmodial activities against *Plasmodium falciparum* lines 3D7 and Dd2, with IC_{50} values ranging from 1.92 to 36.03 μM [36].

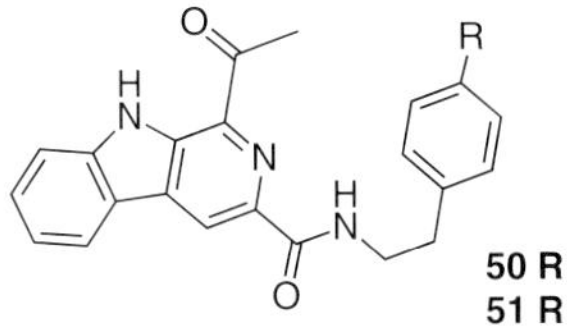

50 R = OCH3
51 R = OH
52 R = H 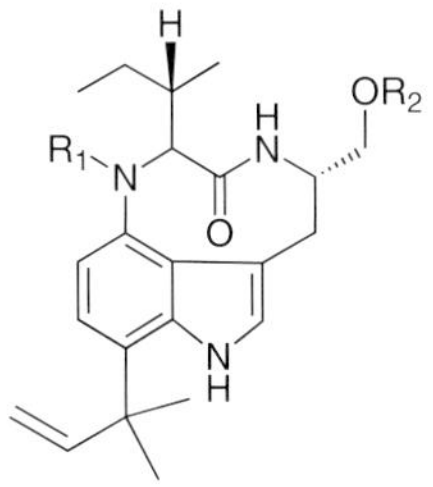 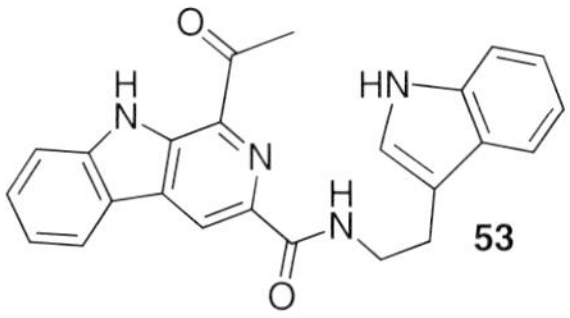
53
54 R_1 = H R_2 = H
55 R_1 = Me R_2 = (glucoside)

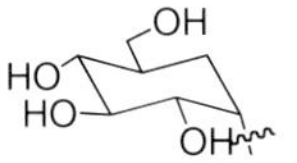

23) Three new α-pyrones, nocapyrones E–G (**56–58**), and three new diketopiperazine derivatives, nocazines A–C (compounds **59–61**), together with a new oxazoline compound, nocazoline A (**62**), were isolated from the marine-derived actinomycete *Nocardiopsis dassonvillei* HR10-5. Compounds **56–58** showed modest antimicrobial activity against *B. subtilis* with MIC values of 26, 14, and 12 μM, respectively [37].

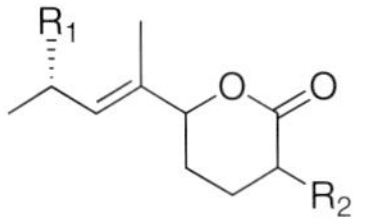

56 R_1 = H $R_2 = CH_2CH_3$
57 R_1 = H $R_2 = CH_3$
58 R_1 = OH $R_2 = CH_2CH_3$
59 R = OCH_3 **60** R = H 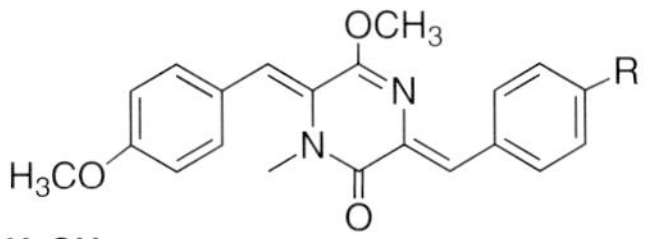
62 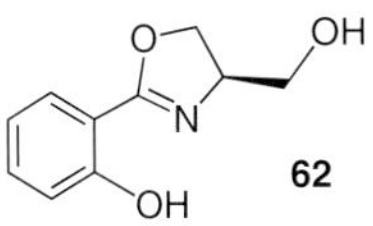
61

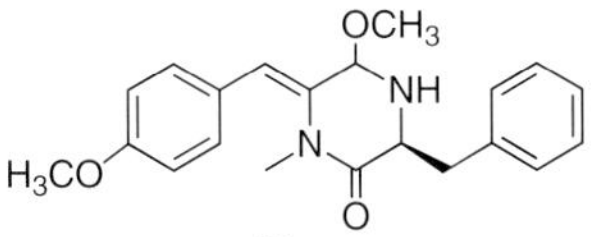

24) A marine cyanobacterial assemblage, composed of *Leptolyngbya* cf. and *Oscillatoria* spp. collected from Palmyra Atoll, yielded a new neuroactive macrolide palmyrolide A (compound **63**), featuring a rare *N*-methyl enamide and an intriguing *t*-butyl branch. Compound **63** showed a significant suppression of calcium influx in cerebrocortical neurons (IC_{50} 3.70 μM), and potent sodium channel blocking activity in neuro-2a cells (IC_{50} 5.2 μM) [38].

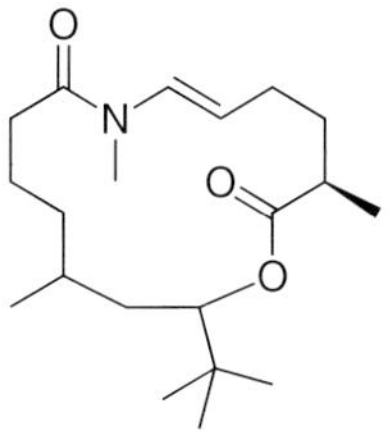

63

25) The marine-derived actinomycete *Actinoalloteichus cyanogriseus* WH1–2216-6 yielded five new bipyridine alkaloids (compounds **64–68**) and a new phenylpyridine alkaloid (compound **69**), which were named caerulomycins F-K. All of the six compounds showed cytotoxicity against several cell lines [39].

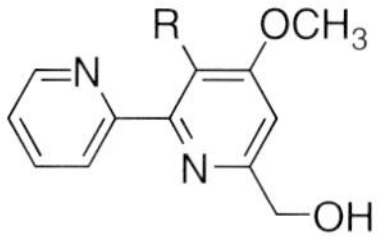

64 R = H **65** R = OCH_3

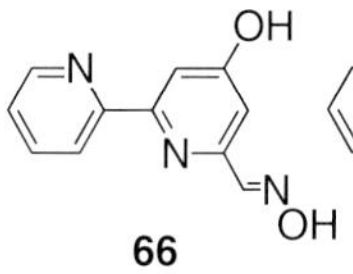

66 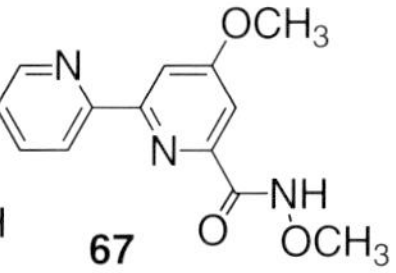 **67**

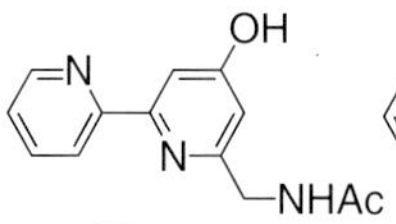

68 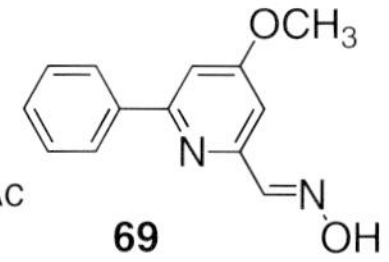 **69**

26) Four new antitumor pyranones, PM050511, PM050463, PM060054, and PM060431 (compounds **70–73**), were isolated from the cell extract of the marine-derived *Streptomyces albus* POR-04–15–053. Compounds **70** and **73** displayed strong cytotoxicity against three human tumor cell lines, whereas **71** showed subnanomolar activity as an inhibitor of EGFRMAPK-AP1-mediated mitogenic signaling, causing inhibition of EGF-mediated AP1 *trans*-activation and EGF-mediated ERK activation and slight inhibition of EGF-mediated JNK activation [40].

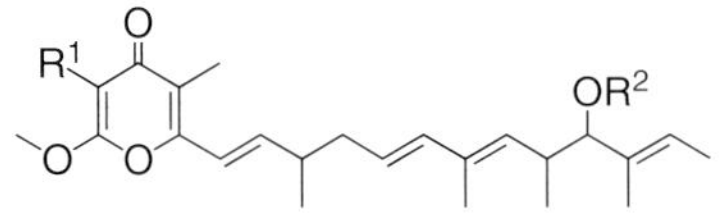

70 R^1 = Me R^2 = X
71 R^1 = Me R^2 = H
72 R^1 = ET R^2 = H
73 R^1 = ET R^2 = X

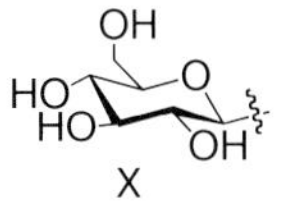

The chemical structures and activities described above are only a small portion of the novel compounds produced by marine microorganisms. It can be seen from the aforementioned descriptions of the above-mentioned active MNPs that the novel MNPs have potential not only in anticancer therapies and antimicrobial treatments but also in the development of antiviral and antiplasmodial drugs [15,36]. These naturally occurring novel compounds not only serve as lead compounds for drug discovery but can also provide basis for studies on its biosynthetic pathway and creation of "unnatural" natural compounds by genetic approaches, which is discussed in the following section.

17.3 Gene-Based Studies of MNPs

Nowadays there is an increasing trend to combine chemical studies with biological studies in the discovery of novel natural products. Traditionally, genetic studies start following the elucidation of chemical structures, and prior to gene-based studies, genetic probes must be generated for the screening of genomic library of the producer. The probes were designed based on the knowledge of the characteristic chemical structures and corresponding enzymes. However, identification of biosynthetic genes and creation of new structures using this strategy have problems in that (i) some chemicals are produced only in special media and under specific conditions; therefore, in the screening process, such compounds may be lost due to the silent expression of biosynthetic genes; (ii) some compounds may have specific activity that cannot be easily detected by the bioassay-based screening; therefore, it is necessary to establish various sensitive assay methods; (iii) some compounds may have complex structures that cannot be easily elucidated, and therefore impedes the identification of genetic probes; (iv) many if not all natural products are produced in low quantity in the natural isolates, and therefore are difficult to be purified for structure elucidation.

Due to the above-mentioned problems, studies on some marine natural products may be limited using the "chemistry to genetics" approach. However, studies of the "from genetics to chemistry" route, which uses gene-based screening strategy to study the biosynthetic potential of the microbial strains, followed by molecular cloning experiments and chemical purifications, have made great progress [41–43]. It has been observed that the genetic elements responsible for the biosynthesis of many, if not all, secondary metabolites are clustered in the bacterial or fungal genomes to form gene clusters, which include the genes encoding biosynthetic enzymes as well as genes responsible for regulation and self-resistance [44]. The organization of the gene clusters has not only greatly enabled the identification and characterization of the biosynthetic machinery but also facilitated the generation of novel compounds through mutagenesis biosynthesis and/or precursor-directed biosynthesis [45–48]. Studies on the biosynthesis of natural products have, therefore, contributed not only to the elucidation of molecular mechanisms and enzymatic machineries but also to the creation of novel molecules with better properties. Below are some of the recent examples on the harnessing of genetic diversity of marine microorganisms for natural product discovery.

$FADH_2$-dependent halogenases are widely distributed in the biosynthetic gene clusters of various antibiotics. Previous work reported by German researchers quickly identified potential producers of halogenated compounds from 550 randomly selected actinomycetes based on the conserved sequences of diverse $FADH_2$-dependent halogenases [42]. Phylogenetic analysis of the PCR products showed close relationship of these halogenases in the biosynthesis of structurally same or similar group of halogenated compounds. Using the degenerate primers in this study, new halogenase genes and novel halometabolites were isolated by heterologous expression of the gene cluster [42]. In the authors' laboratory, we also designed different pairs of degenerate primers of halogenases for screening the marine actinobacteria strain library, and isolated gene clusters putatively involved in the biosynthesis of glycopeptide antibiotics (unpublished data). In a recent report, $FADH_2$-dependent halogenases from marine sponge-associated microbial consortia were explored, and new clades of novel halogenases were discovered, including novel deltaproteobacterial symbiont-derived halogenase, poribacterial symbiont group of halogenases [49], and the studies also suggested that there exists sponge-specific symbiotic microbial consortium.

It is well known that dTDP-glucose-4,6-dehydratase (dTGD) is responsible for the biosynthesis of the key intermediate, namely, dTDP-4-keto-6-deoxy-D-glucose, for most deoxysugars in many antibiotics in different chemical groups, including peptides, polyketides, aminoglycosides, and terpenoids. In a recent study [50], degenerate primers of dTGD were designed and used to screen 91 marine sediment-derived bacteria. High percentage of positive strains was revealed, suggesting that the deoxysugar glycosylated compounds are widely distributed in these marine bacteria. Grouping the dTGD sequences showed that these genes were proposed to synthesize polyketide, aminoglycoside, indolocarbazole, terpenoid, liponucleotide, and enediyne antibiotics, and the production of medermycin and chromomycin A3 was confirmed, which is consistent with the sequence prediction of the corresponding dTGD genes [50].

In addition to the halogenase and dTGD genes, degenerate primers of epoxidase were also designed to probe new polyether gene clusters from microbial producers [51]. The employment of such genome-based screening strategy rapidly revealed potential producers of certain types of antibiotics prior to the isolation of the active compounds, and also had wide applications in the discovery of MNPs.

Marine environment is rich in various genetic resources. Shotgun sequencing of the environmental genome in the seawater samples collected from Sargasso Sea near Bermuda has revealed tremendous diversity [52], which contains 148 previously unknown bacterial phylotypes and 1.2 million novel genes. However, the majority of the marine microbial population is difficult to cultivate [53]. To fully access the genetic elements of all the microbial communities and explore the chemical diversity of uncultured microorganisms, the environmental DNA (eDNA) is extracted directly and packed into vectors, followed by transformation to different microbial hosts, such as *E. coli* and *S. lividans*. Clone libraries are then subjected to sequence-based or function-based screening for the discovery of new genes or new enzymes [54]. This cultivation-independent approach is termed metagenomics.

Marine invertebrates, such as sponges, tunicates, and bryozoans, are rich resources of valuable natural products, many of which have exhibited cytotoxic, antimicrobial, antiviral, or anti-inflammatory activity [55]. However, drug development from these marine macroorgainsms has been seriously limited by supply problem. It was confirmed by metagenomic studies that many valuable marine natural products that were previously thought to be produced by sponges, tunicates, and bryozoans are produced or obtained from marine microorganisms associated with the marine invertebrates [1,17,56–61].

Onnamides and theopederins are antitumor polyketides isolated from marine sponge *Theonella swinhoei*. Biosynthetic genes of these two compounds isolated from the metagenome of the marine sponge strongly indicated a bacterial origin, suggesting that these potent antitumor agents may be produced by the symbiotic bacteria [57]. Similarly, bryostatin has been evaluated for the treatment of various leukemias, ovarian and prostate cancers, and the bioinformatic analysis of bryostatin suggested that it was produced by a symbiotic bacterium *Candidatus Endobugula sertula* associated with the marine byrozoan [61]. Microbial origin of the potent antitumor agent psymberin was also confirmed by analysis of the data from the metagenomic library of marine sponge *Psammocinia* aff. *bulbosa* [59]. Overexpression of the biosynthetic gene cluster in the heterologous host will be of great help in solving the supply problems of these valuable natural products [8]. From chemistry point of view, development of chemical synthesis method is also an alternative way, and combining it with biosynthesis studies, more diversity in novel chemical structures can be obtained. The gene-based strategies for MNPs discovery are depicted in Figure 17.1.

17.4
MNPs Discovery Using Genome Mining

Although gene-based screening has facilitated rapid discovery of MNPs with certain special features in chemical structures, gene-based MNPs discovery has the limitation in that:

1) Gene-based screening relies on the conserved region of biosynthetic enzymes of special class of natural compounds, and therefore restricts the diversity of MNPs discovery.

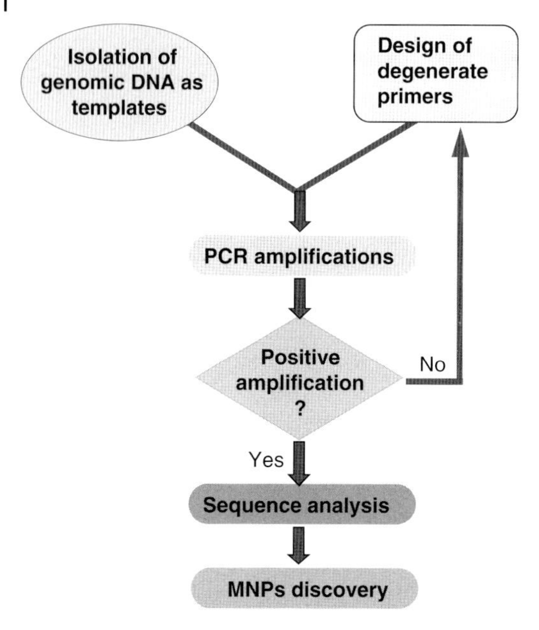

Figure 17.1 Gene-based studies of marine natural products (MNPs).

2) Quality of degenerate primers. Some biosynthetic enzymes do not harbor a well-defined conserved region, thus making the primer design complex; and in the meantime, primers obtained from well-conserved region only result in a limited diversity of biosynthetic genes.
3) PCR amplifications are affected by multiple conditions, including temperature, template DNA amount, inhibitory agents in the reaction mixture, and so on.
4) Sometimes not all genes involved in the biosynthesis of certain natural products are clustered in the same gene cluster [62], therefore resulting in an unsuccessful expression of full set of biosynthesis genes.

On the other hand, the biosynthesis of natural products requires different precursors and cofactors derived from the primary metabolism of the host cells [63]; therefore, knowledge on the balanced interaction of metabolic network of both primary metabolism and secondary metabolism is helpful in achieving optimized production. Understanding the genomic context of MNPs is of great importance to harness the diversity of both chemical structures and genetic elements, and is especially helpful for those compounds that have supply problem. Although omics studies of MNPs are still in their early stage, systems biological approaches, such as transcriptome studies, proteomic studies, and metabolome studies, in the model *Streptomyces* strain *S. coelicolor* have provided valuable information for related studies [64–67].

With the development of high-throughput sequencing technology and reduction of the sequencing cost, a vast variety of microbial genomic sequencing projects were achieved. The genomic sequencing data of various marine microorganisms are also accumulating rapidly in the recent years. According to the Genomes On Line Database (GOLD), there are at least 13 652 genome sequencing projects that were registered in GOLD, while according to the last update, 3173 complete genome projects are documented currently in GOLD (www.genomesonline.org) of which 2847 complete genomes are from bacterial strains. And in the meantime, 10 479 genome projects are ongoing, which are threefolds more than those reported in 2009 [68]. It can be anticipated that more marine microbial genomic sequences will be available in the near future.

As mentioned above, *Streptomyces* strains are the major producers of antibiotics that find a wide range of utilities in medical treatments. The first genomic sequencing of the model streptomycete, *S. coelicolor* A3(2), was released in 2002 [69], and analysis of the genome revealed about 31 biosynthetic gene clusters that were predicted to be involved in the biosynthesis of small-molecule metabolites, which are the major source of bioactive natural products. This was a great surprise to many researchers because by that time, only four antibiotics in *S. coelicolor* had been extensively studied, and the identification of previously unknown clusters indicated that microbial cells have the potential to produce much more molecules than previously anticipated. Later on, the genome sequence of *Streptomyces avermitilis* was also reported [70], followed by the blooming uncovering of genomic sequences of various antibiotic producers; a few examples are listed in this chapter [71–79].

Analysis of the genomic sequence of marine microorganisms not only provides basis to fully explore the biosynthetic potential of these microorganisms but also the starting point for further genetic engineering of the natural product biosynthesis for drug discovery. Studies on the biodiscovery of natural products using genomic sequences information were defined as "genome mining," and the strategies for the discovery of novel compounds by genome mining have been the subject of many review articles [7,48,80–91]. The early genome-mining studies can be dated back to the period right after the release of the genomic sequence of *S. coelicolor*, and the isolation and characterization of coelichelin, which is a new tris-hydroxamate tetrapeptide iron chelator from *S. coelicolor*, is one of the best examples [92], which showed that the chemical structures of putative biosynthetic gene clusters could be predicted accurately by analysis of the biosynthetic enzymes.

The genome sequence of the first marine obligate actinomycete *Salinispora tropica* was reported [71], and bioinformatic analysis revealed that it owns a large proportion of genes (about 9.9%) responsible for MNPs biosynthesis. In contrast, the corresponding numbers in *S. coelicolor* and *S. avermitilis* are 4.5 and 6%, respectively [93] indicating its great potential in MNPs discovery. The genome of *S. tropica* contains very large portion of genes (516 kb) dedicated to polyketide synthase (PKS) and/or nonribosomal peptide synthetase (NRPS) biosynthesis, which are megasynthases involved in the biosynthesis of many active natural product biosynthesis, and have been extensively studied for the creation of novel compounds by combinational biosynthesis [94–96]. Subsequently, the comparative genomic studies of *S. pacifica* and *S. tropica* resulted in the discovery of

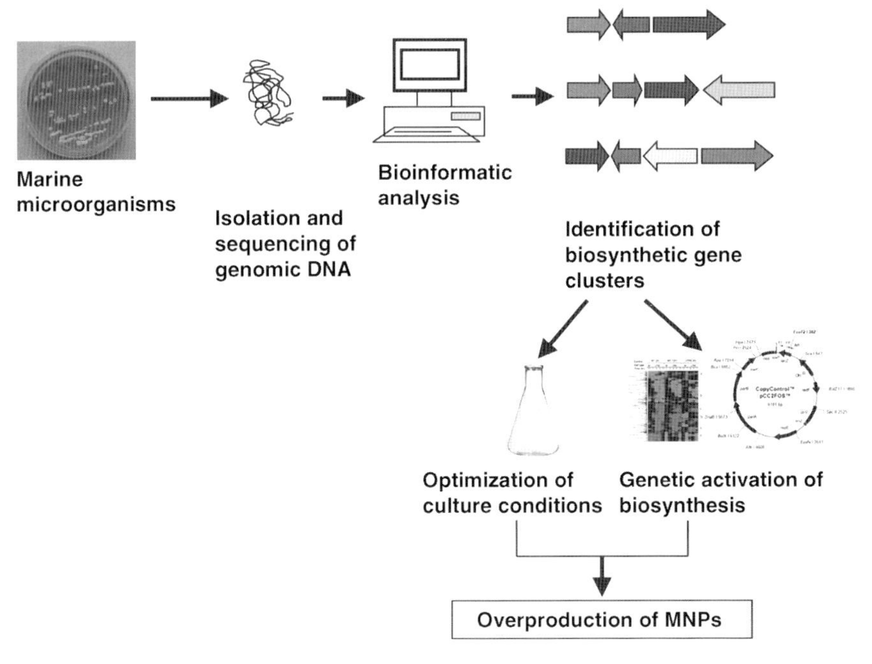

Figure 17.2 Genome-mining studies of MNPs.

salinosporamide K [97]. A truncated biosynthetic gene cluster was identified in *S. pacifica*, which resembled the 41 kb gene cluster in *S. tropica* for salinosporamide A biosynthesis, but the genes coding for chloroethylmalonyl-CoA pathway enzymes were missing. This information guided the isolation of salinosporamide K, which structurally resembles salinosporamide A [77]. More emerging examples include the studies on the biosynthesis of thiopeptide antibiotic TP-1161 and prenylated indole alkaloids stephacidin as well as notoamide metabolites by analysis of the genomes of marine *Nocardiopsis* sp. strain TFS65-07 and marine-derived fungus strain *Aspergillus* sp. MF297-2 [46,98]. The general procedure of genome mining for MNPs discovery was illustrated in Figure 17.2. The strategies of genome mining have been reviewed previously [74,90], and several key technologies with the recent reports are summarized in the following section.

17.4.1
Genome Mining of Peptide Products Using MALDI-TOF Mass Spectrometry-Based Approaches

Peptide fragment ions generated by MALDI-TOF mass spectrometry analysis were utilized in a new genome-mining approach for the discovery of peptide natural products. Imaging mass spectrometry (IMS) technology has been developed in Dr. Pieter Dorrestein's group for the studies of microbial intraspecies communications and natural product biosynthesis during microbial interactions [100,101]. Recently,

IMS was combined with a short sequence tagging (SST)-based genome-mining approach with only one or two amino acid residues for the discovery of peptide antibiotics [102]. Peptide antibiotics are mainly composed of nonribosomal peptides and ribosomal peptides, and are ubiquitous in many marine microorganisms. Using IMS, it was observed that a cluster of ions produced by *S. roseosporus* correlated with the inhibition of *S. aureus* and *S. epidermidis*, and subsequent analysis retrieved a sequence tag of glycine and alanine, which corresponds to a gene cluster with six amino acids encoding a nonribosomal peptide in the genome of *S. roseosporus*. Combining with the search of the NORINE database that contains more than 1000 NRPS-derived molecules, the ions discovered by IMS were determined to be arylomycins, which was indeed detected in the subsequent analysis [102]. Later on, natural product peptidogenomics (NPP) concept was proposed that takes advantages of the knowledge of peptide natural product biosynthesis and combines the knowledge with the MS^n data analysis, which generates fragment ions. NPP studies not only revealed novel peptide natural products from various streptomycetes but also resulted in the identification of 14 peptides from well-studied *Streptomyces* strains of *S. griseus* and *S. coelicolor* [47], which strongly suggest the potential of microbial producers in natural compounds production. Employment of this NPP approaches in the genome mining of marine microorganisms will rapidly enlarge our vision on the chemical diversities of MNPs.

17.4.2
Comparative Metabolic Profiling for the Discovery of Novel Compounds

Analysis of the microbial genomic sequences resulted in the identification of many biosynthetic gene clusters that are not related to the production of any known metabolites, and such gene clusters are often called "orphan" gene clusters [85]. Such gene clusters can be activated by changing the culture conditions as well as by genetic approaches (deletion or overexpression of regulatory genes) followed by the comparison of the metabolic profiling using HPLC or LC–MS of the mutants [103–105]. The fungus of *A. nidulans* contains a lot of genes involved in the biosynthesis of polyketides. However, under normal growth conditions, many gene clusters involved in polyketide biosynthesis are silent. Transcriptome and metabolome analysis were performed by comparing the samples in chemostat cultures under various nutrient limitation conditions (N-, P-, and C-limitation), and a new prenylated benzophenone derivative, preshamixanthone, was identified [105] that represents a proposed intermediate in the biosynthesis of xanthone. Comparative metabolic profiling can also be performed after the biosynthetic gene clusters of natural products are heterologously expressed in a well-characterized host strain (e.g., *E. coli*, *S. lividans*, *Saccharomyces cerevisiae*), and the comparison of metabolic profiling of the mutant, with the host cells harboring the empty plasmid, can then be performed to identify the novel products [83,106]. Development of robust cell factory of marine microorganisms and production of MNPs using synthetic biology approaches will be promising for further development of MNPs for drug discovery

[9,107]. Although so far not so many marine-derived natural products have been heterologously expressed [108,109], the development of heterologous expression technology will greatly facilitate MNPs discovery for medical treatments in the near future.

17.4.3
Identification of the Biosynthetic Genes by Single-Cell Genome Amplification

Despite the progress in the genome mining of microorganism pure cultures, genome exploration of complex microbial population is less advanced. For example, marine filamentous *Lyngbya bouillonii* coexists with abundant proteobacteria imbedded in and surrounding its polysaccharide sheath, and is thus difficult to be purified. Single-cell genome amplification was recently developed to identify the biosynthetic gene cluster of apratoxin A, which is a potent cancer cell cytotoxin [110]. Single cell of *L. bouillonii* was isolated and the genomic DNA was amplified by multiple displacement amplification. Isolation of the entire gene cluster for apratoxin A was performed by screening of the genomic DNA library using the probes designed based on the information of partial gene cluster from the single-cell genome sequencing. The single-cell genome amplification technology greatly facilitates the genome mining of the yet-uncultured marine microorganisms, and will find applications in further studies on the discovery of novel MNPs.

17.5
Conclusion and Prospects

The discovery of active natural products with medical treatment potential, from marine environment, is now rapidly developing; there has been growing recognition in the recent years that marine microorganisms, including marine actinomycetes, marine fungi, marine cyanobacteria, and others, are actually the most important source of novel compounds. Based on the knowledge of the biosynthetic pathways for natural products and genetic approaches developed in the last few decades, significant progress has been made in the elucidation of the chemical and genetic diversity of marine microorganisms. Genetic and biochemical studies are now playing a significant role in the discovery and creation of novel chemical structures and scaffolds for lead development.

Recent years have witnessed the exponentially increasing genomic sequencing information that enables gene-based studies and discovery of novel compounds by genome mining. In contrast to traditional strategies that can only discover chemical diversity in restricted variation of culture conditions, exploration of the genomic basis of natural products biosynthesis can fully take advantage of the biosynthetic potential of the microbial producers, and thus enable the maximized utilization of microbial resources, including those yet-uncultured microorganisms associated with marine plants and animals.

On the other hand, despite the success in the mining of natural products using gene-based and genome-mining studies, there are limitations in developing powerful tools in bioinformatic and chemical analyses. The poor quality of genomic sequence and annotation may lead to incomplete data for genetic studies, and sometimes even lead to wrong conclusion. Mass spectrometry tools nowadays are only successful in the mining of peptide natural compounds, and implementation of similar genome-mining approaches will facilitate the discovery of novel compounds with diverse structures. On the other hand, although examples of creation of novel MNPs using synthetic biology concept are still not reported, developing robust cell factories and design of novel molecules with improved activities are beneficial for drug discovery from MNPs. We are confident that there is a bright future for harnessing both the chemical and genetic diversities of marine microorganisms for full exploration of MNPs for drug discovery and medical treatments.

Acknowledgments

The authors are grateful for the financial support from the Next-Generation BioGreen 21 295 Program (PJ0080932011), Rural Development Administration, Republic of Korea. We appreciate helpful discussions with Professor Sergey B. Zotchev in Norwegian University of Science and Technology on marine natural product research.

References

1 Piel, J. (2009) Metabolites from symbiotic bacteria. *Nat. Prod. Rep.*, **26**, 338–362.

2 Xu, M.J., Liu, X.J., Zhao, Y.L., Liu, D., Xu, Z.H., Lang, X.M., Ao, P., Lin, W.H., Yang, S.L., Zhang, Z.G., and Xu, J. (2012) Identification and characterization of an anti-fibrotic benzopyran compound isolated from mangrove-derived *Streptomyces xiamenensis*. *Mar. Drugs*, **10** (3), 639–654.

3 Gulder, T.A. and Moore, B.S. (2009) Chasing the treasures of the sea-bacterial marine natural products. *Curr. Opin. Microbiol.*, **12**, 252–260.

4 Bull, A.T. and Stach, J.E. (2007) Marine actinobacteria: new opportunities for natural product search and discovery. *Trends. Microbiol.*, **15**, 491–499.

5 Williams, P.G. (2009) Panning for chemical gold: marine bacteria as a source of new therapeutics. *Trends. Biotechnol.*, **27**, 45–52.

6 Waters, A.L., Hill, R.T., Place, A.R., and Hamann, M.T. (2010) The expanding role of marine microbes in pharmaceutical development. *Curr. Opin. Biotechnol.*, **21**, 780–786.

7 Zotchev, S.B., Sekurova, O.N., and Katz, L. (2012) Genome-based bioprospecting of microbes for new therapeutics. *Curr. Opin. Biotechnol.*, **23** (6), 941–947.

8 Zhang, H., Boghigian, B.A., Armando, J., and Pfeifer, B.A. (2011) Methods and options for the heterologous production of complex natural products. *Nat. Prod. Rep.*, **28**, 125–151.

9 Zotchev, S.B. (2012) Marine actinomycetes as an emerging resource for the drug development pipelines. *J. Biotechol.*, **158**, 168–175.

10 Imhoff, J.F., Labes, A., and Wiese, J. (2011) Bio-mining the microbial treasures of the ocean: new natural products. *Biotechnol. Adv.*, **29**, 468–482.

11 Rateb, M.E. and Ebel, R. (2011) Secondary metabolites of fungi from marine habitats. *Nat. Prod. Rep.*, **28**, 290–344.
12 Blunt, J.W., Copp, B.R., Keyzers, R.A., Munro, M.H., and Prinsep, M.R. (2012) Marine natural products. *Nat. Prod. Rep.*, **29**, 144–222.
13 Williams, D.E., Dalisay, D.S., Patrick, B.O., Matainaho, T., Andrusiak, K., Deshpande, R., Myers, C.L., Piotrowski, J.S., Boone, C., Yoshida, M., and Andersen, R.J. (2011) Padanamides A and B, highly modified linear tetrapeptides produced in culture by a *Streptomyces* sp. isolated from a marine sediment. *Org. Lett.*, **13** (15), 3936–3939.
14 Raju, R., Piggott, A.M., Huang, X.C., and Capon, R.J. (2011) Nocardioazines: a novel bridged diketopiperazine scaffold from a marine-derived bacterium inhibits P-glycoprotein. *Org. Lett.*, **13** (10), 2770–2773.
15 Shushni, M.A.M., Singh, R., Mentel, R., and Lindequist, U. (2011) Balticolid: a new 12-membered macrolide with antiviral activity from an *Ascomycetous* fungus of marine origin. *Mar. Drugs*, **9** (5), 844–851.
16 Xie, Z., Liu, B., Wang, H., Yang, S., Zhang, H., Wang, Y., Ji, N., Qin, S., and Laatsch, H. (2012) Kiamycin, a unique cytotoxic angucyclinone derivative from a marine *Streptomyces* sp. *Mar. Drugs*, **10** (3), 551–558.
17 Xu, Y., Kersten, R.D., Nam, S.J., Lu, L., Al-Suwailem, A.M., Zheng, H., Fenical, W., Dorrestein, P.C., Moore, B.S., and Qian, P.Y. (2012) Bacterial biosynthesis and maturation of the didemnin anti-cancer agents. *J. Am. Chem. Soc.*, **134**, 8625–8632.
18 Han, Z., Xu, Y., McConnell, O., Liu, L., Li, Y., Qi, S., Huang, X., and Qian, P. (2012) Two antimycin A analogues from marine-derived actinomycete *Streptomyces lusitanus*. *Mar. Drugs*, **10** (3), 668–676.
19 Raju, R., Piggott, A.M., Barrientos Diaz, L.X., Khalil, Z., and Capon, R.J. (2010) Heronapyrroles A–C: farnesylated 2-nitropyrroles from an Australian marine-derived *Streptomyces* sp. *Org. Lett.*, **12** (22), 5158–5161.
20 Fu, P., Yang, C., Wang, Y., Liu, P., Ma, Y., Xu, L., Su, M., Hong, K., and Zhu, W. (2012) Streptocarbazoles A and B, two novel indolocarbazoles from the marine-derived actinomycete strain *Streptomyces* sp. FMA. *Org. Lett.*, **14** (9), 2422–2425.
21 Le, G.G., Martin, M.T., Servy, C., Cortial, S., Lopes, P., Bialecki, A., Smadja, J., and Ouazzani, J. (2012) Isolation and characterization of α,β-unsaturated γ-lactono-hydrazides from *Streptomyces* sp. *J. Nat. Prod.*, **75**, 915–919.
22 Mevers, E., Liu, W.-T., Engene, N., Mohimani, H., Byrum, T., Pevzner, P.A., Dorrestein, P.C., Spadafora, C., and Gerwick, W.H. (2011) Cytotoxic veraguamides, alkynyl bromide-containing cyclic depsipeptides from the marine cyanobacterium cf. *Oscillatoria margaritifera*. *J. Nat. Prod.*, **74** (5), 928–936.
23 Kim, D.G., Moon, K., Kim, S.H., Park, S.H., Park, S., Lee, S.K., Oh, K.B., Shin, J., and Oh, D.C. (2012) Bahamaolides A and B, antifungal polyene polyol macrolides from the marine actinomycete *Streptomyces* sp. *J. Nat. Prod.*, **75**, 959–967.
24 Gärtner, A., Ohlendorf, B., Schulz, D., Zinecker, H., Wiese, J., and Imhoff, J.F. (2011) Levantilides A and B, 20-membered macrolides from a micromonospora strain isolated from the Mediterranean deep sea sediment. *Mar. Drugs*, **9** (1), 98–108.
25 Wyche, T.P., Hou, Y., Vazquez-Rivera, E., Braun, D., and Bugni, T.S. (2012) Peptidolipins B–F, antibacterial lipopeptides from an ascidian-derived *Nocardia* sp. *J. Nat. Prod.*, **75** (4), 735–740.
26 Sousa, T.d.S., Jimenez, P.C., Ferreira, E.G., Silveira, E.R., Braz-Filho, R., Pessoa, O.D. L., and Costa-Lotufo, L.V. (2012) Anthracyclinones from *micromonospora* sp. *J. Nat. Prod.*, **75** (3), 489–493.
27 Huang, H., Yang, T., Ren, X., Liu, J., Song, Y., Sun, A., Ma, J., Wang, B., Zhang, Y., Huang, C., Zhang, C., and Ju, J. (2012) Cytotoxic angucycline class glycosides from the deep sea actinomycete *Streptomyces lusitanus* SCSIO LR32. *J. Nat. Prod.*, **75** (2), 202–208.

28 Hosoya, T., Hirokawa, T., Takagi, M., and Shin-ya, K. (2012) Trichostatin analogues JBIR-109, JBIR-110, and JBIR-111 from the marine sponge-derived *Streptomyces* sp. RM72. *J. Nat. Prod.*, **75** (2), 285–289.

29 Li, S., Tian, X., Niu, S., Zhang, W., Chen, Y., Zhang, H., Yang, X., Zhang, W., Li, W., Zhang, S., Ju, J., and Zhang, C. (2011) Pseudonocardians A–C, new diazaanthraquinone derivatives from a deep-sea actinomycete *Pseudonocardia* sp. SCSIO 01299. *Mar. Drugs*, **9** (8), 1428–1439.

30 Sun, Y., Takada, K., Takemoto, Y., Yoshida, M., Nogi, Y., Okada, S., and Matsunaga, S. (2012) Gliotoxin analogues from a marine-derived fungus, *Penicillium* sp., and their cytotoxic and histone methyltransferase inhibitory activities. *J. Nat. Prod.*, **75** (1), 111–114.

31 Kondratyuk, T.P., Park, E.-J., Yu, R., vanBreemen, R.B., Asolkar, R.N., Murphy, B.T., Fenical, W., and Pezzuto, J.M. (2012) Novel marine phenazines as potential cancer chemopreventive and anti-inflammatory agents. *Mar. Drugs*, **10** (2), 451–464.

32 Tsukimoto, M., Nagaoka, M., Shishido, Y., Fujimoto, J., Nishisaka, F., Matsumoto, S., Harunari, E., Imada, C., and Matsuzaki, T. (2011) Bacterial production of the tunicate-derived antitumor cyclic depsipeptide didemnin B. *J. Nat. Prod.*, **74** (11), 2329–2331.

33 Pan, E., Jamison, M., Yousufuddin, M., and MacMillan, J.B. (2012) Ammosamide D, an oxidatively ring opened ammosamide analog from a marine-derived *Streptomyces variabilis*. *Org. Lett.*, **14** (9), 2390–2393.

34 Tian, X.P., Tang, S.K., Dong, J.D., Zhang, Y.Q., Xu, L.H., Zhang, S., and Li, W.J. (2009) *Marinactinospora thermotolerans* gen. nov., sp. nov., a marine actinomycete isolated from a sediment in the northern South China Sea. *Int. J. Syst. Evol. Microbiol.*, **59** (5), 948–952.

35 Tian, X.P., Zhi, X.Y., Qiu, Y.Q., Zhang, Y. Q., Tang, S.K., Xu, L.H., Zhang, S., and Li, W.J. (2009) *Sciscionella marina* gen. nov., sp. nov., a marine actinomycete isolated from a sediment in the northern South China Sea. *Int. J. Syst. Evol. Microbiol.*, **59** (2), 222–228.

36 Huang, H., Yao, Y., He, Z., Yang, T., Ma, J., Tian, X., Li, Y., Huang, C., Chen, X., Li, W., Zhang, S., Zhang, C., and Ju, J. (2011) Antimalarial β-carboline and indolactam alkaloids from *Marinactinospora thermotolerans*, a deep sea isolate. *J. Nat. Prod.*, **74** (10), 2122–2127.

37 Fu, P., Liu, P., Qu, H., Wang, Y., Chen, D., Wang, H., Li, J., and Zhu, W. (2011) α-Pyrones and diketopiperazine derivatives from the marine-derived actinomycete *Nocardiopsis dassonvillei* HR10-5. *J. Nat. Prod.*, **74** (10), 2219–2223.

38 Pereira, A.R., Cao, Z., Engene, N., Soria-Mercado, I.E., Murray, T.F., and Gerwick, W.H. (2010) Palmyrolide A, an unusually stabilized neuroactive macrolide from Palmyra Atoll cyanobacteria. *Org. Lett.*, **12** (20), 4490–4493.

39 Fu, P., Wang, S., Hong, K., Li, X., Liu, P., Wang, Y., and Zhu, W. (2011) Cytotoxic bipyridines from the marine-derived actinomycete *Actinoalloteichus cyanogriseus* WH1–2216-6. *J. Nat. Prod.*, **74** (8), 1751–1756.

40 Schleissner, C., Pérez, M., Losada, A., Rodríguez, P., Crespo, C., Zúñiga, P., Fernández, R., Reyes, F., and de la Calle, F. (2011) Antitumor actinopyranones produced by *Streptomyces albus* POR-04-15-053 isolated from a marine sediment. *J. Nat. Prod.*, **74** (7), 1590–1596.

41 Zazopoulos, W., Huang, K., Staffa, A., Liu, W., Bachmann, B.O., Nonaka, K., Ahlert, J., Thorson, J.S., Shen, B., and Farnet, C.M. (2003) A genomics-guided approach for discovering and expressing cryptic metabolic pathways. *Nat. Biotechnol.*, **21**, 187–190.

42 Hornung, A., Bertazzo, M., Dziarnowski, A., Schneider, K., Welzel, K., Wohlert, S.E., Holzenkämpfer, M., Nicholson, G.J., Bechthold, A., Süssmuth, R.D., Vente, A., and Pelzer, S. (2007) A genomic screening approach to the structure-guided identification of drug candidates from natural sources. *ChemBioChem*, **8** (7), 757–766.

43 Toyomasu, T., Kaneko, A., Tokiwano, T., Kanno, Y., Kanno, Y., Niida, R., Miura, S., Nishioka, T., Ikeda, C., Mitsuhashi, W.,

Dairi, T., Kawano, T., Oikawa, H., Kato, N., and Sassa, T. (2009) Biosynthetic gene-based secondary metabolite screening: a new diterpene, methyl homopsenonate, from the fungus *Phomopsis amygdali*. *J. Org. Chem.*, **74**, 1541–1548.

44 Martin, J.F. (1992) Clusters of gene for the biosynthesis of antibiotics: regulatory genes and overproduction of pharmaceuticals. *J. Ind. Microbiol.*, **9**, 73–90.

45 McGlinchey, R.P., Nett, M., Eustáquio, A.S., Asolkar, R.N., Fenical, W., and Moore, B.S. (2008) Engineered biosynthesis of antiprotealide and other unnatural salinosporamide proteasome inhibitors. *J. Am. Chem. Soc.*, **130** (25), 7822–7823.

46 Engelhardt, K., Degnes, K.F., and Zotchev, S.B. (2010) Isolation and characterization of the gene cluster for biosynthesis of the thiopeptide antibiotic TP-1161. *Appl. Environ. Microbiol.*, **76**, 7093–7101.

47 Kersten, R.D., Yang, Y.L., Xu, Y., Cimermancic, P., Nam, S.J., Fenical, W., Fischbach, M.A., Moore, B.S., and Dorrestein, P.C. (2011) A mass spectrometry-guided genome mining approach for natural product *peptidogenomics*. *Nat. Chem. Biol.*, **7** (11), 794–802.

48 Lane, A.L. and Moore, B.S. (2011) A sea of biosynthesis: marine natural products meet the molecular age. *Nat. Prod. Rep.*, **28**, 411–428.

49 Bayer, K., Scheuermayer, M., Fieseler, L., and Hentschel, U. (2013) Genomic mining for novel $FADH_2$-dependent halogenases in marine sponge-associated microbial consortia. *Mar. Biotechnol.*, **15** (1), 63–72.

50 Chen, F., Lin, L., Wang, L., Tan, Y., Zhou, H., Wang, Y., Wang, Y., and He, W. (2011) Distribution of dTDP-glucose-4,6-dehydratase gene and diversity of potential glycosylated natural products in marine sediment-derived bacteria. *Appl. Microbiol. Biotechnol.*, **90** (4), 1347–1359.

51 Liu, G., Cane, D.E., and Deng, Z.X. (2009) The enzymology of polyether biosynthesis. *Methods Enzymol.*, **459**, 187–214.

52 Venter, J.C., Remington, K., Heidelberg, J.F., Halpern, A.L., Rusch, D., Eisen, J.A., Wu, D., Paulsen, I., Nelson, K.E., Nelson, W., Fouts, D.E., Levy, S., Knap, A.H., Lomas, M.W., Nealson, K., White, O., Peterson, J., Hoffman, J., Parsons, R., Baden, T.H., Pfannkoch, C., Rogers, Y.H., and Smith, H.O. (2004) Environmental genome shotgun sequencing of the Sargasso Sea. *Science*, **304**66–74.

53 Ogura, A., Lin, M., Shigenobu, Y., Fujiwara, A., Ikeo, K., and Nagai, S. (2011) Effective gene collection from the metatranscriptome of marine microorganisms. *BMC. Genomics*, **3**, 15.

54 Ouyang, Y., Dai, S., Xie, L., Ravi Kumar, M.S., Sun, W., Sun, H., Tang, D., and Li, X. (2010) Isolation of high molecular weight DNA from marine sponge bacteria for BAC library construction. *Mar. Biotechnol.*, **12** (3), 318–325.

55 Piel, J., Hui, D., Wen, G., Butzke, D., Platzer, M., Fusetani, N., and Matsunaga, S. (2004) Antitumor polyketide biosynthesis by an uncultivated bacterial symbiont of the marine sponge *Theonella swinhoei*. *Proc. Natl. Acad. Sci. USA*, **101**, 16222–16227.

56 Schirmer, A., Gadkari, R., Reeves, C.D., Ibrahim, F., DeLong, E.F., and Hutchinson, C.R. (2005) Metagenomic analysis reveals diverse polyketide synthase gene clusters in microorganisms associated with the marine sponge *Discodermia dissoluta*. *Appl. Environ. Microbiol.*, **71** (8), 4840–4849.

57 Piel, J. (2004) Metabolites from symbiotic bacteria. *Nat. Prod. Rep.*, **21**, 519–538.

58 Piel, J. (2011) Approaches to capturing and designing biologically active small molecules produced by uncultured microbes. *Annu. Rev. Microbiol.*, **65**, 431–453.

59 Fisch, K.M., Gurgui, C., Heycke, N., van derSar, S.A., Anderson, S.A., Webb, V.L., Taudien, S., Platzer, M., Rubio, B.K., Robinson, S.J., Crews, P., and Piel, J. (2009) Polyketide assembly lines of uncultivated sponge symbionts from structure-based gene targeting. *Nat. Chem. Biol.*, **5**, 494–501.

60 Schmidt, E.W., Nelson, J.T., Rasko, D.A., Sudek, S., Eisen, J.A., Haygood, M.G., and Ravel, J. (2005) Patellamide A and C biosynthesis by a microcinlike pathway in *Prochloron didemni*, the cyanobacterial symbiont of *Lissoclinum patella*. *Proc. Natl. Acad. Sci. USA*, **102**, 7315–7320.

61 Sudek, S., Lopanik, N.B., Waggoner, L.E., Hildebrand, M., Anderson, C., Liu, H., Patel, A., Sherman, D.H., and Haygood, M.G. (2007) Identification of the putative bryostatin polyketide synthase gene cluster from "*Candidatus Endobugula sertula*", the uncultivated microbial symbiont of the marine bryozoan *Bugula neritina*. *J. Nat. Prod.*, **70**, 67–74.

62 Sanchez, J.F., Entwistle, R., Hung, J.H., Yaegashi, J., Jain, S., Chiang, Y.M., Wang, C.C., and Oakley, B.R. (2011) Genome-based deletion analysis reveals the prenyl xanthone biosynthesis pathway in *Aspergillus nidulans*. *J. Am. Chem. Soc.*, **133**, 4010–4017.

63 Borodina, I., Siebring, J., Zhang, J., Smith, C.P., Keulen, G., Dijkhuizen, L., and Nielsen, J. (2008) Antibiotic overproduction in *Streptomyces coelicolor* A3(2) mediated by phosphofructokinase deletion. *J. Biol. Chem.*, **283**, 25186–25199.

64 Borodina, I., Krabben, P., and Nielsen, J. (2005) Genome-scale analysis of *Streptomyces coelicolor* A3(2) metabolism. *Genome Res.*, **15** (6), 820–829.

65 Castro, M.M., Charaniya, S., Karypis, G., Takano, E., and Hu, W.S. (2010) Genome-wide inference of regulatory networks in *Streptomyces coelicolor*. *BMC Genomics*, **11**, 578.

66 Jankevics, A., Merlo, M.E., deVries, M., Vonk, R.J., Takano, E., and Breitling, R. (2011) Metabolomic analysis of a synthetic metabolic switch in *Streptomyces coelicolor* A3(2). *Proteomics*, **11** (24), 4622–4631.

67 Martín, J.F., Santos-Beneit, F., Rodríguez-García, A., Sola-Landa, A., Smith, M.C., Ellingsen, T.E., Nieselt, K., Burroughs, N.J., and Wellington, E.M. (2012) Transcriptomic studies of phosphate control of primary and secondary metabolism in *Streptomyces coelicolor*. *Appl. Microbiol. Biotechnol.*, **95**, 61–75.

68 Liolios, K., Chen, I.M., Mavromatis, K., Tavernarakis, N., Hugenholtz, P., Markowitz, V.M., and Kyrpides, N.C. (2010) The Genomes On Line Database (GOLD) in 2009: status of genomic and metagenomic projects and their associated metadata. *Nucleic Acids Res.*, **38**, 346–354.

69 Bentley, S.D., Chater, K.F., Cerdeño-Tárraga, A.M., Challis, G.L., Thomson, N.R., James, K.D., Harris, D.E., Quail, M.A., Kieser, H., Harper, D., Bateman, A., Brown, S., Chandra, G., Chen, C.W., Collins, M., Cronin, A., Fraser, A., Goble, A., Hidalgo, J., Hornsby, T., Howarth, S., Huang, C.H., Kiese, T., Larke, L., Murphy, L., Oliver, K., O'Neil, S., Rabbinowitsch, E., Rajandream, M. A., Rutherford, K., Rutter, S., Seeger, K., Saunders, D., Sharp, S., Squares, R., Squares, S., Taylor, K., Warren, T., Wietzorrek, A., Woodward, J., Barrell, B. G., Parkhill, J., and Hopwood, D.A. (2002) Complete genome sequence of the model actinomycete *Streptomyces coelicolor* A3(2). *Nature*, **417**, 141–147.

70 Ikeda, H., Ishikawa, J., Hanamoto, A., Shinose, M., Kikuchi, H., Shiba, T., Sakaki, Y., Hattori, M., and Omura, S. (2003) Complete genome sequence and comparative analysis of the industrial microorganism *Streptomyces* avermitilis. *Nat. Biotechnol.*, **21**, 526–531.

71 Oliynyk, M., Samborskyy, M., Lester, J. B., Mironenko, T., Scott, N., Dickens, S., Haydock, S.F., and Leadlay, P.F. (2007) Complete genome sequence of the erythromycin-producing bacterium *Saccharopolyspora erythraea* NRRL23338. *Nat. Biotechnol.*, **25**, 447–453.

72 Udwary, D.W., Zeigler, L., Asolkar, R.N., Singan, V., Lapidus, A., Fenical, W., Jensen, P.R., and Moore, B.S. (2007) Genome sequencing reveals complex secondary metabolome in the marine actinomycete *Salinispora tropica*. *Proc. Natl. Acad. Sci. USA*, **104**, 10376–10381.

73 Ichikawa, N., Oguchi, A., Ikeda, H., Ishikawa, J., Kitani, S., Watanabe, Y., Nakamura, S., Katano, Y., Kishi, E., Sasagawa, M., Ankai, A., Fukui, S.,

Hashimoto, Y., Kamata, S., Otoguro, M., Tanikawa, S., Nihira, T., Horinouchi, S., Ohnishi, Y., Hayakawa, M., Kuzuyama, T., Arisawa, A., Nomoto, F., Miura, H., Takahashi, Y., and Fujita, N. (2010) Genome sequence of kitasatospora setae NBRC 14216T: an evolutionary snapshot of the family Streptomycetaceae. *DNA Res.*, **17**, 393–406.

74 Zhao, X.Q. and Yang, T.H. (2011) Draft genome sequence of the marine derived actinomycete *Streptomyces xinghaiensis* NRRL B24674^{T}. *J. Bacteriol.*, **193**, 5543.

75 Qin, S., Zhang, H., Li, F., Zhu, B., and Zheng, H. (2012) Draft genome sequence of marine *Streptomyces* sp. strain W007, which produces angucyclinone antibiotics with a benz[*a*] anthracene skeleton. *J. Bacteriol.*, **194** (6), 1628–1629.

76 Ohnishi, Y., Ishikawa, J., Hara, H., Suzuki, H., Ikenoya, M., Ikeda, H., Yamashita, A., Hattori, M., and Horinouchi, S. (2008) Genome sequence of the streptomycin-producing microorganism *Streptomyces griseus* IFO 13350. *J. Bacteriol.*, **190**, 4050–4060.

77 Song, J.Y., Jeong, H., Yu, D.S., Fischbach, M.A., Park, H.S., Kim, J.J., Seo, J.S., Jensen, S.E., Oh, T.K., Lee, K.J., and Kim, J.F. (2010) Draft genome sequence of *Streptomyces clavuligerus* NRRL 3585, a producer of diverse secondary metabolites. *J. Bacteriol.*, **192**, 6317–6318.

78 Wang, J., Li, Y., Bian, J., Tang, S.K., Ren, B., Chen, M., Li, W.J., and Zhang, L.X. (2010) *Prauserella marina* sp. nov., isolated from ocean sediment of the South China Sea. *Int. J. Syst. Evol. Microbiol.*, **60** (4), 985–989.

79 Wang, X.J., Yan, Y.J., Zhang, B., An, J., Wang, J.J., Tian, J., Jiang, L., Chen, Y.H., Huang, S.X., Yin, M., Zhang, J., Gao, A.L., Liu, C.X., Zhu, Z.X., and Xiang, W.S. (2010) Genome sequence of the milbemycin-producing bacterium *Streptomyces bingchenggensis*. *J. Bacteriol.*, **192**, 4526–4527.

80 VanLanen, S.G. and Shen, B. (2006) Microbial genomics for the improvement of natural product discovery. *Curr. Opin. Microbiol.*, **9**, 252–260.

81 Baltz, R.H. (2008) Renaissance in antibacterial discovery from actinomycetes. *Curr. Opin. Pharmacol.*, **8**, 557–563.

82 Velásquez, J.E. and Donk, W.A. (2011) Genome mining for ribosomally synthesized natural products. *Curr. Opin. Chem. Biol.*, **15** (1), 11–21.

83 Corre, C., Song, L., O'Rourke, S., Chater, K.F., and Challis, G.L. (2008) 2-Alkyl-4-hydroxymethylfuran-3-carboxylic acids, antibiotic production inducers discovered by *Streptomyces coelicolor* genome mining. *Proc. Natl. Acad. Sci. USA*, **105**, 17510–17515.

84 Gross, H. (2009) Genomic mining – a concept for the discovery of new bioactive natural products. *Curr. Opin. Drug Discov. Dev.*, **12**, 207–219.

85 Gross, H. (2007) Strategies to unravel the function of orphan biosynthesis pathways: recent examples and future prospects. *Appl. Microbiol. Biotechnol.*, **75**, 267–277.

86 Gross, H., Stockwell, V.O., Henkels, M.D., Nowak, T., Loper, B.J.E., and Gerwick, W. H. (2007) The genomisotopic approach: a systematic method to isolate products of orphan biosynthetic gene clusters. *Chem. Biol.*, **14**, 53–63.

87 McAlpine, J.B. (2009) Advances in the understanding and use of the genomic base of microbial secondary metabolite biosynthesis for the discovery of new natural products. *J. Nat. Prod.*, **72**, 566–572.

88 Zhao, X.Q. (2011) Genome-based studies of marine microorganisms to maximize the diversity of natural products discovery for medical treatments. *Evid. Based Compl. Alt. Med.* doi: 10.1155/2011/384572.

89 Gerwick, W.H. and Moore, B.S. (2012) Lessons from the past and charting the future of marine natural products drug discovery and chemical biology. *Chem. Biol.*, **19** (1), 85–98.

90 Challis, G.L. (2008) Genome mining for novel natural product discovery. *J. Med. Chem.*, **51**, 2618–2628.

91 Zerikly, M. and Challis, G.L. (2009) Strategies for the discovery of new natural products by genome mining. *ChemBioChem*, **10**, 625–633.

92 Lautru, S., Deeth, R.J., Bailey, L.M., and Challis, G.L. (2005) Discovery of a new peptide natural product by *Streptomyces coelicolor* genome mining. *Nat. Chem. Biol.*, **1**, 265–269.

93 Nett, M., Ikeda, H., and Moore, B.S. (2009) Genomic basis for natural product biosynthetic diversity in the actinomycetes. *Nat. Prod. Rep.*, **26**, 1362–1384.

94 Cane, D.E., Walsh, C.T., and Khosla, C. (1998) Harnessing the biosynthetic code: combinations, permutations, and mutations. *Science*, **282**, 63–68.

95 Floss, H.G. (2006) Combinatorial biosynthesis-potential and problems. *J. Biotechnol.*, **124**, 242–257.

96 Du, L., Sánchez, C., and Shen, B. (2001) Hybrid peptide-polyketide natural products: biosynthesis and prospects toward engineering novel molecules. *Metab. Eng.*, **3**, 78–95.

97 Eustáquio, A.S., Nam, S.J., Penn, K., Lechner, A., Wilson, M.C., Fenical, W., Jensen, P.R., and Moore, B.S. (2011) The discovery of salinosporamide K from the marine bacterium "*Salinispora pacifica*" by genome mining gives insight into pathway evolution. *ChemBioChem*, **12**, 61–64.

98 Ding, Y., Wet, J.R., Cavalcoli, J., Li, S., Greshock, T.J., Miller, K.A., Finefield, J.M., Sunderhaus, J.D., McAfoos, T.J., Tsukamoto, S., Williams, R.M., and Sherman, D.H. (2010) Genome-based characterization of two prenylation steps in the assembly of the stephacidin and notoamide anticancer agents in a marine-derived *Aspergillus* sp. *J. Am. Chem. Soc.*, **132**, 12733–12740.

99 Engelhardt, K., Degnes, K.F., Kemmler, M., Bredholt, H., Fjaervik, E., Klinkenberg, G., Sletta, H., Ellingsen, T.E., and Zotchev, S.B. (2010) Production of a new thiopeptide antibiotic, TP-1161, by a marine *Nocardiopsis* species. *Appl. Environ. Microbiol.*, **76** (15), 4969–4976.

100 Esquenazi, E., Dorrestein, P.C., and Gerwick, W.H. (2009) Probing marine natural product defenses with DESI-imaging mass spectrometry. *Proc. Natl. Acad. Sci. USA*, **106** (18), 7269–7270.

101 Yang, Y.L., Xu, Y., Kersten, R.D., Liu, W.T., Meehan, M.J., Moore, B.S., Bandeira, N., and Dorrestein, P.C. (2011) Connecting chemotypes and phenotypes of cultured marine microbial assemblages by imaging mass spectrometry. *Angew. Chem. Int. Ed. Engl.*, **50** (26), 5839–5842.

102 Liu, W.T., Kersten, R.D., Yang, Y.L., Moore, B.S., and Dorrestein, P.C. (2011) Imaging mass spectrometry and genome mining via short sequence tagging identified the anti-infective agent arylomycin in *Streptomyces roseosporus*. *J. Am. Chem. Soc.*, **133** (45), 18010–21813.

103 Laureti, L., Song, L., Huang, S., Corre, C., Leblond, P., Challis, GL., and Aigle, B. (2011) Identification of a bioactive 51-membered macrolide complex by activation of a silent polyketide synthase in *Streptomyces ambofaciens*. *Proc. Natl. Acad. Sci. USA*, **108** (15), 6258–6263.

104 Bergmann, S., Funk, AN., Scherlach, K., Schroeckh, V., Shelest, E., Horn, U., Hertweck, C., and Brakhage, AA. (2010) Activation of a silent fungal polyketide biosynthesis pathway through regulatory cross talk with a cryptic nonribosomal peptide synthetase gene cluster. *Appl. Environ. Microbiol.*, **76** (24), 8143–8149.

105 Sarkar, A., Funk, A.N., Scherlach, K., Horn, F., Schroeckh, V., Chankhamjon, P., Westermann, M., Roth, M., Brakhage, A.A., Hertweck, C., and Horn, U. (2012) Differential expression of silent polyketide biosynthesis gene clusters in chemostat cultures of *Aspergillus nidulans*. *J. Biotechnol.*, **31** (160), 64–71.

106 Blasiak, L.C. and Clardy, J. (2010) Discovery of 3-formyl-tyrosine metabolites from *Pseudoalteromonas tunicata* through heterologous expression. *J. Am. Chem. Soc.*, **132**, 926–927.

107 Mitchell, W. (2011) Natural products from synthetic biology. *Curr. Chem. Biol.*, **15**, 505–515.

108 Lombó, F., Velasco, A., Castro, A., de laCalle, F., Braña, A.F., Sánchez-Puelles, J.M., Méndez, C., and Salas, J.A. (2006) Deciphering the biosynthesis pathway of the antitumor thiocoraline from a marine actinomycete and its expression in two *Streptomyces* species. *ChemBioChem*, **7**, 366–367.

109 Winter, J.M., Moffitt, M.C., Zazopoulos, E., McAlpine, J.B., Dorrestein, P.C., and Moore, B.S. (2007) Molecular basis for chloronium-mediated meroterpene cyclization: cloning, sequencing, and heterologous expression of the napyradiomycin biosynthetic gene cluster. *J. Biol. Chem.*, **282**, 16362–16368.

110 Grindberg, R.V., Ishoey, T., Brinza, D., Esquenazi, E., Coates, R.C., Liu, W.T., Gerwick, L., Dorrestein, P.C., Pevzner, P., Lasken, R., and Gerwick, W.H. (2011) Single cell genome amplification accelerates identification of the apratoxin biosynthetic pathway from a complex microbial assemblage. *PLoS One*, **6** (4), 18565.

18
Marine Symbiotic Microorganisms: A New Dimension in Natural Product Research

S.W.A. Himaya and Se-Kwon Kim

18.1
Introduction

Marine environment covers over 70% of the earth surface and represents great biological diversity compared to terrestrial counterpart. Majority of animals and plants in the marine environment are soft-bodied sessile organisms and highly depend on a chemical defense mechanism. This chemical defense mechanism seems to be a generous gift for these organisms to survive in an extremely competitive marine environment. Moreover, prevailing extreme physiological conditions in the environment also have resulted in making marine organisms a rich source of chemically diverse group. However, limited accessibility to these sources kept the marine chemical diversity underneath the ocean till recent years. Even though some of the chemical constituents of marine origin were used, relatively easy accessible sources were subjected to exploration [1]. Werner Bergmann's landmark exploration of Caribbean marine sponge *Cryptotethya crypta* in 1945 eventually led to the development of ara-A (vidarabine) and ara-C (cytarabine): two nucleosides with significant anticancer and antiviral activity that were approved for clinical use as the first marine-derived natural product. This finding brought a considerable attention on marine sources for exploration of lead compounds for pharmaceuticals and cosmeceuticals [2]. The knowledge and technological developments in natural product explorations contributed to the present understanding of chemical diversity of the marine environment. As the data suggest, nearly 20 000 new marine natural products have been isolated during the last few decades. These isolated metabolites offer diverse chemical and biological properties. Especially, halogenated compounds have been considered as one of the specialties of marine-derived natural products. Several marine-derived natural products are currently used as pharmaceutical and cosmeceutical products. Moreover, many more marine-derived compounds are in clinical and preclinical trials as drug candidates [3].

Although marine flora and fauna are ideal sources of chemical compounds for improving health and well-being of humans, marine natural products have the

Marine Microbiology: Bioactive Compounds and Biotechnological Applications, First Edition.
Edited by Se-Kwon Kim

so-called supply problem. The supply problem has hampered the development of these chemical constituents as commercial products. Likewise, wild harvest of marine organisms for natural product extraction is not advisable as it threatens the marine biodiversity. As an alternative to these drawbacks, marine microorganisms have gained much attention. It is well known that marine microbial diversity is tremendously high and marine microorganisms are capable of producing a wide range of chemically diverse compounds. Recent reports have highlighted that symbiotic microorganisms isolated from marine sources produce chemical compounds, previously ascribed to host organism [4]. This finding has given a new insight into marine natural product exploration. In this chapter, marine-derived symbiotic microorganisms and their metabolites are discussed in detail with special emphasis on biological activities.

18.2 Marine Microorganisms and Their Symbiotic Relationships

Marine microorganisms represent a large ecological group rather than a taxonomical classification. Current estimates are reaching over millions of different microbial species, which account for more than 90% of the ocean's biomass. Till date a greater part of this diversity remains unknown because of difficulties in culturing them under laboratory culture conditions. So far, most probably, less than 0.1% of available microbial species in marine environment have been studied.

This tiny creature inhabits all kinds of the marine ecosystem, from ice-covered polar regions to the boiling hydrothermal vents in the deep sea. Even though the exact ecological role of marine microbes has not yet been fully understood, they play a pivotal role in marine food web and recycling nutrients. Recently, marine microorganisms, bacteria, archaea, microbial eukaryotes, and their associated viruses have gained increasing attention due to their great potential in natural product research [5]. Remarkably higher diversity of marine microorganisms has made them a treasure box for chemists to explore novel chemical entities. Moreover, both new and known chemical compounds of marine microbes with a potential reproducibility give a great capacity to study them as sources of lead compounds for pharmaceutical use as well as sources of other valuable products such as supplements, neutraceuticals, and cosmeceuticals [6].

As scientists assume, marine microbial species live in close association with sessile marine organisms and plants to survive in extreme conditions in marine environment. In particular, many bacteria and fungi live on the surface, or within the body or cells of macroorganisms such as sponge, coral, algae, and fish, because any surface immersed in sea is covered by a biofilm, and this biofilm provides good carbon source for microbial species. The relationship between microbe and higher organism in the ocean could be either mutualism or symbiosis. In many cases, marine microbes make some beneficial association with host organism. Several examples are available in literature to explain this beneficial association of marine microorganisms with host organism. Epibiotic bacteria on the larvae of some crustaceans provide protection against fungal infection by the production of simple antimicrobial compounds [7]. Many algal species use surface bacteria to prevent heavy biofouling. Some marine sponges, such as bacteriosponges, comprise up to 40% of microbial cells per milliliter of sponge tissue volume. This close and often long-term interaction provides many advantages to both parties. Most of the time, microbial species do fulfill their nutritional requirements (except photosynthetic microbes) from host organisms while providing protection to them.

It has been identified that marine microorganisms strengthen the host defense mechanism by producing various kinds of secondary metabolites. Especially antimicrobial and antifouling compounds serve as best protectors in the competitive marine environment. Because of the growing recognition that many of these defense metabolites are rich in chemical and biological properties, many researches have been done to culture these microbes in order to discover biologically active compounds. Moreover, compelling evidences prove that symbiotic microorganism is a true producer of some metabolites, previously ascribed to host organism [8]. A recent discovery revealed that a polyketide synthase (PKS) of bacterial symbiont *Candidatus Endobugula sertula* is responsible for the biosynthesis of potent anticancer drug bryostatin. For the first time bryostatin, a cyclic polyketide, was isolated from marine bryozoan. Further, the genomic analysis of cynobacterial symbiont *Prochloron* sp. clearly showed that patellamides A and C are metabolites of the microorganism not the product of the tunicate *Lissoclinum patella* [9]. All these findings highlighted that symbiotic microorganisms are fascinating sources of biologically active metabolites. Novel molecular approaches facilitated the identification of symbiotic marine microorganisms, and the development of culture techniques strengthened the journey.

18.3 Biologically Active Metabolites of Marine Symbiotic Microbes

Chemically diverse and biologically significant metabolites of marine symbiotic microbes gained considerable popularity among natural product researchers within a short period of time. As a result, various kinds of marine sources were explored for culturable microorganisms that have ability to produce bioactive metabolites. Marine algae, sponges, coral, and many marine invertebrates are a popular source for symbiotic microorganisms. Careful observations and isolations resulted in dozens of marine symbiotic microorganism, which have unique characteristics. These symbiotic microorganisms are either surface associated or endophytic. Among them bacteria, fungi, and microalgae have been identified as promising sources of biologically active metabolites. Both surface method and submerged cultivation are used in culturing of these symbionts for secondary metabolites extraction. Hildebrand *et al.* [10] comprehensively discussed approaches to identify marine symbionts for natural product extraction. Potent bioactive metabolites isolated from marine bacteria, fungi, and microalgae that have been isolated from different sources and their potentials are discussed hereafter.

18.3.1 Microorganisms Isolated from Marine Algae

The association of marine microorganisms and algae has shown some beneficial symbiotic relationship as well as harmful parasitic interactions. Symbiosis that strengthens them is more prominent since both of them struggle to survive in harsh marine environment [11]. Many marine algal species provide shelter for microorganisms as one of protective mechanisms against heavy biofouling. These epiphytic microorganisms get plenty of organic substances secreted by living or dead algae cells while producing various kinds of antifouling compounds. Some close observations have proved that many marine macroalgae are growing well in the presence of associated microorganisms. It is also proved that symbiotic microorganisms secrete many nutrients such as vitamins and growth factors that facilitate growth of host in addition to antifouling compounds [12]. Moreover, there are some microorganisms live as endophytics in marine algae. Endophytic interaction of marine algae and microorganisms has been observed long time before. Seaweeds provide an interesting biotic environment for these bacterial communities [13]. There are ample evidences to prove that both endophytic and epiphytic microorganisms of marine algae are rich sources of bioactive metabolites.

Since marine environment is full of various kinds of microorganisms, marine plants and animals are continually exposed to high concentration of pathogenic microorganisms. Thus, marine algae that lack cell-based immune system gain fitness through a chemical defense. In order to strengthen the chemical defense, many macroalgae provide shelters for symbiotic microorganisms. One of the pioneer studies conducted by Boyd *et al.* [14] isolated 280 pure epiphytic strains from marine algae, and these microbes showed significant potential to control the growth of fouling bacteria. These defense metabolites also elicit strong antimicrobial activity against clinically important microorganisms. The extracts obtained from the bacteria isolated from seaweed *Egregia menziesii, Codium fragile, Sargassum muticum, Endarachne binghamiae, Centroceras clavulatum,* and *Laurencia pacifica,* collected from Todos Santos Bay México, inhibited the growth of the Gram-negative bacterium *Proteus mirabilis.* Further investigations of bacteria associated with these seaweed surfaces revealed that bacteria of the family Firmicutes, Proteobacteria, and Actinobacteria produce compounds capable of inhibiting the growth of HCT-116 colorectal cancer cells [15]. In a detailed chemical study of the secondary metabolites of seaweed-associated myxobacterium revealed that at 2–3% NaCl concentration the microorganism produces haliangicin, which shows potent antifungal activity against important pathogenic strain *Phytophthora capsici* [16]. Kubanek *et al.* isolated 22-membered cyclic lactone, lobophorolide from seaweed *Lobophora variegata,* which shows activity against pathogenic and saprophytic marine fungi. Since the compound has structural characteristics of polyketide origin finally speculated that microbial symbiont of the seaweed could be responsible for the production of active compound [17]. In an investigation of epiphytic bacteria associated with surface of several brown algae species, it was found that 20% of the isolated strains show potent antibacterial activity. Primary disk-diffusion screening was conducted for screening the basic activity. The research team deeply investigated the antibacterial activity with a panel of bacterial species including some pathogenic bacteria. In conclusion, authors suggested that genus *Bacillus* associated with brown algae is a promising source for the isolation of antibacterial metabolite [18]. Moreover, marine fungal strains associated with algae have shown array of biological actives. Lateff *et al.* [19] isolated new antioxidant hydoquinone derivatives from an algicolous marine fungal strain. Another study conducted to screen fungal isolate associated with algae collected from Abou-keer, Alexanderia resulted in 13 secondary metabolites that shows antimicrobial and anticancer activities against several tumor cell lines [20]. All these literatures stress that marine alga-associated microorganisms are good candidates for lead compound explorations.

18.3.2
Marine Sponge-Associated Microorganism

Marine sponge harbors numerous microbial species throughout the body. In some cases population densities of sponge-associated microbes exceed those of seawater. Mesophyl matrix of sponge's body is a good shelter of extracellular heterotrophic and autotrophic microbes, while the outer layer is occupied by photosynthetic microbes. Thus in some sponges microbial communities contribute over 40–60% of the animal biomass [21]. Nevertheless, marine sponge-associated microbes gained interest recently with the main constraint of sponge-derived active metabolites, sustainable production of active metabolites. Isolation of marine *Micrococcus* sp. from sponge *Tedania ignis* that produces metabolites previously ascribed to the sponge steepened the interest of sponge-associated microbes for the production of biologically active secondary metabolites. If compounds are derived from a symbiotic microorganism, culturing the microorganism would provide an improved source of the bioactive compound [22]. Waters *et al.* [23] clearly indicated that there is considerable amount of evidence to prove that marine microbes are the true producers of many marine natural products used in clinical trials (Table 18.1). Thus, marine sponge-associated microbes are a fascinating source to explore

Table 18.1 Marine-derived compounds currently in clinical trials that have evidences of microbial origin [23].

Clinical status	Trade mark	Compound	Biological activity	Microorganism
Phase III	Soblidotin	Peptide	Anticancer	Bacteria
Phase II	Plinabulin	Diketopiperazine	Anticancer	Fungi
	Isokahalide	Depsipeptide	Anticancer	Bacteria
	Tasidotin	Peptide	Anticancer	Bacteria
Phase I	Bryostatin 1	Bryostatin 1	Anticancer	Bacteria
	Marizomib	Beta-lactone–gamma-lactam	Anticancer	Bacteria

biologically active metabolites because marine sponges have already proved their richness in chemically diverse metabolites. Followings are only few recent findings to prove the potency of sponge-associated microbes to produce a wide array of biologically active metabolites.

Microbial communities associated with marine sponges are well-known producers of novel bioactive compounds. Both marine fungi and bacteria have been reported as producers of these metabolites. Eurocristatine is a recently identified novel diketopiperazine from marine sponge-associated fungus *Eurotium cristatum*. This species is a sexual state of *Aspergillus* species. In this experiment another eight known, biologically active compounds were identified in addition to the new compound [24]. Lee *et al.* [25] discovered a new cytotoxic peptide from sponge-derived fungus *Aspergillus versicolor.* The ethyl acetate extract of fungal strain isolated from a marine sponge *Petrosia* sp. has been subjected to bioactive-guided fractionation. The potent cytotoxic fraction among the others resulted in the discovery of a new derivative of Fellutamide that has significant cytotoxicity against solid tumor cell lines. Efrapeptins Eα, H-, and N-methylated octapeptides RHM3 and RHM4 are other examples of marine fungi-derived bioactive peptides. These peptides exhibit potent anticancer activity against HCT-116 colon cancer cells at nanomolar concentrations [26]. Small peptides are among the highest ranked candidates to be developed as lead compounds due to their specificity in biological activity and simplicity in chemical synthesis. Further, two novel cyclodepsipeptides, scopularides A and B, were obtained from fungus *Scopulariopsis brevicaulis*, which was isolated from the marine sponge *Tethya aurantium*. The structural novelty of this peptide is sufficient to explain the potential of marine sponge-derived fungal strain to be a good candidate of structurally diverse compounds [27].

Marine bacterial species have received considerable attention for marine microbial research due to their immense capability to produce novel bioactive compounds. The best example is salinosporamide A that has recently been approved for phase I clinical trials for the treatment of multiple myeloma. The compound was approved for the clinical trials in less than 3 years after its discovery from marine actinobacterial strain *Salinispora tropica* [28]. Sponge-associated bacterial community has shown great potential as a source of antibiotics. Especially, symbiotic actinomycetes are noteworthy producers of potent antimicrobial compounds. For example, Kim *et al.* [29] discovered that marine sponge-associated actinobacterial *Salinispora* strains have PKS gene sequence that is most closely related to rifamycin B synthase, which is responsible for the production of clinically important antibiotics rifamycins. Till date rifamycin classes are semisynthesized, and the effectiveness of some has been threatened by the development of resistance. Thus, the isolated strain and its host would be a potential source to develop as new source for the production of rifamycins as well as new compounds with improved antibiotics properties. Moreover, studies of Abdelmohsen *et al.* [30] and Selvin *et al.* [31] have given some information about diversity of Actinobacteria, having ability to produce novel antimicrobial agents, associated with marine sponge.

Interestingly, sponge-associated bacteria also have been proved as potent anticancer metabolites producers. Bendigoles D–F are recently identified anticancer sterols, produced by sponge-derived *Actinomadura* sp. For initial screening of the active compound, novel high-content screen for NF-κB and glucocorticoid receptor (GR) activity have been used. Final result proved that the bendigoles D is a potent cytotoxic compound (with IC_{50} value of 30 μM) and it blocks the nuclear translocation of NF-κB [32]. Further, actinobacterial species *Micromonospora* isolated from marine sponges *Clathrina coriacea* have produced strong antitumor indolocarbazole alkaloids. Detailed biological activity studies suggested that the sugar moiety of compounds is the important part for their strongest activity [33]. Another striking analysis of marine Actinobacteria associated with the sponge *Halichondria panicea* revealed that 30 strains out of 46 isolated strains consist of biosynthesis genes encoding PKS and nonribosomal peptide synthetases (NRPSs), which provide good indication of the pharmaceutical potential of these strains. This result was confirmed by isolating potent cytotoxic alkaloids, polyketides, and macrolide from culture extracts [34].

18.3.3
Marine Invertebrate-Associated Microorganisms

Even though marine sponge belongs to the group of marine invertebrates, it was separately discussed above to elaborate the interesting association of marine sponge and symbiotic microorganism. Here we discuss the diversity of other marine invertebrate-associated microorganisms and the potentials of their metabolic products in lead compound exploration.

All three domains of life, bacteria, archaea, and eukaryotes, reside within marine invertebrates. Now it is not a surprise that these microbial symbionts are responsible for the high biosynthetic virtuosity of marine invertebrates. Further, most phyla of marine invertebrates, Porifera, Cnidaria, Bryozoa, Mollusca, Pogonophora, Echinodermata, Urochordata, and Crustacea have been investigated for microbial symbiosis. In particularly, bryozoans and sponges have become model organisms to study symbiotic relationships of marine invertebrates and explain why most of the sessile invertebrates are rich in unusual chemical metabolites [35].

Bryozoans are well-known producers of various kinds of alkaloids. Sponge's bryozoans-associated microbial symbionts also have been implicated in biosynthesis of active metabolites. As mentioned previously, the best available example is bryostatin 1, a potent modulator of protein kinase C (PKC), which was first isolated in the 1960s by George Pettit from extracts of a species of bryozoan. Recently this compound has passed several clinical trials as anticancer agent and memory enhancer. However, recent molecular evidences prove that this potent bioactive compound is a metabolic product of a microbial symbiont associated with the bryozoa [36]. Moreover, there is enough experimental evidence to reveal the chemical diversity of bryozoan-associated microbes [37].

Ascidians, well known as tunicates, are another popular source of bioactive natural products. Ecteinascidin 743, which is an approved anticancer agent in Europe, and few more compounds of ascidian origin are now in advanced clinical trials. Hence, this small organism has gained much attention in natural product research. Ascidian's tough outer polysaccharide cover provides a good shelter for symbiotic microbes. The symbiotic relationship between didemnid ascidians and photosynthetic cyanobacterial sp. *Prochloron* symbiotic is a well-studied interaction of this group [38]. Interestingly, it has been shown that some bacterial symbionts of ascidians produce some bioactive compounds, previously isolated from whole-organism extracts. Compelling evidences show that a series of ascidian's polyketides, small peptides, and alkaloids have microbial origin. Structural similarities, genome sequencing, protein profiling, and pure culture experiments have brought enough facts to prove the strength of ascidian-associated microbial communities in producing potent bioactive metabolites [39]. Thus, available literatures highlight that many potently bioactive natural products have microbial origin, and isolation and culturing of these microbes would add new value to the marine natural product researches.

In addition, several other marine invertebrates including organisms with hard exoskeleton have been studied for their chemical potentials as well as interactions with microorganisms. In most cases, symbiotic microbes have been recognized as true producers of biologically active metabolites. Thus, it is noteworthy to hypothesize that detailed study of marine invertebrates such as corals, jellyfish, sea anemones, sea warms, and so on is useful in marine biotechnology.

18.3.4
Microbial Association with Mangrove

Mangrove, a highly productive ecosystem that covers roughly 60–75% of the world's tropical coastlines, provides unique ecological space to various microbial species. Especially, the mutualistic bacterial interaction with mangroves has been highlighted due to the important services of bacteria such as N-fixation. In some cases fungal species also have shown similar interactions with mangrove. All these interactions have been described in order to explain the ecological role of mangrove-associated microbes [40]. Nevertheless, limited amount of nutrient content in mangrove has created considerable competition among these microbial species. In this regard, mangrove-associated microbial communities produce diverse chemical compounds for their survival. Thus, these microbial species have earned significant attention in natural product research as sources of novel metabolites.

A few studies, with special attention to chemical metabolites of microbial species of mangrove, have been conducted, and a wide range of biologically active secondary metabolites have been identified. Among the identified metabolites, anticancer potentials of mangrove-associated fungal species have taken great interest. Tao *et al.* [41] described 87 different metabolites of mangrove-associated fungi. Finally, the author suggested that these metabolites show great potential as anticancer drug leads for patients with multidrug resistance.

Moreover, antimicrobial compounds isolated from mangrove-derived microbial stains have given new insight into antimicrobial research of pathogenic strains of animals as well as plants [42].

18.4 Concluding Remarks

Symbiotic microbial communities of marine origin undoubtedly brought a new wave to marine natural product research. With respect to lead compound exploration, this is an indispensable source of potent biologically active compounds and could be a sustainable approach to explore marine chemical diversity. However, till date only very few available species have been fully explored for their biotechnological potentials. Some unique properties of these symbionts, especially in culture-dependent experiments, have hindered their potential applications not only in drug development but even in explorations. In this regard, advanced biotechnological approaches would be required to explore true potentials of these organisms. Improved culture techniques will bring great value to marine microbial researches. Further, identification of genetic relationships of biosynthesis of potent bioactive compounds will pave another way to hidden potentials of symbiotic microbes. It will facilitate identification, cloning, and expressing of bioactive metabolite genes of symbionts. Interestingly, polyketide synthase and nonribosomal peptide synthetase gene probes have already taken a part of identification of biotechnologically important microbial symbionts. Thus, we strongly believe that marine symbiotic microbes will be a new paradigm of marine natural product research.

References

1 Kornprobst, J.M. (2010) *Encyclopedia of Marine Natural Product*, Wiley-VCH Verlag GmbH, Weinheim.
2 Baby, J. and Sujatha, S. (2010) Pharmacologically important natural products from marine sponges. *J. Nat. Prod.*, **4**, 5–12.
3 Blunt, J.W., Copp, B.R., Munro, M.H.G., Northcote, P.T., and Prinsep, M.R. (2011) Marine natural products. *Nat. Prod. Rep.*, **28**, 196–268.
4 Waters, A.L., Hill, R.T., Place, A.R., and Hamann, M.T. (2010) The expanding role of marine microbes in pharmaceutical development. *Curr. Opin. Biotechnol.*, **21**, 780–786.
5 Imhoff, F.J., Labes, A., and Wiese, J. (2011) Bio-mining the microbial treasures of the ocean: new natural products. *Biotechnol. Adv.*, **29**, 468–482.
6 Penesyan, A., Kjelleberg, S., and Egan, S. (2010) Development of novel drugs from marine surface associated microorganisms. *Mar. Drugs*, **8**, 438–459.
7 Armstrong, E., Yan, L., Boyd, K.G., Wright, P.C., and Burgess, J.G. (2001) The symbiotic role of marine microbes on living surfaces. *Hydrobiologia*, **461**, 37–40.
8 Egan, S., Thomas, T., and Kjelleberg, S. (2008) Unlocking the diversity and biotechnological potential of marine surface associated microbial communities. *Curr. Opin. Microbiol.*, **11**, 219–225.
9 Schmidt, E.W., Nelson, J.T., Rasko, D.A., Sudek, S., Eisen, J.A., Haygood, M.G., and Ravel, J. (2005) Patellamide A and C biosynthesis by a microcin-like pathway in *Prochloron didemni*, the cyanobacterial symbiont of *Lissoclinum patella*. *Proc. Natl. Acad. Sci. USA*, **102**, 7315–7320.

10 Hildebrand, M., Waggoner, L.E., Lim, G.E., Sharp, K.H., Ridley, C.P., and Haygood, M.G. (2003) Approaches to identify, clone, and express symbiont bioactive metabolite genes. *Nat. Prod. Rep.*, **21**, 122–142.

11 Goecke, F., Labes, A., Wiese, J., and Imhoff, J.F. (2010) Chemical interactions between marine macroalgae and bacteria. *Mar. Ecol. Prog. Ser.*, **409**, 267–299.

12 Harder, T. (2009) Marine epibiosis: concepts, ecological consequences and host defence, in *Marine and Industrial Biofouling* (ed. J.W. Costerton), Springer-Verlag, Berlin, pp. 219–231.

13 Dawes, C.J. and Lohr, C.A. (1978) Cytoplasmic organization and endosymbiotic bacteria in the growing points of *Caulerpa prolifera*. *Rev. Algol.*, **13**, 309–314.

14 Boyd, K.G., Adams, D.R., and Burgess, J.G. (1999) Antibacterial and repellent activities of marine bacteria associated with algal surfaces. *Biofouling*, **14**, 227–236.

15 Gomez, L.J.V., Mercado, I.E.S., Rivas, G.G., and Sanchez, N.E.A. (2010) Antibacterial and anticancer activity of seaweeds and bacteria associated with their surface. *Rev. Biol. Mar. Oceanogr.*, **45**, 267–275.

16 Kundim, B.A., Itou, Y., Sakagami, Y., Fudou, R., Iizuka, T., Yamanaka, S., and Ojika, M. (2003) New haliangicin isomers, potent antifungal metabolites produced by a marine myxobacterium. *J. Antibiot. (Tokyo)*, **56**, 630–638.

17 Kubanek, J., Jensen, P.R., Keifer, P.A., Sullards, M.C., Collins, D.O., and Fenical, W. (2003) Seaweed resistance to microbial attack: a targeted chemical defence against marine fungi. *Proc. Natl. Acad. Sci. USA*, **100**, 6916–6921.

18 Kanagasabhapathy, M., Sasaki, H., Haldar, S., Yamasaki, S., and Nagata, S. (2006) Antibacterial activities of marine epibiotic bacteria isolated from brown algae of Japan. *Ann. Microbiol.*, **56**, 167–173.

19 Lateff, A.A., Konig, G.M., Fisch, K.M., Holler, U., Jones, P.G., and Wright, A.D. (2002) New antioxidant hydroquinone derivatives from the algicolous marine fungus *Acremonium* sp. *Nat. Prod. Rep.*, **65**, 1605–1611.

20 Mabrouk, A.M., Kheiralla, Z.H., Hamed, E.R., Youssry, A.A., and Aty, A.A.A.E. (2008) Production of some biologically active secondary metabolites from marine-derived fungus *Varicosporina ramulosa*. *Malaysian J. Microbiol.*, **4**, 14–24.

21 Grozdanov, L. and Hentschel, U. (2005) An environmental genomics perspective on the diversity and function of marine sponge-associated microbiota. *Curr. Opin. Microbiol.*, **10**, 215–220.

22 Stierle, A.A., Cardellina, H., and Singleton, F.L. (1991) Benzothiazoles from a putative bacterial symbiont of the marine sponge. *Tetrahedron Lett.*, **32**, 4847–4848.

23 Waters, A.L., Hill, R.T., Place, A.R., and Hamann, M.T. (2010) The expanding role of marine microbes in pharmaceutical development. *Curr. Opin. Biotechnol.*, **21**, 780–786.

24 Gomes, N.M., Dethoup, T., Singburaudom, N., Gales, L., Silva, A.M.S., and Kijjoa, A. (2012) Eurocristatine, a new diketopiperazine dimer from the marine sponge-associated fungus *Eurotium cristatum*. *Photochem. Lett.* doi. org/10.1016/j.phytol.2012.07.010.

25 Lee, Y.M., Dang, H.T., Hong, J., Lee, C., Bae, K.S., Kim, D.K., and Jung, J.H. (2010) A cytotoxic lipopeptide from the sponge-derived fungus *Aspergillus versicolor*. *Bull. Korean Chem. Soc.*, **31**, 205–208.

26 Boota, C.M., Amagataa, T., Tenneya, K., Comptona, J.E., Pietraszkiewiczb, H., Valerioteb, F.A., and Crewsa, C.P. (2007) Four classes of structurally unusual peptides from two marine-derived fungi: structures and bioactivities. *Tetrahedron*, **63**, 9903–9914.

27 Yu, Z., Lang, G., Kajahn, I., Schmaljohann, R., and Imhoff, J.F. (2008) Scopularides A and B, cyclodepsipeptides from a marine sponge-derived fungus, *Scopulariopsis brevicaulis*. *J. Nat. Prod.*, **71**, 1052–1054.

28 Bull, A.T. and Stach, J.E.M. (2007) Marine actinobacteria: new opportunities for natural product search and discovery. *Trends Microbiol.*, **15**, 491–499.

29 Kim, T.K., Hewavitharana, A.K., Shaw, P.N., and Fuerst, J.A. (2006) Discovery of a new source of rifamycin antibiotics in marine sponge actinobacteria by phylogenetic

prediction. *Appl. Environ. Microbiol.*, **72**, 2118–2125.

30 Abdelmohsen, U.A., Elardo, S.M.P., Hanora, A., Radwan, M., Ela, S.H.A.E., Ahmed, S., and Hentschel, U. (2010) Isolation, phylogenetic analysis and anti-infective activity screening of marine sponge-associated actinomycetes. *Mar. Drugs*, **8**, 399–412.

31 Selvin, S., Shanmughapriya, S., Gandhimath, R., Kiran, G.S., Ravji, T.R., Natarajaseenivasan, K., and Hema, T.A. (2009) Optimization and production of novel antimicrobial agents from sponge associated marine actinomycetes *Nocardiopsis dassonvillei* MAD08. *Appl. Microbiol. Biotechnol.*, **83**, 435–445.

32 Simmons, L., Kaufmann, K., Garcia, R., Schwar, A., Huch, V., and Muller, R. (2011) Bendigoles D–F, bioactive sterols from the marine sponge-derived *Actinomadura* sp. SBMs009. *Bioorg. Med. Chem.*, **19**, 6570–6575.

33 Hernandez, L.M.C., Blanco, J.L.D.L.F., Baz, J.P., Puentes, J.L.F., Millan, F.R., Vazquez, F.E., and Chimeno, R.I.F. (2000) 4′-7V-Methyl-5′-hydroxystaurosporine and 5′hydroxystaurosporine, new indolocarbazole alkaloids from a marine *Micromonospora* sp. strain. *J. Antibiot. (Tokyo)*, **53**, 895–902.

34 Schneemann, I., Nagel, K., Kajahn, I., Labes, A., Wiese, J., and Imhoff, J.F. (2010) Comprehensive investigation of marine Actinobacteria associated with the sponge *Halichondria panicea*. *J. Appl. Microbiol.*, **76**, 3702–3714.

35 Haygood, M.G., Schmidt, E.W., Davidson, S.K., and Faulkner, D.J. (1999) Microbial symbionts of marine invertebrates: opportunities for microbial biotechnology. *J. Mol. Microbiol. Biotechnol.*, **1**, 33–43.

36 Keck, G.E., Poudel, Y.B., Cummins, T.J., Rudra, A., and Covel, J.A. (2011) Total synthesis of bryostatin 1. *J. Am. Chem. Soc.*, **133**, 744–747.

37 Heindl, H., Wiese, J., Thiel, V., and Imhoff, J.F. (2010) Phylogenetic diversity and antimicrobial activities of bryozoan-associated bacteria isolated from Mediterranean and Baltic Sea habitats. *Syst. Appl. Microbiol.*, **33**, 94–104.

38 Gault, P.M. and Marler, H.J. (2009) *Handbook on Cyanobacteria: Biochemistry, Biotechnology and Applications*, Nova Science Publishers, New York.

39 Schmidt, E.W. and Donia, M.S. (2010) Life in cellulose houses: symbiotic bacterial biosynthesis of ascidian drugs and drug leads. *Curr. Opin. Biotechnol.*, **21**, 827–833.

40 Sahoo, K. and Dhal, N.K. (2009) Potential microbial diversity of mangrove ecosystems: a review. *Indian J. Mar. Sci.*, **38**, 249–256.

41 Tao, L.Y., Zhang, J.Y., Liang, Y.J., Chen, L.M., Zhen, L.S., Wang, F., Mi, Y.J., She, Z.G., To, K.K.W., Lin, Y.C., and Fu, L.W. (2010) Anticancer effect and structure–activity analysis of marine products isolated from metabolites of mangrove fungi in the south china sea. *Mar. Drugs*, **8**, 1094–1105.

42 Hu, H.Q., Li, X.S., and He, H. (2010) Characterization of an antimicrobial material from a newly isolated *Bacillus amyloliquefaciens* from mangrove for biocontrol of *Capsicum* bacterial wilt. *Biol. Control*, **54**, 359–365.

19
Application of Probiotics from Marine Microbes for Sustainable Marine Aquaculture Development

Nguyen Van Duy, Le Minh Hoang, and Trang Sy Trung

19.1
Introduction

19.1.1
The Concept of Sustainable Development in Aquaculture

Aquaculture is currently one of the fastest growing food production systems in the world. Most of the global aquaculture output is produced in developing countries and significantly in low-income food-deficit countries [1]. According to the Food and Agricultural Organization (FAO) of the United Nations, Aquaculture is "Farming of aquatic organisms including fish, mollusks, crustacean and plants, with some form of intervention in the rearing process to enhance production, such as regular stocking, feeding, protection from predators." Aquaculture consists of cultivating freshwater or seawater organisms under controlled conditions by humans, and can be contrasted with commercial fishing, which is the harvesting of natural (wild) fish.

Sustainable development is "Development which meets the present without compromising the ability of future generations to meet their own needs" [2]. According to this definition of sustainable development, a sustainable aquaculture should be environmentally acceptable, economically viable, and socially equitable. However, even if these principles are clear, their application is not so straightforward.

Environmental acceptability is the most difficult component of the definition. Generally, aquaculture as a human activity has to take into consideration other human activities occurring in the same area, as well as acceptability being linked to the participation of all stakeholders. Furthermore, in order to understand what would be environmentally acceptable, the ecosystem where the activity takes place has to be identified and understood to the greatest possible degree.

Economic viability is the most obvious element of the definition, but this concept is deeply linked with the economic system of the country where the development

Marine Microbiology: Bioactive Compounds and Biotechnological Applications, First Edition.
Edited by Se-Kwon Kim

takes place. The former is the process by which an economic activity obtains all the tools and knowledge necessary to operate successfully and reach an adequate level of maturity; the latter is the process that can be indefinite in a finite world.

Social equity or fairness is the most viable aspect of the definition. It depends greatly on the social parameters of the society where the activity takes place. It is the most difficult to use because of its intrinsic variability.

19.1.2 Freshwater and Marine Culture for a Sustainable Development

Sustainable freshwater aquaculture is a mean to profitable production, efficient use of national resources, the best aquatic environmental practices, and a way forward to keep pace with aquaculture consumption without compromising the overall ecological integrity of our ecosystem. This will be considered for diversification of cultured species, ecological filtration techniques, ecological methods of fish disease control, and breeding and rearing methods [3].

Marine aquaculture holds great promise throughout the world. However, a host of physical, environmental, social, economic, legal, and institutional factors must be addressed if marine aquaculture is to realize this potential. Applying the principles of sustainable development to aquaculture presents many challenges to the practitioner. There are no easy solutions to the issues related to sustainable development of marine aquaculture [2,4].

19.1.3 Current Approaches for Sustainable Marine Aquaculture Development

Classical antibiotics appeared ineffective due to overuse in aquaculture. It not only increased disease resistance by bacteria, damaged normal microflora, and caused microdysbiosis as double pollution but also made antibiotic residue, accumulated in aquatic products, harmful for human consumption. Therefore, scientific communities have proposed friendly alternatives such as vaccines, antibiotic substitutes, or use of prebiotics and probiotics.

19.1.3.1 Vaccine

Vaccination is an easy, effective, and preventive method of protecting fish from diseases. Vaccination is a process by which a protective immune response is induced in an animal by administration of vaccines. Vaccines are preparations of antigens derived from pathogenic organisms, rendered nonpathogenic by various means, which will stimulate immune system of the animal to increase the resistance to the disease on natural encounter with pathogens. Once stimulated by a vaccine, the antibody-producing cells, called B lymphocytes, remain sensitized and ready to respond to the agent should it ever gain entry to the body. During the last two decades vaccination has become established as an important method for prevention of infectious diseases in farmed fish, mainly salmonid species. So far, most commercial vaccines have been inactivated vaccines, administered by injection or immersion. Bacterial infections caused by Gram-negative bacteria such as *Vibrio* sp., *Aeromonas* sp., and *Yersinia* sp. have been effectively controlled by vaccination. With furunculosis, the success is attributed to the use of injectable vaccines containing adjuvants. Vaccines against virus infections, including infectious pancreatic necrosis, have also been used in commercial fish farming. Vaccines against several other bacterial and viral infections have been studied and found to be technically feasible. The overall positive effect of vaccination in farmed fish is reduced mortality. However, for the future of the fish farming industry it is also important that vaccination contributes to a sustainable biological production with negligible consumption of antibiotics [5].

19.1.3.2 **Antibiotic Alternatives**

The wide and frequent use of antibiotics in aquaculture has resulted in the development and spread of antibiotic resistance. Because of the health risks associated with the use of antibiotics in animal production, there is a growing awareness that antibiotics should be used with more care. This is reflected in the recent implementation of more strict regulations on the prophylactic use of antibiotics and the presence of antibiotic residues in aquaculture products. For the sustainable further development of the aquaculture industry, novel strategies to control bacterial infections are needed. This review evaluates several alternative biocontrol measures that have emerged recently. Most of these methods are still in research phase; few have been tested in real aquaculture settings. It is important to further develop different strategies that may be combined or used in rotation in order to maximize the chance of successfully protecting the animals and to prevent resistance development [6]. To keep a sustainable growth pattern, health management strategies must go beyond antibiotics and chemotherapeutics that create resistance in the bacteria and immunosuppression in the host. Besides development of drug-resistant bacteria and pathogens, the adverse effect of antibiotics is caused by their influence on the aquatic microflora and the retention of harmful residues in aquatic animals. Because of this, certain antibiotics such as chloramphenicol have been banned in many countries [7].

19.1.3.3 **Prebiotics**

Prebiotics defined by Gibson and Roberfroid [8] are "nondigestible ingredients that beneficially affect the host by selectively stimulating growth and/or activity of one or a limited number of bacteria in the colon, and thus improves the host health." Inulin, fructooligosaccharides (FOS), mannan oligosaccharides (MOS), galactooligosaccharides (GOS), xylooligosaccharides (XOS), arabinoxylooligosaccharides (AXOS), and isomaltooligosaccharides (IMO) have been considered as prebiotics and used in aquaculture [9,10]. Ringo *et al.* [10] reviewed the role of prebiotics in fish and shellfish: effect on growth, feed conversion, gut microbiota, cell damage/morphology, resistance against pathogenic bacteria, and innate immune parameters. Common prebiotics have been extracted from plants, macroalgae, and microbes but data on prebiotics extracted from marine microbes are very limited.

More research efforts are needed to exploit the potential of marine prebiotics, especially from tropical marine regions.

19.1.3.4 **Probiotics**

The term probiotics was coined by Parker [11] to describe organisms and substances that contribute to intestinal microbial balance. Probiotics are viable cultures of bacteria and fungi that, when introduced through feed, have a positive effect on health. According to the currently adopted definition by FAO/WHO: probiotics are "Live microorganisms which when administered in adequate amounts confer a health benefit on the host." Use of probiotics has been highlighted and reviewed by many workers [7]. Improved resistance against infectious diseases can be achieved by the use of probiotics.

Compared to probiotic administration, the use of vaccines is more directly effective but often laborious, costly, and highly stressful to the animals. Currently, Zokaeifar *et al.* [12] isolated and screened two potential probiotics *Bacillus subtilis* L10 and G1 with antagonistic activity against *Vibrio harveyi* and *Vibrio parahaemolyticus*, and nonpathogenic effect in juvenile shrimp *Litopenaeus vannamei*. The result also showed proteinaceous nature of the antibacterial substances from these strains, which demonstrated that they can produce bacteriocins or bacteriocin-like inhibitory substances (BLISs). Bacteriocinogenic bacterial strains could be an excellent candidate with dual role because bacteriocin would be used as a human-safe and environmentally friendly antibiotic substitute, whereas bacteria would be a potential probiotic [13]. This chapter reviews the application of probiotics in the marine culture of fishes, crustaceans, and mollusks with a particular discussion on cultured species, administrated popularly by probiotics.

19.2 The Application of Probiotics for Marine Fishes

19.2.1 Turbot (*Scophthalmus maximus*)

In 1989, Gatesoupe and coworkers started their long-term studies on the development of probiotics for improving the growth and survival of turbot (*S. maximus*) larvae with rotifers (*Brachionus plicatilis*) used as live food. Typical examples of developed probiotics include *Bacillus toyoi*, *Bacillus* sp. IP5832 spores, *Vibrio alginolyticus*, *Vibrio* sp. E, *Streptococcus thermophilus*, *Lactobacillus helveticus*, *Lactobacillus plantarum*, and *Carnobacterium* sp., which have been shown to enhance growth and/or survival rate after challenging with *Vibrio* spp. [14,15] (Table 19.1). The mode of action for these probionts was proposed to be antagonistic activity and/or improving the nutritional value of the rotifers, or competition for iron. After two decades of research, Gatesoupe appreciates the role of lactic acid bacteria in improving the turbot larvae health [16]. For example, the addition of *Streptococcus lactis* and *Lactobacillus bulgaricus* to rotifers, and *Artemia* used in turbot larva feeding, enhanced the survival rate of the turbot larvae [17].

Table 19.1 The application of probiotics for turbot (*S. maximus*).

Probiotic	Origin	Treatment method	Effect on host animal	Pathogen	Mode of action	References
B. toyoi		Added to rotifers	Enhanced larval growth			[14]
V. alginolyticus			Enhanced survival rate			[18]
Bacillus sp. IP5832 spores			Enhanced growth and survival rate after challenging with *Vibrio*	*Vibrio* sp.	Antagonism	[19]
					Nutrition support for rotifers	
S. thermophilus, *L. helveticus*, and *L. plantarum*		Added to rotifers	Enhanced survival rate			[20]
S. lactis and *L. bulgaricus*		Added to rotifers and *Artemia*	Enhanced survival rate			[17]
L. plantarum and *Carnobacterium* sp.	Rotifers	Added to rotifers	Enhanced survival rate after challenge with *Vibrio* sp.	*Vibrio* sp.	Antagonism	[21]
					Nutrition support for rotifers	
Vibrio sp. E	Healthy turbot larvae	Added to rotifers	Enhanced survival and growth rate after challenge with *Vibrio splendidus*	*V. splendidus*	Competition for iron	[15]
V. pelagius	Rearing water, healthy turbot larvae	Added to rearing water	Enhanced larval survival when added alone or in combination with *A. caviae*			[22]

(*continued*)

Table 19.1 (*Continued*)

Probiotic	Origin	Treatment method	Effect on host animal	Pathogen	Mode of action	References
C. divergens and *V. pelagius*	Rearing water, healthy turbot larvae	Added to rearing water	Beneficial effect of *C. divergens* on survival inconclusive from *in vivo* study			[23]
Vibrio mediterranei Q40	Sea bream larvae	Added to rotifers	Reduced colonization of gut opportunistic microflora		Mucosal colonization	[24]
Roseobacter sp. 27-4, *Roseobacter* spp.	Spanish turbot larvae rearing farms		Enhanced survival rate	*V. anguillarium*	Antagonism	[25–27]
			Not lethal to egg yolk sac turbot larvae			
Phaeobacter spp.	Danish turbot larval farms		Antibacterial activity under a broader range of growth conditions than did *Phaeobacter* strain 27-4	*V. anguillarum*	Antagonism	[28]
					Tropodithietic acid production	
					Brown pigment production	

The hypothesis was that the lactic acid bacteria would act as antagonists to pathogens or produce bacteriocins as antibacterial peptides. Recent evidences are provided that lactic acid bacteria can stimulate the immune system in fish [16], provoking a regained interest in allochthonous strains used as probiotics in marine aquaculture.

In contrast to Gatesoupe's approach, Ringo *et al.* have focused their research on the intestinal microflora of turbot larvae and fry to screen suitable probiotics. The establishment of a balanced gut flora in fish larvae is assumed to be influenced by the microflora of the egg, the live feed, and the bacteria present in the rearing water. Although very few bacteria can be detected in the intestine of newly hatched larvae, it is rapidly colonized during the first few days. Therefore, seeding the gut with harmless bacteria that occupy the attachment sites may prevent pathogen infection [23]. The introduction of *Vibrio pelagius* seemed to improve larval survival as compared to fish exposed to *Aeromonas caviae* and with the control group [22,23]; thus, perhaps this promising probiotic had to compete with the microbiota already present in the larval gut for suitable attachment sites. However, a new strain of *V. pelagius* (Hq 222) was then found to be very virulent for larvae and postlarvae [29]. Clearly, an *in vivo* assessment to evaluate the positive or negative effect of a potential probiotic should be carried out carefully instead of bacterial identification only.

In 2004, the research group of Hjelm [25] developed a long-term, large-scale screening based on *in vitro* antibacterial activity against *Vibrio anguillarum*, resulting in the introduction of a promising probiotic candidate, *Roseobacter* strains. *Roseobacter* sp. strain 27-4 was found to be nonlethal to egg yolk sac turbot larvae, and to reduce the mortality of turbot larvae [25,26]. Interestingly, this strain produces acylated homoserine lactone in culture media; thus, the production of antibacterial compounds in this strain may be quorum regulated [27]. Recently, Porsby *et al.* [28] have shown that bacteria from the *Roseobacter clade* appear to be universal colonizers of marine larval rearing units. Roseobacters isolated from a Spanish turbot larval unit with antagonistic activity against *V. anguillarum* also colonize a Danish turbot larval farm that relies on a very different water source, in which the Danish *Phaeobacter* spp. displayed antibacterial activity under a broader range of growth conditions than did *Phaeobacter* strain 27-4. Thus, *Phaeobacter* strains are expected to be used as future probiotics for turbot larvae.

19.2.2 Atlantic Cod (*Gadus morhua*)

Viral and bacterial diseases often occur in the cultivation of Atlantic cod (*G. morhua)*, resulting in heavy losses, especially at the larval and juvenile stages. Vibriosis has long been the most important bacterial disease in cod, with *V. anguillarum* dominant among pathogenic isolates [30]. Hansen and Olafsen [31] first attempted to manipulate the egg microbiota of Atlantic cod by bathing gnotobiotic eggs in the suspension of some bacterial strains isolated from eggs of cod and halibut (Table 19.2). Although these bacteria failed to prevent the adherence of environmental bacteria to cod eggs, this is a good approach for preselection that requires a further step to confirm the

Table 19.2 The application of probiotics for Atlantic cod (*G. morhua*).

Probiotic	Origin	Treatment method	Effect on host animal	Pathogen	Mode of action	References
Several bacterial strains	Eggs of cod and halibut	Bathing in bacterial suspension	Failed to prevent the adhesion of environmental bacteria to cod eggs		Antagonism	[31]
L. plantarum		Added into water	Reduced opportunistic colonization by 70%		Mucosal colonization	[33]
C. divergens			Improved survival after challenge with *V. anguillarum*	*V. anguillarum*		
Four strains of lactic acid bacteria	Cod rearing environment			*V. anguillarum*, *A. salmonicida*, and *V. salmonicida*	Antagonism	[36]
					Metabolite production	
					Adhesion to fish cells	
Probiotics			Repressed some proteins related to stress and immune responses			[37]
			Induced proteins related to growth and development			
Enterococcus sp.	Cod rearing environment	Added to rearing water	Enhanced larval growth, performance and microflora control			[35,38]
Probiotics (GP21 and GP12)	Cod gut		Enhanced the adhesion to cod gut and competition exclusion with pathogens	*V. anguillarum* and *A. salmonicida*	Adhesion to mucus,	[39]
					Competition exclusion	
Five strains: *Vibrio*, *Microbacterium* spp., *Ruegeria*, *Pseudoalteromonas*	*Codlarvae*		Enhanced the survival of larvae	*V. anguillarum*	Diverse mechanisms	[40]

effect of selected bacteria [32]. Based on this approach, later studies demonstrated the probiotic effect of two strains of lactic acid bacteria, *L. plantarum*, reducing opportunistic colonization by 70% in cod larvae [33], and *Carnobacterium divergens*, improving the survival rate of cod fry after challenge with *V. anguillarum* [34]. Moreover, another lactic acid bacterium, *Enterococcus* sp., was shown to enhance larval growth, survival rate, hypersalinity tolerance, and physiological measurements but lower *Vibrio* levels in the larvae [35]. These results indicate the important potential of lactic acid bacteria in cod larvicultivation, and the administration of these probiotics perhaps promotes larval growth, performance, development, stress tolerance, and pathogen control at early stages, thus increasing production yield in intensive culture.

Fjellheim *et al.* [40] showed that the introduction of three dominant bacteria (*Vibrio* sp. and *Microbacterium* spp.) and two strains with high antagonistic activity (*Ruegeria* and *Pseudoalteromonas*) significantly improved the survival of larvae. Thus, the authors suggest that the selection of both *in vitro* and *in vivo* strategies should be applied for the development of multistrain probiotics with complementary modes of action. In addition, the effects of probiotic treatment on the global protein expression in early Atlantic cod larvae have recently been analyzed using proteomic techniques [41]. The results demonstrated that the administration of probiotic bacteria repressed some proteins related to stress and immune responses, and induced proteins related to growth and development. A reduction in environmental stress responses caused by the probiotic bacteria mixture could result from inhibiting the growth of pathogenic bacteria. Clearly, proteome analysis allows us to get a deeper understanding about the effects of probiotic administration on larval global protein expression and posttranslational modifications. All important information about several environmental factors interfering with the protein expression in larvae and affecting larval quality, such as growth and survival rate, could be controlled for a normal development to adult stages. Therefore, challenges in marine fish hatcheries, such as the production of good quality larvae, may be overcome.

19.2.3 Rainbow Trout (*Oncorhynchus mykiss*)

Joborn *et al.* [42] first reported a potential probiotic *Carnobacterium* sp. K1, isolated from the gastrointestinal tract of Atlantic salmon, *Salmo salar*, which was shown by *in vitro* tests to produce inhibitory substances against two common fish pathogens *V. anguillarum* and *Aeromonas salmonicida*, by *in vivo* experiments to remain viable in the gastrointestinal tract of salmonides including rainbow trout for several days and give no virulence to the fish (Table 19.3). Similarly, another *Carnobacterium* strain isolated from Atlantic salmon gut was also evaluated for potential use as a probiotic for salmonids by *in vitro* and *in vivo* experiments. The results indicated that this bacterium remained viable in the gut and reduced disease caused by *A. salmonicida, Vibrio ordalii, and Yersinia ruckeri* but not *V. anguillarum* [43].

Table 19.3 The application of probiotics for rainbow trout (*O. mykiss*).

Probiotic	Origin	Treatment method	Effect on host animal	Pathogen	Mode of action	References
Carnobacterium sp. K1	Salmon gut		Remained viable in the gut	*V. anguillarum* and *A. salmonicida*	Antagonism	[42]
			No detrimental effect to fish			
S. cerevisiae CBS 7764	Rainbow trout gut		Induced heat shock proteins related to mucosal colonization		Mucosal colonization	[44]
P. fluorescens AH2			Reduced accumulated mortality by 46% when challenged with *V. anguillarum*	*V. anguillarum*	Antagonism	[45]
					Competition for iron	
Carnobacterium sp.	Salmon gut	Added to diets	Remained viable in the gut	*A. salmonicida*, *V. ordalii*, and *Y. ruckeri*	Antagonism	[43]
			Reduced disease caused by *A. salmonicida*, *V. ordalii*, and *Y. ruckeri* but not *V. anguillarum*			
Pseudomonas spp.	Rainbow trout	Added to rearing water	Improved survival by 13–43%	*V. anguillarum*	Antagonism	[46]
					Competition for iron	
B. subtilis and *B. licheniformis*	BioPlus2B	Added to diets	Enhanced resistance against *Y. ruckeri*	*Y. ruckeri*	Antagonism	[47]
L. rhamnosus (ATCC 53103)	ATCC	Added to diets	Remained viable in the gut within 1 week		Mucosal colonization	[48]
			Stimulated innate immunity response		Immune modulation	

L. rhamnosus JCM 1136		Added to diets	Increased with the feeding duration in the intestine but not in the stomach		Mucosal colonization	[49]
			Stimulated innate immunity response		Immune modulation	
L. rhamnosus, *E. faecium*, and *B. subtilis*		Added to diets	Stimulated innate immunity response		Immune modulation	[50]
A. sobria GC2	Ghost carp (*Cyprinus* sp.) gut	Added to diets	Stimulated innate immunity response	*A. salmonicida*, *L. garvieae*, *S. iniae*, *V. anguillarum*, *V. ordalii*, and *Y. ruckeri*	Antagonism	[51–54]
			Remained healthy with total mortalities of only 0–16% after pathogen infection		Immune modulation	
					Siderophore and chitinase production	
A. sobria GC2, *B. thermosphacta* BA211		Added to diets	Enhanced survival when challenged with *Aeromonas bestiarum* and *Ichthyophthirius multifiliis*	*A. bestiarum* and *I. multifiliis*	Antagonism	[55]
			Enhanced innate immunity response		Immune modulation	
Bacillus sp. JB-1	Rainbow trout gut	Added to diets	Enhanced innate immunity response	*A. salmonicida*, *L. garvieae*, *S. iniae*, *V. anguillarum*, *V. ordalii*, and *Y. ruckeri*	Antagonism	[52–54]
			Remained healthy with total mortalities of only		Siderophore and chitinase production	

(continued)

Table 19.3 (*Continued*)

Probiotic	Origin	Treatment method	Effect on host animal	Pathogen	Mode of action	References
			0–13% after pathogen infection			
					Immune modulation	
C. maltaromaticum B26 and *C. divergens* B33	Rainbow trout gut		Enhanced expression ratios of IL-1beta and TNF-alpha of head kidney leucocytes		Immune modulation	[56]
C. maltaromaticum B26 and *C. divergens* B33	Rainbow trout gut	Added to diets	Broad antibacterial spectrum against many pathogens	*A. salmonicida*, *A. hydrophila*, *S. iniae*, and *V. anguillarum*	Antagonism	[57,58]
			Remained viable in the gut for up to 3 weeks		Mucosal colonization	
			Enhanced the cellular and humoral immune responses		Immunity modulation	
Kocuria sp. SM1	Rainbow trout gut	Added to diets	Reduced mortalities when challenged with *V. anguillarum* and *V. ordalii*	*V. anguillarum* and *V. ordalii*	Antagonism	[59–62]
Subcellular components of *Kocuria* sp. SM1 and *Rhodococcus* sp. SM2			Stimulated innate immunity		Immune modulation	
Lactococcus lactis sp. *lactis* CLFP 100, *Leuconostoc mesenteroides* CLFP 196, and *L. sakei* CLFP 202		Added to diets	Enhanced survival after challenged with *A. salmonicida* and *L. garvieae*	*A. salmonicida*, *L. garvieae*	Antagonism	[63–65]
			Enhanced humoral and cellular innate immune response		Immune modulation	

B. subtilis AB1	Rainbow trout gut	Added to diets	Enhanced survival against *Aeromonas* sp.	*Aeromonas* sp.	Antagonism	[66]
			Stimulated innate immunity response		Immune modulation	
Two bacterial strains, A3-47 and A3-51			Produced cross-reactive antibodies against *V. harveyi*	*V. harveyi*	Antagonism	[67]
			Enhanced survival		Immune modulation	
Enterobacter cloacae and *Bacillus mojavensis*	Rainbow trout gut	Added to diets	Enhanced survival by 64% after challenge with *Y. ruckeri*	*Y. ruckeri*	Antagonism	[68]
			Affected white blood cells, hemoglobin and weight gains		Immune modulation	
			Enhanced digestibility and utilization of feed			
L. plantarum	Rainbow trout		Improved protection against *L. garvieae*	*L. garvieae*	Antagonism	[69,70]
			Stimulated innate immunity		Immune modulation	
Pseudomonas sp. MSB1				*Flavobacterium psychrophilum*	Antagonism	[71]
					Competition for iron	
Pseudomonas sp. M174		Added to diets	Enhanced survival after challenged with *F. psychrophilum*	*F. psychrophilum*	Antagonism	[72]
			Stimulated innate immunity		Siderophore production	
					Immune modulation	

(*continued*)

Table 19.3 (*Continued*)

Probiotic	Origin	Treatment method	Effect on host animal	Pathogen	Mode of action	References
Pseudomonas sp. M162			Enhanced survival by 39.2% after challenged with *F. psychrophilum* Stimulated innate immunity	*F. psychrophilum*	Antagonism Siderophore production Mucozal colonization Immune modulation	[73]
Strains of lactic acid bacteria	Estuarine source			*Y. ruckeri* and *A. salmonicida*	Adhesion to mucus Survival in bile Antagonism	[74]

Andlid *et al.* [44] introduced a wild-type *Saccharomyces cerevisiae* CBS 7764 isolated from the intestine of rainbow trout, which was shown to induce heat shock proteins related to mucosal colonization. Gram *et al.* [45] evaluated the *in vitro* and *in vivo* antagonism of antibacterial strain *Pseudomonas fluorescens* AH2 against the fish-pathogenic bacterium *V. anguillarum*. This probiotic was found to reduce calculated accumulated mortality by 46% after challenged with *V. anguillarum*. Also, because iron is important in virulence and bacterial interactions, the effect of *P. fluorescens* AH2 was studied under iron-rich and iron-limited conditions. The results showed that *P. fluorescens* AH2 inhibited the growth of *V. anguillarum* during coculture, independently of the iron concentration. Similarly, many other strains of *Pseudomonas* from the indigenous microflora of rainbow trout were also shown to improve the survival of rainbow trout by 13–43% after challenged with *V. anguillarum* [46]. This effect was proposed to be based on the antagonistic activity and iron competition of putative probiotics; however, *in vitro* antagonism could not completely predict an *in vivo* effect. Therefore, further studies on the underlying mechanism of activity are required to screen probiotic bacteria in the cultivation of rainbow trout.

Since 2003, the mechanism of immune modulation has started to be used as one of important criteria to screen probiotics for rainbow trout. The addition of *Lactobacillus rhamnosus*, strain ATCC 53103 [48] or strain JCM 1136 [49] to diets stimulated innate immunity responses after feeding. Panigrahi *et al.* [50] also studied immune modulation including the expression of cytokine genes following dietary administration of three selected probiotic bacteria *L. rhamnosus*, *Enterococcus faeceium*, and *B. subtilis* to this fish, indicating that probiotic bacteria delivered in feed exerted their influence on the immune system of fish, at both cellular and molecular levels. Based on similar approaches, the research group of Austin successfully developed seven probiotics, including *Aeromonas sobria* GC2 isolated from the gut of ghost carp (*Cyprinus* sp.), *Bacillus* sp. JB-1 [51–54], *Brochothrix thermosphacta* BA211 [55], *Carnobacterium maltaromaticum* B26, *C. divergens* B33 [56–58], and *Kocuria* sp. SM1 and *Rhodococcus* sp. SM2 from rainbow trout intestine [59]. All strains were found to promote innate immune responses; however, each probiotic seemed to stimulate different pathways within the innate immune system. Also, these seven probiotics showed diverse antibacterial spectra against different pathogens. Many other probiotics developed later showed similar tendencies, indicating the advance in the development of probiotics acting as immunostimulants in rainbow trout.

19.2.4 Gilthead Sea Bream (*Sparus auratu*)

In 1997, Maeda *et al.* [75] first introduced the probiotic candidate *Pseudoalteromonas undina* in the cultivation of gilthead sea bream (*S. auratu*) because it was found to decrease infection with various pathogenic viruses (Table 19.4). Also, six bacterial strains belonging to different genera, *Cytophaga*, *Roseobacter*, *Ruergeria*, *Paracoccus*, *Aeromonas*, and *Shewanella* isolated from live food cultures were found to enhance survival of gilthead sea bream larvae when compared with filtered seawater [76].

Table 19.4 The application of probiotics for gilthead sea bream (*S. auratu*).

Probiotic	Origin	Treatment method	Effect on host animal	Pathogen	Mode of action	References
P. undina			Reduced infection with various pathogenic viruses	Viruses		[75]
L. fructivorans (AS17B); *L. plantarum* (906)	Adult sea bream gut; human feces	Delivered via rotifers and/or *Artemia* and dry feed	Enhanced survival of larvae		Immune modulation	[77,78]
			Stimulated innate immune response			
Cytophaga, Roseobacter, Ruergeria, Paracoccus, Aeromonas, and *Shewanella*	Live food cultures	Added to rearing water	Enhanced survival of larvae when compared with filtered seawater			[76]
Two killed bacteria (Pdp11 and 51M6) of Vibrionaceae family; two killed bacteria (*L. delbrueckii* and *B. subtilis*)	Gilthead sea bream skin; human and cattle source	Added to diets	Enhanced humoral and cellular innate immune response		Immune modulation	[79–81]
D. hansenii CBS 8339		Added to diets	Enhanced cellular innate immune response		Immune modulation	[82]
L. fructivorans + *L. plantarum*	Human feces	Delivered via rotifers and *Artemia*	Increased body weight, decreased cortisol levels, and improved stress response		Immune modulation	[83]
			Enhanced innate immune Response			

Similarly, two bacterial strains *Lactobacillus fructivorans* (AS17B) and *L. plantarum* (906) administered during sea bream development using rotifers and/or *Artemia salina* and dry feed as vectors reduced larval mortality between days 39 and 66 post hatch [77]. Immunohistochemical studies show that early feeding with probiotic-supplemented diets increased the number of Ig(+) cells and acidophilic granulocytes, indicating a stimulatory effect of probiotics on the gut immune system that correlates with improvement of fish survival as shown before [78].

Such innate immune responses after the administration of heat-inactivated probiotics have been found in gilthead sea bream also. For example, two bacteria (Pdp11 and 51M6) from the Vibrionaceae family, which were obtained from the skin of gilthead sea bream, induced the serum peroxidase content, the natural hemolytic complement activity, the phagocytic activity, and the cytotoxic activity (for 51M6 only) [79]. Moreover, the advantages provided by administration of killed probiotic bacteria as well as multispecies versus monospecies formulations are indicated, which contribute to their application in aquaculture [80]. Similarly, the multispecies probiotic formulation of *L. fructivorans* and *L. plantarum* from human feces delivered via rotifers and *Artemia* increased body weight, decreased cortisol levels, and improved stress response of sea bream [83], whereas an enhancement in innate immune responses via intestinal immunoglobulin (Ig(+)) cells and acidophilic granulocytes confirmed stimulatory actions of probiotics on the gut immune system that correlated with improvement of fry survival.

While two killed bacteria Pdp11 and 51M6 stimulated innate immune response at both humoral and cellular levels, the live yeast *Debaryomyces hansenii* CBS 8339 strongly increased the innate immune parameters in sea bream at the cellular level rather than at humoral level [82]. *D. hansenii* administration significantly enhanced leukocyte peroxidase, respiratory burst, phagocyte, cytotoxin, and liver superoxide dismutase activity. Moreover, the probiotic repressed the expression of most sea bream genes, except C3, in liver and intestine, and induced all of them in head–kidney, revealing the tissue-specific regulation of gene expression affected by the probiotic.

19.2.5 Other Marine Fishes

19.2.5.1 Sea Bass (*Dicentrarchus labrax*)

Frouel *et al.* [84] reported a preliminary study of the effects of the commercial probiotic product (SORBIAL, France) containing two strains of lactobacilli on digestive metabolism of juvenile sea bass (*D. labrax*). Added to diets, the probiotics improved survival rate and were accompanied by the disappearance of the ventral, dorsal, and operculum malformations that usually occur in juveniles. Further studies should be carried out for a comprehensive understanding of the complex mechanisms of these probiotics. Immune modulation is one of the mechanisms increasingly considered for the development of probiotics in marine fish. *Lactobacillus delbrueckii* was shown to increase body weight, decrease cortisol levels, and improve stress response of sea bass, whereas innate immune responses were promoted and immune-related genes were differentially regulated after the probiotic treatment [85]. Recently, Sorroza *et al.* [86] screened a new strain, *Vagococcus fluvialis*, which improved survival rate of sea bass by 42.3% after challenged with *V. anguillarum*, indicating a promising probiotic in sea bass culture in the future.

19.2.5.2 Atlantic Halibut (*Hippoglossus hippoglossus*)

In 1993, the research group of Vadstein and Skjermo suggested a strategy so-called microbial maturation to obtain microbial control during larval development of marine fish [87,88]. Microbial maturation was shown to increase the survival of halibut yolk sac larvae [87]. These results support the hypothesis that microbial maturation screens nonopportunistic bacteria to protect the marine larvae from the proliferation of detrimental opportunistic bacteria. That is why the introduction of the mixture of Pseudomonas and Cytophaga/Flavobacterium improves larval survival and reproducibility of halibut larvae [88]. The incubation of *Vibrio salmonicida*-like strains and *L. plantarum* enhanced the survival of halibut larvae by 14% in the first 2 weeks after hatching, whereas the presence of *Vibrio iliopiscarius* reduced survival to 18% [32]. Moreover, *V. salmonicida* increased halibut larval survival nearly 15% after 32 days posthatch [89]. As broodstock is expensive to maintain and obtain, an increase in larval survival means that fewer broodstock animals are required for production. In contrast, bacterial strains PB 52 and 4:44, in addition to water and attachment to rotifers, enabled to colonize the gut of halibut larvae without improving survival rate, whereas the bioencapsulation of *Vibrio* strains PB 1–11 and PB 6-1 in *Artemia franciscana metanauplii* failed to make any changes in the numbers of bacteria in the intestine or growth rate of halibut larvae [90].

19.2.5.3 Olive Flounder (*Paralichthys olivaceus*)

When the candidate probiotic *Lactobacillus* sp. DS-12 was added to the diet, olive flounder or Bastard halibut (*P. olivaceus*) had a higher growth than the control group, while fewer bacteria were counted in the intestine of the experimental group [91]. Few other lactic acid bacteria were tested as probiotics. For example, *Weisella hellenica* DS-12 isolated from the Japanese flounder was antagonistic to fish pathogens such as *Edwarsiella tarda*, *Pasteurella piscicida*, *Aeromonas hydrophila*, and *V. anguillarum*, but not *Streptococcus faecalis* [92].

With the development of intensive fish farming, periodic and epidemic outbreaks of *Lactococcosis*, *Streptococcosis*, or *Scuticociliatosis* have occurred worldwide. Infections resulting from *Lactococcus* and *Streptococcus* have been noticed in a variety of wild and farmed animals including olive flounder. Kim *et al.* have studied the development of probiotics that promote immune system to protect olive flounder from these diseases [93,94]. For example, the putative probiotic *Zooshikella* sp. JE-34 enhanced the growth and innate immune response, and reduced mortality of *P. olivaceus* against *Streptococcus iniae* [93]. Similar results were shown by *E. faecium* for protecting olive flounder against *Lactococcus garvieae* [94]. Scuticociliatosis is another severe disease caused by about 20 species belonging to the phylum Ciliophora. It is recognized as an emerging problem, inflicting significant economic loss in the aquaculture industry worldwide. Among the 20 species *Philasterides dicentrarchi*, *Miamiensis avidus*, and *Uronema marinum* are the three species responsible for scuticociliatosis in olive flounder farms of South Korea [95]. Till now diverse probiotics and herbal mixtures have been used against freshwater bacterial and fungal diseases, but little information is available regarding the different prophylactic measures against marine scuticociliatosis [95]. Interestingly, the mixtures of five probiotics, *L. plantarum*, *L. acidophilus*, *Lactobacillus sakei*, *B. subtilis*, and *S. cerevisiae* were found to promote the growth, feed efficiency, blood biochemistry, survival rate, and nonspecific immune response in *U. marinum* infected olive flounder [95].

Besides having antibacterial activity, some probiotics were shown to reduce virus infection in marine fish by immune modulation. Lymphocystis disease virus (LCDV) is a typical example in olive flounder, causing the chronic, self-limiting clinical disease, lymphocystis. Fish with lymphocystis exhibit grossly visible papilloma-like skin lesions that substantially reduce their commercial value. No vaccines are currently available for lymphocystis viruses. However, Harikrishnan *et al.* [96] have demonstrated that two probiotics Lactobacil and Sporolac, as separate or mixed diets, act as immunostimulants that enhance the innate immune response and disease resistance in olive flounder.

19.2.5.4 Clownfish (*Amphiprion* spp.)

The selection of probiotics for aquaculture is usually based on their antagonism toward pathogens. However, Vine *et al.* [97] suggested other criteria such as growth, attachment to intestinal mucus, and production of beneficial compounds to isolate and screen bacteria from the gut of the common clownfish, *Amphiprion percula*, for probiotic development. The results indicated that only *Pseudoalteromonas* sp. AP5 reduced *Vibrio* spp. in the culture water microflora and *Kocuria* sp. AP4 improved clownfish larval survival [98]. Also, the probiotic *L. rhamnosus* IMC 501 delivered to the false percula clownfish *Amphiprion ocellaris* via live prey and added to the rearing water increased the body weight twofold in both clownfish larvae and juveniles, with accelerated metamorphosis occurring 3 days earlier in fingerlings, and significantly induced gene expression of factors involved in growth and development [99]. The findings confirmed that the introduction of lactic acid bacteria modulated the development of larval clownfish.

19.2.5.5 Common Snook (*Centropomus undecimalis*)

Kennedy *et al.* [100] suggested that accurate decisions about microbial health of cultured fish require determination of normal and pathogenic microbes inside. Most bacterial infections of marine fish larvae perhaps start in the gut, whereas the presence of some strains of *Bacillus* sp. and Gram-negative bacteria in the intestine of common snook seems responsible for the exclusion of pathogens, thus conferring general fish health. The administration of the putative *Bacillus* no. 48 with salinity-reduced culture water was shown to significantly remove *Vibrio* spp. from juvenile snook, contributing to better survival rate of the fish.

19.2.5.6 Pollack (*Pollachius pollachius*)

Gatesoupe [101] introduced a formaldehyde treatment to disinfect *Artemia* cysts and nauplii, followed by the administration of two commercial probiotics, Bactocell (*Pediococcus acidilactici*) and Levucell (*S. cerevisiae*), to the nauplii that was fed to pollack (*P. pollachius*) larvae. The results showed that the probiotics enhanced the mean weight of pollack, suggesting that they should be used for fish larvae. However, resistant strains had emerged after 3 months of daily cyst incubation with formaldehyde, leading to a distribution in some genotypic characteristics between new and reference strains. Clearly, formaldehyde treatment should be limited in aquaculture to avoid long-term risks caused by the wide distribution of resistance.

19.2.5.7 Goldfish (*Carassius auratus*)

Although there has been an increasing interest in the use of probiotics in aquaculture to control fish diseases, the possible use of probiotics to control diseases of ornamental fish has been neglected. In 2003, Irianto *et al.* [102] first introduced the oral administration of killed *A. hydrophila* A3-51 for the control of the infection by atypical *A. salmonicida* in goldfish, *C. auratus*. Also, the administration of diets supplemented with either probiotics, triherbal extract or azadirachtin improved survival rate of goldfish after *A. hydrophila* infection and enhanced innate immune responses [103]. Currently, Chu *et al.* [104] isolated the first bacterium, *Bacillus* sp. QSI-1, from the goldfish gut that can degrade acyl-homoserine lactones (AHLs). The coculture of QSI-1 with fish pathogen *A. hydrophila* YJ-1 significantly lowered the amount of AHLs and the extracellular proteases activity of the pathogen. QSI-1 was then shown to enhance survival rate of goldfish after *A. hydrophila* infection, indicating good probiotic activities for its use in aquaculture.

19.3
The Application of Probiotics for Marine Crustaceans

19.3.1
Black Tiger Shrimp (*Penaeus monodon*)

Maeda and Liao [105] reported the use of a soil bacterial strain PM-4 as a food source for increasing the growth of *P. monodon* nauplii or as an antagonist for decreasing *V. anguillarum* densities (Table 19.5). When added to tanks inoculated with diatoms and rotifers, the strain resulted in 57% survival of the larvae after 13 days, while without the bacterium all the larvae died after 5 days [106]. This strain was identified as *Thalassobacter utilis* [107] and also found to promote the growth of Japanese blue crab (*Portunus trituberculatus*), and repress the growth of the fungus *Haliphthoros* sp.

Later studies screened many *Bacillus* strains that mainly expressed *in vitro* antagonistic activity against *vibrios*, and an *in vivo* positive effect on the growth rate of *P. monodon* [108,110,112,114,116]. The application of *Bacillus* strains as

Table 19.5 The application of probiotics for black tiger shrimp (*P. monodon*).

Probiotic	Origin	Treatment method	Effect on host animal	Pathogen	Mode of action	References
T. utilis PM-4	Soil	Added to rearing water	Enhanced survival rate	*V. anguillarum*	Antagonism	[105,106]
			Reduced *Vibrio* densities		Food source for larvae	
Bacillus sp.		Added to rearing water	Enhanced survival rate	Luminous *Vibrio*	Antagonism	[108,109]
			Reduced *V. harveyi* densities			
Bacillus spp.	Commercial source	Added to rearing water	Enhanced total heterotrophic bacteria, *Bacillus* spp. and reduced vibrios in water and sediment of the pond	Vibrios	Water quality improvement	[110]
Bacillus sp. S11	Tiger shrimp ponds	Added to diets	Enhanced growth and survival of larvae and postlarvae after challenge with the pathogen	*V. harveyi* 1526	Antagonism	[111]
B. subtilis BT23	Tiger shrimp ponds	Added to rearing water	Enhanced survival after challenge with the pathogen	*V. harveyi*, causing black gill disease	Antagonism	[112]
Pseudomonas sp. PM-11 and *V. fluvialis* PM-17	Subadult shrimp gut	Added to rearing water	Induced immunity indicators: hemocyte counts, phenol oxidase, antibacterial activity		Immune modulation	[113]

(*continued*)

Table 19.5 (*Continued*)

Probiotic	Origin	Treatment method	Effect on host animal	Pathogen	Mode of action	References
Paenibacillus sp. Q, *B. cereus* Q1	Marine sediment	Added to rearing water	Enhanced survival after challenge with the pathogen	*V. harveyi*, *V. vulnificus*, and *Vibrio* spp.	Antagonism	[114]
Synechocystis spp. MCCB 114 and 115	Seawater	Added to diets	Enhanced intestinal bacterial flora but reduced total *Vibrio*	*V. harveyi*	Nutrient support	[115]
			Enhanced survival rate when challenge with *V. harveyi*		Antagonism	
B. pumilus	Healthy tiger shrimp mid-gut	Added to rearing water		*V. alginolyticus*, *V. mimicus*, and *V. harveyi*	Antagonism	[116]
C. haemulonii S27, *Bacillus* sp. MCCB101	Marine source		Enhanced the production of crustin-like AMP and confer protection to shrimps against WSSV infection	WSSV	Immune modulation	[117]
					Antiviral activity	

probiotics in shrimp aquaculture was evaluated using economic effect analysis, which showed a clear benefit to the producer [109]. Typically, the strain *Bacillus* S11 promoted the growth and survival rate, and enhanced resistance of black tiger shrimp to *Vibrio* challenge. Annual yields were 49% greater for shrimps fed with this probiotic [111]. These studies indicate that pathogenic *vibrios* were controlled by *Bacillus* strains under *in vitro* and *in vivo* conditions and that probiotic treatment offers a promising alternative to the use of antibiotics in *P. monodon* culture.

However, Alavandi *et al.* [113] reported that the criteria used for the selection of putative probiotic strains based on *in vitro* assays did not always bring about the desired effect *in vivo* and improve the immune system in shrimp. For example, two strains of *Pseudomonas* sp. PM-11 and *Vibrio fluvialis* PM-17 were selected as candidate probionts from the tiger shrimp gut microbiota and tested for their effect on the shrimp immunity indicators such as hemocyte counts, phenol oxidase, and antibacterial activity. All three immunity indicators showed declining trends after 45 days of the experiment. Therefore, the authors suggested that new protocols be evolved for selection of probiotics, in which the evaluation of immunity promotion of microbes on the host animal may provide a clue on the screening procedure. According to this approach, Antony *et al.* [117] have currently found a crustin-like antimicrobial peptide of 99 amino acids from the hemocytes of *P. monodon*, which is shown to be divergent from other shrimp crustins using phylogenetic and sequence analysis. The differential expression of the crustin-like AMP in *P. monodon*, in response to the administration of the marine yeast *C. haemulonii* S27 and the probiotic bacteria *Bacillus* sp. MCCB101, indicated that these microbes enhanced the production of crustin-like AMP and conferred significant protection to *P. monodon* against the infection of White Spot Syndrome Virus (WSSV).

Besides antagonistic activity and immunity promotion effect, putative probiotics could be used as a valuable nutrient source for tiger shimps. Among them, cyanobacteria were suggested due to high levels of protein (34–43%), in addition to carotenoids. Therefore, they can become candidates of multiple-role probiotics when combined to other activities. On feeding *P. monodon* postlarvae with the cyanobacteria *Synechocystis* spp. MCCB 114 and 115, survival rate of shrimp was shown to be enhanced when challenged with *V. harveyi*, and the generic diversity of the intestinal bacterial flora also increased with substantial reduction or total absence of *Vibrio* spp. [115].

19.3.2
White Shrimp (*L. vannamei*)

Garriques and Arevalo [118] first reported the use of probiotics in the cultivation of *L. vannamei* (Table 19.6). A nonpathogen recognized *V. alginolyticus* strain was shown to enhance the average survival rate and wet weight of shrimps compared to oxytetracycline treatment and the control. Also, no presence of *V. parahaemolyticus* was detected in any probiotic treated shrimp, while control and antibiotic treated shrimps had nearly 10% of this potentially pathogenic bacterium [118].

Table 19.6 The application of probiotics for white shrimp (*L. vannamei*).

Probiotic	Origin	Treatment method	Effect on host animal	Pathogen	Mode of action	References
V. alginolyticus	Pacific Ocean seawater	Added to rearing water	Enhanced survival rate and weight of postlarvae Reduced *V. parahaemolyticus* in shrimps	*V. parahaemolyticus*	Antagonism	[118]
Probiotics		Added to rearing water	Enhanced the accumulation of nutrients and improved the bacterial flora in pond sediment		Water quality improvement Antagonism	[119]
L. plantarum		Added to diets	Enhanced survival rate after challenge with the pathogen Enhanced innate humoral immune responses but reduced total hemocyte counts	*V. alginolyticus*	Immune modulation Antagonism	[120]
V. alginolyticus UTM 102, *B. subtilis* UTM 126, *R. gallaeciensis* SLV03, and *P. aestumarina* SLV22	Adult white shrimp gut	Added to diets	Reduced feed conversion ratio and mortality when challenged with the pathogen	*V. parahaemolyticus* PS-017	Antagonism	[121]
B. subtilis UTM 126	Adult white shrimp gut	Added to diets	Reduced mortality when challenged with the pathogens	*V. alginolyticus*, *V. parahaemolyticus*, and *V. harveyi*	Antagonism	[121]
B. licheniformis		Added to rearing water	Reduced total vibrios counts		Immune modulation	[122]

			Enhanced hemocyte counts, phenoloxidase and superoxide dismutase activities		Antagonism	
B. subtilis E20	Natto (fermented soybeans)	Added to diets	Enhanced growth rate		Protease activity	[123]
		Added to diets	Enhanced phenoloxidase activity, phagocytic activity, and clearance efficiency	*A. hydrophila*	Immune modulation	[124]
					Antagonism	
		Added to rearing water	Promoted larval development		Development modulation	[125]
			Weakly reduced total bacterial count and *Vibrio* count		Antagonism	
			Enhanced tolerance to salt and nitrite stress		Stress tolerance	
			Induced the expression of immune-related genes		Immune modulation	
Live or killed *V. gazogenes* (NCIMB 2250)	Commercial product	Added with chitin into diets	Enhanced nutritional and immunological health	*V. harveyi*, *V. anguillarum*, and *V. alginolyticus*	Immune modulation	[126]
			Changed gut microbial flora		Antagonism	

(*continued*)

Table 19.6 (*Continued*)

Probiotic	Origin	Treatment method	Effect on host animal	Pathogen	Mode of action	References
B. subtilis	Commercial product	Added with soybean meal into diets	Enhanced survival, growth, and food conversion ratio, hemolymph metabolites and stress tolerance of juveniles		Enzyme activity for feed digestion and assimilation	[127]
L. plantarum MRO3.12	Wild shrimp gut	Added to diets	Enhanced growth rate Reduced total bacterial and nonfermenting vibrios counts Enhanced survival rate when challenged with the pathogen	*V. harveyi*	Antagonism	[128]
B. subtilis L10 and G1	Fermented pickles			*V. harveyi* and *V. parahaemolyticus*	Antagonism (bacteriocin activity)	[12]

From 1995 to 2005, no reports on probiotics in *L. vannamei* culture were published. From 2006 to present, the extended cultivation of this shrimp in the world led to an increasing consideration in the development of probiotics. Besides *Lactobacillus, Vibrio, Pseudomonas,* and *Roseobacter,* strains of *Bacillus* genus have become the most popular candidates of probiotics for *L. vannamei*, which is similar to the case of *P. monodon* as mentioned above. These probiotics were proposed to express diverse modes of action such as microbial antagonism or competitive exclusion, immune modulation, and enzyme activity for feed digestion and assimilation, larvae metamorphose, stress tolerance, and water quality improvement.

Competitive exclusion approaches for probiotic development have been used as one method to control pathogens in poultry and now in aquaculture with ubiquitous pathogens belonging to the genus *Vibrio*. For instance, Balcazar and Rojas-Luna [121] reported the inhibitory activity of probiotic *B. subtilis* UTM 126 against pathogenic *Vibrio* species. Then juveniles of white shrimp treated with the probiotic were found to have reduced mortality rates after an immersion challenge with *V. harveyi*, which confers protection against vibriosis in this shrimp species. Moreover, feeding shrimp with diets containing this probiotic bacterium along with three other probiotics *V. alginolyticus* UTM 102, *Roseobacter gallaeciensis* SLV03, and *Pseudomonas aestumarina* SLV22 reduced feed conversion ratio and mortality when challenged with *V. parahaemolyticus* PS-017 [129]. Currently, Thompson *et al.* [126] showed the promise of *Vibrio gazogenes* NCIMB 2250 (live or dead cells) together with chitin to improve the health and welfare of white shrimp, whereas the growth, feed conversion ratio, and survival of white shrimp when challenged with *V. harveyi* were provoked by *L. plantarum* MRO3.12 [128].

The administration of probiotics also influences the development of the immune response in white shrimp. The exact mechanisms that mediate the immune modulation of probiotics are not clear. However, it has been shown that the probiotics often promote phenoloxidase activity in white shrimps [120,124,130]. Among them, *L. plantarum* was found to reduce total hemocyte counts [120], whereas these numbers were enhanced when shrimps were fed with the probiotic *Bacillus licheniformis* [122] and there were no significant differences after the administration to *B. subtilis* E20 [124]. Moreover, *L. plantarum* was shown to enhance innate immune responses and the survival rate after challenging with *V. alginolyticus* [120]. Thus, administration of the probiotics can improve the white shrimp's intestinal microflora and its immune response. This may facilitate the development of alternative strategies, including novel prebiotics and synbiotics [122], for shrimp growth and health management.

The above results suggest that *B. subtilis* E20 would be a potential candidate for use as a probiotic to improve shrimp growth performance, and consequently reduce feed costs. Moreover, this probiotic was then demonstrated to enhance the larvae growth rate through protease activity [123], accelerate larval metamorphose, enhance tolerance to salt and nitrite stress, as well as induce the expression of some immune-related genes in white shrimp larvae [125]. Similar results were also reported on another *B. subtilis* strain with its enzyme activity for feed digestion and assimilation, leading to an increase in the survival and growth rate, food conversion ratio, hemolymph metabolites, and stress tolerance of juveniles [127]. Therefore, the probiotics *B. subtilis* was expected to be used for white shrimp larvae breeding to improve survival and development of larvae as well as increase stress tolerance and immune status of post larvae.

19.3.3 Other Crustaceans

19.3.3.1 Blue Shrimp (*Litopenaeus stylirostris*)

The blue shrimp (*L. stylirostris*) immersed with pathogenic *Vibrio nigripulchritudo* SFn1 led to maximal infection level in the hemolymph at 24 h postinfection, preceding the mortality peak recorded at 48 h postinfection. Moreover, the antioxidant defenses in the digestive gland significantly reduced from 24 h postinfection, leading to enhanced oxidative stress level and tissue damage. However, shrimps fed with the probiotic *P. acidilactici* MA18/5M showed lower infection and mortality rate. Interestingly, shrimps fed with the probiotic sustained higher antioxidant defenses and lower oxidative stress level. The findings indicate that bacterial infection perhaps induces oxidative stress in *L. stylirostris*, and the probiotic treatment can overcome this status by some modes of action such as an antagonistic activity or a competitive exclusion effect leading to a reduction of infection level, and/or an induction of the antioxidant defenses of the blue shrimp [131].

19.3.3.2 Western King Prawn (*Penaeus latisulcatus*)

Van Hai *et al.* [132] used customized probiotics, *Pseudomonas synxantha* and *Pseudomonas aeruginosa*, in the cultivation of western king prawns (*P. latisulcatus* Kishinouye, 1896). These probiotics were shown to repress the growth of *vibrios* isolated from western king prawns and other aquatic animals. Also, they conclusively met all the essential requirements for appropriate probiotics such as suitability and safety in the cultivation of western king prawn. *P. aeruginosa* was more effective for improving prawn health than *P. synxantha*, but the combined probiotics were the best. Compared to the common prebiotics, Bio-Mos and beta-1,3-D-glucan, these probiotics were found to have similar beneficial effects on the growth, survival, and immune responses of the prawns. Therefore, they were suggested to be used as appropriate probiotics and as a suitable replacement of antibiotics, for disease control in western king prawn aquaculture.

19.3.3.3 Tropical Rock Lobster (*Panulirus ornatus*)

Larval rearing of the tropical rock lobster, *P. ornatus* is now facing high mortality during early larval stages, particularly on initial stocking and around molt periods. *Vibrio* sp. and other pathogens have been implicated as a potential cause of larval mortalities. The colonization by a dominant organism *Thiothrix* sp. suggested that this bacterium perhaps contributes to phyllosoma survival [133]. Also, isolates affiliated with the *Roseobacter* group were obtained from the rearing water, whereas

Bacillus strains were isolated from phyllosoma [134]. *Bacillus* species are often used as probiotics within aquaculture systems; certain members of the *R. clade* are believed to play an important role in oceanic sulfur and carbon cycling. Although the ecological role of *Roseobacter* detected on wild phyllosoma is currently unknown, their dominance in the bacterial community suggests that they may possess attributes that promote phyllosoma health [135]. Further work should be carried out to identify potential probiotic strains of the phyllosoma rearing system.

19.3.3.4 Crabs

In contrast to the increasingly popular application of probiotics in cultured penaeid shrimps, relatively few studies have been published on the development of probiotics in crabs. It seems to be that only the bacterial strain PM-4 isolated from a crustacean culturing pond has been considered as an in-depth studied probiotic in this area. This bacterium improved the growth of Japanese blue crab (*P. trituberculatus*) larvae and repressed the growth of *V. anguillarum* in seawater [106,136,137]. PM-4 was identified as *T. utilis* and also shown to repress the growth of a fungus *Haliphthoros* sp. Cultured and added daily to seawater during the first to third zoeae growth stage of the crab with diatoms and rotifers, PM-4 decreased in cell numbers during the first 3 days, because of feeding by the first zoeae stage of larvae. The survival rate of PM-4 added crab larvae was 28.3% whereas only 15.6% of crab larvae without probiotic treatment survived, which indicated that the bacterial strain PM-4 is effective as a biocontrol agent for this crab [107].

To develop new probiotics in crab cultivation, phylogenetic analysis of crab intestinal bacteria has currently become one of the most popular approaches to define the dominant bacteria, which may be a promising candidate for probiotics. For example, Li *et al.* [138] aimed to identify the dominant intestinal bacteria in the Chinese mitten crab (*Eriocheir sinensis*) and investigate the differences in the intestinal bacteria between pond-raised and wild crabs using molecular techniques. Results have shown that the intestinal bacteria of pond-raised crabs expressed higher intersubject variation, total diversity, and abundance than that observed in wild crabs. Also, Proteobacteria and Bacteroidetes may be the dominant bacteria in the gut of this crab, which supports an important clue for the development of new probiotics in crab farming.

19.4 The Application of Probiotics for Marine Mollusks

The production of bivalve mollusks has been increasing to follow the even more rapidly increasing demand of worldwide consumers. However, disease outbreaks caused by bacterial pathogens led to heavy loss of complete batches, compromising the regular production and the economic viability of the industry [139]. Scientists focus on nutritional and environmental studies more often than on the control of microbiota, whereas in bivalve larvae the filter-feeding behavior increases the modification of these bacterial populations. Larvae and bacteria, both beneficial and potentially pathogenic strains, share a common habitat. The classical treatments are directed toward to the complete elimination of total bacteria from culture seawater, while some beneficial bacteria can supply food source for larvae or inhibit the growth of pathogens. Moreover, the use of antimicrobial agents may lead to the rapid development of pathogen populations resistant to antibiotics [140], the elimination of beneficial organisms, and the emergence of other microbial pathogens of the bivalves. So the use of probiotics was introduced as an alternative method in aquaculture, but mainly on fish and crustaceans rather than on mollusks.

Lodeiros *et al.* [141] published a study on the effect of antibiotic-producing marine bacteria on the larval survival of the zigzac scallop *Euvola ziczac* (formerly *Pecten ziczac*). Till now only some mollusk species, such as Chilean scallop (*Argopecten purpuratus*), great scallop (*Pecten maximus*), Pacific oyster (*Crassostrea gigas*), European flat oyster *(Ostrea edulis)*, and Cortez oyster (*Crassostrea corteziensis*), have been assessed and incorporated for potential marine probiotics (Table 19.7).

19.4.1
Chilean Scallop (*A. purpuratus*)

In the 1990s, the mass culture of Chilean scallop *A. purpuratus* (Lamarck, 1819) faced serious problems because of high larval mortalities. The main cause of mortality was the presence of pathogenic bacteria belonging to the genus *Vibrio*. Therefore, Riquelme *et al.* studied the application of probiotics to this mollusk species. In 1996, a native bacterial strain identified as *Pseudoalteromonas haloplanktis* (formerly *Alteromonas haloplanktis*), was shown to clearly suppress the growth of *V. alginolyticus* and *V. anguillarum*, two strains that cause severe mortalities in larval cultures of *A. purpuratus* [140]. In 1997, a total of 506 bacterial isolates from laboratory and hatchery sources were evaluated *in vitro* for the antagonism activity on a *V. anguillarum*-related larval pathogen, of which 11 (2.2%) were found to be positive. One of these strains (*Vibrio* sp.), when used as a pretreatment, protected the scallop larvae against subsequent experimental infection with the pathogen [142].

In 1999, the same group [143] studied the use of axenic microalgal cultures (*Isochrysis galbana*) as a vector for transmitting putative probiotics into cultures of larval bivalves as antagonists of pathogenic bacteria. Among three selected strains including *Pseudomonas* sp. 11, *Vibrio* sp. C33, and *Arthrobacter* sp. 77, only cells of strains 11 and 77 were found to be ingested by *A. purpuratus* larvae. While comparing bacterial incorporation among these strains, the 77 became the dominant bacteria of the larval microflora, causing no differences in larval survival at different bacterial concentrations. Such results suggested that *Arthrobacter* sp. 77 could be used as a putative probiotic for scallop larvae, and hence as a promising method to control and prevent infections in hatcheries systems.

Finally, they studied the incorporation of the probiotic bacteria to massive larval cultures [144], inoculating periodically a mixture of antibiotic-producing strains to

Table 19.7 The application of probiotics for marine mollusks.

Probiotic	Origin	Treatment method	Effect on host animal	Pathogen	Mode of action	References
Zigzac scallop (*E. ziczac*)						
Probiotics	Marine source		Enhanced survival rate		Antibiotic production	[141]
Chilean scallop (A. *purpuratus*)						
P. haloplanktis INH	Scallop hatchery		Enhanced protection against infection of the pathogen	*V. alginolyticus* and *V. anguillarum*	Antagonism	[140]
Vibrio sp.	Scallop hatchery		Enhanced protection against infection of the pathogen	*V. anguillarum*-like	Antagonism	[142]
Pseudomonas sp. 11	Scallop hatchery	Fed with algae	Enhanced protection against infection of the pathogen	*V. anguillarum*-like	Antagonism	[143]
Vibrio sp. C33	Scallop hatchery	Fed with algae	Enhanced survival after challenge with the pathogen	*V. anguillarum*-like	Antagonism	[143]
Vibrio sp. C33, *Pseudomonas* sp. 11, *Bacillus* sp. B2	Scallop hatchery		Completed larval phase without antibiotic treatment		Development modulation	[144]
Arthrobacter sp. strain 77	Scallop hatchery		Antibacterial activity		Antagonism	[145]
			Replaced resident microflora within 24 h		Mucosal colonization	
Great scallop (*P. maximus*)						
P. gallaeciensis BS107	Marine source		Enhanced survival rate after challenge with the pathogen	*V. pectenicida*	Antagonism	[146]

(*continued*)

Table 19.7 (*Continued*)

Probiotic	Origin	Treatment method	Effect on host animal	Pathogen	Mode of action	References
					Thermostable bacteriocin production	
Pseudoalteromonas sp. X153	Pebble		Enhanced survival rate after challenge with the pathogen	Ichthyopathogenic *Vibrio*	Antagonism	[147]
			Reduced growth		Bacteriocin production	
Pacific oyster (*C. gigas*)						
Alteromonas sp. CA2	Marine source	Fed with algae	Enhanced growth		Food source for larvae	[148,149]
			Enhanced survival rate			
A. media A 199		Added to rearing water	Enhanced survival rate when challenged with pathogen *V. tubiashii*	*A. caviae, A. hydrophila, A. salmonicida, A. veronii, V. anguillarum, P. damsella,* and *Y. ruckeri*	Bacteriocin production	[150]
					Antagonism	
Gram-negative short rod, strain S21	Oyster rearing seawater		Enhanced survival rate when challenged with pathogen *V. alginolyticus*	*V. alginolyticus* and *V. tubiashii*	Antagonism	[151]
European flat oyster (*O. edulis*)						
P. gallaeciensis 154	Mollusk hatcheries		Enhanced survival rate	*V. anguillarum, V. neptunius,* and *Vibrio* spp.	Antagonism	[139,152]
Cortez oyster (*C. corteziensis*)						
Lactic acid bacteria (strain NS61)	Lions-paw scallop	Added to rearing water	Enhanced survival rate			[153]
P. aeruginosa YC58	White shrimp		No increase of growth rate			
B. cepacia Y021	Cortez oyster					

the system and demonstrating that the addition of the selected bacterial cultures such as *Vibrio* sp. C33, *Pseudomonas* sp. 11, and a newly identified *Bacillus* sp. B2 did not adversely affect the scallop larvae. The larval phase could be completed without the use of antibiotic treatment. Such controls changed the bacterial flora associated with the larvae and reduced potential pathogens.

19.4.2 Pacific Oyster (*C. gigas*)

Douillet and Langdon [148,149] tested bacteria-free oyster larvae (*C. gigas*) cultured under aseptic conditions, fed axenic algae (*I. galbana*), and inoculated with the marine bacterium *Alteromonas* sp. CA2 in the rearing water. Results demonstrated that addition of strain CA2 enhanced larval survival (21–22%) and growth rate (16–21%), contributed to normal distributions in size–frequency of larvae populations, and reduced the proportion of slow-growing larvae. Such results suggest a bacterial nutritional contribution to larval growth.

Gibson [150] tested *in vitro* antagonism activity of three strains of *Aeromonas media* (A161, A164, and A199) against a variety of fish/shellfish pathogens. Among these strains, the strain A199 displayed a bacteriocin activity of broad inhibitory spectrum to all strains tested of *A. caviae, A. hydrophila, A. salmonicida, A. veronii, Listonella anguillarum, Photobacterium damsella, Y. ruckeri,* and eight species of *Vibrio*. However, an important question is whether the activity observed *in vitro* could be repeated *in vivo*. Therefore, field trials were carried out, showing that the introduction of the strain A199 to the host animal *C. gigas* protected the larvae when challenged with *Vibrio tubiashii*. These findings may perhaps contribute to reduction in heavy losses in the oyster-producing industry where infectious outbreaks can often be faced.

Similarly, Nakamura *et al.* [151] isolated marine bacterial strains from the rearing seawater of oyster brood stock and screened the strains having suppressive activity for the growth of pathogenic vibrios (*V. alginolyticus, V. tubiashii,* and *Vibrio* sp.). Among screened isolates, the Gram-negative short rod bacterium strain S21 was found to express the highest vibriostatic activity and did not have adverse effect on the survival of larval oysters. The addition of this bacterium into the oyster rearing water enhanced about 70% of the larval survival rate after challenge with *V. alginolyticus*, indicating that strain S21 protected larvae from this pathogen and could be used as an effective biocontrol agent for vibriosis in the larval oyster culture.

19.4.3 Other Mollusks

19.4.3.1 Great Scallop (*P. maximus*)

The marine bacterium *Phaeobacter gallaeciensis* BS107 (formerly *R. gallaeciensis*) was shown to express antagonistic activity to *Vibrio* species on agar plates thanks to the production of a proteinaceous antibacterial substance that was sensitive to trypsin but stable to heat [146]. At the concentration of 10^6 cells/ml this bacterium significantly enhanced scallop larval survival, thus being a putative probiotic for the rearing process. Another antimicrobial protein purified from the marine bacterium *Pseudoalteromonas* sp. X153 with a molecular mass of 87 kDa was found to induce this strain highly active against human pathogenic strains involved in dermatologic diseases, and marine bacteria including various ichthyopathogenic *Vibrio* strains. In the scallop rearing condition, the X153 bacterium protected the larvae against mortality when challenging with ichthyopathogenic *Vibrio*, so it could be a promising probiotic candidate in marine aquaculture [147].

19.4.3.2 **European Flat Oyster (*O. edulis*)**

Prado *et al.* [152] isolated a total of 523 bacterial strains during a 4-year period from the hatcheries of flat oysters and clams in Galicia, Spain. All of the strains were tested for their antibacterial activity against three larval pathogens (*V. anguillarum* USC-72, *V. neptunius* PP-145.98, and *Vibrio* sp. PP-203). Of the isolates, four similar strains belonging to the genus *Phaeobacter* showed the strongest activity. Strain PP-154, selected as a representative of this group, displayed a wide spectrum of inhibitory activity against aquaculture pathogens, especially against members of the genus *Vibrio*, which is responsible for most larval deaths. The inhibitory ability of such strain on solid medium was confirmed in seawater experiments, and the optimal conditions for antibacterial activity were established. These strains are promising probiotics for aquaculture facilities. Their potential benefit is based on the capacity to control the proliferation of a variety of aquaculture bacterial pathogens in mollusk larval cultures.

19.4.3.3 **Cortez Oyster (*C. corteziensis*)**

Recently, Campa-Cordova *et al.* [153] studied the effects of three probiotic bacteria (lactic acid bacteria strain NS61, *P. aeruginosa* YC58, *Burkholderia cepacia* Y021) isolated from different marine animals on the survival and growth of Cortez oyster larvae, *C. corteziensis*. The results demonstrated that the introduction of lactic acid bacteria or mixed bacilli, directly into culture tanks, enhanced survival rate but did not increase the growth rate of oyster larvae.

References

1 FAO (2012) Technical guidelines for responsible fisheries aquaculture development, http://www.fao.org/DOCREP/003/W4493E/w4493e03.htm (accessed June 2012).

2 Shakouri, B. and Yazdi, S.K. (2012) The sustainable marine aquaculture. *Adv. Environ. Biol.*, **6**, 18–23.

3 Romanowski, N. (2006) *Sustainable Freshwater Aquaculture: the Complete Guide from Backyard to Investor*, University of New South Wales Press.

4 Stickney, R.R. and Mcvey, J.P. (2004) *Responsible Marine Aquaculture*, World Aquaculture Society.

5 Arun, S.S., Ferose, K., and Linga, P.D. (2012) Fish vaccination: a health management tool for aquaculture, http://aquafind.com/articles/Vaccination.php (accessed June 2012).

6 Defoirdt, T., Sorgeloos, P., and Bossier, P. (2011) Alternatives to antibiotics for the control of bacterial disease in aquaculture. *Curr. Opin. Microbiol.*, **14**, 251–258.

7 Ram, C.S. and Sharma, P. (2012) Probiotics: the new ecofriendly alternative measures of disease control for sustainable aquaculture. *J. Fish. Aquat. Sci.*, **7**, 72–103.

8 Gibson, G.R. and Roberfroid, M.B. (1995) Dietary modulation of the human colonic microbiota: introducing the concept of prebiotics. *J. Nutr.*, **125**, 1401–1412.

9 Merrifield, D.L., Dimitroglou, A., Foey, A., Davies, S.J., Baker, R.T.M., Bøgwald, J., Castex, M., and Ringø, E. (2010) The current status and future focus of probiotic and prebiotic applications for salmonids. *Aquaculture*, **302**, 1–18.

10 Ringo, E., Olsen, R.E., Gifstad, T.O., Dalmo, R.A., Amlund, H., Hemre, G.I., and Bakke, A.M. (2010) Prebiotics in aquaculture: a review. *Aquacult. Nutr.*, **16**, 117–136.

11 Parker, G.A. (1974) Assessment strategy and the evolution of animal conflicts. *J. Theor. Biol.*, **47**, 223–243.

12 Zokaeifar, H., Luis Balcazar, J., Kamarudin, M.S., Sijam, K., Arshad, A., and Saad, C.R. (2012) Selection and identification of non-pathogenic bacteria isolated from fermented pickles with antagonistic properties against two shrimp pathogens. *J. Antibiot. (Tokyo)*, **65** (6), 289–294.

13 Desriac, F., Defer, D., Bourgougnon, N., Brillet, B., LeCHevalier, P., and Fleury, Y. (2010) Bacteriocin as weapons in the marine animal-associated bacteria warfare: inventory and potential applications as an aquaculture probiotic. *Mar. Drugs*, **8**, 1153–1177.

14 Gatesoupe, F.J. (1989) Further advances in the nutritional and antibacterial treatments of rotifers as food for turbot larvae, Scophthalmus maximus, in *Aquaculture – A Biotechnology in Progress* (ed. N. de Pauw), European Aquaculture Society, Bredene, pp. 721–730.

15 Gatesoupe, F.J. (1997) Sidephore production and probiotic effect of *Vibrio* sp. associated with turbot larvae *Scophthalmus maximus*. *Aquat. Liv. Resour.*, **10**, 239–246.

16 Gatesoupe, F.J. (2008) Updating the importance of lactic acid bacteria in fish farming: natural occurrence and probiotic treatments. *J. Mol. Microbiol. Biotechnol.*, **14**, 107–114.

17 García de la Banda, I., Chereguini, O., and Rasines, I. (1992) Influenciade la adición de bacteria lácticas en el cultivo larvario delrodaballo (*Scophthalmus maximus* L.). *Bol. Inst. Esp. Oceanogr.*, **8**, 247–254.

18 Gatesoupe, F.J. (1990) The continuous feeding of turbot larvae *Scophthalmus maximus*, and control of the bacterial environment of rotifers. *Aquaculture*, **89**, 139–148.

19 Gatesoupe, F.J. (1991) Bacillus sp. spores as food additive for the rotifer Brachionus plicatilis: improvement of their bacterial environment and their dietary value for larval turbot, Scopthalamus maximus L. Fish Nutrition in Practice, in *Proceedings of the 4th International Symposium on Fish Nutrition and Feeding* (ed. S. Kaushik), Institut National de la Recherche Agronomique, Paris, pp. 561–568.

20 Gatesoupe, F.J. (1991) The effect of three strains of lactic bacteria on the production rate of rotifers, *Brachionus plicatilis*, and their dietary value for larval turbot, *Scophthalmus maximus*. *Aquaculture*, **96**, 335–342.

21 Gatesoupe, F.J. (1994) Lactic acid bacteria increase the resistance of turbot larvae, *Scophthalmus maximus*, against pathogenic *vibrio*. *Aquat. Liv. Resour.*, **7**, 277–282.

22 Ringo, E. and Vadstein, O. (1998) Colonization of *Vibrio pelagius* and *Aeromonas caviae* in early developing turbot (*Scophthalmus maximus* L.) larvae. *J. Appl. Microbiol.*, **84**, 227–233.

23 Ringo, E. and Birkbeck, T.H. (1999) Intestinal microflora of fish larvae and fry. *Aquacult. Res.*, **30**, 73–93.

24 Huys, L., Dhert, P.H., Robles, R., Ollevier, F., Sorgeloos, P., and Swings, J. (2001) Search for beneficial bacterial strains for turbot (*Scophthalmus maximus* L.) larviculture. *Aquaculture*, **193**, 25–37.

25 Hjelm, M., Bergh, O., Riaza, A., Nielsen, J., Melchiorsen, J., Jensen, S., Duncan, H., Ahrens, P., Birkbeck, H., and Gram, L. (2004) Selection and identification of autochthonous potential probiotic bacteria from turbot larvae (*Scophthalmus maximus*) rearing units. *Syst. Appl. Microbiol.*, **27**, 360–371.

26 Hjelm, M., Riaza, A., Formoso, F., Melchiorsen, J., and Gram, L. (2004) Seasonal incidence of autochthonous antagonistic *Roseobacter* spp. and Vibrionaceae strains in a turbot larva (*Scophthalmus maximus*) rearing system. *Appl. Environ. Microbiol.*, **70**, 7288–7294.
27 Bruhn, J.B., Nielsen, K.F., Hjelm, M., Hansen, M., Bresciani, J., Schulz, S., and Gram, L. (2005) Ecology, inhibitory activity, and morphogenesis of a marine antagonistic bacterium belonging to the *Roseobacter clade*. *Appl. Environ. Microbiol.*, **71**, 7263–7270.
28 Porsby, C.H., Nielsen, K.F., and Gram, L. (2008) *Phaeobacter* and *Ruegeria* species of the *Roseobacter clade* colonize separate niches in a Danish Turbot (*Scophthalmus maximus*)-rearing farm and antagonize *Vibrio anguillarum* under different growth conditions. *Appl. Environ. Microbiol.*, **74**, 7356–7364.
29 Villamil, L., Figueras, A., Toranzo, A.E., Planas, M., and Novoa, B. (2003) Isolation of a highly pathogenic *Vibrio pelagius* strain associated with mass mortalities of turbot, *Scophthalmus maximus* (L.), larvae. *J. Fish. Dis.*, **26**, 293–303.
30 Samuelsen, O.B., Nerland, A.H., Jorgensen, T., Schroder, M.B., Svasand, T., and Bergh, O. (2006) Viral and bacterial diseases of Atlantic cod *Gadus morhua*, their prophylaxis and treatment: a review. *Dis. Aquat. Organ.*, **71**, 239–254.
31 Hansen, G.H. and Olafsen, J.A. (1989) Bacterial colonization of cod (*Gadus morhua* L.) and halibut (*Hippoglossus hippoglossus*) eggs in marine aquaculture. *Appl. Environ. Microbiol.*, **55**, 1435–1446.
32 Olafsen, J.A. (1998) In Interactions between hosts and bacteria in aquaculture. Proceedings from the US-EC Workshop on Marine Microorganisms: Research Issues for Biotechnology, European Commission, Brussels, Belgium, European Commission, Brussels, Belgium, pp. 127–145.
33 Strom, E. and Ring, E. (1993) Changes in the bacterial composition of early developing cod, *Gadus morhua* (L.) larvae following inoculation of *Lactobacillus plantarum* into the water, in *Physiology and Biochemical Aspects of Fish Development* (eds B. Walther and H.J. Fyhn), University of Bergen, Bergen, Norway, pp. 226–228.
34 Gildberg, A, Mikkelsen, H., Sandaker, E., and Ring, E. (1997) Probiotic effect of lactic acid bacteria in the feed on growth and survival of fry of Atlantic cod (*Gadus morhua*). *Hydrobiologia*, **352**, 279–285.
35 Lauzon, H.L., Gudmundsdottir, S., Steinarsson, A., Oddgeirsson, M., Petursdottir, S.K., Reynisson, E., Bjornsdottir, R., and Gudmundsdottir, B.K. (2010) Effects of bacterial treatment at early stages of Atlantic cod (*Gadus morhua* L.) on larval survival and development. *J. Appl. Microbiol.*, **108**, 624–632.
36 Lauzon, H.L., Gudmundsdottir, S., Pedersen, M.H., Budde, B.B., and Gudmundsdottir, B.K. (2008) Isolation of putative probionts from cod rearing environment. *Vet. Microbiol.*, **132**, 328–339.
37 Sveinsdóttir, H., Steinarsson, A., and Gudmundsdóttir, Á. (2009) Differential protein expression in early Atlantic cod larvae (*Gadus morhua*) in response to treatment with probiotic bacteria. *Comp. Biochem. Physiol.*, **4**, 249–254.
38 Lauzon, H.L., Gudmundsdottir, S., Petursdottir, S.K., Reynisson, E., Steinarsson, A., Oddgeirsson, M., Bjornsdottir, R., and Gudmundsdottir, B.K. (2010) Microbiota of Atlantic cod (*Gadus morhua* L.) rearing systems at pre- and posthatch stages and the effect of different treatments. *J. Appl. Microbiol.*, **109**, 1775–1789.
39 Lazado, C.C., Caipang, C.M., Brinchmann, M.F., and Kiron, V. (2011) *In vitro* adherence of two candidate probiotics from Atlantic cod and their interference with the adhesion of two pathogenic bacteria. *Vet. Microbiol.*, **148**, 252–259.
40 Fjellheim, A.J., Klinkenberg, G., Skjermo, J., Aasen, I.M., and Vadstein, O. (2010) Selection of candidate probionts by two different screening strategies from Atlantic cod (*Gadus morhua* L.) larvae. *Vet. Microbiol.*, **144**, 153–159.
41 Sveinsdottir, H., Steinarsson, A., and Gudmundsdottir, A. (2009) Differential protein expression in early Atlantic cod larvae (*Gadus morhua*) in response to

treatment with probiotic bacteria. *Comp. Biochem. Physiol.*, **4**, 249–254.

42 Joborn, A., Olsson, J.C., Westerdahl, A., Conway, P.L., and Kjelleberg, S. (1997) Colonization in the fish intestinal tract and production of inhibitory substances in intestinal mucus and faecal extracts by *Carnobacterium sp.* strain K1. *J. Fish. Dis.*, **20**, 383–392.

43 Robertson, P.A.W., O'Dowd, C., Burrells, C., Williams, P., and Austin, B. (2000) Use of *Carnobacterium* sp. as a probiotic for Atlantic salmon (*Salmo salar* L.) and rainbow trout (*Oncorhynchus mykiss*, Walbaum). *Aquaculture*, **185**, 235–243.

44 Andlid, T., Blomberg, L., Gustafsson, L., and Blomberg, A. (1999) Characterization of *Saccharomyces cerevisiae* CBS 7764 isolated from rainbow trout intestine. *Syst. Appl. Microbiol.*, **22**, 145–155.

45 Gram, L., Melchiorsen, J., Spanggaard, B., Huber, I., and Nielsen, T.F. (1999) Inhibition of *vibrio anguillarum* by *Pseudomonas fluorescens* AH2, a possible probiotic treatment of fish. *Appl. Environ. Microbiol.*, **65**, 969–973.

46 Spanggaard, B., Huber, I., Nielsen, J., Sick, E.B., Pipper, C.B., Martinussen, T., Slierendrecht, W.J., and Gram, L. (2001) The probiotic potential against vibriosis of the indigenous microflora of rainbow trout. *Environ. Microbiol.*, **3**, 755–765.

47 Raida, M.K., Larsen, J.L., Nielsen, M.E., and Buchmann, K. (2003) Enhanced resistance of rainbow trout, *Oncorhynchus mykiss* (Walbaum), against *Yersinia ruckeri* challenge following oral administration of *Bacillus subtilis* and *B. licheniformis* (BioPlus2B). *J. Fish. Dis.*, **26**, 495–498.

48 Nikoskelainen, S., Ouwehand, A.C., Bylund, G., Salminen, S., and Lilius, E.M. (2003) Immune enhancement in rainbow trout (*Oncorhynchus mykiss*) by potential probiotic bacteria (*Lactobacillus rhamnosus*). *Fish Shellfish Immunol.*, **15**, 443–452.

49 Panigrahi, A., Kiron, V., Kobayashi, T., Puangkaew, J., Satoh, S., and Sugita, H. (2004) Immune responses in rainbow trout *Oncorhynchus mykiss* induced by a potential probiotic bacteria *Lactobacillus rhamnosus* JCM 1136. *Vet. Immunol. Immunopathol.*, **102**, 379–388.

50 Panigrahi, A., Kiron, V., Satoh, S., Hirono, I., Kobayashi, T., Sugita, H., Puangkaew, J., and Aoki, T. (2007) Immune modulation and expression of cytokine genes in rainbow trout *Oncorhynchus mykiss* upon probiotic feeding. *Dev. Comp. Immunol.*, **31**, 372–382.

51 Brunt, J. and Austin, B. (2005) Use of a probiotic to control lactococcosis and streptococcosis in rainbow trout, *Oncorhynchus mykiss* (Walbaum). *J. Fish. Dis.*, **28**, 693–701.

52 Brunt, J., Newaj-Fyzul, A., and Austin, B. (2007) The development of probiotics for the control of multiple bacterial diseases of rainbow trout, *Oncorhynchus mykiss* (Walbaum). *J. Fish. Dis.*, **30**, 573–579.

53 Brunt, J., Hansen, R., Jamieson, D.J., and Austin, B. (2008) Proteomic analysis of rainbow trout (*Oncorhynchus mykiss*, Walbaum) serum after administration of probiotics in diets. *Vet. Immunol. Immunopathol.*, **121**, 199–205.

54 Abbass, A., Sharifuzzaman, S.M., and Austin, B. (2010) Cellular components of probiotics control *Yersinia ruckeri* infection in rainbow trout, *Oncorhynchus mykiss* (Walbaum). *J. Fish. Dis.*, **33**, 31–37.

55 Pieters, N., Brunt, J., Austin, B., and Lyndon, A.R. (2008) Efficacy of in-feed probiotics against *Aeromonas bestiarum* and *Ichthyophthirius multifiliis* skin infections in rainbow trout (*Oncorhynchus mykiss*, Walbaum). *J. Appl. Microbiol.*, **105**, 723–732.

56 Kim, D.H. and Austin, B. (2006a) Cytokine expression in leucocytes and gut cells of rainbow trout, *Oncorhynchus mykiss* Walbaum, induced by probiotics. *Vet. Immunol. Immunopathol.*, **114**, 297–304.

57 Kim, D.H. and Austin, B. (2006) Innate immune responses in rainbow trout (*Oncorhynchus mykiss*, Walbaum) induced by probiotics. *Fish Shellfish Immunol.*, **21**, 513–524.

58 Kim, D.H. and Austin, B. (2008) Characterization of probiotic carnobacteria isolated from rainbow trout (*Oncorhynchus mykiss*) intestine. *Lett. Appl. Microbiol.*, **47**, 141–147.

59 Sharifuzzaman, S.M. and Austin, B. (2010a) Development of protection in

rainbow trout (*Oncorhynchus mykiss*, Walbaum) to *Vibrio anguillarum* following use of the probiotic Kocuria SM1. *Fish Shellfish Immunol.*, **29**, 212–216.

60 Sharifuzzaman, S.M. and Austin, B. (2009) Influence of probiotic feeding duration on disease resistance and immune parameters in rainbow trout. *Fish Shellfish Immunol.*, **27**, 440–445.

61 Sharifuzzaman, S.M. and Austin, B. (2010b) Kocuria SM1 controls vibriosis in rainbow trout (*Oncorhynchus mykiss*, Walbaum). *J. Appl. Microbiol.*, **108**, 2162–2170.

62 Sharifuzzaman, S.M., Abbass, A., Tinsley, J.W., and Austin, B. (2011) Subcellular components of probiotics Kocuria SM1 and Rhodococcus SM2 induce protective immunity in rainbow trout (*Oncorhynchus mykiss*, Walbaum) against *Vibrio anguillarum. Fish Shellfish Immunol.*, **30**, 347–353.

63 Balcazar, J.L., Vendrell, D., de Blas, I., Ruiz-Zarzuela, I., Girones, O., and Muzquiz, J.L. (2006) Immune modulation by probiotic strains: quantification of phagocytosis of *Aeromonas salmonicida* by leukocytes isolated from gut of rainbow trout (*Oncorhynchus mykiss*) using a radiolabelling assay. *Comp. Immunol. Microbiol. Infect. Dis.*, **29**, 335–343.

64 Balcazar, J.L., de Blas, I., Ruiz-Zarzuela, I., Vendrell, D., Girones, O., and Muzquiz, J.L. (2007) Enhancement of the immune response and protection induced by probiotic lactic acid bacteria against furunculosis in rainbow trout (*Oncorhynchus mykiss*). *FEMS Immunol. Med. Microbiol.*, **51**, 185–193.

65 Vendrell, D., Balcazar, J.L., de Blas, I., Ruiz-Zarzuela, I., Girones, O., and Luis Muzquiz, J. (2008) Protection of rainbow trout (*Oncorhynchus mykiss*) from lactococcosis by probiotic bacteria. *Comp. Immunol. Microbiol. Infect. Dis.*, **31**, 337–345.

66 Newaj-Fyzul, A., Adesiyun, A.A., Mutani, A., Ramsubhag, A., Brunt, J., and Austin, B. (2007) *Bacillus subtilis* AB1 controls Aeromonas infection in rainbow trout (*Oncorhynchus mykiss*, Walbaum). *J. Appl. Microbiol.*, **103**, 1699–1706.

67 Arijo, S., Brunt, J., Chabrillon, M., Diaz-Rosales, P., and Austin, B. (2008) Subcellular components of *Vibrio harveyi* and probiotics induce immune responses in rainbow trout, *Oncorhynchus mykiss* (Walbaum), against *V. harveyi*. *J. Fish. Dis.*, **31**, 579–590.

68 Capkin, E. and Altinok, I. (2009) Effects of dietary probiotic supplementations on prevention/treatment of Yersiniosis disease. *J. Appl. Microbiol.*, **106**, 1147–1153.

69 Perez-Sanchez, T., Balcazar, J.L., Garcia, Y., Halaihel, N., Vendrell, D., de Blas, I., Merrifield, D.L., and Ruiz-Zarzuela, I. (2011a) Identification and characterization of lactic acid bacteria isolated from rainbow trout, *Oncorhynchus mykiss* (Walbaum), with inhibitory activity against *Lactococcus garvieae*. *J. Fish. Dis.*, **34**, 499–507.

70 Perez-Sanchez, T., Balcazar, J.L., Merrifield, D.L., Carnevali, O., Gioacchini, G., de Blas, I., and Ruiz-Zarzuela, I. (2011) Expression of immune-related genes in rainbow trout (*Oncorhynchus mykiss*) induced by probiotic bacteria during *Lactococcus garvieae* infection. *Fish Shellfish Immunol.*, **31**, 196–201.

71 Strom-Bestor, M. and Wiklund, T. (2011) Inhibitory activity of *Pseudomonas* sp. on *Flavobacterium psychrophilum, in vitro*. *J. Fish. Dis.*, **34**, 255–264.

72 Korkea-aho, T.L., Heikkinen, J., Thompson, K.D., von Wright, A., and Austin, B. (2011) *Pseudomonas* sp. M174 inhibits the fish pathogen *Flavobacterium psychrophilum*. *J. Appl. Microbiol.*, **111**, 266–277.

73 Korkea-Aho, T.L., Papadopoulou, A., Heikkinen, J., von Wright, A., Adams, A., Austin, B., and Thompson, K.D. (2012) *Pseudomonas* M162 confers protection against rainbow trout fry syndrome by stimulating immunity. *J. Appl. Microbiol.*, **113** (1), 24–35

74 Sica, M.G., Brugnoni, L.I., Marucci, P.L., and Cubitto, M.A. (2012) Characterization of probiotic properties of lactic acid bacteria isolated from an estuarine environment for application in rainbow trout (*Oncorhynchus mykiss*, Walbaum) farming. *Antonie Van Leeuwenhoek.*, **101** (4), 869–879.

75 Maeda, M., Nogami, K., Kanematsu, M., and Hirayama, K. (1997) The concept of biological control methods in aquaculture. *Hydrobiologia*, **358**, 285–290.

76 Makridis, P., Martins, S., Vercauteren, T., Van Driessche, K., Decamp, O., and Dinis, M.T. (2005) Evaluation of candidate probiotic strains for gilthead sea bream larvae (*Sparus aurata*) using an *in vivo* approach. *Lett. Appl. Microbiol.*, **40**, 274–277.

77 Carnevali, O., Zamponi, M.C., Sulpizio, R., Rollo, A., Nardi, M., Orpianesi, C., Silvi, S., Caggiano, M., Polzonetti, A.M., and Cresci, A. (2004) Administration of a probiotic strain to improve sea bream wellness during development. *Aquacult. Int.*, **12**, 377–386.

78 Picchietti, S., Mazzini, M., Taddei, A.R., Renna, R., Fausto, A.M., Mulero, V., Carnevali, O., Cresci, A., and Abelli, L. (2007) Effects of administration of probiotic strains on GALT of larval gilthead seabream: immunohistochemical and ultrastructural studies. *Fish Shellfish Immunol.*, **22**, 57–67.

79 Diaz-Rosales, P., Salinas, I., Rodriguez, A., Cuesta, A., Chabrillon, M., Balebona, M.C., Morinigo, M.A., Esteban, M.A., and Meseguer, J. (2006) Gilthead seabream (*Sparus aurata* L.) innate immune response after dietary administration of heat-inactivated potential probiotics. *Fish Shellfish Immunol.*, **20**, 482–492.

80 Salinas, I., Abelli, L., Bertoni, F., Picchietti, S., Roque, A., Furones, D., Cuesta, A., Meseguer, J., and Esteban, M.A. (2008) Monospecies and multispecies probiotic formulations produce different systemic and local immunostimulatory effects in the gilthead seabream (*Sparus aurata* L.). *Fish Shellfish Immunol.*, **25**, 114–123.

81 Salinas, I., Diaz-Rosales, P., Cuesta, A., Meseguer, J., Chabrillon, M., Morinigo, M.A., and Esteban, M.A. (2006) Effect of heat-inactivated fish and non-fish derived probiotics on the innate immune parameters of a teleost fish (*Sparus aurata* L.). *Vet. Immunol. Immunopathol.*, **111**, 279–286.

82 Reyes-Becerril, M., Salinas, I., Cuesta, A., Meseguer, J., Tovar-Ramirez, D., Ascencio-Valle, F., and Esteban, M.A. (2008) Oral delivery of live yeast *Debaryomyces hansenii* modulates the main innate immune parameters and the expression of immune-relevant genes in the gilthead seabream (*Sparus aurata* L.). *Fish Shellfish Immunol.*, **25**, 731–739.

83 Abelli, L., Randelli, E., Carnevali, O., and Picchietti, S. (2009) Stimulation of gut immune system by early administration of probiotic strains in *Dicentrarchus labrax* and *Sparus aurata*. *Ann. N. Y. Acad. Sci.*, **1163**, 340–342.

84 Frouel, S., Le Bihan, E., Serpentini, A., Lebel, J.M., Koueta, N., and Nicolas, J.L. (2008) Preliminary study of the effects of commercial lactobacilli preparations on digestive metabolism of juvenile sea bass (*Dicentrarchus labrax*). *J. Mol. Microbiol. Biotechnol.*, **14**, 100–106.

85 Picchietti, S., Fausto, A.M., Randelli, E., Carnevali, O., Taddei, A.R., Buonocore, F., Scapigliati, G., and Abelli, L. (2009) Early treatment with *Lactobacillus delbrueckii* strain induces an increase in intestinal T-cells and granulocytes and modulates immune-related genes of larval *Dicentrarchus labrax* (L.). *Fish Shellfish Immunol.*, **26**, 368–376.

86 Sorroza, L., Padilla, D., Acosta, F., Roman, L., Grasso, V., Vega, J., and Real, F. (2012) Characterization of the probiotic strain *Vagococcus fluvialis* in the protection of European sea bass (*Dicentrarchus labrax*) against vibriosis by *Vibrio anguillarum*. *Vet. Microbiol.*, **155**, 369–373.

87 Vadstein, O., Oie, G., Olsen, Y., Salvesen, I., Skjermo, J., and Skjak-Brak, G. (1993) A strategy to obtain microbial control during larval development of marine fish. Proceedings of the First International Conference of on Fish Farming Technology, Balkema, Rotterdam, pp. 69–75.

88 Skjermo, J., Salvesen, I., Oie, G., Olsen, Y., and Vadstein, O. (1997) Microbially matured water: a technique for selection of non opportunistic bacterial flora in water may improve performance of marine larvae. *Aquacult. Int.*, **5**, 13–28.

89 Ottesen, O.H. and Olafsen, J.A. (2000) Effects on survival and mucous cell proliferation of Atlantic halibut, *Hippoglossus hippoglossus* L, larvae

following microflora manipulation. *Aquaculture*, **187**, 225–238.

90 Makridis, P., Bergh, O., Skjermo, J., and Vadstein, O. (2001) Addition of bacteria bioencapsulated in *Artemia metanauplii* to a rearing system for halibut larvae. *Aquacult. Int.*, **9**, 225–235.

91 Byun, J.W., Park, S.C., Benno, Y., and Oh, T.K. (1997) Probiotic effect of *Lactobacillus* sp. DS-12 in flounder (*Paralichthys olivaceus*). *J. Gen. Appl. Microbiol.*, **43**, 305–308.

92 Cai, Y., Benno, Y., Nakase, T., and Oh, T.K. (1998) Specific probiotic characterization of *Weissella hellenica* DS-12 isolated from flounder intestine. *J. Gen. Appl. Microbiol.*, **44**, 311–316.

93 Kim, J.S., Harikrishnan, R., Kim, M.C., Balasundaram, C., and Heo, M.S. (2010) Dietary administration of *Zooshikella sp.* enhance the innate immune response and disease resistance of *Paralichthys olivaceus* against *Sreptococcus iniae*. *Fish Shellfish Immunol.*, **29**, 104–110.

94 Kim, Y.R., Kim, E.Y., Choi, S.Y., Hossain, M.T., Oh, R., Heo, W.S., Lee, J.M., Cho, Y. C., and Kong, I.S. (2012) Effect of a probiotic strain, *Enterococcus faecium*, on the immune responses of Olive Flounder (*Paralichthys olivaceus*). *J. Microbiol. Biotechnol.*, **22**, 526–529.

95 Harikrishnan, R., Balasundaram, C., and Heo, M.S. (2010) Scuticociliatosis and its recent prophylactic measures in aquaculture with special reference to South Korea Taxonomy, diversity and diagnosis of scuticociliatosis: Part I. Control strategies of scuticociliatosis: Part II. *Fish Shellfish Immunol.*, **29**, 15–31.

96 Harikrishnan, R., Balasundaram, C., and Heo, M.S. (2010c) Effect of probiotics enriched diet on *Paralichthys olivaceus* infected with lymphocystis disease virus (LCDV). *Fish Shellfish Immunol.*, **29**, 868–874.

97 Vine, N.G., Leukes, W.D., and Kaiser, H. (2004) *In vitro* growth characteristics of five candidate aquaculture probiotics and two fish pathogens grown in fish intestinal mucus. *FEMS Microbiol. Lett.*, **231**, 145–152.

98 Vine, N.G., Leukes, W.D., Kaiser, H., Daya, S., Baxter, J., and Hecht, T. (2004) Competition for attachment of aquaculture candidate probiotic and pathogenic bacteria on fish intestinal mucus. *J. Fish. Dis.*, **27**, 319–326.

99 Avella, M.A., Olivotto, I., Silvi, S., Place, A. R., and Carnevali, O. (2010) Effect of dietary probiotics on clownfish: a molecular approach to define how lactic acid bacteria modulate development in a marine fish. *Am. J. Physiol.*, **298**, 359–371.

100 Kennedy, S.B., Tucker, J.W.J., Neidig, C.L., Vermeer, G.K., Cooper, V.R., Jarrell, J.L., and Sennett, D.G. (1998) Bacterial management strategies for stock enhancement of warm water marine fish: a case study with common snook (*Centropomus undecimalis*). *Bull. Mar. Sci.*, **62**, 573–588.

101 Gatesoupe, F.J. (2002) Probiotic and formaldehyde treatments of *Artemia* nauplii as food for larval pollack. *Pollachius pollachius*. *Aquaculture*, **212**, 347–360.

102 Irianto, A., Robertson, P.A., and Austin, B. (2003) Oral administration of formalin-inactivated cells of *Aeromonas hydrophila* A3–51 controls infection by atypical *A. salmonicida* in goldfish, *Carassius auratus* (L.). *J. Fish. Dis.*, **26**, 117–120.

103 Harikrishnan, R., Balasundaram, C., and Heo, M.S. (2009) Effect of chemotherapy, vaccines and immunostimulants on innate immunity of goldfish infected with *Aeromonas hydrophila*. *Dis. Aquat. Organ.*, **88**, 45–54.

104 Chu, W., Lu, F., Zhu, W., and Kang, C. (2011) Isolation and characterization of new potential probiotic bacteria based on quorum-sensing system. *J. Appl. Microbiol.*, **110**, 202–208.

105 Maeda, M. and Liao, I.C. (1992) Effect of bacterial population on the growth of a prawn larva, *Penaeus monodon*. *Bull. Natl. Res. Inst. Aquacult.*, **21**, 25–29.

106 Maeda, M. (1994) Biocontrol of the larvae rearing biotope in aquaculture. *Bull. Natl. Res. Inst. Aquacult.*, **1**, 71–74.

107 Nogami, K., Hamasaki, K., Maeda, M., and Hirayama, K. (1997) Biocontrol method in aquaculture for rearing the swimming crab larvae *Portunus trituberculatus*. *Hydrobiologia*, **358**, 291–295.

108 Moriarty, D.J.W. and Body, A.G.C. (1995) Modifying microbial ecology in ponds: the

key to sustainable aquaculture. Proceedings of Fish Asia '95 Conference: 2nd Asian Aquaculture and Fisheries Exhibition and Conference, RAI Exhibitions, Singapore, Singapore, pp. 1–10.

109 Moriarty, D.J.W. (1998) Control of luminous *Vibrio* species in penaeid aquaculture ponds. *Aquaculture*, **164**, 351–358.

110 Dalmin, G., Kathiresan, K., and Purushothaman, A. (2001) Effect of probiotics on bacterial population and health status of shrimp in culture pond ecosystem. *Indian J. Exp. Biol.*, **39**, 939–942.

111 Rengpipat, S., Tunyanun, A., Fast, A.W., Piyatiratitivorakul, S., and Menasveta, P. (2003) Enhanced growth and resistance to *Vibrio* challenge in pond-reared black tiger shrimp *Penaeus monodon* fed a *Bacillus* probiotic. *Dis. Aquat. Organ.*, **55**, 169–173.

112 Vaseeharan, B. and Ramasamy, P. (2003) Control of pathogenic *Vibrio* spp. by *Bacillus subtilis* BT23, a possible probiotic treatment for black tiger shrimp *Penaeus monodon*. *Lett. Appl. Microbiol.*, **36**, 83–87.

113 Alavandi, S.V., Vijayan, K.K., Santiago, T. C., Poornima, M., Jithendran, K.P., Ali, S.A., and Rajan, J.J. (2004) Evaluation of *Pseudomonas* sp. PM 11 and *Vibrio fluvialis* PM 17 on immune indices of tiger shrimp, *Penaeus monodon*. *Fish Shellfish Immunol.*, **17**, 115–120.

114 Ravi, A.V., Musthafa, K.S., Jegathammbal, G., Kathiresan, K., and Pandian, S.K. (2007) Screening and evaluation of probiotics as a biocontrol agent against pathogenic *Vibrios* in marine aquaculture. *Lett. Appl. Microbiol.*, **45**, 219–223.

115 Preetha, R., Jayaprakash, N.S., and Singh, I.S. (2007) Synechocystis MCCB 114 and 115 as putative probionts for *Penaeus monodon* post-larvae. *Dis. Aquat. Organ.*, **74**, 243–247.

116 Hill, J.E., Baiano, J.C., and Barnes, A.C. (2009) Isolation of a novel strain of *Bacillus pumilus* from penaeid shrimp that is inhibitory against marine pathogens. *J. Fish. Dis.*, **32**, 1007–1016.

117 Antony, S.P., Singh, I.S., Sudheer, N.S., Vrinda, S., Priyaja, P., and Philip, R. (2011) Molecular characterization of a crustin-like antimicrobial peptide in the giant tiger shrimp, *Penaeus monodon*, and its expression profile in response to various immunostimulants and challenge with WSSV. *Immunobiology*, **216**, 184–194.

118 Garriques, D. and Arevalo, G. (1995) An evaluation of the production and use of a live bacterial isolate to manipulate the microbial flora in the commercial production of *Penaeus vannamei* postlarvae in Ecuador, in *Swimming Through Troubled Water. Proceedings of the Special Session on Shrimp Farming, Aquaculture '95* (eds C.L. Browdy and J.S. Hopkins), World Aquaculture Society, Baton Rouge, LA, pp. 53–59.

119 Wang, Y., Zha, L., and Xu, Z. (2006) Effects of probiotics on *Penaeus vannamei* pond sediments. *Ying Yong Sheng Tai Xue Bao*, **17** (9), 1765–1767.

120 Chiu, C.H., Guu, Y.K., Liu, C.H., Pan, T.M., and Cheng, W. (2007) Immune responses and gene expression in white shrimp, *Litopenaeus vannamei*, induced by *Lactobacillus plantarum*. *Fish Shellfish Immunol.*, **23**, 364–377.

121 Balcazar, J.L and Rojas-Luna, T. (2007) Inhibitory activity of probiotic *Bacillus subtilis* UTM 126 against vibrio species confers protection against vibriosis in juvenile shrimp (*Litopenaeus vannamei*). *Curr. Microbiol.*, **55**, 409–412.

122 Li, K., Zheng, T., Tian, Y., Xi, F., Yuan, J., Zhang, G., and Hong, H. (2007) Beneficial effects of *Bacillus licheniformis* on the intestinal microflora and immunity of the white shrimp, *Litopenaeus vannamei*. *Biotechnol. Lett.*, **29**, 525–530.

123 Liu, C.H., Chiu, C.S., Ho, P.L., and Wang, S.W. (2009) Improvement in the growth performance of white shrimp, *Litopenaeus vannamei*, by a protease-producing probiotic, *Bacillus subtilis* E20, from natto. *J. Appl. Microbiol.*, **107**, 1031–1041.

124 Tseng, D.Y., Ho, P.L., Huang, S.Y., Cheng, S.C., Shiu, Y.L., Chiu, C.S., and Liu, C.H. (2009) Enhancement of immunity and disease resistance in the white shrimp, *Litopenaeus vannamei*, by the probiotic, *Bacillus subtilis* E20. *Fish Shellfish Immunol.*, **26**, 339–344.

125 Liu, K.F., Chiu, C.H., Shiu, Y.L., Cheng, W., and Liu, C.H. (2010) Effects of the probiotic, *Bacillus subtilis* E20, on the survival, development, stress tolerance, and immune status of white shrimp, *Litopenaeus vannamei* larvae. *Fish Shellfish Immunol.*, **28**, 837–844.

126 Thompson, J., Gregory, S., Plummer, S., Shields, R.J., and Rowley, A.F. (2010) An *in vitro* and *in vivo* assessment of the potential of *Vibrio* spp. as probiotics for the Pacific white shrimp, *Litopenaeus vannamei*. *J. Appl. Microbiol.*, **109**, 1177–1187.

127 Olmos, J., Ochoa, L., Paniagua-Michel, J., and Contreras, R. (2011) Functional feed assessment on *Litopenaeus vannamei* using 100% fish meal replacement by soybean meal, high levels of complex carbohydrates and *Bacillus* probiotic strains. *Mar. Drugs*, **9**, 1119–1132.

128 Kongnum, K. and Hongpattarakere, T. (2012) Effect of *Lactobacillus plantarum* isolated from digestive tract of wild shrimp on growth and survival of white shrimp (*Litopenaeus vannamei*) challenged with *Vibrio harveyi*. *Fish Shellfish Immunol.*, **32**, 170–177.

129 Balcazar, J.L., Rojas-Luna, T., and Cunningham, D.P. (2007) Effect of the addition of four potential probiotic strains on the survival of pacific white shrimp (*Litopenaeus vannamei*) following immersion challenge with *Vibrio parahaemolyticus*. *J. Invertebr. Pathol.*, **96**, 147–150.

130 Li, P., Burr, G.S., Gatlin, D.M., Hume, M.E., Patnaik, S., Castille, F.L., and Lawrence, A.L. (2007) Dietary supplementation of short-chain fructooligosaccharides influences gastrointestinal microbiota composition and immunity characteristics of Pacific white shrimp, *Litopenaeus vannamei*, cultured in a recirculating system. *J. Nutr.*, **137**, 2763–2768.

131 Castex, M., Lemaire, P., Wabete, N., and Chim, L. (2010) Effect of probiotic *Pediococcus acidilactici* on antioxidant defences and oxidative stress of *Litopenaeus stylirostris* under *Vibrio nigripulchritudo* challenge. *Fish Shellfish Immunol.*, **28**, 622–631.

132 Van Hai, N., Buller, N., and Fotedar, R. (2009) The use of customised probiotics in the cultivation of western king prawns (*Penaeus latisulcatus* Kishinouye, 1896). *Fish Shellfish Immunol.*, **27**, 100–104.

133 Bourne, D., Hoj, L., Webstera, N., Paynea, M., Skindersoeb, M., Givskovb, M., and Halla, M. (2007) Microbiological aspects of phyllosoma rearing of the ornate rock lobster *Panulirus ornatus*. *Aquaculture*, **268**, 274–287.

134 Payne, M.S., Hall, M.R., Sly, L., and Bourne, D.G. (2007) Microbial diversity within early-stage cultured *Panulirus ornatus* phyllosomas. *Appl. Environ. Microbiol.*, **73**, 1940–1951.

135 Payne, M.S., Hoj, L., Wietz, M., Hall, M.R., Sly, L., and Bourne, D.G. (2008) Microbial diversity of mid-stage palinurid phyllosoma from great barrier reef waters. *J. Appl. Microbiol.*, **105**, 340–350.

136 Nogami, K. and Maeda, M. (1992) Bacteria as biocontrol agents for rearing larvae of the crab *Portunus trituberculatus*. *Can. J. Fish. Aquat. Sci.*, **49**, 2373–2376.

137 Maeda, M. and Liao, I.C. (1994) Microbial processes in aquaculture environment and their importance for increasing crustacean production. *Jpn. Int. Res. Center Agricult. Sci.*, **28**, 283–288.

138 Li, K., Guan, W., Wei, G., Liu, B., Xu, J., Zhao, L., and Zhang, Y. (2007) Phylogenetic analysis of intestinal bacteria in the Chinese mitten crab (*Eriocheir sinensis*). *J. Appl. Microbiol.*, **103**, 675–682.

139 Prado, S., Romalde, J.L., and Barja, J.L. (2010) Review of probiotics for use in bivalve hatcheries. *Vet. Microbiol.*, **145**, 187–197.

140 Riquelme, C.E., Hayashida, G., Araya, R., Uchida, A., Satomi, M., and Ishida, Y. (1996) Isolation of a native bacterial strain from the scallop *Argopecten purpuratus* with inhibitory effects against pathogenic vibrios. *J. Shellfish Res.*, **15**, 369–374.

141 Lodeiros, C., Freites, L., Fernández, E., Vélez, A., and Bastardo, J. (1989) Efecto antibiótico de tres bacterias marinas en la supervivencia de larvas de la vieira Pecten ziczac infectadas con el germen *Vibrio anguillarum*. *Bol. Inst. Oceanogr. Venezuela Univ. Oriente*, **28**, 165–169.

142 Riquelme, C.E., Araya, R., Vergara, N., Rojas, A., Guaita, M., and Candia, M. (1997) Potential probiotic strains in the culture of the Chilean scallop *Argopecten* *purpuratus* (Lamarck, 1819). *Aquaculture*, **154**, 17–26.

143 Avendano, R.E. and Riquelme, C.E. (1999) Establishment of mixed-culture probiotics and microalgae as food for bivalve larvae. *Aquacult. Res.*, **30**, 893–900.

144 Riquelme, C.E., Jorquera, M.A., Rosas, A.I., Avendano, R.E., and Reyes, N. (2001) Addition of inhibitor-producing bacteria to mass cultures of *Argopecten purpuratus* larvae (Lamarck, 1819). *Aquaculture*, **192**, 111–119.

145 Riquelme, C.E., Araya, R., and Escribano, R. (2000) Selective incorporation of bacteria by *Argopecten purpuratus* larvae: implications for the use of probiotics in culturing systems of the Chilean scallop. *Aquaculture*, **181**, 25–36.

146 Ruiz-Ponte, C., Samain, J.F., Sanchez, J.L., and Nicolas, J.L. (1999) The benefit of a *Roseobacter* species on the survival of scallop larvae. *Mar. Biotechnol. (NY)*, **1**, 52–59.

147 Longeon, A., Peduzzi, J., Barthelemy, M., Corre, S., Nicolas, J.L., and Guyot, M. (2004) Purification and partial identification of novel antimicrobial protein from marine bacterium *Pseudoalteromonas* species strain X153. *Mar. Biotechnol.*, **6**, 633–641.

148 Douillet, P. and Langdon, C.J. (1993) Effects of marine bacteria on the culture of axenic oyster *Crassostrea gigas* (Thunberg) larvae. *Biol. Bull.*, **184**, 36–51.

149 Douillet, P.A. and Langdon, C.J. (1994) Use of a probiotic for the culture of larvae of the pacific oyster (*Crassostrea gigas*). *Aquaculture*, **119**, 25–40.

150 Gibson, L.F. (1998) Bacteriocin activity and probiotic activity of *Aeromonas media*. *J. Appl. Microbiol.*, **85** (Suppl. 1), 243S–248S.

151 Nakamura, A., Takahashi, K.G., and Mori, K. (1999) Vibriostatic bacteria isolated from rearing seawater of oyster brood stock: potentiality as biocontrol agents for vibriosis in oyster larvae. *Fish Pathol.*, **34**, 139–144.

152 Prado, S., Montes, J., Romalde, J.L., and Barja, J.L. (2009) Inhibitory activity of Phaeobacter strains against aquaculture pathogenic bacteria. *Int. Microbiol.*, **12**, 107–114.

153 Campa-Cordova, A.I., Luna-Gonzalez, A., Mazon-Suastegui, J.M., Aguirre-Guzman, G., Ascencio, F., and Gonzalez-Ocampo, H.A. (2011) Effect of probiotic bacteria on survival and growth of Cortez oyster larvae, *Crassostrea corteziensis* (Bivalvia: Ostreidae)]. *Rev. Biol. Trop.*, **59**, 183–191.

20
Antimicrobial Properties of Eicosapentaenoic Acid (C20 : 5*n*−3)

Andrew P. Desbois

20.1
Introduction

20.1.1
Potential of Eicosapentaenoic Acid as an Antimicrobial Agent

Eicosapentaenoic acid (EPA) (C20 : 5$n-3$) is a long-chain polyunsaturated fatty acid (PUFA) that serves a multitude of functions in living cells. Interestingly, this fatty acid demonstrates antagonistic activity against a broad range of microorganisms, including bacteria, fungi, cyanobacteria, microalgae, protozoans, and viruses (Table 20.1). At present, there exists an immediate need to develop new antimicrobials because of the increasing prevalence of antibiotic resistance and the reduced efficacy of contemporary agents; fatty acids such as EPA warrant further investigation to address this global problem [1,2]. However, despite the relative abundance and low cost of obtaining EPA and other fatty acids, the commercial potential of these natural antimicrobial compounds is yet to be realized [2]. Thus, this chapter aims to present and discuss the antimicrobial actions of EPA, which may encourage its exploitation in new biotechnological applications.

20.1.2
Biochemistry, Biological Functions, and Natural Sources

EPA consists of a straight chain of 20 carbon atoms that terminates with a methyl (—CH_3) group at one end and a carboxyl (—COOH) group at the other (Figure 20.1a). Unsaturated C=C bonds in *cis* orientation are found at bond positions 5, 8, 11, 14, and 17 from the carboxyl end (Figure 20.1a). EPA is largely considered to be a marine-derived fatty acid because by far the most important producers of this compound are the algae, especially diatoms [23]. In these primary producers, fatty acids are formed in the plastids and then typically esterified to form glycolipids, which are particularly rich in EPA. Like other fatty acids, EPA is found in phospholipids that serve as the major structural constituents of cell membranes.

Marine Microbiology: Bioactive Compounds and Biotechnological Applications, First Edition.
Edited by Se-Kwon Kim

Table 20.1 Spectrum of antimicrobial activities exerted by eicosapentaenoic acid.

Microorganism	Susceptibility	References
Gram-positive bacteria		
Bacillus cereus	+	[3]
Bacillus subtilis	+	[4]
Bacillus weihenstephanensis	+	[3]
Bacillus fibrisolvens[a)]	+	[5,6]
Butyrivibrio hungatei[a)]	+	[5]
Clostridium aminophilum	+	[5]
Clostridium proteoclasticum	+	[5]
Eubacterium pyruvativorans	+	[5]
Eubacterium ruminantium	+	[5]
Lachnospira multipara	+	[5]
Lactococcus garvieae	+	[7]
Listeria monocytogenes	+	[4]
Micrococcus luteus	+	[3]
Peptostreptococcus anaerobius	−	[5]
Planococcus citreus	+	[3]
Pseudobutyrivibrio ruminis[a)]	+	[5]
Pseudobutyrivibrio xylanovorans[a)]	+	[5]
Ruminococcus albus	+	[5]
Ruminococcus flavefaciens	+	[5]
Staphylococcus aureus	+	[3, 4, 8–11]
Staphylococcus epidermidis	+	[3]
Streptococcus bovis	+	[5]
Streptococcus mutans	+	[12]
Gram-negative bacteria		
Aeromonas hydrophila	−	[3]
Aggregatibacter actinomycetemcomitans	+	[12]
Anaerovibrio lipolytica	−	[5]
Enterobacter aerogenes	−	[4]
Escherichia coli	−	[3,4]
Fibrobacter succinogenes	+	[5]
Fusobacterium nucleatum	+	[12]
Helicobacter pylori	+	[13]
Megasphaera elsdenii	−	[5]
Mitsuokella multiacidus	−	[5]
Photobacterium phosphoreum	+	[14]
Photobacterium sp.	+	[3]
Porphyromonas gingivalis	+	[12]
Prevotella albensis	+	[5]
Prevotella brevis	+	[5]
Prevotella bryantii	−	[5]
Prevotella ruminicola	−	[5]
Pseudoalteromonas haloplanktis	−	[3]
Pseudomonas aeruginosa NCIMB 10775	−	[3]
Pseudomonas aeruginosa KCTC2004	+	[4]
Psychrobacter immobilis	−	[3]
Ruminobacter amylophilus	−	[5]

Salmonella enteritidis	−	[4]
Salmonella typhimurium	−	[4]
Selenomonas ruminantium	−	[5]
Veillonella parvula	−	[5]
Vibrio anguillarum	+	[3,7]
Vibrio harveyi	+	[7]
Vibrio alginolyticus	+	[7]
Mycobacteria		
Mycobacterium aurum	+	[15]
Mycobacterium chelonae spp. *abscessus*	+	[15]
Mycobacterium chelonae spp. *chelonae*	+	[15]
Mycobacterium chitae	+	[15]
Mycobacterium diernhoferi	+	[15]
Mycobacterium duvalii	+	[15]
Mycobacterium flavescens	+	[15]
Mycobacterium fortuitum	+	[15]
Mycobacterium gilvum	+	[15]
Mycobacterium neoaurum	+	[15]
Mycobacterium parafortuitum	+	[15]
Mycobacterium phlei	+	[15]
Mycobacterium rhodesiae	+	[15]
Mycobacterium smegmatis	+	[15]
Mycobacterium thermoresistible	+	[15]
Mycobacterium vaccae	+	[15]
Fungi		
Candida albicans	+	[12,16]
Candida dubliniensis	+	[16]
Candida glabrata	−	[3]
Candida neoformans	−	[3]
Candida sp.	−	[3]
Saccharomyces cerevisiae	−	[3]
Cyanobacteria		
Anabaena sp.	+	[17]
Microalgae		
Chaetoceros gracile (Bacillariophyceae)	+	[14]
Chattonella marina (Raphidophyceae)	+	[18]
Chlorella vulgaris (Chlorophyceae)	+	[17]
Heterosigma akashiwo (Raphidophyceae)	+	[18,19]
Monoraphidium contortum (Chlorophyceae)	+	[17]
Protozoa		
Plasmodium falciparum	+	[20]
Viruses		
Hepatitis C	+	[21]

a) *Butyrivibrio* spp. and *Pseudobutyrivibrio* spp. stain Gram-negative but are ultrastructurally Gram-positive [22].

(a)

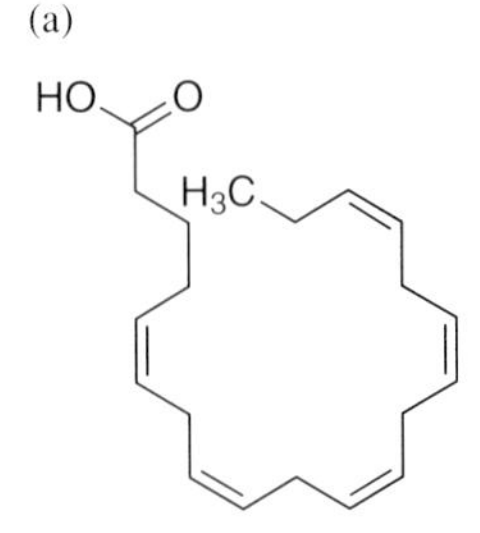

(b)

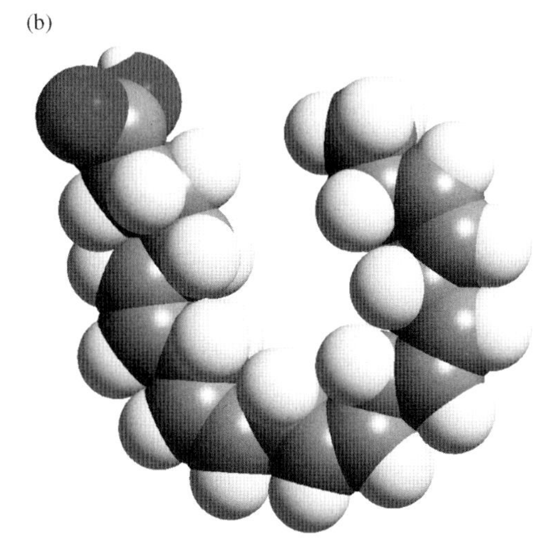

Figure 20.1 Molecular structure (a) and space-filled representation (b) of eicosapentaenoic acid.

Membranes containing high contents of PUFAs tend to demonstrate greater fluidity and flexibility. Greater proportions of these fatty acids are incorporated into membrane lipids at cold temperatures to maintain correct membrane functioning [24–26]. EPA can also be esterified to glycerol, forming compounds such as triglycerides that function as energy stores. In mammals, EPA exerts anti-inflammatory properties [27] and serves as the precursor of eicosanoid hormones and other bioactive compounds that play important roles in controlling inflammation and other critical immune processes [27–29]. Besides autotrophic microbes, certain marine heterotrophic bacteria can produce high quantities of EPA, particularly *Shewanella* spp. and species within the *Cytophaga–Flavobacterium–Bacteroides* group, and therefore have attracted attention as alternative sources of PUFAs [23]. EPA is accumulated through food chains; humans obtain much of this important fatty acid mainly through the consumption of seafood and freshwater fish. Consumption of EPA is associated with beneficial effects on human health and well-being, particularly with respect to cardiovascular health and inflammatory disorders [30]. It is possible that certain favorable effects may derive directly from its antimicrobial actions.

20.2
Spectrum of Antimicrobial Activity and Potency

20.2.1
Isolation and Confirmation of Antimicrobial Activity

Perhaps the first indication of the antimicrobial actions of EPA came in a report by Pesando [10]. A bioassay-guided fractionation approach was used to isolate and identify compounds responsible for the anti-*Staphylococcus aureus* activity of solvent extracts from the marine diatom *Asterionella japonica* [10]. The most potent antimicrobial fraction contained EPA; however, the activity of this fraction was not ascribed to this compound [10]. Instead, the action was attributed to photooxidative products of EPA because extracts prepared under dark conditions only became antibacterial once they had been exposed to light for 2 h [10]. Desbois *et al.* [3] isolated an antibacterial fraction from another marine diatom, *Phaeodactylum tricornutum*, which contained almost pure EPA. Subsequently, commercially available EPA (>99%) was used to demonstrate growth inhibitory activity against a range of bacteria, including *S. aureus* [3]. In this case, the antibacterial activity was attributed to EPA itself because the fatty acid remained active even when tested in the dark by disc diffusion assay [3]. EPA was re-extracted from the disk at the end of the assay, and there was no evidence for the presence of any other compounds, including photooxidized derivatives [3]. There are now numerous studies that have confirmed that EPA is effective against various bacteria, including human pathogens, gut bacteria, and marine and environmental isolates (Table 20.1). As often observed with long-chain PUFAs, EPA generally exerts more potent activity against Gram-positive species compared with Gram-negative strains [3–5], which is likely due to the structure of the Gram-negative outer cell membrane that can provide protection against the detrimental actions of free fatty acids [31,32].

20.2.2
Antibacterial Activities

Shin *et al.* [4] showed that EPA was effective at inhibiting the growth of important foodborne pathogens, including three Gram-positive species (*Bacillus subtilis*, *Listeria monocytogenes*, and *S. aureus*) and *Pseudomonas aeruginosa* (Gram-negative). The minimal inhibitory concentrations (MICs) against these strains ranged from 500 to 1350 mg/l [4], but these values are far greater than the concentrations reported in similar studies [3,11]. This may be due to the high concentration of bacterial inocula used by Shin *et al.* [4]. Indeed, Desbois *et al.* [3] showed that EPA inhibited *S. aureus* growth by >50% at just 6 mg/l (with a minimum bactericidal concentration (MBC) of 24 mg/l), while Zhang *et al.* [11] determined the MIC to be 125 mg/l. Moreover, EPA is effective against methicillin-resistant strains of *S. aureus* (MRSA), which are a major cause of patient mortality in healthcare institutions across the world [3,9]. EPA also exerts potent antibacterial activity against oral human pathogens; at 2.5 mg/l the growth of *S. mutans* (Gram-positive) was reduced by 93% during 16 h at 37 °C [12]. This PUFA was less effective against three oral Gram-negative species, but at 25 mg/l, it still inhibited bacterial growth by 46–72% [12] (Table 20.1). Furthermore, of six C18 and C20 unsaturated fatty acids examined, EPA, C18 : 3*n* − 3, and C18 : 3*n* − 6 showed the greatest potency against the etiological agent of peptic ulcers, *Helicobacter pylori* (Gram-negative), and at 75 mg/l, these fatty acids abolished the growth of this pathogen [13]. This finding prompted Thompson *et al.* [13] to speculate that EPA may provide natural protection against the development of peptic ulcers caused by *H. pylori*. Finally, in a study against 15 species of group IV mycobacteria, including the potential human pathogens *Mycobacterium fortuitum* and *Mycobacterium chelonae*, EPA prevented the growth of 80% of the isolates of each species (MIC_{80}) during 7 days at concentrations ranging from 12.5 to 100 mg/l [15] (Table 20.1 and Figure 20.2).

The effects of EPA have been evaluated against 24 strains of bacteria isolated originally from the guts of ruminants [5]. As expected, the Gram-positive species were by far the most sensitive to the actions of EPA, and antimicrobial potency (defined as an increase in lag phase) was of the order EPA > 22 : 6*n* − 3 > C18 : 3*n* − 3 > C18 : 2*n* − 6 [5]. A subsequent study confirmed the potency of EPA; at 50 mg/l, this PUFA was as effective as C22 : 6*n* − 3 and C18 : 3*n* − 6 at inhibiting

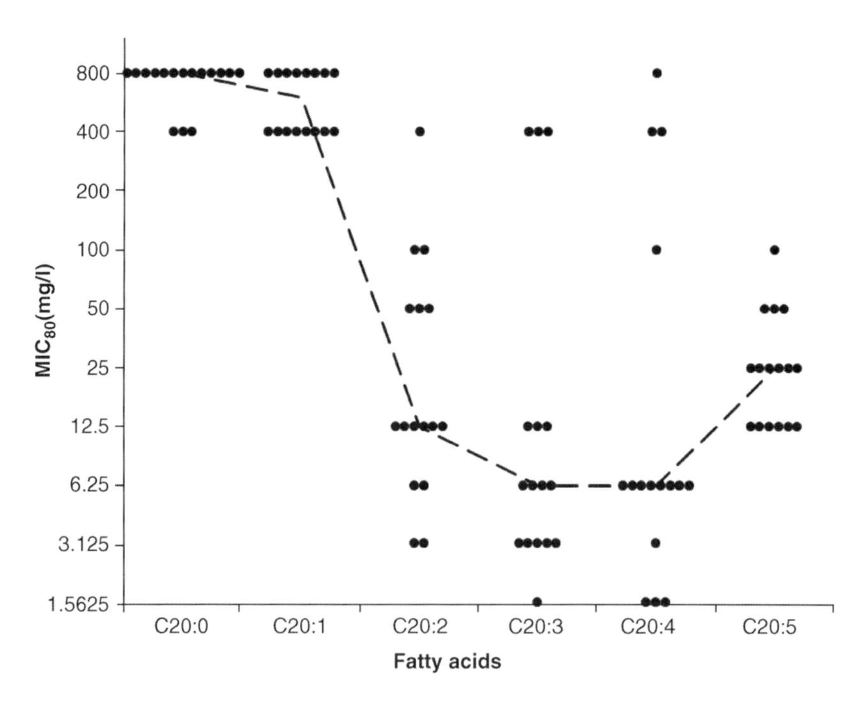

Figure 20.2 Minimum inhibitory concentrations to inhibit 80% of isolates (MIC_{80}) from 16 *Mycobacterium* spp. of six C20 fatty acids containing zero to five double bonds in the carbon chain. Each dot indicates the MIC_{80} for a single *Mycobacterium* spp., while the dashed line indicates the median MIC_{80} for each fatty acid tested. Data adapted from Ref. [15].

the growth of the ruminant bacterium *Butyrivibrio fibrisolvens* (Gram-positive) but more effective than C18 : 3n − 3, C18 : 2n − 6, C18 : 2n − 7t,9, C18 : 1n − 9, C18 : 1n − 7t, and C18 : 0 [6]. Marine bacteria are also susceptible to the actions of EPA. Benkendorff *et al.* [7] evaluated the activity of 11 medium- and long-chain saturated and unsaturated fatty acids against three *Vibrio* spp. (Gram-negative) and *Lactococcus garvieae* (Gram-positive). EPA was one of the more potent fatty acids examined, and at 10 mg/l EPA, it reduced the viability of exponential phase preparations of each of the four bacterial strains by >99% [7]. In addition, the effects of 11 long-chain saturated and unsaturated (C18 : 0, C18 : 1n − 9, C18 : 2n − 6, C18 : 3n − 3, C18 : 3n − 6, C18 : 4n − 3, C18 : 5n − 3, C20 : 0, C20 : 4n − 6, EPA, and C22 : 6n − 3) fatty acids have been tested against *Photobacterium phosphoreum* (Gram-negative) [14]. EPA was the most potent of the fatty acids examined, and bioluminescence production by *P. phosphoreum* (a surrogate measure of cellular toxicity) was inhibited by 50% at just 12.1 μg/l [14]. EPA is also effective against other marine bacteria, including *Vibrio anguillarum* (Gram-negative), *Planococcus citreus* (Gram-positive), and *Micrococcus luteus* (Gram-positive) [3].

20.2.3
Antifungal Activities

There are conflicting reports of antifungal activity attributable to EPA. Although EPA was ineffective against three *Candida* spp. and an isolate of *Saccharomyces cerevisiae* when assessed by disk diffusion assay [3], Huang and Ebersole [12] found that EPA at 25 mg/l did cause partial growth inhibition of *Candida albicans*. Moreover, although Thibane *et al.* [16] demonstrated that EPA could inhibit mitochondrial function almost completely in biofilms of *C. albicans*, there was no concomitant effect on biomass production. Still, in an identical assay, EPA did reduce mitochondrial function and biomass accumulation in biofilms of the human opportunistic pathogen, *Candida dubliniensis* [16]. These contrasting reports of antifungal activity may stem from differences in the methodologies and strains used to determine activity. Further investigation of detrimental effects of EPA on fungi is warranted.

20.2.4
Antagonistic Activities Against Microalgae

Fatty acids can kill or hinder the growth of eukaryotic microalgae [14, 17–19]. Arzul *et al.* [14] demonstrated that EPA inhibited the growth of the diatom *Chaetoceros gracile* (3.75×10^4 cells/ml) almost completely at just 1.5 mg/l. Of eight long-chain unsaturated fatty acids tested, EPA was the third most active and only C18 : 5 and C22 : 6 showed greater potencies [14]. In another study, EPA was the least active of five unsaturated fatty acids against two chlorophytes, *Monoraphidium contortum* and *Chlorella vulgaris* (the four other fatty acids tested were C18 : 1n − 9, C18 : 2n − 6, C18 : 3n − 3, and C22 : 6n − 3), but even so EPA reduced the growth of these species by 50% at 100 and 125 mg/l, respectively [17]. While investigating the effects of 16 medium- and long-chain saturated and unsaturated fatty acids on the growth of single-cell raphidophytes, Alamsjah *et al.* [18] calculated that EPA killed 50% of inocula of *Heterosigma akashiwo* and *Chattonella marina* during 4 h at 2.1 and 6.8 mg/l, respectively, and this PUFA was the third most active of the fatty acids examined. In agreement, Kakisawa *et al.* [19] showed that EPA at 2 mg/l killed a 7×10^4 cell/ml inoculum of *H. akashiwo* within just 30 min.

20.2.5
Antagonistic Activities Against Protozoa and Viruses

Interestingly, EPA exerts detrimental effects on two intracellular pathogens, specifically the protozoan *Plasmodium falciparum* (a causative agent of malaria) and the hepatitis C virus [20,21]. After 2 h at 37 °C, EPA at 40 mg/l reduced the proliferation of *P. falciparum* in human red blood cells (RBCs) *in vitro* by 60% compared with untreated controls; only C22 : $6n-3$ was more effective (76% inhibition), while four other fatty acids (C20 : $4-6$, C18 : $2n-6$, C18 : $1n-9$, and C22 : 0) showed lower antimalarial activities [20]. Intriguingly, although EPA was not pursued in *in vivo* studies, C22 : $6n-3$ was highly effective in a murine infection model against another malarial parasite, *Plasmodium berghei* [20]. Finally, EPA reduced viral genome replication in human Ava5 cells infected with hepatitis C, although more studies are required to confirm and further characterize this activity [21].

20.3
Structure Relationship with Activity

20.3.1
Importance of a Free Carboxyl Group

The presence of a free carboxyl group at the terminus of EPA is important for its antibacterial activity and potency, and EPA is up to eight times more effective against MRSA and *S. aureus* than methyl-EPA [9]. Moreover, against four oral pathogenic bacteria (one Gram-positive, three Gram-negative), EPA exerted greater activity than its methyl and ethyl esters [12]. Similarly, the lag phase of *B. fibrisolvens* was extended to >72 h by 50 mg/l EPA, but the lag phase of cultures exposed to methyl-EPA was <16 h [6]. That methylated fatty acids demonstrate lower antibacterial activity compared with fatty acids having a free carboxyl group confirms previous observations with other fatty acids [9,33]. However, methyl-EPA demonstrates greater potency against *C. albicans* than EPA itself, and even ethyl-EPA is more effective than EPA [12]. Thus, the status of the carboxyl group plays an important role in the antimicrobial activity of EPA, but the target microbe is just as crucial in each interaction.

20.3.2
Importance of Unsaturated Double Bonds

The presence of unsaturated double bonds within the carbon chain of EPA also plays a critical role in the antimicrobial potency of this molecule [15,8]. Knapp and Melly [8] tested the anti-*S. aureus* activity of a series of C20 fatty acids containing zero to five double bonds and determined a direct correlation between antibacterial potency and the number of double bonds. Thus, EPA was the most active of the fatty acids followed by (in decreasing order of potency) C20 : 4 > C20 : 3 > C20 : 2 > C20 : 1 > C20 : 0 [8]. Saito *et al.* [15] performed a similar experiment with C20 fatty acids against 71 strains of group IV mycobacteria. Again, EPA showed the greatest activity when spectrum and potency of activity were considered, as this was the only fatty acid for which the MIC_{80} values were $\leq$100 mg/l against each of the species tested [15] (Figure 20.2). However, when median MIC_{80} values against the 15 species are used for comparison, antimicrobial potency was of the order C20 : 4 = C20 : 3 > C20 : 2 > EPA > C20 : 1 > C20 : 0, and the trend of greater activity with increasing unsaturation in the carbon chain is not so evident [15] (Figure 20.2). Nevertheless, the C20 fatty acids with multiple double bonds still showed considerably greater antimicrobial activity, which is thought to be due to the effect that these bonds have on the shape of the molecule [15]. Zhang *et al.* [11] examined the effects of various molecular parameters of fatty acids and close derivatives on the ability of these compounds to inhibit the growth of *S. aureus*. Interestingly, a particular spatial distribution of atoms is associated with greater antibacterial activity, and compounds exhibiting denser atom distributions and smaller space volumes, such as EPA and certain other PUFAs, exert more potent antibacterial actions [11]. Essentially, the evenly distributed double bonds in the structure distort the molecule, making it more spherical, particularly in aqueous milieus, which results in a smaller molecular space volume and greater antibacterial activity compared with long-chain saturated fatty acids or those containing just one double bond [11] (Figure 20.1b). The electronic properties of the overall structure of EPA may also contribute to its antimicrobial activity [11].

20.4
Mechanism of Antimicrobial Action

20.4.1
Microbial Growth Inhibition and Cell Killing

Many studies have investigated the mechanisms and targets of antimicrobial fatty acids, and these compounds can affect a variety of cellular targets in different organisms (for an overview see [34]). It is known that EPA can exert both growth inhibitory and cidal activities on microorganisms [3]. Knapp and Melly [8] showed that EPA was cidal at 30 mg/l against *S. aureus*, as a

10^5colony-forming units (CFU) inoculum was reduced by 99.99% during 1 h at room temperature in 0.01 mM phosphate buffer solution (PBS) (pH 7). At 250 mg/l, Shin *et al.* [4] showed that EPA completely killed a 10^8 CFU inoculum of *S. aureus* within just 15 min at 37 °C. Lower concentrations, specifically 62.5 and 125 mg/l, also killed the inoculum during 30 and 180 min, respectively, thus demonstrating dose- and time-dependent killing [4]. Interestingly, under identical conditions, EPA at 31.25–125 mg/l reduced 10^6 CFU inocula of *P. aeruginosa* by 4–5 $\log_{10}$ CFU within 15 min in each case; however, complete killing was avoided and, instead, the effect was bacteriostatic, meaning that dose-dependent effects were not observed against this bacterium under these test conditions [4]. Benkendorff *et al.* [7] showed that EPA was cidal against four Gram-positive and Gram-negative marine bacteria at 1–10 mg/l. At 75 mg/l, EPA prevented the growth of *H. pylori* during 24 h at 37 °C, and there was even a reduction in optical density of the culture, perhaps indicative of cidal action, although this was not confirmed by performing CFU counts [13]. With regard to the actions of EPA against eukaryotic microbes, this PUFA at $\leq$6.8 mg/l was cidal against two species of microalgae in as little as 30 min [18,19].

20.4.2
Effects on the Cell Membrane

The vast diversity of microbes that are affected by EPA perhaps suggests a conserved nonspecific mechanism of action such as membrane disruption. The fast kill kinetics observed in some interactions certainly points to cell lysis brought about by membrane perturbation [4,8,18]. Indeed, PUFAs are thought to exert many of their detrimental effects on susceptible microbes by acting on the cell membrane [34]. Some studies have provided evidence to suggest that the cell membrane is a major target of EPA [4,17,19]. For example, after 6 h exposure to 750 mg/l EPA, the outer cell surface of *S. aureus* exhibited large surface collapses when observed by scanning electron microscopy, which is indicative of the breakdown of the cell wall and loss of cellular constituents [4]. Similarly, *P. aeruginosa* cells incubated with 500 mg/l EPA for 6 h showed small clefts and wrinkled surfaces, which again is suggestive of severe damage to the cell membrane and a disruption of its integrity [4]. In further support, radiolabel experiments have shown that long-chain PUFAs can accumulate into bacterial cell membranes [13]; the incorporation of these PUFAs may alter membrane fluidity leading to increases in permeability [26], with concomitant losses of cellular constituents and metabolites, that may ultimately cause cell lysis [31]. Moreover, there is evidence that EPA exerts its toxic effects against microalgae by disrupting the cell membrane [17,19]. Indeed, Wu *et al.* [17] determined that EPA and various other fatty acids caused the leakage of potassium ions from *M. contortum* and *C. vulgaris* cells, which is indicative of membrane perturbation. Notably, there was a direct correlation between the growth inhibitory activities of the fatty acids and the magnitude of potassium ion release [17]. In addition, Kakisawa *et al.* [19] reported rapid and complete lysis of a eukaryotic microalga at just 2 mg/l EPA.

20.4.3
Effects on Other Cell Targets

Despite much evidence to suggest that EPA exerts much of its antimicrobial effects via membrane perturbation, other specific mechanisms may result in the demise of certain microbes. EPA can inhibit mitochondrial function [16] as well as key enzymes such as glucosyltransferase (GTF), which is an important virulence factor that allows the formation of dental plaque leading to dental caries [35]. GTF from *Streptococcus sobrinus* is inhibited by 50% at 25.7 mg/l EPA [35], but interestingly, GTF from *S. mutans* is unaffected by EPA at 50 mg/l, indicating the importance of characterizing the effects of this PUFA against similar targets in different circumstances [12]. Finally, it has been suggested that EPA might kill cells by a mechanism involving the generation of toxic lipid peroxidation products [8,20,36]. While this may be important in many cases [8,20,36], this process is unlikely to be important against *S. mutans*, as this bacterium is killed in anaerobic conditions where oxygen needed for lipid peroxidation is unavailable [12]. In summary, the literature implicates the involvement of multiple mechanisms of antimicrobial action by EPA, but disruption of the cell membrane seems to be the most common process observed during cellular killing.

20.5
Safety, Delivery, and Biotechnological Application

20.5.1
Toxicity *In Vitro*

The broad spectrum of antimicrobial activity of EPA makes this compound suitable for a variety of potential applications that make use of this biological attribute. Nevertheless, consideration must be given to the selectiveness of its action against the target microbe compared to any toxic effects on nontarget cells, tissues, and structures. The therapeutic window is the difference between an effective antimicrobial dose and the dose causing nontarget toxicity, and intuitively, a larger therapeutic window offers greater opportunity for subsequent application. Various studies have evaluated the safety of EPA, and the literature reveals a variation in reported toxicities, tolerances, and responses [14,20,36,37]. Initial toxicity assessments are usually performed *in vitro*, and perhaps one of the simplest measures of acute effects is the hemolysis assay that indicates the propensity of a compound to lyse RBCs. Kumaratilake *et al.* [20] determined that EPA at 40 mg/l caused no hemolysis to human RBCs during 20 h; however, Fu *et al.* [38] showed that EPA at 10 mg/l lysed 90% of human RBCs (4.5×10^6 cells/ml) during 24 h at 15 °C. Moreover, EPA at 4.8 mg/l caused approximately 45% lysis of sheep RBCs during 90 min at 18 °C, while complete lysis was observed at 12 mg/l [14]. Cell culture experiments can provide insight into likely chronic toxicity, and EPA was nontoxic to human neutrophils at 4.84 mg/l during 30 min [36]. Against a murine macrophage cell line (J774), EPA at 45.4 mg/l induced a significant loss in membrane integrity (37%) during 24 h, while DNA fragmentation and loss of mitochondrial transmembrane potential were also reported [39]. The effects of various fatty acids on membrane integrity of J774 cells were (in decreasing order of potency) $C16:0 > C22:6n-3 > C18:0 = C20:4n-6 = EPA > C18:1n-9 > C18:2n-6$ [39]. In contrast, Gorjão *et al.* [26] reported low toxicity of EPA to murine RINm5F cells even at 121 mg/l, although few details were given. Similarly, EPA at 30 mg/l did not significantly affect the viability of human THC-1 macrophages during 24 h [40].

20.5.2
Toxicity *In Vivo*

Still, whole organism *in vivo* assessments are far more indicative of likely toxicity. Against the brine shrimp (*Artemia salina*), an indicator species often used to assess aquatic toxicity, EPA showed no effect on viability at 10 mg/l during 24 h, but at 72 h just 1 mg/l caused a slight, but nonetheless significant, reduction in viability [41]. In tests on another invertebrate, the anostracan *Thamnocephalus platyurus*, the EPA concentration required to kill 50% of the population during 24 h was 10.3 mg/l [42]. EPA at 167 mg/l killed Japanese killifish, *Oryzias latipes* (300–350 mg; 3–3.5 cm in length), within 160 ± 49 min and, of six long-chain unsaturated fatty acids tested, only $C18:3n-3$ was more toxic [37]. Damselfish (*Acanthochromis polyacanthus*; 600–1200 mg) are far more susceptible to EPA, and just 2.7 mg/l EPA was sufficient to kill 50% of its population in 155 min [43]. Takagi *et al.* [37] determined that 6–12 mg EPA given by intraperitoneal injection per 16–20 g mouse (300–750 mg/kg) was sufficient to cause death within 24 h. EPA was as toxic as $C18:3n-3$ and $C20:4n-6$ but less toxic than $C18:1n-9$, $C18:2n-6$, and $C22:6n-3$ [37]. In summary, there appears to be an exploitable therapeutic window to administer EPA in certain situations, although in some cases this may be relatively narrow.

20.5.3
Delivery and Biotechnological Exploitation

There are a number of ways in which toxicity of a compound such as EPA can be reduced to more acceptable levels, and this can be achieved by altering its mechanism of delivery, chemical form, and concentration in sensitive areas. For example, dietary consumption of 3 g/day (≈40 mg/kg) is generally regarded as safe [30], and the fatty acid is most often delivered bound to glycerol as triglycerides, which are far less bioactive compared with free EPA [44]. Nevertheless, EPA can be released from the triglyceride glycerol backbone by enzyme action, which then allows the fatty acid to exert its bioactivity [44]. Hence, a safer form of the fatty acid can be delivered, which is then released *in situ* to become antimicrobial, and this approach may be used to reduce nontarget toxicity, especially if the location and rate of EPA release are controlled [2,44]. These principles are incorporated into at least one medical product that makes use of the bioactive properties of EPA, including its antimicrobial actions [45]. Recently, an artificial mesh used by surgeons to repair hernias has been made available, and this structure is covered in a fish oil-containing gel, consisting of a mixture of cross-linked fatty acids and glycerides [45,46]. EPA is a major constituent of fish oil and free fatty acids are suggested to be released slowly from the gel coating *in vivo* by natural degradation processes [46,47]. Reportedly, the coating releases fatty acids into the vicinity of the surgical area for a period of weeks to months [46], and such slow, relatively controlled release increases the safety of administering antimicrobial fatty acids. This mesh aids healing and reduces inflammation in animal models due to the release of anti-inflammatory fatty acids, such as EPA. Adverse effects have not been reported in animal studies or human patients to date [45,46,48]. Moreover, the antimicrobial properties of the fatty acids released from the mesh may reduce the opportunity for microbial infection; although direct evidence must still be gathered, there is a single clinical observation in a patient where the surgical area became infected but the fish oil mesh was not colonized by the bacterium [49,50]. In a similar way, EPA may help in the topical treatment of wounds, as a prophylactic or against established infections [51]. A key benefit of using EPA is the broad spectrum of activity against diverse pathogens, including drug-resistant strains, as wound infections are typically polymicrobial [3,52]. The anti-inflammatory properties of EPA and its derivatives provide additional benefit; crucially, hydrogel-coated dressings containing large quantities of EPA can promote wound healing *in vivo* [51]. Slow, natural release of EPA from inactive triglycerides delivered directly to the wound is expected to limit toxicity problems. Finally, EPA is also being developed as an antimicrobial coating on the surface of titanium-implanted medical devices [53–55]. This fatty acid is selected for its antimicrobial actions because it has anti-inflammatory attributes and can improve bone healing [53]. The titanium surface of an implant is coated in EPA and then irradiated with ultraviolet light, which leads to the covalent attachment of various peroxidative fatty acid derivatives [54]. These photooxidation products of EPA prevent surface colonization by opportunistic pathogens such as *Staphylococcus epidermidis* and reduce the viability of these bacteria, but still promote bone formation [53,55]. These implants have been demonstrated as promising in animal studies [53], and controlled trials in human patients are eagerly anticipated.

20.6 Concluding Remarks

Given the need for new antimicrobial agents for a variety of applications, the investigation of fatty acids such as EPA, which are known to exert potent, broad-spectrum anti-infective capabilities, is warranted [2]. Indeed, EPA is being developed and employed for antimicrobial applications as coatings for implanted medical devices, and further clinical uses for this PUFA are anticipated. In addition, there are prospects for exploiting this PUFA in other applications, considering its activity against oral pathogens, food spoilage microbes, and ruminant bacteria, perhaps indicating its potential as an active agent in healthcare and cosmetic products, food preservation, and for controlling digestion in farm animals. Nevertheless, further investigations are required to realize the full potential of this PUFA as an antimicrobial agent. Although fatty acids are thought to be unaffected by classical microbial drug resistance mechanisms, more studies that investigate and characterize these mechanisms are warranted. Such studies may also provide insight into mechanisms of action, which appear varied and dependent on multiple components, not just the target microbe. Moreover, as there is growing clinical preference for combination therapies that act to prevent the emergence of microbial resistance [56], studies to investigate interactions between EPA and other antibiotics would be of value, particularly for medical applications.

There are a multitude of applications in which EPA may find future use as an antimicrobial agent; encouragingly, this PUFA is already being used as a bioactive coating on medical devices to aid healing and reduce opportunity for microbial colonization [46,53,55]. Still, further work is required to confirm the safety and efficacy of this compound, which may involve the development of improved methods of delivery that will enhance and optimize these parameters and enlarge the therapeutic window.

Abbreviations

C16 : 0	palmitic acid
C18 : 0	stearic acid
C18 : 1*n* − 7*t*	*trans*-vaccenic acid
C18 : 1*n* − 9	oleic acid
C18 : 2*n* − 6	linoleic acid
C18 : 2*n* − 7*t*,9	bovinic acid
C18 : 3*n* − 3	α-linolenic acid
C18 : 3*n* − 6	γ-linolenic acid
C18 : 4*n* − 3	stearidonic acid
C18 : 5*n* − 3	octadecapentaenoic acid
C20 : 0	arachidic acid
C20 : 1	eicosenoic acid
C20 : 2	eicosadienoic acid
C20 : 3	eicosatrienoic acid
C20 : 4*n* − 6	arachidonic acid
C20 : 5*n* − 3	eicosapentaenoic acid
C22 : 0	behenic acid
C22 : 6*n* − 3	docosahexaenoic acid

Acknowledgment

Sincere thanks to Dr. Valerie Smith (University of St. Andrews, Scotland) for helpful discussions and constructive comments during the preparation of this chapter.

References

1 Payne, D.J. (2004) Antimicrobials – where next? *Microbiol. Today*, **31**, 55–57.

2 Desbois, A.P. (2012) Potential applications of antimicrobial fatty acids in medicine, agriculture and other industries. *Recent Pat. Antiinfect. Drug Discov.*, **7**, 111–122.

3 Desbois, A.P., Mearns-Spragg, A., and Smith, V.J. (2009) A fatty acid from the diatom *Phaeodactylum tricornutum* is antibacterial against diverse bacteria including multi-resistant *Staphylococcus aureus* (MRSA). *Mar. Biotechnol.*, **11**, 45–52.

4 Shin, S.Y., Bajpai, V.K., Kim, H.R., and Kang, S.C. (2007) Antibacterial activity of eicosapentaenoic acid (EPA) against foodborne and food spoilage microorganisms. *LWT – Food Sci. Technol.*, **40**, 1515–1519.

5 Maia, M.R., Chaudhary, L.C., Figueres, L., and Wallace, R.J. (2007) Metabolism of polyunsaturated fatty acids and their toxicity to the microflora of the rumen. *Antonie Van Leeuwenhoek*, **91**, 303–314.

6 Maia, M.R., Chaudhary, L.C., Bestwick, C.S., Richardson, A.J., McKain, N., Larson, T.R., Graham, I.A., and Wallace, R.J. (2010) Toxicity of unsaturated fatty acids to the biohydrogenating ruminal bacterium, *Butyrivibrio fibrisolvens. BMC Microbiol.*, **10**, 52.

7 Benkendorff, K., Davis, A.R., Rogers, C.N., and Bremner, J.B. (2005) Free fatty acids and sterols in the benthic spawn of aquatic molluscs, and their associated antimicrobial properties. *J. Exp. Mar. Biol. Ecol.*, **316**, 29–44.

8 Knapp, H.R. and Melly, M.A. (1986) Bactericidal effects of polyunsaturated fatty acids. *J. Infect. Dis.*, **154**, 84–94.

9 Ohta, S., Shiomi, Y., Kawashima, A., Aozasa, O., Nakao, I.T., Nagate, T., Kitamura, K., and Miyata, H. (1995) Antibiotic effect of linolenic acid from *Chlorococcum* strain HS-101 and *Dunaliella primolecta* on methicillin-resistant *Staphylococcus aureus. J. Appl. Phycol.*, **7**, 121–127.

10 Pesando, D. (1972) Étude chimique et structurale d'une substance lipidique antibiotique produite par une diatomée marine: *Asterionella japonica. Rev. Intern. Océanogr. Méd.*, **25**, 49–69.

11 Zhang, H., Zhang, L., Peng, L.J., Dong, X. W., Wu, D., Wu, V.C., and Feng, F.Q. (2012) Quantitative structure–activity relationships of antimicrobial fatty acids and derivatives against *Staphylococcus aureus. J. Zhejiang Univ. Sci. B.*, **13**, 83–93.

12 Huang, C.B. and Ebersole, J.L. (2010) A novel bioactivity of omega-3 polyunsaturated fatty acids and their ester derivatives. *Mol. Oral. Microbiol.*, **25**, 75–80.

13 Thompson, L., Cockayne, A., and Spiller, R.C. (1994) Inhibitory effect of polyunsaturated fatty acids on the growth of *Helicobacter pylori*: a possible explanation of the effect of diet on peptic ulceration. *Gut*, **35**, 1557–1561.

14 Arzul, G., Gentien, P., Bodennec, G., Toularastel, F., Youenou, A., and Crassous, M.P. (1995) Comparison of toxic effects in Gymnodinium cf. nagasakiense polyunsaturated fatty acids, in *Harmful Marine Algal Blooms* (eds P. Lassus *et al.*), Intercept, Andover, pp. 395–400.

15 Saito, H., Tomioka, H., and Yoneyama, T. (1984) Growth of group IV mycobacteria on medium containing various saturated and unsaturated fatty acids. *Antimicrob. Agents Chemother.*, **26**, 164–169.

16 Thibane, V.S., Kock, J.L.F., Ells, R., van Wyk, P.W.J., and Pohl, C.H. (2010) Effect of marine polyunsaturated fatty acids on biofilm formation of *Candida albicans* and *Candida dubliniensis. Mar. Drugs*, **8**, 2597–2604.

17 Wu, J.-T., Chiang, Y.-R., Huang, W.-Y., and Jane, W.-N. (2006) Cytotoxic effects of free fatty acids on phytoplankton algae and cyanobacteria. *Aquat. Toxicol.*, **80**, 338–345.

18 Alamsjah, M.A., Hirao, S., Ishibashi, F., Oda, T., and Fujita, Y. (2008) Algicidal activity of polyunsaturated fatty acids derived from *Ulva fasciata* and *U. pertusa* (Ulvaceae, Chlorophyta) on phytoplankton. *J. Appl. Phycol.*, **20**, 713–720.

19 Kakisawa, H., Asari, F., Kusumi, T., Toma, T., Sakurai, T., Oohusa, T., Hara, Y., and Chiharai, M. (1988) An allelopathic fatty acid from the brown alga *Cladosiphon okamuranus. Phytochemistry*, **27**, 731–735.

20 Kumaratilake, L.M., Robinson, B.S., Ferrante, A., and Poulos, A. (1992) Antimalarial properties of $n-3$ and $n-6$ polyunsaturated fatty acids: *in vitro* effects on *Plasmodium falciparum* and *in vivo* effects on *P. berghei*. *J. Clin. Invest.*, **89**, 961–967.
21 Leu, G.Z., Lin, T.Y., and Hsu, J.T. (2004) Anti-HCV activities of selective polyunsaturated fatty acids. *Biochem. Biophys. Res. Commun.*, **318**, 275–280.
22 Kopečný, J., Zorec, M., Mrázek, J., Kobayashi, Y., and Marinšek-Logar, R. (2003) *Butyrivibrio hungatei* sp. nov. and *Pseudobutyrivibrio xylanivorans* sp. nov., butyrate-producing bacteria from the rumen. *Int. J. Syst. Evol. Microbiol.*, **53**, 201–209.
23 Bergé, J.P. and Barnathan, G. (2005) Fatty acids from lipids of marine organisms: molecular biodiversity, roles as biomarkers, biologically active compounds, and economical aspects. *Adv. Biochem. Eng. Biotechnol.*, **96**, 49–125.
24 Somerville, C. and Browse, J. (1991) Plant lipids: metabolism, mutants, and membranes. *Science*, **252**, 80–87.
25 Cybulski, L.E., Albanesi, D., Mansilla, M.C., Altabe, S., Aguilar, P.S., and de Mendoza, D. (2002) Mechanism of membrane fluidity optimization: isothermal control of the *Bacillus subtilis* acyl-lipid desaturase. *Mol. Microbiol.*, **45**, 1379–1388.
26 Gorjão, R., Azevedo-Martins, A.K., Rodrigues, H.G., Abdulkader, F., Arcisio-Miranda, M., Procopio, J., and Curi, R. (2009) Comparative effects of DHA and EPA on cell function. *Pharmacol. Ther.*, **122**, 56–64.
27 Mayer, K., Kiessling, A., Ott, J., Schaefer, M. B., Hecker, M., Henneke, I., Schulz, R., Günther, A., Wang, J., Wu, L., Roth, J., Seeger, W., and Kang, J.X. (2009) Acute lung injury is reduced in fat-1 mice endogenously synthesizing $n-3$ fatty acids. *Am. J. Respir. Crit. Care Med.*, **179**, 474–483.
28 Schmitz, G. and Ecker, J. (2008) The opposing effects of $n-3$ and $n-6$ fatty acids. *Prog. Lipid Res.*, **47**, 147–155.
29 Seki, H., Fukunaga, K., Arita, M., Arai, H., Nakanishi, H., Taguchi, R., Miyasho, T., Takamiya, R., Asano, K., Ishizaka, A., Takeda, J., and Levy, B.D. (2010) The anti-inflammatory and proresolving mediator resolvin E1 protects mice from bacterial pneumonia and acute lung injury. *J. Immunol.*, **184**, 836–843.
30 Kris-Etherton, P.M., Harris, W.S., and Appel, L.J. (2002) Fish consumption, fish oil, omega-3 fatty acids, and cardiovascular disease. *Circulation*, **106**, 2747–2757.
31 Galbraith, H. and Miller, T.B. (1973) Physicochemical effects of long chain fatty acids on bacterial cells and their protoplasts. *J. Appl. Bacteriol.*, **36**, 647–658.
32 Miller, R.D., Brown, K.E., and Morse, S.A. (1977) Inhibitory action of fatty acids on the growth of *Neisseria gonorrhoeae*. *Infect. Immun.*, **17**, 303–312.
33 Zheng, C.J., Yoo, J.S., Lee, T.G., Cho, H.Y., Kim, Y.H., and Kim, W.G. (2005) Fatty acid synthesis is a target for antibacterial activity of unsaturated fatty acids. *FEBS Lett.*, **579**, 5157–5162.
34 Desbois, A.P. and Smith, V.J. (2010) Antibacterial free fatty acids: activities, mechanisms of action and biotechnological potential. *Appl. Microbiol. Biotechnol.*, **85**, 1629–1642.
35 Kurihara, H., Goto, Y., Aida, M., Hosokawa, M., and Takahashi, K. (1999) Antibacterial activity against cariogenic bacteria and the inhibition of insoluble glucan production by free fatty acids obtained from dried *Gloiopeltis furcata*. *Fish. Sci.*, **65**, 129–132.
36 Kumaratilake, L.M., Ferrante, A., Robinson, B.S., Jaeger, T., and Poulos, A. (1997) Enhancement of neutrophil-mediated killing of *Plasmodium falciparum* asexual blood forms by fatty acids: importance of fatty acid structure. *Infect. Immun.*, **65**, 4152–4157.
37 Takagi, T., Hayashi, K., and Itabashi, Y. (1982) Toxic effect of free polyenoic acids: a fat-soluble marine toxin. *Bull. Fac. Fish. Hokkaido Univ.*, **33**, 255–262.
38 Fu, M., Koulman, A., van Rijssel, M., Lützen, A., de Boer, M.K., Tyl, M.R., and Liebezeit, G. (2004) Chemical characterisation of three haemolytic compounds from the microalgal species *Fibrocapsa japonica* (Raphidophyceae). *Toxicon*, **43**, 355–363.
39 Martins de Lima, T., Cury-Boaventura, M.F., Giannocco, G., Nunes, M.T., and Curi, R. (2006) Comparative toxicity of fatty acids on

a macrophage cell line (J774). *Clin. Sci. (Lond.)*, **111**, 307–317.

40 Weldon, S.M., Mullen, A.C., Loscher, C.E., Hurley, L.A., and Roche, H.M. (2007) Docosahexaenoic acid induces an anti-inflammatory profile in lipopolysaccharide-stimulated human THP-1 macrophages more effectively than eicosapentaenoic acid. *J. Nutr. Biochem.*, **18**, 250–258.

41 Caldwell, G.S., Bentley, M.G., and Olive, P.J. (2003) The use of a brine shrimp (*Artemia salina*) bioassay to assess the toxicity of diatom extracts and short chain aldehydes. *Toxicon*, **42**, 301–306.

42 Jüttner, F. (2001) Liberation of 5,8,11,14,17-eicosapentaenoic acid and other polyunsaturated fatty acids from lipids as a grazer defense reaction in epilithic diatom biofilms. *J. Phycol.*, **37**, 744–755.

43 Marshall, J.A., Nichols, P.D., Hamilton, B., Lewis, R.J., and Hallegraeff, G.M. (2003) Ichthyotoxicity of *Chattonella marina* (Raphidophyceae) to damselfish (*Acanthochromis polycanthus*): the synergistic role of reactive oxygen species and free fatty acids. *Harmful Algae*, **2**, 273–281.

44 Decuypere, J.A. and Dierick, N.A. (2003) The combined use of triacylglycerols containing medium-chain fatty acids and exogenous lipolytic enzymes as an alternative to in-feed antibiotics in piglets: concept, possibilities and limitations: an overview. *Nut. Res. Rev.*, **16**, 193–210.

45 Franklin, M.E., Matthews, B.D., Voeller, G., and Earle, D.B. (2010) The benefits of omega-3 fatty acid-coated mesh in ventral hernia repair. General Surgery News 6, http://www.generalsurgerynews.com/download/SR104_Atrium_Hernia_WM.pdf (28 June 2012).

46 Pierce, R.A., Perrone, J.M., Nimeri, A., Sexton, J.A., Walcutt, J., Frisella, M.M., and Matthews, B.D. (2009) 120-day comparative analysis of adhesion grade and quantity, mesh contraction, and tissue response to a novel omega-3 fatty acid bioabsorbable barrier macroporous mesh after intraperitoneal placement. *Surg. Innov.*, **16**, 46–54.

47 Atrium Medical Corporation (2007) Technical data report no. 10, http://www.atriummed.com/PDF/CQUR-TechData010.pdf (5 July 2012).

48 Atrium Medical Corporation (2007) Technical data report no. 11, http://www.atriummed.com/PDF/CQUR-TechData011.pdf (5 July 2012).

49 Atrium Medical Corporation (2010) Technical data report no. 12, http://www.atriummed.com/PDF/CQUR-TechData012.pdf (5 July 2012).

50 Atrium Medical Corporation (2010) Technical data report no. 13, http://www.atriummed.com/PDF/CQUR-TechData013.pdf (5 July 2012).

51 Shingel, K.I., Faure, M.P., Azoulay, L., Roberge, C., and Deckelbaum, R.J. (2008) Solid emulsion gel as a vehicle for delivery of polyunsaturated fatty acids: implications for tissue repair, dermal angiogenesis and wound healing. *J. Tissue Eng. Regen. Med.*, **2**, 383–393.

52 James, G.A., Swogger, E., Wolcott, R., de Lancey Pulcini, E., Secor, P., Sestrich, J., Costerton, J.W., and Stewart, P.S. (2008) Biofilms in chronic wounds. *Wound Repair Regen.*, **16**, 37–44.

53 Petzold, C., Rubert, M., Lyngstadaas, S.P., Ellingsen, J.E., and Monjo, M. (2011) *In vivo* performance of titanium implants functionalized with eicosapentaenoic acid and UV irradiation. *J. Biomed. Mater. Res. A.*, **96**, 83–92.

54 Petzold, C., Lyngstadaas, S.P., Rubert, M., and Monjo, M. (2008) UV-induced chemical coating of titanium surfaces with eicosapentaenoic acid. *J. Mater. Chem.*, **18**, 5502–5510.

55 Petzold, C., Gomez-Florit, M., Lyngstadaas, S.P., and Monjo, M. (2012) EPA covalently bound to smooth titanium surfaces decreases viability and biofilm formation of *Staphylococcus epidermidis in vitro*. *J. Orthop. Res.*, **30**, 1384–1390.

56 Cottarel, G. and Wierzbowski, J. (2007) Combination drugs, an emerging option for antibacterial therapy. *Trends Biotechnol.*, **25**, 547–555.

21
Bioprospecting of Marine Microbial Symbionts: Exploitation of Underexplored Marine Microorganisms

Ocky K. Radjasa

21.1
Introduction

Bioprospecting is defined as the collection of small samples of biological material for screening in the search for commercially exploitable biologically active compounds or attributes such as genetic information. While the focus is frequently on the design and development of pharmaceuticals, other types of commercial products sourced from biological resources include agrochemicals, industrial chemicals, construction materials, crops, cosmetics, food, and flavorings [1]. Although oceans represent a center of biological prospecting, the biotechnological use of marine resources, particularly in drug discovery, is a relatively recent activity. Unlike bioprospecting on land, marine bioprospecting, widely described as a systematic search for valuable compounds in marine organisms, is a relatively new phenomenon. Despite the fact that many marine plants and organisms harbor complex microbial communities, this chapter deals mainly with microbial symbionts of reef invertebrates as the source of sustainable marine natural products.

Marine invertebrates are the main components of coral reefs that have pronounced pharmacological activities. However, one of the most serious bottlenecks in developing natural products from coral reefs has been the availability of biomass to gain sufficient amounts of substances for preclinical and clinical studies. Exploitation is further complicated by the fact that most of these metabolites possess highly complex structures, making them difficult to be produced economically via chemical synthesis.

There is substantial evidence that many natural products extracted from marine invertebrates are in fact the products of associated microorganisms [2]. However, there is still a general neglect of this highly important field of research and development.

Secondary metabolites of marine microbial symbionts with a wide spectrum of biological activities are known, and these can be used within a wide field of applications. It is very unfortunate that studies on the biological role of marine

Marine Microbiology: Bioactive Compounds and Biotechnological Applications, First Edition.
Edited by Se-Kwon Kim

microbial symbionts have largely focused on their ecology, pathogenicity, taxonomic relationships, and exploitable hydrolytic enzymes, with limited exploration of secondary metabolites.

The availability of novel bacteria and fungi from marine habitats is still low as compared to their estimated biodiversity. Therefore, one key aspect is the isolation of new microbial symbionts from intriguing habitats such as coral reefs for the recovery of new marine bacteria and fungal isolates. A large number of new microbial symbionts capable of producing a variety of different secondary metabolites currently awaiting discovery from the oceans can be obtained using this approach.

It has been very well established for more than half a century [3] that terrestrial bacteria and fungi are sources of valuable bioactive metabolites. It has also been noted that the rate at which new compounds are being discovered from traditional microbial resources, however, has diminished significantly in recent decades as exhaustive studies of soil microorganisms repeatedly yield the same species that in turn produce an unacceptably large number of previously described compounds [4]. Therefore, it is reasonable to expect that the exploration of untapped marine microbial diversity and resources will improve the rates at which new classes of secondary metabolites are discovered. In particular, the studies regarding screening of secondary metabolite producing microbial symbionts are important for understanding their biotechnological potential [5].

21.2 Marine Microbial Symbionts

Biologists often use the word "symbiosis" in the sense of mutualism, that is, living together for the benefit of both partners. Symbiosis is also generally defined as a situation in which two different organisms live together in close association [6]. Symbiotic systems in which there is a strong likelihood of microbial bioactive metabolite synthesis offer attractive alternatives to chemical synthesis or extraction from natural sources. Symbionts that can be cultivated in the laboratory and still produce the bioactive metabolite can then be subjected to fermentation technology to produce large amounts of the targeted compound [7].

Marine bacteria and other microorganisms often live in close association with higher organisms. These associations attracted much interest during recent years because of the potential production of some bioactive compounds by the microbial symbionts.

The study of symbiotic microorganisms is a rapidly growing field, as some reports from the recent past suspect that a number of metabolites obtained from algae and invertebrates may be produced by their associated microorganisms. Many marine invertebrate-associated microbes are known for their tremendous activities covering a wide range of biological functions.

In almost all cases, the development and production of reef invertebrate-derived drugs is seriously hampered by the environmental and technical problems associated with collecting or cultivating large amounts of animals (or organisms).

The existence of secondary metabolite producing microbial symbionts is therefore especially intriguing because a sustainable source of invertebrate-derived drug candidates can be generated by establishing symbionts in culture or by transferring symbiont biosynthetic genes into culturable bacteria.

The supply of marine metabolites tested preclinically and in the clinic can be provided by several methods, including open aquaculture of the invertebrates, total synthesis, semisynthesis, and fermentation of the producing microbes. It is likely that fermentation is the most appropriate method for the production of natural products [8].

The increasing need for new marine natural products for the treatment of clinical diseases coupled with the recognition of marine microbial symbionts of reef invertebrates as a rich source of suitable substances for these purposes provides a strong rationale for focusing on marine organisms in the search for novel marine natural products. Efforts have been mounted to enforce research on marine natural products in particular to promote the use of microorganisms that enable large-scale production of marine natural products in laboratory cultures, thereby avoiding any harm to coral reefs.

21.3 Bioethical and Supply Issues in Utilizing Marine Invertebrates

The collection of marine organisms for the discovery and development of pharmaceuticals has been perceived variously as both sustaining and threatening conservation, especially regarding issues of overcollection of target marine organisms for bioprospecting [9]. A further discussion of the protection and sustainability of coral reefs in regard to the use of reef invertebrates as sources of bioactive compounds was published by Sukarmi and Radjasa [10]. It was noted that ecological ethics must be taken into account, especially in light of the importance of coral reefs for humans in tropical communities. Less attention has been given to the widening gap between the exploitation of marine organisms and its ethical implications.

A significant threat regarding the development of drugs and perhaps the most significant problem that has hampered the investigation of secondary metabolites produced by reef invertebrates is their low concentration. In marine invertebrates, many highly active compounds contribute to $<10^{-6}\%$ of the body wet weight. Hence, providing sufficient amounts of these biologically active substances may be a difficult task (Table 21.1).

Significant case examples are the bryostatins, halichondrins, and other anti-tumoral or anti-inflammatory active substances from marine invertebrates. In these cases, the biologically active secondary metabolite content in the animals was very low and it was not possible to harvest such large amounts of organisms from nature without destroying the habitats, nor was it possible to cultivate the organisms or cell cultures thereof in sufficient scale and time [13–15]. As for the sponge *Lissodendoryx* sp., to obtain sufficient halichondrins to carry out trials up to

Table 21.1 Minute concentrations of described natural products from reef invertebrates.

No.	Source of biota	Compound	Biological activity	Concentration	Reference
1	Sponge *Lisso-dendoryx* sp.	Halicondrins	Anticancer	300 mg/t	Hart *et al.* [11]
2	Soft coral *P. elisabethae*	Pseudopterosins	Anti-inflamma-tory	1 g/t	Mayer *et al.* [12]

the clinical level, at least 15 t would be required. It is unlikely that permission would be secured to undertake such a collection due to the limited biomass established for this sponge [11]. Furthermore, the structural complexity displayed by many natural product compounds often limits the ability of chemical synthesis to access the compounds of interest and analogues thereof. Thus, considering the bioethical perspective and finding alternative solutions to the supply problem for marine natural products produced by reef invertebrates must be given high priority.

21.4 Marine Fungal Symbionts of Corals as Sustainable Sources of Marine Natural Products

The number of novel fungi isolated from marine habitats is still low as compared to their estimated high biodiversity [16]. Very little is known about the global diversity and distribution of marine fungi. The probability of isolating fungal strains belonging to new taxonomic groups from selected marine habitats and from marine samples in general remains high [12,17]. Overall, the number of strains available from marine sources is limited and the knowledge of marine fungi in general is scarce.

Research on marine fungi has suffered neglect, although the fungi are extremely potent producers of secondary metabolites and bioactive substances [13,18,19]. In addition, marine-derived fungi have contributed an important proportion of the important bioactive molecules discovered.

Marine-derived fungal strains have been isolated, screened, and reported to produce novel antimicrobial compounds belonging to the alkaloids, macrolides, terpenoids, peptide derivatives, and other structure types. A review covering more than 23 000 bioactive microbial products, that is, antifungal, antibacterial, antiviral, cytotoxic, and immunosuppressive agents, shows that the prolific producing organisms are mainly from the fungal kingdom. Hence, fungi represent one of the most promising sources of bioactive compounds [20,21].

The majority of novel fungal secondary metabolites originate from inorganic matter such as sediments, soil, sandy habitats, and artificial substrates, whereas marine invertebrate-derived fungi contribute less than the marine plants such as algae, sea grasses, mangrove plants, and woody habitats. Research on marine-

Table 21.2 Selected example of marine natural products from marine fungi.

No.	Product	Activity	Fungus	Source
1	Asperazine	Anticancer	*Aspergillus niger*	Sponge *Hyrtios* sp.
2	Trichodenones	Anticancer	*Trichoderma harzianum*	Sponge *Halichondria okadai*
3	Exophilin A	Antibacterial	*Exophiala pisciphila*	Sponge *Mycale adhaerens*
4	Evariquinone	Antiproliferative	*Emericella variecolor*	Sponge *Haliclona valliculata*
5	Flavicerebroside A	Cytotoxic	*Aspergillus flavipes*	Sea anemone *Anthopleura xanthogrammica*
6	Yanuthone A	Antimicrobial	*A. niger*	Tunicate *Aplidium* sp.
7	Gymnastatin F	Cytotoxic	*Gymnascella dankaliensis*	Sponge *Halichondria japonica*
8	Aspergillitine	Antibacterial	*Aspergillus versicolor*	Sponge *Xestospongia exigua*
9	Microsphaeropsin	Antifungal	*Microsphaeropsis* sp.	Sponge *Myxilla incrustans*
10	Sorbicillacton A	Cytotoxic	*Penicillium chrysogenum*	Sponge *Ircinia fasciculate*

derived fungi up to 2002 has led to the discovery of some 272 new natural products and another 240 new structures were discovered between 2002 and 2004. This provides significant evidence that marine-derived fungi have high potential to be a rich source of pharmaceutical leads [22].

In contrast to the existing knowledge of symbiotic or otherwise associated bacteria and cyanobacteria, whose presence in invertebrates is well documented, our knowledge of the occurrence and function of fungi in marine invertebrates is relatively scarce. Until now, most of the isolation and screening of marine fungi reported was from marine environments; however, fungal symbionts of reef invertebrates have been relatively ignored. However, a positive trend has been observed in the bioactive metabolites that have been documented so far, as shown in Table 21.2.

21.5 Marine Actinomycete Symbionts as Prolific Marine Natural Products

Microbial natural products remain an important resource for drug discovery. Among the potential sources of natural products, bacteria have proven to be particularly prolific resources with a surprisingly small group accounting for most of the compounds discovered [4].

Only five bacterial phyla are reported to produce anti-infective agents [23], and the class Actinobacteria, more specifically those belonging to the order of Actinomycetales (commonly called actinomycetes) that consist of a diverse range of Gram-positive bacteria with high G + C DNA content [24], account for approximately 7000 of the compounds reported in the Dictionary of Natural Products. The genus *Streptomyces* alone accounts for 80% of the actinomycete natural products, a biosynthetic capacity that remains unrivaled in the secondary metabolite producing microbial world.

The fact that there has been a decline in the discovery of new compounds from common soil-derived actinomycetes in the past two decades led to the attempted cultivation of rare or novel actinomycete taxa [25].

Feling *et al.* [26] mentioned that a logical extension of the search for new actinomycete natural products is the study of marine-derived strains. Very few natural product studies have assessed the taxonomic novelty of marine-derived actinomycetes, especially those that have yielded new structures, suggesting that targeting marine actinomycetes represents a productive and rational approach to marine natural product discovery. It is reasonable that new groups of actinomycetes from unexplored or underexploited habitats be targeted as sources of novel bioactive secondary metabolites, such as those associated with marine invertebrates. The uniqueness of the particular habitat as in the marine invertebrates may be reflected in the presence of uncommon diversity of marine actinomycetes with specific genetic and metabolite diversity.

The exploitation of marine actinomycetes as a source for novel secondary metabolites is considered to be in its infancy; however, the discovery rate of novel active metabolites from marine actinomycetes has recently surpassed that of their terrestrial counterparts [1,4]. Unfortunately, less attention has been given to this area since the majority of previous studies focused on marine sediment-derived actinomycetes [1,4]. Hence, further systematic investigation of symbiotic marine actinomycetes is necessary because it will provide us with very useful ecological information and a path to the discovery of new bioactive natural products with a higher hit rate. The slow growing accumulation of the bioactive molecules from marine actinomycetes has prompted the bioprospecting of actinomycete symbionts of reef invertebrates. A list of selected bioactive molecules produced by actinomycete symbionts is illustrated in Table 21.3.

Table 21.3 Selected natural products from actinomycete symbionts of invertebrates.

No.	Product	Activity	Actinomycetes	Source
1	Thiocoraline	Antibacterial	*Micromonospora* sp.	Unidentified soft coral
2	Octalasins	Anticancer	*Streptomyces* sp.	Soft coral *Pacificorgia* sp.
3	Urauchimycins	Antifungal	*Streptomyces* sp.	Unidentified sponge
4	Salinamides	Anti-inflammatory	*Streptomyces* sp.	Jellyfish *Cassiopiea xamachana*
5	5′-Hydroxystaurosporine	Antitumor	*Micromonospora* sp.	Sponge *Clathrina coriacea*
6	Metacycloprodigiosin	Anticancer	*Saccharopolyspora* sp.	Sponge *Mycale plumose*

The distribution of actinomycetes in the marine invertebrates is largely unexplored and their presence remains elusive. This is partly due to the lack of effort spent in exploring marine actinomycetes associated with marine invertebrates. The marine invertebrates represent an untapped source of actinomycetes that may indeed produce novel bioactive compounds with therapeutically relevant biological activities. The success of exploring this unexploited group of marine microorganisms relies on the availability of isolation techniques, media, and an in-depth understanding of the chemical ecology of this particular class of symbionts.

21.6 New Avenue of Research: Marine Natural Products from Fungal Symbionts of Corals

Coral microbiology is a new, immature field, which is driven largely by a desire of a marine microbiologist to understand the interactions between corals and their symbiotic microorganisms. The concept of coral holobiont, which contains the host corals plus all of its associated microorganisms, is a complex system containing microbial representatives of all three domains: Eukarya, Bacteria, and Archaea, as well as numerous viruses [27]. Research into the fungi associated with coral reef ecosystems has predominantly been concerned with pathogenic disease symptoms in various 105 host organisms, usually in the form of tissue necrosis or biomineralization [28]. Less information has been documented on the potential of fungal symbionts of corals as the producer of marine natural products.

In a recent work, we have successfully isolated and screened fungal symbionts of hard corals such as *Acropora* spp. from Indonesian coral reefs, which were found to be active against multidrug-resistant (MDR) *Escherichia coli* and *Staphylococcus aureus* collected from Kariadi Hospital in Semarang, Central Java, Indonesia. The biological activity of selected fungal symbionts against MDR bacteria is shown in Figure 21.1 and Table 21.4.

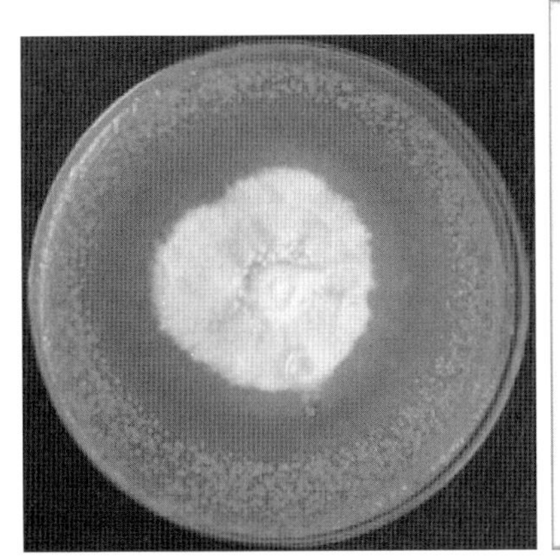

Figure 21.1 Selected anti-MDR activities of marine fungal symbionts of Indonesian corals.

Table 21.4 Antibacterial activity of fungal symbionts of hard corals from Raja Ampat.

No.	Fungal isolate	MDR pathogen *E. coli*	*S. aureus*
1	*Penicillium citrinum*	+	+
2	*Geosmithia pallida*	–	+
3	*P. citrinum*	–	+
4	*P. citrinum*	+	–

Table 21.5 Cytotoxic and antiproliferative activities of fungal symbionts of hard corals.

No.	Isolate	Biological activity Cytotoxicity[a)]	Antiproliferative[b)]
1	*P. citrinum*	+	–
2	*G. pallida*	+	–
3	*Rigidiporus* sp.	+	–
4	*Trametes maxima*	+	–
5	*P. citrinum*	–	+

a) Against three cell lines: M-14, MCF-7, and HL-60.
b) Against cell line M-14.

Another screening effort on cytotoxity and antiproliferative activities on fungal symbionts from hard corals from Indonesian coral reefs also shows the potential of these fungal symbionts as seen in Table 21.5.

The above-mentioned results revealed the potential of marine fungi for the treatment of cancer. An ongoing research is now being implemented to perform extraction, fractionation, and bioassay-guided purification to obtain the promising marine anticancer compounds.

21.7 Concluding Remarks

Bioprospecting of reef invertebrates in an environmentally sustainable manner must be continued as part of exploring novel bioactive molecules with potential applications in health and industrial sectors. The bioethical aspect, however, must be given careful consideration in conjunction with conservation efforts of one of the most productive ecosystem in coastal areas.

The field study of marine microbial natural products from fungal symbionts of corals is immature, but the growing and accumulating results have prompted the development of the underutilized group of marine fungi, a sustainable source of novel marine natural products with various applications.

References

1 Fiedler, H.P., Bruntner, C., Bull, A.T., Ward, A.C., Goodfellow, M., Potterat, O., Puder, C., and Mihm, G. (2005) Marine actinomycetes as a source of novel secondary metabolites. *Antonie Van Leeuwenhoek*, **87**, 37–42.

2 Radjasa, O.K., Vaske, Y.M., Navarro, G., Vervoort, H.C., Tenney, K., Linington, R.G., and Crews, P. (2011) Highlights of marine invertebrate-derived biosynthetic products: their biomedical potential and possible production by microbial associants. *Bioorg. Med. Chem.*, **19**, 6658–6674.

3 Kelecom, A. (2002) Secondary metabolites from marine microorganisms. *Ann. Braz. Acad. Sci.*, **74** (1), 151–170.

4 Jensen, P.R. and Fenical, W. (2000) Marine microorganisms and drug discovery: current status and future potential, in *Drugs from the Sea* (ed. N. Fusetani), Karger, Basel, pp. 6–29.

5 Radjasa, O.K., Martens, T., Grossart, H.-P., Brinkoff, T., Sabdono, A., and Simon, M. (2007) Antagonistic activity of a marine bacterium *Pseudoalteromonas luteoviolacea* TAB4.2 associated with coral *Acropora* sp. *J. Biol. Sci.*, **7** (2), 239–246.

6 Hoffmeister, M. and Martin, W. (2003) Interspecific evolution: microbial symbiosis, endosymbiosis and gene transfer. *Environ. Microbiol.*, **5** (8), 641–649.

7 Hildebrand, M., Waggoner, L.E., Liu, H., Sudek, S., Allen, S., Anderson, C., Sherman, D.H., and Haygood, M. (2004) *bryA*: an unusual modular polyketide synthase gene from the uncultivated bacterial symbiont of the marine bryozoan *Bugula neritina*. *Chem. Biol.*, **11**, 1543–1552.

8 Salomon, C.E., Magarvey, N.A., and Sherman, D.H. (2004) Merging the potential of microbial genetics with biological and chemical diversity: an even brighter future for marine natural product drug discovery. *Nat. Prod. Rep.*, **21**, 105–121.

9 Hunt, B. and Vincent, A.C.J. (2006) Scale and sustainability of marine bioprospecting for pharmaceuticals. *Ambio*, **35**, 57–64.

10 Sukarmi, R. and Radjasa, O.K. (2007) Bioethical consideration in the search for bioactive compounds from reef invertebrates. *J. Appl. Sci.*, **7** (8), 1235–1238.

11 Hart, J.B., Lill, R.E., Hickford, S.J.H., Blunt, J.W., and Munro, M.H.G. (2000) The halichondrins: chemistry, biology, supply and delivery, in *Drugs from the Sea* (ed. N. Fusetani), Karger, Basel, pp. 134–153.

12 Mayer, K.M., Ford, J., Macpherson, G.R., Padgett, D., Volkmann-Kohlmeyer, B., Kohlmeyer, J., Murphy, C., Douglas, S.E., Wright, J.M., and Wright, J.L. (2007) Exploring the diversity of marine-derived fungal polyketide synthases. *Can. J. Microbiol.*, **53**, 291–302.

13 Raghukumar, C. (2008) Marine fungal biotechnology: an ecological perspective. *Fungal Divers.*, **31**, 19–35.

14 Pan, J.H., Jones, E.B.G., She, Z.G., Pang, J.Y., and Lin, Y.C. (2008) Review of bioactive compounds from fungi in the South China Sea. *Bot. Mar.*, **51**, 179–190.

15 Schulz, B., Draeger, S., de la Cruz, T.E., Rheinheimer, J., Siems, K., Loesgen, S., Bitzer, J., Schloerke, O., Zeeck, A., Kock, I., Hussain, H., Dai, J.Q., and Krohn, K. (2008) Screening strategies for obtaining novel, biologically active, fungal secondary metabolites from marine habitats. *Bot. Mar.*, **51**, 219–234.

16 Sponga, F., Cavaletti, L., Lazzarini, A., Borghi, A., Ciciliato, I., Losi, D., and Marinelli, F. (1999) Biodiversity and potentials of marine-derived microorganisms. *J. Biotechnol.*, **70**, 65–69.

17 Burgaud, G., Le Calvez, T., Arzur, D., Vandenkoornhuyse, P., and Barbier, G. (2009) Diversity of culturable marine filamentous fungi from deep-sea hydrothermal vents. *Environ. Microbiol.*, **11**, 1588–1600.

18 Lang, G., Wiese, J., Schmaljohann, R., and Imhoff, J.F. (2007) New pentaenes from the sponge-derived marine fungus *Penicillium rugulosum*: structure determination and biosynthetic studies. *Tetrahedron*, **63**, 11844–11849.

19 Zhiguo, Y., Lang, G., Kajahn, I., Schmaljohann, R., and Imhoff, J.F. (2008) Scopularides A and B, cyclodepsipeptides

from a marine sponge-derived fungus *Scopulariopsis brevicaulis*. *J. Nat. Prod.*, **71**, 1052–1054.

20 Brakhage, A.A., Spröte, P., Al-Abdallah, W., Gehrke, A., Plattner, H., and Tüncher, A. (2004) Regulation of penicillin biosynthesis in filamentous fungi, in *Molecular Biotechnolgy of Fungal Beta-Lactam Antibiotics and Related Peptide Synthetases* (ed. A. A. Brakhage), Springer, pp. 45–90.

21 Saleem, M., Ali, M.S., Hussain, S., Jabbar, A., Ashraf, M., and Lee, Y.S. (2007) Marine natural products of fungal origin. *Nat. Prod. Rep.*, **24**, 1142–1152.

22 Ebel, R. (2006) Secondary metabolites from marine-derived fungi, in *Frontiers in Marine Biotechnology* (eds P. Proksch and W.E.G. Müller), Horizon Bioscience, England, pp. 73–143.

23 Keller, M. and Zengler, K. (2004) Tapping into microbial diversity. *Nat. Rev. Microbiol.*, **2**, 141–150.

24 Montalvo, N.F., Mohamed, N.M., Enticknap, J.J., and Hill, R.T. (2005) Novel actinobacteria from marine sponges. *Antonie Van Leeuwenhoek*, **87**, 29–36.

25 Mincer, T.J., Jensen, P.R., Kauffman, C.A., and Fenical, W. (2002) Widespread and persistent population of a major new marine actinomycete taxon in ocean sediments. *Appl. Environ. Microbiol.*, **68**, 5005–5011.

26 Feling, R.H., Buchanan, G.O., Mincer, T.J., Kauffman, C.A., Jensen, P.R., and Fenical, W. (2003) Salinosporamide A: a highly cytotoxic proteasome inhibitor from a novel microbial source, a marine bacterium of the new genus *Salinospora*. *Angew. Chem., Int. Ed.*, **42**, 355–357.

27 Rosenberg, E., Kellogg, C.A., and Rohwer, F. (2007) Coral microbiology. *Oceanography*, **20**, 146–154.

28 Rand, T.G., Bunkley-Williams, L., and Williams, E.H. (2000) A hyphomycete fungus, *Paecilomyces lilacinus*, associated with wasting disease in two species of *Tilapia* from Puerto Rico. *J. Aquat. Anim. Health*, **12**, 149–156.

22
Marine Microorganisms and Their Versatile Applications in Bioactive Compounds

Sougata Jana, Arijit Gandhi, Samrat Chakraborty, Kalyan K. Sen, and Sanat K. Basu

22.1
Introduction

The hydrosphere marine environment represents the major component of Earth's biosphere. The oceans represent a vast and exhaustive source of natural products on Earth, harboring the most diverse groups of flora and fauna. The marine microorganisms have developed unique metabolic and physiological capabilities to thrive in extreme habitats and produce novel metabolites [1]. Therefore, this rich marine habitat provides a magnificent opportunity to discover newer compounds such as antibiotics, enzymes, vitamins, drugs, biosurfactants, bioemulsifiers, and other valuable compounds of commercial importance [2–5]. The marine environment contains about 80% of the world's plant and animal species. In recent years, many bioactive compounds have been extracted from various marine animals like tunicates, sponges, soft corals, bryozoans, sea slugs, and other marine organisms. The marine environment covers wide thermal, pressure, and nutrient ranges and has extensive photic and nonphotic zones. This extensive variability has facilitated extensive specification at all phylogenetic levels, from microorganisms to mammals [6]. The potency of bioactive compounds from marine life is mainly due to the intensive ecological pressure from the stronger predators. Investigations in their chemical ecology have revealed that the secondary metabolites play various roles not only in the metabolism of the producer but also in their strategies in the given environment. The study on marine chemical compounds produced by different organisms showed the strategies for their use for human benefit [7–9].

22.2
Separation and Isolation Techniques of Bioactive Compounds from Marine Organisms

Marine organisms produce a variety of bioactive secondary metabolites. Chemically, these bioactive metabolites can be divided into amino acids, peptides,

Marine Microbiology: Bioactive Compounds and Biotechnological Applications, First Edition.
Edited by Se-Kwon Kim

nucleosides, alkaloids, terpenoids, sterols, saponins, polycyclic ethers, and so on. The ethanolic/methanolic extracts of marine organisms exhibiting biological activities can be a mixture of several classes of compounds. Since the chemical nature of bioactive compounds of the complex mixture is not known, it is not possible to follow any specific technique for the separation of the constituents of the complex mixture. However, a broad separation of the mixture can be achieved by fractionation with organic solvents. The ethanolic/methanolic extract is successively extracted with hexane, chloroform, and ethyl acetate and then divided into water-soluble and water-insoluble fractions. Each of these fractions is then subjected to a biological assay. If the separation is good, the biological activity may concentrate in a particular fraction. The biological activity may sometimes be in more than one fraction. Generally, lipophilic compounds are present in hexane- and chloroform-soluble fractions. The isolation of a pure compound from a hexane- and chloroform-soluble fraction is comparatively easier than the isolation from a water-soluble fraction. The nonpolar compounds that are extracted in hexane, benzene, and chloroform are generally esters, hydrocarbons of terpenoids, sterols, fatty acids, and so on. The mixtures of these compounds are resolved by standard chromatographic techniques over SiO_2, Al_2O_3, HPLC, and so on. The techniques like ion-exchange chromatography, reverse-phase columns, high/medium pressure chromatography on porous materials, combination of ion exchange and size-exclusion chromatography, and bioassay directed fractionation are used [10–16].

22.3 Different Bioactive Compounds from Marine Organisms

22.3.1 Bioemulsifiers and Biosurfactants

Surface-active molecules-producing microorganisms are ubiquitous, inhabiting both water (sea, freshwater, and groundwater) and land (soil, sediment, and sludge) as well as extreme environments (hypersaline sites, oil reservoirs, etc.), and thriving at a wide range of temperatures, pH values, and salinity (Figure 22.1). These microorganisms produce biosurfactants (BS)/bioemulsifiers (BE) to mediate solubilization of hydrophobic compounds in their environment to be able to utilize them as substrates [17]; however, this fact may not be always true. Few microbes produce BS/BE on water-soluble substrates (Table 22.1) [18].

BS/BE produced by microbes may be either extracellularly released into the environment or localized on the surfaces, that is, become associated with the cell membrane. When BS/BE are associated with the cell, the organism itself behaves as a BS/BE in controlling the adherence property to water-insoluble substrates [26].

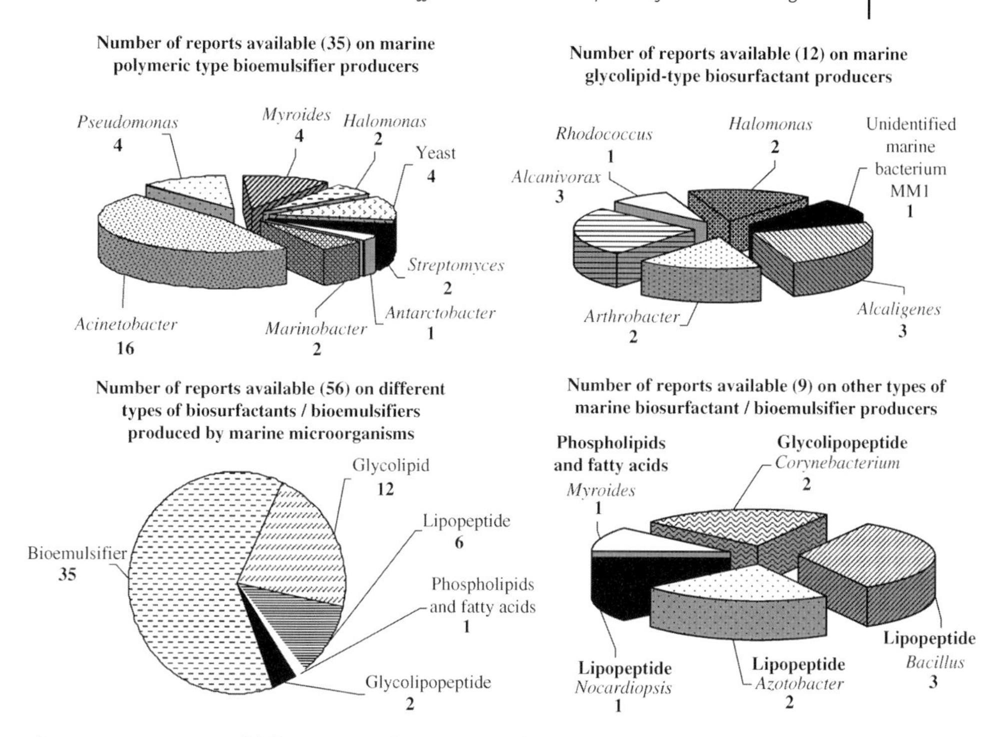

Figure 22.1 Summary of different types of marine biosurfactant—bioemulsifier-producing microorganisms [19].

22.3.2 Bioactive Metabolites of Therapeutic Value

Marine organisms comprise approximately half of the total biodiversity on Earth, and marine ecosystem is the greatest source to discover useful therapeutics. The sessile marine invertebrates such as sponges, bryozoans, and tunicates, mostly lacking morphological defense structures, have developed the largest number of

Table 22.1 Important types of marine microbial emulsifiers.

Bioemulsifiers	Producing microorganisms
BD4 emulsan [20]	*Acinetobacter calcoaceticus* BD4
Polypeptide [21]	*Acinetobactor* sp. A3
Emulsan [22]	*A. calcoaceticus* RAG-1
Biodispersan [23]	*A. calcoaceticus* A2
HE39 [24]	*Halomonas* TG39
HE67 [24]	*Halomonas* TG67
Yansan [25]	*Yarrowia lipolytica* IMUFRJ 50682

Table 22.2 Potential antimicrobial/anticancer compounds from marine organisms [27–39].

Metabolites	Sources	Applications	Reference
Enniatin	*Fusarium* sp.	Antibacterial	[27]
Modiolides A–B	*Paraphaeospheria* sp.N-119	Antibacterial	[28]
SS-228 Y	*Chainia* sp.	Antibacterial	[29]
Hypoxysordarin	*Hypoxylon croceum*	Antifungal	[30]
Keisslone	*Keissleriella* sp.	Antifungal	[31]
Artemisinic acid	*Saccharomyces cerevisiae*	Antiparasitic	[32]
Hypothemycin	*Aigialus parvus*	Antiparasitic	[33]
Shikimic acid	*E. coli*	Antiviral	[34]
Halovirs A–E	*Scytidium* sp.	Antiviral	[35]
Salinosporamide A	*Salinispora* sp.	Anticancer	[36]
Bryostatin	*Candida* sp.	Anticancer	[37]
Apratoxin	*Lyngbya majuscula*	Antitumor	[38]
Tylactone	*Streptomyces* sp.	Antibiotics	[39]

marine-derived secondary metabolites, including some of the most interesting drug candidates (Table 22.2).

In recent years, a significant number of novel metabolites with potent pharmacological properties have been discovered from marine organisms.

Although there are only a few marine-derived products currently on the market, several marine natural products are now in the clinical pipeline, with more undergoing development [40]. Similar work has been conducted targeting uncultivable microbes of marine sediments and sponges using metagenomic-based techniques to develop recombinant secondary metabolites [41]. Marine bacteria are emerging as an exciting resource for the discovery of new classes of therapeutics. The promising anticancer clinical candidates such as salinosporamide A and bryostatin only hint at the incredible wealth of drug leads hidden just beneath the ocean surface. Salinosporamide A, isolated from marine bacteria, is currently in several phase I clinical trials for the treatment of drug-resistant multiple myelomas and three other types of cancers [42].

Microbes generally lack an active means of defense and thus have resulted in developing chemical warfare to protect them from attack. In addition, many invertebrates (including sponges, tunicates, bivalves, and so on) are filter feeders, resulting in high concentrations of marine viruses and bacteria in their systems. For their survival, potent antivirals and antibacterials had to be developed to combat any opportunistic infectious organisms (Table 22.1). It is hoped that many of these chemicals can be used as the basis for future generations of antimicrobials usable in humans.

22.3.3 Enzymes

With the recent advent of biotechnology, there has been a growing interest and demand for enzymes with novel properties. When compared with terrestrial environment, the marine environment gives marine microorganisms, unique genetic structures and life habitats [43].

22.3.3.1 Protease

In 1960, Dane first isolated alkaline protease from *Bacillus licheniformis*. So far, it has been found that microorganisms are still the most suitable resources for protease production. In 1972, Nobou Kato isolated a new type of alkaline protease from marine *Psychrobacter*, and since then quite a few proteases have been continually obtained from marine microorganisms. An alkaline protease, previously isolated from a symbiotic bacterium found in the gland of Deshayes of a marine shipworm, was evaluated as a cleansing additive [44]. A yeast strain (*Aureobasidium pullulans*) with a high yield of alkaline protease was isolated from sea saltern of the China Yellow Sea by Chi *et al.* (2007), and the maximum production of enzyme was 623.1 U/mg protein (7.2 U/ml) [45]. In 2009, *Bacillus mojavensis* A21 producing alkaline proteases were isolated from seawater by Haddar *et al.*, [46] and they purified two detergent-stable alkaline serine-proteases (BM1 and BM2) from this strain. Both proteases showed high stability toward nonionic surfactants. In addition, both of them showed excellent stability and compatibility with a wide range of commercial liquid and solid detergents.

22.3.3.2 Lipases

Lipases have received much attention recently, as evidenced by the increasing amount of information about lipases in the current literature. Also, many microbial lipases are available as commercial products, majority of which being used in detergents, paper production, cosmetic production, food flavoring, organic synthesis, and some other industrial applications. The enzyme detergent market share has currently reached 90% in Europe and around 80% in Japan. Lipases are valuable biocatalysts because they act under mild conditions, are highly stable in organic solvents, and show broad substrate specificity. Microbial lipase was first found from *Penicillium oxalicum* and *Aspergillus flavus* in 1935 by David [47]. Feller *et al.* [48] screened four cold-adapted lipases secreted by *Moraxella*. These *Moraxella* were obtained from the Antarctic seawater with the optimum growth temperature of 25 °C, and the maximum secretion of lipases was supposed to occur at lower temperature conditions; the lowest secretion temperature can reach 3 °C [48]. Wang *et al.* screened out 9 lipase-producing strains from a total of 427 yeast strains. They belonged to *Candida intermedia* YA01a, *Pichia guilliermondii* N12c, *Candida parapsilosis* 3eA2, *Lodderomyces elongisporus* YF12c, *Candida quercitrusa* JHSb, *Candida rugosa* wl8, *Yarrowia lipolytica* N9a, *Rhodotorula mucilaginosa* L10-2, and *Aureobasidium pullulans* HN2–3 strain. Some lipases could actively hydrolyze different oils, indicating that they may have potential applications in industry [49]. In 2009, a novel extracellular phospholipase C was purified from a marine *streptomycete*, which was selected from approximately 400 marine bacteria by Mo *et al* [50]. Its enzyme activity was optimal at pH 8.0 at 45 °C, and it hydrolyzed only phosphatidylcholine [50].

22.3.3.3 Chitinase

Osawa *et al.* found chitinase from six species of marine bacteria: *Vibrio fluvialis*, *Vibrio parahaemolyticus*, *Vibrio mimicus*, *Vibrio alginolyticus*, *Listonella anguillarum*, and *Aeromonas hydrophila* [51]. In addition, a variety of chitinase genes were already cloned from marine bacteria and fungi. Suolow and Jones inserted two chitinase genes (ChiA and ChiB) into *Escherichia coli*, and subsequently these genes were transferred into *Pseudomonas*; finally, they acquired four high-yielding chitinase strains [52]. Meanwhile, Roberts and Cabib put the chitinase gene into tobacco plant cells and were able to develop a new tobacco plant with a strong disease resistance to the pathogen *Alternaria Longipes* [53].

22.3.3.4 Alginate Lyases

Brown alga is one of the largest marine biomass resources. Alginate has a wide range of applications; furthermore, the degraded low molecular weight fragment shows more potential. Alginate lyases, characterized as either mannuronate or guluronate lyases, are a complex copolymer of α-L-guluronate and its C5 epimer β-D-mannuronate. They have been isolated from a wide range of organisms, including algae, marine invertebrates, and marine and terrestrial microorganisms. In recent years, the marine microbial alginate lyases have been greatly developed. Discovering and characterizing alginate lyases will enhance and expand the use of these enzymes to engineer novel alginate polymers for applications in various industrial, agricultural, and medical fields [54–57].

22.3.3.5 Agarase

Agar-degrading microorganisms can be divided into two groups: bacteria that soften the agar, and other bacteria that violently liquefy the agar. In 1902, Gran isolated agar-degrading *Pseudomonas galatica* from seawater. Until now, researchers have found the presence of agarase from species within the genus *Cytophaga*, *Bacillus*, *Vibrio*, *Alteromonas*, *Pseudoalteromonas*, and *Streptomyces* [58–60]. Susgano *et al.* reported a marine bacterium *Vibrio* sp. (JT0107), which can hydrolyze the α-l,3-glycosidic bond of agar by α-Neoagaro-oligosaccharides [61]. Several agarase genes have been cloned and sequenced. In 1987, Buttner *et al.* found the *Streptomyces* agarase gene (dagA) [62].

22.4 Polysaccharides

The field of natural polysaccharides of marine origin is already large and expanding. Seaweeds are the most abundant source of polysaccharides such as alginates, agar, and agarose as well as carrageenans.

Recently, microalgae have become particularly interesting because of the possibility to easily control the growth conditions in a bioreactor together with the demonstrated biochemical diversity of these organisms. Greater screening have been developed and selection efforts made for biologically active compounds, including polysaccharides

[63]. Examples of microalgae with commercial value are the unicellular red algae *Porphyridium cruentum* and *P. aerugineum*, because of the large quantities of extracellular polysaccharides they produce [64,65]. Lewis [66] screened a number of *Chlamidomonas* spp. for extracellular polysaccharide production. The most useful of these is *Ceratozamia mexicana*, which yields up to 25% of its total organic production as polysaccharides. Moore and Tischer [67] have also reported high extracellular production levels for a number of green and blue-green algae. A number of patents have been issued concerning the production methods and applications for the *Porphyridium* polysaccharide [68]. The *Porphyridium* polysaccharide can also replace existing polysaccharide polymers such as carrageenan in biomedical applications.

Sodium alginate, chitosan, agar, and carrageenan, in combination with polyacrylics such as poly(acrylic acid) and/or poly(acrylamide), form interpenetrating networks that give rise to superabsorbent and superporous hydrogels of enhanced elasticity [69–71]. Hybrid hydrogels are multifunctional as their properties depend on cross-linking density and medium pH, and their potential for controlled release is under investigation [72]. Alternatively, polysaccharides have been cross-linked by diacrylates also leading to superabsorbent and/or superporous full-polysaccharide hydrogels [73].

Natural bioadhesives are also obtained from marine sources. Many actives can be released via bioadhesives, such as steroids, anti-inflammatory agents, pH-sensitive peptides, and small proteins such as insulin, and local treatments to alleviate pain in the buccal cavity. A bioadhesive system based exclusively on polysaccharides and potentially useful for bone glue has been recently proposed by Hoffmann *et al.* [74]. The authors developed a two-component system based on chitosan and oxidized dextran or starch. The bonding mechanism employs the reaction of aldehyde groups with amino groups in the presence of water, which covalently bind to each other in a Schiff base reaction. Chitosan was chosen as the amino carrier, and was previously partially depolymerized with acid treatment to obtain a higher ratio between amino and aminoacetyl groups. Aldehyde groups on starch or dextran are generated by oxidation with periodates. In addition, L-DOPA, an important element of mussel adhesives [75–77], was first conjugated to oxidized dextran or starch in analogy to the gluing mechanism of mussels and then oxidized to quinone. The quinone structure of L-DOPA, which is covalently bound to the aldehydes on dextran/starch, can also react with the amino groups of chitosan by an imine formation or a Michael adduct formation. All these reactions result in a strong adhesive force within the glue. With respect to fibrin glue and cyanoacrylate adhesives, which are currently used in clinical practice, biomechanical studies revealed that the new glue is superior to fibrin glue, but has less adhesive strength than cyanoacrylates. Nonetheless, cyanoacrylates, besides having toxic side effects [78], are not resorbable and thus inhibit endogenous bone repair. In conclusion, because both components are natural, biodegradable polysaccharides, and without any cytotoxic effects, this bioadhesive seems to be a good candidate for bone or soft tissue gluing applications in surgery.

In recent years, there has been a growing interest in isolating new exopolysaccharides (EPSs)-producing bacteria from marine environments, particularly from various extreme marine environments [79]. Many new marine microbial EPSs with

Table 22.3 Biologically active pigmented compounds isolated from marine bacteria.

Pigment	Activity	Bacterial strains
Undecylprodigiosin [81]	Anticancer	*Streptomyces ruber*
Cycloprodigiosin [82]	Immunosuppressant; anticancer; antimalarial	*Pseudoalteromonas denitrificans*
Heptyl prodigiosin [83]	Antiplasmodial	Alphaproteobacteria
Prodigiosin [84]	Antibacterial	*Pseudomonas* sp.
Astaxanthin [85]	Antioxidation	*Agrobacterium aurantiacum*
Violacein [86]	Antibiotic; Antiprotozoan	*Collimonas* C
Methyl saphenate (phenazine derivative) [87]	Antibiotic	*Pseudonocardia* sp. B6273
Phenazine [88]	Cytotoxic	*Bacillus* sp.
Pyocyanin [89]	Antibacterial	*Pseudomonas aeruginosa*

novel chemical compositions, properties, and structures have been found to have potential applications in fields such as adhesives, textiles, pharmaceuticals, and medicine for anticancer, food additives, oil recovery, and metal removal in mining and industrial waste treatments, and so on. General information about the EPSs produced by marine bacteria, including their chemical compositions, properties, and structures, together with their potential applications in industry, is widely reported [80].

22.5 Pigments

A number of bacterial species, including those inhabiting the vast marine environment, produce a wide variety of pigments that are important to cellular physiology and survival. Many of these natural metabolites were found to have antibiotic, anticancer, and immunosuppressive activities. These secondary metabolites, mostly produced by microorganisms via the quorum sensing mechanism, have the ability to inhibit the growth of or even kill bacteria and other microorganisms at very low concentrations. Due to such diverse and promising activities against different kinds of diseases, these compounds can play an important role in both pharmaceutical and agricultural research. Bioactive pigments from marine bacteria are summarized in Table 22.3.

22.6 Conclusions

Natural products have played a significant role in drug discovery. Over the past 75 years, natural product-derived compounds have led to the discovery of many drugs to treat human diseases. Drugs developed from marine sources give us this hope and also give us novel mechanisms to fight some of the most debilitating diseases encountered today, including HIV, osteoporosis, Alzheimer's disease, and cancer. Although the costs associated with developing drugs from marine sources have been prohibitive in the past, the development of new technology and a greater understanding of marine organisms and their ecosystem allow us to further develop our research into this area of drug development. This chapter is based on several research reports and their outcomes have been cited here in a concise manner. We hope this chapter will contribute to further investigations by the new researchers.

References

1 Fenical, W. (1993) Chemical studies of marine bacteria: developing a new resource. *Chem. Rev.*, **93**, 1673–1683.

2 Jensen, P.R. and Fenical, W. (1994) Strategies for the discovery of secondary metabolites from marine bacteria: ecological perspectives. *Annu. Rev. Microbiol.*, **48**, 559–584.

3 Austin, B. (1989) Novel pharmaceutical compounds from marine bacteria. *J. Appl. Bacteriol.*, **67**, 461–470.

4 Lang, S. and Wagner, F. (1993) Biosurfactants from marine microorganisms, in *Biosurfactants: Production, Properties, Applications, Surfactant Science Series*, vol. **48** (ed. N. Kosaric), Marcel Dekker, New York, pp. 391–417.

5 Romanenko, L.A., Kalinovskaya, N.I., and Mikhailov, V.V. (2001) Taxonomic composition and biological activity of microorganisms associated with a marine ascidian *Halocynthia aurantium*. *Russ. J. Mar. Biol.*, **27**, 291–295.

6 Harvey, A. (2000) Strategies for discovering drug from previously unexplored natural products. *Drug Discov. Today*, **5**, 294–300.

7 Munro, M.H.G., Blunt, J.W., Dumdei, E.J., Hickford, S.J.H., Lill, R.E., Li, S., Battershill, C.N. *et al.* (1999) The discovery and development of marine compounds with pharmaceutical potential. *J. Biotechnol.*, **70**, 15–25.

8 Mayer, A.M.S. (1999) Marine pharmacology in 1998: antitumor and cytotoxic compounds. *The Pharmacologist*, **41**, 159–164.

9 Muller, W.E.G., Brummer, F., Batel, R., Muller, I.M., and Schroder, H.C. (2003) Molecular biodiversity. Case study: Porifera (sponges). *Naturwissenschaften.*, **90**, 103–120.

10 Uemura, D., Hirata, Y., Iwashita, T., and Naoki, H. (1985) Studies on palytoxins. *Tetrahedron*, **41**, 1007.

11 Moore, R.E. and Scheuer, P.J. (1971) Palytoxin: new marine toxin from a coelenterate. *Science*, **172**, 495.

12 Hashimoto, Y., Fusetani, N., and Kimura, S. (1969) Aluterin: a toxin of filefish, *Alutera scripta*, probably originating from a zoantharian, *Palythoa tuberculosa*. *Bull. Jpn. Soc. Sci. Fish.*, **35**, 1095.

13 Satake, M., Murata, M., and Yasumoto, T. (1993) Gambierol: a new toxic polyether compound isolated from the marine dinoflagellate *Gambierdiscus toxicus*. *J. Am. Chem. Soc.*, **115**, 361.

14 Tachibana, K., Scheuer, P.J., Tsukitani, Y., Kikuchi, H., Engen, D.V., Clardy, J., Gopichand, Y., and Schmitz, F.J. (1981) Okadaic acid, a cytotoxic polyether from two marine sponges of the genus *Halichondria*. *J. Am. Chem. Soc.*, **103**, 2469.

15 Kobayashi, J., Ishibashi, M., Wälchli, M.R., Nakamura, H., Hirata, Y., Sasaki, T., and Ohizuni, Y. (1988) Amphidinolide C, the first 25-membered macrocyclic lactone with potent antineoplastic activity from the cultured dinoflagellate *Amphidinium* sp. *J. Am. Chem. Soc.*, **110**, 490.

16 Satake, M., Murata, M., Yasumoto, T., Fujita, T., and Naoki, H. (1991) Amphidinol, a polyhydroxypolyene antifungal agent with an unprecedented structure, from a marine

dinoflagellate, *Amphidnium klebsii*. *J. Am. Chem. Soc.*, **113**, 9859.

17 Margesin, R. and Schinner, F. (2001) Bioremediation (natural attenuation and biostimulation) of diesel-oil contaminated soil in an alpine glacier skiing area. *Appl. Environ. Microbiol.*, **67**, 3127–3133.

18 Gauthier, M.J., Lafay, B., Christen, R., Fernandez, L., Acquaviva, M., Bonin, P. *et al.* (1992) *Marinobacter hydrocarbonoclasticus* gen. nov., sp. nov., a new, extremely halotolerant, hydrocarbon-degrading marine bacterium. *Int. J. Syst. Bacteriol.*, **42**, 568–576.

19 Satpute, S.K., Banat, I.M., Dhakephalkar, P. K., Banpurkar, A.G., and Chopade, B.A. (2010) Biosurfactants, bioemulsifiers and exopolysaccharides from marine microorganisms. *Biotechnol. Adv.*, **28**, 436–450.

20 Kaplan, N., Zosim, Z., and Rosenberg, E. (1987) Reconstitution of emulsifying activity of *Acinetobacter calcoaceticus* BD4 emulsan by using pure polysaccharide and protein. *Appl. Environ. Microbiol.*, **53**, 440–446.

21 Hanson, K.G., Kale, V.C., and Desai, A.J. (1994) The possible involvement of cell surface and outer membrane proteins of *Acinetobacter* sp. A3 in crude oil degradation. *FEMS Microbiol. Lett.*, **122**, 275–279.

22 Belsky, I., Gutnick, D.L., and Rosenberg, E. (1979) Emulsifier of *Arthrobacter* RAG-1: determination of emulsifier-bound fatty acids. *FEBS Lett.*, **10**, 175–178.

23 Rosenberg, E. and Kaplan, N. (1979) Surface active properties of *Acinetobacter* exopolysaccharides, in *Bacterial Outer Membranes as Model Systems* (ed. M. Inouye), John Wiley & Sons, Inc., New York, pp. 311–342.

24 Gutiérrez, T., Mulloy, B., Black, K., and Green, D.H. (2007) Glycoprotein emulsifiers from two marine *Halomonas* species: chemical and physical characterization. *J. Appl. Microbiol.*, **103**, 1716–1727.

25 Trindade, J.R., Freire, M.G., Amaral, P.F.F., Coelho, M.A.Z., Coutinho, J.A.P., and Marrucho, I.M. (2008) Aging mechanisms of oil-in-water emulsions based on a bioemulsifier produced by *Yarrowia lipolytica*. *Colloids Surf. A*, **324**, 149–154.

26 Maneerat, S. and Dikit, P. (2007) Characterization of cell-associated bioemulsifier from *Myroides* sp. SM1, a marine bacterium. *Songklanakarin J. Sci. Technol.*, **29**, 769–779.

27 Meca, G., Sospedra, I., Valero, M.A., Mañes, J., Font, G., and Ruiz, M.J. (2011) Antibacterial activity of the enniatin B, produced by Fusarium tricinctum in liquid culture, and cytotoxic effects on Caco-2 cells. *T. Mech. Methods.*, **21**, 503–512.

28 Tsuda, M., Mugishima, T., Komatsu, K., Sone, T., Tanaka, M., Mikami, Y., and Kobayashi, J. (2003) Modiolides A and B, two new 10-membered macrolides from a marine-derived fungus. *J. Nat. Prod.*, **66**, 412–415.

29 Okazaki, T., Kitahara, T., and Okami, Y. (1975) Studies on marine microorganisms: IV. A new antibiotic SS-228 Y produced by Chainia isolated from shallow sea mud. *J. Antibiot. (Tokyo)*, **28**, 176–184.

30 Christie, S.N., McCaughey, C., McBride, M., and Coyle, P.V. (1997) Herpes simplex type 1 and genital herpes in northern Ireland. *Int. J. STD AIDS*, **8**, 68–69.

31 Isaka, M., Suyarnsestakorn, C., Tanticharoen, M., Kongsaeree, P., and Thebtaranonth, Y. (2002) Aigialomycins A–E, new resorcylic macrolides from the marine mangrove fungus *Aigialus parvus*. *J. Org. Chem.*, **67**, 1561–1566.

32 Martin, V.J.J., Pitera, D.J., Withers, S.T., Newman, J.D., and Keasling, J.D. (2003) Engineering a mevalonate pathway in *Escherichia coli* for production of terpenoids. *Nat. Biotech.*, **21**, 796–802.

33 Chandran, S.S., Yi, J., Draths, K.M., vonDaeniken, R., Weber, W., and Frost, J.W. (2003) Phosphoenolpyruvate availability and the biosynthesis of shikimic acid. *Biotechnol. Prog.*, **19**, 808–814.

34 Rowley, D.C., Kelly, S., Kauffman, C.A., Jensen, P.R., and Fenical, W. (2003) Halovirs-E, new antiviral agents from a marine-derived fungus of the genus *Scytalidium*. *Bioorg. Med. Chem.*, **11**, 4263–4274.

35 Feling, R.H., Buchanan, G.O., Mincer, T.J., Kauffman, C.A., Jensen, P.R., and Fenical, W. (2003) Salinosporamide A: a highly cytotoxic proteasome inhibitor from a novel microbial source, a marine bacterium of the

new genus *Salinospora*. *Angew. Chem., Int. Ed.*, **42**, 355–357.

36 Sudek, S., Lopanik, N.B., Waggoner, L.E., Hildebrand, M., Anderson, C. *et al.* (2007) Identification of the putative bryostatin polyketide synthase gene clusters from *Candidatus endobugula* sertula, the uncultivated microbial symbiont of the marine byrozoan *Bugula neritina*. *J. Nat. Prod.*, **70**, 67–74.

37 Luesch, H., Yoshida, W.Y., Moore, R.E., Paul, V.J., and Corbett, T.H. (2001) Total structure determination of apratoxin A, a potent novel cytotoxin from the marine cyanobacterium *Lyngbya majuscula*. *J. Am. Chem. Soc.*, **123**, 5418–5423.

38 Jung, W.S., Lee, S.K., Hong, J.S.J., Park, S. R., Jeong, S.J. *et al.* (2006) Heterologous expression of tylosin polyketide synthase and production of a hybrid bioactive macrolide in *Streptomyces venezuelae*. *Appl. Microbiol. Biotechnol.*, **72**, 763–769.

39 Thakur, N.L., Jain, R., Natalio, F., Hamer, B., Thakur, A.N., and Muller, W.E.G. (2008) Marine molecular biology: an emerging field of biological sciences. *Biotechnol. Adv.*, **26**, 233–245.

40 Rawat, D.S., Joshi, M.C., Joshi, P., and Atheaya, A. (2006) Marine peptides and related compounds in clinical trial. *Anticancer Agents Med. Chem.*, **6**, 33–40.

41 Moreira, D., Rodriguez-Valera, F., and Lopez-Garcia, P. (2004) Analysis of a genome fragment of a deep-sea uncultivated Group II euryarchaeote containing 16S rDNA, a spectinomycin-like operon and several energy metabolism genes. *Environ. Microbiol.*, **6**, 959–969.

42 Ahn, K.S., Sethi, G., Chao, T.H., Neuteboom, S.T., Chaturvedi, M.M. *et al.* (2007) Salinosporamide A (NPI-0052) potentiates apoptosis, suppresses osteoclastogenesis and inhibits invasion through down-modulation of NF-κB-regulated gene products. *Blood*, **10**, 2286–2295.

43 Stach, J.E.M., Maldonado, L.A., Ward, A.C., Goodfellow, M., and Bull, A.T. (2003) New primers for the class Actinobacteria: application to marine and terrestrial environments. *Environ. Microbiol.*, **5**, 828–841.

44 Greene, R.V., Griffin, H.L., and Cotta, M.A. (1996) Utility of alkaline protease from marine shipworm bacterium in industrial cleansing applications. *Biotechnol. Lett.*, **18**, 759–764.

45 Chi, Z.M., Ma, C., Wang, P., and Li, H.F. (2007) Optimization of medium and cultivation conditions for alkaline protease production by the marine yeast *Aureobasidium pullulans*. *Bioresour. Technol.*, **98**, 534–538.

46 Haddar, A., Agrebi, R., Bougatef, A., Hmidet, N., Sellami-Kamoun, A., Nasri, M. (2009) Two detergent stable alkaline serine-proteases from *Bacillus mojavensis* A21: Purification, characterization and potential application as a laundry detergent additive. *Bioresour. Technol.*, **100**, 3366–3373.

47 David, K. (1935) Lipase production by *Penicillium oxalicum* and *Aspergillus flavus*. *Bot. Gaz.*, **97**, 321.

48 Feller, G., Thiry, M., Arpigy, J.L., Mergeay, M., and Gerday, C. (1990) Lipases from psychrotrophic Antarctic bacteria. *FEMS Microbiol. Lett.*, **66**, 239–244.

49 Wang, L., Chi, Z.M., Wang, X.H., Liu, Z.Q., and Li, J. (2007) Diversity of lipase-producing yeasts from marine environments and oil hydrolysis by their crude enzymes. *Ann. Microbiol.*, **57**, 495–501.

50 Mo, S.J., Kim, J.H., and Cho, K.W. (2009) Enzymatic properties of an extracellular phospholipase C purified from a marine *Streptomycete*. *Biosci. Biotechnol. Biochem.*, **73**, 2136–2137.

51 Osawa, R. and Koga, T. (1995) An investigation of aquatic bacteria capable of utilizing chitin as the sole source of nutrients. *Lett. Appl. Microbiol.*, **21**, 288–291.

52 Suolow, T.V. and Jones, J. (1988) Chitinase-producing bacteria. U.S. Patent No. 4751081.

53 Roberts, R.L. and Cabib, E. (1982) Serratia marcescens chitinase: one-step purification and use for the determination of chitin. *Anal. Biochem.*, **127**, 402–412.

54 Wong, T.Y., Preston, L.A., and Schiller, N.L. (2000) Alginate lyase: review of major sources and enzyme characteristics, structure–function analysis, biological roles, and applications. *Ann. Rev. Microbiol.*, **54**, 289–340.

55 Xiao, L., Han, F., Yang, Z., Lu, X.Z., and Yu, W.G. (2006) A novel alginate lyase with

high activity on acetylated alginate of *Pseudomonas aeruginosa* FRD1 from *Pseudomonas* sp. QD03. *World J. Microbiol. Biotechnol.*, **22**, 81–88.

56 Alkawash, M.A., Soothill, J.S., and Schiller, N.L. (2006) Alginate lyase enhances antibiotic killing of mucoid *Pseudomonas aeruginosa* in biofilms. *APMIS*, **114**, 131–138.

57 Gacesa, P. (1988) Alginates. *Carbohydr. Polym.*, **8**, 161–182.

58 Aoki, T., Araki, T., and Kitamikado, M. (1990) Purification and characterization of a novel β-agarase from *Vibrio* sp. AP-2. *Eur. J. Biochem.*, **187**, 461–465.

59 Leon, O., Quintana, L., Peruzzo, G., and Slebe, J.C. (1992) Purification and properties of an extracellular agarase from *Alteromonas* sp. strain C-1. *Appl. Environ. Microbiol.*, **58**, 4060–4063.

60 Hosoda, A., Sakai, M., and Kanazawa, S. (2003) Isolation and characterization of agar-degrading *Paenibacillus* spp. associated with the rhizosphere of spinach. *Biosci. Biotechnol. Biochem.*, **67**, 1048–1055.

61 Sugano, Y., Terada, I., Arita, M., and Noma, M. (1993) Purification and characterization of a new agarase from a marine bacterium, *Vibrio* sp. strain JT0107. *Appl. Environ. Microbiol.*, **59**, 1549–1554.

62 Buttner, M.J., Fearnleiy, I.M., and Bibb, M.J. (1987) The agarase gene (dagA) of *Streptomyces coelicolor* A3(2): nucleotide sequence and transcriptional analysis. *Mol. Genet. Genomics*, **209**, 101–109.

63 Carlsson, A.S., vanBeilen, J., Möller, R., and Clayton, D. (2007) Micro- and macro-algae: utility for industrial applications, in *Outputs from the EPOBIO Project* (ed. D. Bowles), CPL Science, Newbury, UK, pp. 1–86.

64 DePauw, N. and Persoone, G. (1988) Microalgae for aquaculture, in *Micro-algal Biotechnology* (eds M.A. Borowitzka and L.J. Borowitzka), Cambridge University, Cambridge, UK, pp. 197–221.

65 Ramus, J.S. (1972) The production of extracellular polysaccharides by the unicellular red alga *Porphyridium eurugineum*. *J. Phycol.*, **8**, 97–111.

66 Lewis, R.A. (1956) Extracellular polysaccharides of green algae. *Can. J. Microbiol.*, **2**, 665–672.

67 Moore, B.G. and Tischer, R.G. (1964) Extracellular polysaccharides of algae: effects on life-support system. *Science*, **145**, 586–587.

68 Ramus, J.S. (1980) Algae biopolymer production. U.S. Patent No. 4,236,349.

69 Guilherme, M.R., Reis, A.V., Paulino, A.T., Fajardo, A.R., Muniz, E.C., and Tambourgi, E.B. (2007) Superabsorbent hydrogel based on modified polysaccharide for removal of Pb^{2+} and Cu^{2+} from water with excellent performance. *J. Appl. Polym. Sci.*, **105**, 2903–2909.

70 Omidian, H., Rocca, J.G., and Park, K. (2006) Elastic, superporous hydrogel hybrids of polyacrylamide and sodium alginate. *Macromol. Biosci.*, **6**, 703–710.

71 Pourjavadi, A., Soleyman, R., Bardajee, G.R., and Ghavami, S. (2009) Novel superabsorbent hydrogel based on natural hybrid backbone: optimized synthesis and its swelling behavior. *Bull. Korean Chem. Soc.*, **30**, 2680–2686.

72 Pourjavadi, A., Farhadpour, B., and Seidi, F. (2009) Synthesis and investigation of swelling behavior of new agar based superabsorbent hydrogel as a candidate for agrochemical delivery. *J. Polym. Res.*, **16**, 655–665.

73 Pourjavadi, A., Barzegar, Sh., and Mahdavinia, G.R. (2006) MBA-crosslinked Na–Alg/CMC as a smart full-polysaccharide superabsorbent hydrogel. *Carbohydr. Polym.*, **66**, 386–395.

74 Hoffmann, B., Volkmer, E., Kokott, A., Augat, P., Ohnmacht, M., Sedlmayr, N., Schieker, M., Claes, L., Mutschle, W., and Ziegler, G. (2009) Characterisation of a new bioadhesives system based on polysaccharides with the potential to be used as bone glue. *J. Mater. Sci. Mater. Med.*, **20**, 2001–2009.

75 Sever, M.J., Weisser, J.T., Monahan, J., Srinivasan, S., and Wilker, J.J. (2004) Metal-mediated cross-linking in the generation of a marine-mussel adhesive. *Angew. Chem., Int. Ed. Engl.*, **43**, 448–450.

76 Yu, M. and Deming, T.J. (1998) Synthetic polypeptide mimics of marine adhesives. *Macromolecules*, **31**, 4739–4745.

77 Deming, T.J. (1999) Mussel byssus and biomolecular materials. *Curr. Opin. Chem. Biol.*, **3**, 100–105.

78 Montanaro, L., Arciola, C.R., Cenni, E., Ciapetti, G., Ravioli, F., Filippini, F., and

Barsanti, L.A. (2001) Cytotoxicity, blood compatibility and antimicrobial activity of two cyanoacrylate glues for surgical use. *Biomaterials*, **22**, 59–66.

79 Nichols, C.A., Guezennec, J., and Bowman, J.P. (2005) Bacterial exopolysaccharides from extreme marine environments with special consideration of the southern ocean, sea ice and deep-sea hydrothermal vents: a review. *Mar. Biotechnol.*, **7**, 253–271.

80 Weiner, R., Langille, S., and Quintero, E. (1995) Structure, function and immunochemistry of bacterial exopolysaccharides. *J. Ind. Microbiol.*, **15**, 339–346.

81 Gerber, N.N. (1975) Prodigiosin-like pigments. *CRC Crit. Rev. Microbiol.*, **3** (4), 469–485.

82 Yamamoto, C., Takemoto, H., Kuno, K. *et al.* (1999) Cycloprodigiosin hydrochloride, a new H^+/Cl^- symporter, induces apoptosis in human and rat hepatocellular cancer cell lines *in vitro* and inhibits the growth of hepatocellular carcinoma xenografts in nude mice. *Hepatology*, **30** (4), 894–902.

83 Lazaro, J.E., Nitcheu, J., Predicala, R.J. *et al.* (2002) Heptyl prodigiosin, a bacterial metabolite, is antimalarial *in vivo* and nonmutagenic *in vitro*. *J. Nat. Toxins*, **11** (4), 367–377.

84 Gerber, M. and Gauthier, J. (1979) New prodigiosin-like pigment from *Alteromonas rubra*. *Appl. Environ. Microbiol.*, **37** (6), 1176–1179.

85 Misawa, N., Satomi, Y., Kondo, K. *et al.* (1995) Structure and functional analysis of a marine bacterial carotenoid biosynthesis gene cluster and astaxanthin biosynthetic pathway proposed at the gene level. *J. Bacteriol.*, **177** (22), 6575–6584.

86 Gauthier, M.J. (1976) Morphological, physiological, and biochemical characteristics of some violet-pigmented bacteria isolated from seawater. *Can. J. Microbiol.*, **22** (2), 138–149.

87 Maskey, R.P., Kock, I., Helmke, E., and Laatsch, H. (2003) Isolation and structure determination of phenazostatin D, a new phenazine from a marine actinomycete isolate *Pseudonocardia* sp. B6273. *Z. Naturforsch*, **58b** (7), 692–694.

88 Li, D., Wang, F., Xiao, X., Zeng, X., Gu, Q.Q., and Zhu, W. (2007) A new cytotoxic phenazine derivative from a deep sea bacterium *Bacillus* sp. *Arch. Pharm. Res.*, **30** (5), 552–555.

89 Saha, S., Thavasi, R., and Jayalakshmi, S. (2008) Phenazine pigments from *Pseudomonas aeruginosa* and their application as antibacterial agent and food colourants. *Res. J. Microbiol.*, **3** (3), 122–128.

23
Metabolites of Marine Microorganisms and Their Pharmacological Activities

Kustiariyah Tarman, Ulrike Lindequist, and Sabine Mundt

23.1
Introduction

The search for bioactive compounds of marine origin has been increasing, as indicated by the rising number of reviews on marine natural products. Approximately 1000 reviews of novel marine natural products are published annually, of which 16–18% are of microbiological origin [1]. The first novel anticancer drug from the sea was approved by the US Food and Drug Administration about 8 years ago; nevertheless, the global marine preclinical pharmaceutical pipeline remains very active [2]. Currently, 13 marine-derived compounds are in phases I–III of clinical development [3]. Some pharmaceutical companies such as PharmaMar, Bedford, Enzon, Eisai Inc., Novartis, Aventis, Eli Lilly, Abbott Inflazyme, Pfizer, and Taiho Pharmaceuticals Co. are presently developing therapeutic compounds of marine origin [4].

Marine environment, with its distinct characteristics, is responsible for the production of diverse metabolites. Among marine microorganisms, fungi (Ascomycota), actinomycetes (Actinobacteria), and cyanobacteria have recently attracted great attention as considerable sources of bioactive metabolites [5]. In this chapter, therefore, we focus on the recently described bioactive metabolites from these organisms. The earlier published compounds have been summarized, for example, in the annually edited papers about marine natural products in *Natural Product Reports* [1,5,6].

23.2
Marine Fungi

The term marine fungi was defined in 1979 by Kohlmeyer: "Obligate marine fungi are those which grow and sporulate exclusively in a marine or estuarine habitat; facultative marine fungi are those from freshwater or terrestrial

Marine Microbiology: Bioactive Compounds and Biotechnological Applications, First Edition.
Edited by Se-Kwon Kim

milieus able to grow and possibly also to sporulate in the marine environment" [7].

More intense studies on marine fungi have been done in the past three decades. These continuous studies revealed that marine fungi are rich sources of bioactive natural products [8–12].

23.2.1
Biological Sources of Marine Fungi

Fungi for chemical studies have been isolated from various organisms in the marine environment, for example, from sponges, algae, mangroves, mollusks, and so on. Since the 1960s, sponges have been the most investigated marine organisms [13]. Correspondingly, the studies of natural products from sponge-associated microorganisms, particularly bacteria and fungi, are also plenty. Fungal associates in sponges contribute 65.71% of the compounds, almost double compared to the compounds produced by sponge-associated bacteria. In more detail, Ascomycota dominate the proportion of fungal producer by division [14].

Sponge-associated fungus *Stachylidium* sp. prolifically produces marilones A–C and stachylines A–D [15,16]. This fungus was isolated from the marine sponge *Callyspongia* cf. *Carduelis flammea* collected at Bear Island, Sydney, Australia. Other species are also reported as biological sources of bioactive compound-producing fungi, such as *Niphates olemda* [17], *Halichondria panicea* [18], *Halichondria japonica* [19], *Halichondria okadai* [20], *Agelas dispar* [21], *Petrosia* sp. [22,23,24], *Tethya aurantium* [25,26], *Suberites domuncula* [27], *Myxilla incrustans* [28,29], *Ectyplasia perox* [28,30], *Xestospongia testudinaria* [31,32], *Ircinia fasciculata* [33], and *Pseudoceratina purpurea* [34].

Algal–fungal relationships have been intensively investigated. There is a symbiotic association of fungi with algae in lichens where both partners benefit. On the extreme end, this association is called mycophycobiosis where an obligate symbiotic association exists between fungi and marine macroalgae [35].

In the last decade, secondary metabolites obtained from marine algicolous fungi have shown significant increase in the number of marine natural products and their diversity [1]. Some algae reported as biological sources of bioactive compound-producing fungi are red algae *Liagora viscida* [36], *Plocamium* sp. [37], *Acanthophora spicifera* [38], *Heterosiphonia japonica* [39], and *Laurencia* sp. [40]; green algae *Ulva* sp. [41,42], *Ulva pertusa* [43], and *Enteromorpha* sp. [44]; and brown algae *Colpomenia sinuosa* [45], *Rosenvingea* sp. [46], *Fucus vesiculosus* [47], *Sargassum horneri* [48], *Sargassum kjellmanianum* [43,49], and *Undaria pinnatifida* [50].

Some marine-derived fungi are of mangrove origin. These fungi are also known as manglicolous fungi. The mangrove species known as biological sources of marine fungi are *Hibiscus tiliaceus* [51,52], *Kandelia candel* [53–55], *Acanthus ilicifolius* [56,57], *Bruguiera gymnorrhiza* [58], *Bruguiera sexangula* [59], *Lumnitzera racemosa* [60], *Excoecaria agallocha* [61,62], *Ceriops tagal* [63], *Pongamia pinnata* [64], *Clerodendrum inerme* [65], and *Avicennia marina* [66].

23.2.2
Marine Fungal Metabolites and Their Pharmacological Activities

Valuable reviews dealing with bioactive compounds of fungal origin [5,6,8,9,11,67–71] cover new biologically active natural products of marine-derived fungi, published until 2010. In this chapter, we list selected new bioactive compounds of marine fungi having considerable pharmacological activities, published between 2011 and mid-2012.

Based on their putative biogenetic origin, natural products isolated from marine fungi are generally classified into seven classes: polyketides, alkaloids, peptides, terpenoids, prenylated polyketides/meroterpenoids, shikimate-derived metabolites, and lipids. Several review papers mention that polyketides dominate marine natural products of fungal origin [1,11,71].

Pharmacological activities of fungal metabolites are very diverse. Three new phthalide derivatives, marilones A–C, were isolated from sponge-derived *Stachylidium* sp. Marilone A exhibited antiplasmodial activity against *Plasmodium berghei* liver stages, while marilone B showed selective antagonistic activity toward the serotonin receptor 5-HT_{2B} [15]. Cyclo(L-Trp-L-Phe) exhibits biological functions such as plant growth regulation and moderate cytotoxicity and accordingly has the application potential in pharmaceutical and agricultural biotechnologies. It was isolated from the sponge *Holoxea* sp.-associated fungus *Aspergillus versicolor* strain TS08 [72].

Chloctanspirones A and B, two novel chlorinated polyketides with an unprecedented skeleton, were isolated from marine sediment-derived fungus *Penicillium terrestre*. The chloctanspirone A was active against both HL-60 and A549, while chloctanspirone B showed weaker activity only against HL-60 cells [73]. Helicascolides isolated from fungus were found for the second time [74]. A new lactone, helicascolide C was isolated from *Daldinia eschscholzii* strain KT32. The fungus was isolated from the red alga *Kappaphycus alvarezii* collected in South Sulawesi, Indonesia (Figure 23.1) [75]. Further selected marine fungal metabolites and their biological activities are listed in Table 23.1 .

23.2.3
Recent Focus on Marine Fungal Research

Developments in nanotechnology have led to the production of bioactive compounds for drug discovery using organisms. Recent publications demonstrate the increasing use of organisms in this technology due to the ease of nanoparticle formation [87]. In addition, this technology is believed to be environment-friendly. Some researchers have studied its application using fungi as the producer [88–91].

Molecular method such as sequencing rDNA is also applied for rapid taxonomic identification of fungal strains. The identification of the strain would be useful for drug discovery because it may help trace the known substances produced by the related strains. Therefore, it might avoid replication in isolation of known compounds.

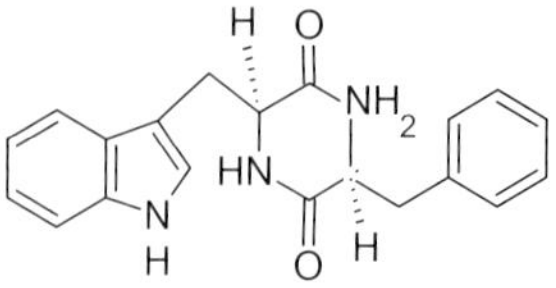

Cyclo(L-Trp-L-Phe)

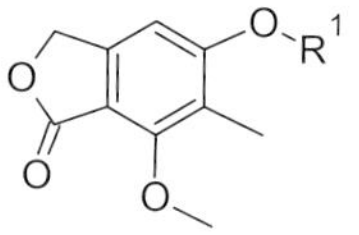

Marilone A, R^1 =

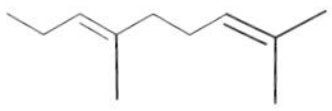

Marilone B, R^1 =H

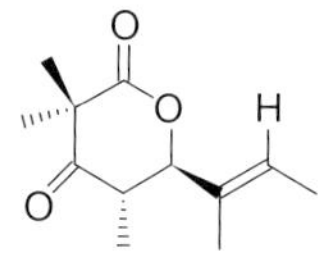

Helicascolide C

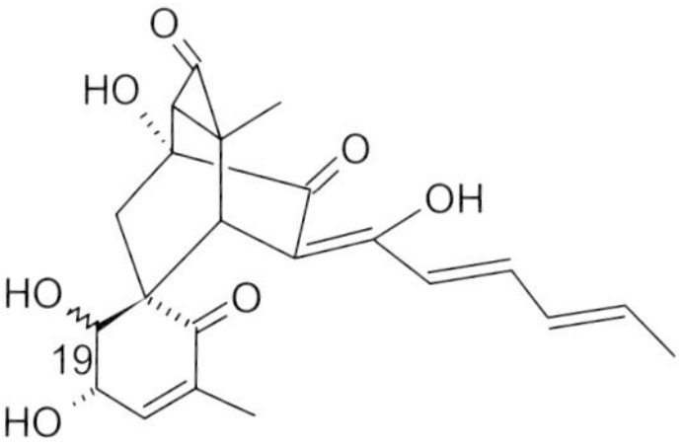

Chloctanspirone A: (19*R*)

Chloctanspirone B: (19*S*)

Figure 23.1 New compounds isolated from marine fungi.

Table 23.1 Selected metabolites from marine fungi (published since 2011).

Substance name	Organism	Chemistry/ biogenesis	Pharmacological activity	Reference
Acremolin	*Acremonium strictum* (associated with Choristida sponge from the coast of Korea)	1*H*-Azirine metabolite	Weak cytotoxicity against A549 cell line	[76]
Acremostrictin		Tricyclic lactone	Antibacterial and antioxidant	[77]
Aspergiterpenoid A	*Aspergillus* sp. (sponge *X. testudinaria*, South China Sea)	Bisabolane-type sesquiterpenes	Antibacterial activity	[32]
Aspochalasin U	*Aspergillus* sp. F00685 (Dongshi Saltern, Fujian, China)	Cytochalasan	TNF-α inhibitor	[78]
Atroviridetide	*Trichoderma atroviride* G20-12 (sediment		Cytotoxic	[79]

	on the root of *C. tagal*, South China Sea)			
Butyrolactone VI	*Aspergillus* sp. (Pacific Sea, Los Molles, IV Coquimbo Región, Chile)	Butyrolactone	Cytotoxic against crown gall tumors	[80]
Emerimidines A and B; emeriphenolicins A and D	*Emericella* sp. (HaiKou, the People's Republic of China)	Isoindolones derivatives	Anti-influenza A virus (H1N1)	[81]
Penicacids A–E	*Penicillium* sp. SOF07 (sediment of the South China Sea)	Mycophenolic acid derivatives	Immunosuppressive activity	[82]
Penicillone A, penicillactam	*Penicillium* sp. F11 (sediment of the South China Sea)		Cytotoxic against HT1080, CNE2, and BEL-7402 cell lines	[83]
Sclerodin derivatives	*Coniothyrium cereale* (green alga *Enteromorpha* sp., the Baltic Sea)	Phenalenone derivatives	Inhibition of human leukocyte elastase antibacterial activity	[84]
Scopararanes C–G	*Eutypella scoparia* FS26 (sediment of the South China Sea)	Oxygenated pimarane diterpenes	Cytotoxic activities against MCF-7, NCI-H460, and SF-268 tumor cell lines	[85]
Terremides A and B; terrelactone A	*Aspergillus terreus* (sediment of the Putian Sea Saltern, Fujian, China)	Alkaloids, lactone	Antibacterial activity against *P. aeruginosa* and *Enterobacter aerogenes*	[86]

23.3 Marine Actinomycetes

Actinomycetes are generally Gram-positive anaerobic bacteria. Their filamentous and branching growth pattern results, in most forms, in an extensive colony or mycelium. Many species also form spores like sporangia (www.britannica.com/EBchecked/topic/4401/actinomycete). Important genera are *Streptomyces, Micromonospora, Nocardia,* and *Pseudonocardia*. Members of these genera also occur in a marine environment.

23.3.1
Biological Sources of Marine Actinomycetes

In marine environment, actinomycetes occur in sediments or are associated with invertebrates. Sponges are the most popular biological sources of marine actinomycetes in the literatures. Some species reported as actinomycete associates are *H. panicea* [92,93], *Haliclona* sp. [94,95], *Cinachyra* sp. [96], *Aplysina aerophoba, Callyspongia* sp., *Dysidea avara, Dysidea tupha, Hemimycale columella, Hyrtios erecta, I. fasciculata* [97], *Dendrilla nigra* [98], *Axinella polypoides, Tethya* sp. [99], and *Tedania* sp. [99,100].

However, actinomycetes are also associated with mangrove plants such as *Aegiceras corniculatum* [101,102] and *B. gymnorrhiza* [103,104].

23.3.2
Metabolites of Marine Actinomycetes and Their Pharmacological Activities

To date, nearly 400 new compounds with cytotoxicity and/or antimicrobial activity have been isolated from marine actinomycetes [105]. The enterocins and wailupemycins isolated from the marine strain *Streptomyces maritimus* are the first actinomycete natural products for which a complete gene cluster was identified, sequenced, and verified. Members of the genus *Salinispora* are intriguing model organisms for whole genome sequencing [106]. *Salinispora tropica* is the producer of the clinically promising salinosporamide A (marizomib (NPI-0052), Nereus Pharmaceuticals). This orally active β-lactone inhibits chymotrypsin-, caspase-, and trypsin-like activities of purified human erythrocyte 20S proteasomes by irreversibly binding to it. In various tumor xenograft models, it is well tolerated and prolongs survival, with significantly reduced tumor recurrence [107–109]. It is used in clinical development for the treatment of various hematological malignancies and solid tumors, especially of multiple myeloma [110]. Halogenated analogues of salinosporamide A possess, for example, trypanocidal activity [111].

Novel compounds antimycins B1 and B2 were isolated from *Streptomyces lusitanus* (rhizosphere of the mangrove plant *A. marina* in Fujian province, China). Antimycin B2 showed antibacterial activity against *Staphylococcus aureus* and *Loktanella hongkongensis* [112]. The new meroterpenoid, merochlorin A was isolated from *Streptomyces* sp. strain CNH-189 (near-shore marine sediments collected off Oceanside, California). This compound was reported to be active against Gram-positive bacteria, including *Clostridium difficile*, and also demonstrated rapid bactericidal activity against MRSA. Its activity was lost in the presence of 20% serum [113].

Two bipyridine alkaloids, caerulomycins F and G, were isolated from *Actinoalloteichus cyanogriseus* WH1–2216-6 (sediment, seashore of Weihai, China). The compounds showed cytotoxic activity against HL-60, K562, KB, and A549 cell lines [114]. The angucyclinone derivate, kiamycin, was cytotoxic to HL-60, A549, and BEL-7402 cell lines. It was isolated from *Streptomyces* strain M268 (sediment, Kiaochow Bay, near Qingdao, China) [115]. Two new 20-membered macrolides, levantilides A and B, were isolated from *Micromonospora* M71-A77 (deep sea of the Eastern Mediterranean). Levantilide A displayed moderate antiproliferative activity against several human tumor cell lines (gastric, lung, pancreas, mammary, melanoma, and renal) [116]. Nocardioazine A, a new prenylated diketopiperazine, was isolated together with nocardioazine B from *Nocardiopsis* sp. (coastal sediment, South Molle Island, near Brisbane, Australia). The compound inhibits P-glycoprotein and reverses doxorubicin resistance [117]. Pseudonocardians A–C, three diazaanthraquinone derivatives from a deep sea actinomycete *Pseudonocardia* sp. SCSIO 01299, exhibited potent cytotoxic activities against three tumor cell lines of SF-268, MCF-7, and NCI-H460. It also showed antibacterial activities against *S. aureus*, *Enterococcus faecalis*, and *Bacillus thuringiensis* [118].

Furthermore, novel compounds exhibiting activities other than antimicrobial and cytotoxic have been reported. A new antifibrotic benzopyran compound was isolated from *Streptomyces xiamenensis* (mangrove sediment, Fujian province, China). The compound showed multiple inhibiting effects on human lung fibroblasts (WI-26) [119]. New β-carbolines, marinacarbolines A–D, and indolactam alkaloids, pendolmycin derivatives, exhibited antiplasmodial activity. The compounds were isolated from *Marinactinospora thermotolerans* (deep sea sediment, the South China Sea) (Figure 23.2) [120].

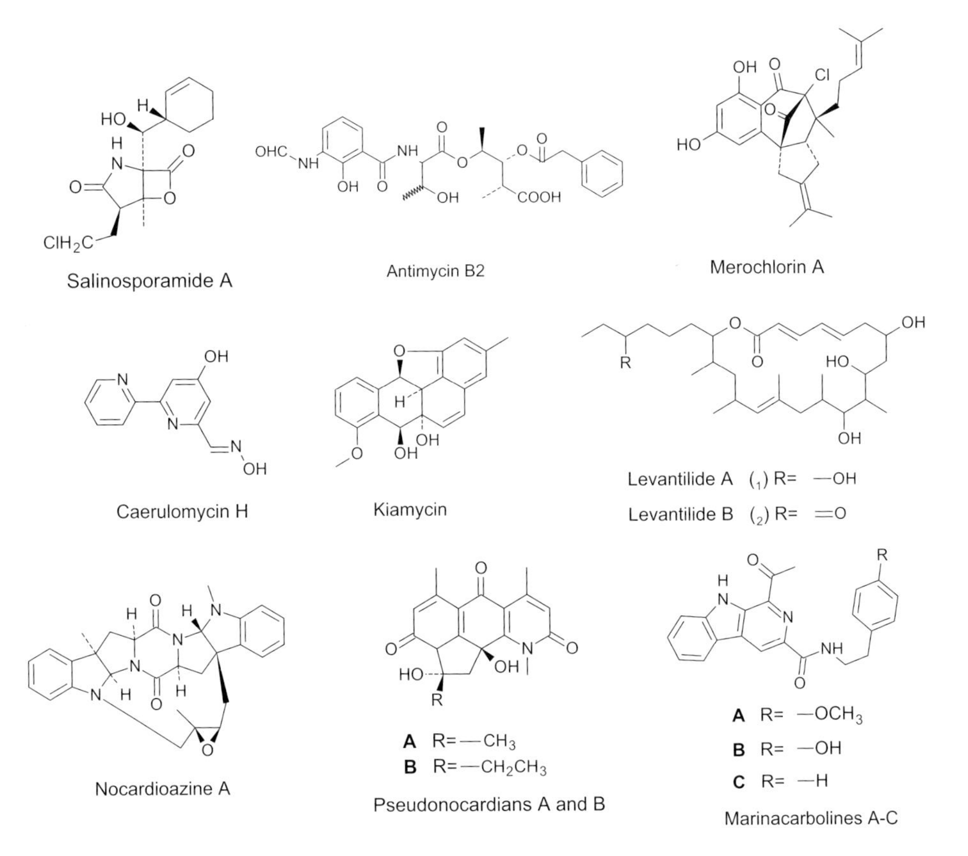

Figure 23.2 New compounds isolated from marine actinomycetes.

Table 23.2 further sums up selected metabolites from marine actinomycete strains, published since 2011, according to their described biological activity.

The examples show that as in the past years, the main focus has been toward the detection of compounds with cytostatic and antimicrobial activities. To an increasing degree, information about cellular target, mode of action, and substances with other activities is given. The newly described compounds cover a wide variety of chemical structures and biogenetic pathways. There are small molecular weight compounds like sterols and alkaloids as well as higher molecular weight compounds like cyclic peptides. Although some compounds represent novel structures, many others are derivatives of known basic structures. From a regional point of view, an accumulation of investigations of organisms from the South China Sea can be observed.

Table 23.2 Selected metabolites from marine actinomycetes (published since 2011).

Substance name	Organism	Chemistry/biogenesis	Activity	Reference
Antibiotic A201A	*M. thermotolerans* SCSIO 00652 (deep sea sediment, the South China Sea)	Nucleoside; efficient metabolic engineering for improvement of antibiotic titer	Antibacterial	[121]
Antimycins A (19) and A(20)	*Streptomyces antibioticus* H74-18	Macrolides, antimycin A biosynthesis catalyzed by NRPS/PKS	Antifungal against *Candida albicans*	[122]
Bendigoles D, E, and F	*Actinomadura* sp. SBMs009 (from tissue sections of the sponge *Suberites japonicus*)	Sterols	Inhibition of NF-κB and glucocorticoid receptor translocation. Only bendigole D mild cytotoxic against L929 mouse fibroblast cell line	[123]
Caerulomycins H, I, J, and K (phenylpyridine alkaloid)	*A. cyanogriseus* WH1–2216-6 (sediment, seashore of Weihai, China)	Bipyridine alkaloids	Cytotoxic against HL-60, K562, KB, and A549 cell lines	[114]
Chromomycin SA analogues	*Streptomyces* sp. (sediment from the hypersaline lake, East Plana Cay, Bahamas)	Glycosylated polyketides; Chromomycin SA_2 and SA_3 first analogues with truncated side chains	Cytotoxic against HCC44 and A549 cell lines	[124]

Cyanogrisides A, B, C, and D	*A. cyanogriseus* WH1–2216-6 (sediment, seashore of Weihai, China)	Cyclic glycosides, N-containing compound	A and C moderate cytotoxic against K562 and KB cells; B reverses multiple drug resistance of several tumor cell lines	[125]
Lobophorins C and D	*Streptomyces rimosus* AZS17 (associated with the marine sponge *Hymeniacidon* sp., the East China Sea)	Nitroaminoglycosides (kijanimicin derivatives)	Cytotoxic against human liver cancer cell line BEL-7402	[126]
Marinactinones A, B, and C	*M. thermotolerans* SCSIO 00606	γ-Pyrones	Cytotoxic against SW1990, HepG2, and SMCC-7721 cell lines, marinactinone B: weak inhibition of DNA topoisomerase II	[127]
Streptobactin, benarthin, dibenarthin, tribenarthin	*Streptomyces* sp. YM5–799 (isolated from surface of algae collected from Hokkaido, Japan)	Tricatechol	Iron chelating (siderophore)	[128]
Thiochondrilline and other analogues of thiocoraline A	*Verrucosispora* sp. (sponge *Chondrilla caribensis*, the Florida Keys, USA)	Thiodepsipeptides	Cytotoxic against A549 cells	[129]
Usabamycins A, B, and C	*Streptomyces* sp. NPS853 (marine sediment, 20 m depth, USA Bay, Kochi Prefecture, Japan	Pyrrolo[1,4]benzodiazepines from anthramycin-type analogues	Selective inhibition of serotonin uptake (5-HT_{2B})	[130]

23.3.3
Recent Focus on Marine Actinomycete Research

Increasing attention is being given to novel genomic approaches. They demonstrate the high biosynthetic potential of actinomycetes and show that a large fraction of genomes still remains unexploited. Activating silent pathways might therefore reveal production of novel metabolites [106,131]. Methods of functional genome analysis, including transcriptome, proteome, and metabolome analysis, would lead to a better understanding of physiological and biochemical processes in bacteria. They would also promote the cultivation of microorganisms uncultivable until now [132]. From the reported about 22 500 bioactive metabolites produced by microorganisms, over 10 000 (about 45%) are produced by actinomycetes. The *Streptomyces* species account for about 7600 of these metabolites [133]. It can be expected that the exploration of the biosynthetic potential of marine actinomycetes would increase this number considerably.

23.4 Marine Cyanobacteria

Cyanobacteria (cyanoprokaryotes, cyanophytes, and blue-green bacteria) are Gram-negative photoautotrophic prokaryotes, capable of performing oxygenic photosynthesis. They show a typical prokaryotic cell organization, but have an elaborate system of internal membranes responsible for both respiratory and photosynthetic electron transport. They accomplish oxygenic photosynthesis because they possess photosystems I and II, but under anaerobic conditions they undergo anoxygenic photosynthesis, only using photosystem I.

The morphology of cyanobacteria varies from unicellular to filamentous or colonial forms. The most recent botanical classification divides cyanobacteria into six orders (http://www.algaebase.org), but more and more cyanobacteria are being classified by molecular approaches to taxonomy, based on nitrogenase (*nifH*) and 16S rRNA gene sequences and random amplified polymorphic DNA analysis, revealing an enormous diversity not indicated by morphological classifications (http://www.ncbi.nlm.nih.gov/taxonomy).

23.4.1 Biological Sources of Marine Cyanobacteria

Marine cyanobacteria used as sources of novel structures are free-living filamentous species, but they can also live in invertebrates. Analogues of dolastatins, originally isolated from *Dolabella auricularia*, such as symplostatins, somamides A and B, and pitiprolamide have been isolated from cyanobacteria [134–136]. Current investigations of extracts from cyanobacterial strains isolated from the Mediterranean sponge *Petrosia ficiformis* revealed strong cytotoxic activity, so that the symbionts could often be the real producers of active compounds [137].

23.4.2 Metabolites of Marine Cyanobacteria and Their Pharmacological Activities

As in the past years, the main focus has been on the detection of compounds with cytostatic/antitumoral, antimicrobial, and enzyme-inhibiting activities, but the test systems used today are more specific and information about cellular target and mode of action is published. Substances isolated earlier are investigated in other test systems and activities such as anti-inflammatory, antiparasitic (antitrypanosomal, antileishmanial, and antimalarial), molluscicidal, insecticidal, and cannabinomimetic or inhibiting effects on neuronal processes are described. Recently, metabolites exhibiting inhibitory activity on quorum sensing have been identified.

From the structural point of view, cyclic and linear peptides often containing lipophilic side chains, macrolides, fatty acids and their amides, and alkaloids have been identified. Very often, swarms of derivatives of known basic structures are presented, such as apratoxins A–E, grassypeptolides A–E, and veraguamides A–L. Genome sequencing studies of cyanobacteria show significant diversity and novelty of genes responsible for bioactive proteins, ribosomal and nonribosomal peptides, and peptide–polyketide hybrid molecules [138]. Most peptides are of mixed polyketide synthase/nonribosomal peptide synthetase origin; peptides synthesized ribosomal and modified posttranslational, such as prochlorosins are rarely described.

New compounds isolated from marine cyanobacteria with various biological activities are presented. Cyclic depsipeptides, veraguamides A–C and H–L, isolated from *Oscillatoria* cf. *margaritifera* (Coiba National Park, Panama) displayed cytotoxic activity to H460 cells [139]. Related compounds, veraguamides A–G, from *Symploca* cf. *hydnoides* (Cetti Bay, Guam) exhibited moderate to weak cytotoxicity to HT29 colorectal adenocarcinoma and HeLa cervical carcinoma cell lines [140].

Marine cyanobacteria are also prolific sources of enzyme inhibitors. Symplocin A, a linear peptide isolated from *Symploca* sp. (San Salvador Island, Bahamas), was reported to inhibit cathepsin E [141]. Biselyngbyaside, a macrolide glycoside, inhibited osteoclast differentiation in RAW-264 cells by inhibition of transcription factors and suppression of bone resorption via inhibition of osteoclastogenesis and induction of apoptosis. The compound was isolated from *Lyngbya* sp. collected in Okinawa, Japan [142,143]. Two alkylamides, credneramides A and B, were isolated from cf. *Trichodesmium* sp. nov. (Credner Islands, Papua New Guinea). They inhibited calcium oscillations in mouse cerebrocortical neurons [144]. Ethyl tumonoate A, an acylproline derivative isolated from cf. *O. margaritifera* (Caracas Baii, Curaçao) exhibited anti-inflammatory activity in murine macrophages and inhibitory activity on calcium oscillations in neocortical neurons [145]. The lipoamide, janthielamide A, isolated from a consortium of unknown cyanobacteria/*Symploca* sp. (Jan Thiel Bay, Curaçao) showed sodium channel-blocking activity in murine neuro-2a cells and antagonized veratridine-induced sodium influx in murine cerebrocortical neurons. Other lipoamide derivatives, kimbeamides A–C, also obtained from a consortium of unknown cyanobacteria/*Symploca* sp. collected at Kimbe Bay (north coast of New Britain) and Papua New Guinea displayed the same activity in murine neuro-2a cells. From the same collection, kimbelactone A, a polyketide, was isolated [146]. The cyclic depsipeptide, lagunamide C, isolated from *Lyngbya majuscula* (Pulau Hantu Besar, Singapore) exhibited significant antimalarial activity against *Plasmodium falciparum* and cytotoxic activity toward P388, A549, PC3, HCT8, and SKOV-3 cell lines [147]. Lyngbyoic acid, a cyclopropane-containing fatty acid, isolated from *L. majuscula* and collected near Fort Pierce, Florida, disrupted quorum sensing in *Pseudomonas aeruginosa* [148]. The cyclic ketones, palmyrrolinone and thiopalmyrone, isolated from an assemblage cf. *Oscillatoria* and *Hormoscilla* spp. (North Beach, Palmyra Atoll) exhibited molluscicidal activity against *Biomphalaria glabrata*, the snail vector of *Schistosoma* species. A slight enhancement in molluscicidal effect was observed when these two compounds were utilized as an equimolar binary mixture [149]. The alkyl amides and propenediester isolated from *Oscillatoria* sp. (Isla Canales de Afuera, Coiba National Park, Panama) and serinolamide A isolated from *L. majuscula* (New Island, Papua New Guinea; Isla Canales, Panama) exhibited cannabinomimetic activity (Figure 23.3) [150].

Table 23.3 summarizes selected metabolites from marine cyanobacteria, published since 2011, according to their described biological activity.

23.4.3 Recent Focus of Research on Marine Cyanobacteria

The inclusion of phylogenetic analysis in the classification of cyanobacteria has improved the efforts on natural products-based drug discovery; the evolutionary relationships of cyanobacteria, as inferred by their 16S rRNA genes can be used as predictors of their potential to produce varied secondary metabolites [145]. To discover and evaluate factors influencing natural product biosynthesis, Gerwick's group used an approach that permits monitoring of *in vivo* natural product synthesis and turnover by mass spectrometry and stable isotope ^{15}N feeding experiments. The temporal comparison of the amount of *in vivo* ^{15}N labeling of nitrogen-containing metabolites permits identification of the timing of specific steps in metabolite assembly and estimation of the turnover rates of natural products from small amounts of biomass [157].

Especially, cyanobacteria belonging to the genus *Lyngbya* are described as very important producers of secondary metabolites (see Table 23.1–23.3) [158]. One reason for the rich secondary metabolite capacity of the genus *Lyngbya* may be that it is a polyphyletic group and bioactive secondary metabolites attributed to it are actually produced by morphologically similar but phylogenetically distant lineages. Thus, taxonomic clarification and revision of polyphyletic cyanobacterial lineages are essential for an accurate understanding of the distribution of bioactive secondary metabolites [159,160]. Cyanobacteria are accepted as an emerging source of drug discovery [134]; of the total marine natural products of prokaryotic origin evaluated for their medical potential, approximately 40% are of cyanobacterial origin [13]. None of these substances have reached the market or are undergoing clinical trials; only a few are in the preclinical pipeline [2,3]. To ensure the substance supply for *in vivo* and other detailed biological studies, synthetic strategies [161] and heterologous expression systems using bacteria such as *Streptomyces coelicolor* or *Escherichia coli* for the production of cyanobacterial metabolites are under research [162,163].

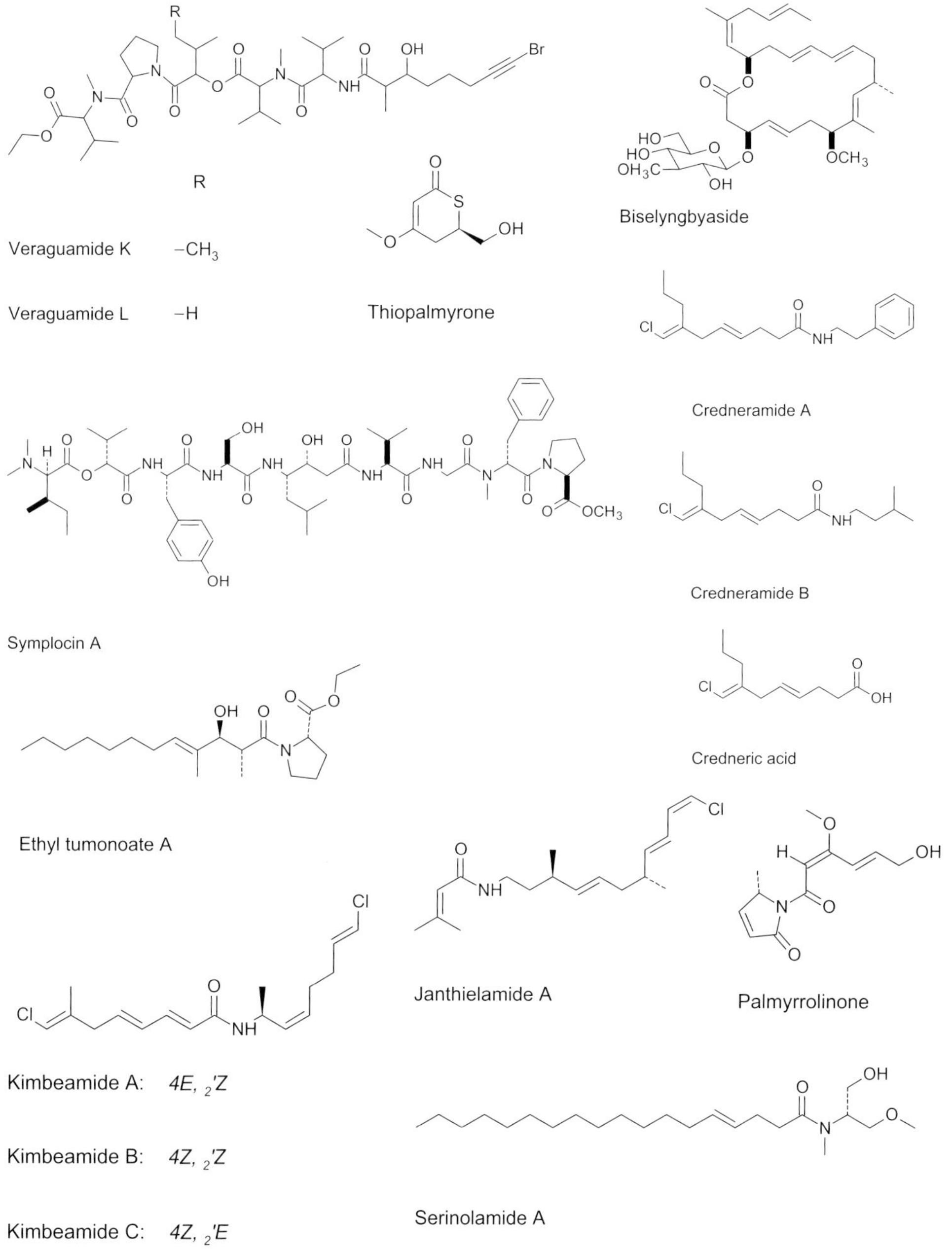

	R
Veraguamide K	$-CH_3$
Veraguamide L	$-H$

Thiopalmyrone

Biselyngbyaside

Credneramide A

Credneramide B

Credneric acid

Symplocin A

Ethyl tumonoate A

Janthielamide A

Palmyrrolinone

Kimbeamide A: *4E, $_2$'Z*

Kimbeamide B: *4Z, $_2$'Z*

Kimbeamide C: *4Z, $_2$'E*

Serinolamide A

Figure 23.3 New compounds isolated from marine cyanobacteria.

Table 23.3 Selected metabolites from marine cyanobacteria (published since 2011).

Substance name	Organism	Chemistry/ biogenetic group	Pharmacological activity	Reference
Grassypeptolides D and E	*Leptolyngbya* sp. (SS Thistlegorm shipwreck, Red Sea)	Cyclic depsipeptides	Cytotoxic to HeLa and mouse neuro-2a blastoma cells	[151]
Grassypeptolides F and G	*L. majuscula* (Ngerderrak Reef, Palau)	Cyclic depsipeptides	Moderate inhibitory activity against the transcription factor AP-1	[152]
Guineamide G	*L. majuscula* (Alotau Bay, Papua New Guinea)	Cyclic depsipeptide	Cytotoxic to mouse neuroblastoma cell lines	[153]
Hoiamide D	*Symploca* sp. (Kape Point, Papua New Guinea)	Linear peptide	Inhibitory activity against p53/MDM2 interaction	[144]
Ibu-epi-demethoxy-lyngbyastatin 3	*Leptolyngbya* sp. (SS Thistlegorm shipwreck, Red Sea)	Cyclic depsipeptide	No cytotoxicity to neuro-2a cells	[151]
Malyngamide 2	cf. *Lyngbya sordida* (Papua New Guinea)	Linear lipopeptides	Inhibitor of NO synthase in RAW cells; anti-inflammatory activity	[154]
Pitipeptolides C–F	*L. majuscula* (Piti Bomb Holes, Guam)	Cyclic depsipeptides	Moderate antibacterial activities against *Mycobacterium tuberculosis*; low cytotoxic effects on HT29 colon adenocarcinoma and MCF-7 breast cancer cells	[155]
Pitiprolamide	*L. majuscula* (Piti Bomb Holes, Guam)	Cyclic depsipeptide	Weak antibacterial activities against *M. tuberculosis* and *Bacillus cereus*; weak cytotoxic activity against HCT116 colon cancer and MCF-7 breast cancer cell lines	[136]
Wewakamide A	*Lyngbya semiplena* (Wewak Bay, Papua New Guinea)	Cyclic depsipeptide	Brine shrimp toxicity	[156]

23.5 Conclusions

Novel metabolites from marine fungi, actinomycetes, and cyanobacteria have been thoroughly explored. Biologically active and chemically diverse compounds, such as lactones, sterols, alkaloids, polyketides, diketopiperazines, and peptides, that could be cytotoxic, antibacterial, antifungal, antiplasmodial, or antioxidant are produced by marine fungi, actinomycetes, and cyanobacteria. New bioactive compounds isolated from cyanobacteria are dominated by peptides, particularly cyclic depsipeptides. However, *in vivo* and additional bioavailability tests of the most potent compounds are still unpublished. The most reported bioactive compound-producing marine fungi are still dominated by *Aspergillus* and *Penicillium*. *Streptomyces* sp. is the most productive marine actinomycete, while for cyanobacteria it is *Lyngbya* sp. Majority of the samples are isolated from marine organisms collected from the South China Sea, South Korea, Japan, Western Europe, and South America. Tropical regions with their high levels of biodiversity, however, remain less explored. Genetic or molecular taxonomic methods are commonly used for identifying marine microorganisms. The molecular approach is applied more frequently for producing bioactive compounds. Recently, nanoparticle biosynthesis is commonly applied to marine microorganisms for producing more potent bioactive compounds.

References

1 Blunt, J.W., Copp, B.R., Munro, M.H.G., Northcote, P.T., and Prinsep, M.R. (2010) Marine natural products. *Nat. Prod. Rep.*, **27**, 165–237.

2 Mayer, A.M.S., Rodríguez, A.D., Berlinck, R.G.S., and Fusetani, N. (2011) Marine pharmacology in 2007–2008: marine compounds with antibacterial, anticoagulant, antifungal, anti-inflammatory, antimalarial, antiprotozoal, antituberculosis, and antiviral activities; affecting the immune and nervous system, and other miscellaneous mechanisms of action. *Comp. Biochem. Physiol. C*, **153**, 191–222.

3 Mayer, A.M.S., Glaser, K.B., Cuevas, C., Jacobs, R.S., Kem, W., Little, R.D., McIntosh, J.M., Newman, D.J., Potts, B.C., and Shuster, D.E. (2010) The odyssey of marine pharmaceuticals: a current pipeline perspective. *Trends Pharmacol. Sci.*, **31**, 255–265.

4 Schumacher, M., Kelkel, M., Dicato, M., and Diederich, M. (2011) Gold from the sea: marine compounds as inhibitors of the hallmarks of cancer. *Biotechnol. Adv.*, **29**, 531–547.

5 Blunt, J.W., Copp, B.R., Munro, M.H.G., Northcote, P.T., and Prinsep, M.R. (2011) Marine natural products. *Nat. Prod. Rep.*, **28**, 196–268.

6 Blunt, J.W., Copp, B.R., Keyzers, R.A., Munro, M.H.G., and Prinsep, M.R. (2012) Marine natural products. *Nat. Prod. Rep.*, **29**, 144–222.

7 Kohlmeyer, J. and Kohlmeyer, E. (1979) *Marine Mycology: The Higher Fungi*, Academic Press, New York.

8 Liberra, K. and Lindequist, U. (1995) Marine fungi: a prolific resource of biologically active natural products? *Pharmazie.*, **50**, 583–588.

9 Pietra, F. (1997) Secondary metabolites from marine microorganisms: bacteria, protozoa, algae and fungi. *Nat. Prod. Rep.*, **14**, 453–464.

10 Jensen, P.R. and Fenical, W. (2002) Secondary metabolites from marine fungi, in *Fungi in Marine Environments*, *Fungal Diversity Research Series*, vol. 7

(ed. K.D. Hyde), Fungal Diversity Press, Hong Kong, pp. 293–315.

11 Ebel, R. (2010) Natural product diversity from marine fungi, in *Comprehensive Natural Products II: Chemistry and Biology*, vol. **2** (eds L. Mander*et al.*), Elsevier, Oxford, pp. 223–262.

12 Tarman, K., Lindequist, U., Wende, K., Porzel, A., Arnold, N., and Wessjohann, L.A. (2011) Isolation of a new natural product and cytotoxic and antimicrobial activities of extracts from fungi of Indonesian marine habitats. *Mar. Drugs*, **9** (3), 294–306.

13 Blunt, J.W., Copp, B.R., Hu, W.-P., Munro, M.H.G., Northcote, P.T., and Prinsep, M. R. (2009) Marine natural products. *Nat. Prod. Rep.*, **26**, 170–244.

14 Thomas, T.R.A., Kavlekar, D.P., and LokaBharathi, P.A. (2010) Marine drugs from sponge-microbe association: a review. *Mar. Drugs*, **8** (4), 1417–1468.

15 Almeida, C., Kehraus, S., Prudêncio, M., and König, G.M. (2011) Marilones A–C, phthalides from the sponge-derived fungus *Stachylidium* sp. *Beilstein J. Org. Chem.*, **7**, 1636–1642.

16 Almeida, C., Part, N., Bouhired, S., Kehraus, S., and König, G.M. (2011) Stachylines A–D from the sponge derived fungus *Stachylidium* sp. *J. Nat. Prod.*, **74**, 21–25.

17 Jadulco, R., Brauers, G., Edrada, R.A., Ebel, R., Wray, V., Sudarsono, S., and Proksch P. (2002) New metabolites from sponge-derived fungi *Curvularia lunata* and *Cladosporium herbarum*. *J. Nat. Prod.*, **65** (5), 730–733.

18 Bringmann, G., Gulder, T.A.M., Lang, G., Schmitt, S., Stöhr, R., Wiese, J., Nagel, K., and Imhoff, J.F. (2007) Large-scale biotechnological production of the antileukemic marine natural product sorbicillactone A. *Mar. Drugs*, **5** (2), 23–30.

19 Amagata, T., Tanaka, M., Yamada, T., Minoura, K., and Numata, A. (2008) Gymnastatins and dankastatins, growth inhibitory metabolites of *Gymnacella* species from a *Halichondria* sponge. *J. Nat. Prod.*, **71**, 340–345.

20 Sugiyama, Y., Ito, Y., Suzuki, M., and Hirota, A. (2009) Indole derivatives from a marine sponge-derived yeast as DPPH radical scavengers. *J. Nat. Prod.*, **72**, 2069–2071.

21 Abdel-Lateff, A., Fisch, K., and Wright, A.D. (2009) Trichopyrone and other constituents from the marine sponge-derived fungus *Trichoderma* sp. *Z. Naturforsch. C*, **64**, 186–192.

22 Lee, Y.M., Mansoor, T.A., Hong, J., Lee, C.-O., Bae, K.S., and Jung, J.H. (2007) Polyketides from a sponge-derived fungus, *Aspergillus versicolor*. *Nat. Prod. Sci.*, **13** (1), 90–96.

23 Elbandy, M., Shinde, P.B., Hong, J., Bae, K.S., Kim, M.A., Lee, S.M., and Jung, J.H. (2009) α-Pyrones and yellow pigments from the sponge-derived fungus *Paecilomyces lilacinus*. *Bull. Korean Chem. Soc.*, **30**, 188–192.

24 Lee, Y.M., Dang, H.T., Li, J., Zhang, P., Hong, J., Lee, C.-O., and Jung, J.H. (2011) A cytotoxic fellutamide analogue from the sponge-derived fungus *Aspergillus versicolor*. *Bull. Korean Chem. Soc.*, **32** (10), 3817–3820.

25 Yu, Z., Lang, G., Kajahn, I., Schmaljohann, R., and Imhoff, J.F. (2008) Scopularides A and B, cyclodepsipeptides from a marine sponge-derived fungus, *Scopulariopsis brevicaulis*. *J. Nat. Prod.*, **71** (6), 1052–1054.

26 Wiese, J., Ohlendorf, B., Blümel, M., Schmaljohann, R., and Imhoff, J.F. (2011) Phylogenetic identification of fungi isolated from the marine sponge *Tethya aurantium* and identification of their secondary metabolites. *Mar. Drugs*, **9** (4), 561–585.

27 Liu, H., Edrada-Ebel, R., Ebel, R., Wang, Y., Schulz, B., Draeger, S., Müller, W.E.G., Wray, V., Lin, W., and Proksch, P. (2009) Drimane sesquiterpenoids from the fungus *Aspergillus ustus* isolated from the marine sponge *Suberites domuncula*. *J. Nat. Prod.*, **72** (9), 1585–1588.

28 Höller, U., König, G.M., and Wright, A.D. (1999) Three new metabolites from marine-derived fungi of the genera *Coniothyrium* and *Microsphaeropsis*. *J. Nat. Prod.*, **62** (1), 114–118.

29 Neumann, K., Kehraus, S., Gütschow, M., and König, G.M. (2009) Cytotoxic and HLE-inhibitory tetramic acid derivatives from marine-derived fungi. *Nat. Prod. Commun.*, **4**, 347–354.

30 Mohamed, I.E., Gross, H., Pontius, A., Kehraus, S., Krick, A., Kelter, G., Maier, A., Fiebig, H.-H., and König, G.M. (2009) Epoxyphomalin A and B, prenylated polyketides with potent cytotoxicity from the marine-derived fungus *Phoma* sp. *Org. Lett.*, **11** (21), 5014–5017.
31 Sun, L.-L., Shao, C.-L., Chen, J.-F., Guo, Z.-Y., Fu, X.-M., Chen, M., Chen, Y.-Y., Li, R., deVoogd, N.J., She, Z.-G., Lin, Y.-C., and Wang, C.-Y. (2012) New bisabolane sesquiterpenoids from a marine-derived fungus *Aspergillus* sp. isolated from the sponge *Xestospongia testudinaria*. *Bioorg. Med. Chem. Lett.*, **22**, 1326–1329.
32 Li, D., Xu, Y., Shao, C.-L., Yang, R.-Y., Zheng, C.-J., Chen, Y.-Y., Fu, X.-M., Qian, P.-Y., She, Z.-G., deVoogd, N.J., and Wang, C.-Y. (2012) Antibacterial bisabolane-type sesquiterpenoids from the sponge-derived fungus *Aspergillus* sp. *Mar. Drugs*, **10** (1), 234–241.
33 Bringmann, G., Lang, G., Bruhn, T., Schäffler, K., Steffens, S., Schmaljohann, R., Wiese, J., and Imhoff, J.F. (2010) Sorbifuranones A–C, sorbicillinoid metabolites from *Penicillium* strains isolated from Mediterranean sponges. *Tetrahedron*, **66**, 9894–9901.
34 Boot, C.M., Amagata, T., Tenney, K., Compton, J.E., Pietraszkiewicz, H., Valeriote, F.A., and Crews, P. (2007) Four classes of structurally unusual peptides from two marine-derived fungi: structures and bioactivities. *Tetrahedron*, **63** (39), 9903–9914.
35 Raghukumar, C. (2006) *Algal–Fungal Interactions in the Marine Ecosystem: Symbiosis to Parasitism*, Central Salt and Marine Chemicals Research Institute.
36 Osterhage, C., König, G.M., Höller, U., and Wright, A.D. (2002) Rare sesquiterpenes from the algicolous fungus *Drechslera dematioidea*. *J. Nat. Prod.*, **65** (3), 306–313.
37 Pontius, A., Mohamed, I., Krick, A., Kehraus, S., and König, G.M. (2008) Aromatic polyketides from marine algicolous fungi. *J. Nat. Prod.*, **71**, 272–274.
38 Greve, H., Schupp, P.J., Eguereva, E., Kehraus, S., Kelter, G., Maier, A., Fiebig, H.-H., and König, G.M. (2008) Apralactone A and a new stereochemical class of curvularins from the marine fungus *Curvularia* sp. *Eur. J. Org. Chem.*, **2008** (30), 5085–5092.
39 Qiao, M.-F., Ji, N.-Y., Liu, X.-H., Li, K., Zhu, Q.-M., and Xue, Q.-Z. (2010) Indoloditerpenes from an algicolous isolate of *Aspergillus oryzae*. *Bioorg. Med. Chem. Lett.*, **20**, 5677–5680.
40 Meyer, S.W., Mordhorst, T.F., Lee, C., Jensen, P.R., Fenical, W., and Köck, M. (2010) Penilumamide, a novel lumazine peptide isolated from the marine-derived fungus, *Penicillium* sp. CNL-338. *Org. Biomol. Chem.*, **8**, 2158–2163.
41 Osterhage, C., Kaminsky, R., König, G.M., and Wright, A.D. (2000) Ascosalipyrrolidinone A, an antimicrobial alkaloid, from the obligate marine fungus *Ascochyta salicorniae*. *J. Org. Chem.*, **65** (20), 6412–6417.
42 Gamal-Eldeen, A.M., Abdel-Lateff, A., and Okino, T. (2009) Modulation of carcinogen metabolizing enzymes by chromanone A; a new chromone derivative from algicolous marine fungus *Penicillium* sp. *Environ. Toxicol. Pharmacol.*, **28** (3), 317–322.
43 Cui, C.-M., Li, X.-M., Li, C.-S., Proksch, P., and Wang, B.-G. (2010) Cytoglobosins A–G, cytochalasans from a marine-derived endophytic fungus, *Chaetomium globosum* QEN-14. *J. Nat. Prod.*, **73** (4), 729–733.
44 Almeida, C., Eguereva, E., Kehraus, S., Siering, C., and König, G.M. (2010) Hydroxylated sclerosporin derivatives from the marine-derived fungus *Cadophora malorum*. *J. Nat. Prod.*, **73** (3), 476–478.
45 Zhang, Y., Li, X.-M., and Wang, B.-G. (2007) Nigerasperones A–C, new monomeric and dimeric naphtho-γ-pyrones from a marine alga-derived endophytic fungus *Aspergillus niger* EN-13. *J. Antibiot. (Tokyo)*, **60**, 204–210.
46 Cueto, M., Jensen, P.R., Kauffman, C., Fenical, W., Lobkovsky, E., and Clardy, J. (2001) Pestalone, a new antibiotic produced by a marine fungus in response to bacterial challenge. *J. Nat. Prod.*, **64** (11), 1444–1446.
47 Abdel-Lateff, A., Fisch, K.M., Wright, A.D., and König, G.M. (2003) A new antioxidant isobenzofuranone derivative from the algicolous marine fungus *Epicoccum* sp. *Planta Med.*, **69** (9), 831–834.

48 Nguyen, H.P., Zhang, D., Lee, U., Kang, J.S., Choi, H.D., and Son, B.W. (2007) Dehydroxychlorofusarielin B, an antibacterial polyoxygenated decalin derivative from the marine-derived fungus *Aspergillus* sp. *J. Nat. Prod.*, **70** (7), 1188–1190.

49 Cui, C.-M., Li, X.-M., Li, C.-S., Sun, H.-F., Gao, S.-S., and Wang, B.-G. (2009) Benzodiazepine alkaloids from marine-derived endophytic fungus *Aspergillus ochraceus*. *Helv. Chim. Acta*, **92** (7), 1366–1370.

50 Wang, F.-W. (2012) Bioactive metabolites from *Guignardia* sp., an endophytic fungus residing in *Undaria pinnatifida*. *Chin. J. Nat. Med.*, **10** (1), 72–76.

51 Wang, W., Zhu, T., Tao, H., Lu, Z., Fang, Y., Gu, Q., and Zhu, W. (2007) Two new cytotoxic quinone type compounds from the halotolerant fungus *Aspergillus variecolor*. *J. Antibiot. (Tokyo)*, **60**, 603–607.

52 Li, D.-L., Li, X.-M., Proksch, P., and Wang, B.-G. (2010) 7-*O*-Methylvariecolortide A, a new spirocyclic diketopiperazine alkaloid from a marine mangrove derived endophytic fungus, *Eurotium rubrum*. *Nat. Prod. Commun.*, **5** (10), 1583–1586.

53 Liu, F., Cai, X.-L., Yang, H., Xia, X.-K., Guo, Z.-Y., Yuan, J., Li, M.-F., She, Z.-G., and Lin, Y.-C. (2010) The bioactive metabolites of the mangrove endophytic fungus *Talaromyces* sp. ZH-154 isolated from *Kandelia candel* (L.) Druce. *Planta Med.*, **76** (2), 185–189.

54 Wen, L., Cai, X., Xu, F., She, Z., Chan, W. L., Vrijmoed, L.L.P., Jones, E.B.G., and Lin, Y. (2009) Three metabolites from the mangrove endophytic fungus *Sporothrix* sp. (#4335) from the South China Sea. *J. Org. Chem.*, **74** (3), 1093–1098.

55 Huang, H., She, Z., Lin, Y., Vrijmoed, L.L. P., and Lin, W. (2007) Cyclic peptides from an endophytic fungus obtained from a mangrove leaf (*Kandelia candel*). *J. Nat. Prod.*, **70** (11), 1696–1699.

56 Lin, Z.-J., Zhang, G.-J., Zhu, T.-J., Liu, R., Wei, H.-J., and Gu, Q.-Q. (2009) Bioactive cytochalasins from *Aspergillus flavipes*, an endophytic fungus associated with the mangrove plant *Acanthus ilicifolius*. *Helv. Chim. Acta*, **92** (8), 1538–1544.

57 Maria, G.L., Sridhar, K.R., and Raviraja, N.S. (2005) Antimicrobial and enzyme activity of mangrove endophytic fungi of southwest coast of India. *J. Agric. Technol.*, **1**, 67–80.

58 Xiaoling, C., Xiaoli, L., Shining, Z., Junping, G., Shuiping, W., Xiaoming, L., Zhigang, S., and Yongcheng, L. (2010) Cytotoxic and topoisomerase I inhibitory activities from extracts of endophytic fungi isolated from mangrove plants in Zhuhai, China. *J. Ecol. Nat. Environ.*, **2** (2), 17–24.

59 Han, Z., Mei, W., Zhao, Y., Deng, Y., and Dai, H. (2009) A new cytotoxic isocoumarin from endophytic fungus *Penicillium* SP. 091402 of the mangrove plant *Bruguiera sexangula*. *Chem. Nat. Compd.*, **45** (6), 805–807.

60 Chaeprasert, S., Piapukiew, J., Whalley, A.J.S., and Sihanonth, P. (2010) Endophytic fungi from mangrove plant species of Thailand: their antimicrobial and anticancer potentials. *Bot. Mar.*, **53** (6), 555–564.

61 Huang, Z., Guo, Z., Yang, R., Yin, X., Li, X., Luo, W., She, Z., and Lin, Y. (2009) Chemistry and cytotoxic activities of polyketides produced by the mangrove endophytic fungus *Phomopsis* SP. ZSU-H76 *Chem. Nat. Compd.*, **45** (5), 625–628.

62 Lu, Z., Zhu, H., Fu, P., Wang, Y., Zhang, Z., Lin, H., Liu, P., Zhuang, Y., Hong, K., and Zhu, W. (2010) Cytotoxic polyphenols from the marine-derived fungus *Penicillium expansum*. *J. Nat. Prod.*, **73** (5), 911–914.

63 Sun, S., Tian, L., Wu, Z.-H., Chen, G., Wu, H.-H., Wang, Y.-N., and Pei, Y.-H. (2009) Two new compounds from fermentation liquid of the marine fungus *Trichoderma atroviride* G20-12. *J. Asian Nat. Prod. Res.*, **11** (10), 898–903.

64 Huang, H.-B., Feng, X.-J., Liu, L., Chen, B., Lu, Y.-J., Ma, L., She, Z.-G., and Lin, Y.-C. (2010) Three dimeric naphtho-γ-pyrones from the mangrove endophytic fungus *Aspergillus tubingensis* isolated from *Pongamia pinnata*. *Planta Med.*, **76** (16), 1888–1891.

65 Wu, H.-H., Tian, L., Feng, B.-M., Li, Z.-F., Zhang, Q.-H., and Pei, Y.-H. (2010) Three new compounds from the marine fungus

Penicillium sp. *J. Asian Nat. Prod. Res.*, **12** (1), 15–19.

66 Pan, J.-H., Deng, J.-J., Chen, Y.-G., Gao, J.-P., Lin, Y.-C., She, Z.G., and Gu, Y.-C. (2010) New lactone and xanthone derivatives produced by a mangrove endophytic fungus *Phoma* sp. SK3RW1M from the South China Sea. *Helv. Chim. Acta*, **93** (7), 1369–1374.

67 Bugni, T.S. and Ireland, C.M. (2004) Marine-derived fungi: a chemically and biologically diverse group of microorganisms. *Nat. Prod. Rep.*, **21**, 143–163.

68 Bhadury, P., Mohammad, B.T., and Wright, P.C. (2006) The current status of natural products from marine fungi and their potential as anti-infective agents. *J. Ind. Microbiol. Biotechnol.*, **33**, 325–337.

69 Saleem, M., Ali, M.S., Hussain, S., Jabbar, A., Ashraf, M., and Lee, Y.S. (2007) Marine natural products of fungal origin. *Nat. Prod. Rep.*, **24** (5), 1142–1152.

70 Debbab, A., Aly, A.H., Lin, W.H., and Proksch, P. (2010) Bioactive compounds from marine bacteria and fungi. *Microb. Biotechnol.*, **3** (5), 544–563.

71 Rateb, M.E. and Ebel, R. (2011) Secondary metabolites of fungi from marine habitats. *Nat. Prod. Rep.*, **28**, 290–344.

72 Chu, D., Peng, C., Ding, B., Liu, F., Zhang, F., Lin, H., and Li, Z. (2011) Biological active metabolite cyclo(L-Trp-L-Phe) produced by South China Sea sponge *Holoxea* sp. associated fungus *Aspergillus versicolor* strain TS08. *Bioprocess Biosyst. Eng.*, **34** (2), 223–229.

73 Li, D., Chen, L., Zhu, T., Kurtán, T., Mándi, A., Zhao, Z., Li, J., and Gu, Q. (2011) Chloctanspirones A and B, novel chlorinated polyketides with an unprecedented skeleton, from marine sediment derived fungus *Penicillium terrestre*. *Tetrahedron*, **67** (41), 7913–7918.

74 Poch, G.K. and Gloer, J.B. (1989) Helicascolides A and B: new lactones from the marine fungus *Helicascus kanaloanus*. *J. Nat. Prod.*, **52** (2), 257–260.

75 Tarman, K., Palm, G.J., Porzel, A., Merzweiler, K., Arnold, N., Wessjohann, L.A., Unterseher, M., and Lindequist, U. (2012) Helicascolide C, a new lactone from an Indonesian marine algicolous strain of *Daldinia eschscholzii* (Xylariaceae, Ascomycota). *Phytochem. Lett.*, **5** (1), 83–86.

76 Julianti, E., Oh, H., Lee, H.S., Oh, D.-C., Oh, K.-B., and Shin, J. (2012) Acremolin, a new 1*H*-azirine metabolite from the marine-derived fungus *Acremonium strictum*. *Tetrahedron Lett.*, **53** (23), 2885–2886.

77 Julianti, E., Oh, H., Jang, K.H., Lee, J.K., Lee, S.K., Oh, D.-C., Oh, K.-B., and Shin, J. (2011) Acremostrictin, a highly oxygenated metabolite from the marine fungus *Acremonium strictum*. *J. Nat. Prod.*, **74** (12), 2592–2594.

78 Liu, J., Hu, Z., Huang, H., Zheng, Z., and Xu, Q. (2012) Aspochalasin U, a moderate TNF-α inhibitor from *Aspergillus* sp. *J. Antibiot. (Tokyo)*, **65** (1), 49–52.

79 Lu, X., Tian, L., Chen, G., Xu, Y., Wang, H.-F., Li, Z.-Q., and Pei, Y.-H. (2012) Three new compounds from the marine-derived fungus *Trichoderma atroviride* G20-12. *J. Asian Nat. Prod. Res.*, **14** (7), 647–651.

80 San-Martín, A., Rovirosa, J., Vaca, I., Vergara, K., Acevedo, L., Viña, D., Orallo, F., and Chamy, M.C. (2011) New butyrolactone from a marine-derived fungus *Aspergillus* sp. *J. Chil. Chem. Soc.*, **56** (1), 625–627.

81 Zhang, G., Sun, S., Zhu, T., Lin, Z., Gu, J., Li, D., and Gu, Q. (2011) Antiviral isoindolone derivatives from an endophytic fungus *Emericella* sp. associated with *Aegiceras corniculatum*. *Phytochemistry*, **72**, 1436–1442.

82 Chen, Z., Zheng, Z., Huang, H., Song, Y., Zhang, X., Ma, J., Wang, B., Zhang, C., and Ju, J. (2012) Penicacids A–C, three new mycophenolic acid derivatives and immunosuppressive activities from the marine-derived fungus *Penicillium* sp. SOF07. *Bioorg. Med. Chem. Lett.*, **22** (9), 3332–3335.

83 Zhuang, P., Tang, X.X., Yi, Z.W., Qiu, Y.K., and Wu, Z. (2012) Two new compounds from marine-derived fungus *Penicillium* sp. F11. *J. Asian Nat. Prod. Res.*, **14** (3), 197–203.

84 Elsebai, M.F., Kehraus, S., Lindequist, U., Sasse, F., Shaaban, S., Gütschow, M., Josten, M., Sahl, H.-G., and König, G.M.

(2011) Antimicrobial phenalenone derivatives from the marine-derived fungus *Coniothyrium cereale*. *Org. Biomol. Chem.*, **9**, 802–808.

85 Sun, L., Li, D., Tao, M., Chen, Y., Dan, F., and Zhang, W. (2012) Scopararanes C–G: new oxygenated pimarane diterpenes from the marine sediment-derived fungus *Eutypella scoparia* FS26. *Mar. Drugs*, **10**, 539–550.

86 Wang, Y., Zheng, J., Liu, P., Wang, W., and Zhu, W. (2011) Three new compounds from *Aspergillus terreus* PT06-2 grown in a high salt medium. *Mar. Drugs*, **9** (8), 1368–1378.

87 Korbekandi, H., Iravani, S., and Abbasi, S. (2009) Production of nanoparticles using organisms. *Crit. Rev. Biotechnol.*, **29** (4), 279–306.

88 Mukherjee, P., Senapati, S., Mandal, D., Ahmad, A., Khan, M.I., Kumar, R., and Sastry, M. (2002) Extracellular synthesis of gold nanoparticles by the fungus *Fusarium oxysporum*. *ChemBioChem.*, **3** (5), 461–463.

89 Li, G., He, D., Qian, Y., Guan, B., Gao, S., Cui, Y., Yokoyama, K., and Wang, L. (2012) Fungus-mediated green synthesis of silver nanoparticles using *Aspergillus terreus*. *Int. J. Mol. Sci.*, **13**, 466–476.

90 Shankar, S.S., Ahmad, A., Pasricha, R., and Sastry, M. (2003) Bioreduction of chloroaurate ions by geranium leaves and its endophytic fungus yields gold nanoparticles of different shapes. *J. Mater. Chem.*, **13**, 1822–1826.

91 Devi, L.S. and Joshi, S.R. (2012) Antimicrobial and synergistic effects of silver nanoparticles synthesized using soil fungi of high altitudes of Eastern Himalaya. *Mycobiology*, **40** (1), 27–34.

92 Mitova, M.I., Lang, G., Wiese, J., and Imhoff, J.F. (2008) Subinhibitory concentrations of antibiotics induce phenazine production in a marine *Streptomyces* sp. *J. Nat. Prod.*, **71**, 824–827.

93 Schneemann, I., Ohlendorf, B., Zinecker, H., Nagel, K., Wiese, J., and Imhoff, J.F. (2010) Nocapyrones A–D, gamma-pyrones from a *Nocardiopsis* strain isolated from the marine sponge *Halichondria panacea*. *J. Nat. Prod.*, **73**, 1444–1447.

94 Pimentel-Elardo, S.M., Gulder, T.A.M., Hentschel, U., and Bringmann, G. (2008) Cebulactams A1 and A2, new macrolactams isolated from *Saccharopolyspora cebuensis*, the first obligate marine strain of the genus *Saccharopolyspora*. *Tetrahedron Lett.*, **49**, 6889–6892.

95 Khan, S.T., Komaki, H., Motohashi, K., Kozone, I., Mukai, A., Takagi, M., and Shin-ya, K. (2011) *Streptomyces* associated with a marine sponge *Haliclona* sp.; biosynthetic genes for secondary metabolites and products. *Environ. Microbiol.*, **13** (2), 391–403.

96 Izumikawa, M., Khan, S.T., Takagi, M., and Shin-ya, K. (2010) Sponge-derived *Streptomyces* producing isoprenoids via the mevalonate pathway. *J. Nat. Prod.*, **73**, 208–212.

97 Abdelmohsen, U.R., Pimentel-Elardo, S. M., Hanora, A., Radwan, M., Abou-El-Ela, S.H., Ahmed, S., and Hentschel, U. (2010) Isolation, phylogenetic analysis and anti-infective activity screening of marine sponge-associated actinomycetes. *Mar. Drugs*, **8** (3), 399–412.

98 Selvin, J. (2009) Exploring the antagonistic producer *Streptomyces* MSI051: implications of polyketide synthase gene type II and a ubiquitous defense enzyme phospholipase A2 in host sponge *Dendrilla nigra*. *Curr. Microbiol.*, **58** (5), 459–463.

99 Pimentel-Elardo, S.M., Kozytska, S., Bugni, T.S., Ireland, C.M., Moll, H., and Hentschel, U. (2010) Anti-parasitic compounds from *Streptomyces* sp. strains isolated from Mediterranean sponges. *Mar. Drugs*, **8**, 373–380.

100 Selvakumar, D., Arun, K., Suguna, S., Kumar, D., and Dhevendaran, K. (2010) Bioactive potential of *Streptomyces* against fish and shellfish pathogens. *Iran. J. Microbiol.*, **2** (3), 157–164.

101 Lin, C., Lu, C., and Shen, Y. (2010) Three new 2-pyranone derivatives from mangrove endophytic actinomycete strain *Nocardiopsis* sp. A00203. *Rec. Nat. Prod.*, **4** (4), 176–179.

102 Wang, F., Xu, M., Li, Q., Sattler, I., and Lin, W. (2010) *p*-Aminoacetophenonic acids produced by a mangrove endophyte

Streptomyces sp. (strain HK10552). *Molecules*, **15**, 2782–2790.

103 Ding, L., Münch, J., Goerls, H., Maier, A., Fiebig, H.-H., Lin, W.-H., and Hertweck, C. (2010) Xiamycin, a pentacyclic indolosesquiterpene with selective anti-HIV activity from a bacterial mangrove endophyte. *Bioorg. Med. Chem. Lett.*, **20**, 6685–6687.

104 Hong, K., Gao, A.-H., Xie, Q.-Y., Gao, H., Zhuang, L., Lin, H.-P., Yu, H.-P., Li, J., Yao, X.-S., Goodfellow, M., and Ruan, J.-S. (2009) Actinomycetes for marine drug discovery isolated from mangrove soils and plants in China. *Mar. Drugs*, **7** (1), 24–44.

105 Fu, P., Liu, P.P., Qu, H.J., Wang, Y., Chen, D.F., Wang, H., Li, J., and Zhu, W.M. (2011) α-Pyrones and diketopiperazine derivatives from the marine-derived actinomycete *Nocardiopsis dassonvillei* HR10-5. *J. Nat. Prod.*, **74** (10), 2219–2223.

106 Lane, A.L. and Moore, B.S. (2011) A sea of biosynthesis: marine natural products meet the molecular age. *Nat. Prod. Rep.*, **28**, 411–428.

107 Chauhan, D., Hideshima, T., and Anderson, K.C. (2006) A novel proteasome inhibitor NPI-0052 as an anticancer therapy. *Br. J. Cancer*, **95** (8), 961–965.

108 Fenical, W. and Jensen, P.R. (2006) Developing a new resource for drug discovery: marine actinomycete bacteria. *Nat. Chem. Biol.*, **2**, 666–673.

109 Fenical, W., Jensen, P.R., Palladino, M.A., Lam, K.S., Lloyd, G.K., and Potts, B.C. (2009) Discovery and development of the anticancer agent salinosporamide A (NPI-0052). *Bioorg. Med. Chem.*, **17** (6), 2175–2180.

110 Dick, L.R. and Fleming, P.E. (2010) Building on bortezomib: second-generation proteasome inhibitors as anti-cancer therapy. *Drug Discov. Today*, **15**, 243–249.

111 Steverding, D., Wang, X., Potts, B.C., and Palladino, M.A. (2012) Trypanocidal activity of beta-lactone-gamma-lactam proteasome inhibitors. *Planta Med.*, **78** (2), 131–134.

112 Han, Z., Xu, Y., McConnell, O., Liu, L., Li, Y., Qi, S., Huang, X., and Qian, P. (2012) Two antimycin A analogues from marine-derived actinomycete *Streptomyces lusitanus*. *Mar. Drugs*, **10**, 668–676.

113 Sakoulas, G., Nam, S.J., Loesgen, S., Fenical, W., Jensen, P.R., Nizet, V., and Hensler, M. (2012) Novel bacterial metabolite merochlorin A demonstrates *in vitro* activity against multi-drug resistant methicillin-resistant *Staphylococcus aureus*. *PLoS One*, **7** (1), e29439-1–e29439-6.

114 Fu, P., Wang, S.X., Hong, K., Li, X., Liu, P.P., Wang, Y., and Zhu, W.M. (2011) Cytotoxic bipyridines from the marine-derived actinomycete *Actinoalloteichus cyanogriseus* WH1–2216-6. *J. Nat. Prod.*, **74** (8), 1751–1756.

115 Xie, Z., Liu, B., Wang, H., Yang, S., Zhang, H., Wang, Y., Ji, N., Qin, S., and Laatsch, H. (2012) Kiamycin, a unique cytotoxic angucyclinone derivative from a marine *Streptomyces* sp. *Mar. Drugs*, **10**, 551–558.

116 Gärtner, A., Ohlendorf, B., Schulz, D., Zinecker, H., Wiese, J., and Imhoff, J.F. (2011) Levantilides A and B, 20-membered macrolides from a *Micromonospora* strain isolated from the Mediterranean deep sea sediment. *Mar. Drugs*, **9** (1), 98–108.

117 Raju, R., Piggott, A.M., Huang, X.C., and Capon, R.J. (2011) Nocardioazines: a novel bridged diketopiperazine scaffold from a marine-derived bacterium inhibits P-glycoprotein. *Org. Lett.*, **13** (10), 2770–2773.

118 Li, S.M., Tian, X.P., Niu, S.W., Zhang, W.J., Chen, Y.C., Zhang, H.B., Yang, X.W., Zhang, W.M., Li, W.J., Zhang, S., Ju, J.H., and Zhang, C.S. (2011) Pseudonocardians A–C, new diazaanthraquinone derivatives from a deep-sea actinomycete *Pseudonocardia* sp. SCSIO 01299. *Mar. Drugs*, **9** (8), 1428–1439.

119 Xu, M.J., Liu, X.J., Zhao, Y.L., Liu, D., Xu, Z.H., Lang, X.M., Ao, P., Lin, W.H., Yang, S.L., Zhang, Z.G., and Xu, J. (2012) Identification and characterization of an anti-fibrotic benzopyran compound isolated from mangrove-derived *Streptomyces xiamenensis*. *Mar. Drugs*, **10**, 639–654.

120 Huang, H.B., Yao, Y.L., He, Z.X., Yang, T.T., Ma, J.Y., Tian, X.P., Li, Y.Y., Huang, C.G., Chen, X.P., Li, W.J., Zhang, S., Zhang, C.S., and Ju, J.H. (2011)

Antimalarial beta-carboline and indolactam alkaloids from *Marinactinospora thermotolerans*, a deep sea isolate. *J. Nat. Prod.*, **74** (10), 2122–2127.

121 Zhu, Q.H., Li, J., Ma, J.Y., Luo, M.H., Wang, B., Huang, H.B., Tian, X.P., Li, W.J., Zhang, S., Zhang, C.S., and Ju, J.H. (2012) Discovery and engineered overproduction of antimicrobial nucleoside antibiotic A201A from the deep-sea marine actinomycete *Marinactinospora thermotolerans* SCSIO 00652. *Antimicrob. Agents Chemother.*, **56** (1), 110–114.

122 Xu, L.Y., Quan, X.S., Wang, C., Sheng, H.F., Zhou, G.X., Lin, B.R., Jiang, R.W., and Yao, X.S. (2011) Antimycins A(19) and A (20), two new antimycins produced by marine actinomycete *Streptomyces antibioticus* H74-18. *J. Antibiot. (Tokyo)*, **64** (10), 661–665.

123 Simmons, L., Kaufmann, K., Garcia, R., Schwär, G., Huch, V., and Müller, R. (2011) Bendigoles D–F, bioactive sterols from the marine sponge-derived *Actinomadura* sp. SBMs009. *Bioorg. Med. Chem.*, **19** (22), 6570–6575.

124 Hu, Y.C., Espindola, A.P.D.M., Stewart, N.A., Wei, S.G., Posner, B.A., and MacMillan, J.B. (2011) Chromomycin SA analogs from a marine-derived *Streptomyces* sp. *Bioorg. Med. Chem.*, **19** (17), 5183–5189.

125 Fu, P., Liu, PP., Li, X., Wang, Y., Wang, S., Hong, K., and Zhu, W. (2011) Cyclic bipyridine glycosides from the marine-derived actinomycete *Actinoalloteichus cyanogriseus* WH1–2216-6. *Org. Lett.*, **13** (22), 5948–5951.

126 Wei, R.B., Xi, T., Li, J., Wang, P., Li, F.C., Lin, Y.C., and Qin, S. (2011) Lobophorin C and D, new kijanimicin derivatives from a marine-sponge-associated actinomycetal strain AZS17. *Mar. Drugs*, **9** (3), 359–368.

127 Wang, F.Z., Tian, X.P., Huang, C.G., Li, Q. X., and Zhang, S. (2011) Marinactinones A–C, new gamma-pyrones from marine actinomycete *Marinactinospora thermotolerans* SCSIO 00606. *J Antibiot. (Tokyo)*, **64** (2), 189–192.

128 Matsuo, Y., Kanoh, K., Jang, J.H., Adachi, K., Matsuda, S., Miki, O., Kato, T., and Shizuri, Y. (2011) Streptobactin, a tricatechol-type siderophore from marine-derived *Streptomyces* sp. YM5–799. *J. Nat. Prod.*, **74** (11), 2371–2376.

129 Wyche, T.P., Hou, Y.P., Braun, D., Cohen, H.C., Xiong, M.P., and Bugni, T.S. (2011) First natural analogs of the cytotoxic thiodepsipeptide thiocoraline A from a marine *Verrucosispora* sp. *J. Org. Chem.*, **76** (16), 6542–6547.

130 Sato, S., Iwata, F., Yamada, S., Kawahara, H., and Katayama, M. (2011) Usabamycins A–C: new anthramycin-type analogues from a marine-derived actinomycete. *Bioorg. Med. Chem. Lett.*, **21** (23), 7099–7101.

131 Genilloud, O., González, I., Salazar, O., Martín, J., Tormo, J.R., and Vicente, F. (2011) Current approaches to exploit actinomycetes as a source of novel natural products. *J. Ind. Microbial. Biotechnol.*, **38**, 375–389.

132 Schweder, T., Lindequist, U., and Lalk, M. (2005) Screening for new metabolites from marine microorganisms. *Adv. Biochem. Eng. Biotechnol.*, **96**, 1–48.

133 Berdy, J. (2005) Bioactive microbial metabolites. *J. Antibiot. (Tokyo)*, **58**, 1–26.

134 Singh, R.K., Tiwari, S.P., Rai, A.K., and Mohapatra, T.M. (2011) Cyanobacteria: an emerging source for drug discovery. *J Antibiot. (Tokyo)*, **64** (6), 401–412.

135 Luesch, H., Moore, R.E., Paul, V.J., Mooberry, S.L., and Corbett, T.H. (2001) Isolation of dolastatin 10 from the marine cyanobacterium *Symploca* species VP642 and total stereochemistry and biological evaluation of its analogue symplostatin 1. *J. Nat. Prod.*, **64** (7), 907–910.

136 Montaser, R., Abboud, K.A., Paul, V.J., and Luesch, H. (2011) Pitiprolamide, a proline-rich dolastatin 16 analogue from the marine cyanobacterium *Lyngbya majuscula* from Guam. *J. Nat. Prod.*, **74**, 109–112.

137 Pagliara, P. and Caroppo, C. (2011) Cytotoxic and antimitotic assessment of aqueous extracts from eight cyanobacterial strains isolated from the marine sponge *Petrosia ficiformis*. *Toxicon*, **57**, 889–896.

138 Sivonen, K. and Börner, T. (2008) Bioactive compounds produced by cyanobacteria, in *The Cyanobacteria: Molecular Biology, Genomics and Evolution* (eds A. Herrero *et al.*), Caister Academic Press, Hethersett, UK.

139 Mevers, E., Liu, W.-T., Engene, N., Mohimani, H., Byrum, T., Pevzner, P.A., Dorrestein, P.C., Spadafora, C., and Gerwick, W.H. (2011) Cytotoxic veraguamides, alkynyl bromide-containing cyclic depsipeptides from the marine cyanobacterium cf. *Oscillatoria margaritifera*. *J. Nat. Prod.*, **74**, 928–936.

140 Salvador, L.A., Biggs, J.S., Paul, V.J., and Luesch, H. (2011) Veraguamides A–G, cyclic hexadepsipeptides from a dolastatin 16-producing cyanobacterium *Symploca* cf. *hydnoides* from Guam. *J. Nat. Prod.*, **74**, 917–927.

141 Molinski, T.F., Reynolds, K.A., and Morinaka, B.I. (2012) Symplocin A, a linear peptide from the bahamian cyanobacterium *Symploca* sp. configurational analysis of *N,N*-dimethylamino acids by chiral-phase HPLC of naphthacyl esters. *J. Nat. Prod.*, **75**, 425–431.

142 Teruya, T., Sasaki, H., Kitamura, K., Nakayama, T., and Suenaga, K. (2009) Biselyngbyaside, a macrolide glycoside from the marine cyanobacterium *Lyngbya* sp. *Org. Lett.*, **11**, 2421–2424.

143 Yonezawa, T., Mase, N., Sasaki, H., Teruya, T., Hasegawa, S.I., Cha, B.-Y., Yagasaki, K., Suenaga, K., Nagai, K., and Woo, J.-T. (2012) Biselyngbyaside, isolated from marine cyanobacteria, inhibits osteoclastogenesis and induces apoptosis in mature osteoclasts. *J. Cell Biochem.*, **113**, 440–448.

144 Malloy, K.L., Choi, H., Fiorilla, C., Valeriote, F.A., Matainaho, T., and Gerwick, W.H. (2012) Hoiamide D, a marine cyanobacteria-derived inhibitor of p53/MDM2 interaction. *Bioorg. Med. Chem. Lett.*, **22**, 683–688.

145 Engene, H., Choi, H., Esquenazi, E., Rottacker, E.C., Ellisman, M.H., Dorrestein, P.C., and Gerwick, W.H. (2011) Underestimated biodiversity as a major explanation for the perceived rich secondary metabolite capacity of the cyanobacterial genus *Lyngbya*. *Environ. Microbiol.*, **13**, 1601–1610.

146 Nunnery, J.K., Engene, N., Byrum, T., Cao, Z., Jabba, S.V., Pereira, A.R., Matainaho, T., Murray, T.F., and Gerwick, W.H. (2012) Biosynthetically intriguing chlorinated lipophilic metabolites from geographically distant tropical marine cyanobacteria. *J. Org. Chem.*, **77** (9), 4198–4208.

147 Tripathi, A., Puddick, J., Prinsep, M.R., Rottmann, M., Chan, K.P., Chen, D.Y.-K., and Tan, L.T. (2011) Lagunamide C, a cytotoxic cyclodepsipeptide from the marine cyanobacterium *Lyngbya majuscula*. *Phytochemistry*, **72**, 2369–2375.

148 Kwan, J.C., Meickle, T., Ladwa, D., Teplitski, M., Paul, V., and Luesch, H. (2011) Lyngbyoic acid, a “tagged” fatty acid from a marine cyanobacterium, disrupts quorum sensing in *Pseudomonas aeruginosa*. *Mol. Biosyst.*, **7**, 1205–1216.

149 Pereira, A.R., Etzbach, L., Engene, N., Müller, R., and Gerwick, W.H. (2011) Molluscicidal metabolites from an assemblage of Palmyra Atoll cyanobacteria. *J. Nat. Prod.*, **74**, 1175–1181.

150 Gutiérrez, M., Pereira, A.R., Debonsi, H.M., Ligresti, A., DiMarzo, V., and Gerwick, W.H. (2011) Cannabinomimetic lipid from a marine cyanobacterium. *J. Nat. Prod.*, **74**, 2313–2317.

151 Thornburg, C.C., Thimmaiah, M., Shaala, L.A., Hau, A.M., Malmo, J.M., Ishmael, J.E., Youssef, D.T.A., and McPhail, K.L. (2011) Cyclic depsipeptides, grassypeptolides D and E and Ibu-epidemethoxylyngbyastatin 3, from a Red Sea *Leptolyngbya* cyanobacterium. *J. Nat. Prod.*, **74**, 1677–1685.

152 Popplewell, W.L., Ratnayake, R., Wilson, J.A., Beutler, J.A., Colburn, N.H., Henrich, C.J., McMahon, J.B., and McKee, T.C. (2011) Grassypeptolides F and G, cyanobacterial peptides from *Lyngbya majuscula*. *J. Nat. Prod.*, **74**, 1686–1691.

153 Han, B., Gross, H., McPhail, K.L., Goeger, D., Maier, C.S., and Gerwick, W.H. (2011) Wewakamide A and guineamide G, cyclic depsipeptides from the marine cyanobacteria *Lyngbya semiplena* and *Lyngbya majuscula*. *J. Microbiol. Biotechnol.*, **21**, 930–936.

154 Malloy, K.L., Villa, F.A., Engene, N., Matainaho, T., Gerwick, L., and Gerwick, W.H. (2011) Malyngamide 2, an oxidized lipopeptide with nitric oxide inhibiting activity from a Papua New Guinea marine cyanobacterium. *J. Nat. Prod.*, **74**, 95–98.

155 Montaser, R., Paul, V.J., and Luesch, H. (2011) Pitipeptolides C–F, antimycobacterial cyclodepsipeptides from the marine cyanobacterium *Lyngbya majuscula* from Guam. *Phytochemistry*, **72**, 2068–2074.

156 Han, B., Reinscheid, U.M., Gerwick, W.H., and Gross, H. (2011) The structure elucidation of isomalyngamide K from the marine cyanobacterium *Lyngbya majuscula* by experimental and DFT computational methods. *J. Mol. Struct.*, **989**, 109–113.

157 Esquenazi, E., Jones, A.C., Byrum, T., Dorrestein, P.C., and Gerwick, W.H. (2011) Temporal dynamics of natural product biosynthesis in marine cyanobacteria. *Proc. Natl. Acad. Sci. USA*, **108**, 5226–5231.

158 Liu, L. and Rein, K.S. (2010) New peptides isolated from *Lyngbya* species: a review. *Mar. Drugs*, **8**, 1817–1837.

159 Engene, N., Choi, H., Esquenazi, E., Byrum, T., Villa, F.A., Cao, Z., Murray, T.F., Dorrestein, P.C., Gerwick, L., and Gerwick, W.H. (2011) Phylogeny-guided isolation of ethyl tumonoate A from the marine cyanobacterium cf. *Oscillatoria margaritifera*. *J. Nat. Prod.*, **74**, 1737–1743.

160 Engene, N., Rottacker, E.C., Kaštovský, J., Byrum, T., Choi, H., Ellisman, M.H., Komárek, J., and Gerwick, W.H. (2012) *Moorea producens* gen. nov., sp. nov. and *Moorea bouillonii* comb. nov., tropical marine cyanobacteria rich in bioactive secondary metabolites. *Int. J. Syst. Evol. Microbiol.*, **62**, 1171–1178.

161 Li, S., Yao, H., Xu, J., and Jiang, S. (2011) Synthetic routes and biological evaluation of largazole and its analogues as potent histone deacetylase inhibitors. *Molecules*, **16**, 4681–4694.

162 Jones, A.C., Ottilie, S., Eustáquio, A.S., Edwards, D.J., Gerwick, L., Moore, B.S., and Gerwick, W.H. (2012) Evaluation of *Streptomyces coelicolor* A3(2) as a heterologous expression host for the cyanobacterial protein kinase C activator lyngbyatoxin A. *FEBS J.*, **279**, 1243–1251.

163 Tang, W. and van derDonk, W.A. (2012) Structural characterization of four prochlorosins: a novel class of lantipeptides produced by planktonic marine cyanobacteria. *Biochemistry*, **51** (21), 4271–4279.

24
Sponges: A Reservoir for Microorganism-Derived Bioactive Metabolites

Visamsetti Amarendra, Ramachandran S. Santhosh, and Kandasamy Dhevendaran

24.1
Introduction

Diseases are caused by newly identified, multidrug-resistant microbes and/or by the reemergence of preexisting microbes. This situation led World Health Organization to take initiatives for discovering novel bioactive compounds from new sources that possess antibacterial, anticancer, antidiabetic, antifungal, anti-HIV, anti-inflammatory, antiparasitic, and antipain properties. Marine habitat is a treasure trove of natural products. Main sources for these natural products are corals, seaweeds, sponges, snails, mollusks, and so on. Marine microorganisms started inhabiting metazoans in symbiotic association from the Precambrian period. Because of this, the microbes are protected from evolutionary changes and it becomes important to characterize bioactive metabolites synthesized by them.

Sponges are sessile, benthic, and most primitive filter feeders among metazoans. Through the porocytes, the water containing minerals, cellular debris, and microorganisms will be taken up into the mesohyl region where it further passes to the archaeocytes for digestion. Filter-feeding habitat of sponges leads to the accumulation of few microbes, which is beneficial; the released chemically potent defense molecules by the microbe in the surrounding environment protect the sponge from predators. Diverse classes of bacteria and fungi have been reported as associated with marine sponges. The bioactive metabolites from these symbionts are used in treating some of the dreadful diseases. But the major drawback lies in procuring these natural products in limited quantity. This chapter aims to discuss various microbes associated with sponges, isolated compounds, their structure, and clinical significance. Table 24.1 provides the chemical names of various compounds synthesized by microbes.

Marine Microbiology: Bioactive Compounds and Biotechnological Applications, First Edition.
Edited by Se-Kwon Kim

Table 24.1 Chemical names of marine microbe-derived compounds associated with different sponges.

No.	Compound name
1	Acetic acid, -butyl ester
2	Ethanol, 2-(octyloxy)-
3	Oxalic acid, allyl nonyl ester
4	2-Isopropyl-5-methyl-1-heptanol
5	Butylated hydroxytoluene
6	Cyclohexanecarboxylic acid, hexyl ester
7	Diethyl phthalate
8	Pentadecanal
9	1-Tridecanol
10	9-Octadecenamide (*Z*)-
11	9-Octadecenal
12	9-Octadecenoicacid (*Z*)-, methyl ester
13	Oleic acid
14	(*E*)-9-Octadecenoicacid ethyl ester
15	*n*-Hexadecanoic acid
16	Hexadecanoic acid, methyl ester
17	Hexadecanoic acid, ethyl ester
18	Manzamine A
19	GGL.2, {1-*O*-acyl-3-[α-glucopyranosyl-(1–3)-(6-*O*-acyl-α-mannopyranosyl)] glycerol} with 14-methyl hexadecanoic acid and 12-methyl tetradecanoic acid positioned at C-6 of the mannose unit and at glycerol moieties, respectively
20 and 21	Urauchimycins A and B
22	2,4,4′-Trichloro-2′-hydroxydiphenyl ether (Triclosan)
23	Acyl-1-(acyl-6′-mannobiosyl)-3-glycerol (Lutoside)
24	Rifamycin B
25	Rifamycin SV
26	Metacycloprodigiosin
27	Undecylprodigiosin
28	GGL11, {1, 2-*O*-diacyl-3-[β-glucopyranosyl-(1–6)-β-glucopyranosyl] glycerol} with 14-methyl hexadecanoic acid and 12-methyl tetradecanoic acid as the main fatty acid moieties
29	Surfactin
30	Iturin
31	Fengycin
32	YM-266183
33	YM-266184
34	Cyclo-(L-Pro-L-Phe)
35	Norharman
36	Cyclo-(L-Pro-L-Met)
37	Alteramide A
38 and 39	Bromoalterochromides A and A′
40	Tetrabromodiphenyl ether
41	Trisindoline
42	2-Undecyl-4-quinolone
43	2-Undecen-1′-yl-4-quinolone
44	2-Nonyl-4-hydroxyquinoline *N*-oxide
45	Andrimid

46	Theopalauamide
47	Theonegramide
48	Leucamide A
49	Majusculamide C
50	2-(2′,4′-Dibromophenyl)-4,6-dibromophenol
51	Onnamide A
52	Swinholide A
53	2-Methylthio-1, 4-naphthoquinone
54	Exophilin A
55 and **56**	Diaporthein A and B
57	1,3,8-Trihydroxy-6-methoxy anthraquinone (lunatin)
58	Cytoskyrin-A
59	(3*S*)-(3′,5′-Dihydroxyphenyl)butan-2-one
60	2-(1′(*E*)-propenyl)-octa-4(*E*),6(*Z*)-diene-1,2-diol
61	(3*R*)-6-Methoxymellein
62	(3*R*)-6-Methoxy-7-chloromellein
63	Cryptosporiopsinol
64	Microsphaeropsisin
65	(*R*)-mellein
66	(3*R*,4*S*)-Hydroxymellein
67	(3*R*,4*R*)-Hydroxymellein
68	4,8-Dihydroxy-3,4-dihydro-2*H*-naphthalen-1-one
69	Sumiki's acid
70	Acetyl Sumiki's acid
71	YM-202204
72	Epoxyphomalin A
73	Ulocladol
74	1-Hydroxy-6-methyl-8-(hydroxymethyl)xanthone
75	Oxaline
76–78	Communesin B–D
79	Griseofulvin
80	Dechlorogriseofulvin
81	Evariquinone
82	3, 3′-Bicoumarin (bicoumanigrin A)
83	Aspernigrin B
84	Aspergillitine
85	Asperazine
86	Malformin C
87 and **88**	Aurantiomides B and C
89	(*S*)-2, 4-Dihydroxy-1-butyl (4-hydroxy)benzoate
90	Fructigenine-A
91	Mycophenolic acid
92	8-Chloro-9-hydroxy-8,9-deoxyasperlactone
93	9-Chloro-8-hydroxy-8,9-deoxyasperlactone
94	9-Chloro-8-hydroxy-8,9-deoxyaspyrone
95	Penicillic acid
96–98	Aspergillides A–C
99	Aspinonene
100	Dihydroaspyrone
101	Xestodecalactone B
102	Fellutamide C
103	Decumbenone A
104	Terretonine E

(continued)

Table 24.1 (*Continued*)

No.	Compound name
105	Aurantiamine
106	Varixanthone
107	Varitriol
108–114	Gymnastatin A–G
115 and **116**	Dankastatin A and B
117–120	Gymnasterone A–D
121	Dankasterone A
122–128	Prugosenes A1–A3, B1, B2, C1, C2
129–131	Brocaenol A–C
132	Sorbicillactone A
133	Isocyclocitrinol A
134	22-Acetylisocyclocitrinol A
135–137	Trichodenone A–C
138 and **139**	Destruxin A and B
140	Destruxin B2
141	Destruxin E chlorohydrin
142–144	Cathestatin A–C
145–147	Roridin A, D, and M
148 and **149**	Verrucarin A and M
150	Isororidin A
151	Epiroridin E
152 and **153**	Trichoverrin A and B
154	Trichodermol
155	8-Deoxy-trichothecin
156	RHM 1
157–161	Efrapeptin E, Eα, and F–H
162	14,15-Secocurvularin
163	Chloriolin B
164	Hirsutanol A
165	*ent*-gloeosteretriol

For convenience, the compound names are indicated using the serial number in bold type.

24.2 Collection of Sponges and Associated Microbes

Sponges are collected from their natural habitats by methods such as kagava, harpoon, skafandro, nargiles, hand collecting, artificial reef matrix structures, and self-contained underwater breathing apparatus.

24.2.1 Identification of Sponges and Extraction of Microbes

Collected sponges are transported aseptically to the laboratory in the minimum possible time by storing them in iceboxes and are identified based on spicule morphology.

24.2.2
Isolation of Microbes

The sponge extract is serially diluted and plated onto isolation media such as starch–yeast extract–peptone–seawater, marine agar, modified marine agar, actinomycete isolation seawater agar, actinomycete isolation agar, chloroplast extraction medium, potato dextrose agar, glycerol asparagine agar, malt extract agar, and so on. The purified isolates are then identified using Bergey's Manual of taxonomy, and at the molecular level using 16S rRNA gene, Denaturing Gradient Gel Electrophoresis (DGGE), and fluorescence in situ hybridization (FISH).

24.2.3
Extraction of Metabolites

Purified strains are cultured in a species-specific medium and the production of metabolites is optimized by using different carbon sources, nitrogen sources, vitamins, micronutrients, mineral salts, and various precursor molecules. Metabolites are extracted from the culture supernatants using different polar or nonpolar solvents such as water, methanol, ethanol, *n*-butanol, acetone, and ethyl acetate or hexane, petroleum ether, benzene, diethyl ether, chloroform, dichloromethane, and dichloroethane, respectively, by batch or continuous method. The metabolites dissolved in the solvent are further concentrated either by distillation or by vacuum rotary evaporator.

24.2.4
Characterization of Biologically Active Compound

The extracts are primarily analyzed for the presence of different groups of compounds such as alkaloids, aromatic compounds, flavanoids, terpenoids, phenolic compounds, quinones, peptides, and steroids. They are also analyzed for antimicrobial, antioxidant, anticancer, hemolytic, dye degrading, enzymatic activities besides some general activity. Preparative techniques are carried out to separate the active compounds from the extracted mixture. Using mass spectrometry analysis, the relative weights of the purified compounds are calculated and if there is no compound corresponding to the mass calculated in database, fractions are further analyzed. UV–visible spectrophotometer and Fourier transform infrared spectroscopy (FTIR) are used to calculate wavelength maximum and detect the presence of functional groups in compound, respectively. Nuclear magnetic resonance (NMR) spectroscopy and X-ray crystallography enable complete structure determination.

24.3
Bacteria

Bacteria represent the largest group associated with sponges that synthesizes bioactive compounds. Phyla Actinobacteria, Firmicutes, Proteobacteria, Verrucomicrobia, and Cyanobacteria represent the major microbes.

24.3.1 Actinobacteria

Actinobacteria are Gram-positive with high guanosine–cytosine (GC) content, ubiquitous, and well known for their ability to produce secondary metabolites, and hence regarded as the superior source for exploitation of novel bioactive compounds. *Nocardiopsis dassonvillei* MAD08 was 1 of the 11 heterotrophic strains that were screened from *Dendrilla nigra,* isolated from the southwest coast of India. Among the 26 compounds present in ethyl acetate extract of the strain, compounds **1–11** had antimicrobial property; **12–14** showed anticancer and anti-inflammatory properties; and **15–17** showed antioxidant, hemolytic, cholesterolemic, nematicide, and antiandrogenic properties [1]. *Micromonospora* sp. isolated from the deep-water Indonesian sponge *Acanthostrongylophora* sp. was found to contain an anti-infectious and antitumor β-carboline alkaloid (**18**) [2]. Recently, it was shown that the compound also has antimalarial activity against rodent malaria parasite, *Plasmodium berghei* [3]. An antifungal enzyme, chitinase was isolated from the *Streptomyces* sp. DA11 associated with the South China sponge, *Craniella australiensis,* collected from the Sanya Island. Purified chitinase was used as an antifungal agent against *Aspergillus niger* and *Candida albicans* [4]. Four unusual glycerolipids and one diphosphatidyl glycerol were produced by marine actinobacterium, *Microbacterium* sp., isolated from the Adriatic sponge *Halichondria panacea,* among which one of the glycerolipids (**19**) showed antitumor activity [5]. Phospholipase A2, a ubiquitous antibacterial enzyme in higher mammals, was detected in high levels in the sample extracts of host sponge *D. nigra* that was associated with the bacterium *Streptomyces dendra* sp. nov. MSI051 [6]. Extracts of *Tedania ignis* isolated from Bermudian waters were found to contain diketopiperazines produced by associated *Micrococcus* sp. [7]. The authors also detected diketopiperazines from marine microbes associated with *Tedania anhelans,* collected from southeast coast of India. The first antimycin macrolides having branched side chains (**20** and **21**) and exhibiting inhibitory activity against morphological differentiation of *C. albicans* were isolated from *Streptomyces* sp. Ni-80 from an unidentified sponge [8]. Two new phenyl ethers with antimicrobial properties (**22** and **23**) were derived from *Micrococcus luteus* R-1588-10 associated with sponge *Xestospongia* sp. (Off Noumea, New Caledonia, Southwest Pacific) [9]. The Great Barrier Reef marine sponge *Suberea clavata* was associated with 10 strains of *Salinispora* group, which were previously noticed for compounds containing cytotoxic, antifungal, and antibacterial activities. Kim *et al.* found the compounds **24** and **25** from the mycelium extracted with absolute ethanol [10]. *Saccharopolyspora* sp. nov., associated with *Mycale plumose* from Qingdao Coast, China, showed the presence of two tripyrrole pigments with potent cytotoxic activities. Chloroform extract from the fermentation broth of *Saccharopolyspora* sp. nov., had two known prodigiosin analogues (**26** and **27**), which showed significant cytotoxic effects on different cancer cell lines [11]. Extracellular products from the *Streptomyces* sp. BTL7 isolated from *D. nigra* contained antibacterial agents that are effective against both types of Gram-staining bacteria [12].

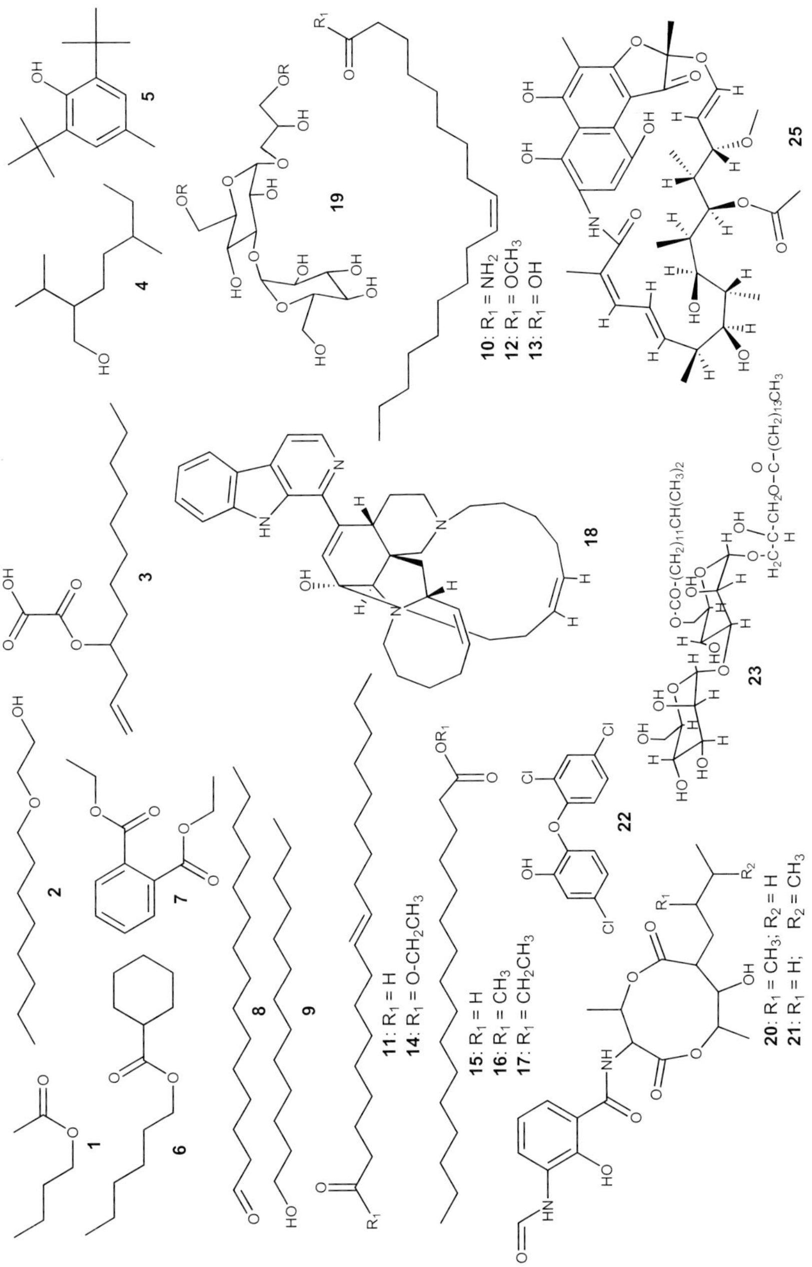

10: $R_1 = NH_2$
12: $R_1 = OCH_3$
13: $R_1 = OH$

11: $R_1 = H$
14: $R_1 = O\text{-}CH_2CH_3$

15: $R_1 = H$
16: $R_1 = CH_3$
17: $R_1 = CH_2CH_3$

20: $R_1 = CH_3$; $R_2 = H$
21: $R_1 = H$; $R_2 = CH_3$

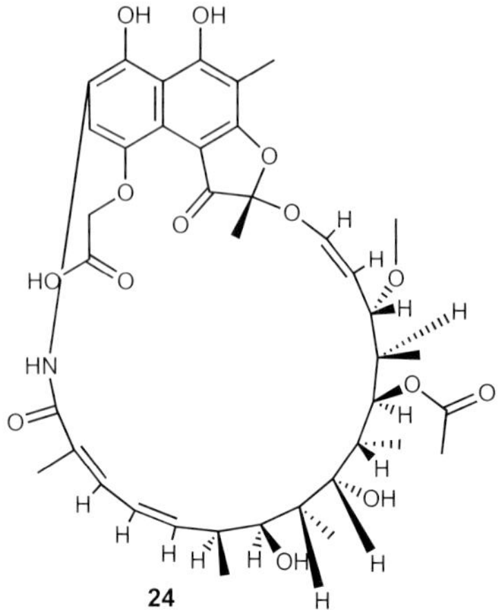

24

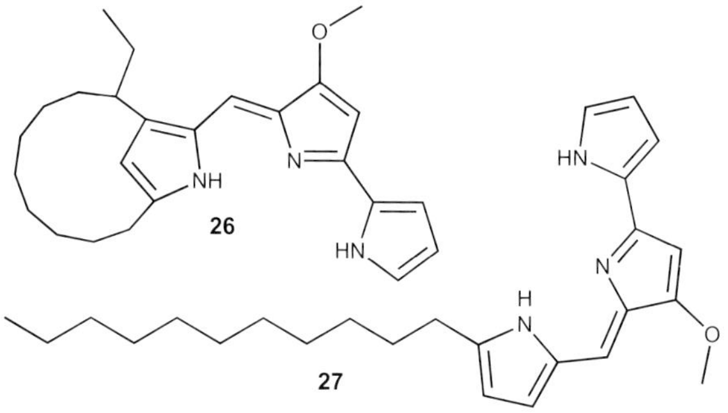

26

27

24.3.2
Firmicutes

Firmicutes are Gram-positive and low in GC content compared to Actinobacteria and found associated with sponges in equal proportions, such as Actinobacteria and Proteobacteria. The marine sponge *Acanthella acuta*-derived bacterial strain *Bacillus pumilus* AAS3 produced an anticancer diglucosyl-glycerolipid **28** [13]. A Mediterranean sponge *Aplysina aerophoba* harboring *B. pumilus* A586 was demonstrated to produce a potent antibacterial compound pumilacidin against *Staphylococcus aureus* [14]. The metabolites from *Bacillus subtilis* strains A184, A190, and A202 associated with same sponge *A. aerophoba* were characterized to find antifungal and hemolytic compounds against *C. albicans.* Mass spectroscopy results for the extracts showed various lipopeptides. Strain A184 was found to contain compounds **29–31**; A190 showed the presence of **29**; and A202 contained the compound **30** [14]. Two novel peptide antibiotics **32** and **33** were isolated from *Bacillus cereus* QN03323 living as a symbiont in marine sponge *Halichondria japonica* picked up from the Hoshizuna Beach, Iriomote Island, Japan [15]. The metabolites exhibited potent antibacterial activities against *Staphylococci* and *Enterococci,* including multidrug-resistant strains, whereas ineffective activities toward Gram-negative microbes [16].

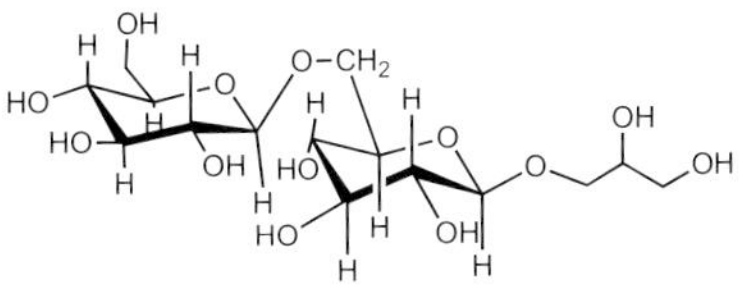

24.3.3
Proteobacteria

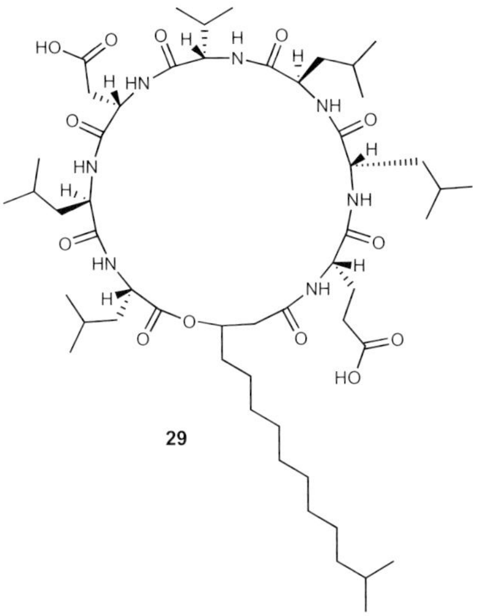

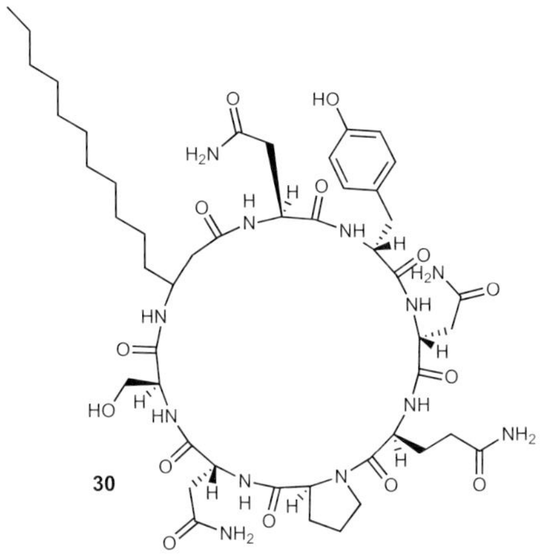

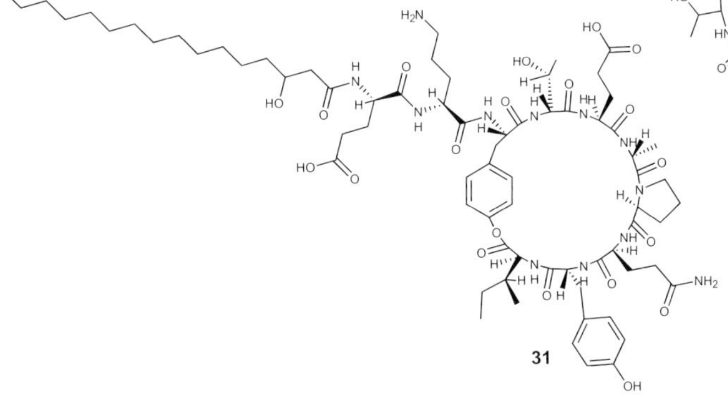

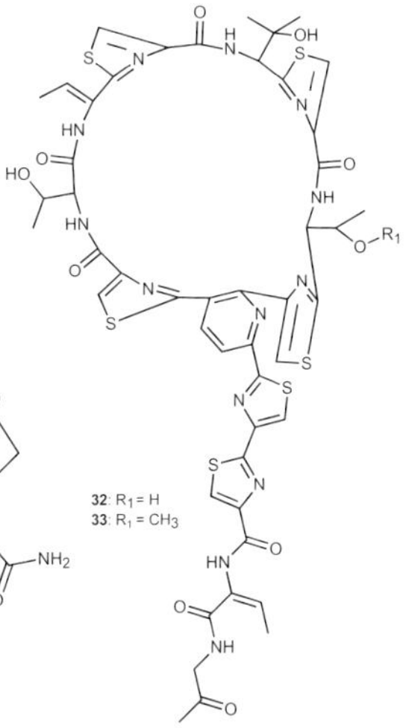

Proteobacteria constitutes Gram-negative, facultative and obligate, aerobic and anaerobic, chemoautotrophs and heterotrophs, pathogenic and symbiotic bacteria that are medically and environmentally important. Furthermore, it is subdivided into six classes: alpha-, beta-, gamma-, delta-, epsilon-, and zetaproteobacteria. Most of the classes represent a major group in the sponge-associated microbial flora, among which gammaproteobacteria was found in large numbers.

24.3.3.1 **Alphaproteobacteria**

The two isolates SB1 and SB2 associated with *Suberites domuncula* collected from the Northern Adriatic Sea displayed more similarity with alphaproteobacteria MBIC3368 and are known to produce unidentified compounds [17].

24.3.3.2 **Betaproteobacteria**

A diketopiperazine compound **34** showed moderate antimicrobial activity against *S. aureus*. The betaproteobacteria *Alcaligenes faecalis* A72 was purified from the South China Sea sponge *Stellata tenius* [18].

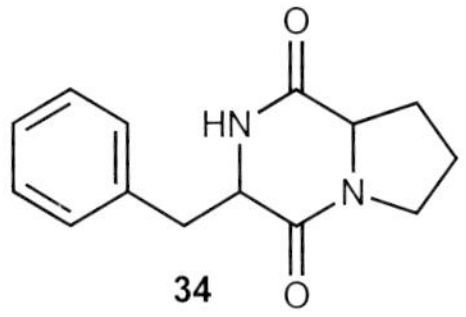

24.3.3.3 **Gammaproteobacteria**

A β-carboline alkaloid **35** was characterized from the isolates of *Pseudoalteromonas piscicida* NJ6-3-1 and showed antimicrobial activity that was associated with *Hymeniacidon perleve* at Nanji Island (East China Sea) [19]. Diketopiperazine class of compounds having antimicrobial activity, **36**, were isolated from the culture broth of *Pseudomonas aeruginosa*, a symbiotic microbe associated with *Isodictya setifera* collected from Ross Island, Antarctica [20]. A tetracyclic alkaloid **37** was found to have cytotoxic activity against carcinoma cell lines that was identified from the extracts of *Alteromonas* sp. associated with the sponge *Halichondria okadai* [2]. A marine strain *Pseudoalteromonas maricaloris* KMM 636^T was found as an epibiont of the Australian sponge *Fascaplysinopsis reticulata* that promisingly demonstrated to produce two stereo chromopeptide isomers **38** and **39** having cytotoxic effects on the developing eggs of sea urchin *Strongylocentrotus intermedius* [21]. *Vibrio* sp. associated with *Dysidea* sp. was prominent in synthesizing bioactive brominated diphenyl ether compound, **40** that showed cytotoxic and antibacterial properties [2]. A new trisindole derivative antibiotic, which is an indole dimer, **41** was found in the extracts of *Vibrio* sp. living as a habitant in the marine Okinawan sponge, *Hyrtios altum* [22]. The compound showed a potent bactericidal property against *Escherichia coli, B. subtilis*, and *S. aureus* and cytotoxic activity against carcinoma cell lines [23]. The bacterial strain *Pseudomonas* sp. 1537-E7 identified in the new Caledonian sponge *Homophymia* sp. was found to synthesize bioactive compounds. A quinolone derivative **42** is active against malaria parasite *Plasmodium falciparum* and HIV-1; **43** showed mild cytotoxicity over human carcinoma cells; and **44** showed cytotoxicity and antimicrobial activity against *S. aureus* [24]. *Vibrio* M22-1 obtained from the homogenate of *Hyatella* sp. was identified to produce an antibacillus peptide antibiotic **45** [2]. The same compound was previously identified to have inhibitory activity against *Xanthomonas campestris* pv. *oryzae* [2]. Marine *Pseudomonas fluorescens* was also found to synthesize this active compound and was active against methicillin-resistant *S. aureus*.

35

36

37

38: $R_1 = CH_3$; $R_2 = H$

39: $R_1 = H$; $R_2 = CH_3$

40

41

42

44

43

45

24.3.3.4 Deltaproteobacteria

Theonella swinhoei isolated from Palau and Philippines harbor promising bioactive compound-producing microorganisms. The bicyclic glycopeptides **46** from *Candidatus Entotheonella palauensis* [25] and **47** from *E. palauenis* exhibited antifungal properties [26].

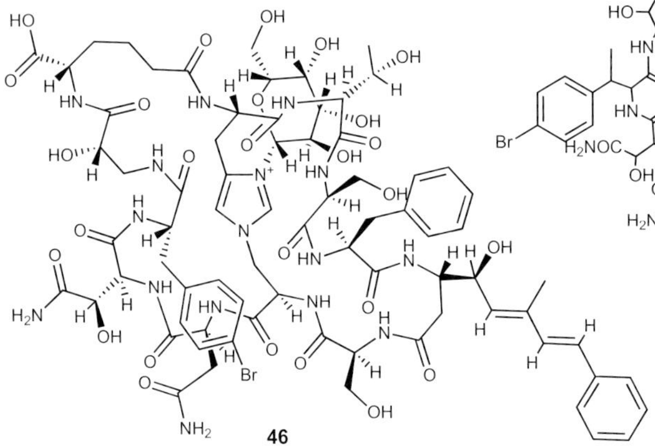

24.3.4 Verrucomicrobia

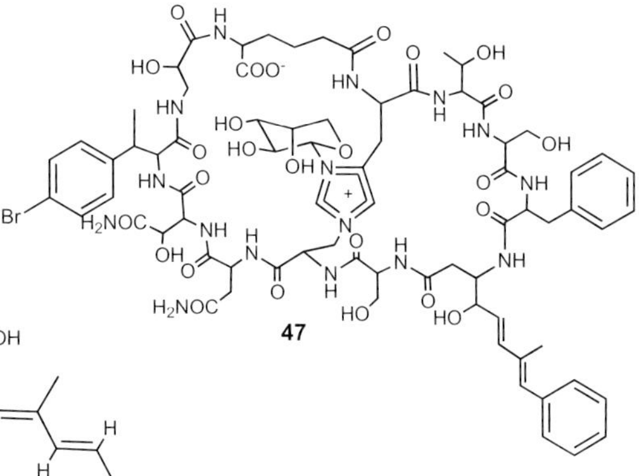

Three novel antioxidants were discovered in the extracts of *Rubritalea squalenifasciens* HOact23^{T}, a marine bacterium from *H. okadai*, as red pigments. Spectroscopic analysis revealed the three structures as acyl glyco-carotenoic acid derivatives having moderate antioxidant activity [2].

24.3.5 Cyanobacteria

Majority of the photosynthetic bacteria belong to this phylum, which live as symbionts in fresh and marine water organisms. The photosynthetic activity of the cyanobacteria provides carbon and nitrogen and gets shelter from the sponges. In addition, the secondary metabolites from cyanobacteria protect sponges from predators.

Sponge and cyanobacterial interactions involve peptides such as a nonribosomal cyclic peptide **48**, an antitumor agent from the Great Barrier Reef sponge *Leucetta microraphis* [27]. A cynaobacterium *Lyngbya majuscula*-derived cyclic depsipeptide **49** exhibited antifungal properties against pathogens of crop plants that have been isolated from sponge *Ptilocaulis trachys* collected at Enewetak Atoll, Marshall Island (Pacific Ocean) [2]. A polybrominated biphenyl compound **50** was synthesized by cyanobacterial cell *Oscillatoria spongeliae* separated from *Lamellodysidea herbacea* (formerly *Dysidea herbacea*), the tropical marine shallow sponge collected from the Republic of Palau, Caroline Island (Western Pacific Ocean). The compound formed as a crystal on the sponge mesohyl region as a result of its insolubility in aqueous medium showed antibacterial activity against Gram-positive, Gram-negative, and other unicellular marine cyanobacteria [28,29].

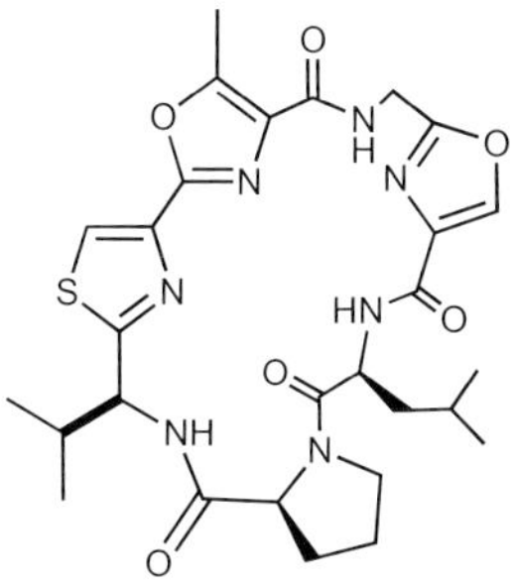

48

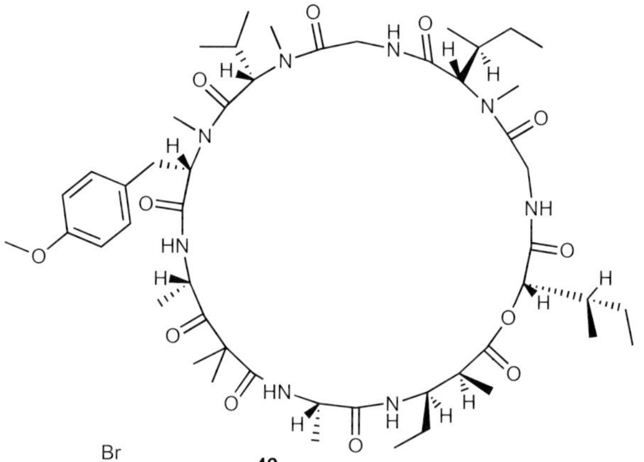

49

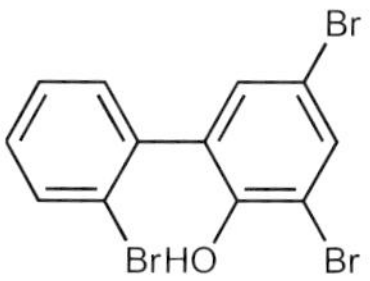

50

24.4 Unidentified Bacteria

Two polyketide genes coding for **51** and theopederin A were characterized in the microbial metagenome of Hachijō-jima Island sponge *T. swinhoei* and they showed antitumor activity [30]. *T. swinhoei* is well noticed for the presence of diverse peptides and polyketide secondary metabolites. Two bacterial cell fractions were separated from it, in which one fraction composed of a mixture of unicellular bacteria containing a macrocyclic, potent cytotoxin **52** [22] and another fraction composed of single morphophyte of filamentous bacteria containing bicyclic glycopeptides **46**. In the same way, an unidentified bacterium associated with *Dysidea avara* from the Adriatic Sea showed to produce **53**, having antiangiogenic and antimicrobial properties [2].

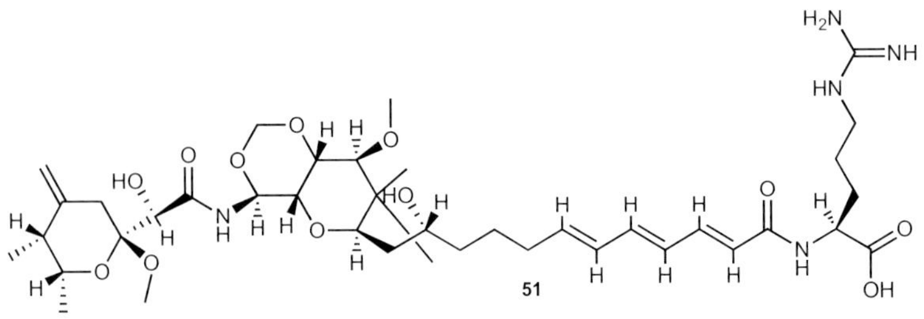

24.5
Fungi

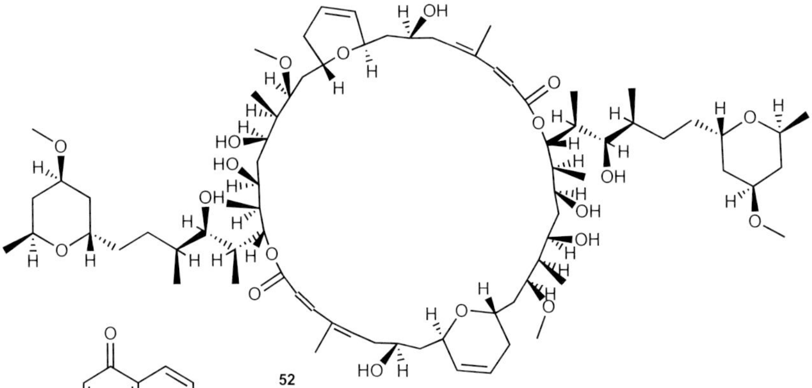

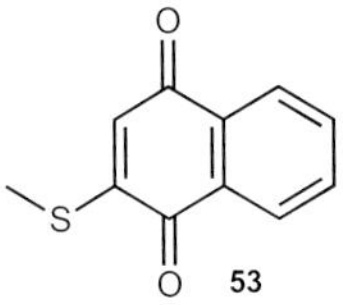

The main focus on secondary metabolites produced by fungi is due to their diverse biological activities having both toxic and curative effects in maintaining human and veterinary health. Division Ascomycota are the major inhabitants widely distributed in sponges. Class Ascomycetes, Dothideomycetes, Eurotiomycetes, and Sordariomycetes were primarily associated with sponges.

24.5.1
Ascomycetes

Little dragon sculpin *Blepsias cirrhosus* lays egg in the marine sponge *Mycale adhaerens* to protect them from predators and infections. Doshida *et al.* [31] revealed the presence of **54**, a new decanoate antibacterial compound from the extracts of *Exophiala pisciphila* NI1012 isolated from *M. adhaerens* that is responsible for the prevention of predators and infections to the eggs of sculpin [31]. *Cryptosphaeria eunomia* obtained from an unidentified sponge collected at Pohnpei was found to secrete a new pimarane type of diterpene antibacterial compounds **55** and **56** that were previously isolated from terrestrial fungus *Diaporthe* sp. BCC 6140 [32].

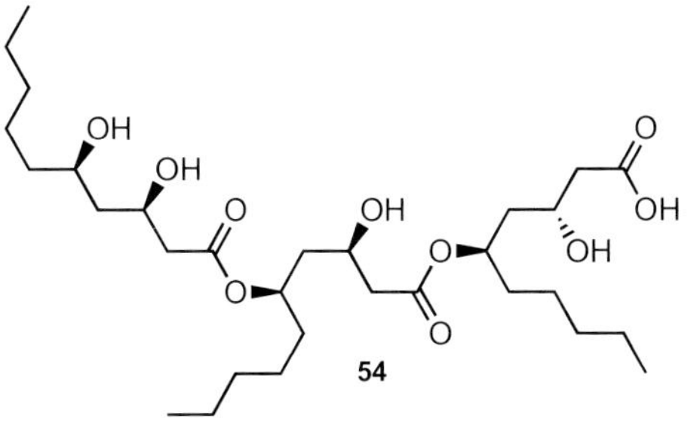

54

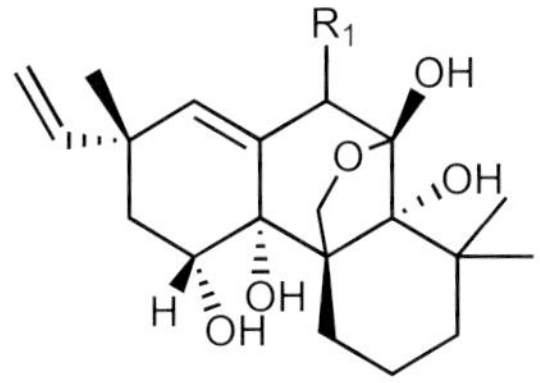

55: R_1 = OH
56: R_1 = O

24.5.2 Dothideomycetes

It is a group of anthraquinones in which a new compound **57** and a modified bisanthraquinone **58** were discovered from the fungus *Curvularia lunata* isolated from the Indonesian sponge *Niphates olemda*. Ethyl acetate extract of *C. lunata* inhibited the growth of *B. subtilis, S. aureus,* and *E. coli* [33]. A new benozenoid derivative, 10-hydroxy-18-ethoxyl-betaenone, from *Microsphaeropsis* sp. was isolated from the Mediterranean sponge *A. aerophoba*, which was shown as a potent inhibitor for protein kinase C – a major target in the development of novel anticancer drugs [34]. A marine sponge *Ectyoplasia perox* from the waters of Caribbean Island, Dominica, was investigated for the associated fungal strains, for example, *Coniothyrium* sp. 193H77 for its antimicrobial agents. Two novel compounds **59** and **60**, along with known fungal metabolites **61–63** were characterized using NMR, among which all, except **63**, demonstrated significant antibacterial activity [35]. An eremophilane derivative and antifungal sesquiterpenoid **64**, derived from an anamorphic fungus *Microsphaeropsis* sp. H5–50, was found to have carbon at positions 14 and 15 in *trans*-configuration, with its counterparts being in *cis*-configuration, derived from *Penicillium roqueforti* and marine *Dendryphiella salina*. Along with **64**, other known benzopyrans **65–68** were found in the fungal strain *Microsphaeropsis* sp. H5–50, seen as an inhabitant in *Myxilla incrustans* sponge collected from Helgoland, Germany [35]. Two furan carboxylic acids, **69** and their derivative **70**, were showed to possess active antimicrobial activity against *B. subtilis* and *S. aureus* that were obtained from the organic extracts of *Cladosporium herbarum*, consorted with marine sponge *Callyspongia aerizusa* [36]. Many symbiotic bacteria, actinomycetes, and fungal strains were associated with *H. japonica* from Japan. A new lactone compound **71** was identified in the culture broth of fungal strain *Phoma* sp. Q60596 associated with *H. japonica*, found at Hoshizuna beach, Iriomote Island, Japan. The compound exhibited potent antifungal activity against *C. albicans, Cryptococcus neoformans,* and *Aspergillus fumigates* and also inhibited glycosylphosphatidylinositol anchoring in yeast cells [37]. A prenylated polyketide **72** from the lipid extracts of *Phoma* sp., associated with *E. perox* collected from Dominica (Caribbean Island), showed potent cytotoxicity against tumor cell lines at nanomolar concentrations [38]. The lipophilic extracts of *Ulocladium botrytis* 193A4 associated with marine sponge *Callyspongia vaginalis* exhibited inhibitory activity against tyrosine protein kinase and it was exhibited by a new polyketide **73**, whereas the antifungal compound was identified as xanthone derivative **74** [39].

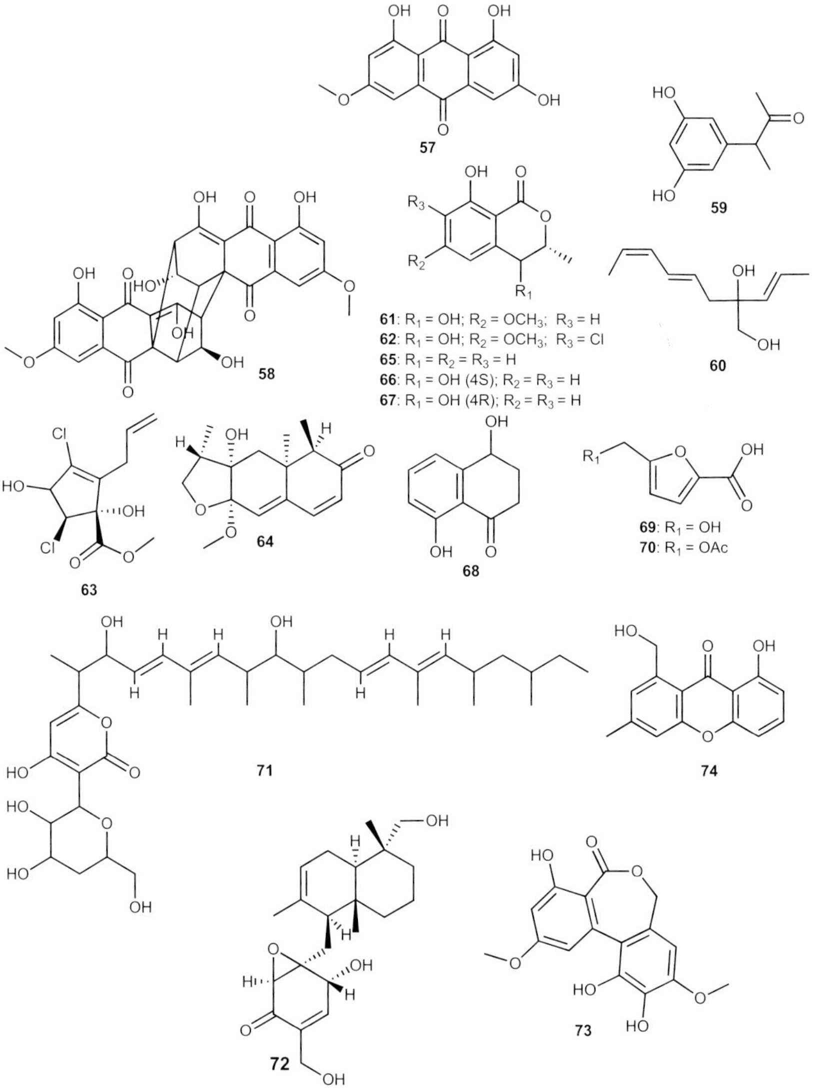

57 **58** **59** **60**

61: R_1 = OH; R_2 = OCH_3; R_3 = H
62: R_1 = OH; R_2 = OCH_3; R_3 = Cl
65: R_1 = R_2 = R_3 = H
66: R_1 = OH (4S); R_2 = R_3 = H
67: R_1 = OH (4R); R_2 = R_3 = H

63 **64** **68**

69: R_1 = OH
70: R_1 = OAc

71 **74** **72** **73**

24.5.3
Eurotiomycetes

Penicillium sp. associated with the Mediterranean sponge *Axinella verrucosa* yielded new polycyclic indole alkaloid compounds **77** and **78** along with the known **75**, **76**, **79**, and **80** [40]. Studies of an antiproliferative agent **75** on human cell lines showed that the cell cycle arrest was done by inhibiting the polymerization of microtubule protein [41]. Indole alkaloid derivatives **76–78** exhibited moderate antiproliferative activity against different leukemia cell lines [40]. A new anthraquinone **81** was isolated and structurally elucidated from the fungal species *Emericella variecolor* derived from *Haliclona valliculata*, which showed a strong antiproliferative activity. Furthermore, a novel anthraquinone isoemericellin and three known metabolites – 7-hydroxyemodin, shamixanthone, and stromemycin – were also characterized from *E. variecolor*. Among these, C-glycosidic depside stromemycin showed metalloproteinase-inhibiting activity; acting as an antimetastatic factor [42]. Ethyl acetate fractions of the *A. niger* culture collected from the Mediterranean sponge *Axinella damicornis* yielded eight natural products. A new coumarin compound **82** showed moderate cytotoxicity against human leukemia and carcinoma cell lines *in vitro*. Another new dihydropyridine from the same isolate **83** displayed potent neuroprotective activity, preventing glutamic acid-induced neuronal cell death [43]. An antimicrobial chromone derivative **84** was purified and characterized from ethyl acetate extracts of *Aspergillus versicolor*, a marine fungus isolated from the Indonesian sponge *Xestospongia exigua*, that showed antimicrobial activity against *B. subtilis* [44]. Among the five compounds belonging to different biosynthetic classes isolated from culture extract of *A. niger* collected from *Hyrtios proteus* (Dry Tortugas National Park, Florida), diketopiperazine (**85**) and cyclicpeptide (**86**) showed cytotoxicity against selective leukemia cells [45]. Synthetic asperazine was tested against L1210 leukemia, Colon 38, H116 colon cancer, and H125 lung cancer cell lines, but it showed a low cytotoxicity against all cancer cell lines, demonstrating difference in relative and absolute configuration of the synthesized asperazine to the natural **85** [46]. A novel strain *Penicillium aurantio-griseum* SP0–19 from *M. plumose* was reported with three new quinazoline alkaloids: aurantiomides A, B (**87**), and C (**88**), among which **87** and **88** showed moderate cytotoxicity against human and murine leukemia cell lines [47]. Previously, a new compound **89** and a known compound **90** were reported from the same strain by spectroscopic and chemical methods [48]. Immunosuppressive agents are widely used in transplantations and in treating autoimmune disorders. One of the well-known drugs of this class, **91** was discovered in the extracts of *Penicillium brevicompactum* isolated from *Petrosia ficiformis* [2]. Three new chlorine-containing antibacterial compounds **92–94** along with **95** were derived from the marine fungus strain *Aspergillus ostianus* 01F313 associated with an unidentified sponge collected from Pohnpei, Micronesia. These compounds exhibited effective antibacterial activity against *Ruegeria atlantica*, *E. coli*, and *S. aureus* [49]. In another study, the same strain cultured in bromine-modified artificial seawater produced three new pentaketides, aspinotriols A, B, and aspinonediol and three 14-membered aspergillides **96–98** along with known **99** and **100**. Among them, aspinotriol, **99** and **100** and all aspergillides showed cytotoxic activity [50,51]. Three novel decalactones, xestodecalactone A, B **(101)**, and C were structurally elucidated from the isolates of *Penicillium* cf. *montanense* collected from the Bali Sea marine sponge *Xestospongia exigua*. Xestodecalactone B **(101)** and C showed similarities with biologically active metabolites found in terrestrial fungal strains. Among the three decalactones, only **101** showed antifungal activity against *C. albicans* [52]. The compound **102** along with the three known polyketides – decumbenones A **(103)**, B, and versiol – was isolated from *A. versicolor* associated with *Petrosia* sp. (Jeju Island, Korea) [53,54]. The cytotoxic lipopeptide **102** is structurally similar to the fellutamides A and B identified in the fish-derived fungus *Penicillium fellutanum*. The compound **102** was evaluated for its cytotoxicity against human solid tumor cell lines [54]. *Aspergillus insuetus* yielded two meroterpenoid compounds – Terretonine E (**104**) and F – and a piperazine compound **105** in association with *Petrosia ficiformis*, which were used as potent inhibitors of mammalian mitochondrial respiratory chain [55]. Along with the known compounds, **106** and **107** were also isolated from an unidentified sponge (collected from Venezuela)-derived marine fungus *E. variecolor*. Among these compounds, **106** showed antibacterial property and **107** showed a potent cytotoxicity toward renal cancer, breast cancer, and central nervous system lymphoma cell lines [56]. A wide range of cytotoxic compounds belonging to diverse classes was isolated from an ascomycete fungal strain *Gymnascella dankaliensis* OUPS-N134, associated with the Osaka Bay (Japan) sponge *H. japonica* [57]. Gymnastatins, gymnasterones, dankastatins, and dankasterones were found in the extracts of the isolate. Gymnastatin A–G **(108–114)**, Q, and R and **115–121** showed significant cytotoxic activity against lymphomatic leukemia test system [58–63]. Compounds **111** and **112** lacked ketone moiety, which showed a weak cytotoxicity effect [60]. Interestingly, gymnastatin Q showed growth inhibition against human breast and stomach cancer cell lines [63]. Lang group found seven new fungal polyketides in ethyl alcohol extracts of *Penicillium rugulosum* associated with marine sponge *Chondrosia reniformis* collected from the island of Elba, Italy. The compounds **122–128** showed a common linear pentaene structure with cyclic moieties. These polyketides can be used as precursors for new anti-infectives [64]. Along with the known diketopiperazines, *Penicillium brocae* F97S76 isolated from the Fijian *Zyzzya* sp. sponge yielded three novel polyketides, **129–131**, containing uncommon enolized oxepine lactones in their structures, and exhibited moderate cytotoxicity against human colon carcinoma cell line [65]. A unique bicyclic lactone **132** seemingly derived from sorbicillin was identified in the methanol extract of strain *Penicillium chrysogenum* cultured from the Mediterranean sponge *Ircinia fasciculate* and showed elevated effects on mammalian and viral test systems. The compound **132** also showed highly selective cytostatic activities against murine leukemic lymphoblasts and ability in protecting human T cells against cytopathic effects of HIV-1 [66]. *Penicillium citrinum*, a marine-derived fungus characterized from the marine sponge *Axinella* sp., collected in Papua New Guinea, was found to synthesize structurally unique steroids **133** and **134**. Both of these compounds exhibited weak antibacterial activity against *Staphylococcus epidermidis* and *Enterococcus durans* [2].

75

79: R_1 = Cl
80: R_1 = H

81

82

83

84

76: R_1 = H
77: R_1 = CH_3
78: R_1 = CHO

85

86

87

88

89

90

91

92: R_1 = Cl; R_2 = OH
93: R_1 = OH; R_2 = Cl

94

95

96: H1 = α, $CH_3$1 = β
97: H1 = β, $CH_3$1 = α

98

99

100

101

102

103

104

105

106

107

108

109

110

111

112

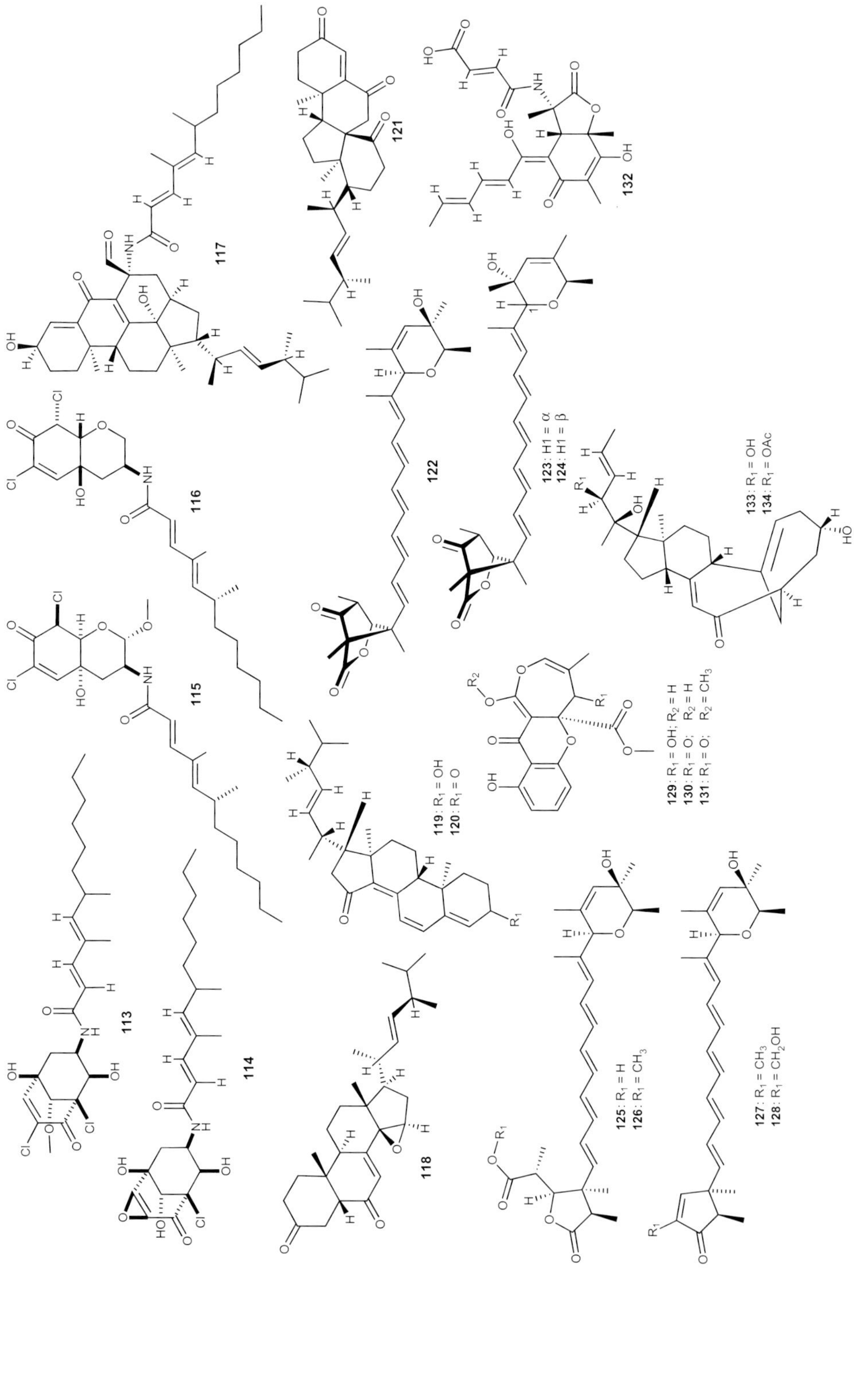

113

114

115

116

117

118

119: R_1 = OH
120: R_1 = O

121

122

123: H1 = α
124: H1 = β

125: R_1 = H
126: R_1 = CH_3

127: R_1 = CH_3
128: R_1 = CH_2OH

129: R_1 = OH; R_2 = H
130: R_1 = O; R_2 = H
131: R_1 = O; R_2 = CH_3

132

133: R_1 = OH
134: R_1 = OAc

24.5.4
Sordariomycetes

Chemically synthesized novel cyclopentenones **135–137** were naturally found in *Trichoderma harzianum* OUPS-N115 associated with *H. okadai*, showing promising cytotoxicity against human cell lines [67]. Six known *N*-methylated cyclic depsipeptides of the Destruxin family were identified in the extracts of *Metarhizium* sp. 001103 separated from *Pseudoceratina purpurea* that were collected at Fiji. The peptides include Destruxin A **(138)**, B **(139)**, B2 **(140)**, desmethyl B, E chlorohydrin **(141)**, and E2 chlorohydrin. The compounds showed selective cytotoxicity on human tumor cell lines. Among these, Destruxin E2 chlorohydrin showed 90% tumor cell death in *in vitro* experiments [68]. Selective cysteine inhibitors **142–144** are found in the isolates of marine fungus *Microascus longirostris* SF-73 screened from an unidentified marine sponge that was collected at Harington Point, Otago Harbor, New Zealand. Protease inhibitors from the above strain can be targeted to selective proteases involved in pathogenesis of human diseases such as AIDS, arthritis, cancer, emphysema, high blood pressure, muscular dystrophy, pancreatitis, thrombosis, and so on [69]. Ethyl acetate extract of fungal isolate *Myrothecium* sp. JS9 from the South China Sea sponge *Axinella* sp. was found to produce two macrocyclic trichothecenes **145** and **146** against plant fungal pathogen *Sclerotinia sclerotiorum* [70]. Culture extracts of *Myrothecium verrucaria* 973023 separated from the Hawaii sponge *Spongia* sp. showed potent cytotoxic properties against cancer cell lines. The 1D and 2D NMR spectral data reveal the presence of three new macrocyclic trichothecenes – 3-hydroxyroridin E, 13′-acetyltrichoverrin B, and miophytocen C – and nine known compounds such as roridin A **(145)**, L, M **(147)**, **148–153**. Except miophytocen C, all other metabolites showed significant cytotoxicity against human and murine cell lines [71]. Another microbial associate *Spicellum roseum* 193H15 from the *E. perox* sponge revealed the presence of two sesquiterpenes **154** and **155** involved in the inhibition of lactosylceramide synthase activity in neural cells by effecting sphingolipid metabolism [72]. A linear octapeptide compound **156** containing five *N*-methyl groups, having $\sim$1000 Da molecular weight, extracted from *Acremonium* sp. 021172cKZ derived from a marine sponge *Teichaxinella* sp., displayed weak cytotoxicity against murine cancer cell lines and also demonstrated to have antibacterial activity. Similarly, RHMs 2–4 and **157–161** were also accompanied in the extracts of *Acremonium* sp. [73]. *Hymeniacidon perleve* from the intertidal zone of Fujiazhuang Coast, China, was identified to harbor a large amount of culturable and active epiphytic and endophytic fungi. *Fusarium oxysporum* DLFP2008005 strain is one of the screened isolates that exhibited effective antibacterial and antifungal activities against Gram-positive (*S. epidermidis* and *B. subtilis*) and Gram-negative (*P. fluorescens, P. aeruginosa*, and *C. albicans*) microbes [2].

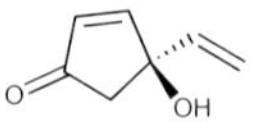

135

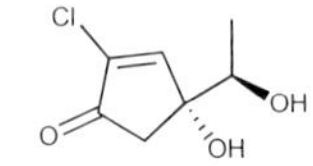

136

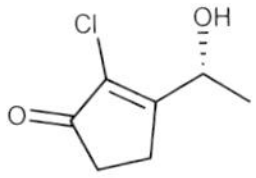

137

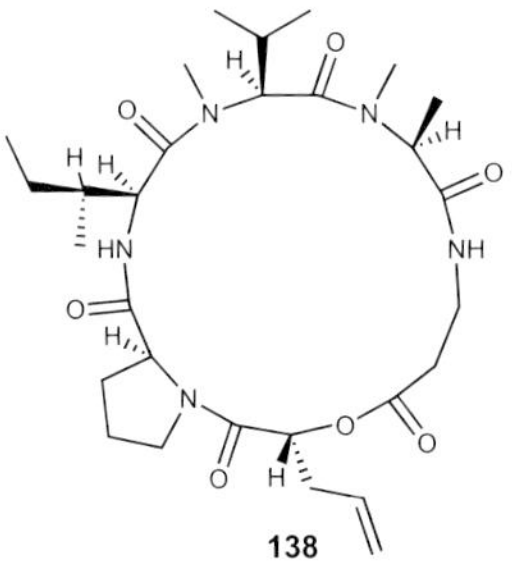

138

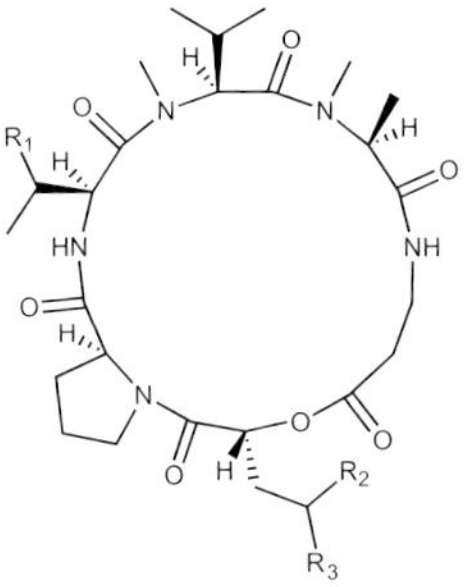

139: $R_1 = CH_2CH_3$; $R_2 = R_3 = CH_3$
140: $R_1 = R2 = R3 = CH_3$
141: $R_1 = CH_2CH_3$; $R_2 = OH$; $R_3 = CH_2Cl$

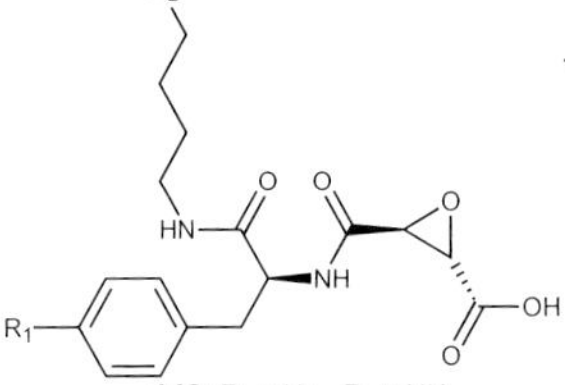

142: $R_1 = H$; $R_2 = NH_2$
143: $R_1 = OH$; $R_2 = NH_2$
144: $R_1 = OH$; $R_2 = CH_2NH_2$

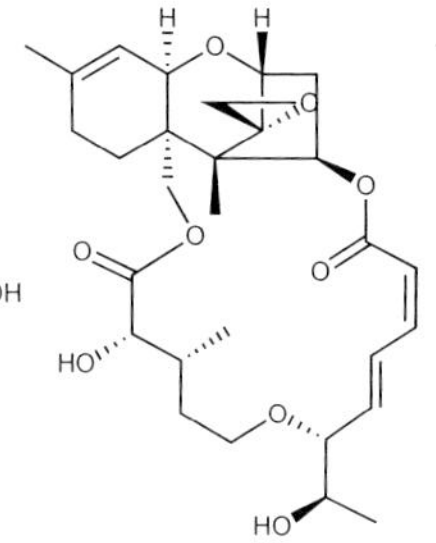

145

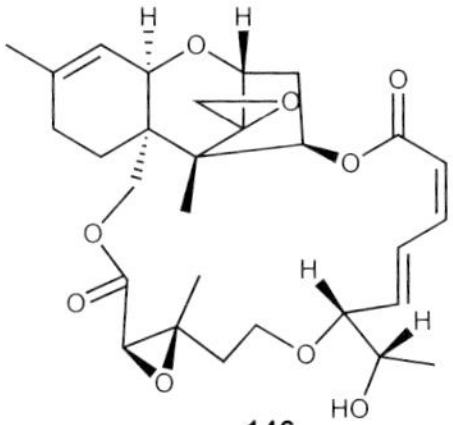

146

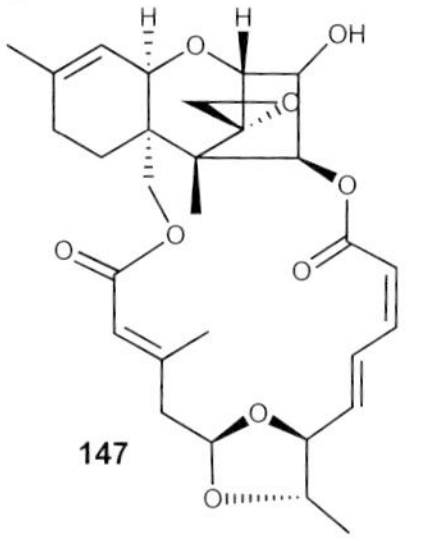

147

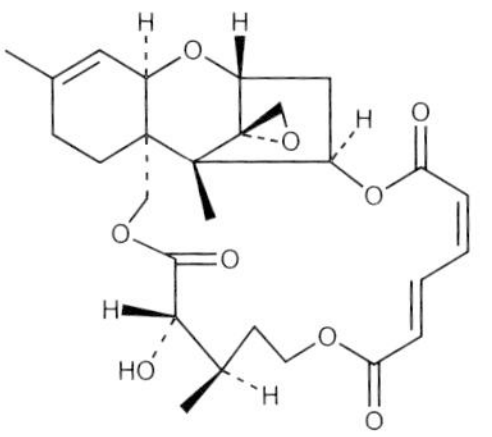

148

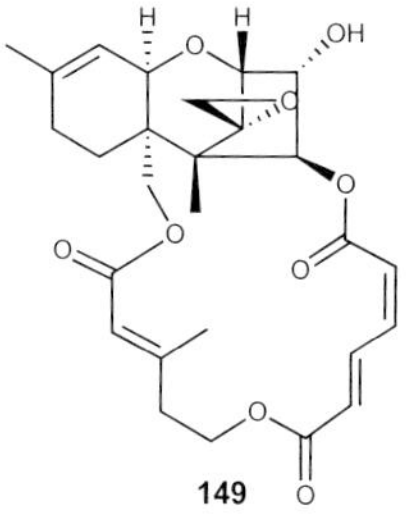

149

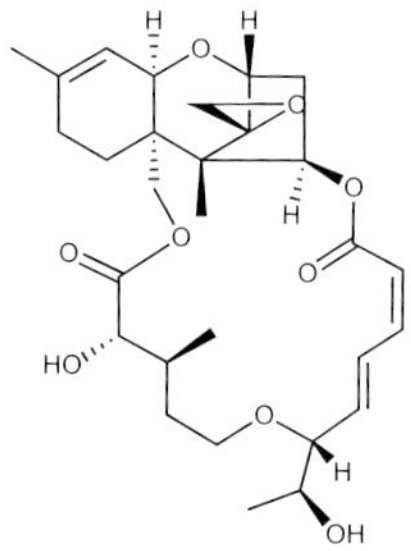

150

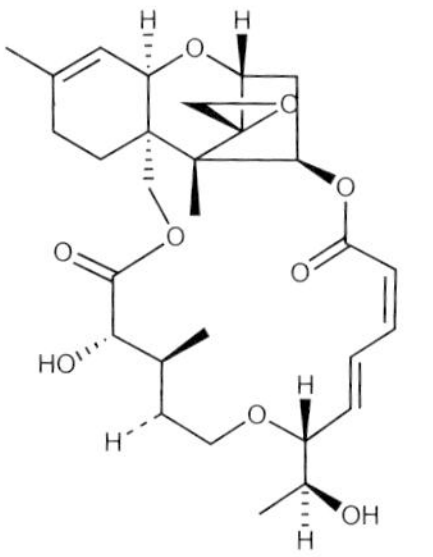

151

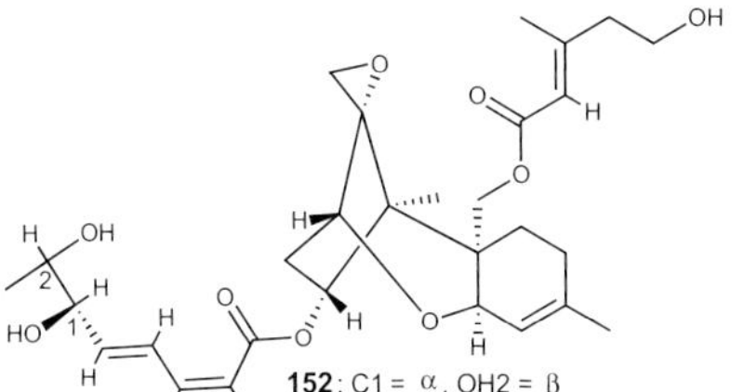

152: C1 = α, OH2 = β
153: C1 = β, OH2 = α

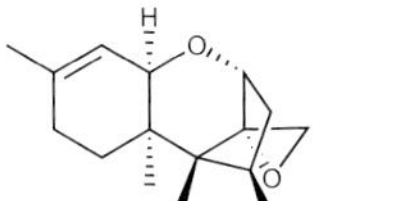

154: R1 = H
155: R1 = COCH=CHCH$_3$

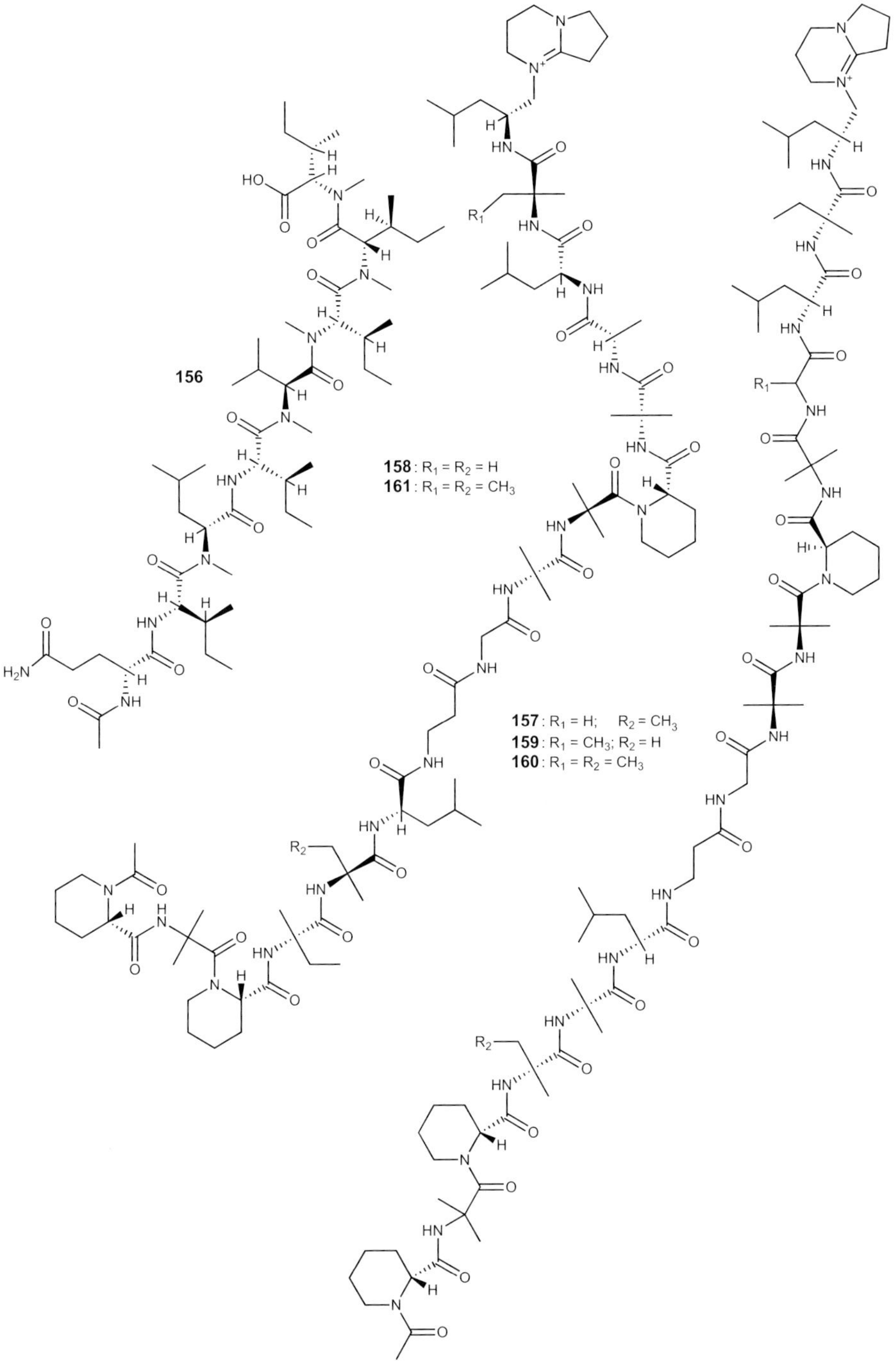

156

158: $R_1 = R_2 = H$
161: $R_1 = R_2 = CH_3$

157: $R_1 = H$; $R_2 = CH_3$
159: $R_1 = CH_3$; $R_2 = H$
160: $R_1 = R_2 = CH_3$

24.6
Unidentified Fungal Strains from Sponges

An unidentified fungus obtained from the Indonesian encrusting sponge *Spirastrella vagabunda* was found to contain **162**, belonging to the class of polyketides, having mild antibiotic activity against *B. subtilis* [2]. An unidentified fungus belonging to the class Hypomycetes isolated from the Indonesian–Pacific sponge *Jaspis* cf. *Johnstoni* was reported to synthesize five compounds belonging to class Chloriolines. Three novel Chloriolines – A, B **(163)**, and C – with known tricyclic sesquiterpenes, coriolin B, and dihydrocoriolin C were detected in the extracts of the fungal isolate [2]. Previously, coriolin B and dihydrocoriolin C were isolated from the terrestrial wood-rotting basidiomycete *Coriolus consors*, among which coriolin B showed potent inhibition of solid tumor cells [74]. Hirsutanols A **(164)**, B, C, and **165** were synthesized by a marine fungal strain obtained from *Haliclona* sp., a marine-derived sponge from Tomini Bay, North Sulawesi, Indonesia [74]. Compounds **64** and **65** possessed mild inhibitory activity against *B. subtilis*.

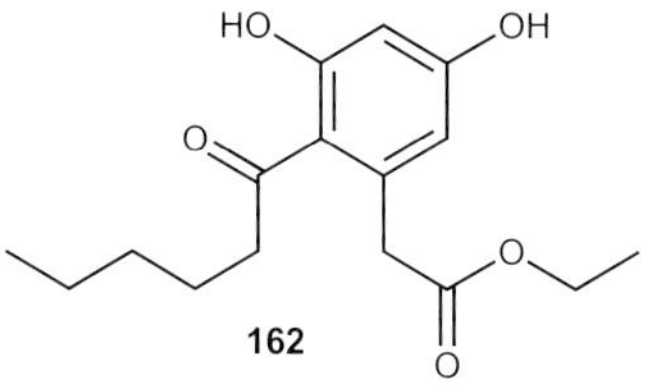

HO OH O O O 162

OH H H Cl O O O OH OH OH H H 163

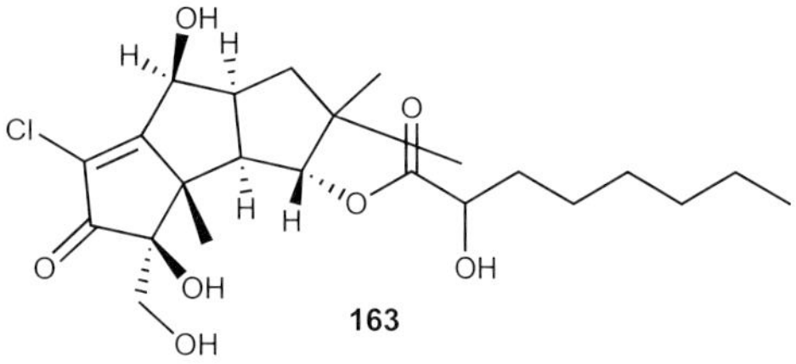

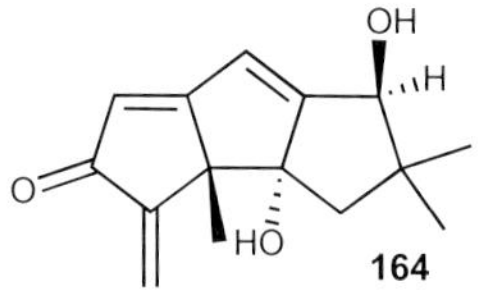

OH H O HO 164

HO OH HO H 165

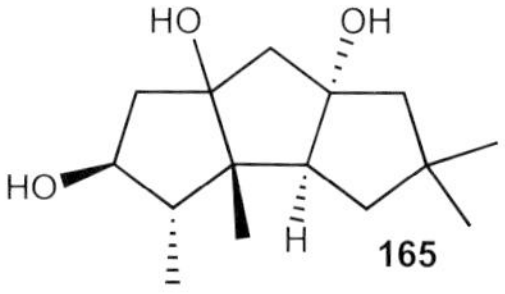

24.7
Compounds in Clinical Trial and Use

The huge numbers of bioactive metabolites characterized from marine microbes can be made available as drug after preclinical testing in laboratory animal models. Evaluation of its biological activity, safety, and formulations is carried out in this process. Phase I clinical trial evaluates the effectiveness of the drug on 20–100 healthy human volunteers and will proceed for its side effects on over 200–500 patient volunteers in the phase II trials. Phase III trials monitor the adverse reaction on long-term usage of the drug on 1000–5000 patient volunteers. Approval of the drug by Food and Drug Administration (FDA) is then followed by reviews of various clinicians. Till now, hundreds of compounds are commercially available. Table 24.2

Table 24.2 Potential therapeutic compounds derived from marine sponges.

Sl. No.	Sponge	Drug name	Disease area	Product name	Company or institution	Clinical status	Ref.
1	*Agelas mauritianus*	KRN700 (α-Galactosylceramide; agelasphin derivative) **(166)** 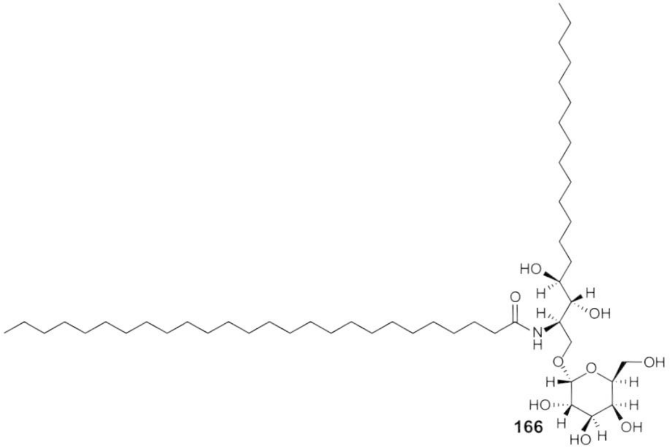	Antitumor; immunostimulatory	α-GalCer	Kirin Brewery Co., Ltd	Phase II	[75,76]
2	*Discodermia dissoluta*	Discodermolide **(167)** 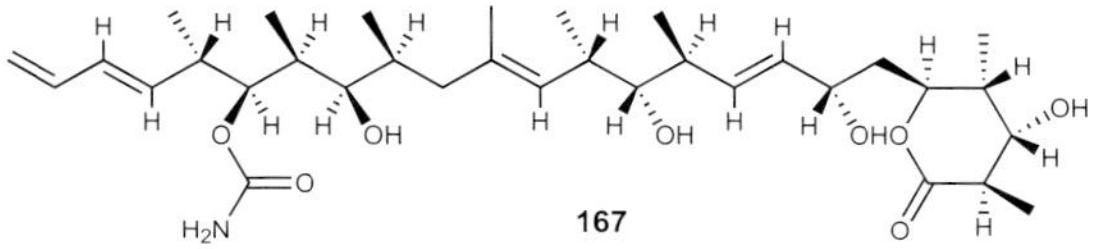	Antitumor	NA	Kosan Pharmaceuticals	Phase II	[77]

(continued)

Table 24.2 (Continued)

Sl. No.	Sponge	Drug name	Disease area	Product name	Company or institution	Clinical status	Ref.
3	*Forcepia* sp.	Lasonolide A **(168)** 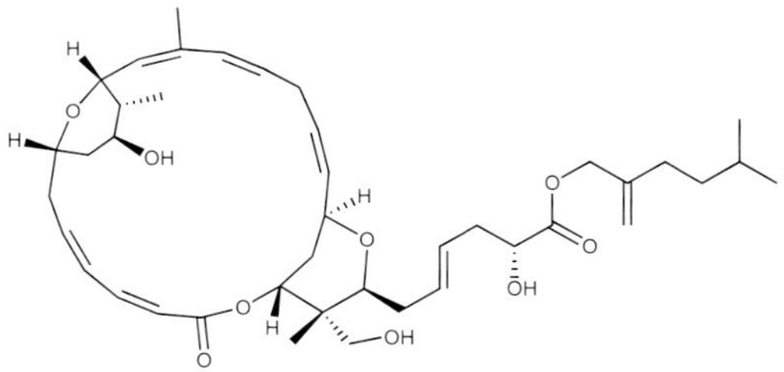 168	Anticancer	NA	Harbor Branch Oceanographic Institute	Preclinical trial	[78]
4	*Halichondria okadai*	Eribulin mesylate (E7389) **(169)** 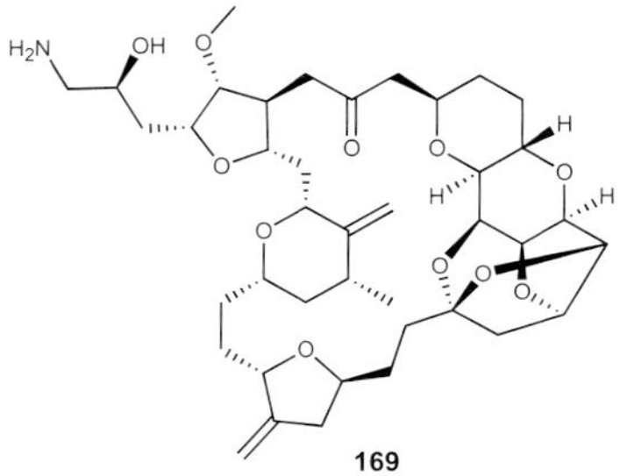 169	Anticancer	Halaven	Eisai Inc.	Approved	[79]

5	*Hemiasterella minor, Cymbastella* sp.; *Siphonochalina* sp.; *Auletta* sp.	Hemiasterlin (E7974) **(170)** 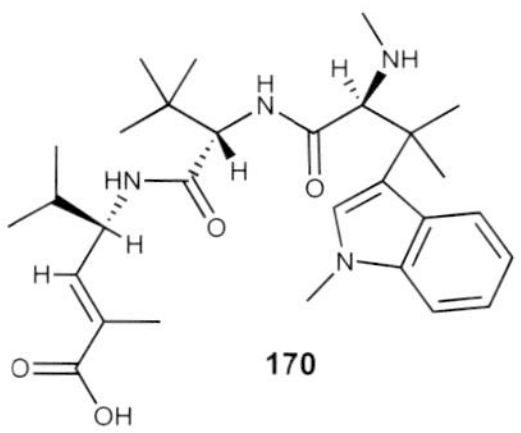	Anticancer	NA	Eisai Inc.	Phase I	[80]
		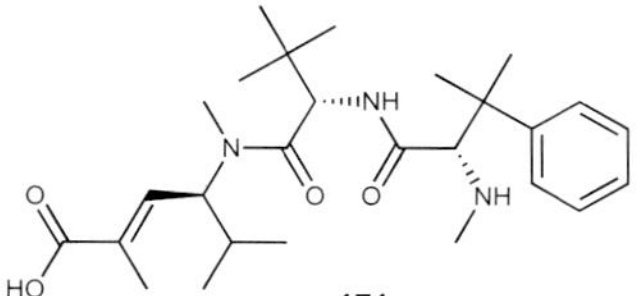 HTI-286 **(171)**	Anticancer	NA	Wyeth	Phase II	[81]
6	*Hexadella* sp.; *Topsentia genitrix; Spongosorites ruetzleri*	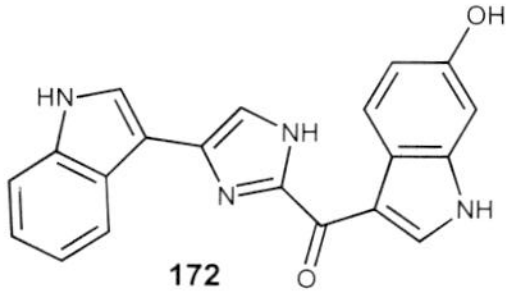 Topsentins **(172)**	Anti-inflammatory	NA	—	Preclinical trial	[77,82]
7	*Jaspis* cf. *coriacea*	Bengamide A and B **(173** and **174)** 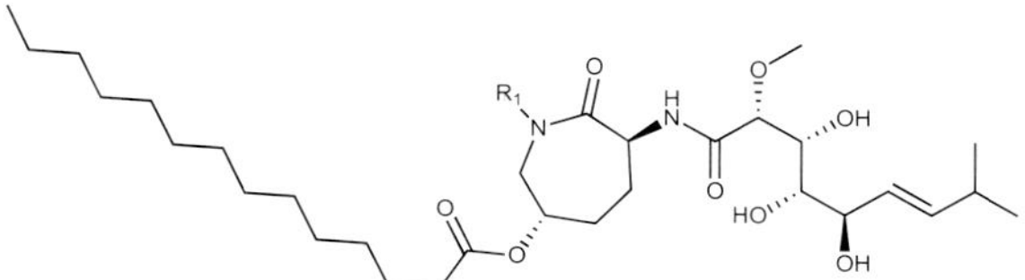 **173**: $R_1 = H$ **174**: $R_1 = CH_3$	Tumor growth inhibitor	NA	Novartis	Synthetic analogue LAF389 withdrawn from phase I clinical trials in 2002	[83]

(*continued*)

Table 24.2 (Continued)

Sl. No.	Sponge	Drug name	Disease area	Product name	Company or institution	Clinical status	Ref.
8	*Latrunculia magnifica*	Latrunculin A and B (**175** and **176**) 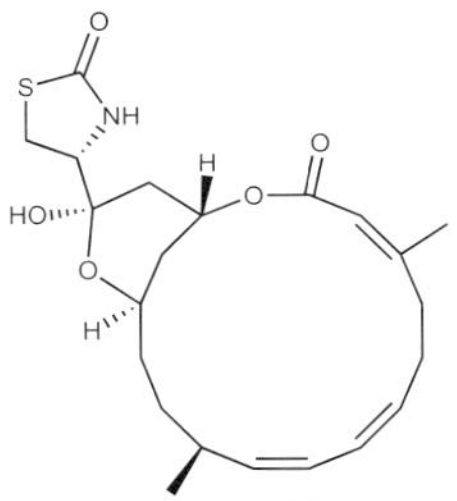 **175** 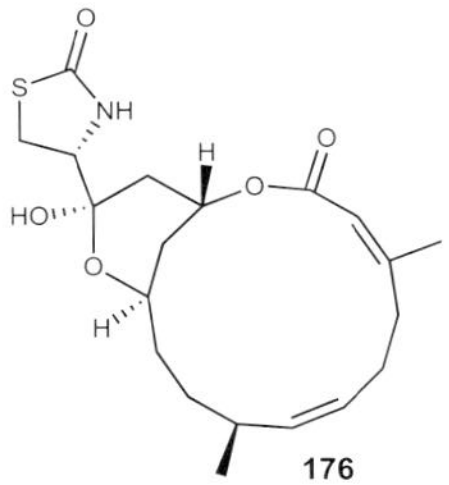 **176**	Anticancer	NA	—	Preclinical trial	[84]
9	*Luffariella variabilis*	Manoalide (**177**) 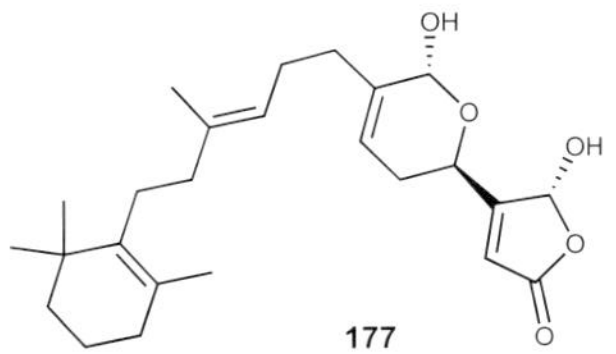 **177**	Anti-inflammatory; analgesia	NA	Allergan pharmaceuticals	Phase II (withdrawn)	[83]
10	*Mycale hentscheli*	Peloruside A (**178**) 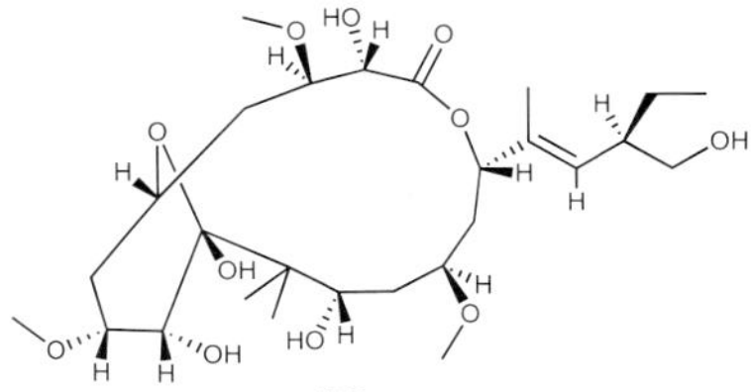 **178**	Anticancer	NA	Reata Pharmaceuticals, Inc.	Preclinical	[85]

11	*Petrosia contignata*	Contignasterol (IZP-94005, IPL576,092) **(179)** 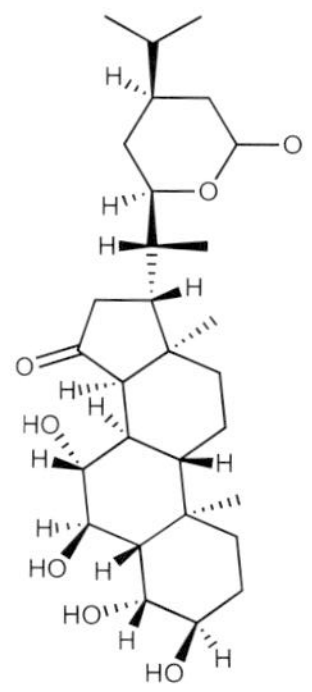 **179**	Anti-asthma	NA	Aventis pharma	Various phases	[83]
12	*Pseudaxinyssa cantharella*	Girolline **(180)** 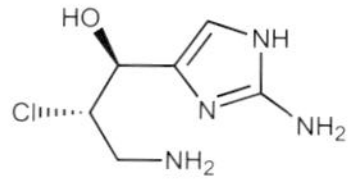 **180**	Anticancer	NA	Rhône-Poulenc	Discontinued	[83]
13	*Spongia* sp.	Dictyostatin **(181)** 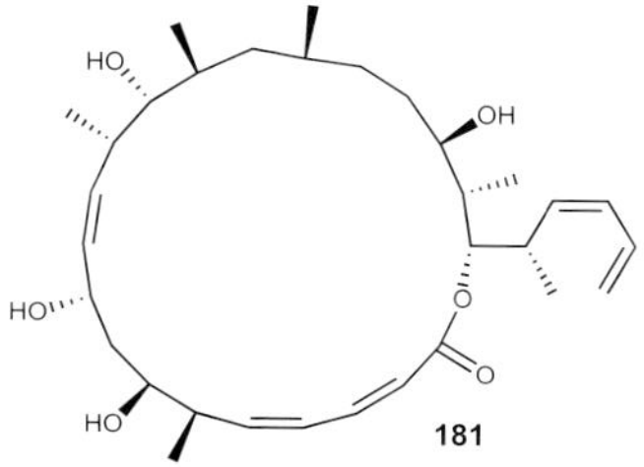 **181**	Anticancer	NA	Harbor Branch Oceanographic Institute	Preclinical trial	[86]

(*continued*)

Table 24.2 (Continued)

Sl. No.	Sponge	Drug name	Disease area	Product name	Company or institution	Clinical status	Ref.
14	*Stylotella aurantium*	Debromohymenialdisine **(182)** 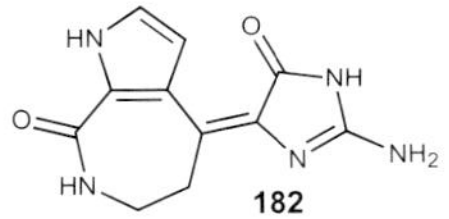	Anti-Alzheimer; anti-osteoarthritis	NA	Genzyme tissue repair for commercial development	Phase I	[77]
15	*Tethya crypta*	Cytarabine (Ara-C) **(183)** 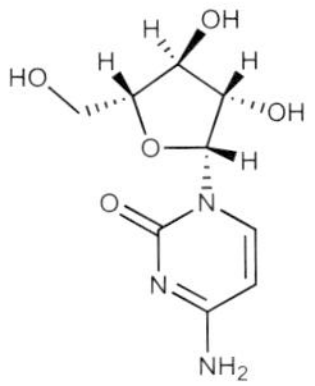	Anticancer	Cytosar-U®	Bedford, Enzon	Approved	[87]
16	*Tethya crypta*	Vidarabine (Ara-A) **(184)** 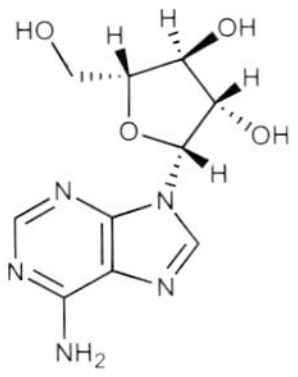	Antiviral	Vira-A®	King Pharmaceuticals	Approved	[87]

describes the various compounds that are in different phases of clinical trials and available on the market.

24.8 Conclusions

Sponges are considered as natural fermenters. A great deal of isolation, characterization, purification, and synthesis of the natural products from them for industrial use is the important objective of the present-day research. This section highlights the abundance of marine sponge-derived metabolites with diverse activities for the biomedical and biotechnological domains. In most of the cases, culture-dependent characterization of the metabolites may not reach up to purification, because it is time-consuming and sometimes compound may be unstable during *in vivo* studies. Metagenomics, which deals with the culture-independent characterization of genes, is now involved in drug synthesis. Many complex biosynthetic pathways have been exploited for the synthesis of principal compounds. The increasing demand for the candidate drug in abundant quantity led to the development of novel strategies for their synthesis, either by cloning of the genes responsible for metabolite synthesis in genetically modified organisms in bioreactors or by formulating them by chemical synthesis. Clinical trials form an important phase in formulating the drug and to study its effectiveness over a group of animal models. This chapter mainly highlights ocean as a new source in discovering novel drugs.

Acknowledgments

The authors acknowledge Indian Council for Medical Research (ICMR), New Delhi, for funding the project (No. 5/8/3(3)/2010-ECD-I) and SASTRA University, Research and Modernization Fund for infrastructure facilities. We thank Prof. V.K. Balasubramanian, Professor of English, SASTRA University, for his constructive suggestions.

References

1 Selvin, J., Shanmughapriya, S., Gandhimathi, R., Kiran, G.S., Ravji, T.R., Natarajaseenivasan, K., and Hema, T.A. (2009) Optimization and production of novel anti-microbial agents from sponge associated marine actinomycetes *Nocardiopsis dassonvillei* MAD08. *Appl. Microbiol. Biotechnol.*, **83** (3), 435–445.

2 Thomas, T.R.A., Kavlekar, D.P., and LokaBharathi, P.A. (2010) Marine drugs from sponge–microbe association – a review. *Mar. Drugs*, **8** (4), 1417–1468.

3 Ang, K.K.H., Holmes, M.J., Higa, T., Hamann, M.T., and Kara, U.A.K. (2000) *In vivo* antimalarial activity of the beta-carboline alkaloid manzamine A. *Antimicrob. Agents Chemother.*, **44** (6), 1645–1649.

4 Han, Y., Yang, B., Zhang, F., Miao, X., and Li, Z. (2009) Characterization of anti-fungal chitinase from marine *Streptomyces* sp. DA11 associated with South China sea sponge *Craniella australiensis*. *Mar. Biotechnol. (NY)*, **11** (1), 132–140.

5 Wicke, C., Hüners, M., Wray, V., Nimtz, M., Bilitewski, U., and Lang, S. (2000) Production and structure elucidation of glycoglycerolipids from a marine sponge-associated *Microbacterium* species. *J. Nat. Prod.*, **63** (5), 621–626.

6 Selvin, J. (2009) Exploring the antagonistic producer *Streptomyces* MSI051: implications of polyketide synthase gene type II and a ubiquitous defense enzyme phospholipase A2 in the host sponge *Dendrilla nigra*. *Curr. Microbiol.*, **58** (5), 459–463.

7 Stierle, A.C., Cardellina, J.H., II, and Singleton, F.L. (1988) A marine *Micrococcus* produces metabolites ascribed to the sponge *Tedania ignis*. *Experientia*, **44** (11–12), 1021.

8 Imamura, N., Nishijima, M., Adachi, K., and Sano, H. (1993) Novel antimycin antibiotics, urauchimycins A and B, produced by marine actinomycete. *J. Antibiot. (Tokyo)*, **46** (2), 241–246.

9 Bultel-Poncé, V., Debitus, C., Berge, J.P., Cerceau, C., and Guyot, M. (1998) Metabolites from the sponge-associated bacterium *Micrococcus luteus*. *J. Mar. Biotechnol.*, **6** (4), 233–236.

10 Kim, T.K., Hewavitharana, A.K., Shaw, P.N., and Fuerst, J.A. (2006) Discovery of a new source of rifamycin antibiotics in marine sponge actinobacteria by phylogenetic prediction. *Appl. Environ. Microbiol.*, **72** (3), 2118–2125.

11 Li, Z. (2009) Advances in marine microbial symbionts in the China Sea and related pharmaceutical metabolites. *Mar. Drugs*, **7** (2), 113–129.

12 Selvin, J., Joseph, S., Asha, K.R.T., Manjusha, W.A., Sangeetha, V.S., Jayaseema, D.M., Antony, M.C., and Vinitha, A.J.D. (2004) Antibacterial potential of antagonistic *Streptomyces* sp. isolated from marine sponge *Dendrilla nigra*. *FEMS Microbiol. Ecol.*, **50** (2), 117–122.

13 Ramm, W., Schatton, W., Wagner-Döbler, I., Wray, V., Nimtz, M., Tokuda, H., Enjyo, F., Nishino, H., Beil, W., Heckmann, R., Lurtz, V., and Lang, S. (2004) Diglucosyl-glycerolipids from the marine sponge-associated *Bacillus pumilus* strain AAS3: their production, enzymatic modification and properties. *Appl. Microbiol. Biotechnol.*, **64** (4), 497–504.

14 Pabel, C.T., Vater, J., Wilde, C., Franke, P., Hofemeister, J., Adler, B., Bringmann, G., Hacker., J., and Hentschel, U. (2003) Antimicrobial activities and matrix-assisted laser desorption/ionization mass spectrometry of *Bacillus* isolates from the marine sponge *Aplysina aerophoba*. *Mar. Biotechnol. (NY)*, **5** (5), 424–434.

15 Nagai, K., Kamigiri, K., Arao, N., Suzumura, K.I., Kawano, Y., Yamaoka, M., Zhang, H., Watanabe, M., and Suzuki, K. (2003) YM-266183 and YM-266184, novel thiopeptide antibiotics produced by *Bacillus cereus* isolated from a marine sponge. I. Taxonomy, fermentation, isolation, physico-chemical properties and biological properties. *J. Antibiot. (Tokyo)*, **56** (2), 123–128.

16 Suzumura, K.I., Yokoi, T., Funatsu, M., Nagai, K., Tanaka, K., Zhang, H., and Suzuki, K. (2003) YM-266183 and YM-266184, novel thiopeptide antibiotics produced by *Bacillus cereus* isolated from a marine sponge. II. Structure elucidation. *J. Antibiot. (Tokyo)*, **56** (2), 129–134.

17 Thakur, N.L., Hentschel, U., Krasko, A., Pabel, C.T., Anil, A.C., and Müller, W.E.G. (2003) Antibacterial activity of the sponge *Suberites domuncula* and its primmorphs: potential basis for epibacterial chemical defense. *Aquat. Microb. Ecol.*, **31** (1), 77–83.

18 Li, Z., Peng, C., Shen, Y., Miao, X., Zhang, H., and Lin, H. (2008) L,L-Diketopiperazines from *Alcaligenes faecalis* A72 associated with South China Sea sponge *Stelletta tenuis*. *Biochem. Syst. Ecol.*, **36** (3), 230–234.

19 Zheng, L., Chen, H., Han, X., Lin, W., and Yan, X. (2005) Antimicrobial screening and active compound isolation from marine bacterium NJ6-3-1 associated with the sponge *Hymeniacidon perleve*. *World J. Microbiol. Biotechnol.*, **21** (2), 201–206.

20 Jayatilake, G.S., Thornton, M.P., Leonard, A.C., Grimwade, J.E., and Baker, B.J. (1996) Metabolites from an Antarctic sponge-

associated bacterium, *Pseudomonas aeruginosa*. *J. Nat. Prod.*, **59** (3), 293–296.

21 Speitling, M., Smetanina, O.F., Kuznetsova, T.A., and Laatsch, H. (2007) Bromoalterochromides A and A′, unprecedented chromopeptides from a marine *Pseudoalteromonas maricaloris* strain KMM 636^T. *J. Antibiot. (Tokyo)*, **60** (1), 36–42.

22 Kobayashi, M. and Kitagawa, I. (1994) Bioactive substances isolated from marine sponge, a miniature conglomerate of various organisms. *Pure Appl. Chem.*, **66** (4), 819–826.

23 Yoo, M., Choi, S.U., Choi, K.Y., Yon, G.H., Chae, J.C., Kim, D., Zylstra, G.J., and Kim, E. (2008) Trisindoline synthesis and anticancer activity. *Biochem. Biophys. Res. Commun.*, **376** (1), 96–99.

24 Bultel-Poncé, V., Berge, J.P., Debitus, C., Nicolas, J.L., and Guyot, M. (1999) Metabolites from the sponge-associated bacterium *Pseudomonas* species. *Mar. Biotechnol. (NY)*, **1** (4), 384–390.

25 Schmidt, E.W., Obraztsova, A.Y., Davidson, S.K., Faulkner, D.J., and Haygood, M.G. (2000) Identification of the antifungal peptide-containing symbiont of the marine sponge *Theonella swinhoei* as a novel δ-proteobacterium, "*Candidatus* Entotheonella palauensis". *Mar. Biol.*, **136** (6), 969–977.

26 Bewley, C.A., Holland, N.D., and Faulkner, D.J. (1996) Two classes of metabolites from *Theonella swinhoei* are localized in distinct populations of bacterial symbionts. *Experientia*, **52** (7), 716–722.

27 König, G.M., Kehraus, S., Seibert, S.F., Abdel-Lateff, A., and Müller, D. (2006) Natural products from marine organisms and their associated microbes. *ChemBioChem*, **7** (2), 229–238.

28 Hinde, R., Pironet, F., and Borowitzka, M.A. (1994) Isolation of *Oscillatoria spongeliae*, the filamentous cyanobacterial symbiont of the marine sponge *Dysidea herbacea*. *Mar. Biol.*, **119** (1), 99–104.

29 Unson, M.D., Holland, N.D., and Faulkner, D.J. (1994) A brominated secondary metabolite synthesized by the cyanobacterial symbiont of a marine sponge and accumulation of the crystalline metabolite in the sponge tissue. *Mar. Biol.*, **119** (1), 1–11.

30 Piel, J., Hui, D., Wen, G., Butzke, D., Platzer, M., Fusetani, N., and Matsunaga, S. (2004) Antitumor polyketide biosynthesis by an uncultivated bacterial symbiont of the marine sponge *Theonella swinhoei*. *Proc. Natl. Acad. Sci. USA*, **101** (46), 16222–16227.

31 Doshida, J., Hasegawa, H., Onuki, H., and Shimidzu, N. (1996) Exophilin A, a new antibiotic from a marine microorganism *Exophiala pisciphila*. *J. Antibiot. (Tokyo)*, **49** (11), 1105–1109.

32 Yoshida, S., Kito, K., Ooi, T., Kanoh, K., Shizuri, Y., and Kusumi, T. (2007) Four pimarane diterpenes from marine fungus: chloroform incorporated in crystal lattice for absolute configuration analysis by X-ray. *Chem. Lett.*, **36** (11), 1386–1387.

33 Jadulco, R., Brauers, G., Edrada, R.A., Ebel, R., Wray, V., Sudarsono, S., and Proksch, P. (2002) New metabolites from sponge-derived fungi *Curvularia lunata* and *Cladosporium herbarum*. *J. Nat. Prod.*, **65** (5), 730–733.

34 Brauers, G., Edrada, R.A., Ebel, R., Proksch, P., Wray, V., Berg, A., Gräfe, U., Schächtele, C., Torzke, F., Finkenzeller, G., Marme, D., Kraus, J., Münchbach, M., Michel, M., Bringmann, G., and Schaumann, K. (2000) Anthraquinones and betaenone derivatives from the sponge-associated fungus *Microsphaeropsis* species: novel inhibitors of protein kinases. *J. Nat. Prod.*, **63** (6), 739–745.

35 Höller, U., König, G.M., and Wright, A.D. (1999) Three new metabolites from marine-derived fungi of the genera *Coniothyrium* and *Microsphaeropsis*. *J. Nat. Prod.*, **62** (1), 114–118.

36 Jadulco, R., Proksch, P., Wray, V., Sudarsono, S., Berg, A., and Gräfe, U. (2001) New macrolides and furan carboxylic acid derivative from the sponge-derived fungus *Cladosporium herbarum*. *J. Nat. Prod.*, **64** (4), 527–530.

37 Nagai, K., Kamigiri, K., Matsumoto, H., Kawano, Y., Yamaoka, M., Shimoi, H., Watanabe, M., and Suzuki, K. (2002) YM-202204, a new antifungal antibiotic produced by marine fungus *Phoma* sp. *J. Antibiot. (Tokyo)*, **55** (12), 1036–1041.

38 Mohamed, I.E., Gross, H., Pontius, A., Kehraus, S., Krick, A., Kelter, G., Maier, A., Fiebig, H.H., and König, G.M. (2009) Epoxyphomalin A and B, prenylated polyketides with potent cytotoxicity from the marine-derived fungus *Phoma* sp. *Org. Lett.*, **11** (21), 5014–5017.

39 Höller, U., König, G.M., and Wright, A.D. (1999) A new tyrosine kinase inhibitor from a marine isolate of *Ulocladium botrytis* and new metabolites from the marine fungi *Asteromyces cruciatus* and *Varicosporina ramulosa*. *Eur. J. Org. Chem.*, **1999** (11), 2949–2955.

40 Jadulco, R., Edrada, R.A., Ebel, R., Berg, A., Schaumann, K., Wray, V., Steube, K., and Proksch, P. (2004) New communesin derivatives from the fungus *Penicillium* sp. derived from the Mediterranean sponge *Axinella verrucosa*. *J. Nat. Prod.*, **67** (1), 78–81.

41 Koizumi, Y., Arai, M., Tomoda, H., and Ômura, S. (2004) Oxaline, a fungal alkaloid, arrests the cell cycle in M phase by inhibition of tubulin polymerization. *Biochim. Biophys. Acta*, **1693** (1), 47–55.

42 Bringmann, G., Lang, G., Steffens, S., Günther, E., and Schaumann, K. (2003) Evariquinone, isoemericellin, and stromemycin from a sponge derived strain of the fungus *Emericella variecolor*. *Phytochemistry*, **63** (4), 437–443.

43 Hiort, J., Maksimenka, K., Reichert, M., Perović-Ottstadt, S., Lin, W.H., Wray, V., Steube, K., Schaumann, K., Weber, H., Proksch, P., Ebel, R., Müller, W.E.G., and Bringmann, G. (2004) New natural products from the sponge-derived fungus *A. niger*. *J. Nat. Prod.*, **67** (9), 1532–1543.

44 Lin, W., Brauers, G., Ebel, R., Wray, V., Berg, A., Sudarsono, S., and Proksch, P. (2003) Novel chromone derivatives from the fungus *Aspergillus versicolor* isolated from the marine sponge *Xestospongia exigua*. *J. Nat. Prod.*, **66** (1), 57–61.

45 Varoglu, M. and Crews, P. (2000) Biosynthetically diverse compounds from a saltwater culture of sponge-derived *A. niger*. *J. Nat. Prod.*, **63** (1), 41–43.

46 Govek, S.P. and Overman, L.E. (2001) Total synthesis of (+)-asperazine. *J. Am. Chem. Soc.*, **123** (38), 9468–9469.

47 Xin, Z.H., Fang, Y., Du, L., Zhu, T., Duan, L., Chen, J., Gu, Q.Q., and Zhu, W.M. (2007) Aurantiomides A–C, quinazoline alkaloids from the sponge-derived fungus *Penicillium aurantiogriseum* SP0–19. *J. Nat. Prod.*, **70** (5), 853–855.

48 Xin, Z.H., Zhu, W.M., Gu, Q.Q., Fang, Y.C., Duan, L., and Cui, C.B. (2005) A new cytotoxic compound from *Penicillium auratiogriseum*, symbiotic or epiphytic fungus of sponge *Mycale plumose*. *Chin. Chem. Lett.*, **16** (9), 1227–1229.

49 Namikoshi, M., Negishi, R., Nagai, H., Dmitrenok, A., and Kobayashi, H. (2003) Three new chlorine containing antibiotics from a marine-derived fungus *Aspergillus ostianus* collected in Pohnpei. *J. Antibiot. (Tokyo)*, **56** (9), 755–761.

50 Kito, K., Ookura, R., Yoshida, S., Namikoshi, M., Ooi, T., and Kusumi, T. (2007) Pentaketides relating to aspinonene and dihydroaspyrone from a marine-derived fungus, *Aspergillus ostianus*. *J. Nat. Prod.*, **70** (12), 2022–2025.

51 Kito, K., Ookura, R., Yoshida, S., Namikoshi, M., Ooi, T., and Kusumi, T. (2008) New cytotoxic 14-membered macrolides from marine-derived fungus *A. ostianus*. *Org. Lett.*, **10** (2), 225–228.

52 Edrada, R.A., Heubes, M., Brauers, G., Wray, V., Berg, A., Gräfe, U., Wohlfarth, M., Mühlbacher, J., Schaumann, K., Sudarsono, S., Bringmann, G., and Proksch, P. (2002) Online analysis of xestodecalactones A–C, novel bioactive metabolites from the fungus *Penicillium* cf. *montanense* and their subsequent isolation from the sponge *Xestospongia exigua*. *J. Nat. Prod.*, **65** (11), 1598–1604.

53 Lee, Y.M., Mansoor, T.A., Hong, J., Lee, C.O., Bae, K.S., and Jung, J.H. (2007) Polyketides from a sponge-derived fungus, *A. versicolor*. *Nat. Prod. Sci.*, **13** (1), 90–96.

54 Lee, Y.M., Dang, H.T., Hong, J., Lee, C.O., Bae, K.S., Kim, D.K., and Jung, J.H. (2010) A cytotoxic lipopeptide from the sponge-derived fungus *Aspergillus versicolor*. *Bull. Korean Chem. Soc.*, **31** (1), 205–208.

55 López-Gresa, M.P., Cabedo, N., González-Mas, M.C., Ciavatta, M.L., Avila, C., and Primo, J. (2009) Terretonins E and F, inhibitors of the mitochondrial respiratory chain from the marine-derived fungus

Aspergillus insuetus. *J. Nat. Prod.*, **72** (7), 1348–1351.

56 Malmstrøm, J., Christophersen, C., Barrero, A.F., Oltra, J.E., Justicia, J., and Rosales, A. (2002) Bioactive metabolites from a marine-derived strain of the fungus *Emericella variecolor*. *J. Nat. Prod.*, **65** (3), 364–367.

57 Amagata, T., Tanaka, M., Yamada, T., Doi, M., Minoura, K., Ohishi, H., Yamori, T., and Numata, A. (2007) Variation in cytostatic constituents of a sponge-derived *Gymnascella dankaliensis* by manipulating the carbon source. *J. Nat. Prod.*, **70** (11), 1731–1740.

58 Numata, A., Amagata, T., Minoura, K., and Ito, T. (1997) Gymnastatins, novel cytotoxic metabolites produced by a fungal strain from a sponge. *Tetrahedron Lett.*, **38** (32), 5675–5678.

59 Amagata, T., Minoura, K., and Numata, A. (1998) Gymnasterones, novel cytotoxic metabolites produced by a fungal strain from a sponge. *Tetrahedron Lett.*, **39** (22), 3773–3774.

60 Amagata, T., Doi, M., Ohta, T., Minoura, K., and Numata, A. (1998) Absolute stereostructures of novel cytotoxic metabolites, gymnastatins A–E, from a *Gymnascella* species separated from a *Halichondria* sponge. *J. Chem. Soc., Perkin Trans. I*, **21**, 3585–3599.

61 Amagata, T., Doi, M., Tohgo, M., Minoura, K., and Numata, A. (1999) Dankasterone, a new class of cytotoxic steroid produced by a *Gymnascella* species from a marine sponge. *Chem. Commun.*, **14**, 1321–1322.

62 Amagata, T., Minoura, K., and Numata, A. (2006) Gymnastatins F–H, cytostatic metabolites from the sponge-derived fungus *Gymnascella dankaliensis*. *J. Nat. Prod.*, **69** (10), 1384–1388.

63 Amagata, T., Tanaka, M., Yamada, T., Minoura, K., and Numata, A. (2008) Gymnastatins and dankastatins, growth inhibitory metabolites of a *Gymnascella* species from a *Halichondria sponge*. *J. Nat. Prod.*, **71** (3), 340–345.

64 Lang, G., Wiese, J., Schmaljohann, R., and Imhoff, J.F. (2007) New pentaenes from the sponge-derived marine fungus *Penicillium rugulosum*: structure determination and biosynthetic studies. *Tetrahedron*, **63** (48), 11844–11849.

65 Bugni, T.S., Bernan, V.S., Greenstein, M., Janso, J.E., Maiese, W.M., Mayne, C.L., and Ireland, C.M. (2003) Brocaenols A–C: novel polyketides from a marine derived *Penicillium brocae*. *J. Org. Chem.*, **68** (5), 2014–2017.

66 Bringmann, G., Gulder, T.A.M., Lang, G., Schmitt, S., Stöhr, R., Wiese, J., Nagel, K., and Imhoff, J.F. (2007) Large-scale biotechnological production of the antileukemic marine natural product sorbicillactone A. *Mar. Drugs*, **5** (2), 23–30.

67 Usami, Y., Ikura, T., Amagata, T., and Numata, A. (2000) First total syntheses and configurational assignments of cytotoxic trichodenones A–C. *Tetrahedron: Asymmetry*, **11** (18), 3711–3725.

68 Boot, C.M., Amagata, T., Tenney, K., Compton, J.E., Pietraszkiewicz, H., Valeriote, F.A., and Crews, P. (2007) Four classes of structurally unusual peptides from two marine-derived fungi: structures and bioactivities. *Tetrahedron*, **63** (39), 9903–9914.

69 Yu, C.M., Curtis, J.M., Walter, J.A., Wright, J.L.C., Ayer, S.W., Kaleta, J., Querengesser, L., and Fathi-Afshar, Z.R. (1996) Potent inhibitors of cysteine proteases from the marine fungus *Microascus longirostris*. *J. Antibiot. (Tokyo)*, **49** (4), 395–397.

70 Xie, L.W., Jiang, S.M., Zhu, H.H., Sun, W., Ouyang, Y.C., Dai, S.K., and Li, X. (2008) Potential inhibitors against *Sclerotinia sclerotiorum*, produced by the fungus *Myrothecium* sp. associated with the marine sponge *Axinella* sp. *Eur. J. Plant Pathol.*, **122** (4), 571–578.

71 Amagata, T., Rath, C., Rigot, J.F., Tarlov, N., Tenney, K., Valeriote, F.A., and Crews, P. (2003) Structures and cytotoxic properties of trichoverroids and their macrolide analogues produced by saltwater culture of *Myrothecium verrucaria*. *J. Med. Chem.*, **46** (20), 4342–4350.

72 Kralj, A., Gurgui, M., König, G.M., and van Echten-Deckert, G. (2007) Trichothecenes induce accumulation of glucosylceramide in neural cells by interfering with lactosylceramide synthase activity. *Toxicol. Appl. Pharmacol.*, **225** (1), 113–122.

73 Boot, C.M., Tenney, K., Valeriote, F.A., and Crews, P. (2006) Highly *N*-methylated linear peptides produced by an atypical sponge-derived *Acremonium* sp. *J. Nat. Prod.*, **69** (1), 83–92.

74 Wang, G.Y.S., Abrell, L.M., Avelar, A., Borgeson, B.M., and Crews, P. (1998) New hirsutane based sesquiterpenes from salt water cultures of a marine sponge-derived fungus and the terrestrial fungus *Coriolus consors*. *Tetrahedron*, **54** (26), 7335–7342.

75 Veldt, B.J., van der Vliet, H.J.J., von Blomberg, B.M.E., van Vlierberghe, H., Gerken, G., Nishi, N., Hayashi, K., Scheper, R.J., de Knegt, R.J., van den Eertwegh, A.J.M., Janssen, H.L.A., and van Nieuwkerk, C.M.J. (2007) Randomized placebo controlled phase I/II trial of α-galactosylceramide for the treatment of chronic hepatitis C. *J. Hepatol.*, **47** (3), 356–365.

76 Motohashi, S., Nagato, K., Kunii, N., Yamamoto, H., Yamasaki, K., Okita, K., Hanaoka, H., Shimizu, N., Suzuki, M., Yoshino, I., Taniguchi, M., Fujisawa, T., and Nakayama, T. (2009) A phase I–II study of α-galactosylceramide-pulsed IL-2/GM-CSF-cultured peripheral blood mononuclear cells in patients with advanced and recurrent non-small cell lung cancer. *J. Immunol.*, **182** (4), 2492–2501.

77 Faulkner, D.J. (2000) Marine pharmacology. *Antonie van Leeuwenhoek*, **77** (2), 135–145.

78 Isbrucker, R.A., Guzmán, E.A., Pitts, T.P., and Wright, A.E. (2009) Early effects of lasonolide A on pancreatic cancer cells. *J. Pharmacol. Exp. Ther.*, **331** (2), 733–739.

79 Huyck, T.K., Gradishar, W., Manuguid, F., and Kirkpatrick, P. (2011) Eribulin mesylate. *Nat. Rev. Drug Discov.*, **10** (3), 173–174.

80 Kuznetsov, G., TenDyke, K., Towle, M.J., Cheng, H., Liu, J., Marsh, J.P., Schiller, S.E.R., Spyvee, M.R., Yang, H., Seletsky, B.M., Shaffer, C.J., Marceau, V., Yao, Y., Suh, E.M., Campagna, S., Fang, F. G., Kowalczyk, J.J., and Littlefield, B.A. (2009) Tubulin-based antimitotic mechanism of E7974, a novel analogue of the marine sponge natural product hemiasterlin. *Mol. Cancer Ther.*, **8** (10), 2852–2860.

81 Loganzo, F., Hari, M., Annable, T., Tan, X., Morilla, D.B., Musto, S., Zask, A., Kaplan, J., Jr., Minnick, A.A., May, M.K., Ayral-Kaloustian, S., Poruchynsky, M.S., Fojo, T., and Greenberger, L.M. (2004) Cells resistant to HTI-286 do not overexpress P-glycoprotein but have reduced drug accumulation and a point mutation in α-tubulin. *Mol. Cancer Ther.*, **3** (10), 1319–1327.

82 Oh, K.B., Mar, W., Kim, S., Kim, J.Y., Lee, T.H., Kim, J.G., Shin, D., Sim, C.J., and Shin, J. (2006) Antimicrobial activity and cytotoxicity of bis(indole) alkaloids from the sponge *Spongosorites* sp. *Biol. Pharm. Bull.*, **29** (3), 570–573.

83 Newman, D.J. and Cragg, G.M. (2004) Marine natural products and related compounds in clinical and advanced preclinical trials. *J. Nat. Prod.*, **67** (8), 1216–1238.

84 Fürstner, A., Kirk, D., Fenster, M.D.B., Aïssa, C., De Souza, D., Nevado, C., Tuttle, T., Thiel, W., and Müller, O. (2007) Latrunculin analogues with improved biological profiles by "diverted total synthesis": preparation, evaluation, and computational analysis. *Chem. Eur. J.*, **13** (1), 135–149.

85 Miller, J.H., Singh, A.J., and Northcote, P.T. (2010) Microtubule-stabilizing drugs from marine sponges: focus on peloruside A and zampanolide. *Mar. Drugs*, **8** (4), 1059–1079.

86 Eiseman, J.L., Bai, L., Jung, W.H., Moura-Letts, G., Day, B.W., and Curran, D.P. (2008) Improved synthesis of 6-*epi*-dictyostatin and antitumor efficacy in mice bearing MDA-MB231 human breast cancer xenografts. *J. Med. Chem.*, **51** (21), 6650–6653.

87 Thakur, N.L. and Müller, W.E.G. (2004) Biotechnological potential of marine sponges. *Curr. Sci.*, **86** (11), 1506–1512.

25
Bioactive Marine Microorganisms for Biocatalytic Reactions in Organic Compounds

Lenilson C. Rocha, Julieta R. de Oliveira, Bruna Vacondio, Gisele N. Rodrigues, Mirna H. Regali Seleghim, and André L. Meleiro Porto

25.1
Introduction

The aim of this chapter is to explain the importance of marine enzymes from a broad range of sources (microorganisms, algae, fish, mollusks, sponges) that can be used to transform organic compounds, emphasizing the ecofriendly reactions of biocatalysis, biotransformation, and biodegradation. In the literature, there are many studies of enzymatic reactions involving biocatalysts from terrestrial microorganisms, but few reports of reactions catalyzed by biocatalysts from marine organisms (living organisms or purified enzymes). The enzymatic reactions selected as examples are those of interest in important transformations of organic compounds, such as biotransformation of natural products and biodegradation of organic pollutants (e.g., pesticides and toxic munitions/explosives). Reductions of acetophenone derivatives and hydrolytic reactions of epoxides by marine fungi, algae, and bacteria have been reported. Another important area discussed here is the isolation of marine fungi from marine organisms (e.g., sponges) for use as biocatalysts for enzymatic reactions of organic compounds. In addition to the work described here, recent literature reviews on the potential use of marine enzymes to promote transformations of organic compounds can be consulted [1–3]. Chemists, biologists, microbiologists, and other professionals interested in the discovery of marine enzymes for biotechnological process can all contribute to the protection, sustainable exploitation, and conservation of marine ecosystems. Although the marine environment is a rich and largely unexplored source of products of potential interest to the human race, only a few of these marine products have reached the stage of commercial production. This gap between discovery and commercialization can be bridged when biologists, chemists, and engineers join forces and integrate their research to develop feasible bioprocess technologies for the production of marine natural compounds [4]. Among the resources from the marine environment that could be exploited are the enzymes that catalyze the transformation of organic compounds by biocatalysis, biotransformation, and biodegradation processes.

Marine Microbiology: Bioactive Compounds and Biotechnological Applications, First Edition.
Edited by Se-Kwon Kim

The limits between the areas are blurred: biotransformations and enzymatic catalysis by crude extracts or pure enzymes are often summarized under the term biocatalysis. Frequently, "biocatalytic processes" are transformations of a defined substrate to a defined product with one or several enzyme-catalyzed steps [5]. Among other definitions, biocatalysis is the use of enzymes as catalysts to perform chemical transformations of organic compounds. Biocatalyst is an enzyme or enzyme complex consisting of, or derived from, an organism or cell culture (in cell-free or whole-cell forms) that catalyzes metabolic reactions in living organisms and/or substrate conversions in various chemical reactions [6].

Biocatalysis has made great advances in the last two decades, especially for the pharmaceutical industries [7]. It is estimated that by 2020, biocatalytic methods will be applied to the production of up to 20% pharmaceutical products. The discovery of new enzymes in living organisms (plants, animals, microorganisms), allied to the new biotechnological techniques, especially the use of recombinant DNA technology and reactions carried out by whole cells of genetically modified organisms on a large scale, boosted organic chemists to use enzymes in a wide variety of organic reactions to obtain compounds of industrial interest.

The advantage of enzymes for the catalytic transformation of fine chemicals is their generally unsurpassed selectivity. While enzymes are used beneficially to increase chemical selectivity or regioselectivity of a reaction, their biggest advantage lies in differentiating between enantiomeric substrates, a pair of substrates with Gibbs free enthalpy differences between the R and the S enantiomers (ΔG_{RS}) of around 1–3 kJ/mol. Enantioselectivities of >99% ee can be achieved routinely by enzymatic reactions [5]. Factors such as production costs, biological activity, and regulatory issues have led to a growing demand for enantiopure building blocks for chiral drugs [8,9]. Biosynthetic techniques using new enzymes from microorganisms are currently established as useful tools for the production of intermediates by fine chemical processes [10,11].

In view of growing importance of the marine environment in the current context, as we move into the new century, and the low exploitation of marine enzymes in organic reactions, this chapter introduces a brief discussion of potential synthetic and environmental applications of marine biocatalysts in the transformations of organic compounds of interest. Thus, some reactions of reduction, oxidation, hydrolysis, and biodegradation that can be achieved using whole cells of marine organisms, but are difficult to carry out by chemical methods, particularly stereoselective reactions, are described here. Another point introduced in this chapter is the collection and isolation of bioactive marine microorganisms.

25.2 Marine Enzymes

The sea covers more than three quarters of the Earth's surface and provides abundant resources for biotechnological research and development [12]. Marine organisms represent a dramatically different environment for the biosynthesis of molecules than terrestrial organisms, and are a vast, untapped source of enzymes [13]. In recent years, a variety of new enzymes with specific activities were isolated from marine bacteria, fungi, and other marine organisms; moreover, some marine organisms produced a considerable number of molecules with a potential to be transformed into commercial drugs [14,15]. In fact, the marine environment is a very rich source of extremely potent compounds exhibiting significant activities in antitumor, anti-inflammatory, analgesic, immunomodulatory, allergic, and antiviral assays [16,17].

Enzyme bioprospecting is a basic research activity devoted to the search for novel biocatalysts. Marine organisms (fungi, bacteria, algae, sponges, fish, prawns, crustaceous, reptiles) can be rich sources of novel enzymes, but most of the current bioprospecting activity is based on microbial organisms. A marine enzyme is a protein molecule with unique properties derived from an organism whose natural habitat is saline or brackish water [2,4].

Marine enzymes can be novel biocatalysts with properties such as high salt tolerance, hyperthermostability, barophilicity, and cold adaptability. Microorganisms isolated from ocean sediment and seawater have been most widely studied as sources of marine enzymes, especially proteases, carbohydrases, and peroxidases. Research on the bioprocessing of marine-derived enzymes has been scanty, focusing mainly on the application of solid-state fermentation to the production of enzymes from microbial sources [14].

Enzymatic reactions catalyzed by marine fungi can be utilized when the fungi are cultured in the presence of artificial seawater. The filamentous fungi *Aspergillus sydowii* Gc12, *Penicillium raistrickii* Ce16, *Penicillium miczynskii* Gc5, and *Trichoderma* sp. Gc1, grown in artificial seawater, were able to catalyze the hydrolysis of benzyl glycidyl ether [18]. Similar results were observed in a study of ligninolytic enzyme production by marine fungi *Aspergillus sclerotiorum* CBMAI 849, *Cladosporium cladosporioides* CBMAI 857, and *Mucor racemosus* CBMAI 847 [19]. Other studies have shown that marine bacteria and fungi cultured in the laboratory have specific requirement for salts, especially sodium, potassium, magnesium, and chloride ions [18,20–23].

Certainly, marine enzymes have a great potential for use in biocatalytic reactions and production of bioactive compounds, due to the peculiar characteristics of the marine environment. For example, they could be used to transform organic compounds in various industrial processes and produce bioactive natural products for healthcare applications.

25.3 Biotransformation of Natural Products by Marine Biocatalysts

The microbial transformation of natural products is an efficient way of converting them to more useful substances, as this approach allows the ready functionalization of inactive carbon atoms. Metabolites of terrestrial origin have been widely used in biotransformations, but a smaller number of marine secondary metabolites have been recorded [17].

The use of marine organisms as biocatalysts in the synthesis of bioactive natural products has rarely been reported. The symbiotic marine bacteria, *Bacillus* sp. NC5, *Bacillus* sp. NK8, and *Bacillus* sp. NK7, isolated from the Red Sea sponge *Negombata magnifica*, transformed cembranoid into six hydroxylated metabolites (Scheme 25.1). Cembranoid and its products of bioconversion showed anti-invasive activity against the human highly metastatic prostate PC-3M cancer cell line at 10–50 nM doses in Matrigel assay [24].

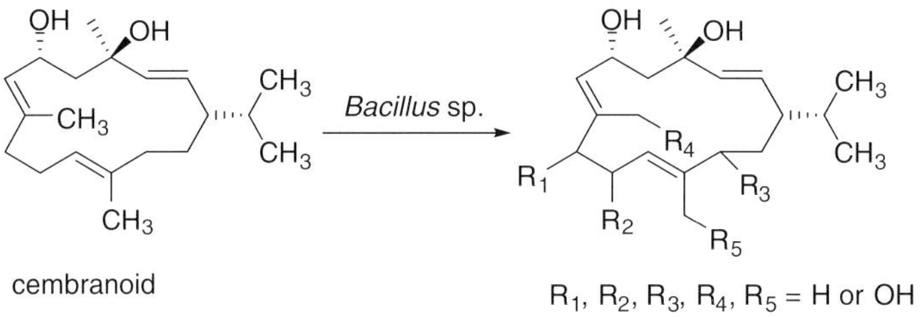

Scheme 25.1 Biotransformation of cembranoid by marine bacteria.

Biotransformation of many organic compounds is hindered or impeded by their low solubility in aqueous media. Organic solvents are generally toxic to microorganisms. However, organic solvent-tolerant bacteria are a group of microorganisms with novel mechanisms of tolerance to organic solvents. Citrinin, a quinone methide, is poorly water soluble but highly soluble in organic solvents. Devi *et al.* [25] described the biotransformation of toxic citrinin to nontoxic decarboxycitrinin by an organic solvent-tolerant marine bacterium, *Moraxella* sp. MB1 (Scheme 25.2). This transformation was catalyzed by the decarboxylase enzyme produced from *Moraxella* sp. MB1 and it was monitored by thin layer chromatography and spectrophotometry. In this study, two experiments were conducted, one in which citrinin was added to a flask culture of *Moraxella* sp. MB1, and the control in which the bacterial culture was not added. There was no transformation of citrinin in the control flasks. Both citrinin and decarboxycitrinin were reported to show antibiotic activity against clinical bacterial pathogens and some multidrug-resistant bacteria. Initially, citrinin was isolated from the fungus *Penicillium citrinum*, but in this study, citrinin was isolated as a secondary metabolite from a marine strain of *Penicillium chrysogenum* [25].

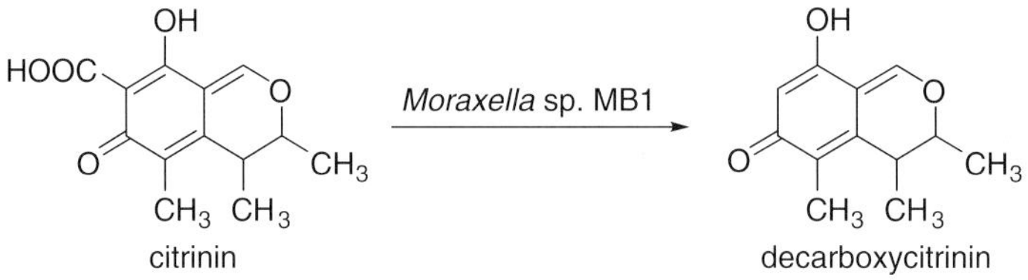

Scheme 25.2 Biotransformation of citrinin by marine bacterium.

Biotransformations of terpenoids have been performed to test the biological activity of the compounds produced. The biotransformation of pacifenol (a polyhalogenated sesquiterpene) to a hydroxylated derivative was catalyzed by *Penicillium brevicompactum*, a facultative halotolerant fungus isolated from the marine sponge *Cliona* sp. The incubation of pacifenol with *P. brevicompactum* for 6 days afforded a 7.3% yield of this product (Scheme 25.3). In order to confirm the hydroxylation reaction and discard a possible nucleophilic substitution reaction, a control reaction was performed in the absence of the growing fungus, in which the transformation did not occur [17]. The biological activity of the hydroxyl derivative was assessed and it exhibited a moderate antibacterial activity against *Staphylococcus enteritidis*. The pacifenol was obtained from the red marine alga *Laurencia claviformis* [26].

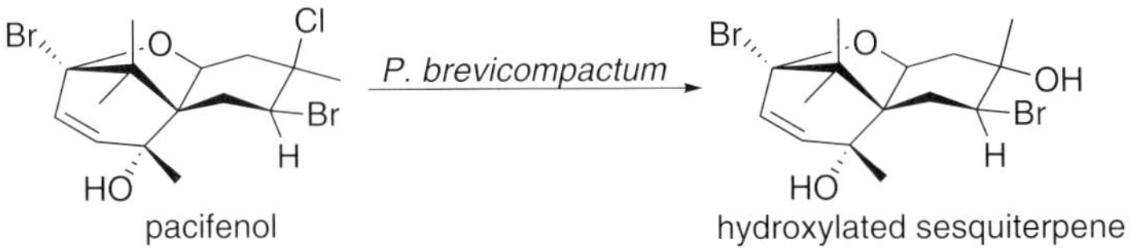

Scheme 25.3 Biotransformation of pacifenol by marine fungus.

Koshimura *et al.* [27] investigated the biotransformation of bromosesquiterpenes with two marine fungi, *Rhinocladiella atrovirens* NRBC 32362 and *Rhinocladiella* sp. K-001, isolated from the marine brown alga *Stypopodium zonale* (Scheme 25.4). Aplysistatin was converted into three compounds, 5α-hydroxyaplysistatin, 5α-hydroxyisoaplysistatin, and 9β-hydroxyaplysistatin, by *R. atrovirens* NRBC 32362, while aplysistatin, palisadin A, and 12-hydroxypalisadin B were biotransformed by the fungus *Rhinocladiella* sp. K-001. Aplysistatin yielded a simple metabolite identified as 3,4-dihydroaplysistatin. Palisadin A gave the corresponding 9,10-dehydrobromopalisadin A. Finally, 12-hydroxypalisadin B yielded palisadin A and 9,10-dehydrobromopalisadin A. The bromosesquiterpene aplysistatin showed antileukemic activity and was biotransformed to novel bioactive compounds [27].

Leutou *et al.* [28] screened various marine microorganisms to determine their ability to transform geraniol. Only the marine fungus *Hypocrea* sp. was capable of catalyzing the conversion of geraniol into its oxidized derivative, 1,7-dihydroxy-3,7-dimethyl-(*E*)-oct-2-ene (Scheme 25.5) [28].

Feng *et al.* [29] described a biological oxidation of the bioactive dihydroisocoumarin, (−)-mellein, to its oxidized metabolite, (3*R*,4*S*)-4-hydroxymellein, by marine bacterium *Stappia* sp. (Scheme 25.6). The dihydroisocoumarin was isolated from the marine fungus *Cladosporium* sp. These compounds are toxic metabolites that display biological activities [29].

Recently, the marine fungus *Aspergillus* sp., isolated from soft coral, *Sarcophyton tortuosum*, was induced by the monoterpene, α-pinene, to produce the mycotoxin penicillic acid (Figure 25.1). Initially, the fungus produced penicillic acid with a yield of 5.5 mg/l in glucose–peptone–yeast (GPY) extract medium and seawater at

aplysistatin → *R. atrovirens* → 5α-hydroxyaplysistatin + 5α-hydroxyisoaplysistatin + 9β-hydroxyaplysistatin

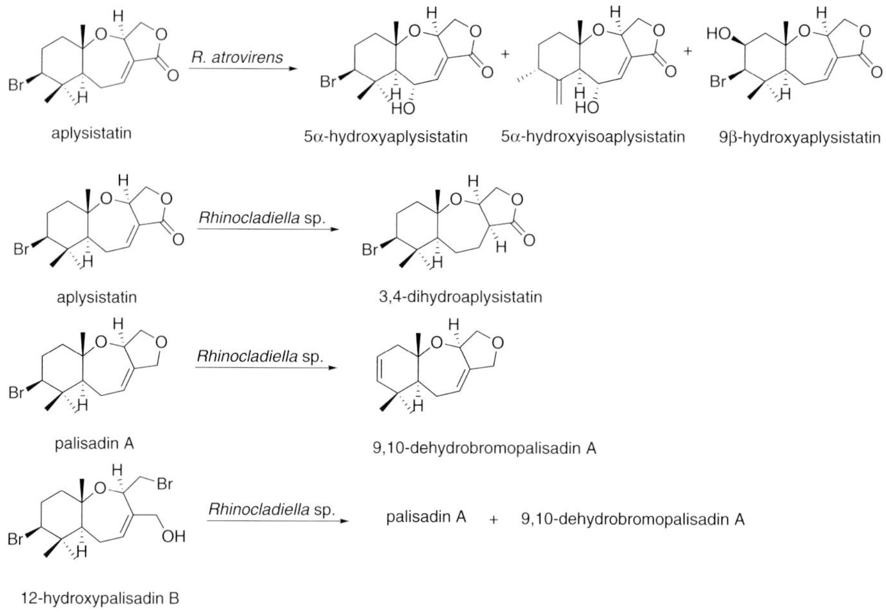

aplysistatin → *Rhinocladiella* sp. → 3,4-dihydroaplysistatin

palisadin A → *Rhinocladiella* sp. → 9,10-dehydrobromopalisadin A

12-hydroxypalisadin B → *Rhinocladiella* sp. → palisadin A + 9,10-dehydrobromopalisadin A

Scheme 25.4 Biotransformation of bromosesquiterpenes by marine fungi.

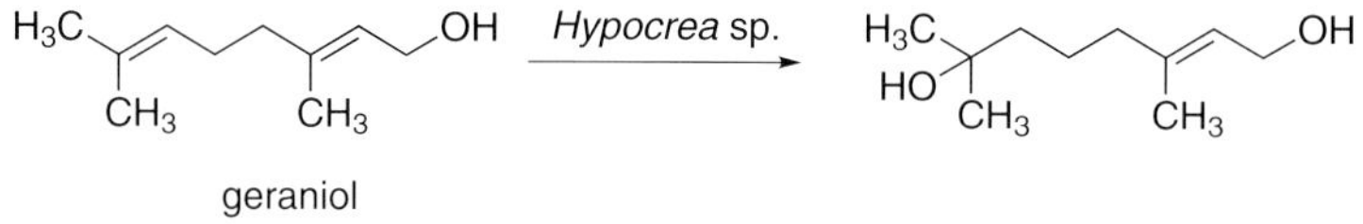

geraniol → *Hypocrea* sp.

Scheme 25.5 Biotransformation of geraniol by marine fungus.

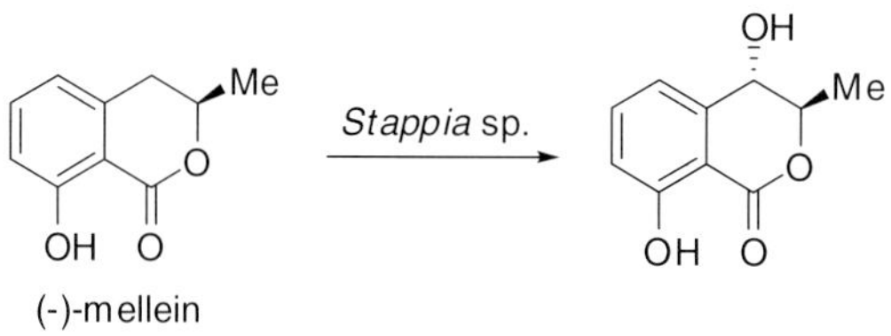

(-)-mellein → *Stappia* sp.

Scheme 25.6 Biotransformation of (−)-mellein by marine bacterium.

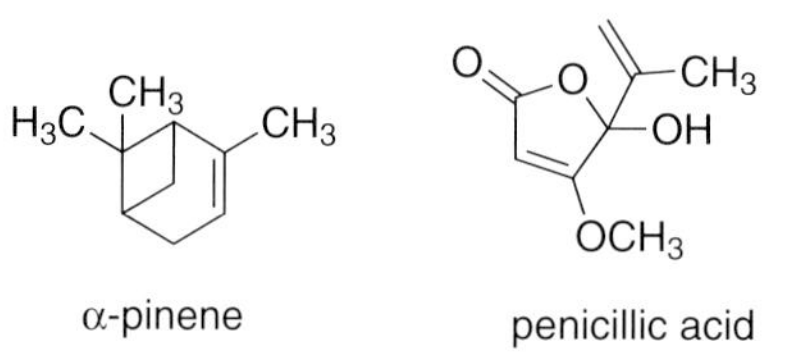

α-pinene

penicillic acid

Figure 25.1 Chemical structures of α-pinene and penicillic acid.

pH 7.5. However, when the GPY medium was supplied with α-pinene at a dose of 200 mg/l, the production of penicillic acid increased dramatically from 5.5 to 29.15 mg/l. The main biotransformed products of α-pinene were characterized as oxygenated monoterpenoids by gas chromatography–mass spectrometry (GC–MS) analysis. This finding suggested that the monoterpene could be acting as an elicitor, changing the oxidase activities, and modulating microbial metabolic pathways for production of penicillic acid [30].

As shown by above examples, marine enzymes have great potential for biotransformation of natural products to produce new molecules. However, there are few published studies of such biotransformations by marine enzymes.

25.4 Biodegradation of Organic Compounds by Marine Biocatalysts

Microbial degradation of chemical compounds in the environment is an important route for the removal of pollutants. A variety of microorganisms possess naturally a sufficiently wide catabolic diversity to degrade, transform, or accumulate a huge range of compounds, including hydrocarbons, polychlorinated biphenyls, polyaromatic hydrocarbons, pharmaceutical substances, and metals.

Despite the enormous concern over the persistence of dichlorodiphenyltrichloroethane (DDT) and other cyclodienes in the environment, very little is known about the biodegradation of these pesticides in the oceans. Patil *et al.* [31] described the first metabolic transformation of DDT, aldrin, and dieldrin by marine microorganisms using radiolabeled insecticides. In their study, 100 microbial cultures of marine isolates from Hawaii, Houston, and Texas were screened to investigate the role of these microorganisms in degrading radiolabeled DDT. Out of the 100 cultures, 35 appeared to be active in degrading DDT to 1,1-dichloro-2,2-bis-(4-chlorophenyl)ethane (DDD), as the predominant metabolite, and 1,1-bis(4-chlorophenyl)ethane (DDNS), 2,2-bis(*p*-chlorophenyl)ethanol (DDOH), and bis(*p*-chlorophenyl)acetic acid (DDA) as minor metabolites (Scheme 25.7) [31].

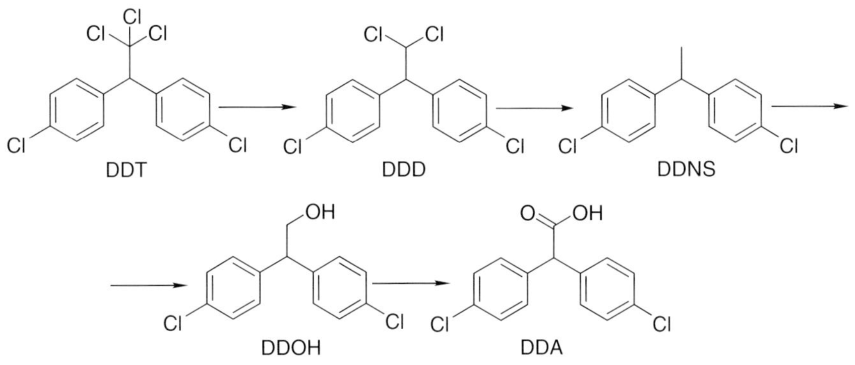

Scheme 25.7 Metabolic pathway of DDT under oceanic conditions.

The radioactive cyclodiene insecticides (dieldrin, aldrin, endrin) were degraded by marine microbial cultures after incubation for 1 month. Photodieldrin was the main metabolite derived from dieldrin and a small amount of diol and other unidentified metabolites were found in some cases. Dieldrin and *trans*-aldrindiol are the metabolic products of aldrin (Scheme 25.8) [31].

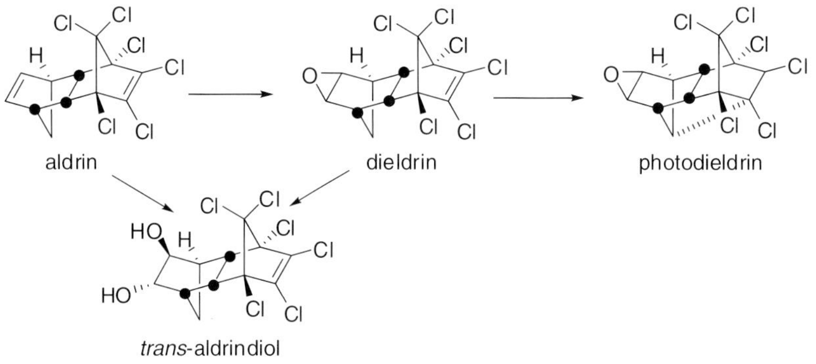

aldrin
dieldrin
photodieldrin
trans-aldrindiol

Scheme 25.8 Metabolic pathway of aldrin and dieldrin under oceanic conditions.

The biodegradation of DDD by marine fungi isolated from the marine sponges *Geodia corticostylifera* and *Chelonaplysylla erecta* has also been investigated. The marine fungi *A. sydowii* Ce15, *A. sydowii* Ce19, *A. sydowii* Gc12, *Bionectria* sp. Ce5, *P. miczynskii* Gc5, *P. raistrickii* Ce16, and *Trichoderma* sp. Gc1 were tested for their ability to grow at a high concentration of DDD pesticide in solid and liquid culture media. Total degradation of DDD was attained in liquid culture medium by *Trichoderma* sp. Gc1, initially cultured for 5 days and then supplemented with DDD in the presence of hydrogen peroxide. However, quantitative analysis showed that DDD was accumulated in the mycelium and the extent of biodegradation reached a maximum of 58% after 14 days [32].

Marine organisms contain arsenic at much higher levels than terrestrials. It has been shown that most of the arsenic in marine organisms is in water-soluble organic form and the main water-soluble organoarsenic compound is arsenobetaine. Arsenobetaine is widely distributed in many marine animals (shark, lobster, shrimp, sole, dab, crab, flat fish, sea cucumber, octopus, shells, and flounder) and is considered to be the final metabolic product of the distribution of arsenic in marine animals. Trimethylarsine oxide was isolated as one of the metabolites of arsenobetaine produced by marine microorganisms from bottom sediments collected from coastal waters at Yoshimi, Shimonoseki, Japan (Figure 25.2) [33,34].

Hanaoka *et al.* [34] described the pathway of arsenobetaine biodegradation after the death of the animals containing this compound; they assumed that arsenobetaine-decomposing microorganisms are common in the marine environment. In their study, the degradation of arsenobetaine to trimethylarsine oxide and/or dimethylarsinic acid by microorganisms associated with marine macroalgae (*Monostroma nitidum* and *Hizikia fusiforme*) was described. The microorganisms

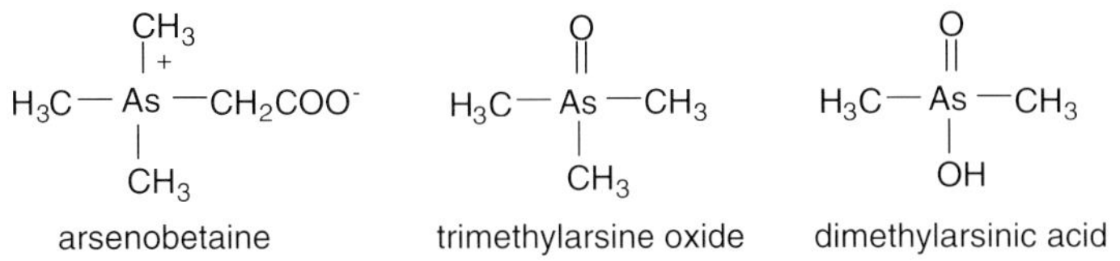

$$H_3C-\overset{CH_3}{\overset{|+}{As}}-CH_2COO^- \quad (\text{with } CH_3 \text{ below As})$$

arsenobetaine

$$H_3C-\overset{O}{\overset{\|}{As}}-CH_3 \quad (\text{with } CH_3 \text{ below As})$$

trimethylarsine oxide

$$H_3C-\overset{O}{\overset{\|}{As}}-CH_3 \quad (\text{with } OH \text{ below As})$$

dimethylarsinic acid

Figure 25.2 Chemical structures of arsenobetaine and its biodegradation products.

isolated from the alga *M. nitidum* catalyzed the conversion of arsenobetaine to trimethylarsine oxide, which was transformed into dimethylarsinic acid. These microorganisms probably used the carboxymethyl moiety of arsenobetaine as a carbon source. Later, as the carbon from arsenobetaine was exhausted, a methyl group in trimethylarsine oxide was cleaved and utilized by the microorganisms generating dimethylarsinic acid. However, the microorganisms associated with the alga *H. fusiforme* converted the arsenobetaine directly to dimethylarsinic acid (Figure 25.2). Subsequently, Hanaoka *et al.* investigated the bioconversion of arsenobetaine to trimethylarsine oxide and dimethylarsinic acid by intestinal bacteria of mollusk *Liolophura japonica* chitons [35]. This process was comparable with the degradation by associated microorganisms from marine macroalgae under aerobic conditions [34,35].

In other study, a similar result was obtained in the biodegradation of arsenobetaine to trimethylarsine oxide by microorganisms occurring in the gill of the clam *Meretrix lusoria* during arsenic circulation in marine ecosystems [36].

Bacterial communities at the surface of sediments are known to be important agents in the transformation of sedimentary organic compounds, including hydrocarbons [37].

Studies performed by Zengler *et al.* [38] demonstrated that anaerobic communities of bacteria and archaea converted the long-chain *n*-alkane, hexadecane (n-$C_{16}H_{34}$), to the simplest hydrocarbon, methane. These microorganisms are assumed to be acetogenic (syntrophic) bacteria that decompose hexadecane to acetate and H_2, a group of archaea that cleave the acetate into CH_4 and CO_2, and another group of archaea converting CO_2 and H_2 into CH_4 (Equations 1–3, Figure 25.3) [38].

The biodegradation of isoprenoid hydrocarbons is of particular interest, since these compounds are often used as "inert biomarkers," especially in studies of oil degradation, due to their relatively long-term preservation, compared to *n*-alkanes, during early diagenesis. The C_{19} acyclic isoprenoid hydrocarbon pristane (2,6,10,14-tetramethylpentadecane) is widely distributed in the biosphere, where it is either directly introduced during oil spills or produced during diagenesis from various

$$4C_{16}H_{34} + 64H_2O \rightarrow 32CH_3COO^- + 32H^+ + 68H_2 \quad (1)$$

$$32CH_3COO^- + 32H^+ \rightarrow 32CH_4 + 32CO_2 \quad (2)$$

$$68H_2 + 17CO_2 \rightarrow 17CH_4 + 34H_2O \quad (3)$$

Figure 25.3 Conversion of hexadecane to CH_4 by anaerobic microorganisms.

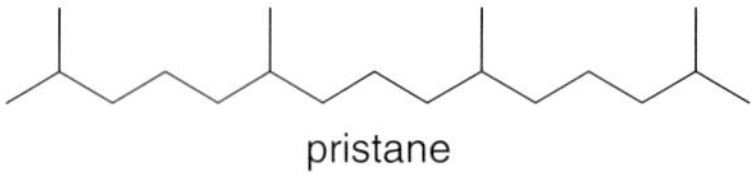

pristane

Figure 25.4 Chemical structure of pristane.

precursors (Figure 25.4). The biodegradation of pristane in anoxic marine sediments by the mixed microbial community (bacterial/archaeal communities) was carried out by incubation of the substrate with the sediment slurry. In this study, the anaerobic conditions were varied, to limit or promote nitrate reduction, in order to examine the impact of distinct microbial metabolisms on pristane degradation. Pristane was significantly degraded only under conditions limiting nitrate reduction, demonstrating that the destiny of sedimentary hydrocarbons in anoxic sediments depends strongly on the environmental conditions and on the microbial populations in place. Although the pathways for the degradation of acyclic isoprenoid hydrocarbons in the absence of oxygen remain unknown, it is clear from this study that such compounds cannot be employed as recalcitrant biomarkers in marine anoxic sediments. The biodegradation of pristane was accompanied by abundant production of methane [37]. The biotransformation/biodegradation of polycyclic aromatic hydrocarbons (PAHs) has been performed by the cells of marine microorganisms. PAHs, such as phenanthrene and anthracene, are mutagenic or carcinogenic environmental pollutants, derived from coal and petroleum. Chun *et al.* [39] described the biotransformation of phenanthrene and 1-methoxynaphthalene with recombinant *Streptomyces lividans* cells expressing a marine bacterial phenanthrene dioxygenase gene cluster. In this study, functional expression of the aromatic compound dioxygenase genes in this actinomycete was used to hydroxylate phenanthrene and 1-methoxynaphthalene. The recombinant *S. lividans* cells converted phenanthrene to *cis*-3,4-dihydroxy-3,4-dihydrophenanthrene and 1-methoxynaphthalene to 8-methoxy-1,2-dihydro-1,2-naphthalenediol (Scheme 25.9). Therefore, such recombinant *Streptomyces* strains may also be promising as a host for the production of *cis*-diols, which are known to be very useful as building blocks for asymmetric synthesis [39].

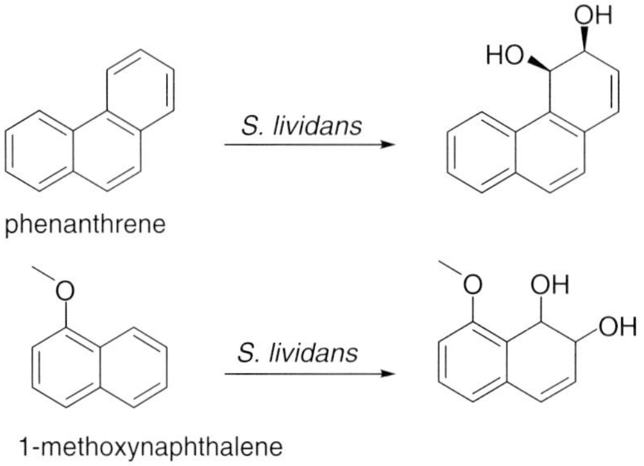

phenanthrene

S. lividans

HO OH

1-methoxynaphthalene

S. lividans

O OH OH

Scheme 25.9 Biodegradation of polycyclic aromatic hydrocarbons by recombinant *S. lividans*.

Nitroaromatic compounds are xenobiotics that have found multiple applications in the synthesis of foams, pharmaceuticals, pesticides, and explosives. These compounds are toxic and recalcitrant as they are degraded relatively slowly in the environment. 2,4,6-Trinitrotoluene (TNT) is the most widely used nitroaromatic compound and is listed as a priority pollutant, urgently recommended for removal from contaminated sites, since it is toxic to living forms [40]. In human beings, chronic exposure to TNT leads to harmful effects such as anemia, abnormal liver function, skin irritation, and development of cataract [41]. There are several reports on the biological transformation of TNT by microbes isolated from explosive-contaminated environments, but studies on the biotransformation of nitroaromatic compounds by microbes isolated from the marine environment are rare [40,41].

Marine yeast *Yarrowia lipolytica* NCIM 3589 was able to degrade TNT. Ring reduction resulted in the formation of the hydride–Meisenheimer complex (H^--TNT) as a transiently accumulating metabolite that was denitrated to 2,4-dinitrotoluene (2,4-DNT) (Scheme 25.10). Additionally, the TNT was reduced to aminodinitrotoluene derivatives. The addition of glucose promoted an increase in biosynthesis of nitroreductases that catalyzed the production of 2,4-DNT and amino derivatives [41].

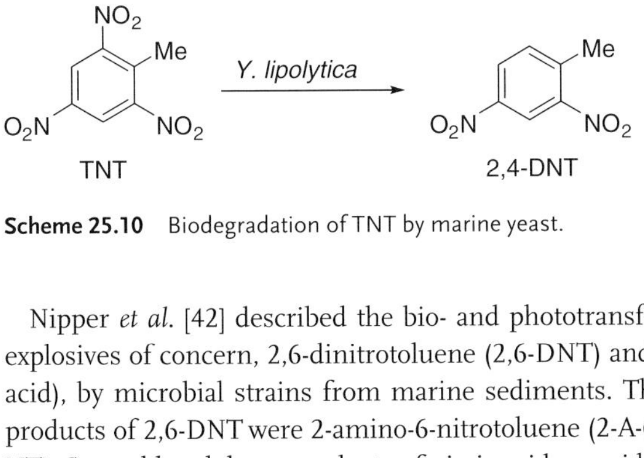

Scheme 25.10 Biodegradation of TNT by marine yeast.

Nipper *et al.* [42] described the bio- and phototransformation of munitions and explosives of concern, 2,6-dinitrotoluene (2,6-DNT) and 2,4,6-trinitrophenol (picric acid), by microbial strains from marine sediments. The major biotransformation products of 2,6-DNT were 2-amino-6-nitrotoluene (2-A-6-NT) and 2-nitrotoluene (2-NT). Several breakdown products of picric acid were identified by GC–MS, such as dinitrophenol, diaminophenol, aminonitrophenol, and nitrodiaminophenol (Scheme 25.11) [42].

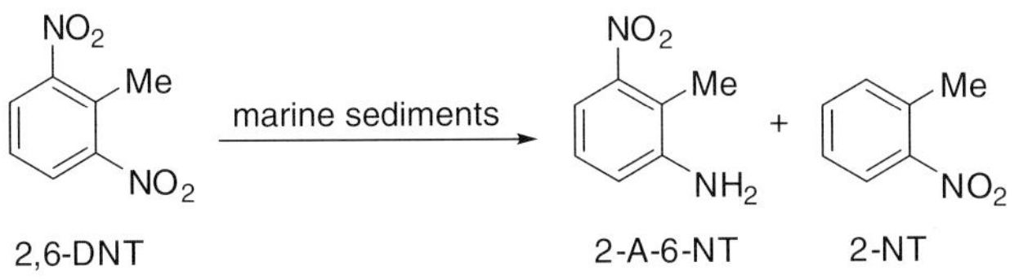

Scheme 25.11 Biodegradation of 2,6-DNT by marine sediments.

Cyclic nitramines, for example, hexahydro-1,3,5-trinitro-1,3,5-triazine (RDX) and octahydro-1,3,5,7-tetranitro-1,3,5,7-tetrazocine (HMX), are powerful and widely used explosives that have resulted in severe contamination of soil and groundwater in various terrestrial and aquatic environments [43,44]. In a report on the phylogeny of cyclic nitramine-degrading psychrophilic (cold-loving) bacteria found in marine sediment, RDX was shown to be degraded to nitroso derivatives, hexahydro-1-nitroso-3,5-dinitro-1,3,5-triazine (MNX) as main product, hexahydro-1-nitro-3,5-dinitroso-1,3,5-triazine (DNX), and the hexahydro-1,3,5-trinitroso-1,3,5-triazine (TNX) (Scheme 25.12). Methylenedinitramine (MEDINA) and 4-nitro-2,4-diazabutanal (NDAB) were detected, suggesting ring cleavage following the denitration of either RDX and/or its initially reduced product (MNX) [43].

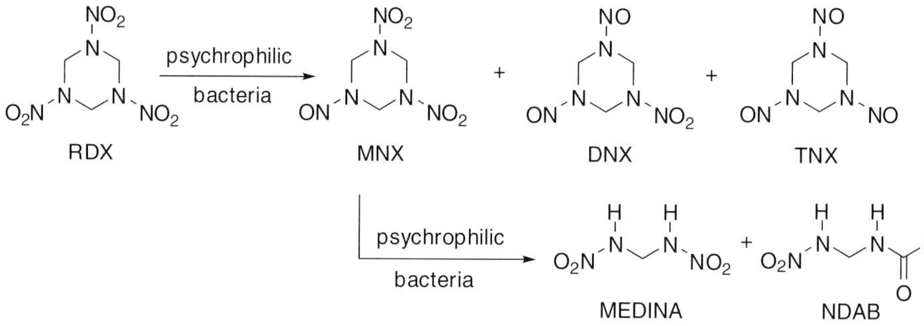

Scheme 25.12 Biodegradation of RDX by psychrophilic bacteria.

Psychrophilic bacteria from marine sediment also removed the HMX, yielding the octahydro-1,3,5-trinitro-7-nitroso-1,3,5,7-tetrazocine (NO-HMX) (Scheme 25.13). These results demonstrate the ability of psychrophilic bacteria to degrade toxic cyclic nitramines present in the marine sediment [43].

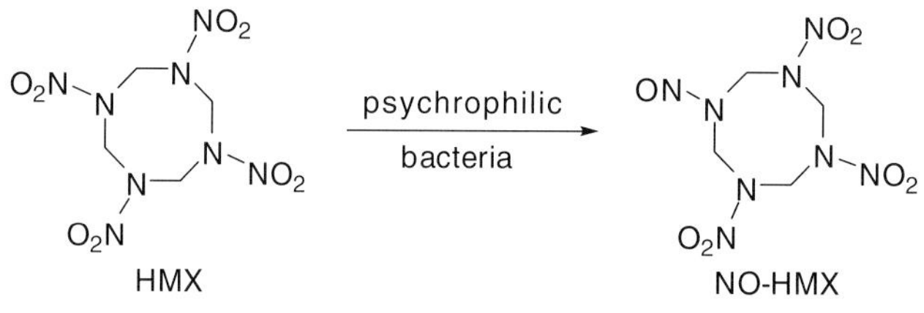

Scheme 25.13 Biodegradation of HMX by psychrophilic bacteria.

Bhatt *et al.* [44] also reported the biodegradation of RDX by novel fungi isolated from unexploded ordnance-contaminated marine sediment. From sediment collected from a coastal area of O'ahu island, Hawaii, four novel RDX-degrading marine aerobic fungi were isolated. These marine fungal isolates, belonging to the genera *Rhodotorula, Bullera, Acremonium,* and *Penicillium,* respectively, degraded 40, 35, 75, and 45% of the RDX in 18 days. *Acremonium* was selected to determine the biotransformation pathway of RDX. When RDX was incubated with resting cells of *Acremonium,* MEDINA, nitrous oxide (N_2O), formaldehyde (HCHO), MNX, DNX, and TNX were detected. Under the same conditions, MNX was metabolized to N_2O and HCHO, together with trace amounts of DNX and TNX (Scheme 25.14) [44].

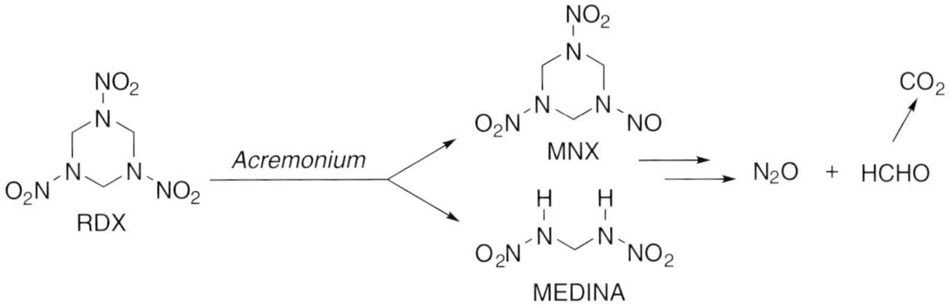

Scheme 25.14 Hypothetical pathways of aerobic RDX degradation by marine fungus.

Bisphenol A is a monomer in polycarbonate plastic and epoxy resins, which exhibit very good physical and chemical properties such as excellent transparency, high mechanical strength, and good thermal stability. However, free bisphenol A is an endocrine disruptor, toxic to several organisms. The biodegradation of polycarbonate plastic was carried out *in vitro* with a mixed marine microbial consortium isolated from the Bay of Bengal (India), over 1 year in controlled conditions. This degradation yielded products such as bisphenol A, by hydrolysis of the carbonate bond, and other compounds such as 4-hydroxyacetophenone, 4-hydroxybenzaldehyde, and 4-hydroxybenzoic acid, by biological oxidations (Scheme 25.15). Both the oxidative and the hydrolytic enzymes seem to play essential parts in this degradation [45].

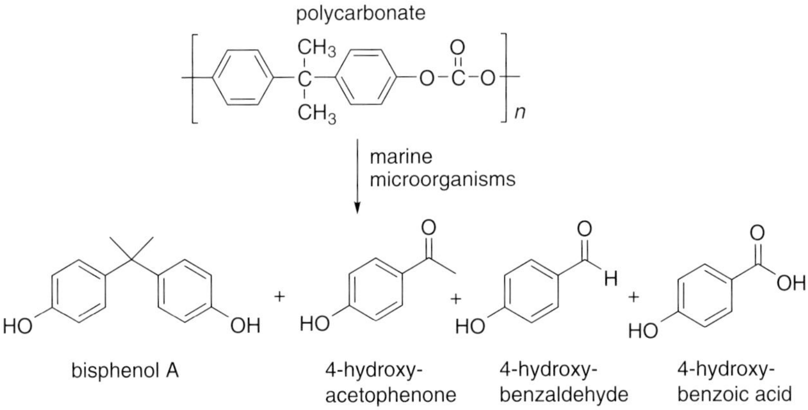

Scheme 25.15 Biodegradation of bisphenol A by marine microbial.

The contamination of air, soil, sediment, groundwater, and surface water by toxic organic compounds has become one of the biggest problems facing the industrialized world. In recent decades, a wide variety of toxic and hazardous substances have been introduced into the environment, in particular, those resulting from the dumping of industrial wastes from accidents involving spills of oil and its derivatives and the highly toxic chemicals used in wars in various parts of the world. The increasing problem of environmental pollution has aroused great concern, leading to a greater awareness of the damage caused by the indiscriminate discharge of restricted pollutants and the need to remediate the polluted sites. The high potential of decomposing microorganisms for use as agents to destroy diverse substances on site, coupled with advanced in biotechnology, promises to be one of the most efficient approaches to reduce the adverse effects of contaminants on the environment [46]. Here, the focus was on the degradation of the contaminants by marine organisms to less toxic compounds, or the removal of pollutant organic compounds by marine biocatalysts.

25.5 Reduction of Carbonyl Groups (Ketones and Keto Esters) by Marine Biocatalysts

The considerable environmental impact of many industrial organic syntheses can be substantially lessened by the use of enzymatic catalysis on biotransformations that require mild conditions and aqueous media. Isolated enzymes and whole-cell biocatalyst systems are increasingly being used to assist in synthetic routes to produce molecules of industrial interest [47]. The microbial reduction of carbonyl compounds is a convenient method to obtain optically pure alcohols. The baker's yeast, *Saccharomyces cerevisiae*, has often been used for the reduction of keto esters to produce enantiopure hydroxy esters [48–50]. The production of enzymes by microorganisms is extremely important in biotechnology, and the advantage of the use of whole living cells is that no cofactors need to be added [51,52].

Selective screenings were carried out for microorganisms capable of catalyzing the ketone reduction and alcohol oxidation. The chosen reactions were the reduction of cyclohexanone and the oxidation of cyclohexanol, both of which depend on coenzymes. The aim of the selection process was to find good biocatalysts for one of the reactions that did not show activity in the reverse reaction. Microorganisms that produced secondary metabolites such as cyclohexanol or cyclohexanone were rejected, for example, those that produce ε-caprolactone via Baeyer–Villiger monooxygenase (MO). The enzymes responsible for these biotransformations are alcohol dehydrogenases (E.C. 1.1.1.245) (Scheme 25.16). The marine fungi selected in this screening catalyzed the reduction of cyclohexanone to cyclohexanol (10 mM, 72 h, 28 °C, 250 rpm) in differing yields: *Ceriosporopsis tubulifera* ATCC 64283 (27%), *Zopfiella latipes* ATCC 26183 (96%), *Buergenerula spartinae* ATCC 62545 (88%), and *Dactylospora haliotrepha* ATCC 66950 (78%). These microorganisms were obtained from American Type Culture Collection [53].

Asymmetric reduction of 2-chloro-1-phenylethanone by marine fungi (*P. miczynskii* Gc5, *Trichoderma* sp. Gc1, *A. sydowii* Gc12, *A. sydowii* Ce19, *A. sydowii* Ce15, *Bionectria* sp. Ce5, and *P. miczynskii* Ce16) has been reported. (*S*)-(−)-2-

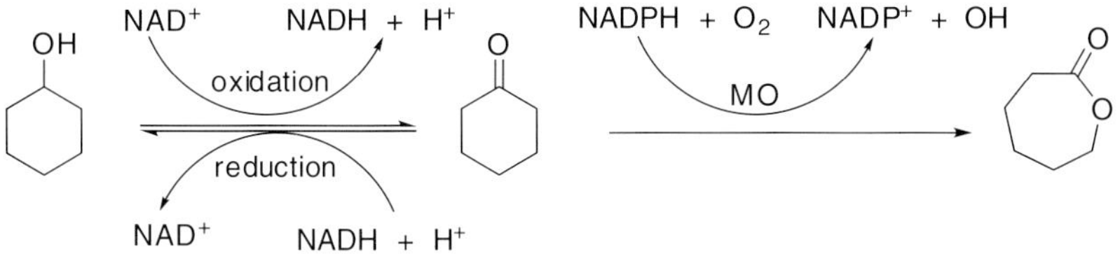

Scheme 25.16 Competitive oxidoreduction reaction by marine fungi.

chloro-1-phenylethanol was produced by *P. miczynskii* Gc5 with 50% ee and an isolated yield of 60% (Scheme 25.17). The ability of marine fungi to catalyze the reduction was directly dependent on growth in an artificial seawater-based medium containing a high concentration of Cl^- (1.2 M). When the marine fungi were grown in the absence of artificial seawater, the reduction of 2-chloro-1-phenylethanone did not occur. The biocatalytic reduction of 2-chloro-1-phenylethanone was more efficient at neutral pH than at acidic pH and in the absence of glucose as cosubstrate [23].

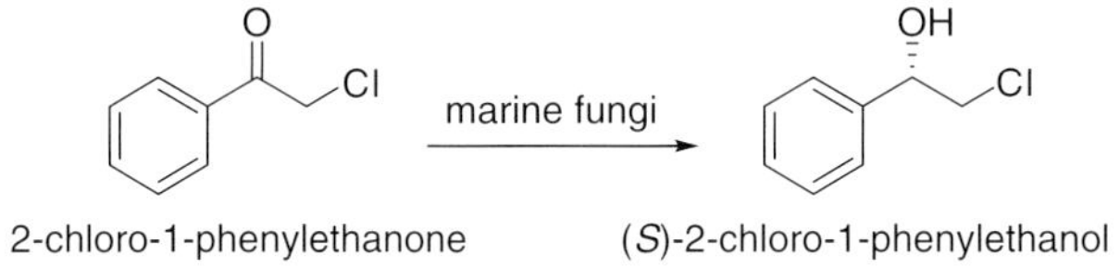

Scheme 25.17 Bioreduction of α-chloroketone by marine fungi.

Whole cells of marine fungi (*A. sydowii* Ce15, *A. sydowii* Ce19, *A. sclerotiorum* CBMAI 849, *Bionectria* sp. Ce5, *Beauveria felina* CBMAI 738, *C. cladosporioides* CBMAI 857, *M. racemosus* CBMAI 847, *P. citrinum* CBMAI 1186, and *P. miczynskii* Gc5) promoted the asymmetric reduction of 1-(4-methoxyphenyl)ethanone. *A. sydowii* Ce15 and *Bionectria* sp. Ce5 produced the enantiopure (*R*)-alcohol (>99% ee) in accordance with the *anti*-Prelog's rule, and *B. felina* CBMAI 738 (>99% ee) and *P. citrinum* CBMAI 1186 (69% ee) yielded the corresponding (*S*)-alcohol in accordance with Prelog's rule (Scheme 25.18). Stereoselective reduction of ketones by whole cells is important for the production and discovery of new reductases from marine fungi [54].

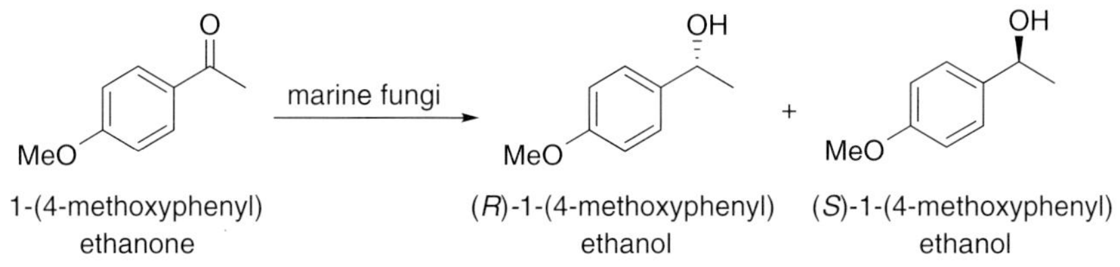

Scheme 25.18 Bioreduction of *p*-methoxyketone by marine fungi.

Marine fungi *B. felina* CBMAI 738, *P. citrinum* CBMAI 1186, *P. miczynskii* Gc5, *Penicillium oxalicum* CBMAI 1185, and *Trichoderma* sp. Gc1 catalyzed the reduction of iodoacetophenones to their corresponding iodophenylethanols. The production of enantiopure iodophenylethanols (up to >98% ee) depended on the strain of the microorganism used and the type of substrate. All marine fungi produced exclusively the (*S*)-*ortho*-iodophenylethanol and (*S*)-*meta*-iodophenylethanol, in accordance with the Prelog's rule, and (*R*)-*para*-iodophenylethanol, by the *anti*-Prelog's rule (Scheme 25.19) [55].

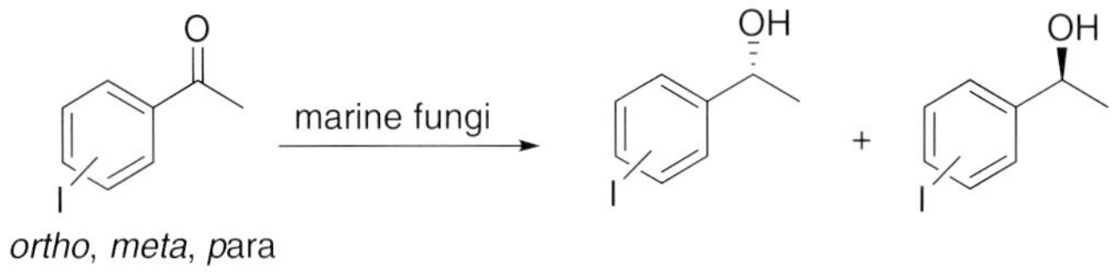

marine fungi

ortho, *meta*, para

Scheme 25.19 Bioreduction of iodoketones by marine fungi.

Red marine algae, *Bostrychia radicans* and *B. tenella*, with their associated marine bacteria (*Bacillus* spp.) catalyzed an efficient reduction of acetophenone derivatives. This algal/bacterial produced the (*S*)-2-phenylethanol derivatives with high enantiomeric excess (>99% ee) (Scheme 25.20) [56].

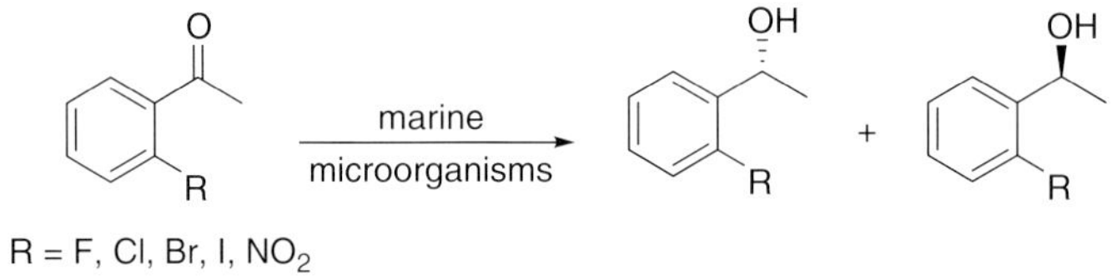

marine microorganisms

R = F, Cl, Br, I, NO_2

Scheme 25.20 Reduction of *ortho*-ketones by marine microorganisms.

In further studies, this group investigated the biocatalytic reduction of acetophenone derivatives by marine fungi isolated from the marine algae *B. radicans* and *Sargassum* sp. (Scheme 25.20). Using conventional and molecular biological approaches, fungi isolated from *B. radicans* were identified as *Botryosphaeria* sp. Br-09, *Eutypella* sp. Br-023, *Hydropisphaera* sp. Br-27, and *Xylaria* sp. Br-61, while the fungi isolated from *Sargassum* sp. were identified as *Pestalotiopsis* sp. SMA2-C, *Penicillium* sp. SMA2–8, and *Arthopyrenia* sp. SGPY-41. In these experiments, the reduction of iodoacetophenones by the marine fungus *Botryosphaeria* sp. Br-09, isolated from the red marine alga *B. radicans*, yielded alcohols in high optical purity (>99% ee) and excellent conversion (>98%) [57].

The marine fungus *A. sydowii* Ce19 was able to catalyze the reduction of 2-bromo-1-phenylethanone to (*R*)-2-bromo-1-phenylethanol, together with other enzymatic and spontaneous reaction products (Scheme 25.21). The product yields were 2-bromo-1-phenylethanol (56%), 2-chloro-1-phenylethanol (9%), 1-phenylethan-1,2-diol (26%), acetophenone (4%), and phenylethanol (5%). In these experiments, the marine fungus showed a potential for the biotransformation and biodegradation of bromoacetophenone [58].

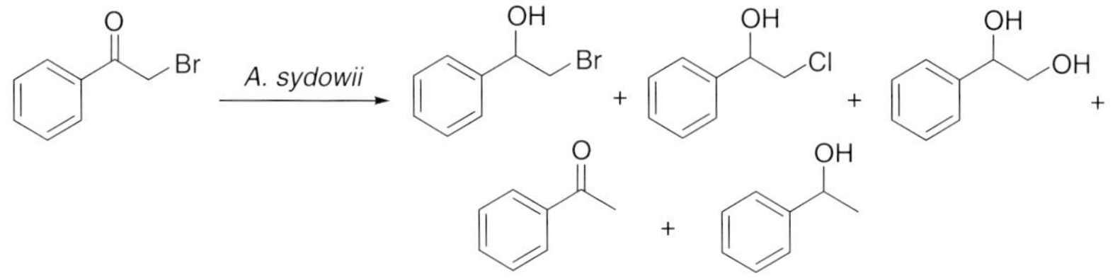

Br
A. sydowii
OH Br + OH Cl + OH OH +
O + OH

Scheme 25.21 Biotransformation of α-bromoacetophenone by marine fungi.

Marine microalgae *Chaetoceros gracilis*, *Chaetoceros* sp., *Nannochloropsis* sp., *Pavlova lutheri*, and *Chlorella* strains were used as biocatalysts for stereoselective reduction of α- and β-keto esters. *C. gracilis* reduced the ethyl benzoylformate to (*S*)-ethyl 2-hydroxy-2-phenylacetate with a conversion of 99% and enantioselectivity 23% ee, and the ethyl 2-oxoheptanoate to (*S*)-ethyl 2-hydroxyheptanoate (89% ee, c = 42%). The ethyl 3-methyl-2-oxobutanoate was reduced by *Nannochloropsis* sp. to (*R*)-ethyl 2-hydroxy-3-methylbutanoate with 98% ee and total conversion. The reduction of ethyl 2-methyl-3-oxobutanoate by microalgae (*C. gracilis, Chaetoceros* sp., *Nannochloropsis* sp., *P. lutheri*) gave the corresponding *anti*-hydroxy ester with conversion ratios of 25–68% (Scheme 25.22). In particular, the ethyl 2-methyl-3-oxobutanoate was reduced by *Nannochloropsis* sp. to the *anti*-hydroxy ester with excellent diastereo- (*syn*:*anti* – 1:99) and enantioselectivity (>99% ee) [59–61].

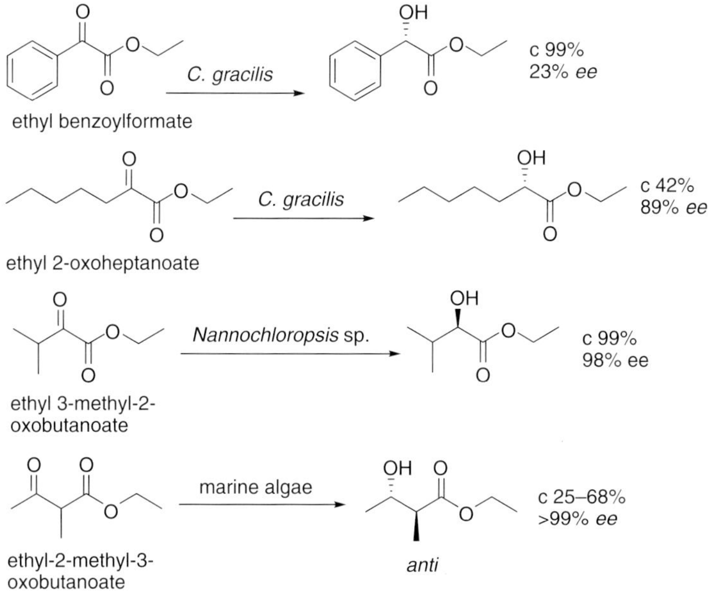

ethyl benzoylformate — *C. gracilis* → c 99%, 23% *ee*

ethyl 2-oxoheptanoate — *C. gracilis* → c 42%, 89% *ee*

ethyl 3-methyl-2-oxobutanoate — *Nannochloropsis* sp. → c 99%, 98% ee

ethyl-2-methyl-3-oxobutanoate — marine algae → *anti*, c 25–68%, >99% *ee*

Scheme 25.22 Bioreduction of α- and β-keto esters by marine algae.

Table 25.1 Effects of additives on the reduction of ethyl 3-methyl-2-oxobutanoate [61].

Additives	*C. gracilis*		*Nannochloropsis* sp.	
	Conversion (%)	ee (%)	Conversion (%)	ee (%)
No additive	66	18 (*S*)	99	98 (*R*)
Glucose	>99	51 (*R*)	84	10 (*R*)
DL-Lactic acid	46	99 (*R*)	64	39 (*R*)
Lithium D-lactate	0	—	0	—
L-Lactic acid	89	99 (*R*)	60	49 (*R*)

Furthermore, the effects of additives on the conversion and stereochemistry of the product ethyl 2-hydroxy-3-methylbutanoate were investigated with marine algae *C. gracilis* and *Nannochloropsis* sp. (Table 25.1, Scheme 25.22). The reduction of ethyl 3-methyl-2-oxobutanoate by *C. gracilis* in the presence of glucose gave the corresponding α-hydroxy ester with high conversion (>99%) and enantioselectivity to (*R*)-hydroxy-3-methylbutanoate (51% ee). In the algal reduction by *C. gracilis* of ethyl 3-methyl-2-oxobutanoate, the addition of DL-lactic acid decreased the conversion, while the enantioselectivity to (*R*)-hydroxy-3-methylbutanoate increased it (99% ee). In the presence of the D-lactate ion, the keto ester was not reduced, suggesting that the D-lactate ion is an inhibitor for (*R*)- and (*S*)-α-hydroxy ester-producing enzyme(s), while the L-lactate ion inhibits only the activity of the (*S*)-enzyme(s). A better result for reduction of ethyl 3-methyl-2-oxobutanoate by marine alga *Nannochloropsis* sp. occurred in the absence of additives (Table 25.1) [60,61].

The marine actinomycetes *Salinispora arenicola* and *Salinispora tropica* produce important bioactive compounds such as arenimycin, an antibiotic against methicillin-resistant *Staphylococcus aureus*, and salinosporamide A, a potent proteasome inhibitor. *Salinispora* strains were used as biocatalysts for stereoselective reduction of α-keto esters and aromatic α-keto amide. *S. arenicola* reduced α-keto esters possessing short alkyl chains (ethyl 2-oxopropanoate, ethyl 2-oxobutanoate, ethyl 2-oxopentanoate, and ethyl 3-methyl-2-oxobutanoate) to the corresponding alcohols, with high conversions and different optical purities: ethyl 2-hydroxypropanoate (c = >99%, 50% ee), ethyl 2-hydroxybutanoate (c = >99%, 77% ee), ethyl 2-hydroxypentanoate (c = >99%, 73% ee), and ethyl 2-hydroxy-3-methylbutanoate (c = 91%, 78% ee) (Scheme 25.23) [62].

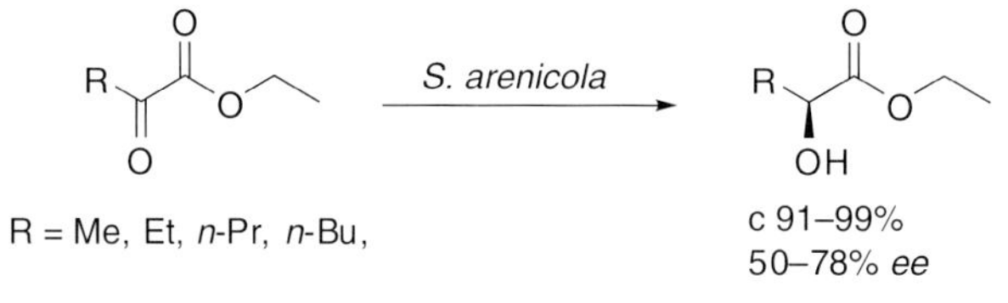

S. arenicola

R = Me, Et, *n*-Pr, *n*-Bu,

c 91–99%
50–78% *ee*

Scheme 25.23 Reduction of α-keto esters by marine actinomycete.

On the other hand, *S. tropica* achieved both high conversions and excellent stereoselectivities in the reduction of ethyl pyruvate, ethyl 2-oxobutanoate, and

2-chlorobenzoylformamide to the corresponding alcohols (Scheme 25.24). From these examples, it can be concluded that some are potentially useful biocatalysts for reduction of carbonyl compounds [62].

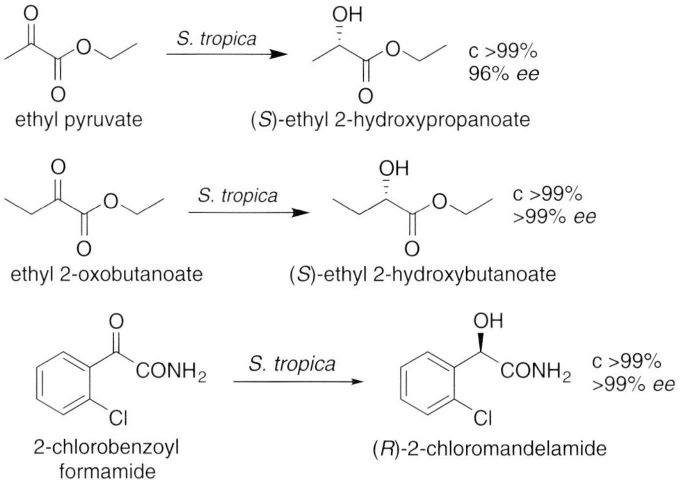

ethyl pyruvate → *S. tropica* → (*S*)-ethyl 2-hydroxypropanoate; c >99%, 96% *ee*

ethyl 2-oxobutanoate → *S. tropica* → (*S*)-ethyl 2-hydroxybutanoate; c >99%, >99% *ee*

2-chlorobenzoyl formamide → *S. tropica* → (*R*)-2-chloromandelamide; c >99%, >99% *ee*

Scheme 25.24 Bioreduction of carbonyl groups by actinomycete strains.

25.6
Hydrolysis of Epoxides by Marine Biocatalysts

Since enantioselective biocatalysts have been discovered in various marine environments, there has been a search for epoxide hydrolase (EH) activities by a combination of screening, conventional molecular engineering, and a genomic approach [63–67]. Epoxides and diols in pure stereochemical forms are of importance in the asymmetric synthesis of drugs and as general synthetic intermediates, so great efforts have been made in the study of catalysis by epoxide hydrolases. Adopting the so-called enantioconvergent process, two different enantioselective enzymes can be used together, each possessing a different specificity for attacking the α- or β-carbon of an asymmetric epoxide ring. In this method, one enzyme acts on one of these carbons while the complementary biocatalyst hits the other, resulting in theoretically 100% yield of a diol with specific stereochemistry [2,68].

As an illustration, enantioconvergent hydrolysis of racemic α-methyl styrene oxide was carried out to prepare enantiopure (*R*)-2-phenylpropane-1,2-diol by using two recombinant EHs, one from bacterium *Caulobacter crescentus* and the other from a marine fish *Mugil cephalus*. The recombinant *C. crescentus* EH primarily attacked the benzylic α-carbon of (*S*)-α-methyl styrene oxide, while the *M. cephalus* EH preferentially attacked the terminal β-carbon of (*R*)-styrene oxide. Thus, the main product was (*R*)-2-phenylpropane-1,2-diol, which was produced with 90% enantiomeric excess and yield as high as 94% from 50 mM racemic styrene oxides in a one-pot process (Scheme 25.25) [2,69,70].

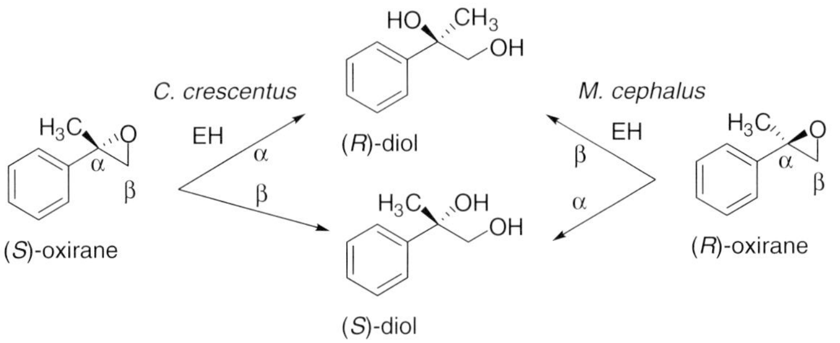

HO CH3 OH
C. crescentus EH α β
(S)-oxirane
(R)-diol
H3C OH OH
(S)-diol
M. cephalus EH β α
(R)-oxirane

Scheme 25.25 Enantioconvergent hydrolysis of styrene oxide by EH from marine organisms.

A cloned epoxide hydrolase from the bacterium *Novosphingobium aromaticivorans* achieved enantioselective hydrolysis of styrene oxide, glycidyl phenyl ether, epoxybutane, and epichlorohydrin (Scheme 25.26). The purified *N. aromaticivorans* enantioselective epoxide hydrolase (NEH) preferentially hydrolyzed the (*R*)-styrene oxide with enantiomeric excess of >99 and 11.7% yield by configuration retention [71]. The hydrolyzing rates of the purified NEH toward epoxide substrates were not affected by concentrations as high as 100 mM of racemic styrene oxide.

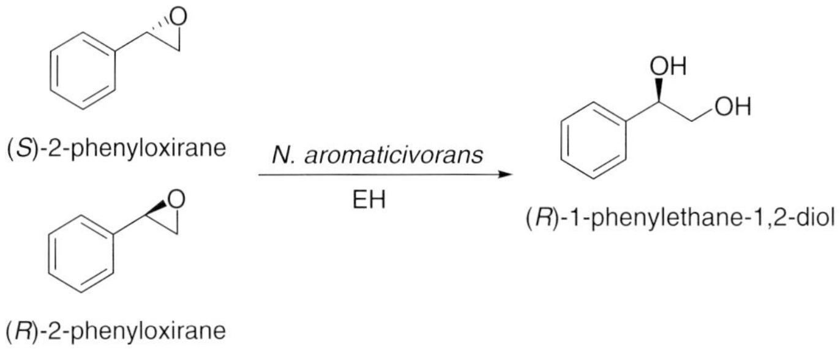

(S)-2-phenyloxirane
(R)-2-phenyloxirane
N. aromaticivorans EH
(R)-1-phenylethane-1,2-diol

Scheme 25.26 Enantioselective hydrolysis of styrene oxide by recombinant EHase.

Novel epoxide hydrolase from a marine bacterium, *Rhodobacterales bacterium* HTCC2654, catalyzed enantioselective biocatalytic kinetic resolution of glycidyl phenyl ether (Scheme 25.27). The purified EHase (REH) hydrolyzed (*S*)-glycidyl phenyl ether preferentially over the (*R*)-enantiomer with enantiopurity of 99.9% ee and 38.4% yield, and an *E*-value of 38.4 [72].

Marine fungi, *A. sydowii* Gc12, *P. raistrickii* Ce16, *P. miczynskii* Gc5, and *Trichoderma* sp. Gc1, isolated from marine sponges, catalyzed the enzymatic hydrolysis of benzyl glycidyl ether. *A. sydowii* Gc12 promoted the kinetic resolution of racemic benzyl glycidyl ether to yield (*R*)-benzyl glycidyl ether with 24–46% ee

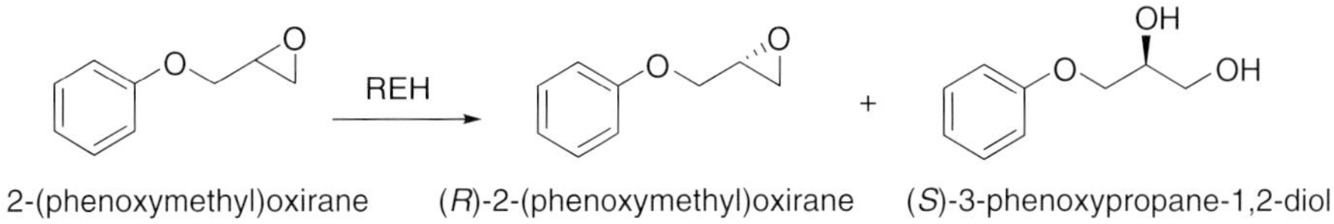

REH

2-(phenoxymethyl)oxirane (*R*)-2-(phenoxymethyl)oxirane (*S*)-3-phenoxypropane-1,2-diol

Scheme 25.27 Hydrolysis of glycidyl phenyl ether by recombinant EHase.

and 3-(benzyloxy)propane-1,2-diol with <10% ee. *Trichoderma* sp. Gc1 afforded (*S*)-oxirane with 60% ee and yields up to 39%, together with (*R*)-3-(benzyloxy)propane-1,2-diol (25% yield, 32% ee). The epoxide hydrolases from marine fungi exhibited complementary regioselectivity in opening the epoxide ring of racemic benzyl glycidyl ether, with *A. sydowii* Gc12 showing (*S*)-preference and *Trichoderma* sp. Gc1 showing (*R*)-preference for the oxirane. Afterwards, the marine fungus *Trichoderma* sp. Gc1 exhibited a preference for the (*R*)-enantiomer of oxirane with attack occurring on the terminal β-carbon yielding (*R*)-diol with retention of configuration (Scheme 25.28) [18].

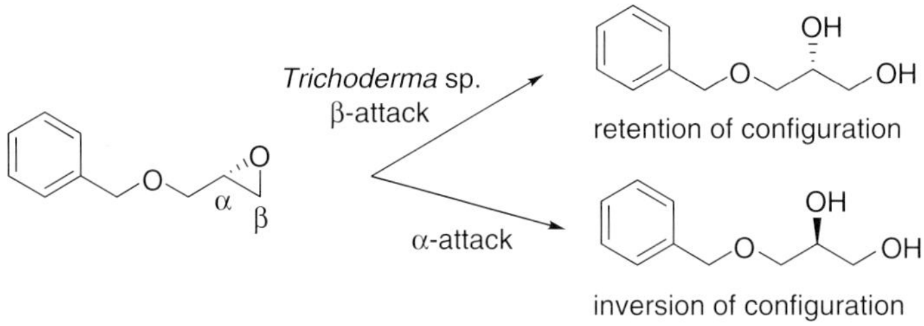

Trichoderma sp. β-attack

α-attack

retention of configuration

inversion of configuration

Scheme 25.28 Hydrolysis of benzyl glycidyl ether by *Trichoderma* sp. Gc1.

Enzymatic hydrolysis of (±)-2-[(allyloxy)methyl]oxirane using whole cells of the marine fungus *Trichoderma* sp. Gc1 grown in artificial seawater produced (*S*)-(+)-2-(allyloxy-methoxyl)oxirane (23% yield, 34% ee) together with (*R*)-(−)-3-(allyloxy)-propane-1,2-diol (yield 60%, 10% ee). The fungal hydrolases exhibited selectivity with preference for (*R*)-(−)-2-[(allyloxy)methyl]oxirane, while the concomitant formation of (*R*)-(−)-3-(allyloxy)propane-1,2-diol indicated that the mechanism involved retention of configuration (Scheme 25.29) [73].

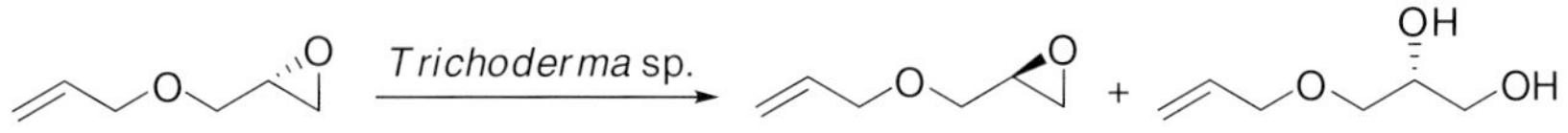

Trichoderma sp.

Scheme 25.29 Hydrolysis of allyl glycidyl ether by *Trichoderma* sp. Gc1.

However, marine epoxide hydrolases may have enantioconvergent applications for the production of interesting diols with specific stereochemistry. Enantioselective epoxide hydrolase activities were found in the microorganism *Sphingomonas echinoides* isolated from seawater [1,64].

Finally, reports of the use of marine fungi, fish, bacteria, and algae as biocatalysts may be rare, but the results reported here suggest that such organisms may be of great value in the biotransformation of organic compounds.

25.7
Collection and Isolation of Bioactive Marine Microorganisms

The oceans are complex systems that support a diverse assemblage of microbial life-forms living in habitats with extreme variations of pressure, salinity, and temperature. Thus, marine microorganisms have developed unique metabolic and physiological capabilities that not only ensure survival in extreme habitats, but also offer a potential for the production of metabolites that would not be obtainable from terrestrial microorganisms [74].

When studying marine microorganisms, it is important to recognize that some microorganisms found in this environment are permanent inhabitants and some have been carried there from terrestrial sources. Isolates can thus be defined as "obligate" and "facultative," respectively, or even as "marine-derived," since it is often hard to give proper classification to a specific microorganism isolated from the sea and definitions are still in debate. As an example, the discovery of the ubiquity and prevalence of the VBNC (viable but nonculturable bacteria) in the oceans puts in check the definitions that classify marine microorganisms as "those that can grow and reproduce in the marine habitat," "those possessing ability to grow only at certain sea-water salt concentrations," or "those possessing physiological adaptations specific to the marine environment."

Despite the problems with definition and cultivation, a great deal of research on marine microorganisms has been carried out. It is now known that microorganisms live in every corner of the oceans. Their habitats are diverse and include open water, sediments, associated with marine macro- and microorganisms (as symbionts or pathogens), estuaries, mangroves, and extreme ones such as hydrothermal vents. The variety of their habitats reflects also their diverse metabolic capabilities, making them very promising for biotechnological applications such as bioremediation and as sources of bioproducts such as enzymes and bioactive molecules. According to Hughes and Fenical [75], habitat characteristics, such as the diluting effect of the ocean, drive the construction of potent molecules that are stable to harsh salty conditions.

Although the discovery of new marine molecules and microorganisms can now be based on molecular "nonculture-dependent methodologies," "culture-dependent methodologies" are still in use and cannot be excluded. According to Hughes and Fenical [75], novel methods for the collection and culture of new species, including bacterial symbionts of invertebrates, are needed [76].

As discussed above, it is important to classify a strain isolated from the sea with respect to its degree of dependence on and tolerance of seawater. To do this, the strain is grown with or without seawater (artificial or natural) and the degree of dependence and tolerance can be assessed from its capacity to grow under these conditions. It is also very important for screening surveys to test production of a natural product or activity with or without seawater, because one strain may, for example, grow better without seawater but be active or produce an antibiotic only when cultured in seawater [23,76]. The same thinking is valid for the choice of the growth medium. The media chosen for isolation purposes need not the same as those used for production purposes.

Many of the methods applied to the isolation of microorganisms from marine habitats are based on the general approaches developed for the nonmarine environment. However, some essential adaptations are described in the various published studies. These depend on the main aims of the research, on the specific organisms being targeted, and on their site of origin (e.g., associated with other marine microorganisms, sediment, rock, or water). If the aim is to isolate truly marine microorganisms, samples should be taken as far away from the coast or islands as possible. It is known that near coastal waters, 95% of the bacteria are salt-tolerant forms and only 5% are true marine forms. In the open ocean and in deep sea, the true marine forms dominate. Regarding culture media, the distilled water should be substituted by artificial or real seawater (the latter must be aged in the dark for about 6 months to allow the organic matter to decompose) and the pH adjusted to 8 (close to that of the sea). Even when these precautions are taken, it cannot be guaranteed that only strictly marine microorganisms will be isolated.

Given that culture media, in general, suffer from the limitation that any chosen medium may select a particular type or group of microorganisms, several different media may be used with the same sample, to broaden the range of isolated strains. The medium should also be chosen to suit the particular goals of the study. For instance, if the intention is to isolate organisms that degrade a specific substance, enriched minimal media are used, in which the target substance must be the only source of carbon, or nitrogen, or some essential nutrient.

The collection of marine samples in the field requires special precautions to avoid cross-contamination between samples, contamination with cutting instruments, air, and so on. Sterile bags or sealed flasks must be used for each sample, which must also be properly identified. Samples must be transported on ice or maintained near the sea temperature until processed. Once in the laboratory, inoculation on culture media can be done in several ways, depending on the organism of interest and the kind of sample. For example, sediment samples have a matrix of particles on which microorganisms are aggregated and they must be released from the particles before inoculation. Often, sediment samples are very diverse and have high microorganism densities, and it is necessary to dilute them with sterile artificial seawater before inoculation. For sediment collection, the most commonly used devices are Grab samplers or dredges (e.g., Peterson, Van Veen, Birge-Ekman samplers) or corers, which remove a cylindrical core of sediment. Sample disturbance or mixing is less frequent in core samplers. For further information on the devices used for sediment sampling, we refer the reader to IAEA-TECDOC-1360 [77]. To access submerged shallow marine habitats, conventional scuba is used and deep-diving submersibles can be used for the deeper sites.

It is well known from previous research that there is a pronounced difference between the microbiological diversity accessible by culture-based methods and that revealed by molecular biological studies based on gene sequences from environmental samples. Aiming to reduce this discrepancy, recently, new and innovative culture techniques have been developed to enable the effective isolation of new microorganisms from the marine environment, those that are not readily cultured by traditional techniques. These new techniques called high-throughput cultivation (HTC) use long incubation times with separated individual microorganisms in low-nutrient media, which mimic the low nutrient concentrations found in marine environments. In the "extinction culturing approach" developed by Connon and Giovannoni [78], based on the ideas and findings reported by Button *et al.* [79] and automated by Bruns *et al.* [80], marine samples are diluted and split into small aliquots, to leave one microorganism in each volume. Recently, Stingl *et al.* [81] made some developments in this approach, which has yielded isolates of many abundant but previously uncultured marine bacterial clades. In the "encapsulation approach," developed by Zengler *et al.* [82], viable cells present in an environmental sample are individually encapsulated in gel microdroplets (20–70 μm diameter) and cultured in a low-nutrient flux. Flow cytometry is used to detect microdroplets containing microcolonies after growth. One of the advantages of this approach is the fact that the encapsulation allows microbial metabolites to cross through the microcapsules, as in their natural environment [81,82]. According to Zengel *et al.* [82], chemical interaction between cells is also an important factor for uncultured microorganisms. Starting with this approach, Kaeberlein *et al.* [83] designed "diffusion chambers" for environmental microorganism culture. These chambers are separated into two parts by 0.03 μm pore-size membranes and placed in the original sampling location or in a simulated natural environment in the laboratory. In one side of the chamber are placed samples diluted serially with seawater blended with agar. New versions of diffusion chambers have been developed and one example is the "isolation chip" (iChip), described by Lewis [84]. This chip is composed of hundreds of miniaturized diffusion chambers assembled in a central unit that is covered by semipermeable membranes allowing nutrient diffusion into the cells and preventing them from escaping from the device. Each chamber contains approximately one viable cell that is isolated from an environmental sample dilution. Recently, Nichols *et al.* [85] found that once a previously uncultured bacterium has been successfully cultured in an isolation chamber, it is then possible to subculture it in conventional Petri dishes. This result increases the possible applications of these techniques.

Although there are several reports of bioactive compounds being isolated from photosynthetic organisms such as cyanobacteria and microalgae, these groups are not discussed here. In the following sections, we give details of specific methodologies used for various marine heterotrophic microbial groups.

25.7.1 Fungi

Fungi are one of the most promising sources of metabolites from marine environments. The chemical diversity of the secondary metabolism of the marine-derived fungi, along with the number of novel strains, makes this group of microorganisms of great interest for the isolation of unusual bioactive natural products [86]. According to Bugni and Ireland [87], the rediscovery of large numbers of previously described metabolites from traditional terrestrial sources led researchers to explore unique habitats, such as the marine environment, for fungi potentially possessing new biosynthetic diversity. In their review, they showed that fungi obtained from sponges, algae, or wood substrates account for 70% of the total chemical compounds in the literature, and that the sponge-derived fungi account for 33%. In a more recent review, Rateb and Ebel [88] pointed that, in the marine environment, algae and sponges are the most important sources of new structures of secondary metabolites reported for fungi.

Research on the biological activities of marine-derived fungi is mainly focused on antibiotic and anticancer properties, but other studied activities include cell-cycle inhibition, antagonism of platelet-activating factor, antiviral activity, neuritogenic activity, phosphatase and kinase inhibition, radical-scavenging activities [87], cytotoxicity to several cancer lineages, insecticidal action, neovascularization effect, and inhibition of the enzyme human leukocyte elastase (HLE) [89]. Recently, products such as proteases [90] and properties, such as effluent detoxification and decolorization [91–93], biocatalysis [18,23,54,58,94], biodegradation of PAHs [95], lignin [96], DDD pesticide [32], and low-density polyethylene [97], have been investigated.

According to Rateb and Ebel [88], marine-derived fungal strains have been isolated from inorganic substrates (soil, sediment, sandy habitats, artificial substrates, and the water column), marine microbial communities, marine plants (algae, sea grasses, driftwood, and other higher plants, especially mangrove plants), marine invertebrates (most notably sponges, but also corals, ascidians, holothurians, bivalves, and crustaceans), and vertebrates (mainly fish). They pointed out that a newly emerging source of marine organisms is the deep sea, where reports exist of new secondary metabolites from fungi derived from deep sea sediments.

When we are dealing with sediment, the sample can be collected and treated as explained later for actinomycetes isolation, except that it is not necessary to carry out the pretreatment step. For fungi associated with marine biological samples, fresh samples must be collected and transported aseptically to the laboratory. If it is not possible to prepare and inoculate samples within a few hours, samples can be frozen. In the laboratory, the sample surface needs to be sterilized by washing with a solution of $HgCl_2$ in EtOH (1 mg/ml) for 1 min, followed by three additional washes with sterilized seawater [98]. For surface sterilization, Sponga *et al.* [99] rinsed sponge samples in sterile marine water, soaked them in ethyl alcohol for 5 min, and then rinsed in sterile water for 5 min. Karnat *et al.* [90] vortexed sample fragments with 0.5% sodium hypochlorite for 10 s and then rinsed them by vortexing with sterile seawater for 10 s, to remove residual sodium hypochlorite. Höller *et al.* [100] tested several techniques for sponge surface sterilization and preferred simply washing the sample several times with sterilized seawater. Surface sterility of the specimen can be checked by plating the water used in the last washing.

Kossuga *et al.* [86] tested three different inoculation procedures for marine-derived fungi associated with biological samples (1. spreading fragments of the inner parts of sample on agar plates; 2. placing the fragments, without spreading, on agar plates; 3. inoculation on agar plates of several dilutions of homogenized fragments in sterilized seawater) and found that the most suitable procedure for recovering a larger number of pure fungal isolates was simply placing 1 cm^3 fragments on agar plates [86]. This method was also used with success by others [99,100].

According to Bugni and Ireland [87], fungal strain isolation from the marine environment is highly dependent on the media used. Some media used for fungal isolation are potato–dextrose agar, Czapek-Dox Agar, Cornmeal Agar, starch casein agar, V8 agar, half nutrient potato–dextrose medium (1/2 PD), Tubaki agar, GPY agar, 2% malt extract agar, 3% malt extract–soy peptone agar, potato–carrot agar, oatmeal agar, cellulose agar, cellulose–yeast extract agar, starch–yeast extract agar, and nutrient agar [86,90,100]. Fremlin *et al.* [101], Kjer *et al.* [102], and Kossuga *et al.* [86] reported the advantages of using malt-based media for the isolation and growth of marine-derived fungal strains for the production of bioactive extracts. According to Kossuga *et al.* [86], 3% malt is a medium free of complex fat mixtures, a useful feature for obtaining crude extracts from microbial growth media.

In culture media used for fungal isolation, antibiotics are frequently added to suppress bacterial growth. The antibacterials used can be benzylpenicillin (250 mg/l) and streptomycin sulfate (250 mg/l), penicillin at 0.2 g/l, or chloramphenicol. The addition of these antibacterials is important because bacteria grow faster than fungi, which need at least 1 week of incubation. Höller *et al.* [100] and Dreyfuss [103] also added 0.5 mg/l cyclosporine A to slow the growth of fast-growing fungi. Plates must be inspected for new fungal colonies during at least 1 month of incubation.

Most of the authors did not mention the pH used in their isolation culture media, but some, like Höller *et al.* [100], used a low pH value of 5.5 that is traditionally used for fungi. Other authors, like Kossuga *et al.* [86], used a pH of 8.0, close to the pH of the marine environment. There is no published assessment of the best culture medium pH for optimized isolation of marine fungal strains and direct comparison of these two authors' data is not possible because they used different isolation methods, culture media, and sponge species in their studies.

Concerning the salt content of the culture media, Bugni and Ireland [87] affirmed that the metabolic production of fungi could be sensitive to seawater concentration. In an experiment using nine fungal strains, they concluded that fungal metabolite production may be enhanced in many species by the addition of higher concentrations of a more diverse array of ions, such as those found in artificial seawater. Thus, this must be taken into account in growth experiments aiming at metabolic production.

25.7.2
Bacteria

The most prevalent bacteria in marine environments are Gram-negative, which comprise approximately 90% of the bacterial flora. The remainder are Gram-positive and, in extreme marine environments, mostly archaebacteria. According to Fenical [74], microhabitats for marine bacteria are sediments, animate and inanimate surfaces, internal spaces of invertebrate animals and marine plants, and animals having symbiotic relationships with numerous microorganisms. Although the Gram-positive actinomycetes are not the most abundant marine bacteria, they are by far the most studied ones when researchers are screening for bioactive products. Thus, actinomycetes are discussed in Section 25.7.3.

Only a small proportion of marine bacteria grow under standard culture conditions and the majority can be classed as "VBNC." New approaches have been tried in recent years with the aim of culturing these bacteria and the new techniques are called HTC. These techniques provide means of studying bacteria that may be abundant in a particular oligotrophic habitat but, because of their adaptation to low nutrient levels, are outcompeted by kinetically more versatile organisms in conventional enrichment methods. These techniques are described above in the introductory text. Examples of the successful use of these techniques are, as yet, few but very important and they are in constant development, improving their capacity to isolate new bacteria. However, as Jensen and Fenical [104] pointed out, our ability to assess the biosynthetic potential of marine bacteria is inevitably coupled with our basic understanding of their biology.

Irrespective of the technique used for bacterial culture (traditional or HTC), the samples collected should be maintained at 4–10 °C and transported within 3 h, and processed soon after arrival in the laboratory. For water and sediment, sample can be spread directly on Petri dishes as the inoculum, or after serial dilution with artificial seawater or autoclaved seawater. To isolate bacteria associated with invertebrate animals and marine plants, the sample surface must be sterilized. The surface sterilization of the specimen and inoculation are often carried out as described for isolation of fungi. Incubation should be at room temperature (or at a temperature similar to that in the sampling station) for 3–7 days.

The most widely used medium for marine heterotrophic bacterial isolation is Zobell marine broth, at pH 8.5. It contains minerals (to simulate seawater), bacteriological peptone (which provides nitrogen, vitamins, trace mineral elements, and amino acids essential for growth), and yeast extract (as a source of vitamins). Others are also used, such as chitin medium and SYP–SW agar, composed of starch, yeast extract, peptone, and artificial sea salts, proposed by Kennedy *et al.* [105]. For isolation of specific groups of bacteria, selective media are required.

As discussed above, new innovative methods are required for the search for new microorganisms and microbial products. In addition to the development of HTC techniques, Jensen and Fenical [104] discussed the need for new culture methods, particularly those that take into account the environmental parameters associated with the habitats sampled. To this end, they suggested the use of marine-derived nutrients (such as pulverized brown alga) that may improve our ability to culture recalcitrant marine bacteria and present additional opportunities for the discovery of novel metabolites.

25.7.3 Actinomycetes

Although actinomycetes (Gram-positive filamentous bacteria pertaining to the order Actinomycetales of the phylum Actinobacteria) can be found in several marine habitats, they are very common in sediments. Densities and types of actinomycetes vary with sediment depth and the distance from the shore. For example, Jensen and Fenical [104] affirmed that actinomycetes groups such as *Actinoplanes* are found in greater numbers as the distance from the shore increases, whereas the opposite occurs with the streptomycetes. In the sediment compartment, the surface layer (0–3 cm deep) has higher microbial densities and should be aseptically removed for analysis.

The methods used for isolation and culture of actinomycetes from marine sediment are very similar to those developed and used for soil. Frequently, the isolation from sediments involves their extraction from particles, coupled with sample decontamination and serial dilution (with phosphate-buffered saline, quarter-strength Ringer solution, sterilized artificial or natural seawater) before plating. The serial dilution can be done by a technique called "stamping" described by Hameş-Kocabaş and Uzel [106]. In this technique, dried sediment samples are grounded with a pestle and mortar. The sample is diluted by a continual stamping process, involving pressing a polyester fiber-tipped sterile swab, small round sponge or foam plug (2 cm diameter) onto the sample and onto the agar. Decontaminated samples (0.5 g) can also be directly sprinkled on agar plates. The release or extraction of the cells from sediment into suspension can be done, for example, using vortexing with glass beads, ULTRA-TURRAX® mixing, blending, sonication, or simply by grinding with a mortar and a pestle. Decontamination is the main problem with isolation of actinomycetes from sediment, because other fast-growing bacteria in the sample tend to dominate the plates. Some actinomycete genera may appear on plates only after 4–5 weeks of incubation, so the isolation procedures and media must be highly selective. The selective methods for actinomycete isolation combine physical or chemical (or both) sample pretreatment with selective media.

The following decontamination techniques are used in the pretreatment of sediment samples: air-drying at room temperature in laminar air flow hood or SpeedVac to reduce the viability of mycelial forms; heat treatment of the samples to eliminate other spore formers (e.g., 120 °C for 1 h in dried samples, according to Bredholdt *et al.* [107]; or 40–60 °C from 1 h to overnight; or 55 °C for 6 min; or lower temperatures for longer periods of time, as suggested by Kokare *et al.* [108], who used 41 °C for 60 days); freezing at −20 °C for 24 h according to Jensen *et al.* [109] or at −18 °C according to Bredholdt *et al.* [110]; UV irradiation at 254 nm for 30 s from a distance of 20 cm, or super high frequency radiation in a microwave oven for 45 s (2460 MHz, 80W), or extremely high frequency radiation (1 kHz within a wavelength band of 8–11.5 mm), according to Bredholdt *et al.* [110]; chemical treatment with substances such as phenol at 0.7–1.5% added to the sample sediment for approximately 30 min and subsequent dilutions to reduce the toxic effects of the phenol, or quaternary salts (e.g., benzethonium chloride); filtration through cellulose ester filters before plating; or baiting techniques (using, for example, pollen).

Combined pretreatments are also found in the literature. According to Bredholdt *et al.* [107], for a dried sample, heat can be combined with a chosen chemical treatment (e.g., dry heat at 120 °C for 60 min and phenol at 1.5% for 30 min at 30 °C or benzethonium chloride at 0.02% for 30 min at 30 °C). Eccleston *et al.* [111], pretreating dried samples at room temperature, used dry heat (55 °C, 30 min) with microwave radiation (80W, 30 s). Discussing the various known pretreatments, Hameş-Kocabaş and Uzel [106], in particular, reported that drying in a laminar air flow hood, dilution with seawater, and heating prior to inoculation are often used.

The selective isolation media for actinomycetes generally have high carbon-to-nitrogen ratios and contain resistant complex carbon and nitrogen sources such as starch, casein, chitin, or humic acid. These media reduce bacterial colonies on isolation plates because, in contrast to actinomycetes, other bacteria grow better in low carbon-to-nitrogen conditions and are usually unable to attack high-molecular-weight polymers. Kokare *et al.* [108] tested eight media (modified starch casein agar, glucose asparagine agar, glycerol asparagine agar, tyrosine agar, yeast–malt extract agar, nutrient agar, maltose yeast extract agar, and glycerol glycine medium, all described in a study by Rathna and Chandrika [112], and found that media with glucose or starch as a carbon source and asparagine or casein as a nitrogen source were suitable for the isolation of actinomycetes from marine sediments). According to Hameş-Kocabaş and Uzel [106], sediment extracts, sponge extracts, and natural seawater are also used alone or as a supplement to mimic natural environmental conditions. These authors, in their review of isolation strategies of marine-derived actinomycetes from sponge and sediment samples, describe the composition of the successful media used for their isolation.

Bredholdt *et al.* [107], using only four different selective agar media, isolated approximately 3200 actinomycete bacteria from the sediment samples collected at various locations and depths (4.5–450 m) in a Norwegian fjord. According to these authors, no reports exist on the effectiveness of selective techniques applied to marine sediments in temperate areas, although these techniques are applied with success for the isolation of groups of actinomycetes from soil. By combining various types of selective treatments commonly used in soil- decontamination techniques (dry heat, phenol treatment, dry heat followed by phenol treatment, dry heat followed by benzethonium chloride treatment, and pollen baiting) combined with selective media, they found that the relative numbers of actinomycetes on the agar plates inoculated with the shallow water (4.5–28 m) near-shore samples were increased. Conversely, for the deep water sediment samples (450 m), these treatments were detrimental, resulting in agar plates free of actinomycetes. In light of these results, they concluded that more effort is required in order to establish methods allowing specific enrichment of marine actinomycetes.

Antibiotics must be added to all isolation media to prevent fast-growing Gram-negative bacteria that dominate sediment samples. Frequently used antibiotics are, for example, rifampicin at 5 μg/ml, penicillin G at 5 μg/ml, nalidixic acid at 10–30 μg/ml, and novobiocin at 25 μg/ml. The antifungals most often used are cycloheximide at 10–70, but often at 40–75 μg/ml, and nystatin at 25–75 μg/ml (frequently in combination). We can also use amphotericin B (30 μg/ml) or $K_2Cr_2O_7$ (50 μg/ml). The antimicrobials incorporated in the culture media can also inhibit some actinomycete species.

Among the innovative HTC approaches to culturing environmental microorganisms discussed above, the "microbial trap," designed recently by Lewis [84], was proposed by Hameş-Kocabaş and Uzel [106] for the isolation of actinomycetes from marine environment. This approach is similar to the diffusion chambers but it is specific to filamentous organisms because one side is sealed with a 0.2 μm pore diameter membrane, enabling actinomycetes to pass through and penetrate the chamber after 2–3 weeks of incubation.

25.7.4 Bacterial Extremophiles

Although representatives of the domain bacteria can be found in extreme environments, those of archaea are more frequently associated with them, especially with very extreme ones. Marine microbial extremophiles can be classified, according to their environmental preferences, as thermophilic, psychrophilic, acidophilic, barophilic, or halophilic, or have, simultaneously, two or three of these environmental preferences. For example, bacteria isolated from deep sea hydrothermal vents are normally anaerobic acidophilic, thermophilic, and barophilic. As these organisms possess physiological adaptations to live in such environmental conditions, they represent opportunities to explore unique enzymes and biotechnological products. Culturing is necessary to describe new microbial representatives thriving in such habitats. The microbial extremophiles must be collected and isolated by appropriate procedures and methods of incubation, taking into account the environmental characteristics of the collection site.

Deep oceans represent the greatest ecosystem on the planet and are the most important marine extreme environment. Most of the deep ocean regions are "nonproductive hydrothermally active," although the "productive hydrothermally active" regions are better studied. Owing to the oligotrophic nature of the "nonproductive hydrothermally active" regions, low-nutrient culture media are used (with longer incubation times – up to 3 months) for bacterial isolation while, on the other hand, for the "productive hydrothermally active" regions, richer culture media are used. At these depths, anaerobic conditions and high hydrostatic pressures prevail and, in "nonhydrothermally active regions," with some exceptions, temperatures range from 2 to 4 °C. In all of these locations, collection is done by means of remotely operated vehicles (ROV) and samples are brought to the surface in decontaminated insulated boxes. Most of the existing studies of bacteria from deep ocean regions were performed at atmospheric pressure and few laboratories have used appropriate *in situ* hydrostatic pressure incubation according to [113]. A system for tube incubation at *in situ* hydrostatic pressure was described by these authors.

Microorganisms from deep sea hydrothermal vent samples are usually cultured by enrichment techniques in flasks or bioreactors [114]. From these samples, Cornec *et al.* [115] carried out bacterial enrichment at temperatures ranging from 60 to 95 °C, on traditional liquid complex organic medium supplemented with colloidal sulfur, and they achieved the purification of strains under anaerobic conditions by incubation on bilayered solid media or monolayered media solidified with agar for about 5 days.

Besides the "nonhydrothermally active" regions of the deep ocean, marine habitats in polar regions (where water and sediments are at approximately −1 °C) are also very important marine environments and sources of psychrophiles. According to Cavicchioli [116], the term psychrophile is effective as a general term for microorganisms that grow in a cold environment, but they classify these as stenopsychrophiles (formerly "true psychrophile") and eurypsychrophiles (formerly "psychrotolerant" or "psychrotroph"). The term stenopsychrophile describes microorganisms with a restricted growth temperature range that cannot tolerate higher temperatures for growth, while eurypsychrophiles are microorganisms that prefer permanently cold environments, but tolerate a wide range of temperatures, including mesophilic. Thus, after microorganisms have been isolated from marine cold regions, their temperature range must be tested, since some may be adapted to cold but grow faster at warmer temperatures, and higher temperatures facilitate massive growth for production purposes.

We know that even the most extreme cold and frozen environments harbor enormously diverse, viable and metabolically active microbial populations representing major phylogenetic groups [117]. Thus, considering the importance and the biotechnological potential of extreme microorganisms, collection and culturing techniques must be improved to explore and exploit thoroughly microbial diversity in extreme marine environments.

25.8 Conclusions and Perspectives

As shown briefly in this chapter, the use of marine enzymes in the transformation of organic compounds shows great potential for development into biotechnological processes.

Reactions catalyzed by marine enzymes may be useful in the synthesis of enantiomerically pure molecules and applied to the synthesis of pharmaceuticals, pesticides, and fragrances, as well as to the formation of achiral molecules that could be used in the synthesis of polymers.

Enzymatic reactions are especially interesting because they are conducted in ecofriendly, mild conditions. The use of marine enzymes has potential in the biotransformation and production of natural products to obtain new molecules that may be used in assays of biological activity and to provide new metabolites that may have anticancer, anti-Alzheimer, antituberculosis, antimalaria, or anti-Chagas disease activity.

Marine enzymes are also promising tools for the biodegradation of organic pollutants such as pesticides, munitions, plastics, and other recalcitrant toxic compounds present in the environment.

In summary, marine enzymes provide new frontiers for research in chemistry, biology, environmental engineering, and biochemistry and these new biocatalysts can contribute not only to the discovery of new products, but also to our knowledge of the marine environment and its preservation.

References

1 Trincone, A. (2011) Marine biocatalysts: enzymatic features and applications. *Mar. Drugs*, **9**, 478–499.

2 Trincone, A. (2010) Potential biocatalysts originating from sea environments. *J. Mol. Catal.: B Enzym.*, **66**, 241–256.

3 Zhang, C. and Kim, S.-K. (2010) Research and application of marine microbial enzymes: status and prospects. *Mar. Drugs*, **8**, 1920–1934.

4 Sarkar, S., Pramanik, A., Mitra, A., and Mukherjee, J. (2010) Bioprocessing data for the production of marine enzymes. *Mar. Drugs*, **8**, 1323–1372.

5 Bommarius, A.S. and Riebel-Bommarius, B.R. (2005) *Biocatalysis: Fundamentals and Applications*, 1st edn, Wiley-VCH Verlag GmbH, Weinheim.

6 Nagel, B., Dellweg, H., and Gierasch, L.M. (1992) Glossary for chemists of terms used in biotechnology (IUPAC Recommendations). *Pure Appl. Chem.*, **64**, 143–168.

7 Meyer, H.-P. and Turner, N.J. (2009) Biotechnological manufacturing options for organic chemistry. *Mini-Rev. Org. Chem.*, **6**, 300–306.

8 Matsuda, T., Yamanaka, R., and Nakamura, K. (2009) Recent progress in biocatalysis for asymmetric oxidation and reduction. *Tetrahedron: Asymmetry*, **20**, 513–557.

9 Wang, Y., Liu, D., Meng, Q., and Zhang, W. (2009) Asymmetric hydrogenation of simple ketones with planar chiral ruthenocenyl phosphinooxazoline ligands. *Tetrahedron: Asymmetry*, **20**, 2510–2512.

10 Valadez-Blanco, R. and Livingston, A.G. (2009) Enantioselective whole-cell biotransformation of acetophenone to (*S*)-phenylethanol by *Rhodotorula glutinis*: Part I. product formation kinetics and feeding strategies in aqueous media. *Biochem. Eng. J.*, **46**, 44–53.

11 Abas, F.Z., Uzir, M.H., and Zahar, M.H.M. (2010) Effect of pH on the biotransformation of (*R*)-1-(4-bromo-phenyl)-ethanol by using *Aspergillus niger* as biocatalyst. *J. Appl. Sci.*, **10**, 3289–3294.

12 Rusch, D.B., Halpern, A.L., Sutton, G., Heidelberg, K.B., Williamson, S., Yooseph, S., Wu, D., Eisen, J.A., Hoffman, J.M., Remington, K., Beeson, K., Tran, B., Smith, H., Baden-Tillson, H., Stewart, C., Thorpe, J., Freeman, J., Andrews-Pfannkoch, C., Venter, J.E., Li, K., Kravitz, S., Heidelberg, J.F., Utterback, T., Rogers, Y.-H., Falcón, L.I., Souza, V., Bonilla-Rosso, G., Eguiarte, L.E., Karl, D.M., Sathyendranath, S., Platt, T., Bermingham, E., Gallardo, V., Tamayo-Castillo, G., Ferrari, M.R., Strausberg, R.L., Nealson, K., Friedman, R., Frazier, M., and Venter, J.C. (2007) The sorcerer II global ocean sampling expedition: northwest

Atlantic through eastern tropical pacific. *PLoS Biol.*, **5**398–431.

13 Venter, J.C., Remington, K., Heidelberg, J.F., Halpern, A.L., Rusch, D., Eisen, J.A., Wu, D., Paulsen, I., Nelson, K.E., Nelson, W., Fouts, D.E., Levy, S., Knap, A.H., Lomas, M.W., Nealson, K., White, O., Peterson, J., Hoffman, J., Parsons, R., Baden-Tillson, H., Pfannkoch, C., Rogers, Y.-H., and Smith, H.O. (2004) Environmental genome shotgun sequencing of the Sargasso sea. *Science*, **304**, 66–74.

14 Ghosh, D., Saha, M., Sana, B., and Mukherjee, J. (2005) Marine enzymes. *Adv. Biochem. Eng. Biotechnol.*, **96**, 189–218.

15 Haefner, B. (2003) Drugs from the deep: marine natural products as drug candidates. *Drug Discov. Today*, **8**, 536–544.

16 Newman, D.J. and Cragg, G.M. (2004) Marine natural products and related compounds in clinical and advanced preclinical trials. *J. Nat. Prod.*, **67**, 1216–1238.

17 San-Martín, A., Rovirosa, J., Astudillo, L., Sepúlveda, B., Ruiz, D., and San-Martín, C. (2008) Biotransformation of the marine sesquiterpene pacifenol by a facultative marine fungus. *Nat. Prod. Res.*, **22**, 1627–1632.

18 Martins, M.P., Mouad, A.M., Boschini, L., Seleghim, M.H.R., Sette, L.D., and Porto, A.L.M. (2011) Marine fungi *Aspergillus sydowii* and *Trichoderma* sp. catalyze the hydrolysis of benzyl glycidyl ether. *Mar. Biotechnol.*, **13**, 314–320.

19 Bonugli-Santos, R.C., Durrant, L.R., DaSilva, M., and Sette, L.D. (2010) Production of laccase, manganese peroxidase and lignin peroxidase by Brazilian marine-derived fungi. *Enzyme Microb. Technol.*, **46**, 32–37.

20 MacLeod, R.A. (1965) The question of the existence of specific marine bacteria. *Bacteriol. Rev.*, **29**, 9–23.

21 Drapeau, G.R., Matula, T.I., and MacLeod, R.A. (1966) Nutrition and metabolism of marine bacteria. XV. Relation of Na^+–activated transport to the Na^+ requirement of a marine pseudomonad for growth. *J. Bacteriol.*, **92**, 63–71.

22 Kogure, K. (1998) Bioenergetics of marine bacteria. *Curr. Opin. Biotechnol.*, **9**, 278–282.

23 Rocha, L.C., Ferreira, H.V., Pimenta, E.F., Berlinck, R.G.S., Seleghim, M.H.R., Javaroti, D.C.D., Sette, L.D., Bonugli, R.C., and Porto, A.L.M. (2009) Bioreduction of α-chloroacetophenone by whole cells of marine fungi. *Biotechnol. Lett.*, **31**, 1559–1563.

24 El Sayed, K.A., Laphookhieo, S., Baraka, H.N., Yousaf, M., Hebert, A., Bagaley, D., Rainey, F.A., Muralidharan, A., Thomas, S., and Shah, G.V. (2008) Biocatalytic and semisynthetic optimization of the anti-invasive tobacco (1*S*,2*E*,4*R*,6*R*,7*E*,11*E*)-2,7,11-cembratriene-4,6-diol. *Bioorg. Med. Chem.*, **16**, 2886–2893.

25 Devi, P., Naik, C., and Rodrigues, C. (2006) Biotransformation of citrinin to decarboxycitrinin using an organic solvent-tolerant marine bacterium, *Moraxella* sp. MB1. *Mar. Biotechnol.*, **8**, 129–138.

26 Rovirosa, J., Astudilo, L., Sanchez, I., Palacios, Y., and San-Martín, A. (1989) Antimicrobial activity of halogenated sesquiterpenes of *Laurencia claviformis* from Easter Island. *Bol. Soc. Chil. Quim.*, **34**, 147–152.

27 Koshimura, M., Utsukihara, T., Kawamoto, M., Saito, M., Horiuchi, C.A., and Kuniyoshi, M. (2009) Biotransformation of bromosesquiterpenes by marine fungi. *Phytochemistry*, **70**, 2023–2026.

28 Leutou, A.S., Yang, G., Nenkep, V.N., Siwe, X.N., Feng, Z., Khong, T.T., Choi, H.D., Kang, J.S., and Son, B.W. (2009) Microbial transformation of a monoterpene, geraniol, by the marine-derived fungus *Hypocrea* sp. *J. Microbiol. Biotechnol.*, **19**, 1150–1152.

29 Feng, Z., Nenkep, V., Yun, K., Zhang, D., Choi, H.D., Kang, J.S., and Son, B.W. (2010) Biotransformation of bioactive (−)-mellein by a marine isolate of bacterium *Stappia* sp. *J. Microbiol. Biotechnol.*, **20**, 985–987.

30 Li, H., Xie, Y., Qiu, X., and Lan, W. (2011) α-Pinene induces marine fungus *Aspergillus* sp. to produce mycotoxin penicillic acid. *Zhongshan Daxue Xuebao, Ziran Kexueban*, **50**, 66–69.

31 Patil, K.C., Matsumura, F., and Boush, M. (1972) Metabolic transformation of DDT, dieldrin, aldrin, and endrin by marine

microorganisms. *Environ. Sci. Technol.*, **6**, 629–632.

32 Ortega, S.N., Nitschke, M., Mouad, A.M., Landgraf, M.D., Rezende, M.O.O., Seleghim, M.H.R., Sette, L.D., and Porto, A.L.M. (2011) Isolation of Brazilian marine fungi capable of growing on DDD pesticide. *Biodegradation*, **22**, 43–50.

33 Kaise, T., Hanaoka, K., and Tagawa, S. (1987) The formation of trimethylarsine oxide from arsenobetaine by biodegradation with marine microorganisms. *Chemosphere*, **16**, 2551–2558.

34 Hanaoka, K., Ueno, K., Torgawa, S., and Kaise, T. (1989) Degradation of arsenobetaine by microorganisms associated with marine macroalgae, *Monostroma nitidum* and *Hizikia fusiforme*. *Comp. Biochem. Physiol.*, **94**, 379–382.

35 Hanaoka, K., Motoya, T., Tagawa, S., and Kaise, T. (1991) Conversion of arsenobetaine by intestinal bacteria of a mollusc *Liolophura japonica* chitons. *Appl. Organomet. Chem.*, **5**, 427–430.

36 Kaise, T., Sakurai, T., Saitoh, T., and Matsubara, C. (1998) Biotransformation of arsenobetaine to trimethylarsine oxide by marine microorganisms in a gill of clam *Meretrix lusoria*. *Chemosphere*, **37**, 443–449.

37 Grossi, V., Raphel, D., Hirschler-Rea, A., Gilewicz, M., Mouzdahir, A., Bertrand, J.C., and Rontani, J.F. (2000) Anaerobic biodegradation of pristane by a marine sedimentary bacterial and/or archaeal community. *Org. Geochem.*, **31**, 769–772.

38 Zengler, K., Richnow, H.H., Rossello-Mora, R., Michaelis, W., and Widdel, F. (1999) Methane formation from long-chain alkanes by anaerobic microorganisms. *Nature*, **401**, 266–269.

39 Chun, H.K., Ohnishi, Y., Misawa, N., Shindo, K., Hayashi, M., Harayama, S., and Horinouchi, S. (2001) Biotransformation of phenanthrene and 1-methoxy naphthalene with *Streptomyces lividans* cells expressing a marine bacterial phenanthrene dioxygenase gene cluster. *Biosci. Biotechnol. Biochem.*, **65**, 1774–1781.

40 Nunez, A.E., Caballero, A., and Ramos, J.L. (2001) Biological degradation of 2,4,6-trinitrotoluene. *Microbiol. Mol. Biol. Rev.*, **65**, 335–352.

41 Jain, M.R., Zinjarde, S.S., Deobagkar, D.D., and Deobagkar, D.N. (2004) 2,4,6-Trinitrotoluene transformation by a tropical marine yeast, *Yarrowia lipolytica* NCIM 3589. *Mar. Pollut. Bull.*, **49**, 783–788.

42 Nipper, M., Qian, Y., Carr, R.S., and Miller, K. (2004) Degradation of picric acid and 2,6-DNT in marine sediments and waters: the role of microbial activity and ultra-violet exposure. *Chemosphere*, **56**, 519–530.

43 Zhao, J.S., Spain, J., Thiboutot, S., Ampleman, G., Greer, C., and Hawari, J. (2004) Phylogeny of cyclic nitramine-degrading psychrophilic bacteria in marine sediment and their potential role in the natural attenuation of explosives. *FEMS Microbiol. Ecol.*, **49**, 349–357.

44 Bhatt, M., Zhao, J.S., Halasz, A., and Hawari, J. (2006) Biodegradation of hexahydro-1,3,5-trinitro-1,3,5-triazine by novel fungi isolated from unexploded ordnance contaminated marine sediment. *J. Ind. Microbiol. Biotechnol.*, **33**, 850–858.

45 Artham, T. and Doble, M. (2012) Bisphenol and metabolites released by biodegradation of polycarbonate in seawater. *Environ. Chem. Lett.*, **10**, 29–34.

46 Cerqueira, V.S. and Costa, J.A.V. (2009) Biodegradação de tolueno e óleo de pescado em solos impactados utilizando surfactantes químico e biológico. *Química Nova*, **32**, 394–400.

47 Pollard, D.J. and Woodley, J.M. (2007) Biocatalysis for pharmaceutical intermediates: the future is now. *Trends Biotechnol.*, **25**, 66–73.

48 Clososki, C.G., Milagre, C.D.F., Moran, P.J.S., and Rodrigues, J.A.R. (2007) Regio- and enantioselective reduction of methyleneketoesters mediated by *Saccharomyces cerevisiae*. *J. Mol. Catal.: B Enzym.*, **48**, 70–76.

49 Nakamura, K., Kawai, Y., Miyai, T., and Ohno, A. (1990) Stereochemical control in diastereoselective reduction with baker's yeast. *Tetrahedron Lett.*, **31**, 2927–2928.

50 Nakamura, K., Kondo, S.-I., Kawai, Y., and Ohno, A. (1993) Stereochemical control in microbial reduction. XXI. Effect of organic solvents on reduction of α-keto esters

mediated by baker's yeast. *Bull. Chem. Soc. Jpn.*, **66**, 2738–2743.

51 Bonugli-Santos, R.C., Durrant, L.R., DaSilva, M., and Sette, L.D. (2010) Production of laccase, manganese peroxidase and lignin peroxidase by Brazilian marine-derived fungi. *Enzyme Microb. Tech.*, **46**, 32–37.

52 Hunter, A.C. (2007) The current chemical utility of marine and terrestrial filamentous fungi in side-chain chemistry. *Curr. Org. Chem.*, **11**, 665–677.

53 Carballeira, J.D., Quezada, M.A., Álvarez, E., and Sinisterra, J.V. (2004) High throughput screening and QSAR-3D/CoMFA: useful tools to design predictive models of substrate specificity for biocatalysts. *Molecules*, **9**, 673–693.

54 Rocha, L.C., Ferreira, H.V., Luiz, R.F., Sette, L.D., and Porto, A.L.M. (2012) Stereoselective bioreduction of 1-(4-methoxyphenyl)ethanone by whole cells of marine-derived fungi. *Mar. Biotechnol.*, **14**, 358–362.

55 Rocha, L.C., Luiz, R.F., Rosset, I.G., Raminelli, C., and Porto, A.L.M. (2012) Bioconversion of iodoacetophenones by marine fungi. *Mar. Biotechnol.*, **14**, 396–401.

56 Mouad, A.M., Martins, M.P., Debonsi, H.M., Oliveira, A.L.L., Felicio, R., Yokoya, N.S., Fujii, M.T., Menezes, C.B.A., Garboggini, F.F., and Porto, A.L.M. (2011) Bioreduction of acetophenone derivatives by red marine algae *Bostrychia radicans* and *B. tenella*, and marine bacteria associated. *Helv. Chim. Acta*, **94**, 1506–1514.

57 Mouad, A.M., Martins, M.P., Romminger, S., Seleghim, M.H.R., De Oliveira, A.L.L., Debonsi, H.M., Yokoya, N.S., Fujii, M.T., Passarini, M.R.Z., Bonugli-Santos, R.C., Sette, L.D., and Porto, A.L.M. (2012) Bioconversion of acetophenones by marine fungi isolated from marine algae *Bostrychia radicans* and *Sargassum* sp. *Curr. Top. Biotechnol.*, **7**, 13–19.

58 Rocha, L.C., Ferreira, H.V., Pimenta, E.F., Berlinck, R.G.S., Rezende, M.O.R., Landgraf, M.D., Seleghim, M.H.R., Sette, L.D., and Porto, A.L.M. (2010) Biotransformation of α-bromoacetophenones by the marine fungus *Aspergillus sydowii*. *Mar. Biotechnol.*, **12**, 552–557.

59 Ishihara, K., Yamaguchi, H., Adachi, N., Hamada, H., and Nakajima, N. (2000) Stereocontrolled reduction of α- and β-keto esters with micro green algae, *Chlorella* strains. *Biosci. Biotechnol. Biochem.*, **64**, 2099–2103.

60 Ishihara, K., Nakajima, N., Yamaguchi, H., Hamada, H., and Uchimura, Y. (2001) Stereoselective reduction of keto esters with marine micro algae. *J. Mol. Catal.: B Enzym.*, **15**, 101–104.

61 Ishihara, K., Yamaguchi, H., and Nakajima, N. (2003) Stereoselective reduction of keto esters: thermophilic bacteria and microalgae as new biocatalysts. *J. Mol. Catal.: B Enzym.*, **23**, 171–189.

62 Ishihara, K., Nagai, H., Takahashi, K., Nishiyama, M., and Nakajima, N. (2011) Stereoselective reduction of α-keto ester and α-keto amide with marine actinomycetes, *Salinispora* strains, as novel biocatalysts. *Biochem. Insights*, **4**, 29–33.

63 Hwang, Y.-O., Kang, S.G., Woo, J.-H., Kwon, K.K., Sato, T., Lee, E.Y., Han, M.S., and Kim, S.-J. (2008) Screening enantioselective epoxide hydrolase activities from marine microorganisms: detection of activities in *Erythrobacter* spp. *Mar. Biotechnol.*, **10**, 366–373.

64 Kim, H.S., Lee, O.K., Lee, S.J., Hwang, S., Kim, S.J., Yang, S.-H., Park, S., and Lee, E.Y. (2006) Enantioselective epoxide hydrolase activity of a newly isolated microorganism *Sphingomonas echinoides* EH-983 from seawater. *J. Mol. Catal.: B Enzym.*, **41**, 130–135.

65 Kim, J.T., Kang, S.G., Woo, J.-H., Lee, J.-H., Jeong, B.C., and Kim, S.-J. (2007) Screening and its potential application of lipolytic activity from a marine environment: characterization of a novel esterase from *Yarrowia lipolytica* CL180. *Appl. Microbiol. Biotechnol.*, **74**, 820–828.

66 Park, S.-Y., Kim, J.-T., Kang, S.G., Woo, J.-H., Lee, J.-H., Choi, H.-T., and Kim, S.-J. (2007) A new esterase showing similarity to putative dienelactone hydrolase from a strict marine bacterium, *Vibrio* sp. GMD509. *Appl. Microbiol. Biotechnol.*, **77**, 107–115.

67 Woo, J.-H., Hwang, O.-K., Kang, S.G., Lee, H.S., Cho, J., and Kim, S.-J. (2007) Cloning and characterization of three novel epoxide hydrolases from a marine bacterium, *Erythrobacter litoralis* HTCC2594. *Appl. Microbiol. Biotechnol.*, **76**, 365–375.

68 Pedragosa-Moreau, S., Archelas, A., and Furstoss, R. (1993) Microbiological transformations. Enantiocomplementary epoxide hydrolyses as a preparative access to both enantiomers of styrene oxide. *J. Org. Chem.*, **58**, 5533–5536.

69 Kim, H.S., Lee, S.J., Lee, E.J., Hwang, J.W., Park, S., Kim, S.J., and Lee, E.Y. (2005) Cloning and characterization of a fish microsomal epoxide hydrolase of *Danio rerio* and application to kinetic resolution of racemic styrene oxide. *J. Mol. Catal.: B Enzym.*, **37**, 30–35.

70 Kim, H.S., Lee, O.K., Hwang, S., Kim, B.J., and Lee, E.Y. (2008) Biosynthesis of (*R*)-phenyl-1,2-ethanediol from racemic styrene oxide by using bacterial and marine fish epoxide hydrolases. *Biotechnol. Lett.*, **30**, 127–133.

71 Woo, J.-H., Kang, J.-H., Kang, S.G., Hwang, Y.-O., and Kim, S.J. (2009) Cloning and characterization of an epoxide hydrolase from *Novosphingobium aromaticivorans*. *Appl. Microbiol. Biotechnol.*, **82**, 873–881.

72 Woo, J.H., Kang, J.H., Hwang, Y.-O., Cho, J.C., Kim, S.J., and Kang, S.G. (2010) Biocatalytic resolution of glycidyl phenyl ether using a novel epoxide hydrolase from a marine bacterium, *Rhodobacterales bacterium* HTCC2654. *J. Biosci. Bioeng.*, **109**, 539–544.

73 Martins, M.P., Mouad, A.M., and Porto, A.L. M. (2012) Hydrolysis of allylglycidyl ether by marine fungus *Trichoderma* sp. Gc1 and the enzymatic resolution of allylchlorohydrin by *Candida antarctica* lipase type B. *Curr. Top. Catal.*, **10**, 27–33.

74 Fenical, W. (1993) Chemical studies of marine bacteria: developing a new resource. *Chem. Rev.*, **93**, 1673–1683.

75 Hughes, C.C. and Fenical, W. (2010) Antibacterials from the sea. *Chem. Eur. J.*, **16**, 12512–12525.

76 Imada, C. (2005) Enzyme inhibitors and other bioactive compounds from marine actinomycetes. *Antonie Van Leeuwenhoek*, **87**, 59–63.

77 IAEA (2003) *Collection and Preparation of Bottom Sediment Samples for Analysis of Radionuclides and Trace Elements*, International Atomic Energy Agency TECDOC 1360, International Atomic Energy Agency, Vienna, p. 130.

78 Connon, S.A. and Giovannoni, S. (2002) High-throughput methods for culturing microorganisms in very-low-nutrient media yield diverse new marine isolates. *Appl. Environ. Microb.*, **68**, 3878–3885.

79 Button, D.K., Schut, F., Quang, P., Martin, R., and Robertson, B.R. (1993) Viability and isolation of marine bacteria by dilution culture: theory, procedures, and initial results. *Appl. Environ. Microb.*, **59**, 881–891.

80 Bruns, A., Hoffelner, H., and Overmann, J. (2003) A novel approach for high throughput cultivation assays and the isolation of planktonic bacteria. *FEMS Microbiol. Ecol.*, **45**, 161–171.

81 Stingl, U., Tripp, H.J., and Giovannoni, S.J. (2007) Improvements of high-throughput culturing yielded novel SAR11 strains and other abundant marine bacteria from the Oregon coast and the Bermuda Atlantic time series study site. *ISME J.*, **1**, 361–371.

82 Zengler, K., Toledo, G., Rappe, M., Elkins, J., Mathur, E.J., Short, J.M., and Keller, M. (2002) Cultivating the uncultured. *Proc. Natl. Acad. Sci. USA*, **99**, 15681–15686.

83 Kaeberlein, T., Lewis, K., and Epstein, S.S. (2002) Isolating ‘uncultivable’ microorganisms in pure culture in a simulated natural environment. *Science*, **296**, 1127–1129.

84 Lewis, K., Epstein, S., D’Onofrio, A., and Ling, L.L. (2010) Uncultured microorganisms as a source of secondary metabolites. *J. Antibiot. (Tokyo)*, **63**, 468–476.

85 Nichols, D., Cahoon, N., Trakhtenberg, E.M., Pham, L., Mehta, A., Belanger, A., Kanigan, T., Lewis, K., and Epstein, S.S. (2010) Use of iChip for high-throughput *in situ* cultivation of “uncultivable” microbial species. *Appl. Environ. Microb.*, **76**, 2445–2450.

86 Kossuga, M.H., Romminger, S., Xavier, C., Milanetto, M.C., Do Valle, M.Z.,

Pimenta, E.F., Morais, R.P., Carvalho, E., Mizuno, C.M., Coradello, L.F.C., Barroso, V.M., Vacondio, B., Javaroti, D.C.D., Seleghim, M.H.R., Cavalcanti, B.C., Pessoa, C., Moraes, M.O., Lima, B.A., Gonçalves, R., Bonugli-Santos, R.C., Sette, L.D., and Berlinck, R.G.S. (2012) Evaluating methods for the isolation of marine-derived fungal strains and production of bioactive secondary metabolites. *Braz. J. Pharmacogn.*, **22**, 257–267.

87 Bugni, T.S. and Ireland, C.M. (2004) Marine-derived fungi: a chemically and biologically diverse group of microorganisms. *Nat. Prod. Rep.*, **21**, 143–163.

88 Rateb, M.E. and Ebel, R. (2011) Secondary metabolites of fungi from marine habitats. *Nat. Prod. Rep.*, **28**, 290–344.

89 Debbab, A., Aly, A.H., and Proksch, P. (2011) Bioactive secondary metabolites from endophytes and associated marine derived fungi. *Fungal Divers.*, **49**, 1–12.

90 Karnat, T., Rodrigues, C., and Nail, C.G. (2008) Marine-derived fungi as a source of proteases. *Indian J. Mar. Sci.*, **37**, 326–328.

91 Raghukumar, C. and Rivonkar, G. (2001) Decolorization of molasses spent wash by the white-rot fungus *Flavodon flavus*, isolated from a marine habitat. *Appl. Microbiol. Biotechnol.*, **55**, 510–514.

92 Verma, A.K., Raghukumar, C., Verma, P., Shouche, Y.S., and Naik, C.G. (2010) Four marine-derived fungi for bioremediation of raw textile mill effluents. *Biodegradation*, **21**, 217–233.

93 Verma, A.K., Raghukumar, C., and Naik, C.G. (2011) A novel hybrid technology for remediation of molasses-based raw effluents. *Bioresour. Technol.*, **102**, 2411–2418.

94 Rocha, L.C., Rosset, I.G., Luiz, R.F., Raminelli, C., and Porto, A.L.M. (2010) Kinetic resolution of iodophenylethanols by *Candida antarctica* lipase and their application for the synthesis of chiral biphenyl compounds. *Tetrahedron: Asymmetry*, **21**, 926–929.

95 Passarini, M.R.Z., Rodrigues, M.V.N., Silva, M., and Sette, L.D. (2011) Marine-derived filamentous fungi and their potential application for polycyclic aromatic hydrocarbon bioremediation. *Mar. Pollut. Bull.*, **62**, 364–370.

96 Chen, H.-Y., Xue, D.-S., Feng, X.-Y., and Yao, S.-J. (2011) Screening and production of ligninolytic enzyme by a marine-derived fungal *Pestalotiopsis* sp. J63. *Appl. Biochem. Biotechnol.*, **165**, 1754–1769.

97 Pramila, R. and Ramesh, K.V. (2011) Biodegradation of low density polyethylene (LDPE) by fungi isolated from marine water – a SEM analysis. *Afr. J. Microbiol. Res.*, **5**, 5013–5018.

98 Newel, S.Y. (1976) Mangrove fungi: the succession in the mycoflora of red mangrove (*Rhizophora mangle* L.), in *Recent Advances in Aquatic Mycology* (ed. E.B.G. Jones), Paul Elek Scientific Books, London, pp. 51–91.

99 Sponga, F., Cavaletti, L., Lazzarini, A., Borghi, A., Ciciliato, I., Losi, D., and Marinelli, F. (1999) Biodiversity and potentials of marine-derived microorganisms. *J. Biotechnol.*, **70**, 65–69.

100 Höller, U., Wright, A.D., Matthe, G.F., Konig, G.M., Draeger, S., Aust, H.-J., and Schulz, B. (2000) Fungi from marine sponges: diversity, biological activity and secondary metabolites. *Mycol. Res.*, **104**, 1354–1365.

101 Fremlin, L.J., Piggott, A.M., Lacey, E., and Capon, R.J. (2009) Cottoquinazoline A and cotteslosins A and B, metabolites from an Australian marine-derived strain of *Aspergillus versicolor*. *J. Nat. Prod.*, **72**, 666–670.

102 Kjer, J., Debbab, A., Aly, A.H., and Proksch, P. (2010) Methods for isolation of marine-derived endophytic fungi and their bioactive secondary products. *Nat. Protoc.*, **5**, 479–490.

103 Dreyfuss, M.M. (1986) Neue Erkenntnisse aus einem pharmakologischen Pilzscreening. *Sydowia*, **39**, 22–36.

104 Jensen, P.R. and Fenical, W. (1994) Strategies for the discovery of secondary metabolites from marine bacteria: ecological perspectives. *Annu. Rev. Microbiol.*, **48**, 559–584.

105 Kennedy, J., Baker, P., Piper, C., Cotter, P.D., Walsh, M., Mooij, M.J., Bourke, M.B., Rea, M.C., O'Connor, P.M., Ross, R.P., Hill, C., O'Gara, F., Marchesi, J.R., and Dobson, A.D.W. (2009) Isolation and analysis of bacteria with antimicrobial activities from the marine sponge *Haliclona simulans* collected from Irish waters. *Mar. Biotechnol.*, **11**, 384–396.

106 Hameş-Kocabaş, E.E. and Uzel, A. (2012) Isolation strategies of marine-derived actinomycetes from sponge and sediment samples. *J. Microbiol. Methods*, **88**, 342–347.

107 Bredholdt, H., Fjaervik, E., Johnsen, G., and Zotchev, S.B. (2008) Actinomycetes from sediments in the Trondheim fjord, Norway: diversity and biological activity. *Mar. Drugs*, **6**, 12–24.

108 Kokare, C.R., Mahadik, K.R., Kadam, S.S., and Chopade, B.A. (2004) Isolation of bioactive marine actinomycetes from sediments isolated from Goa and Maharashtra coastlines (west coast of India). *Indian J. Mar. Sci.*, **33**, 248–256.

109 Jensen, P.R., Gontang, E., Mafnas, C., Mincer, T.J., and Fenical, W. (2005) Culturable marine actinomycete diversity from tropical pacific ocean sediments. *Environ. Microbiol.*, **7**, 1039–1048.

110 Bredholdt, H., Galatenko, O.A., Engelhardt, K., Tjaervik, E., Terekhova, L.P., and Zotchev, S.B. (2007) Rare actinomycete bacteria from the shallow water sediments of the Trondheim fjord, Norway: isolation, diversity and biological activity. *Environ. Microbiol.*, **9**, 2756–2764.

111 Eccleston, G.P., Brooks, P.R., and Kurtböke, D.I. (2008) The occurrence of bioactive micromonosporae in aquatic habitats of the sunshine coast in Australia. *Mar. Drugs*, **6**, 243–261.

112 Rathna, K.R. and Chandrika, V. (1993) Effect of different media for isolation, growth and maintenance of actinomycetes from mangrove sediments. *Indian J. Mar. Sci.*, **22**, 297–299.

113 Gärtner, A., Blümel, M., Wiese, J., and Imhoff, J.F. (2011) Isolation and characterisation of bacteria from the Eastern Mediterranean deep sea. *Antonie Van Leeuwenhoek*, **100**, 421–435.

114 Postec, A., Urios, L., Lesongeur, F., Ollivier, B., Querellou, J.I., and Godfroy, A. (2005) Continuous enrichment culture and molecular monitoring to investigate the microbial diversity of thermophiles inhabiting deep-sea hydrothermal ecosystems. *Curr. Microbiol.*, **50**, 138–144.

115 Cornec, L., Robineau, J., Rolland, J.L., Dietrich, J., and Barbier, G. (1998) Thermostable esterases screened on hyperthermophilic archaeal and bacterial strains isolated from deep-sea hydrothermal vents: characterization of esterase activity of a hyperthermophilic archaeum, *Pyrococcus abyssi*. *J. Mar. Biotechnol.*, **6**, 104–110.

116 Cavicchioli, R. (2006) Cold adapted archaea. *Nat. Rev. Microbiol.*, **4**, 331–343.

117 Margesin, R. and Miteva, V. (2011) Diversity and ecology of psychrophilic microorganisms. *Res. Microbiol.*, **162**, 346–361.

26
Marine Microbial Enzymes: Biotechnological and Biomedical Aspects

Barindra Sana

26.1
Introduction

Enzymes have historical applications in daily life, although it was not known earlier. With the development of science, enzymes became inevitable for a range of industrial processes. Enzymes play a key role in "green" industrial processes – they are renewable, biodegradable, eco-friendly, and nonhazardous for a worker's health. Many traditional chemical processes have been partially or completely replaced by enzymatic methods. Enzymatic reactions are very specific and take place in mild environment, which avoids unnecessary by-product formation and makes enzyme-based industrial processes more energy efficient. Enzymes can even carry out reactions that are impossible with conventional chemistry, and thereby help establish novel industrial processes. Compared with inorganic catalysis, enzymatic processes have better chemical precision, which can lead to more efficient synthesis of single stereoisomer. Therapeutic enzymes are valuable medicines because they avoid side effects due to their unique specificity. Currently, enzymes are largely used in food processing industry, detergent industry, pharmaceutical industry, organic synthesis, and biotechnological research. They have vast potential in emerging fields of biotechnology, biofuels, and biomedicine.

High specific activity, desired substrate specificity, and stability at reaction conditions are basic requirements for an enzyme to qualify for an industrial application. With the recent progress in biotechnology, enzymes are studied for applications in complex reaction environments that pushed the search for novel enzymes with high activity and extraordinary tolerance at unusual physicochemical conditions. Although many important enzymes have been isolated from plants and animals, continuous supply from these resources faces problems associated with large-scale cultivation, disruption of ecosystem, or depletion of resources [1]. Microbial enzymes are excellent alternatives with diverse biochemical properties and have added advantages of easy genetic manipulation and large-scale production. Many important industrial enzymes are isolated from terrestrial microorganisms,

Marine Microbiology: Bioactive Compounds and Biotechnological Applications, First Edition.
Edited by Se-Kwon Kim

but their marine counterpart is poorly explored [2]. In the course of evolution, marine microorganisms have excellently adapted to diverse environmental parameters such as high salt concentration, extreme temperatures, acidic and alkaline pH, extreme barometric pressure, and low nutrient availability. They developed unique metabolic capabilities to ensure survival in diverse habitats and produce a range of metabolites atypical to terrestrial microbial products [3,4]. Marine microbial enzymes are of special interest due to their better stability, activity, and tolerance to extreme conditions that most of the other proteins cannot withstand. These properties are often helpful for industrial process development [5]. So far archaea, extremophiles, and symbiotic microorganisms are reported to be sources of most of the interesting marine enzymes with distinct structure, novel chemical properties, and biocatalytic activity.

Bioprospecting of unexplored environment is very effective for searching novel biocatalysts, but simultaneous engineering of existing enzymes and process development are equally important for their industrial application. Most reports of marine enzymes are concluded with bioprospecting and physicochemical characterization. In spite of immense possibility, scale up and process development is grossly neglected for marine biotechnology. Potential marine microbial enzymes can be industrially exploited in a better way after successful research in this direction.

26.2
Extremozymes: Most Potential Marine Enzymes

Marine microbial enzymes are of special interest due to their habitat-related distinct features. Global marine environment cannot be defined by a set of physical parameters, it is rather more diverse than its terrestrial counterpart. Marine ecosystem includes hot and cold streams, acidic and alkaline water flow, bright surface, dark deep sea with high barometric pressure, hypersaline hydrothermal vents, or cold seeps enriched with different minerals/gases and deep sea volcanoes. There are also diverse flora and fauna in water or at sea floor. Reasonably, this complex environment contains extremely diverse microbial populations, including several extremophiles that can survive in and even may require extreme physicochemical conditions that are detrimental to most living systems on Earth. In order to adapt to the extreme environments, marine microorganisms produce enzymes with high stability and extraordinary functionality under extreme conditions, where most proteins are unstable. Many possible mechanisms are reported to impart distinct structural, physical, chemical, stereochemical, and catalytic features of extremozymes. The higher specific activity of psychrophilic enzymes may be explained by significantly lower activation energy [6]. Thermostability and cold adaptivity of enzymes are often explained respectively by rigidity and flexibility of their molecular structure, but there are little experimental evidences [7,8]. Rigidity of thermostable enzymes can even explain their low specific activity, while higher specific activities of cold-adapted enzymes agree with their molecular flexibility. In the presence of high salt concentration or organic solvents, proteins suffer from lack of hydration/water and become more rigid, which can predict faster inactivation of thermostable enzymes [9]. Recent experimental evidences suggest that the relationship between stability, flexibility/rigidity, and specific activity is much more complex than expected. Enzyme activity is mostly dependent on the structural features of the active site and may not be directly related with the overall structural rigidity or flexibility of the molecule. Similarly, a specific region of protein structure that determines protein stability may be distinct from catalytic site of an enzyme [10,11].

However, extremozymes are very important from the industrial point of view. Many industrial processes are facilitated by higher solubility, lower viscosity, better mixing, faster reaction rate, and decreased risk of microbial contamination at high temperatures. Thermostable enzymes from thermophilic marine microorganisms are capable of catalyzing these reactions. Two thermostable DNA polymerases of marine microbial origin – the *Taq* polymerase from *Thermus aquaticus* and *Pfu* polymerase from *Pyrococcus furiosus* – are inevitable in molecular biology research [12,13]. The use of thermostable DNA polymerases eliminated the need of addition of fresh enzyme after each polymerase chain reaction (PCR) cycle and makes the process feasible for vast applications. Verenium® Corporation revealed the discovery of a thermostable α-amylase from a deep sea microorganism [14]. This enzyme (Fuelzyme®) is active over a broad range of temperatures and can increase fuel ethanol yields due to improved starch hydrolysis at much lower concentrations.

Cold-adapted enzymes catalyze reactions at low temperatures, and thus reduce energy consumption, stabilize thermolabile reactants/products, minimize evaporation, and diminish the likelihood of microbial contamination. In addition, specific activities of cold-adapted enzymes are very high in comparison to their mesophilic counterparts. Cold-adapted enzymes are abundantly reported from deep sea cold streams, cold seeps, and polar seas. Collins *et al.* [15] reported potential application of psychrophilic xylanases from Antarctic microorganisms in baking industry and the enzymes are already patented. Several cold-active lipases were reported from Antarctic deep sea sediments metagenome and microbial isolates, including *Pseudomonas* and *Psychrobacter* species [16,17]. Cold-active lipolytic enzymes can be exploited in specific industrial processes such as bioremediation of fat-contaminated aqueous systems [18].

Several reactions of food processing, paper industry, detergent industry, and textile industry need to be performed at extremely acidic or alkaline conditions. Biocatalysis of these reactions can be considered only after the availability of suitable enzymes that are stable and active at extreme pH. Alkaliphilic proteases and lipases are useful additives in laundry detergents. Alkaline phosphatase is another useful alkaliphilic enzyme frequently reported from marine microorganisms [19,20]. A marine *Streptomyces* is reported to produce a novel α-amylase that is stable in the pH range of 9–11 for 48 h [21]. This enzyme also retains about 50% of its activity at 85 °C, and therefore its potential widespread application in the detergent industry is conceivable.

Several extracellular enzymes from marine microorganisms have optimal activity in the presence of high salt concentrations. These enzymes are a must for the industrial processes, where reaction mixture contains high salt concentrations that inhibit most common industrial enzymes. Halophilic proteolytic enzymes are useful in peptide synthesis and they also have a potential application in fish and meat processing industries. Halo-tolerant enzymes are uniquely adapted to function in low water availability and eventually many of them are reported to be organic solvent tolerant [22,23]. A solvent-tolerant marine *Bacillus aquimaris* was reported to produce organic solvent stable alkaline cellulase that is also activated by the presence of ionic liquids [24]. The organic solvent-tolerant enantioselective alcohol dehydrogenase from *P. furiosus* is being considered for its potential applications in industrial biocatalysis [25]. Organic solvent-tolerant lipases have vast potential application in both biodiesel production and synthesis of chirally pure compounds.

26.3 Biotechnological Aspects

Enzymes are of importance in biotechnology, as being frequently used from biotechnological research to bioprocess industries. Marine microbial enzymes have distinct advantages over others due to several extraordinary characteristics. A few marine-derived enzymes are already in use in a range of industries and many more are being characterized for their potential applications. In the following section, we will discuss the potential industrial enzymes derived from marine microbial resources with reference to currently available biocatalysts.

26.3.1 Detergent Industry

Enzymes are being used in the detergent industry since 1913 and the first bacterial protease containing detergent was available in 1956 [26]. The detergent industry is now a major consumer of industrial enzymes. Proper functioning of detergents depends on their ability to cope with stains from food and other sources of grease – this can be done with the help of proteases and lipases. In addition to removing proteinaceous stains from foods and blood, proteolytic detergents can cope with normal dermal secretions. In addition, currently available detergents also contain some carbohydrate-degrading enzymes such as cellulase and amylase. Ideal enzymes for detergent application should have broad substrate specificity and activity over a broad range of temperatures and at extremely alkaline pH. Also, they should be compatible with other ingredients of detergent formulation and stable in cleaning conditions. For dry cleaning, the enzymes should be active in the presence of organic solvents.

Several marine microbial enzymes are stable at extreme conditions and in the presence of different chemicals such as oxidants and surfactants, which are essential in detergent formulations [27]. Some of them outperform currently available enzymes when tested in cleansing formulations. Alkaline proteases from marine *Bacillus cereus* and *Streptomyces fungicidicus* efficiently remove recalcitrant bloodstains alone or in combination with commercially available detergents [28,29]. Two proteases from a marine *Bacillus* species showed optimum activity at high temperatures and extreme pH [30,31]. Both the enzymes were stable in the presence of surfactants, bleaching agents, and commercial detergents. Another alkaline protease from a marine shipworm bacterium was reported to increase the cleansing power of standard detergents up to twofold [32]. The enzyme can degrade lysozyme more extensively than subtilisin and was effective in the presence of hydrogen peroxide, which is usually used to sterilize contact lenses, thus confirming its potential application in contact lens cleaning. Alkaline serine proteases from marine γ-Proteobacterium and *Engyodontium album* are active in the presence of organic solvents that suggest their potential application in dry cleaning [33,34]. Purified enzyme from the γ-Proteobacterium was reported to remove various proteinaceous stains even in the presence of organic solvents. Solvent stable lipases/esterases were reported to derive from marine microbial resources that have potential application in dry cleaning [35]. Two thermostable alkaliphilic amylases from marine microorganisms showed excellent stability toward surfactants and commercial detergents, suggesting their potential applications in detergent industry [21,36].

26.3.2 Food Processing Industry

Enzymes have multidimensional applications in food processing industries. They are used for preservation of foods and food components, efficient utilization of raw materials, improvement of taste/texture, and reduction of processing costs. Several steps of food processing are carried out at extreme pH or temperature and extremozymes have valuable application in catalyzing these reactions. Thermostable polysaccharolytic enzymes are frequently used in food processing industries for conversion of starch into oligosaccharides, cyclodextrins, and maltose or glucose, while acid-stable enzymes can potentially improve starch liquefaction [37].

Proteases have a long history of application in food processing industry. They are used in cheese manufacturing, baking, meat tendering, and soy protein hydrolysis. In baking and dairy applications, proteases need to be inactivated by moderate temperature, while cheese manufacturing and meat and seafood processing need enzyme activity at high salt concentrations. Cold-adapted and salt-tolerant marine proteases have potential application for these purposes. Several proteases are being used as digestive enzyme. A protease from marine yeast *Metschnikowia reukaufii* W6b was reported to have potential application in cheese, food, and fermentation industries [38]. It showed high skimmed milk coagulability in skimmed milk clotting test. Lipases are important in food processing industry due to their distinct catalytic properties in reactions such as esterification, hydrolysis, and *trans*-esterification [39]. They have numerous applications in the fat and oil industries, such as the production of triglycerides enriched with n-3 long chain polyunsaturated fatty acids. Polysaccharolytic enzymes are used in baking, brewing, and production of natural sweeteners. Amylases and glucosidases are used in glucose or maltose syrup production from starch. Glucose isomerase catalyzes the conversion of glucose to fructose, while pullulanase is used for debranching of amylopectin molecules containing α-1,4- and α-1,6-glucosidic linkages [40]. Several amylases and pullulanases were reported from marine microorganisms that can hydrolyze starch of various food grains and have potential applications in food and beverage industries. Li *et al.* [41] reported a yeast amylase that can digest potato starch and corn starch. Several α-amylases were reported from a marine *Bacillus* that can degrade raw starch from corn, cassaca, sago, potato, rice, and maize to produce simple sugars such as maltose, maltotriose, maltotetraose, maltopentaose, maltohexaose, and glucose [42,43]. Chitin hydrolysate has potential application in food and pharmaceutical industries and the chitinolytic enzymes can be used in the production of commercially valuable oligomeric units from chitin. Currently, the conversion is carried out by acid hydrolysis, but more consistent product formation is possible by the use of chitinase. Several chitinases are reported from marine bacteria and hyperthermophilic archaea [44].

26.3.3
Chemical and Pharmaceutical Synthesis

Enzymes are used as catalysts in many chemical synthesis processes. Marine enzymes are of special interest for their unique catalytic properties, novel stereochemical properties, and solvent stability. Some aqueous hydrolysis reverts to synthesis in organic solvent system, and suitable solvent stable enzymes are useful to catalyze these synthesis reactions. Marine enzymes are reported to produce pure racemic compounds that are not observed in normal catalysis. Several examples are found in the class of lipase, esterase, and oxidoreductase. An organic solvent-tolerant alcohol dehydrogenase from *P. furiosus* catalyzes the reduction of aryl ketones to their corresponding chiral alcohols in an enantiomerically pure form [25]. Marine epoxide hydrolases were reported to have enantioconvergent applications for the synthesis of drug intermediates with specific stereochemistry [18,45,46]. Enantioselective epoxide hydrolases were purified from two marine bacteria, and pure enzymes were reported to perform better than existing enzymes in terms of enantioselective resolution of racemic compounds [45,46].

Novel catalytic properties of enzymes are best exploited in pharmaceutical industry. For many drug molecules, only one stereoisomer is physiologically active – high stereospecificity of enzymatic processes make them favorable over chemical processes in pharmaceutical industry. Lipase B from *Candida antarctica* is a commercial enzyme with high enantioselectivity and is used in enzymatic acylation of Nelarabin for the production of its 5′-monoacetate derivative, which is an antileukemic agent with higher solubility and thus with better bioavailability [47]. This enzymatic process is highly regioselective, almost 99% substrate converts to desired product. This reaction take place in the presence of various organic solvents and the best substrate conversion was achieved in anhydrous dioxane. However, this level of conversion is almost impossible in conventional chemical acetylation processes due to the lack of regioselectivity. *C. antarctica* lipase B and many esterase/lipase are active in the presence of organic solvents that facilitate their use in catalyzing transesterification reactions. *C. antarctica* lipase B is being studied for its potential application for enzymatic synthesis of vitamin E acetate by lipase-catalyzed transesterification.

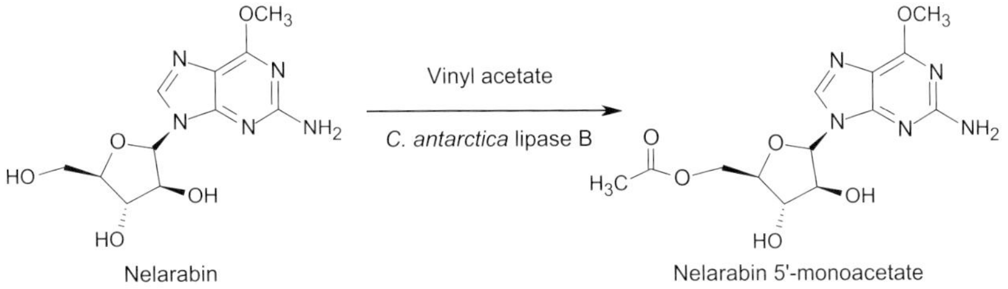

Nelarabin → (Vinyl acetate; *C. antarctica* lipase B) → Nelarabin 5'-monoacetate

The esterase from the marine isolate *Yarrowia lipolytica* CL180 was overexpressed in *Escherichia coli* and the recombinant enzyme was used to produce levofloxacin (*S*-ofloxacin) by preferential hydrolysis of *S*-enantiomer from a racemic mixture of ofloxacin ester [48]. In addition, this enzyme is extremely psychrophilic and retains about 40% of the optimum activity at 10 °C. Another esterase was reported from a marine *Vibrio* that also preferentially produces levofloxacin by the hydrolysis of ofloxacin ester [49].

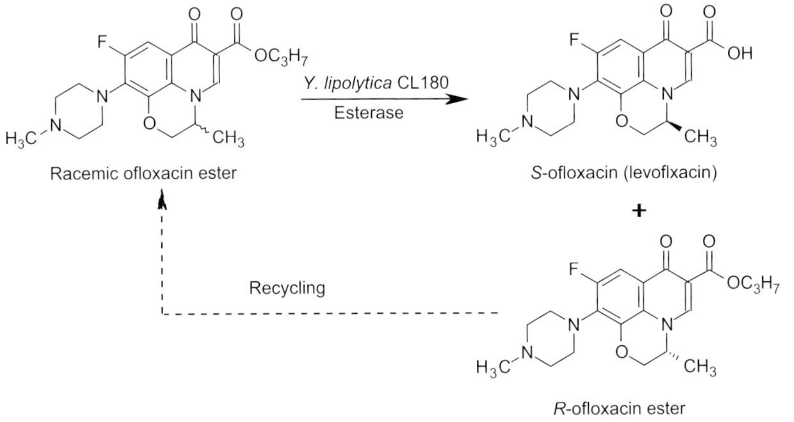

Racemic ofloxacin ester → (*Y. lipolytica* CL180 Esterase) → *S*-ofloxacin (levoflxacin) + *R*-ofloxacin ester (Recycling)

Peptides with high angiotensin-converting enzyme (ACE) inhibitory activity have potential application in the treatment of hypertension. Marine-derived recombinant proteases were reported to hydrolyze various proteins for the production of bioactive peptides having ACE inhibitory activity and antioxidant property [50,51].

26.3.4
Biotechnological Research

Enzymes are well-known tools of biotechnological research. Several enzymes (such as DNA polymerase, restriction endonuclease, ligase, alkaline phosphatase, reverse transcriptase, and kinase) play a very important role in molecular biology and recombinant DNA research. Many biotechnological reactions favor extreme environments and often the corresponding enzymes are marine-derived. Two thermostable DNA polymerases from marine microorganisms *Thermococcus litoralis* (Vent polymerase, New England Biolab) and *Pyrococcus furiosus* (*Pfu* polymerase, Stratagene) are vastly used for DNA amplification by PCR. DNA ligase from the marine isolate *Thermus thermophilus* is another important enzyme in biotechnological research. Several restriction endonuclease, including *Asp*MD1, *Dma*I, *Dpa*I, *Age*I, *Hja*I, *Hac*1, *Hsa*1, and *Hag*1, were isolated from marine bacteria and some of them are already commercially available. All these enzymes are potential tools for double-stranded DNA digestion. DNA topoisomerases play an important role in replication, transcription, and gene expression. Several DNA topoisomerases have been isolated from various marine bacteria and a few of them are characterized for their potential application in biotechnological research [27].

26.3.5
Leather Industry

Traditional leather processing techniques are highly polluting to environment and the worldwide leather industry is under pressure to comply with pollution and discharge legislations [52]. Currently, the industry is replacing harsh chemical processes with eco-friendly biotechnological means. Several proteases and lipases are employed for partial or complete replacement of traditional chemical processes for soaking, bating, dehairing, and degreasing of rawhides in leather industry [53–56]. Most of these processes work at high temperatures and extreme pH or in the presence of high salt/surfactant concentrations, which reduce activity of many currently used enzymes in leather industry [54]. Several thermostable salt-tolerant alkaline proteases/lipases from marine microorganisms fulfill all these requirements and may be better alternatives for this purpose. A keratinolytic enzyme was reported from a marine fungus that has potential application in leather industry. It is stable at extremely alkaline environment and active at high temperatures and extreme pH even in the presence of detergents, oxidizing/reducing agents, and several metal ions [57]. Enzymes are also employed for waste processing in leather industry, but high salinity of tannery wastewater makes it difficult to be treated with many common industrial enzymes. Some salt-tolerant marine bacteria were reported to degrade various organic pollutants of tannery wastage and showed up to 80% chemical oxygen demand (COD) reduction in the presence of high NaCl concentrations [58].

26.3.6 Paper Industry

In paper industry, enzymatic processes are preferred over chemical reactions because of improved fiber quality, least pollution, and less energy consumption. Enzymes have historical applications in pulp and paper industries, but due to limited availability of suitable enzymes, their use was restricted to raw starch treatment. Now microbial enzymes are used to replace harsh chemical processes in paper industry. Xylanases are widely used in the prebleaching of Kraft pulp. Applications of lipases in deresination and pitch control process are gradually increasing. Lignolytic enzymes and laccase have potential application in pulp bleaching. Unlike xylanases, lignolytic enzymes attack lignin directly and are more effective in this process. In addition to giving desired biobleaching effect, the biobleaching enzymes should be highly thermostable and should be completely free from cellulase activity. The xylanase produced by *Bacillus pumilus* strain can efficiently degrade birch wood xylan and has potential application in paper pulp industry [59]. Another highly thermostable xylanase was reported from a marine *Thermotoga* strain that showed 90 min half-life at 95 °C [60]. Luo *et al.* [61] reported xylanolytic and ligninolytic activity in several marine-derived fungal isolates that can produce xylanase or laccase enzymes.

26.3.7 Textile Industry

Traditional raw material processing in textile industry was dependent on hazardous chemical processes, which were operated at high temperatures and required acid, alkali, bleach, and detergent treatment. First enzymatic process used in textile industry was in the field of desizing, that is, removal of size (usually starch-based adhesive) applied to prevent damage due to mechanical strain during weaving. Desizing with acid, alkali, or oxidizing agents sometimes damage or discolor the cotton that can be greatly prevented by using starch-degrading enzymes – the amylases. In 1987, introduction of enzymatic alternatives for stonewashing jeans is considered a major breakthrough in this field, after which a majority of denim finishing laundries replaced pumice stones with enzymes. Today, various polysaccharolytic enzymes are also used for weaving, scouring, impurity reduction, and pretreatment before dying. Verenium Corporation sells an enzyme product Cottonase®, which is a versatile, economically viable, and environment-friendly alternative to chemical scouring in cotton preparation [62]. Colotex Biotechnology sells several enzymes for textile industry applications [63]. Most of these enzymes are derived from terrestrial sources, but several marine microbial enzymes showed similar properties and have potential application in textile industries. The desizing amylases need to be thermostable and scouring enzymes should be active at high pH (8.0–9.0). An amylase from *Rhodothermus marinus* showed maximum activity at 80 °C and has a half-life of 73.7 min at this temperature [64]. A marine *Streptomyces* produced a novel α-amylase that is active at high temperatures and stable in the pH range of 9–11 for 48 h [21]. This enzyme has potential application as a desizing enzyme in the textile industry. A cellulase enzyme produced by the marine isolate *Streptomyces ruber* showed optimum activity at pH 6.0 and 40 °C. This acidic cellulase can be potentially used in polishing for developing denim-finished products [65].

26.3.8
Biofuel Industry

Due to fast depletion of fossil fuels and several environmental reasons, biofuels are emerging as the preferred fuels of tomorrow. Biodiesel, bioalcohols, and biohydrogen are the major biofuels being considered as sustainable/renewable energy. Enzymes may play a vital role in the production of all these kinds of biofuels. In biodiesel industry, lipase and esterase have potential applications in oil degradation and transesterification reactions. Cellulolytic, lignolytic, and other polysaccharolytic enzymes may replace acid hydrolysis procedures used in bioalcohol industries for the production of simple sugars from polysaccharide biomasses. Hydrogenases are proved to be involved in photoevolution of hydrogen by phototropic bacteria and in water splitting by cyanobacteria. Marine algae and cyanobacteria are being studied for biohydrogen production. However, uses of isolated enzymes in photolytic hydrogen production techniques are yet to be established.

Marine microorganisms are well-known sources of highly active esterases and lipases – several of them are active in the presence of organic solvents essential for biodiesel production. The lipase produced by the marine *B. pumilus* B106 is active in extreme saline condition and has potential application in marine organism-based biodiesel production where raw materials have unusually high salt content [66]. The crude lipase from *Aureobasidium pullulans* HN2–3 strain (isolated from marine saltern) showed highest lipolytic activity toward peanut oil and can be used in biodiesel production from waste peanut oil of household and restaurants [67]. Cellulose, xylan, and lignocellulose are major components of almost all biowastes and some industrial wastage. Conversion of these polysaccharides to simple fermentable sugar has a vast potential application in biofuel industry. Several marine microbial xylanases and cellulases were reported to degrade industrial and agricultural wastage. Cellulase-producing marine microorganisms showed preferential degradation of rice straw, rice bran, and cottonseed biomasses, respectively [65,68]. *Saccharophagus degradans* were reported to synthesize a group of enzymes that can degrade several complex polysaccharides from diverse algal, plant, and invertebrate sources [69]. A cold-active halo-tolerant xylanase from *Glaciecola mesophila* KMM 241 was reported to hydrolyze xylooligosaccharides and xylan into xylobiose and xylotriose [70]. Two halo-tolerant xylanases were reported from two different *Bacillus* species isolated from the west coast of India – one of the xylanases preferentially degrades agricultural waste such as wheat bran [59,71]. Lignocellulosic materials are often exposed to marine environment and several marine fungi were reported to have lignocellulose-degrading activity [61,72].

Suitable lignolytic enzymes may be extremely useful in biofuel production from waste wood and other plant biomasses. Salt-tolerant polysaccharolytic enzymes may play an important role in bioalcohol production from (saline) seaweed biomasses. Carrageenolytic enzymes are used for liquefaction of carrageenan in food, pharmaceutical, and cosmetic industries. Salt-tolerant carragenase enzymes can also be used for the production of simple sugar from carrageenan-rich seaweed biomasses. This enzyme is abundant in marine microorganisms in comparison to its terrestrial counterpart. Mou *et al.* [73] isolated 69 κ-carrageenase-producing marine microorganisms from soil and seaweed surfaces from the different areas of China Sea. Several other carrageenolytic enzymes were reported from marine *Pseudomonas* species [74,75].

26.3.9 Bioremediation and Biofouling

Bioremediation is the technique for removal of pollutants by biological means, mainly by the use of microorganisms. Although several terrestrial microorganisms were reported to be highly efficient in the bioremediation of toxic organic pollutants, marine microorganisms have added advantages of adaptability to high salinity, high temperatures, and extreme pH. Although almost all large-scale bioremediation techniques deal with intact microorganisms, several enzymes were tested for this purpose. An extracellular lipase was isolated from the marine fungus *Aspergillus awamori* BTMFW032 that is able to reduce 92% fats and oils in the oil-rich effluent [76]. The enzyme can reduce several commercially used fats and oils that suggest its potential application in effluent treatments. However, large-scale production and immobilization of enzymes may not be feasible for bioremediation applications.

Marine microbial enzymes have potential application in the control of biofouling. Submerged marine structures and ships are affected by biofouling, that is, attachment and growth of marine biomasses on submerged surfaces that increase surface roughness, damage surface coating, and enhance corrosion of surface materials. Historically, biofouling is controlled by the use of antifouling coatings on submerged surfaces that prevent microbial attachment by releasing biocides, which are toxic to marine wildlife and hazardous to marine environment. Eco-friendly antifouling agents became crucial after global and regional ban of highly toxic antifouling paints. Extracellular polymeric matrixes reported to play key role in biofouling, primarily by providing initial adhesion to microorganisms. These matrixes are composed of polysaccharides, proteins, glycoproteins, and phospholipids – hydrolytic enzymes can potentially prevent their deposition and inhibit biofouling at the very initial step. Several environmental parameters are to be considered for selection of antifouling enzymes – they must be stable/active in highly saline marine environments and over a broad range of pH and temperature. Logically, in this environment, marine microbial enzymes should perform better over their terrestrial counterparts. Protease, cellulase, and agarase from marine microorganisms may play lead roles as antifouling enzymes.

26.4 Biomedical Aspects

Apart from their application in pharmaceutical process development, enzymes can be used as therapeutic agents in treating a range of physiological disorders starting from digestive problem to neoplastic disorders. High affinity and specificity of each enzyme to a particular substrate is the major advantage for their therapeutic applications. It decreases incidence of unwanted side effects, and therefore enzyme therapy is preferable over many existing chemotherapeutic treatments.

Proteases are traditionally used as digestive aids. Today, with a broad range of proteolytic enzymes, digestive enzyme preparations also contain cellulases (for fiber digestion), amylases (for carbohydrate digestion), and lipases (for fat digestion). Stability at acidic pH of gastrointestinal tract is the biggest challenge for an enzyme to qualify for digestive applications. Many plant and animal enzymes were classically used as digestive enzymes without any remarkable tolerance to acidic pH, which are now partially replaced by microbial enzymes and recombinant proteins. Acid-stable marine microbial enzymes can be a good replacement for these traditional digestive enzymes.

Most of the cardiovascular diseases are related with blood clotting that blocks arteries, veins, and smaller blood vessels. This problem can be tackled either by preventing blood coagulation or by dissolving the clots. Microbial enzymes have been employed in both ways. Streptokinase, urokinase, and other plasminogen activator enzymes are well known for thrombolytic applications. Blood clots are composed of protein (fibrin) and can be potentially dissolved by injecting fibrinolytic/proteolytic enzymes from external sources. New microbial fibrinolytic enzymes have potential application for this purpose [77–79].

Collagenases have medical applications in wound healing. They are effective for the removal of dead tissue from wounds, burns, and ulcers, which can speed up the growth of new tissues and skin grafts. Collagenase SANTYL® is a debriding ointment available on the market that contains collagenase as active ingredients; the enzyme helps to dissolve dead cells from chronic dermal ulcers and severely burned areas [80]. Collagenase also has the ability to inhibit the growth of some contaminant pathogens and is used in combination with some antimicrobial agents [26,27]. Several collagenases were reported from marine microbial sources, but their application in wound healing is yet to be established [81,82].

Two amino acid hydrolyzing enzymes – asparaginase and glutaminase – have potential application as anticancer agents. L-Asparaginase is a drug of choice in combination therapy for treatment of children lymphoblastic leukemia [83]. A microbial asparaginase was reported to eliminate asparagine from the bloodstream and is effective in the treatment of lymphocytic leukemia [26]. There is growing interest in screening of these enzymes for their exploitation as anticancer drugs. Marine microorganisms are well-known sources of these two enzymes [84–86]. *Pseudomonas fluorescens* is a marine bacterium that produces a salt-tolerant L-glutaminase reported to have antineoplastic activity [27,87]. However, systemic administration of these microbial enzymes is questionable due to their potential antigenic activity to human immune system.

Lysozyme is well known for its antibacterial activity and is used to prevent bacterial contamination in food and personal care products. It exhibits antimicrobial activity by hydrolyzing polysaccharides of bacterial cell walls. Cell walls of some pathogenic fungus, protozoa, and helminths are composed of chitin that could be hydrolyzed by microbial chitinases, and therefore chitinase can be used for the treatment of these microbial infections. RNA is the sole genetic material in several pathogenic viruses, which can be killed by ribonuclease enzymes. However, for hydrolyzing its genetic material, the enzyme must be able to penetrate the virus capsid. Several ribonucleases are being studied for their potential application in treatment of HIV and other viral infections, but mostly from higher organisms. Microbial ribonucleases can be studied, and screening of marine microorganisms may open a potential field of antiviral research.

In spite of many potential applications and without any side effects, success of enzyme therapy is mostly restricted to extracellular applications either for pure topical uses or for treatments of blood circulation disorders. Biomedical applications of enzymes are being restricted by several factors, which need to be overcome for considering their therapeutic applications. First, enzymes are too large to be distributed within the cells and that is why the enzyme therapy cannot be used to treat any disease at genetic level. Second, in most cases, enzymes are foreign proteins to the human body and are susceptible to antigenic reactions that may cause mild allergic reactions to severe life-threatening immune responses. Third, enzymes have very short effective lifetime in blood circulation that may not be enough to complete the enzymatic reactions to treat some disorders. Entrapment of enzymes in nonproteinaceous materials may help in most cases, but many of these materials often cause undesired side effects. Current biomedical researchers are addressing these issues and enzyme therapy would be applicable in much larger area of biomedicine only after answering all these practical problems.

26.5
Concluding Remarks and Perspectives

Successful exploitation of an industrial enzyme is a multidisciplinary task involving systematic application of a range of technologies. Every step from screening to process development faces unique challenge for a particular application and no universal method is practicable to address them. In fact, a single problem can be tackled in different stages of development, but a successful story maker needs to find out the specific problem to be addressed to satisfy the exact requirements of an industrial application. The solution is hidden in the logical application of a variety of techniques from relatively unrelated segments of science; successful researchers need to master them, but their application at appropriate time is equally important.

Hundreds of marine microbial enzymes are already reported to have novel characteristics that would be certainly advantageous for their industrial applications, but most reports are concluded with bioprospecting and physicochemical characterization of the isolated enzymes. A few of them are further studied for cloning and gene overexpression, laboratory-scale product optimization, and structural profile. In spite of immense possibility, scale up of marine microbial enzyme is a rare phenomenon and there is almost no report of selective modification of marine microbial enzymes. Enzyme immobilization is a very important technique for repeated catalysis in industrial processes. It is inevitable in terms of cost- effectiveness of industrial biocatalysis, but it is rarely practiced in marine biotechnology. Future research need to focus on protein engineering, structural profile, and scale-up and downstream processing of marine microbial enzymes – orderly use of the techniques would show the way to their effective biotechnological and biomedical applications.

References

1 Pomponi, S.A. (1999) The bioprocess-technological potential of the sea. *J. Biotechnol.*, **70**, 5–13.

2 Schafer, T., Borchert, T.W., Nielsen, V.S. *et al.* (2007) Industrial enzymes. *Adv. Biochem. Eng. Biotechnol.*, **105**, 59–131.

3 Faulkner, D.J. (2000) Highlights of marine natural products chemistry (1972–1999). *Nat. Prod. Rep.*, **17**, 1–6.

4 Fenical, W. and Jensen, P.R. (1993) Marine microorganisms: a new biomedical resource, in *Marine Biotechnology: Pharmaceutical and Bioactive Natural Products* (eds D.H. Attaway and O.R. Zabrosky), Plenum Press, New York, p. 419.

5 Burg, B.V.D. (2003) Extremophiles as a source of novel enzymes. *Curr. Opin. Microbiol.*, **6**, 213–218.

6 Marx, J.C., Collins, T., D'Amico, S., Feller, G., and Gerday, C. (2007) Cold-adapted enzymes from marine Antarctic microorganisms. *Mar. Biotechnol.*, **9**, 293–304.

7 Aghajari, N., Feller, G., Gerday, C., and Haser, R. (1996) Crystallization and preliminary X-ray diffraction studies of α-amylase from the Antarctic psychrophile *Alteromonas haloplanctis* A23. *Protein Sci.*, **5**, 2128–2129.

8 Vihinen, M. (1987) Relationship of protein flexibility to thermostability. *Protein Eng.*, **1**, 477–480.

9 Partridge, J., Hailing, P.J., and Moore, B.D. (1998) Practical route to high activity enzyme preparations for synthesis in organic media. *Chem. Commun.*, **7**, 841–842.

10 Gerday, C., Aittaleb, M., Bentahir, M. *et al.* (2000) Cold-adapted enzymes: from fundamentals to biotechnology. *Trends Biotechnol.*, **18**, 103–107.

11 Shoichet, B.K., Baase, W.A., Kuroki, R., and Matthews, B.W. (1995) A relationship between protein stability and protein function. *Proc. Natl. Acad. Sci. USA*, **92**, 452–456.

12 Chien, A., Edgar, D.B., and Trela, J.M. (1976) Deoxyribonucleic acid polymerase from the extreme thermophile *Thermus aquaticus*. *J. Bacteriol.*, **127**, 1550–1557.

13 Lundberg, K.S., Shoemaker, D.D., Adams, M.W.W., Short, J.M., Sorge, J.A., and Mathur, E.J. (1991) High-fidelity amplification using a thermostable DNA polymerase isolated from *Pyrococcus furiosus*. *Gene*, **108**, 1–6.

14 Verenium-Fuelzym® (2012) Fuelzyme® enzyme is a next generation alpha amylase for starch liquefaction. Available online http://www.verenium.com/products_fuelzyme.html (accessed March 22, 2012).

15 Collins, T., Hoyoux, A., Dutron, A. *et al.* (2006) Use of glycoside hydrolase family 8 xylanases in baking. *J. Cereal Sci.*, **43**, 79–84.

16 Zhang, J., Lin, S., and Zeng, R. (2007) Cloning, expression, and characterization of a cold-adapted lipase gene from an Antarctic deep-sea psychrotrophic bacterium, *Psychobacter* sp. 7195. *J. Microbiol. Biotechnol.*, **17**, 604–610.

17 Zhang, J. and Zeng, R. (2008) Molecular cloning and expression of a cold-adapted lipase gene from an Antarctic deep sea psychrotrophic bacterium *Pseudomonas* sp. 7323. *Mar. Biotechnol.*, **10**, 612–621.

18 Tricorne, A. (2011) Marine biocatalysts: enzymatic features and applications. *Mar. Drugs*, **9**, 478–499.

19 Plisova, E.Y., Balabanova, L.A., Ivanova, E.P. *et al.* (2005) A highly active alkaline phosphatase from the marine bacterium *Cobetia*. *Mar. Biotechnol.*, **7**, 173–178.

20 Sebastian, M. and Ammerman, J.W. (2009) The alkaline phosphatase PhoX is more widely distributed in marine bacteria than the classical PhoA. *ISME J.*, **3**, 563–572.

21 Chakraborty, S., Khopade, A., Kokare, C., Mahadik, K., and Chopade, B. (2009) Isolation and characterization of novel α-amylase from marine *Streptomyces* sp. D1. *J. Mol. Catal. B*, **58**, 17–23.

22 Marhuenda-Egea, F.C. and Bonete, M.J. (2002) Extreme halophilic enzymes in organic solvents. *Curr. Opin. Biotechnol.*, **13**, 385–389.

23 Sellek, G.A. and Chaudhuri, J.B. (1999) Biocatalysis in organic media using enzymes from extremophiles. *Enzyme Microb. Technol.*, **25**, 471–482.

24 Trivedi, N., Gupta, V., Kumar, M., Kumari, P., Reddy, C.R.K., and Jha, B. (2011) Solvent tolerant marine bacterium *Bacillus aquimaris* secreting organic solvent stable alkaline cellulase. *Chemosphere*, **83**, 706–712.

25 Zhu, D., Malik, H.T., and Hua, L. (2006) Asymmetric ketone reduction by a hyperthermophilic alcohol dehydrogenase. The substrate specificity, enantioselectivity and tolerance of organic solvents. *Tetrahedron: Asymmetry*, **17**, 3010–3014.

26 Rao, M.B., Tanksale, A.M., Ghatge, M.S., and Deshpande, V.V. (1998) Molecular and biotechnological aspects of microbial proteases. *Microbiol. Mol. Biol. Rev.*, **62**, 597–635.

27 Chandrasekaran, M. and Rajeev Kumar, S. (2003) Marine microbial enzymes in Biotechnology, in *Encyclopedia of Life Support Systems (EOLSS)* (eds H.W. Doelle, S. Rokem, and M. Berovic), Eolss Publishers, Oxford, UK 1–35.

28 Abou-Elela, G.M., Ibrahim, H.A.H., Hassan, S.W., Abd-Elnaby, H., and El-Toukhy, N.M.K. (2011) Alkaline protease production by alkaliphilic marine bacteria isolated from Marsa-Matrouh (Egypt) with special emphasis on *Bacillus cereus* purified protease. *Afr. J. Biotechnol.*, **10**, 4631–4642.

29 Ramesh, S., Rajesh, M., and Mathivanan, N. (2009) Characterization of a thermostable alkaline protease produced by marine *Streptomyces fungicidicus* MML1614. *Bioprocess. Biosyst. Eng.*, **32**, 791–800.

30 Gouda, M.K. (2006) Optimization and purification of alkaline proteases produced by marine *Bacillus* sp. MIG newly isolated from eastern harbour of Alexandria. *Pol. J. Microbiol.*, **55**, 119–126.

31 Kumar, C.G., Joo, H.S., Koo, Y.M., Paik, S. R., and Chang, C.S. (2004) Thermostable alkaline protease from a novel marine haloalkalophilic *Bacillus clausii* isolate. *World J. Microbiol. Biotechnol.*, **20**, 351–357.

32 Greene, R.V., Griffin, H.L., and Cotta, M.A. (1996) Utility of alkaline protease from marine shipworm bacterium in industrial cleansing applications. *Biotechnol. Lett.*, **18**, 759–764.

33 Chellappan, S., Jasmin, C., Basheer, S.M. *et al.* (2011) Characterization of an extracellular alkaline serine protease from marine *Engyodontium album* BTMFS10. *J. Ind. Microbiol. Biotechnol.*, **38**, 743–752.

34 Sana, B., Ghosh, D., Saha, M., and Mukherjee, J. (2006) Purification and characterization of a salt, solvent, detergent and bleach tolerant protease from a new gamma-Proteobacterium isolated from the marine environment of the *Sundarbans*. *Process Biochem.*, **41**, 208–215.

35 Sana, B., Ghosh, D., Saha, M., and Mukherjee, J. (2007) Purification and characterization of an extremely dimethylsulfoxide tolerant esterase from a salt-tolerant *Bacillus* species isolated from the marine environment of the Sundarbans. *Process Biochem.*, **42**, 1571–1578.

36 Chakraborty, S., Khopade, A., Biao, R. *et al.* (2011) Characterization and stability studies on surfactant, detergent and oxidant stable α-amylase from marine haloalkaliphilic *Saccharopolyspora* sp. A9. *J. Mol. Catal. B*, **68**, 52–58.

37 Synowiecki, J., Grzybowska, B., and Zdzieblo, A. (2006) Sources, properties and suitability of new thermostable enzymes in food processing. *Crit. Rev. Food Sci. Nutr.*, **46**, 197–205.

38 Li, J., Peng, Y., Wang, X., and Chi, Z. (2010) Optimum production and characterization of an acid protease from marine yeast *Metschnikowia reukaufii* W6. *J. Ocean Univ. China*, **9**, 359–364.

39 Shahidi, F. and Wanasundara, U.N. (1998) Omega-3 fatty acid concentrates: nutritional aspects and production technologies. *Trends Food Sci. Technol.*, **9**, 230–240.

40 Guzman-Maldonado, H. and Paredes-Lopez, O. (1995) Amylolytic enzymes and products derived from starch: a review. *Crit. Rev. Food Sci. Nutr.*, **35**, 373–403.

41 Li, H., Chi, Z., Wang, X., Duan, X., Ma, L., and Gao, L. (2007) Purification and characterization of extracellular amylase from the marine yeast *Aureobasidium pullulans* N13d and its raw potato starch digestion. *Enzyme Microb. Technol.*, **40**, 1006–1012.

42 Puspasari, F., Nurachman, Z., Noer, A.S., Radjasa, O.K., van derMaarel, M.J.E.C., and Natalia, D. (2011) Characteristics of raw starch degrading α-amylase from *Bacillus aquimaris* MKSC 6.2 associated with soft coral *Sinularia* sp. *Starch*, **63**, 462–467.

43 Vidilaseris, K., Hidayat, K., Retnoningrum, D.S., Nurachman, Z., Noer, A.S., and Natalia, D. (2009) Biochemical characterization of a raw starch degrading α-amylase from the Indonesian marine bacterium *Bacillus* sp. ALSHL3. *Biologia*, **64**, 1047–1052.

44 Lee, J.S., Joo, D.S., Cho, S.Y., Ha, J.H., and Lee, E.H. (2000) Purification and characterization of extracellular chitinase produced by marine bacterium *Bacillus* sp. LJ-25. *J. Microbiol. Biotechnol.*, **10**, 307–311.

45 Kim, H.S., Lee, O.K., Lee, S.J. *et al.* (2006) Enantioselective epoxide hydrolase activity of a newly isolated microorganism, *Sphingomonas echinoides* EH-983, from seawater. *J. Mol. Catal. B*, **41**, 130–135.

46 Woo, J.H., Kang, J.H., Hwang, Y.O., Cho, J.C., Kim, S.J., and Kang, S.G. (2010) Biocatalytic resolution of glycidyl phenyl ether using a novel epoxide hydrolase from a marine bacterium, *Rhodobacterales bacterium* HTCC2654. *J. Biosci. Bioeng.*, **109**, 539–544.

47 Mahmoudian, M., Eddy, J., and Dowson, M. (1999) Enzymic acylation of 506U78 (2-amino-9-b-D-arabinofuranosyl-6-methoxy-9*H*-purine), a powerful new anti-leukaemic agent. *Biotechnol. Appl. Biochem.*, **29**, 229–233.

48 Kim, J.T., Kang, S.G., Woo, J.H., Lee, J.H., Jeong, B.C., and Kim, S.J. (2007) Screening and its potential application of lipolytic activity from a marine environment: characterization of a novel esterase from *Yarrowia lipolytica* CL180. *Appl. Microbiol. Biotechnol.*, **74**, 820–828.

49 Park, S.Y., Kim, J.T., Kang, S.G. *et al.* (2007) A new esterase showing similarity to putative dienelactone hydrolase from a strict marine bacterium, *Vibrio* sp. GMD509. *Appl. Microbiol. Biotechnol.*, **77**, 107–115.

50 Ma, C., Ni, X., Chi, Z., Ma, L., and Gao, L. (2007) Purification and characterization of an alkaline protease from the marine yeast *Aureobasidium pullulans* for bioactive peptide production from different sources. *Mar. Biotechnol.*, **9**, 343–351.

51 Ni, X., Yue, L., Chi, Z., Li, J., Wang, X., and Madzak, C. (2009) Alkaline protease gene cloning from the marine yeast *Aureobasidium pullulans* HN2–3 and the protease surface display on *Yarrowia lipolytica* for bioactive peptide production. *Mar. Biotechnol.*, **11**, 81–89.

52 Chowdhary, R.B., Jana, A.K., and Jha, M.K. (2004) Enzyme technology applications in leather industries. *Indian J. Chem. Technol.*, **11**, 659–671.

53 Kamini, N.R., Hemachander, C., Mala, J.G. S., and Puvanakrishnan, R. (1999) Microbial enzyme technology as an alternative to conventional chemicals in leather industry. *Curr. Sci.*, **77**, 80–86.

54 Madhavi, J., Srilakshmi, J., Rao, F.V.R., and Rao, K.R.S.S. (2011) Efficient leather dehairing by bacterial thermostable protease. *Int. J. Biosci. Biotechnol.*, **3**, 11–36.

55 Peper, K.W. and Wyatt, K.G.E. (1989) Enzymatic unhairing of heavy hides. *J. Ind. Leather Technol. Assoc.*, **36**, 214–233.

56 Puvankrishnan, R. and Dhar, S.C. (1986) Recent advances in the enzymatic depilation of hides and skins. *Leather Sci.*, **33**, 177–191.

57 El-Gendy, M.M.A. (2010) Keratinase production by endophytic *Penicillium* spp. Morsy1 under solid-state fermentation using rice straw. *Appl. Biochem. Biotechnol.*, **162**, 780–794.

58 Sivaprakasam, S., Mahadevan, S., Sekar, S., and Rajakumar, S. (2008) Biological treatment of tannery wastewater by using salt-tolerant bacterial strains. *Microb. Cell Fact.*, **7**, 15.

59 Menon, G., Mody, K., Keshri, J., and Jha, B. (2010) Isolation, purification, and characterization of haloalkaline xylanase from a marine *Bacillus pumilus* strain, GESF-1. *Biotechnol. Bioprocess. Eng.*, **15**, 998–1005.

60 Simpson, H.D., Haufler, U.R., and Daniel, R.M. (1991) An extremely thermostable xylanase from the thermophilic eubacterium *Thermotoga*. *Biochem. J.*, **277**, 413–417.

61 Luo, W., Vrijmoed, L.L.P., and Jones, E.B.G. (2005) Screening of marine fungi for lignocellulose-degrading enzyme activities. *Bot. Mar.*, **48**, 379–386.

62 Verenium Corporation (2012) Cottonase® Textile Processing Enzyme for Cotton. Available online http://www.verenium.com/prod_cottonase.html (accessed May 22, 2012).

63 Meng Tai Co. (2012) Bioproducts from Colotex. Available online http://www.mengtaith.com/upFile/2010381350890.pdf (accessed April 29, 2012).

64 Seong-Ae, Y., Ryu, S.I., Lee, S.B., and Moon, T.W. (2008) Purification and characterization of branching specificity of a novel extracellular amylolytic enzyme from marine *Hyperthermophilic Rhodothermus* marinus. *J. Microbiol. Biotechnol.*, **18**, 457–464.

65 El-Sersy, N.A., Abd-Elnaby, H., Abou-Elela, G.M., Ibrahim, H.A.H., and El-Toukhy, N. M.K. (2010) Optimization, economization and characterization of cellulase produced by marine *Streptomyces ruber*. *Afr. J. Biotechnol.*, **9**, 6355–6364.

66 Zhang, H., Zhang, F., and Li, Z. (2009) Gene analysis, optimized production and property of marine lipase from *Bacillus pumilus* B106 associated with South China Sea sponge *Halichondria rugosa*. *World J. Microbiol. Biotechnol.*, **25**, 1267–1274.

67 Liu, Z., Li, X., Chi, Z., Wang, L., Li, J., and Wang, X. (2008) Cloning, characterization and expression of the extracellular lipase gene from *Aureobasidium pullulans* HN2–3 isolated from sea saltern. *Antonie Van Leeuwenhoek*, **94**, 245–255.

68 Ravindran, C., Naveenan, T., and Varatharajan, G.R. (2010) Optimization of alkaline cellulase production by the marine-derived fungus *Chaetomium* sp. using agricultural and industrial wastes as substrates. *Bot. Mar.*, **53**, 275–282.

69 Taylor, L.E., II, Henrissat, B., Coutinho, P.M., Ekborg, N.A., Hutcheson, S.W., and Weiner, R.M. (2006) Complete cellulase system in the marine bacterium *Saccharophagus degradans* strain 2–40T. *J. Bacteriol.*, **188**, 3849–3861.

70 Guo, B., Chen, X.L., Sun, C.Y., Zhou, B.C., and Zhang, Y.Z. (2009) Gene cloning, expression and characterization of a new cold-active and salt-tolerant *endo*-β-1,4-xylanase from marine *Glaciecola mesophila* KMM 241. *Appl. Microbiol. Biotechnol.*, **84**, 1107–1115.

71 Khandeparker, R., Verma, P., and Deobagker, D. (2011) A novel halotolerant xylanase from marine isolate *Bacillus subtilis* cho40: gene cloning and sequencing. *New Biotechnol.*, **28**, 814–821.

72 Pointing, S.B. and Hyde, K.D. (2000) Lignocellulose-degrading marine fungi. *Biofouling*, **15**, 221–229.

73 Mou, H., Jiang, X., Liu, Z., and Guan, H. (2004) Structural analysis of κ-carrageenan oligosaccharides released by carrageenase from marine *Cytophaga* MCA-2. *J. Food Biochem.*, **28**, 245–260.

74 Liu, J.L., Li, Y., Chi, Z., and Chi, Z.M. (2011) Purification and characterization of κ-carrageenase from the marine bacterium *Pseudoalteromonas porphyrae* for hydrolysis of κ-carrageenan. *Process Biochem.*, **46**, 265–271.

75 Ma, Y.S., Dong, S.L., Zing, X.L., Li, J., and Mou, H.J. (2010) Purification and characterization of κ-carragenase from marine bacterium mutant strain *Pseudoalteromonas* sp. AJ5–13 and its degraded products. *J. Food Biochem.*, **34**, 661–678.

76 Basheer, S.M., Chellappan, S., Beena, P.S., Sukumaran, R.K., Elyas, K.K., and Chandrasekaran, M. (2011) Lipase from marine *Aspergillus awamori* BTMFW032: production, partial purification and application in oil effluent treatment. *New Biotechnol.*, **28**, 627–638.

77 Demina, N.S. and Lysenko, S.V. (1991) Microorganisms synthesizing enzymes with thrombolytic action. *Nauchnye doklady vysshei shkoly. Biol. Nauki*, **9**, 136–153.

78 Hassanein, W.A., Kotb, E., Awny, N.M., and El-Zawahry, Y.A. (2011) Fibrinolysis and anticoagulant potential of a metallo protease produced by *Bacillus subtilis* K42. *J. Biosci.*, **36**, 1–7.

79 Wei, X., Luo, M., Xu, El. *et al.* (2011) Production of fibrinolytic enzyme from *Bacillus amyloliquefaciens* by fermentation of chickpeas, with the evaluation of the anticoagulant and antioxidant properties of chickpeas. *J. Agric. Food Chem.*, **59**, 3957–3963.

80 Healthpoint (2012) Understanding Collagenase SANTYL® Ointment. Available online http://www.santyl.com/ (accessed May 10, 2012).

81 Kim, B.J.O., Kim, H.J.U., and Hwang, S.H. (1998) Cloning and expression of a collagenase gene from the marine bacterium *Vibrio vulnificus* CYK279H. *J. Microbiol. Biotechnol.*, **8**, 245–250.

82 Yishan, L., Jiaming, F., Zaohe, W., and Jichang, J. (2011) Genotype analysis of collagenase gene by PCR-SSCP in *Vibrio alginolyticus* and its association with virulence to marine fish. *Curr. Microbiol.*, **62**, 1697–1703.

83 Holcenberg, J.S. (1977) Enzymes as Drugs. *Ann. Rev. Pharmacol. Toxicol.*, **17**, 97–116.

84 Basha, N.S., Rekha, R., Komala, M., and Ruby, S. (2009) Production of extracellular anti-leukaemic enzyme L-asparaginase from marine actinomycetes by solid-state and submerged fermentation: purification and characterisation. *Trop. J. Pharm. Res.*, **8**, 353–360.

85 Dhevagi, P. and Poorani, E. (2006) Isolation and characterization of L-asparaginase from marine actinomycetes. *Indian J. Biotechnol.*, **5**, 514–520.

86 Kumar, S.R. and Chandrasekaran, M. (2003) Continuous production of L-glutaminase by an immobilized marine *Pseudomonas* sp. BTMS-51 in a packed bed reactor. *Process Biochem.*, **38**, 1431–1436.

87 Kondrat'eva, N.A., Dobrynin, I.V., and Merkulov, M.F. (1978) Biological properties of an asparaginase–glutaminase preparation from *Pseudomonas fluorescens* in cell cultures. *Antibiotiki*, **23**, 122–125.

27
Biomedical Applications of Mycosporine-Like Amino Acids

Richa and Rajeshwar P. Sinha

27.1
Introduction

The ultraviolet radiation (UVR) (<400 nm), an important constituent of the solar spectrum, is reaching at an alarming rate on Earth's surface due to continued depletion of the stratospheric ozone layer. UV photons can be absorbed by a number of biomolecules, including nucleic acids and proteins, causing their transition to excited states that are capable of inducing photochemical reactions, resulting in cellular photodamage [1]. On the basis of severity, the solar UVR has been broadly divided into UV-A (315–400 nm), UV-B (280–315 nm), and UV-C (<280 nm). UV-A indirectly affects the DNA by photosensitizing reactions or produces a secondary photoreaction of the existing DNA photoproducts [1,2], whereas UV-B directly denatures the DNA and RNA. It triggers the formation of certain mutagenic and cytotoxic DNA lesions such as cyclobutane pyrimidine dimers and 6-4 photoproducts (6-4 PPs), which can inhibit translation and transcription. Conformational changes and loss of functionality of protein mainly occur due to the direct absorption of UV by the tryptophan residues. Various microbial life processes such as growth, survival, motility, pigmentation, tetrapyrrole synthesis, CO_2 and N_2 fixation, nitrate and ammonium uptake, and O_2 production have been reported to be adversely affected by UVR [3,4]. This has generated concern about the negative impact of highly energetic UV-B radiation on aquatic as well as terrestrial life-forms, ranging from bacteria to humans [3,5].

To counteract the photodamage, various organisms have developed a number of defensive mechanisms such as avoidance, scavenging, screening and repair of damaged DNA, and resynthesis of proteins. Screening is done by a number of UV-absorbing compounds such as phenylpropanoids and flavonoids (in higher plants), melanins (in humans and animals), scytonemins (exclusively in cyanobacteria), mycosporines (in fungi), mycosporine-like amino acids (MAAs) in cyanobacteria, algae, and animals, and several other UV-absorbing substances of unknown chemical structures [6,7]. Mycosporine-like amino acids have received much

Marine Microbiology: Bioactive Compounds and Biotechnological Applications, First Edition.
Edited by Se-Kwon Kim

attention for their putative role in UV photoprotection [8,9] or as antioxidants [10,11], which were originally identified in fungi as having a role in UV-induced sporulation [12]. This chapter describes the structure and physicochemical properties of MAAs, including their biomedical applications.

27.2
Mycosporine-Like Amino Acids (MAAs)

MAAs (Table 27.1) are small (<400 Da), colorless, and water-soluble secondary metabolites consisting of cyclohexenone or cyclohexenimine chromophores conjugated with the nitrogen substituent of amino acids or its imino alcohol [13].

Table 27.1 Chemical structure and the corresponding absorbance maxima of certain known MAAs.

Mycosporine-taurine (309 nm) 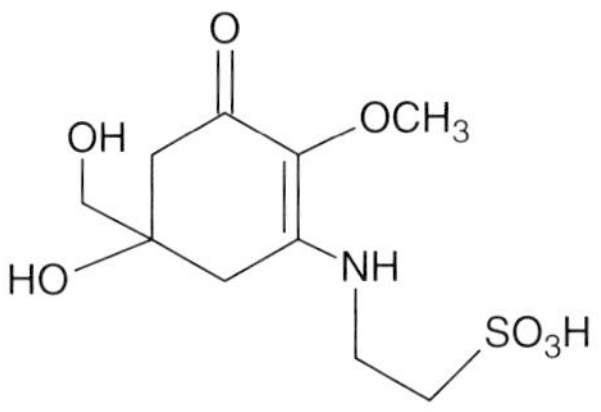	Mycosporine-glycine (310 nm) 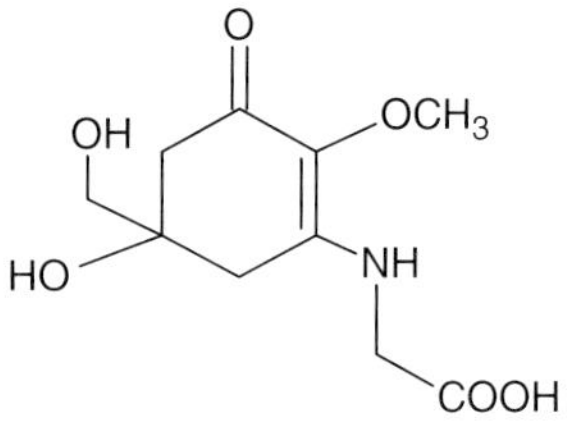
Palythine (320 nm) 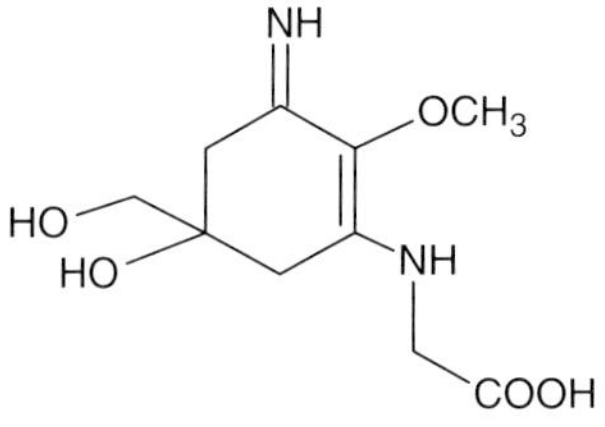	Palythine-serine-sulfate (320 nm) 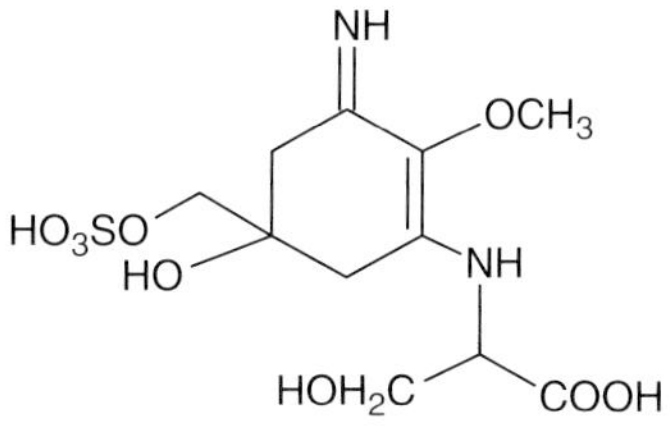
Paythine-serine (320 nm) 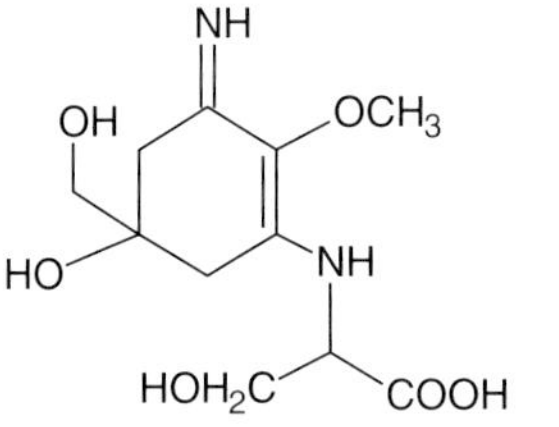	Mycosporine-methylamine-serine (327 nm) 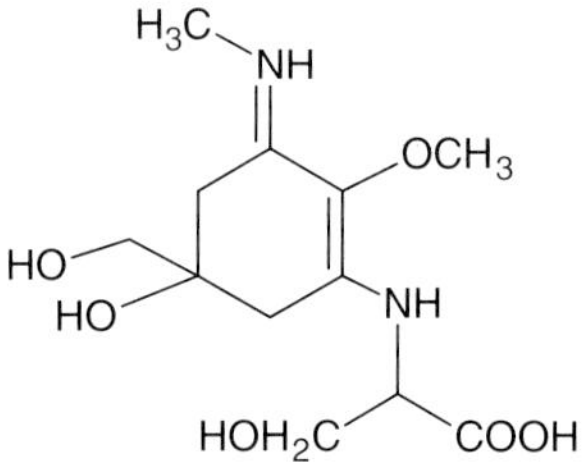

Mycosporine-methylamine-threonine (327 nm)

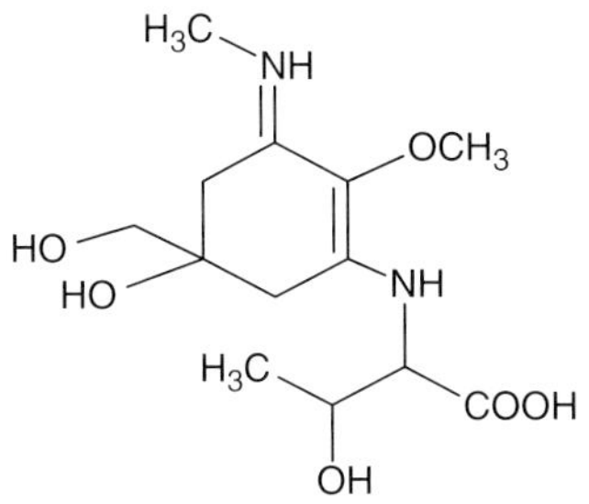

Asterina-330 (330 nm)

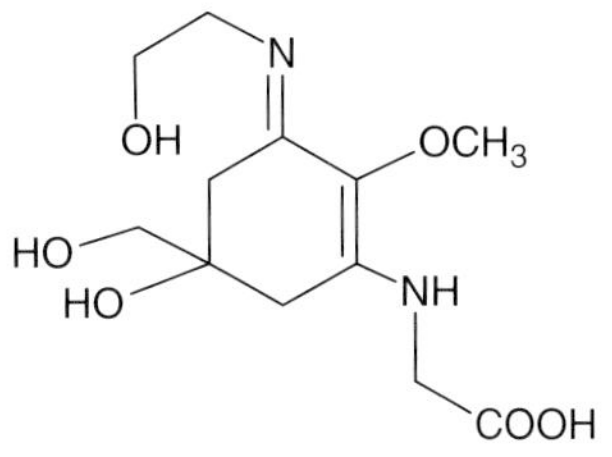

Mycosporine-glutamic acid-glycine (330 nm)

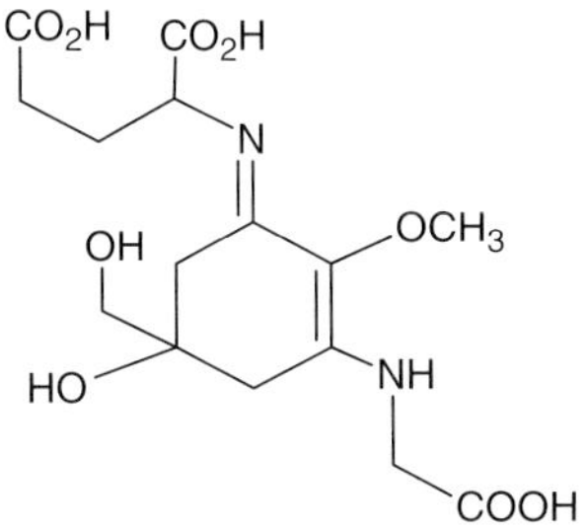

Palythinol (332 nm)

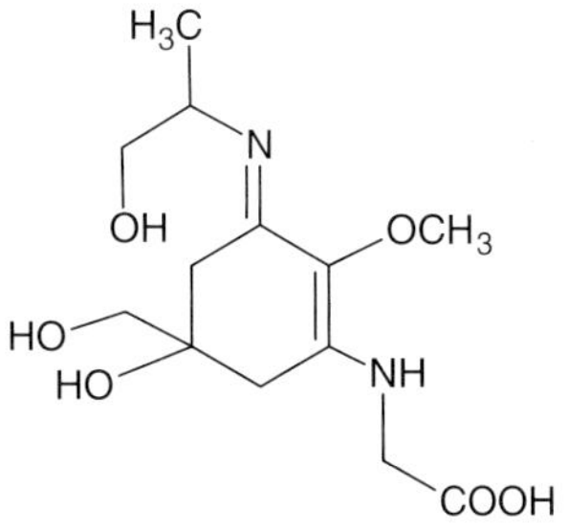

Mycosporine-2-glycine (334 nm)

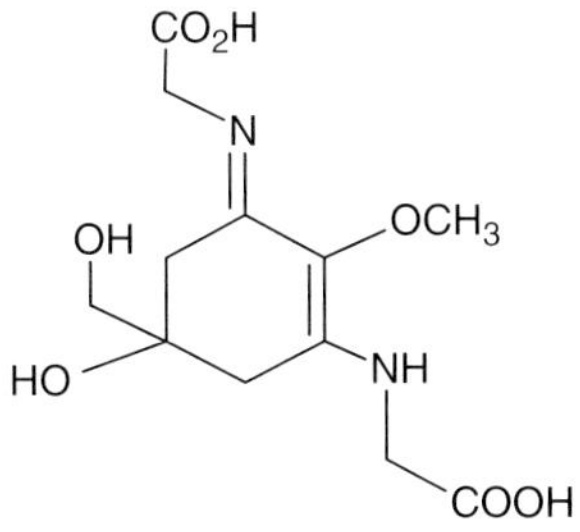

Shinorine (334 nm)

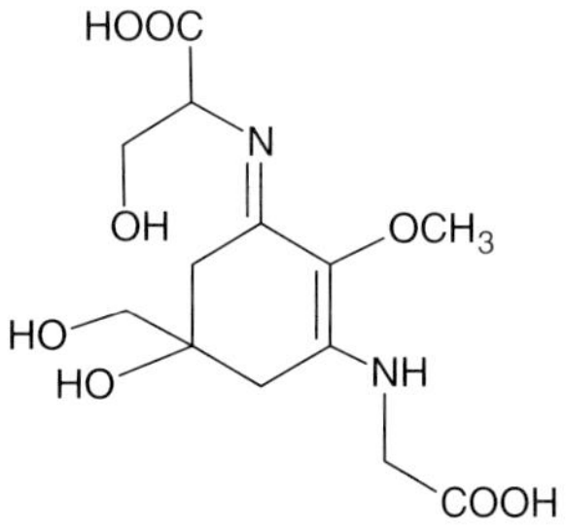

Porphyra-334 (334 nm)

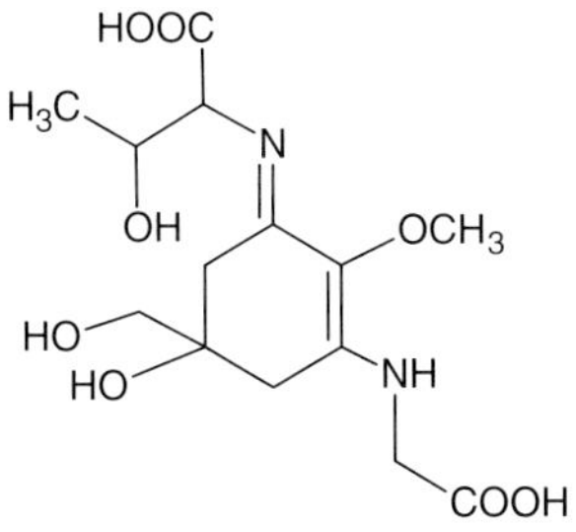

Mycosporine-glycine-valine (335 nm)

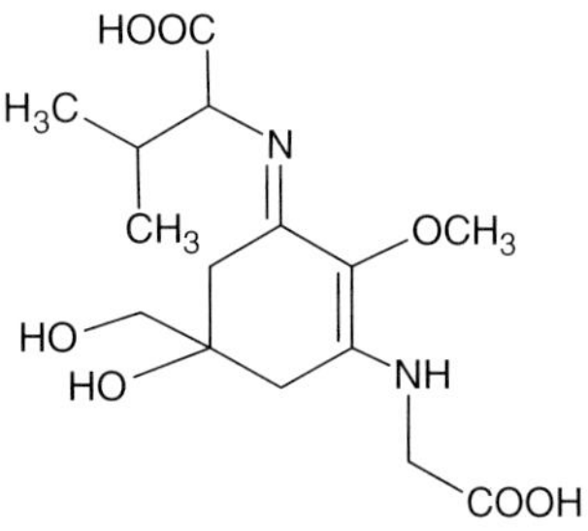

(continued)

Table 27.1 (*Continued*)

Palythenic acid (337 nm)	Usujirene (357 nm)
Palythene (360 nm)	Euhalothece-362 (362 nm)

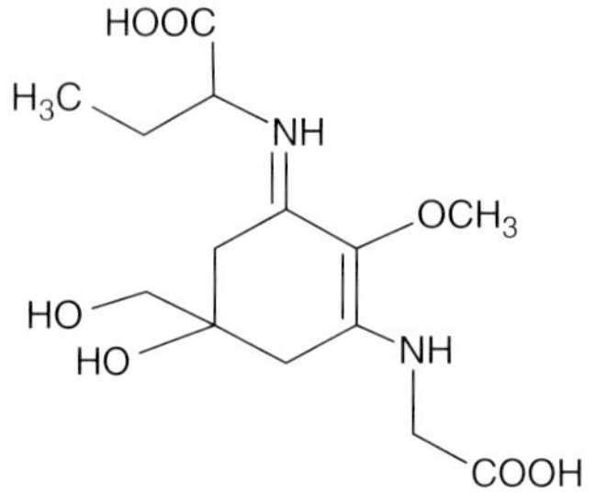

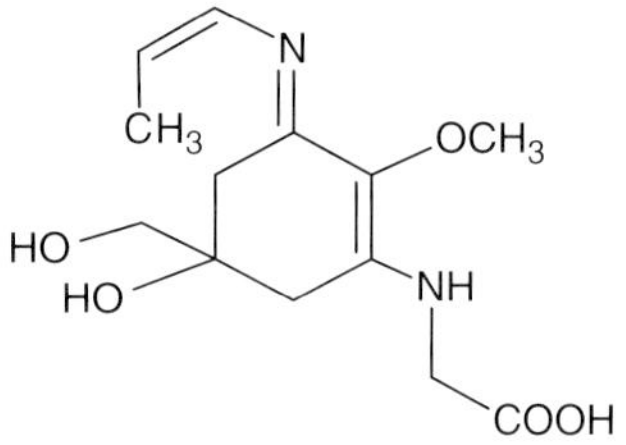

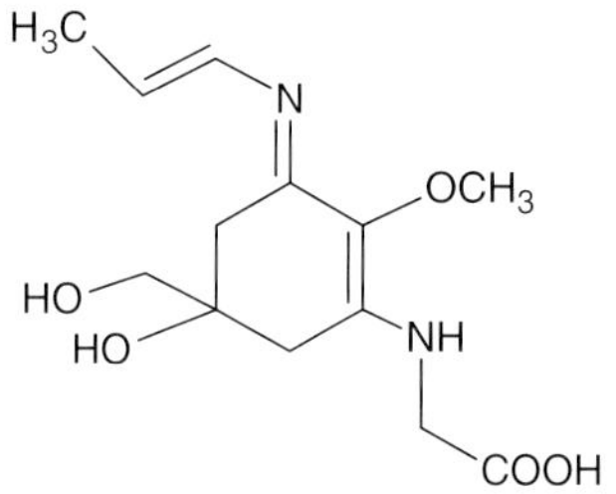

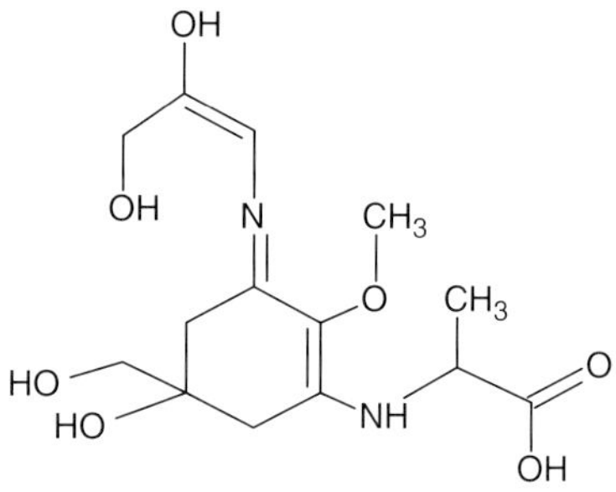

Generally, each MAA has a glycine subunit at the third carbon atom of the cyclohexenimine ring, although some MAAs contain sulfate esters in some corals [14] or glycosidic linkages through the imine substituents [14]. A second amino acid (porphyra-334, shinorine (SH), mycosporine-2-glycine, and mycosporine–glycine–glutamic acid), amino alcohols (palythinol and asterina-330), or an enaminone system (palythene and usujirene) is linked to the C1 (Carreto, J.I. and Carignan, M.O. (2011) Instituto Nacional de Investigación y Desarrollo Pesquero, Mar del Plata, Argentina. Unpublished work.). However, in some corals, glycine has been replaced by methylamine (mycosporine–methylamine–serine and mycosporine–methylamine–threonine) [15] or by an amine group (palythine–serine and palythine–threonine) [15,16]. Recently, a novel MAA, containing the amino acid alanine (2-(*E*)-2,3-dihydroxipro-1-enylimino-mycosporine-alanine), has been isolated from the unicellular cyanobacterium *Euhalothece* sp. [17]. Another unique MAA, tentatively identified as dehydroxylusujirene, has been isolated from the cyanobacterium *Synechocystis* sp. [18]. Each MAA has a unique retention time (Figure 27.1), absorbance between 310 and 362 nm (Figure 27.2), and high molar extinction coefficients ($\varepsilon = 28\,100-50\,000\ M^{-1}\,cm^{-1}$) that favor them as a potent

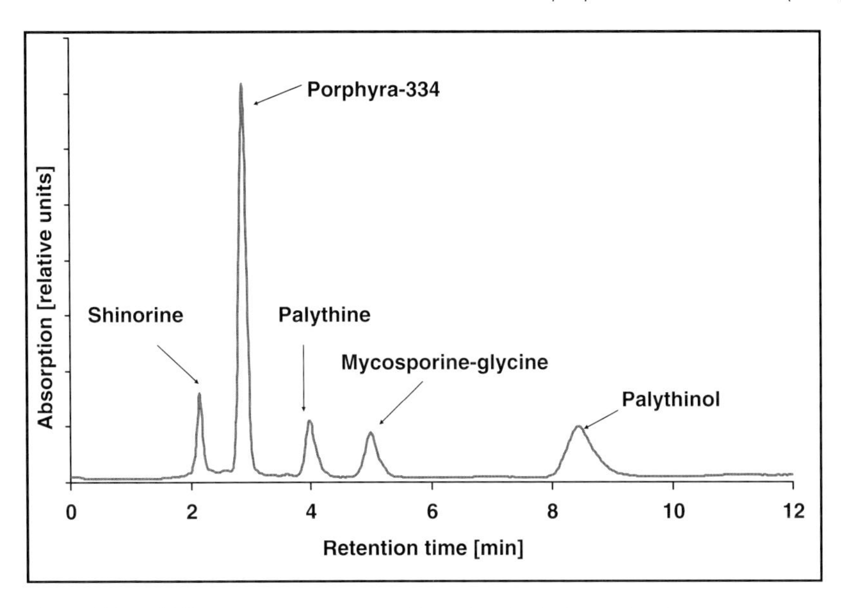

Figure 27.1 HPLC chromatogram of some MAAs showing their retention times.

photoprotectant. They rely on electronic transitions in the conjugated enone or enamine part of the cyclohexene core for absorption. These are highly photostable and resistant against abiotic stressors such as temperature, UV radiation, various solvents, and pH [19,20]. MAAs have been reported extensively from taxonomically diverse organisms, including many terrestrial, freshwater, and marine groups such as heterotrophic bacteria [21], cyanobacteria, and micro/macroalgae. Many animals

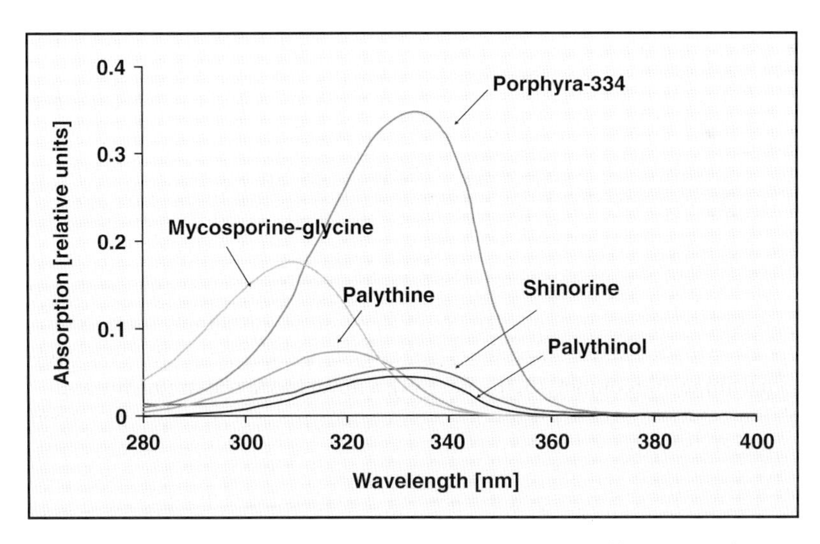

Figure 27.2 Absorption spectra of the corresponding mycosporine-like amino acids (MAAs) as shown in Figure 27.1.

such as arthropods, rotifers, mollusks, fish, cnidarians, tunicates, eubacteriobionts, poriferans, nemerteans, echinodermates, platyhelminthes, polychaetes, bryozoans, and protozoans have also been reported to protect themselves from UVR by MAAs [22]. Till date, about 21 MAAs have been reported. MAAs provide protection against UVR to their producers and consumers through the food chain. The exact location of MAAs is still unclear and is thought to be mostly accumulated as solutes in the cytoplasm, but in the case of *Nostoc commune*, the derivatives are covalently bound to oligosaccharides [23]. MAAs are probably concentrated around UV-sensitive organelles in marine algae [24,25]. In algal–invertebrate symbioses, MAAs are present in soluble form and are not associated with any protein species, whereas in asymbiotic metazoans, they are protein-associated and occur especially in the epidermis [26], in the ocular tissues of many shallow water fishes [26,27], and in some cephalopod mollusk such as *Sepia officinalis* [28]. Asterina-330 and gadusol exist in association with soluble proteins in fish lenses. In the case of certain diatoms, MAAs are found to be in close association with a mineral phase [29]. MAAs protect the cells by absorbing highly energetic UVR and then dissipating this energy in the form of harmless heat radiation to their surroundings [30]. They can also act as antioxidants to prevent damage from reactive oxygen species (ROS) resulting from UVR [31].

27.3
Distribution of MAAs

27.3.1
MAAs in Cyanobacteria

The presence of large amounts of MAAs in cyanobacteria was first reported by Shibata [32] from the Great Barrier Reef. The accumulation of high concentrations of MAAs in the field populations of halotolerant cyanobacteria provides an evidence for its osmotic regulation [33]. The probability that MAAs act as UV-B-absorbing compounds has been derived from the fact that the distribution of MAAs in marine organisms shows a significant correlation with depth, which in turn controls the exposure of UV or photosynthetically active radiation (PAR) [34]. Shinorine and porphyra-334 (P-334) have been found to be the most dominant MAAs in several species of marine cyanobacteria [35–37] such as *Nodularia spumigena*, *Nodularia baltica*, and *Nodularia harveyana* from the Baltic Sea. The filamentous bloom-forming cyanobacterium *Aphanizomenon flos-aquae* is marketed as a food supplement and has been used as a source for the preparation of porphyra-334 [38]. Liu *et al.* [39] reported the presence of MAAs such as porphyra-334 and shinorine in the freshwater bloom-forming cyanobacterium *Microcystis aeruginosa*. The diazotrophic ocean bloom-forming cyanobacterium *Trichodesmium* sp. was also reported to contain high amounts of MAAs. Recently, a rare novel MAA, containing the amino acid alanine (2-(*E*)-2,3-dihydroxipro-1-enylimino-mycosporine-alanine), was isolated from the unicellular cyanobacterium *Euhalothece* sp. inhabiting a hypersaline saltern pond [17]. MAA shinorine has been reported in a number of *Anabaena* sp. [22,40]. Recently four cyanobacteria, for example, *Anabaena variabilis* PCC 7937, *Anabaena* sp. PCC 7120, *Synechocystis* sp. PCC 6803, and *Synechococcus* sp. PCC 6301, were tested for their ability to synthesize MAAs, and genomic and phylogenetic analyses were conducted to identify the possible set of genes that might be involved in the biosynthesis of these compounds. Of the four investigated species, only *A. variabilis* PCC 7937 was able to synthesize MAA [41,42]. However, in the case of *Synechocystis* sp. PCC 6803, Zhang *et al.* [18] have reported the presence of some unusual and new MAAs tentatively identified as mycosporine–taurine and dehydroxylusujirene. A symbiotic cyanobacterium *Prochloron* sp. in association with tropical colonial ascidians has also been reported to contain high amounts of MAAs. Isolated *Prochloron* cells from *Lissoclinum patella* contained shinorine, which was also dominant in the host tunic together with the minor amounts of mycosporine–glycine (MG) and palythine [43]. The same species was reported to contain shinorine and mycosporine–glycine [31] or only mycosporine–glycine [44]. Karsten and Garcia-Pichel [35] reported the presence of shinorine and three unidentified compounds showing absorption maxima at 332, 344, and 346 nm among the *Microcoleus* strains. Recently, Volkmann *et al.* [17] reported the occurrence of mycosporine–alanine, mycosporine–glutaminol, and mycosporine–glutaminol–glucoside in terrestrial cyanobacteria (*Oscillatoriales*), which were previously described in terrestrial fungi. Diazotrophic ocean bloom-forming cyanobacteria *Trichodesmium* sp. was reported to contain high amounts of certain MAAs, identified as asterina-330 and shinorine, but mycosporine–glycine, porphyra-334, and palythene were reported to be present as minor components [45]. Garcia-Pichel *et al.* [46] found a common set of MAAs in 13 unicellular halophilic cyanobacteria. Rastogi *et al.* [47] reported the presence of highly polar and critical class of certain MAAs (shinorine, porphyra-334, and mycosporine–glycine) in the studied hot spring-inhabiting cyanobacteria.

27.3.2 MAAs in Macroalgae

Several species of macroalgae, belonging to Rhodophyceae (red), Phaeophyceae (brown), and Chlorophyceae (green algae), also synthesize and accumulate high concentrations of MAAs as UV sunscreen compounds. Tsujino and Saito [48] were the first to report the presence of UV-absorbing substances in a macroalgal species. Certain red algae such as *Acanthophora, Bangia, Bostrychia, Caloglossa, Catenella, Devaleraea, Ceramium, Chondrus, Corallina, Devaleraea, Gelidiella, Gelidium, Gracilaria, Iridea, Palmaria, Phyllophora, Polysiphonia,* and *Porphyra* produce abundant MAAs. Shinorine and porphyra-334 constitute the most common class of MAAs in macroalgal species collected from tropical to polar waters [22,49,50]. Hoyer *et al.* [51] proposed that the red algae might be divided into three different physiological groups related to MAAs synthesis, among which the first group completely lacks MAAs, second group contains MAAs in variable concentrations, and the third group always contains a stable and high concentration of MAAs. *Palmaria palmata,* edible red algae from New Brunswick, was found to synthesize palythine, shinorine, porphyra-334, asterina-330, palythinol, and usujirene [52]. Coba *et al.* [53] recently reported the presence of the MAAs porphyra-334 and shinorine from the red alga *Porphyra rosengurttii*, asterina-330 and palythine from *Gelidium corneum*, as well as shinorine from *Ahnfeltiopsis devoniensis*. A highly diverse profile of complex MAAs has been reported in certain intertidal red macroalgae such as *Palmaria decipiens*, *Iridea chordata*, and *Curdiea racovitzae* from Antarctic waters [54,55]. UV-B radiation positively induces the synthesis of asterina-330, palythinol, and palythene in the red alga *Chondrus crispus*, but had a negative effect on the accumulation of the major MAAs shinorine and palythine [56]. The imino-MAAs porphyra-334 and shinorine, isolated from *Gracilaria cornea*, have been found to be highly stable against UV and heat stress [57]. Cardozo *et al.* [58] for the first time reported the occurrence of five MAAs in three species of the genus *Gracilaria Greville* (*Gracilaria birdiae*, *Gracilaria domingensis*, and *Gracilaria tenuistipitata*) from Brazilian coast. Zhaohui *et al.* [59] reported prophyra-334 from a thermostable marine alga that was quite stable in water at a temperature of 60 °C. The occurrence of MAAs in brown and green algae is restricted. In an investigation of the occurrence of MAAs in 13 macroalgae collected from the intertidal zone of the tropical island, Hainan and Karsten *et al.* [49] found a significant concentration of photoprotective compounds such as mycosporine–glycine and porphyra-334 only in two green algae: *Boodlea composita* and *Caulerpa racemosa*, respectively. Hoyer *et al.* [51] reported the presence of an UV-absorbing compound of unknown chemical nature with an absorption maximum at 324 nm in the subaerial green macroalgae *Prasiola crispa* spp. *Antarctica*. Later, Gröniger and Häder [60] confirmed the occurrence of this 324 nm MAA in the closely related *Prasiola stipitata* from the supralittoral zone of the rocky island Helgoland (North Sea). UV-B-absorbing compounds with absorption maximum at 294 nm was found to be induced in the green alga *Ulva pertusa* by using different cutoff filters [61]. Phlorotannin, a polyphenolic compound having absorbance between 280–320 nm, has been reported in a number of brown algae such as *Ascophyllum nodosum* that protect them from UV damage [62].

27.3.3 MAAs in Microalgae

MAAs have been reported to provide protection against intense solar UVR in certain phytoplankton such as diatoms, chlorophytes, euglenophytes, eustigmatophytes, rhodophytes, some dinoflagellates, and some prymnesiophytes.

27.3.3.1 Dinoflagellates

All species of symbiotic dinoflagellates do not synthesize MAAs, but it has been reported in the case of *Symbiodinium* [63,64]. Free-living photosynthetic species of dinoflagellates such as *Prorocentrum minimum* [6] and *Woloszynskia* sp. [65] also contain a few MAAs, whereas in *Amphidinium carterae*, only mycosporine–glycine was found [66]. Surface bloom-forming dinoflagellates possess high concentration of MAAs [67–70]. This is made possible either by the endosymbiotic bacteria present in some of these organisms [71] capable of *de novo* synthesis of MAAs [21] or, more probably, they may be responsible for some interconversions between primary and secondary MAAs [68]. Red tide dinoflagellates such as *Akashiwo sanguinea* [72] and *P. minimum* contain several MAAs such as mycosporine–glycine, shinorine, porphyra-334, palythine, palythene, and usujirene. Shinorine methyl ester is exclusively present in *Alexandrium* sp. [67] and *P. minimum* [73]. *Alexandrium* species [68,74–76] has been reported to produce several atypical MAAs. One of them, originally called M-333, was tentatively identified as a monomethyl ester of shinorine [68]. Llewellyn and Aris [76] reported the abundant presence of a similar unknown compound with a λ_{max} at 310 nm in the dinoflagellate *Scrippsiella trochoidea*. Compounds having putative role as MAAs with λ_{max} >360 nm have also been observed in the dinoflagellate *Gymnodinium catenatum* [65], in the euglenophyte *Euglena gracilis*, and in the raphidophytes *Heterosigma akashiwo* and *Porphyridium purpureum* [76]. An UV-induced alteration in the synthesis of amino acids enhances the MAAs accumulation [77]. Elevated levels of PAR (400–700 nm) [67], blue light (400–500 nm) [78], and UV-A [78,79] and UV-B wavelengths [80] have been implicated in the MAA induction response. An increase in MAA concentration was observed in *Prorocentrum micans* only in the presence of UV radiation [80]. Photoinduction of MAA synthesis was also found in the dinoflagellate *Gyrodinium dorsum* only in the presence of PAR [70] and UV-A and UV-B radiation [81].

27.3.3.2 **Prymnesiophytes**

Phaeocystis pouchetii from Antarctica are known to produce MAAs such as shinorine and mycosporine–glycine in abundance [82,83], and palythenic acid and increase its concentration in response to exposure to PAR [83,84] and PAR + UVR [82,84]. The concentration of MAAs was found to be dependent upon the strain, stage in life cycle, and presence of bacteria in the prymnesiophytes [84]. Shinorine was predominant, followed by palythine and porphyra-334 in bloom-forming coccolithophorid *Emiliania huxleyi* [55]. Llewellyn and Airs [76] investigated that *Isochrysis galbana* and *E. huxleyi* contained MAAs at low concentrations and *Phaeocystis globosa* completely lacked it.

27.3.3.3 **Raphidophytes**

The raphidophyte species *Heterosigma carterae* and *Fibrocapsa* sp. contain significantly high levels of secondary MAA asterina-330 [65,76]. Faster growth rate and five times more production of MAAs in an Australian strain of the raphidophyte *Chattonella marina* in comparison to Japanese strain of the same species under inhibiting UV-B radiation [85] provide evidences suggesting that differences between species or algal groups are as important as differences between strains.

27.3.3.4 **Diatoms**

Nine bacillariophyte species were reported to contain predominantly the primary MAAs porphyra-334 (75–95%) with lesser amounts of shinorine (5–20%) and mycosporine–glycine was additionally present in one species, *Porosira pseudodenticulata* [81]. Six different species of *Thalassiosira*, a chain-forming diatom from Antarctica, either lack MAAs or produce shinorine and porphyra-334 [24,54,86], while only mycosporine–glycine was identified in another study [66]. Carreto *et al.* [55] for the first time reported the presence of mycosporine–taurine in *Pseudo-nitzschia multiseries*. The secondary MAAs palythine and palythene have been reported in *Corethron criophilum* [87]. UVR-mediated induction of UV-absorbing compounds, with maximal absorption at 334 nm, has recently been observed in *Skeletonema costatum*, which is a cosmopolitan marine diatom and a major component of most of the red tides in eutrophic regions [88]. Ingalls *et al.* [29] for the first time analyzed the diatom frustules-bound organic matter in opal-rich Southern Ocean plankton and sediments and revealed the presence of several MAAs such as palythine, porphyra-334, and shinorine as well as traces of asterina-330, palythinol, and palythinic acid. The occurrence of MAAs in close association with a mineral phase shows that the mineral matrix can stabilize these compounds and may increase photoprotection against the harmful effects of UVR [29].

27.3.4
MAAs in Lower Invertebrates

Sinha *et al.* [22] have reported the occurrence of MAAs in lower groups such as arthropods, rotifers, mollusks, fishes, cnidarians, tunicates, eubacteriobionts, poriferans, nemerteans, echinodermates, platythelminthes, polychaetes, bryozoans, and protozoans. Symbionts have been inferred to be responsible for the synthesis of MAAs [26,43,64,89–93] in the symbiotic association. Banaszak *et al.* [92] investigated that all clades of *Symbiodinium* can produce MAAs, predominantly the mycosporine–glycine under natural conditions. However, the freshly isolated symbiotic algae can synthesize a maximum of 5 MAAs: mycosporine–glycine, porphyra-334, shinorine, palythine, and mycosporine-2-glycine [64,92], whereas at least 14 different MAAs have been characterized in many coral species [16,64,89,92–95] with up to 12 MAAs identified in a single coral, *Stylophora pistillata* [16,55,64,91]. There is not much information regarding the occurrence of MAAs in marine alga-bearing protists except for coral reefs [26,34]. However, Sommaruga *et al.* [96] reported for the first time some UV-absorbing compounds (shinorine, palythenic acid, palythine, mycosporine-2-glycine, and porphyra-334) in a marine algal-bearing ciliate *Maristentor dinoferus* isolated from coral reefs in Guam, Mariana Islands [97]. Palythine was found as the dominant photoprotective compound in *Sinularia polydactyla* and *Sarcophyton trocheliophorum* [98]. In some coral species such as *Heteroxenia fuscescens* and *Goniastrea retiformis*, it would seem that the source of MAAs is not the algal partner of the symbiosis [99]. Yakovleva and Baird [99] have suggested that MAA synthesis and conversion of MAAs in planulae of the stony coral *G. retiformis* occurred in the absence of zooxanthellae, suggesting a possible contribution of prokaryotes associated with the animal tissue to these processes. Bacteria are also capable of *de novo* synthesis of MAAs [21] and can provide at least some of the intermediate metabolites, or may be responsible for interconversions of MAAs such as the transformation of primary MAAs to secondary MAAs. Seasonal variation and increasing depth primarily affect the MAAs concentration, especially in *Montastraea faveolata* [100]. Recently, a major novel MAA, palythine–threonine, has been reported from the hermatypic coral *Pocillopora capitata* and also from *Pocillopora eydouxi* and *S. pistillata* [16]. Shick *et al.* [101] reported the presence of mycosporine–taurine, shinorine, porphyra-334, and mycosporine-2 glycine in four sea anemones from California in the genus *Anthopleura* (Actiniidae). Genes encoding some enzymes of the shikimic acid pathway are present in a putative, bacterial (Flavobacteriaceae) symbiont of the sea anemone *Nematostella vectensis*, closely related to those of *Tenacibaculum* sp. MED152 [102]. Moreover, a novel bacterial symbiont species in the genus *Tenacibaculum* (*Tenacibaculum aiptasiae* sp. nov.) was recently isolated from the tropical sea anemone *Aiptasia pulchella* [103]. Notably, among invertebrates, mycosporine–taurine was only detected in sea anemones, indicating that MAAs composition, although related with their free amino acid pool composition [101], was dependent on the expression of nonidentified specific enzymes encoded into the host genome, involved in the incorporation of taurine to 4-deoxigadusol. Mycosporine-like amino acids have also been reported in the freshwater ciliate *Stentor amethystinus* Leidy 1880 that hosts *Chlorella* [104]. The nonphotosymbiotic ascidians hosting the algal symbiont *Prochloron* sp. and colonial didemnid ascidians – *L. patella* and *Diplosoma* sp. – that contain the symbiotic photo-oxygenic prokaryote *Prochloron* sp. produce the major MAA (about 94%) as shinorine and the minor MAA as palythine [43,105–107]. *Dysidea herbacea*, a shallow marine sponge common in the Great Barrier Reef containing the symbiotic cyanobacterium *Oscillatoria spongelidae*, contains an unusual MAA, mycosporine–glutamic acid–glycine, as the major component [108]. Mycosporine–glycine, usujirene, and palythene are the minor components [108]. About six different MAAs (shinorine, porphyra-334, palythine, asterina-330, palythinol, and palythenic acid) have been characterized in *Halychondria japonica* from Japan [109], whereas some Antarctic species (i.e., *Inflatella belli*) have been reported to contain only palythine [110], the secondary MAAs most widely distributed among sponges.

27.4 Genetic Control of MAAs Synthesis

The exact biosynthetic pathway for MAAs is still lacking. However, it is considered that 3-dehydroquinate, a six-carbon ring formed in the center of the shikimate pathway (Figure 27.3), serves as a precursor for the synthesis of fungal mycosporines and MAAs via gadusols [13,26]. The primary MAA mycosporine–glycine, synthesized in the shikimate pathway, undergoes chemical or biochemical transformations to yield another secondary MAAs [55,74]. The synthesis of MAAs occurs in fungi, bacteria, cyanobacteria, phytoplankton, and macroalgae (red, brown, and green algae) but not in animals because they lack the shikimate pathway. Studies have shown that in animals, MAAs are derived from

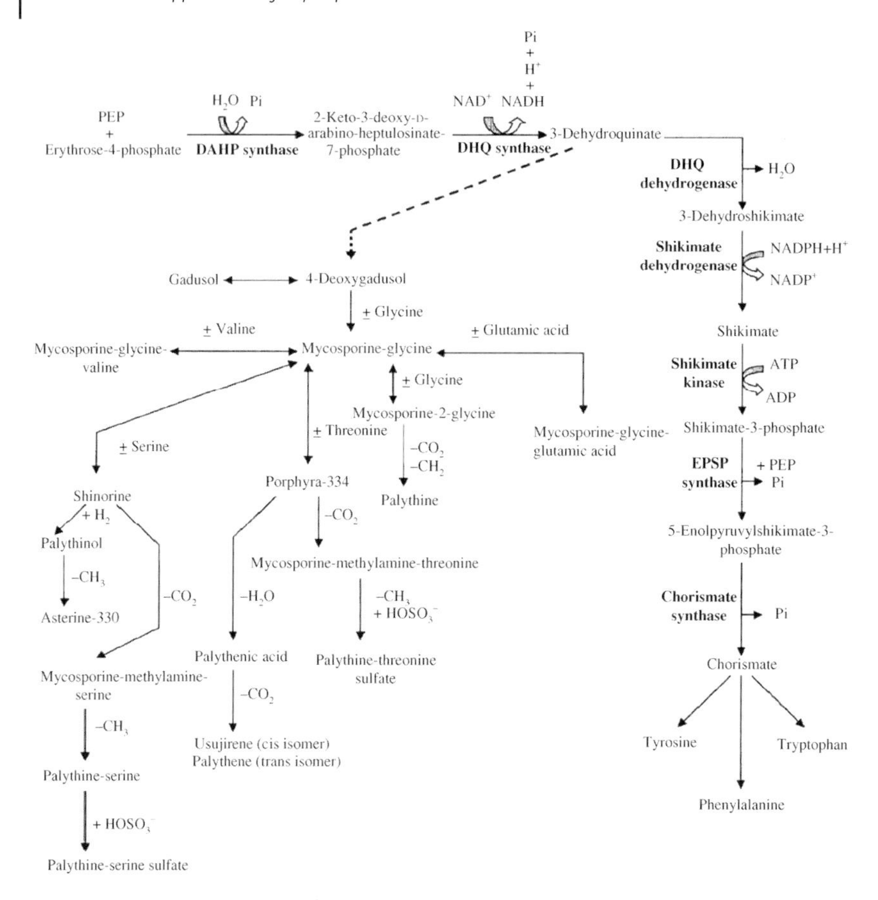

Figure 27.3 Shikimate pathway for MAAs biosynthesis. Adapted from Ref. [13].

their algal diet or symbiotic associations [50,82,111]. The blockage of the MAAs synthesis in the coral *S. pistillata* by the addition of *N*-phosphonomethyl-glycine, a shikimate pathway inhibitor, provided the first direct evidence of the MAAs synthesis via this route in marine organisms [26,91]. Portwich and Garcia-Pichel [112], based on their observations from *Chlorogloeopsis* sp. strain 6912, suggested the similarity of MAAs biosynthesis in both prokaryotes and eukaryotes. Recently, Singh *et al.* [42,113] proposed that in *A. variabilis*, PCC 7937, YP_324358, and YP_324357 gene products are involved in the biosynthesis of deoxygadusol, constituting the common core of all MAAs. The YP_324879 gene product is exclusively involved in the shikimate pathway catalyzing the formation of dehydroquinate, while the YP_324358 gene product in conjunction with the YP_324357 gene product (*O*-methyltransferase) catalyzes the formation of deoxygadusol, which is the core of all MAAs (Figure 27.4). Balskus and Walsh [114] identified a MAA biosynthetic gene cluster in *A. variabilis* ATCC 29413 and discovered the analogous pathways in other sequenced organisms. However, the present assumption that MAA biosynthesis involved a shikimate pathway

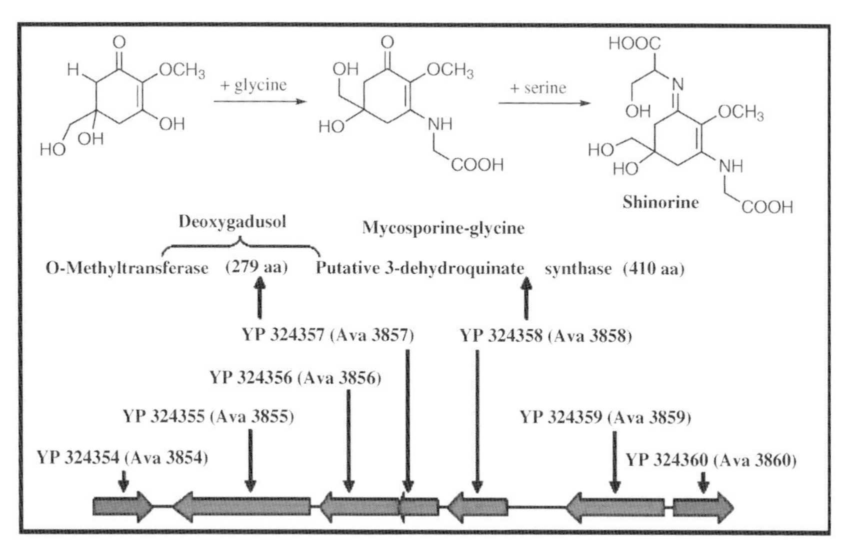

Figure 27.4 Genes involved in MAAs biosynthesis. Modified from Ref. [42].

intermediate was challenged by Balskus and Walsh [114] as these authors failed to observe any 6-deoxygadusol production when the putative substrate 3-dehydroquinate was incubated with the dehydroquinate synthase (DHQS) and *O*-methyltransferase (*O*-MT) homologues (NpR 5600 and NpR 5599, respectively) from *Nostoc punctiforme* (ATCC 29133) and the typical cofactors *S*-adenosylmethionine (SAM), nicotinamide adenine dinucliotide (NAD^+), and Co^{2+}. Instead, sedoheptulose 7-phosphate (SH7-P), an intermediate of the pentose phosphate pathway, produces 6-deoxygadusol with the above treatment. Portwich and Garcia-Pichel [112] observed the incorporation of radioactive ^{14}C-glycine and ^{14}C-serine in the consequent side chains of mycosporine–glycine and shinorine, indicating that these free amino acids are direct precursors. Singh [41] has reported that MAAs biosynthesis depends on photosynthesis for the carbon source, evident by the fact that the addition of fructose [67] reverses the inhibitory effect of (3-(3,4-dichlorophenyl)-1,1-dimethylurea) (DCMU) on MAAs synthesis. Overexpression of gene Ava_3856 from *A. variabilis* was able to convert 6-deoxygadusol and glycine into mycosporine–glycine in the presence of adenosine triphosphate (ATP) and Mg^{2+} [114].

27.5 MAAs Induction

MAAs synthesis is an energy-dependent process and depends on solar energy for its maintenance in natural habitats. Carreto *et al.* [78] reported that increase in PAR, UV-A, and blue light stimulated the synthesis of MAAs in several species of free-living dinoflagellates. However, UV-B wavelengths (280–315 nm) were found to be more effective in inducing MAAs accumulation in the dinoflagellate *G. dorsum*. In Antarctic diatoms, UV-A and blue light were effective to induce MAAs synthesis [66,81,86], while a combination of UV-B + UV-A plays a similar role in prymnesiophytes [66,81]. In cyanobacteria, MAAs can be induced by PAR and UV-A and UV-B radiations [115]. However, UV-B has the most pronounced effect in comparison to the other wavelength ranges [115,116]. Corals needed a combination of UV-B and UV-A in addition to PAR to stimulate the synthesis [91,100]. PAR was shown to induce the UV-absorbing compounds in a marine macroalga *C. crispus* [49] as well as the dinoflagellate *Alexandrium excavatum* when grown at 200 μmol photons/(m^2 s) compared to 20 μmol photons/(m^2 s) [78]. However, cyanobacterium *Chlorogloeopsis* PCC 6912 [117] and the marine bacterium *Micrococcus* sp. AK-334 [21] synthesize MAAs even in the absence of PAR. Among different light qualities, blue light positively influences the accumulation of porphyra-334, palythine, and asterina-330, while shinorine was found to accumulate under white, green, yellow, and red light in *Porphyra leucosticte* [117]. Osmotic stress, alone or in combination with UV-B radiation, can induce the synthesis of MAAs in some cyanobacteria [115,118]. Portwich and Garcia-Pichel [119] have reported the induction of MAA synthesis by salt stress without PAR or UV radiation in the cyanobacterium *Chlorogloeopsis* sp. PCC 6912. Temperature also affects the MAAs synthesis, but there is no available information on MAAs synthesis and accumulation under P-limited conditions and growth at suboptimal temperatures. The MAAs composition is also highly influenced by the nitrogen status as nitrogen limitation significantly decreases the synthesis of MAAs in the dinoflagellates *A. sanguinea* and *Gymnodinium* cf. *instriatum* [120]. Enrichment of medium with ammonium in combination with different radiation treatment has resulted in increased MAAs concentration in *Porphyra columbina* [121]. Singh *et al.* [115] have also reported the induction of MAAs synthesis by salt and ammonium in a concentration-dependent manner without UV stress in the cyanobacterium *A. variabilis* PCC 7937. Recently, Singh *et al.* [122] suggested that the sulfur deficiency plays an important role in the synthesis and bioconversion of MAAs.

27.6 Biomedical Potentials of MAAs

Continuous exposure to UV radiation may lead to a number of complications, including various pathological consequences of the skin damage and sunburn, which occurs when exposure to UV radiation exceeds the protective capacity of an individual's melanin [123,124]. UV irradiation induces photodamage of the skin, resulting in wrinkles, laxity, coarseness, mottled pigmentation, and histological changes that include increased epidermal thickness and connective tissue alteration [125,126]. Imbalance in the major components of connective tissue (collagens, proteoglycans, and glycoproteins) leads to the harmful injurious effects, for example, photoaging in dermal fibroblasts. UV-induced generation of ROS is mainly responsible for the intrinsic aging and photoaging of human skin *in vivo*

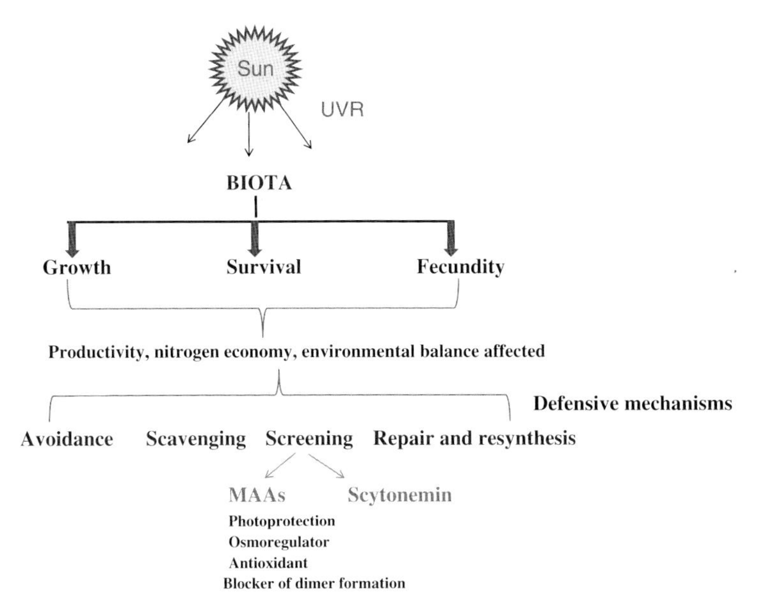

Figure 27.5 Possible UVR effects and biomedical potentials of MAAs.

leading to various skin cancers and cutaneous inflammatory disorders [127,128]. MAAs are secondary metabolites with high absorption in the UV region of the solar spectrum, and thus has the outstanding potentialities for the benefit of humans. The physicochemical properties of MAAs suggest their potential applications as natural photoprotectants and antioxidants (Figure 27.5) in cosmetics, pharmaceuticals, and toiletries as discussed in the following section.

27.7 MAAs as Photoprotectants

MAAs have the capability to efficiently dissipate radiation as heat without producing ROS. MAAs have been reported to reduce the signs of acute UV damage, including erythema, edema, and sunburn. Coba *et al.* [129] reported the cutaneous photoprotective properties of a combination of porphyra-334 and shinorine from the red alga *rosengurttii* applied topically on the skin of female albino hairless mice irradiated with solar stimulated UV radiation. Clinical signs of erythema and edema were found to be lower in photoprotected than the nonphotoprotected margins in the mouse dorsal skin. The production of sunburn cells (SBCs) after UV exposure is believed to be a protective mechanism that eliminates damaged cells, thereby avoiding the risk of malignant transformations. SBCs have been extensively used as a marker of the severity of solar damage and can also be used as a measure of the quantity of energy that a specific filter can absorb efficiently [130,131]. The fact that a sunscreen causes a decrease in the number of SBCs after exposure to UV radiation supports its effectiveness and, theoretically, its anticarcinogenic capacity [132]. Topical application of the MAAs formulation also had a positive influence on the reduction of SBCs. Results from *in vivo* experiment demonstrated that P-334 + SH treatment completely prevented the formation of SBCs in UV-exposed skin supporting the DNA-protective effects of MAAs. Skinfold thickening is considered one of the main markers of actinic damage, since its increase is proportional to the quantity of radiation able to reach different skin layers. UV-B radiation seems to be responsible for the human epidermal and dermal thickening. However, MAAs have been reported to reduce the epidermal and dermal thickening. The role of MAAs in the expression of the heat shock protein HSP70 as a potential biomarker for acute UV damage has also been worked out. Presence of MAAs has also been reported in the black sea cucumber *Holothuria atra* (Jaeger) and their probable role in photoprotection has been hypothesized. It is believed that MAAs could probably function as a broad-spectrum UV absorbers [133]. UV-absorbing compounds have also been isolated from the ovaries of scallop *Patinopecten yessoensis* by Oyamada *et al.* [134]. It was found that the examined MAAs protected the cells from UV-induced cell death and had a protective effect on human cells. It is further expected that these compounds may have potential applications in cosmetics as antiphotoaging/photoprotective agents. Carefoot *et al.* [135] documented the role of MAAs as a sunscreen in the spawn of sea hare *Aplysia dactylomela*. The MAAs such as shinorine (SH), P-334, and mycosporine–glycine have potential to protect the fibroblast cells from UVR-induced cell death [134]. The combined action of P-334 and shinorine, extracted from the red alga *Porphyra umbilicalis*, has been reported to suppress UV-induced aging in human skin [136]. Carefoot *et al.* [135] also investigated the photoprotective role of MAAs in the spawn of sea hare *A. dactylomela*.

27.8
MAAs as an Antioxidant

MAAs exhibit a high antioxidant activity, scavenging superoxide anions and inhibiting lipid peroxidation [137,138] resulting from UV-induced production of ROS [10,139]. Antioxidant activities of these molecules have been related to the presence of a cetonic group in the molecule, while a second group of MAAs with nitrogen in the structure, namely, imino-MAAs, have been scarcely investigated since they were presented as strong antioxidants in lipid medium (phosphatidylcholine peroxidation inhibition assay) [31]. The role of certain MAAs such as P-334 and shinorine in the maintenance of the antioxidant defense system of the skin has been elaborated by Coba *et al.* [129]. Recently, the *in vitro* antioxidative activity of P-334 + SH from *Porphyra* as well as other MAAs from different red algae and a marine lichen has been found out. Coba *et al.* [53,137,138] demonstrated the scavenging of superoxide anions and inhibition of lipid peroxidation by MAAs in a water-soluble medium, confirming their high antioxidant activity. The MAAs glycine and usujilene have also been reported to inhibit lipid peroxidation in aqueous extracts of marine organisms and were able to scavenge singlet oxygen generated from certain endogenous photosensitizers [10,26,140]. The antioxidative activity of certain MAAs such as porphyra-334, shinorine, asterina-330, and palythine in terms of scavenging of hydrosoluble radicals was found to be dose dependent and it increased with the alkalinity of the medium (pH 6–8.5) [53]. Yakovleva *et al.* [141] illustrated the antioxidative role of MG in the thermal stress susceptibility of two scleractinian corals, *Platygyra ryukyuensis* and *S. pistillata*. Their findings strongly suggest that MG is a biological antioxidant in the coral tissue and zooxanthellae and reveal its importance in the survival of reef-building corals under thermal stress.

27.9
MAAs as Blocker of Dimer Formation

Solar UV-B radiation might lead to an enhanced formation of DNA lesions in the form of thymine dimers. UV-B-induced formation of thymine dimers was detected in *A. varabilis* PCC 7937 and *Rivularia* HKAR-4 using immunodot blot and chemiluminescence methods [142,143]. It has been reported by these authors that the frequency of thymine dimer increases with the increase in UVR exposure time in *A. variabilis* PCC 7937 [143]. The UV-absorbing compound MAAs from a marine red alga *Porphyra yezoensis* have been reported to block the formation of certain cytotoxic DNA lesions such as 6-4 photoproduct and cyclobutane pyrimidine dimer (CPD) formation [144]. Rastogi [142] has also demonstrated MAAs response in the elimination of DNA lesions. A steady decrease in thymine dimer formation was recorded in MAAs-induced cells of *A. varabilis* PCC 7937 (Figure 27.6) and *Rivularia* sp. strain HKAR-4.

27.10
MAAs as an Osmoregulator

Oren [33] reported the accumulation of large concentrations of MAAs in a community of unicellular cyanobacteria inhabiting a gypsum crust developed on the bottom of a hypersaline saltern pond, suggesting their role in osmotic stabilization of the cells. Being polar, highly soluble, uncharged, or zwitterionic amino acid derivatives [78], MAAs fulfill at least part of the criteria for osmotic solutes [145].

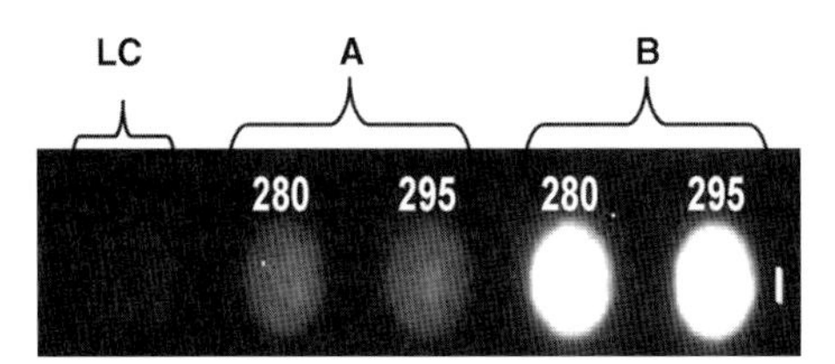

Figure 27.6 Immunodot blot detection of thymine dimer formation in MAAs-induced and noninduced cultures of *A. variabilis* PCC 7937 after UV-B irradiation. LC – light control, A – after MAAs induction, and B – before MAAs induction.

27.11
Conclusions and Future Prospects

The intense UV-B radiation reaching onto Earth's surface has detrimental effects on various vital life processes of all sun-exposed organisms, leading to the reduction in their growth and survival. However, the synthesis of MAAs has been developed as an important defensive mechanism by these organisms that sustain their successful growth and survival in various habitats receiving high solar UVR. MAAs are the imperative and ubiquitous group of sunscreen compounds in a large number of organisms that can potentially reduce the detrimental effects of UVR. They are valuable bioactive compounds having commercial potential for food, cosmetics, and biomedical research as well as in the design of very specific and potent new pharmaceuticals. Although shikimate pathway has been suggested for their synthesis, it is still to be well documented. The loci for synthesis and storage in the cells as well as for the spatial distribution of MAAs have to be ascertained. The biotechnological exploitations of these natural compounds are still very limited and extensive research is needed to enhance the analysis, synthesis, and application of these high-value products for the benefit of living beings.

Acknowledgment

The research work related to MAAs was partially supported by a project (No. SR/WOS-A/LS-140/2011) sanctioned by the Department of Science and Technology, Government of India.

References

1 Gao, Q. and Garcia-Pichel, F. (2011) Microbial ultraviolet sunscreens. *Nat. Rev. Microbiol.*, **9**, 791–802.
2 Hargreaves, A., Taiwo, F.A., Duggan, O., Kirk, S.H., and Ahmad, S.I. (2007) Near-ultraviolet photolysis of β-phenylpyruvic acid generates free radicals and results in DNA damage. *J. Photochem. Photobiol. B: Biol.*, **89**, 110–116.
3 Häder, D.-P., Kumar, H.D., Smith, R.C., and Worrest, R.C. (2007) Effects of solar UV radiation on aquatic ecosystems and interactions with climate change. *Photochem. Photobiol. Sci.*, **6**, 267–285.
4 Sinha, R.P., Kumari, S., and Rastogi, R.P. (2008) Impacts of ultraviolet-B radiation on cyanobacteria: photoprotection and repair. *J. Sci. Res.*, **52**, 125–142.
5 Hansson, L.-A. and Hylander, S. (2009) Effects of ultraviolet radiation on pigmentation, photoenzymatic repair, behavior, and community ecology of zooplankton. *Photochem. Photobiol. Sci.*, **8**, 1266–1275.
6 Sinha, R.P., Klisch, M., Gröniger, A., and Häder, D.-P. (1998) Ultraviolet-absorbing/screening substances in cyanobacteria, phytoplankton and macroalgae. *J. Photochem. Photobiol. B*, **47**, 83–94.
7 Cockell, C.S. and Knowland, J. (1999) Ultraviolet radiation screening compounds. *Biol. Rev.*, **74**, 311–345.
8 Adams, N.L. and Shick, J.M. (2001) Mycosporine-like amino acids prevent UVB-induced abnormalities during early development of the green sea urchin

Strongylocentrotus droebachiensis. *Mar. Biol.*, **138**, 267–280.

9 Klisch, M., Sinha, R.P., Richter, P.E., and Häder, D.-P. (2001) Mycosporine-like amino acids (MAAs) protect against UV damage in *Gyrodinium dorsum* Kofoid. *J. Plant Physiol.*, **158**, 1449–1454.

10 Suh, H.J., Lee, H.W., and Jung, J. (2003) Mycosporine glycine protects biological systems against photodynamic damage by quenching singlet oxygen with a high efficiency. *Photochem. Photobiol.*, **78**, 109–113.

11 Yoshiki, M., Tsuge, K., Tsuruta, Y., Yoshimura, T., Koganemaru, K., Sumi, T., Matsui, T., and Matsumoto, K. (2009) Production of new antioxidant compound from mycosporine-like amino acid, porphyra-334 by heat treatment. *Food Chem.*, **113**, 1127–1132.

12 Leach, C.M. (1965) Ultraviolet-absorbing substances associated with light-induced sporulation in fungi. *Can. J. Bot.*, **43**, 185–200.

13 Singh, S.P., Kumari, S., Rastogi, R.P., Singh, K.L., and Sinha, R.P. (2008) Mycosporine-like amino acids (MAAs): chemical structure, biosynthesis and significance as UV-absorbing/screening compounds. *Indian J. Exp. Biol.*, **46**, 7–17.

14 Wu Won, J.J., Chalker, B.E., and Rideout, J.A. (1997) Two new UV-absorbing compounds from *Stylophora pistillata*: sulfate esters of mycosporine-like amino acids. *Tetrahedron Lett.*, **38**, 2525–2526.

15 Teai, T., Raharivelomanana, P., Bianchini, J.P., Faura, R., Martín, P.M.V., and Cambon, A. (1997) Structure de deux nouvelles iminomycosporines isolées de *Pocillopora eydouxy*. *Tetrahedron Lett.*, **38**, 5799–5800.

16 Carignan, M.O., Cardozo, K.H.M., Oliveira-Silva, D., Colepicolo, P., and Carreto, J.I. (2009) Palythine-treonine, a major novel mycosporine-like amino acid (MAA) isolated from the hermatypic coral *Pocillopora capitata*. *J. Photochem. Photobiol. B*, **94**, 191–200.

17 Volkmann, M., Gorbushina, A.A., Kedar, L., and Oren, A. (2006) Structure of euhalothece-362, a novel red-shifted mycosporine-like amino acid, from a halophilic cyanobacterium (*Euhalothece* sp). *Microbiol. Lett.*, **258**, 50–54.

18 Zhang, L., Li, L., and Wu, Q. (2007) Protective effects of mycosporine-like amino acids of *Synechocystis* sp. PCC 6803 and their partial characterization. *J. Photochem. Photobiol. B*, **86**, 240–245.

19 Gröniger, A. and Häder, D.-P. (2000) Stability of mycosporine-like amino acids. *Recent Res. Dev. Photochem. Photobiol.*, **4**, 247–252.

20 Whitehead, K. and Hedges, J.I. (2005) Photodegradation and photosensitization of mycosporine-like amino acids. *J. Photochem. Photobiol.*, **80**, 115–121.

21 Arai, T., Nishijima, M., Adachi, K., and Sano, H. (1992) *Isolation and Structure of a UV Absorbing Substance from the Marine Bacterium Micrococcus sp. AK-334*, Marine Biotechnology Institute, Tokyo, Japan, pp. 88–94.

22 Sinha, R.P., Singh, S.P., and Häder, D.-P. (2007) Database on mycosporines and mycosporine-like amino acids (MAAs) in fungi, cyanobacteria, macroalgae, phytoplankton and animals. *J. Photochem. Photobiol. B: Biol.*, **89**, 29–35.

23 Böhm, G.A., Pfleiderer, W., Böger, P., and Scherer, S. (1995) Structure of a novel oligosaccharide-mycosporine-amino acid ultraviolet A/B sunscreen pigment from the terrestrial cyanobacterium *Nostoc commune*. *J. Biol. Chem.*, **270**, 8536–8539.

24 Laurion, I., Blouin, F., and Roy, S. (2003) The quantitative filter technique for measuring phytoplankton absorption: interference by MAAs in the UV waveband. *Limnol. Oceanogr. Methods*, **1**, 1–9.

25 Laurion, I., Blouin, F., and Roy, S. (2004) Packaging of mycosporine-like amino acids in dinoflagellates. *Mar. Ecol. Prog. Ser.*, **279**, 297–303.

26 Shick, J.M. and Dunlap, W.C. (2002) Mycosporine-like amino acids and related gadusols: biosynthesis, accumulation, and UV protective function in aquatic organisms. *Annu. Rev. Physiol.*, **64**, 223–262.

27 Dunlap, W.C., Williams, D.M., Chalker, B.E., and Banaszak, A.T. (1989) Biochemical photoadaptations in vision: UV-absorbing

pigments in fish eye tissues. *Comp. Biochem. Physiol.*, **93**, 601–607.

28 Shashar, N., Hárosi, F.I., Banaszak, A.T., and Hanlon, R.T. (1998) UV radiation blocking compounds in the eye of the cuttlefish *Sepia officinalis*. *Biol. Bull.*, **195**, 187–188.

29 Ingalls, A.E., Whitehead, K., and Bridoux, M.C. (2010) Tinted windows: the presence of the UV absorbing compounds called mycosporine-like amino acids embedded in the frustules of marine diatoms. *Geochim. Cosmochim. Acta*, **74**, 104–115.

30 Conde, F.R., Churio, M.S., and Previtali, C.M. (2004) The deactivation pathways of the excited-states of the mycosporine-like amino acids shinorine and porphyra-334 in aqueous solution. *Photochem. Photobiol. Sci.*, **3**, 960–967.

31 Dunlap, W.C. and Yamamoto, Y. (1995) Small-molecule antioxidants in marine organisms: antioxidant activity of mycosporine–glycine. *Comp. Biochem. Physiol.*, **112**, 105–114.

32 Shibata, K. (1969) Pigments and a UV-absorbing substance in coral and a blue-green alga living in the Great Barrier Reef. *Plant Cell Physiol.*, **10**, 325–335.

33 Oren, A. (1997) Mycosporine-like amino acids as osmotic solutes in a community of halophilic cyanobacteria. *Geomicrobiol. J.*, **14**, 231–240.

34 Dunlap, W.C. and Shick, J.M. (1998) Ultraviolet radiation-absorbing mycosporine-like amino acids in coral reef organisms: a biochemical and environmental perspective. *J. Phycol.*, **34**, 418–430.

35 Karsten, U. and García-Pichel, F. (1996) Carotenoids and mycosporine-like amino acids compounds in members of the genus *Microcoleus* (Cyanobacteria): a chemosystematic study. *Syst. Appl. Microbiol.*, **19**, 285–294.

36 Sinha, R.P., Klisch, M., Helbling, E.W., and Häder, D.-P. (2001) Induction of mycosporine-like amino acids (MAAs) in cyanobacteria by solar ultraviolet-B radiation. *J. Photochem. Photobiol. B*, **60**, 129–135.

37 Sinha, R.P., Ambasht, N.K., Sinha, J.P., Klisch, M., and Häder, D.-P. (2003) UVB-induced synthesis of mycosporine-like amino acids in three strains of *Nodularia* (cyanobacteria). *J. Photochem. Photobiol. B*, **71**, 51–58.

38 Torres, A., Enk, C.D., Hochberg, M., and Srebnik, M. (2006) Porphyra-334, a potential natural source for UVA protective sunscreens. *Photochem. Photobiol. Sci.*, **5**, 432–435.

39 Liu, Z., Häder, D.-P., and Sommaruga, R. (2004) Occurrence of mycosporine-like amino acids (MAAs) in the bloom-forming cyanobacterium *Microcystis aeruginosa*. *J. Plankton Res.*, **26**, 963–966.

40 Sinha, R.P. and Häder, D.-P. (2008) UV-protectants in cyanobacteria. *Plant Sci.*, **174**, 278–289.

41 Singh, S.P. (2009) Study on mycosporine-like amino acids (MAAs) in cyanobacteria: a biochemical, bioinformatics and molecular biology approach, PhD Thesis, University of Erlangen-Nürnberg, Erlangen, Germany.

42 Singh, S.P., Klisch, M., Sinha, R.P., and Häder, D.-P. (2010) Genome mining of mycosporine-like amino acid (MAA) synthesizing and non-synthesizing cyanobacteria: a bioinformatics study. *Genomics*, **95**, 120–128.

43 Dionisio-Sese, M.L., Ishikura, M., Maruyama, T., and Miyachi, S. (1997) UV-absorbing substances in the tunic of a colonial ascidian protect its symbiont, *Prochloron* sp., from damage by UV-B radiation. *Mar. Biol.*, **128**, 455–461.

44 Lesser, M.P., Stochaj, W.R., and Tapley, D.W., and Shick, J.M. (1990) Bleaching in coral reef anthozoans: effects of irradiance, ultraviolet radiation and temperature on the activities of protective enzymes against active oxygen. *Coral Reefs*, **8**, 225–232.

45 Subramanian, A., Carpenter, E.J., Karentz, D., and Falkowski, P.G. (1999) Bio-optical properties of the marine diazotrophic cyanobacteria *Trichodesmium* spp: I. Absorption and photosynthetic action spectra. *Limnol. Oceanogr.*, **44**, 608–617.

46 Garcia-Pichel, F., Nübel, U., and Muyzer, G. (1998) The phylogeny of unicellular, extremely halotolerant cyanobacteria. *Arch. Microbiol.*, **169**, 469–482.

47 Rastogi, R.P., Kumari, S., Sinha, R., Han, T., and Sinha, R.P. (2012) Molecular characterization of hot spring

cyanobacteria and evaluation of their photoprotective compounds. *Can. J. Microbiol.*, **58**, 719–727.

48 Tsujino, I. and Saito, T. (1961) Studies on the compounds specific for each group of marine algae. Presence of characteristic ultraviolet absorbing material in *Rhodophyceae. Bull. Fac. Fish. Hokkaido Univ.*, **12**, 49–58.

49 Karsten, U., Sawall, T., and Wiencke, C. (1998) A survey of the distribution of UV-absorbing substances in tropical macroalgae. *Phycol. Res.*, **46**, 271–279.

50 Helbling, E.W., Menchi, C.F., and Villafañe, V.E. (2002) Bioaccumulation and role of UV-absorbing compounds in two marine crustacean species from Patagonia, Argentina. *Photochem. Photobiol. Sci.*, **1**, 820–825.

51 Hoyer, K., Karsten, U., and Wiencke, C. (2002) Induction of sunscreen compounds in Antarctic macroalgae by different radiation conditions. *Mar. Biol.*, **141**, 619–627.

52 Yuan, Y.V., Westcott, N.D., Huc, Ch., and Kitts, D.D. (2009) Mycosporine-like amino acid composition of the edible red alga, *Palmaria palmata* (dulse) harvested from the west and east coasts of Grand Manan Island, New Brunswick. *Food Chem.*, **112**, 321–328.

53 Coba, F.D.L., Aguilera, J., Figueroa, F.L., de Gálvez, M.V., and Herrera, E. (2009) Antioxidant activity of mycosporine-like amino acids isolated from three red macroalgae and one marine lichen. *J. Appl. Phycol.*, **21**, 161–169.

54 Karentz, D., McEuen, E.S., Land, M.C., and Dunlap, W.C. (1991) Survey of mycosporine-like amino acid compounds in Antarctic marine organisms: potential protection from ultraviolet exposure. *Mar. Biol.*, **108**, 157–166.

55 Carreto, J.I., Carignan, M.O., and Montoya, N.G. (2005) A high-resolution reverse-phase liquid chromatography method for the analysis of mycosporine-like amino acids (MAAs) in marine organisms. *Mar. Biol.*, **146**, 237–252.

56 Kräbs, G., Watanabe, M., and Wiencke, C. (2004) A monochromatic action spectrum for the photoinduction of the UV-absorbing mycosporine-like amino acid shinorine in the red alga *Chondrus crispus*. *Photochem. Photobiol.*, **79**, 515–519.

57 Sinha, R.P., Klisch, M., Gröniger, A., and Häder, D.-P. (2000) Mycosporine-like amino acids in the marine red alga *Gracilaria cornea*—effects of UV and heat. *Environ. Exp. Bot.*, **43**, 33–43.

58 Cardozo, K.H.M., Marques, L.G., Carvalho, V.M., Carignan, M.O., Pinto, E., Marinho-Soriano, E., and Colepicolo, P. (2011) Analyses of photoprotective compounds in red algae from the Brazilian coast. *Braz. J. Pharmacogn.*, **21**, 202–208.

59 Zhaohui, Z., Xin, G., Tashiro, Y., Matsukawa, S., and Ogawa, H. (2005) The isolation of prophyra-334 from marine algae and its UV absorption behavior. *Chin. J. Oceanol. Limnol.*, **23**, 400–405.

60 Gröniger, A. and Häder, D.-P. (2002) Induction of the synthesis of an UV-absorbing substance in the green alga *Prasiola stipitata*. *J. Photochem. Photobiol. B*, **66**, 54–59.

61 Han, Y.-S. and Han, T. (2005) UV-B induction of UV-B protection in *Ulva pertusa* (chlorophyta). *J. Phycol.*, **41**, 523–530.

62 Pavia, H., Cervin, G., Lindgren, A., and Åberg, P. (1997) Effects of UVB radiation and simulated herbivory on phlorotannins in the brown alga *Ascophyllum nodosum*. *Mar. Ecol. Prog. Ser.*, **157**, 139–146.

63 Banaszack, A.T. and Trench, R.K. (2001) Ultraviolet sunscreens in dinoflagellates. *Protist*, **152**, 93–101.

64 Shick, J.M. (2004) The continuity and intensity of ultraviolet irradiation affect the kinetics of biosynthesis, accumulation, and conversion of mycosporine-like amino acids (MAAs) in the coral *Stylophora pistillata*. *Limnol. Oceanogr.*, **49**, 442–458.

65 Jeffrey, S.W., MacTavish, H.S., Dunlap, W. C., Vesk, M., and Groenewould, K. (1999) Occurrence of UVA and UVB-absorbing compounds in 152 species (206 strains) of marine microalgae. *Mar. Ecol. Prog. Ser.*, **189**, 35–51.

66 Hannach, G. and Sigleo, A.C. (1998) Photoinduction of UV-absorbing compounds in six species of marine phytoplankton. *Mar. Ecol. Prog. Ser.*, **174**, 207–222.

67 Carreto, J.I., Carignan, M.O., Daleo, G., and DeMarco, S.G. (1990) Occurrence of mycosporine-like amino acids in the red-tide dinoflagellate *Alexandrium excavatum*: UV photoprotective compounds? *J. Plankton Res.*, **12**, 909–921.

68 Carreto, J.I., Carignan, M.O., and Montoya, N.G. (2001) Comparative studies on mycosporine-like amino acids, paralytic shellfish toxins and pigment profiles of the toxic dinoflagellates *Alexandrium tamarense, A. catenella* and *A. minutum*. *Mar. Ecol. Prog. Ser.*, **223**, 49–60.

69 Ekelund, N.G.A. (1994) Influence of UV-B radiation on photosynthetic light response curves, absorption spectra and motility of four phytoplankton species. *Physiol. Plant.*, **91**, 696–702.

70 Klisch, M. and Häder, D.-P. (2000) Mycosporine-like amino acids in the marine dinoflagellate *Gyrodinium dorsum*: induction by ultraviolet irradiation. *J. Photochem. Photobiol. B*, **55**, 178–182.

71 Doucette, G.J. (1995) Assessment of the interaction of prokaryotic cells with harmful algal species, in *Harmful Marine Algae Blooms Technique & Documentation* (eds P. Lassus, G. Arzul, E. Erard, P. Gentien, and C. Marcaillou), Lavoisier Ltd., Paris, France, pp. 385–400.

72 Neale, P.J., Banaszak, A.T., and Jarriel, C.R. (1998) Ultraviolet sunscreens in *Gymnodinium sanguineum* (Dinophyceae): mycosporine-like amino acids protect against inhibition of photosynthesis. *J. Phycol.*, **34**, 928–938.

73 Cardozo, K.H. (2007) Estudos de compostos fotoprotetores da radiaçao ultravioleta em algas: Aminoácidos tipo micosporinas (MAAs), PhD Thesis, Universidade de Sao Paulo, Instituto de Química, Brasil.

74 Callone, A.I., Carignan, M., Montoya, N.G., and Carreto, J.I. (2006) Biotransformation of mycosporine like amino acids (MAAs) in the toxic dinoflagellate *Alexandrium tamarense*. *J. Photochem. Photobiol. B*, **84**, 204–212.

75 Laurion, I. and Roy, S. (2009) Growth and photoprotection in three dinoflagellates (including two strains of *Alexandrium tamarense*) and one diatom exposed to four weeks of natural and enhanced UVB radiation. *J. Phycol.*, **45**, 16–33.

76 Llewellyn, C.A. and Airs, R.L. (2010) Distribution and abundance of MAAs in 33 species of microalgae across 13 classes. *Mar. Drugs*, **8**, 1273–1291.

77 Goes, J.I., Handa, N., Taguchi, S., Harna, T., and Saito, H. (1995) Impact of UV radiation on the production patterns and composition of dissolved free and combined amino acids in marine phytoplankton. *J. Plankton Res.*, **17**, 1337–1362.

78 Carreto, J.I., Lutz, V.A., Dc Marco, S.G., and Carignan, M.O. (1990) Fluence and wavelength dependence of mycosporine-like amino acid synthesis in the dinoflagellate *Alexandrium excavatum*, in *Toxic Marine Phytoplankton* (eds E. Graneli, L. Edler, B. Sundström, and D.M. Anderson), Elsevier, New York, pp. 275–279.

79 Ferreyra, G.A., Schloss, I., Demers, S., and Neale, P.J. (1994) Phytoplankton responses to natural ultraviolet irradiance during early spring in the Weddell-Scotia Confluence: an experimental approach. *Antarct. J. US*, **29**, 268–270.

80 Lesser, M.P. (1996) Acclimation of phytoplankton to UV-B radiation: oxidative stress and photoinhibition of photosynthesis are not prevented by UV-absorbing compounds in the dinoflagellate *Prorocentrum micans*. *Mar. Ecol. Prog. Ser.*, **132**, 287–297.

81 Riegger, L. and Robinson, D. (1997) Photoinduction of UV-absorbing compounds in Antarctic diatoms and *Phaeocystis antarctica*. *Mar. Ecol. Prog. Ser.*, **160**, 13–25.

82 Newman, S.J., Dunlap, W.C., Nicol, S., and Ritz, D. (2000) Antarctic krill (*Euphausia superba*) acquire a UV-absorbing mycosporine-like amino acid from dietary algae. *J. Exp. Mar. Biol. Ecol.*, **255**, 93–110.

83 Moisan, T.A. and Mitchell, B.G. (2001) UV absorption by mycosporine-like amino acids in *Phaeocystis antarctica* induced by photosynthetically available radiation. *Mar. Biol.*, **138**, 217–227.

84 Marchant, H.J., Davidson, A.T., and Kelly, G.J. (1991) UV-B protecting compounds in the marine alga *Phaeocystis*

pouchetii from Antarctica. *Mar. Biol.*, **109**, 391–395.

85 Marshall, J.A. and Newman, S. (2002) Differences in photoprotective pigment production between Japanese and Australian strains of *Chattonella marina* (Raphidophyceae). *J. Exp. Mar. Biol. Ecol.*, **272**, 13–27.

86 Hernando, M., Carreto, J.I., Carignan, M.O., Ferreyra, G.A., and Gross, C. (2002) Effects of solar radiation on growth and mycosporine-like amino acids content in *Thalassiosira* sp., an Antarctic diatom. *Polar Biol.*, **25**, 12–20.

87 Helbling, E.W., Chalker, B.E., Dunlap, W.C., Holm-Hansen, O., and Villafañe, V.E. (1996) Photoacclimation of Antarctic marine diatoms to solar ultraviolet radiation. *J. Exp. Mar. Biol. Ecol.*, **204**, 85–101.

88 Wu, H., Gao, K., and Wu, H. (2009) Responses of a marine red tide alga *Skeletonema costatum* (Bacillariophyceae) to long-term UV radiation exposures. *J. Photochem. Photobiol. B*, **94**, 82–86.

89 Dunlap, W.C. and Chalker, B.E. (1986) Identification and quantification of near-UV absorbing compound (S-320) in a hermatypic scleractinian. *Coral Reefs*, **5**, 1–5.

90 Ishikura, M., Kato, C., and Maruyama, T. (1997) UV-absorbing substances in zooxanthellate and azooxanthellate clams. *Mar. Biol.*, **128**, 649–655.

91 Shick, J.M., Romaine-Lioud, S., Ferrier-Pagès, C., and Gattuso, J.P. (1999) Ultraviolet-B radiation stimulates shikimate pathway-dependent accumulation of mycosporine-like amino acids in the coral *Stylophora pistillata* despite decreases in its population of symbiotic dinoflagellates. *Limnol. Oceanogr.*, **44**, 1667–1682.

92 Banaszak, A.T., Santos, M.G., LaJeunesse, T.C., and Lesser, M.P. (2006) The distribution of mycosporine-like amino acids (MAAs) and the phylogenetic identity of symbiotic dinoflagellates in cnidarian hosts from the Mexican Caribbean. *J. Exp. Mar. Biol. Ecol.*, **337**, 131–146.

93 Ferrier-Pagés, C., Richard, C., Forcioli, D., Allemand, D., Pichon, M., and Shick, J.M. (2007) Effects of temperature and UV radiation increases on the photosynthetic efficiency in four scleractinian coral species. *Biol. Bull.*, **213**, 76–87.

94 Yakovleva, I. and Hidaka, M. (2004) Diel fluctuations of mycosporine-like amino acids (MAAs) in shallow water scleractinian corals. *Mar. Biol.*, **145**, 863–873.

95 Torres-Pérez, J.L. (2005) Responses of two species of Caribbean shallow-water branching corals to changes in ultraviolet radiation, PhD Thesis, University of Puerto Rico, San Juan, Puerto Rico.

96 Sommaruga, R., Whitehead, K., Shick, J.M., and Lobban, C.S. (2006) Mycosporine-like amino acids in the zooxanthella-ciliate symbiosis *Maristentor dinoferus. Protist*, **157**, 185–191.

97 Lobban, C.S., Schefter, M., Simpson, A.G. B., Pochon, X., Pawlowski, J., and Foissner, W. (2002) *Maristentor dinoferus* n. gen., n. sp., a giant heterotrich ciliate (Spirotrichea: Heterotrichida) with zooxanthellae, from coral reefs on Guam, Mariana Island. *Mar. Biol.*, **140**, 411–423.

98 Al-Otaibi, A.A., Al-Sofyani, A., Niaz, G.R., and Al-Lihaibi, S.S. (2006) Temporal and depth variation of photoprotective mycosporine-like amino acids in soft coral species from the eastern red sea coast. *Mar. Sci.*, **17**, 169–180.

99 Yakovleva, I. and Baird, A.H. (2005) Ontogenetic change in the abundance of mycosporine-like amino acids in non-zooxanthellate coral larvae. *Coral Reefs*, **24**, 443–452.

100 Lesser, M.P. (2000) Depth-dependent photoacclimatization to solar ultraviolet radiation in the Caribbean coral *Montastraea faveolata. Mar. Ecol. Prog. Ser.*, **192**, 137–151.

101 Shick, J.M., Dunlap, W.C., Pearse, J.S., and Pearse, V.B. (2002) Mycosporine-like amino acid content in four species of sea anemones in the genus *Anthopleura* reflects phylogenetic but not environmental or symbiotic relationships. *Biol. Bull.*, **203**, 315–330.

102 Starcevic, A., Akthar, S., Dunlap, W.C., Shick, J.M., Hranueli, D., Cullum, J., and Long, P.F. (2008) Enzymes of the shikimic

acid pathway encoded in the genome of a basal metazoan, *Nematostella vectensis*, have microbial origins. *Proc. Natl. Acad. Sci. USA*, **105**, 2533–2537.

103 Wang, J.T., Chou, Y.J., Chou, J.H., Chen, C.A., and Chen, W.M. (2008) *Tenacibaculum aiptasiae* sp. nov., isolated from a sea anemone *Aiptasia pulchella*. *Int. J. Syst. Evol. Microbiol.*, **58**, 761–766.

104 Tartarotti, B., Baffico, G., Temporetti, P., and Zagarese, H.E. (2004) Mycosporine-like amino acids in planktonic organisms living under different UV exposure conditions in Patagonian lakes. *J. Plankton Res.*, **26**, 753–762.

105 Maruyama, T., Hirose, E., and Ishikura, M. (2003) Ultraviolet-light-absorbing tunic cells in didemnid ascidians hosting a symbiotic photo-oxygenic prokaryote, *Prochloron*. *Biol. Bull.*, **204**, 109–113.

106 Hirose, E., Otsuka, K., Ishikura, M., and Maruyama, T. (2004) Ultraviolet absorption in ascidian tunic and ascidian-*Prochloron* symbiosis. *J. Mar. Biol. Assoc. UK*, **84**, 789–794.

107 Hirose, E., Hirabayashi, S., Hori, K., Kasai, F., and Watanabe, M.M. (2006) UV protection in the photosymbiotic ascidian *Didemnum molle* inhabiting different depths. *Zool. Sci.*, **23**, 57–63.

108 Bandaranayake, W.M., Bemis, J.E., and Bourne, D.J. (1996) Ultraviolet absorbing pigments from the marine sponge *Dysidea herbacea*: isolation and structure of a new mycosporine. *Comp. Biochem. Physiol.*, **115**, 281–286.

109 Nakamura, H., Kobayashi, J., and Hirata, Y. (1982) Separation of mycosporine-like amino acids in marine organisms using reversed-phase high-performance liquid chromatography. *J. Chromatogr.*, **250**, 113–118.

110 Mc Clintock, J.B. and Karentz, D. (1997) Mycosporine-like amino acids in 38 species of subtidal marine organisms from Mc Murdo Sound, *Antarctica*. *Antarct. Sci.*, **9**, 392–398.

111 Rastogi, R.P., Richa, Sinha, R.P., Singh, S.P., and Häder, D.-P. (2010) Photoprotective compounds from marine organisms. *J. Ind. Microbiol. Biotechnol.*, **37**, 537–558.

112 Portwich, A. and Garcia-Pichel, F. (2003) Biosynthetic pathway of mycosporines (mycosporine-like amino acids) in the cyanobacterium *Chlorogloeopsis* sp. strain PCC 6912. *Phycologia*, **42**, 384–392.

113 Singh, S.P., Häder, D.-P., and Sinha, R.P. (2012) Bioinformatics evidence for the transfer of mycosporine-like amino acid core (4-deoxygadusol) synthesizing gene from cyanobacteria to dinoflagellates and an attempt to mutate the same gene (YP_324358) in *Anabaena variabilis* PCC 7937. *Gene*, **500**, 155–163.

114 Balskus, E.P. and Walsh, C.T. (2010) The genetic and molecular basis for sunscreen biosynthesis in cyanobacteria. *Science*, **329**, 1653–1656.

115 Singh, S.P., Klisch, M., Sinha, R.P., and Häder, D.-P. (2008) Effects of abiotic stressors on synthesis of the mycosporine-like amino acid shinorine in the cyanobacterium *Anabaena variabilis* PCC 7937. *Photochem. Photobiol.*, **84**, 1500–1505.

116 Castenholz, R.W. and Garcia-Pichel, F. (2000) Cyanobacterial responses to UV-radiation, in *The Ecology of Cyanobacteria: Their Diversity in Time and Space* (eds B.A. Whitton and M. Potts), Kluwer, Dordrecht, The Netherlands, pp. 591–611.

117 Korbee, N., Huovinen, P., Figueroa, F.L., Aguilera, J., and Karsten, U. (2005) Availability of ammonium influences photosynthesis and the accumulation of mycosporine-like amino acids in two *Porphyra* species (Bangiales, Rhodophyta). *Mar. Biol.*, **146**, 645–654.

118 Portwich, A. and Garcia-Pichel, F. (1999) Ultraviolet and osmotic stress induce and regulate the synthesis of mycosporines in the cyanobacterium *Chlorogloeopsis* PCC 6912. *Arch. Microbiol.*, **172**, 187–192.

119 Portwich, A. and Garcia-Pichel, F. (2000) A novel prokaryotic UVB photoreceptor in the cyanobacterium *Chlorogloeopsis* PCC 6912. *Photochem. Photobiol.*, **71**, 493–498.

120 Litchman, E., Neale, P.J., and Banaszak, A.T. (2002) Increased sensitivity to ultraviolet radiation in nitrogen-limited dinoflagellates: photoprotection and repair. *Limnol. Oceanogr.*, **47**, 86–94.

121 Korbee Peinado, N., Abdala Díaz, R.T., Figueroa, F.L., and Helbling, E.W. (2004) Ammonium and UV radiation stimulate the accumulation of mycosporine-like

amino acids in *Porphyra columbina* (Rhodophyta) from Patagonia, Argentina. *J. Phycol.*, **40**, 248–259.

122 Singh, S.P., Klisch, M., Sinha, R.P., and Häder, D.-P. (2010) Sulfur deficiency changes mycosporine-like amino acid (MAA) composition of *Anabaena variabilis* PCC 7937: a possible role of sulfur in MAA bioconversion. *Photochem. Photobiol.*, **86**, 862–870.

123 Agar, N., Halliday, G., Barnetson, R., Ananthaswamy, H., Wheeler, M., and Jones, A. (2004) The basal layer in human squamous tumors harbors more UVA than UVB fingerprint mutations: a role for UVA in human skin carcinogenesis. *Proc. Natl. Acad. Sci. USA*, **101**, 4954–4959.

124 Ryu, B., Qian, Z.J., Kim, M.M., Nam, K.W., and Kim, S.K. (2009) Anti-photoaging activity and inhibition of matrixmetalloproteinase (MMP) by marine red alga, *Corallina pilulifera* methanol extract. *Radiat. Phys. Chem.*, **78**, 98–105.

125 Kondo, S. (2000) The roles of cytokines in photoaging. *J. Dermatol. Sci.*, **23**, 30–36.

126 Rittie, L. and Fisher, G.J. (2002) UV-light-induced signal cascades and skin aging. *Ageing Res. Rev.*, **1**, 705–720.

127 Record, I.R., Dreosti, I.E., Konstantinopoulos, M., and Buckley, R.A. (1991) The influence of topical and systemic vitamin E on ultraviolet light-induced skin damage in hairless mice. *Nutr. Cancer*, **16**, 219–226.

128 Kawaguchi, Y., Tanaka, H., Okada, T., Konishi, H., Takahashi, M., Ito, M., and Asai, J. (1996) The effects of ultraviolet A and reactive oxygen species on the mRNA expression of 72-kDa type IV collagenase and its tissue inhibitor in cultured human dermal fibroblasts. *Arch. Dermatol. Res.*, **288**, 39–44.

129 Coba, F.D.L., Aguilera, J., de Gálvez, M.V., Álvarez, M., Gallego, E., Figueroa, F.L., and Herrera, E. (2009) Prevention of the ultraviolet effects on clinical and histopathological changes, as well as the heat shock protein-70 expression in mouse skin by topical application of algal UV-absorbing compounds. *J. Dermatol. Sci.*, **55**, 161–169.

130 Woodcock, A. and Magnus, I.A. (1976) The sunburn cell in mouse skin: preliminary quantitative studies on its production. *Br. J. Dermatol.*, **95**, 459–468.

131 Baba, T., Katsumi, H., and Hashimoto, I. (1996) The study of ultraviolet B-induced apoptosis in cultured mouse keratinocytes and in mouse skin. *J. Dermatol. Sci.*, **12**, 18–23.

132 Drolet, B.A. and Connor, M.J. (1992) Sunscreens and the prevention of ultraviolet radiation induced. *J. Dermatol. Surg. Oncol.*, **18**, 571–576.

133 Bandaranayake, W.M. and Rocher, A.D. (1999) Role of secondary metabolites and pigments in the epidermal tissues, ripe ovaries, viscera, gut contents and diet of the sea cucumber *Holothuria atra*. *Mar. Biol.*, **133**, 163–169.

134 Oyamada, C., Kaneniwa, M., Ebitani, K., Murata, M., and Ishihara, K. (2008) Mycosporine-like amino acids extracted from Scallop (*Patinopecten yessoensis*) ovaries: UV protection and growth stimulation activities on human cells. *Mar. Biotechnol.*, **10**, 141–150.

135 Carefoot, T.H., Harris, M., Taylor, B.E., Donovan, D., and Karentz, D. (1998) Mycosporine-like amino acids: possible UV protection in eggs of the sea hare *Aplysia dactylomela*. *Mar. Biol.*, **130**, 389–396.

136 Daniel, S., Cornelia, S., and Fred, Z. (2004) UV-A sunscreen from red algae for protection against premature skin aging. *Cosmet. Toilet. Manuf. Worldw.*, 139–143.

137 Coba, F.D.L., Aguilera, J., and Figueroa, F.L. (2007) Use of mycosporine-type aminoacid Porphyra-334 as an antioxidant. Intl Patent WO2007/026035 A2.

138 Coba, F.D.L., Aguilera, J., and Figueroa, F.L. (2007) Use of mycosporine-type aminoacid shinorine as an antioxidant. Intl Patent WO2007/026038 A2.

139 Oren, A. and Gunde-Cimerman, N. (2007) Mycosporines and mycosporine-like amino acids: UV protectants or multipurpose secondary metabolites? *FEMS Microbiol. Lett.*, **269**, 1–10.

140 Nakayama, R., Tamura, Y., Kikuzaki, H., and Nakatani, N. (1999) Antioxidant effect of the constituents of susabinori (*Porphyra yezoensis*). *J. Am. Oil Chem. Soc.*, **76**, 649–653.

141 Yakovleva, I., Bhagooli, R., Takemura, A., and Hidaka, M. (2004) Differential susceptibility to oxidative stress of two scleractinian corals: antioxidant functioning of mycosporine–glycine. *Comp. Biochem. Physiol.*, **139**, 721–730.

142 Rastogi, R.P. (2010) UV-B induced DNA damage and repair in cyanobacteria, PhD Thesis, Banaras Hindu University, Varanasi, India.

143 Rastogi, R.P., Singh, S.P., Häder, D.-P., and Sinha, R.P. (2011) Ultraviolet-B-induced DNA damage and photorepair in the cyanobacterium *Anabaena variabilis* PCC 7937. *Environ. Exp. Bot.*, **74**, 280–288.

144 Misonou, T., Saitoh, J., Oshiba, S., Tokitomo, Y., Maegawa, M., Inoue, Y., Hori, H., and Sakurai, T. (2003) UV-absorbing substance in the red alga *Porphyra yezoensis* (Bangiales, Rhodophyta) block thymine dimer production. *Mar. Biotechnol.*, **5**, 194–200.

145 Galinski, E.A. (1993) Compatible solutes of halophilic cyanobacteria: molecular principles, water-solute interaction, stress protection. *Experimentia*, **49**, 487–496.

Index

Marine Microbiology: Bioactive Compounds and Biotechnological Applications, First Edition.
Edited by Se-Kwon Kim

c

n